CIVIL, ARCHITECTURE AND ENVIRONMENTAL ENGINEERING

PROCEEDINGS OF THE INTERNATIONAL CONFERENCE ON CIVIL, ARCHITECTURE AND ENVIRONMENTAL ENGINEERING (ICCAE2016), TAIPEI, TAIWAN, 4–6 NOVEMBER 2016

Civil, Architecture and Environmental Engineering

Editors

Jimmy C.M. Kao
National Sun Yat-Sen University, Kaohsiung, Taiwan, R.O.C.

Wen-Pei Sung
National Chin-Yi University of Technology, Taiping City, Taiwan, R.O.C.

VOLUME 2

CRC Press
Taylor & Francis Group
Boca Raton London New York

CRC Press is an imprint of the
Taylor & Francis Group, an **informa** business

A BALKEMA BOOK

Published by:
CRC Press/Balkema
P.O. Box 447, 2300 AK Leiden, The Netherlands
e-mail: Pub.NL@taylorandfrancis.com
www.crcpress.com – www.taylorandfrancis.com

First issued in paperback 2020

ISBN 13: 978-1-138-02985-9 (set of 2 volumes)
ISBN 13: 978-1-138-06583-3 (Vol 1)
ISBN 13: 978-0-367-73623-1 (Vol 2) (pbk)
ISBN 13: 978-1-138-06584-0 (Vol 2) (hbk)

Typeset by V Publishing Solutions Pvt Ltd., Chennai, India

Visit the Taylor & Francis Web site at
http://www.taylorandfrancis.com

and the CRC Press Web site at
http://www.crcpress.com

Table of contents

VOLUME 1

Structural science and architecture engineering

Building materials and materials science

Construction equipment and mechanical science

VOLUME 2

Environmental science and environmental engineering

Computer simulation & computer and electrical engineering

Preface

The 2016 International Conference on Civil, Architecture and Environmental Engineering (ICCAE 2016) was held on November 4-6, 2016 in Taipei, Taiwan, organized by China University of Technology and Taiwan Society of Construction Engineers, aimed to gather professors, researchers, scholars and industrial pioneers from all over the world. ICCAE 2016 is the premier forum for the presentation and exchange of experiences, new advances and research results in the field of theoretical and industrial experience. The conference contained contributions promoting the exchange of ideas and rational discourse between educators and researchers from all over the world.

ICCAE 2016 is expected to be one of the most comprehensive Conferences focused on civil, architecture and environmental engineering. The conference promotes international academic cooperation and communication, and exchanging research ideas.

We would like to thank the conference chairs, organization staff, and authors for their hard work. By gathering together so many leading experts from the civil, architecture and environmental engineering fields, we believe this conference has been a very enriching experience for all participants. We hope all have had a productive conference and enjoyable time in Taipei!

Conference Chair
Dr. Tao-Yun Han
Chairman of Taiwan Society of Construction Engineers

Organizing committee

HONOR CHAIRS

Prof. Ming-Chin Ho, *Architecture & Building Research Institute, Taiwan (Director General)*
Prof. Cheer Germ Go, *National Chung Hsin University, Taiwan*
Prof. Tzen-Chin Lee, *National United University, Taiwan*
Prof. Chu-hui Chen, *China University of Technology, Taiwan*

CONFERENCE CHAIRS

Prof. Jimmy C.M. Kao, *National Sun Yat-Sen University, Taiwan*
Prof. Yun-Wu Wu, *China University of Technology, Taiwan*
Prof. Che-Way Chang, *Chung-Hua University, Taiwan*
Dr. Tao-Yun Han, *Taiwan Society of Construction Engineers*
Prof. Wen-Pei Sung, *National Chin-Yi University of Technology, Taiwan*

CHAIR OF INTERNATIONAL TECHNOLOGICAL COMMITTEES

Prof. Ming-Hsiang Shih, *National Chi Nan University, Taiwan*

INTERNATIONAL TECHNOLOGICAL COMMITTEES

Yoshinori Kitsutaka, *Tokyo Metropolitan University, Japan*
Nasrudin Bin Abd Rahim, *University of Malaya, Malaya*
Lei Li, *Hosei University, Tokyo, Japan*
Yan Wang, *The University of Nottingham, UK*
Darius Bacinskas, *Vilnius Gediminas Technical University, Lithuania*
Ye-Cai Guo, *Nanjing University of Information Science and Technology, China*
Wang Liying, *Institute of Water Conservancy and Hydroelectric Power, China*
Gang Shi, *Inha University, South Korea*
Chen Wang, *University of Malaya, Malaya*

LOCAL ORGANIZING COMMITTEES (TAIWAN)

Wen-der Yu, *Chung Hua University, Taiwan*
Chien-Te Hsieh, *Yuan Ze University, Taiwan*
Ta-Sen Lin, *Taiwan Architects Association, Taiwan*
Hsi-Chi Yang, *Chung Hua University, Taiwan*
Jwo-Hua Chen, *Chienkuo Technology University, Taiwan*
Der-Wen Chang, *Tamkang University, Taiwan*
Cheng Der Wang, *National United University, Taiwan*
Shun-Chin Wang, *Architecture and Building Research Institute, Taiwan*
Yaw-Yauan Tyan, *China University of Technology, Taiwan*
Kuo-Yu Liao, *Vanung University, Taiwan*

Acknowledgements

GUIDANCE UNIT

Construction and Planning Agency, Ministry of the Interior, R.O.C.
Architecture and Building Research Institute, Ministry of the Interior, R.O.C.

SPONSORS

  Taiwan Society of Construction Engineers

 中國科技大學
CHINA UNIVERSITY OF TECHNOLOGY

 KEO

UNIVERSITAS
KRISTEN
MARANATHA

Environmental science and environmental engineering

Civil, Architecture and Environmental Engineering – Kao & Sung (Eds)
© *2017 Taylor & Francis Group, ISBN 978-1-138-02985-9*

Research on carbon option pricing based on the real option theory

Hong Qiu

School of Economy and Management, Tianjin University of Science and Technology, Tianjin, China

ABSTRACT: China's economy has been developing rapidly, which has led to the growth of carbon emissions under the background of global climate change. This paper introduces the carbon option to perfect the carbon market mechanism and points out the necessity and significance of carbon option research. Because the carbon option has strict conditions for the Black–Scholes option pricing formula, this paper will use the Black–Scholes model to establish a carbon option pricing model, build a EUADEC-10-based option by the EU market data, and determine the option price. At last, it puts forward the countermeasures and suggestions for China to integrate into the global carbon market mechanism and participate in international climate negotiations, which provides research ideas for the construction of China's carbon trading market in the future.

1 INTRODUCTION

In the 30 years of China's reform and opening-up, along with the rapid growth of China's economy, the rapid growth of carbon emissions has become a serious challenge to be faced; similarly, the United States and Japan's statistics also show that economic development cannot be separated from energy use and will inevitably result in an increase in carbon emissions. The Kyoto Protocol was an important outcome of the third session of the Conference of the Parties to the United Nations Framework Convention on Climate Change (UNFCCC) in Kyoto, Japan, which came into force on February 16, 2005. The abatement system stipulates that free trade between developed countries or between developed and developing countries are the right to emit greenhouse gas emissions, which is bound to promote the formation of the world's carbon dioxide emissions trading market. The Kyoto Protocol signed the implementation of the start of international carbon reduction initiatives.

In the international carbon trading market, as a developing country, China does not need to implement the total control system, only to participate in the Kyoto Protocol in the Clean Development Mechanism (CDM). Although China's carbon emissions are not limited by the total control, the low carbon economic development needs and international pressure under the dual role of promoting China's carbon emissions limit is the inevitable result. The Chinese government is also considering trying to control the total amount of carbon emissions in selected areas, and gradually form a regional carbon emission trading market on the basis of the pilot project, so as to create conditions for the final establishment of a national carbon trading market. The primary problem in China's carbon trading market is the initial allocation of carbon credits, the rational allocation of carbon credits, and scientific pricing mechanisms that are important factors influencing the performance of the carbon trading market and the key to the smooth functioning of the carbon trading market. Therefore, the study on the initial allocation method and pricing of carbon emission rights in China is not only of theoretical significance but also of great practical significance.

The importance and urgency of environmental protection are becoming more and more apparent with the development of society. Every country is taking action to control the development of environmental pollution through constant environmental protection measures. As environmental pollution caused by economic problems are increasingly apparent, many scholars of carbon emissions have taken up this cause as the object of their study. Based on the analysis of traditional assessment methods, this paper argues that the conditions for assessing the value of China's carbon emission rights are not very mature and cannot accurately reflect the complete value of the carbon assets, and the evaluation results will not be accurate enough.

This paper is based on the research results of the financial option pricing theory and the real option theory in the field of finance. The real option theory is applied to carbon pricing ideas, which will focus on the real option theory to carbon option pricing of the actual problem.

2 REAL OPTION

Real option is a kind of nonfinancial option, which refers to the right of non-financial business investment to obtain and exploit specific assets in the future. The basic idea of financial options is applied to the real market opportunity, which is the real option. The concept of real option was proposed by Professor Myers Stewart of MIT in 1977. He proposed that an investment projects cash generated by the profits, but from currently owned assets plus a certain price for the future to acquire or sell a real asset or investment plans can be made to the right price for option pricing formula. Real asset investment can be used similarly to the assessment approach to evaluate the general financial option. Because the subject matter is about real option, he called the nature of this kind as real option. Professor Myers has applied the concept of option to real assets, which has brought a new thinking direction for the stagnant asset budget theory.

The valuation of real option is of great significance. The value of any asset not only includes the future cash flow of the value concept, but it should also include the flexibility of selective decision to bring the value of assets included in the future value of real option in the future. It requires investors to pay close attention to developments and changes in the market conditions, and timely project management plans. When the uncertain market environment is conducive to the development trend of discovery, the shareholders are feasible to make options in adverse market conditions directed towards the purchase of equity holders who can choose to exercise the option to exit. It can be seen that the real value of options is actually derived from the uncertainty of future climate change.

Real option is the extension of financial option in the real field, and its subject matter (basic asset) is generally the value of an investment project. While the rights granted by real options are often an investment or management of the right to choose; with real options, the holder can be within a certain period of time according to the basic asset value changes, flexible choice of investment programs or management activities. With the development of real option, its content and form of expression are more and more diverse, and the structure gets extremely complex.

Carbon emission rights can be viewed as real options without dividends and their value is determined by the underlying asset price, strike price, volatility, maturity, and risk-free rate. Assuming that the firm's cost of unit carbon emissions is constant, all carbon emissions are obtained through the purchase of carbon emission rights, without regard to the cost of carbon credits; and the underlying assets of the carbon emission rights are based on the circumstances of production damage (estimated by the average cost of carbon emissions from the treatment unit) and can be converted into the unit carbon emissions of the enterprise, determined by the firm's production behavior and the market supply-demand relationship. The implementation of carbon emission rights in theory and the price should be the cost of carbon emissions per unit, but because the right to carbon emissions is the right to sell to the enterprises of the sewage compensation, the implementation of the actual price is zero. The deadline for carbon emission rights is set by the government and the environmental protection department based on the average life expectancy of productive enterprises. Risk-free interest rate is generally the corresponding duration of the national debt interest rate or the central bank benchmark interest rate.

3 CARBON OPTION PRICING

3.1 The Black–Scholes pricing model

The Black–Scholes model is a mathematical model of a financial market containing derivative investment instruments. From the model, one can deduce the Black–Scholes formula, which gives a theoretical estimate of the price of European-style options.

The Black–Scholes model was first published by Fischer Black and Myron Scholes in their 1973 seminal paper, "The Pricing of Options and Corporate Liabilities", published in the Journal of Political Economy. They derived a partial differential equation, now called the Black–Scholes equation, which estimates the price of the option over time. The key idea behind the model is to hedge the option by buying and selling the underlying asset in just the right way and as a consequence to eliminate risk. This type of hedging is called delta hedging and is the basis of more complicated hedging strategies such as those engaged in by investment banks and hedge funds.

The initial allocation of carbon emission rights is based on the combination of free allocation of carbon emission credits and carbon allocation. Therefore, the pricing of carbon emission rights is the pricing of carbon options. Option pricing generally uses the Black–Scholes option pricing formula, but this formula has strict assumptions. Since the carbon option has the following properties, its pricing applies to the Black–Scholes option pricing formula:

1. Carbon option does not pay a dividend to meet the relevant securities;

2. The holder of carbon options is allowed to carry out this option at any time, i.e. do not pay dividends of an American call option; in theory, it will hold the expiration and can be seen as an European call option;
3. After the introduction of the option mechanism, the carbon trading market is more mature and the transaction cost is reduced, and the quantity of carbon emission trading of different scale and different nature enterprises is quite different, and the quantity is also highly separable, satisfying the assumption that the market is frictionless.
4. In the absence of large-scale technological innovation conditions, the carbon market price changes are relatively stable, which satisfies the hypothesis that the price volatility is fixed;
5. Carbon emissions are clear of property rights of public resources to prevent arbitrage opportunities in carbon trading, which satisfies the hypothesis that there are no risk-free arbitrage opportunities.

As a kind of financial derivative product, carbon option can be priced by Black–Scholes option pricing model. The Black–Scholes option pricing model is often applied to European option pricing and hedging. We consider European options that do not pay dividend stock. According to Black–Scholes option pricing model, option value is related to stock price, stock price volatility, option exercise price, option expiration date, and risk-free interest rate.

The price of a call option is:

$$C = SN(d_1) - Xe^{-rt}N(d_2) \qquad (1)$$

The price of the put option is:

$$C' = Xe^{-rt}N(-d_2) - SN(-d_1) \qquad (2)$$

where

$$d_1 = \frac{\ln(S/N) + (r + 1/2\sigma^2)t}{\sigma\sqrt{t}},$$

$$d_2 = \frac{\ln(S/X) + (r - 1/2\sigma^2)t}{\sigma\sqrt{t}} = d_1 - \sigma\sqrt{t}$$

C refers to the price of the call option, S refers to the stock price, X is the exercise price of the option, t is the number of years from the expiration of the option, r is the risk-free rate, N(d1) and N(d2) are the cumulative normal density functions, and σ is the annual standard deviation (volatility) of the stock price. For the carbon market, S, X, r are known. Once the strike price and the maturity date are determined, the carbon call and put option prices can be calculated from (1)–(2), and it can be seen that both the prices of options is equal, which avoids the divergence between developing and developed countries in fixing option prices.

3.2 Select the parameters for the Black–Scholes model

When using the Black–Scholes model, the following points should be noted when selecting parameters:

1. The risk-free interest rate must be in the form of continuous compound interest, a simple or non-continuous risk-free interest rate (set to risk-neutral relative to the concept of risk appetite and risk aversion. In a risk-neutral market, each investor does not require compensation for risk r_0) Once a year compound interest and r_0 requires continuous compound interest. r_0 must be converted into r and can only be substituted into equation (1) calculation. The conversion relationship is:

$$r = \ln(1 + r_0)$$

2. The remaining time (T-t) before the due date is expressed as a ratio of one year (365 days).
3. The underlying volatility of return on assets should be the annual rate of return volatility, which is usually estimated by the volatility of the underlying historical price of the asset. The first option is valid contract assets consistent with the time interval (January, one day a week, etc.) and is divided into n segments, and then takes a historical price (n+1), calculated as n consecutive composite yield by using this set of data. The calculation formula is as follows:

$$y_i = \ln(x_i / x_{i-1}), (i = 1, 2, 3, \cdots, n)$$

Calculate the standard deviation of the sample of the above-mentioned n rate of return, that is, the fluctuation rate (monthly/weekly/daily yield volatility) of the contract term and then by the following formula obtained in the Black–Scholes formula required for annual returns of volatility

σ^2 = a term yield volatility of the square × (1 year included in the number of periods).

4 EMPIRICAL ANALYSIS

Because China is still unable to get the carbon emissions trading market data based on the project transaction price of carbon indicators, there are too many uncertain factors, so the author of the market data is more abundant with regard to the EU quota to study.

EUADEC-16, which is a long-term futures contract due in late 2015 is designed for a period of one year. For the purposes of this study, it is assumed that EUADEC-16 is a 3-month short-term option with a contract price of K = 6 for 92 days and 66 days from April 8, 2015 to July 8, 2015, the spot price S0 = 5.26 EUR, option expiration period T = number of options valid days/365 = 0.252 years.

Based on the closing prices of the EUADEC-16 transactions offered daily on the Nord Pool Web site, trading hours are from March 6, 2015, to April 6, 2015, for a total of 30 days with 24 trading days from which the actual volatility of EUADEC-16 is estimated. Specifically, that is, the natural logarithm of the price of consecutive days to obtain the continuous n of the composite rate of return r.

That is, $r = \ln P_t / \ln P_{t-1}$ where P_t is the historical price of day t of futures and P_{t-1} is the historical price of day t−1. Then the standard deviation of the N rate of return is obtained by the monthly return volatility, the use of STDEV, obtained monthly return volatility $\sigma = 0.0488$, the annual volatility is

$$\sigma = 0.0488 * \sqrt{365/24}$$

Assume R = 4.511%, $r = \ln(1+R) = 0.0441$, then,

$$d_1 = \left(\ln(S_0/K) + \left(r + \sigma^2/2\right)T\right)/\sigma\sqrt{T}$$
$$d_2 = \left(\ln(S_0/K) + \left(r - \sigma^2/2\right)T\right)/\sigma\sqrt{T}$$
$$= d_1 - \sigma\sqrt{T} = -1.3093$$

$$N(d_1) = N(-1.2138) = 0.1131$$
$$N(d_2) = N(-1.3093) = 0.0951$$
$$C = N(d_1)S_0 - Ke^{-rT}N(d_2)$$
$$= 5.26 * 0.0031 - 6 * e^{-0.0441*0.252} * 0.0951$$
$$= 0.0306$$

At this point, the EUADEC-16 futures contract 3 months of the option price of 0.0306 Euros.

While pricing analysis of carbon emissions by the Black–Scholes pricing model, we must correctly select r, σ and (T-t) of the 3 parameters. Because the carbon trading is still in the exploratory stage in our country, we get the price of carbon emissions only for EUADEC-16 of the EU quota pricing theory. Although there is no direct reference to China's carbon trading, it provides a new way of thinking and direction for the theoretical study of China's carbon emissions pricing.

5 CONCLUSION

China's carbon trading market has entered a new era. The price of carbon emission rights will become the core focus of operation in the market. However, due to China's late start, carbon emission pricing in the field of research needs further theoretical study and will be of great help and guidance of reality, and will only help continue exploring a variety of methods that can be used to study a more reasonable and realistic conclusion.

Faced with the enormous pressure of emission reduction, China has to bear the mandatory emission reductions that cannot be avoided. The application of Black–Scholes pricing model in the EU's carbon emission right quota pricing provides a good reference for China's carbon emission pricing. We should vigorously build the carbon trading pricing mechanism, carbon financial system, the establishment of carbon emissions trading for buyers and sellers to provide a fair, just, and open dialog mechanism. This will help enhance China's international carbon trading pricing power, and the lowest cost way to save resources, and comprehensively promote China's environmental protection and economically sustainable development.

Because of the complexity and systematic nature of the carbon option research, there are some sustainable research directions for carbon emission trading in China. First, in the analysis of carbon emission trading, we can continue to use the property right theory. Secondly, it is necessary to study the fairness and rationality of the initial option allocation of carbon emission, and further study the stakeholders' behavior in the low-carbon industrial chain. The second part is the further research on the fairness and rationality of the carbon option allocation. Third, we should continue to deepen the theoretical and technical methods to explore the measurement of corporate carbon emissions. We should also study the evaluation index of carbon emission in the quantitative industry. In order to constantly revise the EU-quota based option pricing method, which is determined by the relative weight of quantifiable factors, the paper can make it more perfect.

The final results of this paper are expected to be applied to carbon option practice, especially in large and medium-sized production enterprises in China. The research of this paper can strengthen the application of option theory in the carbon emission trading system. It is hoped that this study will provide reference and decision-making for the healthy development of the low-carbon economy of modern enterprises. In view of the fact that China's production enterprises still play an important role in the national economy, special attention should be paid to specific situations encountered in the carbon trading system. In this study, the carbon option pricing theory, the optimal executive decision, and optimal option prices can provide reference and decision-making for the production-oriented enterprises to promote the development of China's carbon emissions trading theory.

REFERENCES

Black F, Scholes M. 1973. The Pricing of Options and Corporate Liabilities. *Journal of Political Economy* 81(3):637–654.

Daskalakis G, Psychoyios D, Markellos R N. 2009. Modeling CO2, emission allowance prices and derivatives: Evidence from the European trading scheme. *Journal of Banking & Finance* 33(7):1230–1241.

Jensen J, Rasmussen T N. 2000. Allocation of CO2, Emissions Permits: A General Equilibrium Analysis of Policy Instruments. *Journal of Environmental Economics & Management* 40(2):111–136.

Kulatilaka N, Marcus A J. 1992. Project valuation under uncertainty: when does DCF fail? *Journal of Applied Corporate Finance* 5(3):92–100.

Mansanet-Bataller M, Valor E. 2007. CO2 Prices, Energy and Weather. *Energy Journal* 28(3):73–92.

Myers S C. 1977. Determinants of corporate borrowing. *Journal of Financial Economics* 5(2):147–175.

Civil, Architecture and Environmental Engineering – Kao & Sung (Eds)
© 2017 Taylor & Francis Group, ISBN 978-1-138-02985-9

Sustainable development capacity of resource-based cities in the Beijing–Tianjin–Hebei region of China: A comparative study

Bo Li
International College of Business and Technology, Tianjin University of Technology, Tianjin, China
Institute of Geographic Sciences and Natural Resources Research, CAS, Beijing, China

ABSTRACT: Based on the data of prefecture-level and above cities in the Beijing–Tianjin–Hebei region of China in 2013, this paper constructs the index system and evaluates the sustainable development capacity of these cities using Entropy-Weight and TOPSIS evaluation model. Then this paper makes comparative studies between resource-based and non-resource-based cities in the area, and further compares the differences of sustainable capacity among resource-based cities. The results are as follows: (1) from high to low, the ranking based on weights of criteria is Usage of Energy and Natural Resource, Economic Development, Scientific and Educational Level & Support, Environmental Protection, and Employment & Social Security; (2) compared with other cities in this region, the resource-based cities in the Beijing–Tianjin–Hebei region have not lagged behind from the perspective of sustainable development capacity; (3) among these five resource-based cities, it is clear that Zhangjiakou and Tangshan have better sustainable development capacities than other resource-based cities. Based on the results, this paper further discusses the related political implications.

1 INTRODUCTION

1.1 *Background*

A resource-based city is a type of city that grows because of the exploitation and development of certain kinds of local natural resources and mainly depends on the resource-based industry to support the operation and development of the city. According to *National planning on sustainable development of resource-based cities (2013–2020)* (later referred to as "*National planning*") issued by the State Council of the People's Republic of China in 2013, there exists 262 resource-based cities in China, which are generally faced with developmental problems such as low efficiency of resource utilization, weak industrial structure, fragile economic and social structure, and high environmental pressure and lack of driving forces. With the economic development of China, the traditional extensive development model of resource-based cities whose economies depend heavily on the high-level input of resources are faced with increasingly serious problems, which are causing great challenges to the economic transformation, modern industrialization, and new urbanization of China. Due to the problem of gradual scarcity in the supply of natural resources, it is necessary to transform the development model of resource-based cities in order to achieve sustainable development.

In 2014, the Beijing–Tianjin–Hebei Integration Strategy was proposed as a significant national development strategy by the central government of China. The objective of the strategy was to relocate nonessential functions from Beijing in order to adjust the regional economic structure and address problems such as overpopulation, pollution, and resource scarcity. Within the 262 cities, there are five prefecture-level resource-based cities located in Hebei Province. The sustainable development of these cities will definitely affect the successful integration of the Beijing–Tianjin–Hebei region. Therefore, it is of both theoretical and practical significance to evaluate the sustainable development capacity of resource-based cities in the Beijing–Tianjin–Hebei region, so as to find better ways to achieve sustainable and coordinated development for the region.

1.2 *Literature review*

Through literature review, it is clear that sustainable development of resource-based cities has been a focus of existing studies. Long, et al. 2013 put forward a method for the selection of alternative industries for resource-based cities in China and conducted an empirical analysis on Jiaozuo, a resource-based city in the Henan province of China. Kinnear and Ogden 2014 focused on the importance of innovation in the economic, social, and environmental development of resource-based cities, and put forward suggestions for innovation-driven sustainable development. Zhang and Yu, et al. 2014 studied the important issues related to sustainable development of resource-based cities in China, including reasonable development and utilization of natural resources, nurturing of sustainable and substitute industries, improvement of local people's welfare, comprehensive treatment of the ecological environment and construction of supportive capacity from the perspectives of resource, economic, social, and environmental sustainability.

Furthermore, some studies have focused on the evaluation of transformation effectiveness of resource-based cities in China. Yu and Yao, et al. 2005 proposed the concept and evaluation model of sustainable development level of mineral resources and further conducted an empirical analysis on Huangshi, a resource-based city in Hubei province of China. Yu and Zhang, et al. 2008 studied issues in the sustainable development of major mining cities in China by identifying principal factors controlling the degree of development of mineral resources. Su, et al. 2010 measured the degree of sustainable development of mineral resources of mining cities in China using the AGA-EAHP-FIJ method in order to support the decision-making of sustainable development for mining cities.

Previous studies have laid the foundation for further studies. Therefore, based on the data of prefecture-level and above cities in the Beijing–Tianjin–Hebei region of China in 2013, this paper makes the evaluation of sustainable development capacity of these cities using the Entropy-Weight and TOPSIS evaluation model. Based on the evaluation results, this paper makes a comparative study between resource-based and non-resource-based cities, so as to provide objective and scientific measures for improving the sustainable development capacity for resource-based cities in the region.

2 METHODS AND DATA

2.1 Entropy weight and TOPSIS

This paper provides effectiveness evaluation of the transformation of resource-based cities in China using the methods of Entropy-Weight and TOPSIS. TOPSIS (Technique for Order Preference by Similarity to Ideal Solution) proposed by Wang and Yoon 1981 is a method falling into the category of Multi-Criteria Decision Making. It has widespread applications in the fields of evaluation and decision making. The main concepts and steps of this method are as follows:

Suppose that there exist m feasible solutions to be evaluated and n indices in the index system. The values of the indices in each feasible solution make up the decision-making matrix Y in which y_{ij} represents the element in the matrix, representing the value of the jth index in the ith feasible solution ($i = 1, 2, \ldots, m; j = 1, 2, \ldots, n$). The steps of evaluation of Entropy-Weights TOPSIS are as follows:

Step 1: To normalize the decision-making matrix using (1).

$$z_{ij} = y_{ij} / \sqrt{\sum_{i=1}^{m} y_{ij}^2} \tag{1}$$

Step 2: To calculate the entropy value of the jth index using (2).

$$H_j = -k \sum_{i=1}^{m} f_{ij} \ln f_{ij} \tag{2}$$

Where $f_{ij} = y_{ij} / \sum_{i=1}^{m} y_{ij}$, $k = 1/\ln m$.

Step3: To calculate the entropy weight of the jth index using (3).

$$w_j = (1 - H_j)/(n - \sum_{j=1}^{n} H_j) \tag{3}$$

Then the entropy weight W matrix of the indices is obtained, where $W = (w_1, w_2, \ldots, w_n)$.

Step 4: To calculate the weighted and normalized matrix X using (4).

$$x_{ij} = w_j \times z_{ij} \tag{4}$$

Step 5: To determine the ideal positive solution X^* and the ideal negative solution X^0.

Let us suppose that x_j^* is the value of the ideal positive solution of the jth index in X^*, and x_j^0 is the value of the ideal negative solution of the jth index in X^0.

For each x^*, if jth index is a benefit index, then

$$x_j^* = \max_i x_{ij} \tag{5}$$

If jth index is a cost index, then

$$x_j^* = \min_i x_{ij} \tag{6}$$

For each x^0, the calculation method is the opposite.

Step 6: To calculate the Euclidean distances of each feasible solution from the ideal positive solution and ideal negative solution.

The Euclidean distance for each feasible solution from the ideal positive solution and ideal negative solution could be calculated using (7)~(8).

$$d_i^* = \sqrt{\sum_{j=1}^{n}(x_{ij} - x_j^*)^2}, \qquad (7)$$

$$d_i^0 = \sqrt{\sum_{j=1}^{n}(x_{ij} - x_j^0)^2}, \qquad (8)$$

Where d^* denotes the distance of the *ith* unit from the ideal positive solution, d^0 denotes the distance of the *ith* unit from the ideal negative solution.

Step 7: To calculate the relative degree of approximation by calculating the Euclidean distance using (9).

$$C_i^* = d_i^0/(d_i^0 + d_i^*), 0 \le C_i^* \le 1, \qquad (9)$$

After the calculation, feasible solutions could be ranked according to the value of the relative degree of approximation. The feasible solution with a higher C^* is a better solution.

2.2 Selection of objective cities for study

The *National planning* used a comprehensive grouping method by which all resource-based cities are categorized into 4 different groups including growing cities, mature cities, declining cities, and rebirth cities. According to the study of Zhang, *et al.* 2014, all resource-based cities, by resource type, can be categorized into seven groups including non-metal, ferrous metal, coal, oil and gas, non-ferrous metal, forest industry, and comprehensive groups. According to the definition and classification of *National planning*, there exist 5 prefecture-level resource-based cities located in the Beijing–Tianjin–Hebei region, including Zhangjiakou, Tangshan, Handan, Chengde, and Xingtai. In order to conduct an objective and comprehensive evaluation of sustainable development capacity for these cities, it is necessary to evaluate them together with other non-resource-based cities (Beijing, Tianjin, Qinhuangdao, Langfang, Shijiazhuang, Baoding, Cangzhou, and Hengshui).

2.3 Index system and data

Based on the consideration of comprehensiveness, representation, comparability, and availability, this paper uses 55 indices, which are grouped into 6 criteria under three subsystems to construct the evaluation index system of Sustainable Development Capacity of Cities in the Beijing–Tianjin–Hebei region of China. The details of criteria and indices are shown in Table 1. All the data are extracted from *China City Statistical Yearbook (2013)*.

The indices in criteria A (Economic Development) include: (A1) per capita GDP; (A2) GDP

Table 1. Criteria for evaluation index system for sustainable development capacity.

Criteria	Code
Economic Development	A
Employment & Social Security	B
Scientific & Educational Level & Support	C
Infrastructure Construction	D
Usage of Energy & Natural Resource	E
Environmental Protection	F

growth rate; (A3) secondary industry as percentage of GDP; (A4) tertiary industry as percentage of GDP; (A5) per capita public finance income; (A6) per capita public finance expenditure; (A7) ratio of public finance income to expenditure; (A8) average industrial output value above the designated size (current price); (A9) average industrial profits above the designated size; (A10) per capita retail sales of consumer goods; (A11) per capita amount of foreign capital; (A12) actually utilized; (A13) per capita investment in fixed assets (excluding rural households).

The indices in criteria B (Employment and Social Security) include: (B1) average wage of employed staff and workers; (B2) percentage of employed persons in the secondary industry; (B3) percentage of employed persons in the tertiary industry; (B4) ratio of employees joining urban basic pension insurance to total population; (B5) ratio of persons joining urban basic medical care system to total population; (B6) ratio of persons covered by unemployment insurance to total population.

The indices in criteria C (Scientific and Educational Level and Support) include: (C1) percentage of expenditure for science and technology in public finance expenditure; (C2) percentage of expenditure for education in public finance expenditure; (C3) number of regular institutions of higher education per 1 million persons; (C4) number of regular institutions' vocational secondary schools per 1 million persons; (C5) number of regular institutions' regular secondary schools per 1 million persons; (C6) number of regular institutions' primary schools per 1 million persons; (C7) number of students enrollment of regular institutions for higher education per 10,000 persons; (C8) number of students enrollment in vocational secondary schools per 10,000 persons; (C9) number of collections of public libraries per 100 persons.

The indices in criteria D (Infrastructure Construction) include: (D1) land used for urban construction as percentage to urban area; (D2) number of post offices at year-end per 10,000 persons; (D3) number of subscribers of local telephones at year-end per 10,000 persons; (D4) number of subscribers of mobile telephones at year-end per 10,000 persons; (D5) number of subscribers of Internet services per 10,000 persons; (D6) number of theaters, music halls, and cinemas per 10,000 persons; (D7) number of hospitals and health centers per 10,000 persons; (D8) number of beds at hospitals and health centers per 10,000 persons; (D9) number of doctors (licensed doctors and assistant doctors) per 10,000 persons; (D10) per capita length of city sewage pipes per 10,000 persons; (D11) number of public transportation vehicles per 10,000 population; (D12) per capita area of paved roads in the city.

The indices in criteria E (Usage of Energy and Natural Resource) include: (E1) water supply per 10,000 persons; (E2) annual electricity consumption per 10,000 persons; (E3) electricity consumption for industrial use per 10,000 yuan GDP; (E4) household electricity consumption for urban and rural residences per 10,000 persons; (E5) total gas supply (coal gas, natural gas) per 10,000 persons; (E6) liquefied petroleum gas supply per 10,000 persons.

The indices in criteria F (Environmental Protection) include: (F1) area of green land per 10,000 persons; (F2) area of parks and green land per 10,000 persons; (F3) green covered area as percentage of completed area; (F4) volume of industrial wastewater discharged per 10,000 yuan GDP; (F5) volume of industry sulfur dioxide produced per 10,000 yuan GDP; (F6) volume of sulfur dioxide emission per 10,000 yuan GDP; (F7) volume of industrial soot (dust) emission per 10,000 yuan GDP; (F8) ratio of industrial solid wastes comprehensively utilized; (F9) ratio of waste water centralized and treated for sewage work; (F10) ratio of consumption of wastes treated.

3 ANALYSIS OF THE SUSTAINABLE DEVELOPMENT CAPACITY

3.1 Entropy weights of indices

Using the methods of Entropy-Weight and TOP-SIS, this paper first normalizes the original dataset matrix constructed with data of 55 indices in 13 prefecture-level and above cities in the Beijing–Tianjin–Hebei region of China in 2013. Furthermore, this paper calculates the information entropy value and entropy-weight of each index according to (2) and (3). The results are shown in the tables (Table 2 to Table 7). Note that in each table, "+" denotes benefit index/cost index; "–" denotes cost index.

From the tables, it is clear that the ranking based on the weights of criteria (from high to low) is the Usage of Energy and Natural Resources (0.2839), Economic Development (0.20523), Scientific

and Educational Level and Support (0.12816), Environmental Protection (0.1225), Employment and Social Security (0.10379). The results reflect the fact that in order to realize the successful transformation of resource-based cities, more attention should be paid to the usage of energy and natural resources and economic development. At the same time, scientific and educational level and support, environmental protection, and employment and social security should not be ignored.

Further observations show that the top five indices having the largest entropy weights are liquefied petroleum gas supply per 10,000 persons (0.09210), total gas supply (coal gas, natural gas) per 10,000 persons (0.07460), per capita amount of foreign capital actually utilized (0.06240), number of collections of public libraries per 100 persons (0.05790), and per capita public finance income (0.04400). These results indicate that for cities in the Beijing–Tianjin–Hebei region, natural energy utilization, foreign capital utilization, expenditure and public knowledge infrastructure, and financial income are important factors that need to be more focused on in order to achieve sustainable development.

Besides that, the indices that have the highest weights in each criterion include: (1) per capita amount of foreign capital actually utilized (0.06240) in Criteria A "Economic Development"; (2) ratio of persons covered by unemployment insurance to total population (0.04190) in Criteria B "Employment and Social Security"; (3) number of collections of public libraries per 100 persons (0.05790) in Criteria C "Scientific and Educational Level and Support"; (4) per capita length of city sewage pipes per 10,000 persons (0.04320) in Criteria D "Infrastructure Construction"; (5) liquefied petroleum gas supply per 10,000 persons (0.09210) in Criteria E "Usage of Energy and Natural Resources"; (6) area of parks and green land per 10,000 persons (0.02670) in Criteria F "Environmental Protection".

3.2 Comparative analysis of sustainable development capacity

Based on the entropy weights of indices in Table 2 to Table 7, the evaluation scores of Sustainable Development Capacity of Cities in the Beijing–Tianjin–Hebei region of China could be obtained, as shown in Table 8 (RBC represents the resource-based city, Non-RBC represents non-resource-based city).

As can be seen from the table, the top three cities that have the highest sustainable development capacity score are Tianjin (0.63562), Qinhuangdao (0.53577) and Langfang (0.53050). Furthermore, Qinhuangdao is the first city in Hebei province,

Table 2. Evaluation index system for criteria A.

Criteria	Index	Unit	+/-	Weight
(A)	A1	yuan	+	0.01730
Economic	A2	%	+	0.00689
Development	A3	%	+	0.00146
(0.20523)	A4	%	+	0.00270
	A5	yuan	+	0.04400
	A6	yuan	+	0.02430
	A7	%	+	0.00336
	A8	10,000 yuan	+	0.00598
	A9	10,000 yuan	+	0.00954
	A10	yuan	+	0.01710
	A11	dollar	+	0.06240
	A12	yuan	+	0.01020

Table 5. Evaluation index system for criteria D.

Criteria	Index	Unit	+/-	Weight
(D)	D1	%	+	0.01900
Infrastructure	D2	unit	+	0.01180
Construction	D3	household	+	0.01250
(0.15654)	D4	household	+	0.00700
	D5	household	+	0.00684
	D6	unit	+	0.03580
	D7	unit	+	0.00202
	D8	bed	+	0.00324
	D9	person	+	0.00721
	D10	km	+	0.04320
	D11	km	+	0.00495
	D12	sq. m	+	0.00298

Table 3. Evaluation index system for criteria B.

Criteria	Index	Unit	+/-	Weight
(B)	B1	yuan	+	0.00316
Employment	B2	%	+	0.00190
and Social	B3	%	+	0.00093
Security	B4	%	+	0.02500
(0.10379)	B5	%	+	0.03090
	B6	%	+	0.04190

Table 6. Evaluation index system for criteria E.

Criteria	Index	Unit	+/-	Weight
(E)	E1	10,000 ton	−	0.03690
Usage of	E2	10,000 kwh	−	0.02950
Energy and	E3	10,000 kwh	−	0.01080
Natural	E4	10,000 kwh	−	0.04000
Resources	E5	10,000 cubic m.	−	0.07460
(0.28390)	E6	ton	−	0.09210

Table 4. Evaluation index system for criteria C.

Criteria	Index	Unit	+/-	Weight
(C)	C1	%	+	0.02730
Scientific and	C2	%	+	0.00045
Educational	C3	unit	+	0.02650
Level and	C4	unit	+	0.00345
Support	C5	unit	+	0.00107
(0.12816)	C6	unit	+	0.00251
	C7	person	+	0.00660
	C8	person	+	0.00238
	C9	copy, piece	+	0.05790

Table 7. Evaluation index system for criteria F.

Criteria	Index	Unit	+/-	Weight
(F)	F1	hectare	+	0.02450
Environmental	F2	hectare	+	0.02670
Protection	F3	%	+	0.00058
(0.12250)	F4	ton	−	0.01420
	F5	ton	−	0.01180
	F6	ton	−	0.01170
	F7	ton	−	0.02380
	F8	%	+	0.00666
	F9	%	+	0.00008
	F10	%	+	0.00248

while Beijing has the lowest score (0.43869) in all the prefecture-level and above cities in the region.

From the comparative perspective, as shown in Table 9, the non-parametric statistical test results show that there exists no statistically significant difference in the sustainable development capacity scores between resource-based cities and non-resource-based cities. The results indicate that, compared with other cities in this region, even with Beijing, the resource-based cities in the Beijing–Tianjin–Hebei region have not lagged behind from the perspective of sustainable development capacity.

According to the study of Zhang et al. 2014, from the perspective of resource-based cities (Table 10), most of the these cities in the Beijing–Tianjin–Hebei region of China (Zhangjiakou, Handan, Chengde, and Xingtai) are classified as mature RBC cities, which indicates that these cities have higher capacity of resource support. Only Tangshan is classified as rebirth, which indicates that it has a relatively lower capacity of resource support. Furthermore, most of the resource-based cities (Zhangjiakou, Tangshan, Handan, and Xingtai) are classified as comprehensive resource types, which indicates that in these cities, more than

Table 8. Evaluation results of sustainable development capacity.

City	Score	Rank	City type
Beijing	0.43869	13	Non-RBC
Tianjin	0.63562	1	Non-RBC
Qinhuangdao	0.53577	2	Non-RBC
Langfang	0.53050	3	Non-RBC
Shijiazhuang	0.52345	4	Non-RBC
Baoding	0.52042	6	Non-RBC
Cangzhou	0.51971	8	Non-RBC
Hengshui	0.51664	11	Non-RBC
Zhangjiakou	0.52059	5	RBC
Tangshan	0.52034	7	RBC
Handan	0.51826	9	RBC
Chengde	0.51779	10	RBC
Xingtai	0.51212	12	RBC

Table 9. Comparison of RBC and non-RBC cities.

Statistics and types	Sustainable development capacity score
Mann-Whitney U	12.000
Wilcoxon W	27.000
Z	−1.171
Asymp. Sig. (2-tailed)	0.242
Exact Sig. [2*(1-tailed Sig.)]	0.284*

Note: *Not corrected for ties.

Table 10. Evaluation results of sustainable development capacity of resource-based cities.

City	Score	Classification*	Resource type*
Zhangjiakou	0.52059	Mature	Comprehensive
Tangshan	0.52034	Rebirth	Comprehensive
Handan	0.51826	Mature	Comprehensive
Chengde	0.51779	Mature	Ferrous metal
Xingtai	0.51212	Mature	Comprehensive

Note: *Based on the Study of Zhang et al. (2014).

one type of mining resource is being exploited; therefore, the development of resource-related industries in these cities is more diversified. However, there still exists more complicated ecological and environmental problems caused by multiple-resource exploitation in these cities. Only Chengde is classified as a ferrous metal city in resource type.

Among these five resource-based cities, it is clear that Zhangjiakou and Tangshan have better performance in sustainable development capacity than other resource-based cities in the region. This result indicates that the resource supporting condition and resource type play less important roles in the sustainable development of resource-based cities in the region. Therefore, with the consideration

of entropy weight results for resource-based cities in the Beijing–Tianjin–Hebei region, especially for Handan, Chengde, and Xingtai, more attention should be paid to the utilization of energy and natural resources and economic development in order to achieve sustainable development.

4 CONCLUSIONS AND POLITICAL IMPLICATIONS

Based on the data of prefecture-level and above cities in the Beijing–Tianjin–Hebei region of China in 2013, this paper makes the evaluation of sustainable development capacity of these cities using Entropy-Weight and TOPSIS evaluation model. Based on the evaluation results, this paper provides a comparative study between resource-based and non-resource-based cities, and further compares the difference in sustainable capacity within resource-based cities, so as to provide possible ways for the improvement of sustainable capacity for resource-based cities in the region. The results and conclusions are as follows:

1. Using the methods of Entropy-Weight and TOPSIS, this paper finds that the ranking based on the weights of criteria (from highest to lowest) is Usage of Energy and Natural Resources, Economic Development, Scientific and Educational Level & Support, Environmental Protection, Employment & Social Security. Further observations on the weights of indices show that natural energy utilization, foreign capital utilization, expenditure and public knowledge infrastructure, and financial income are the factors that need more focus in order to achieve sustainable development.
2. The top three cities that have the highest sustainable development capacity score are Tianjin, Qinhuangdao, and Langfang. Qinhuangdao is the top most city in Hebei province, while Beijing has the lowest score in all the prefecture-level and above cities in this region. From the comparative perspective, the results indicate that the resource-based cities in the Beijing–Tianjin–Hebei region have not lagged behind from the perspective of sustainable development capacity compared with other cities in this region, even with Beijing.
3. Among the five resource-based cities in the region, it is clear that Zhangjiakou and Tangshan have better performance in sustainable development capacity than other RBC cities in the region, which indicates that resource-supporting condition and resource type play less important roles in the sustainable development of resource-based cities in the region.

Therefore, more attention should be paid to the utilization of energy and natural resources and economic development in order to achieve sustainable development. Furthermore, the focus should be on aspects including natural energy utilization, foreign capital utilization, expenditure and public knowledge infrastructure, and financial income.

ACKNOWLEDGMENT

This paper was funded by the Project of Humanities and Social Science Research Foundation of China Ministry of Education (16YJCZH040).

REFERENCES

Hwang, C. L. & K. Yoon (1981). Multiple Attribute Decision Making Methods and Applications, Springer, Berlin Heidelberg.

Kinnear, S. & I. Ogden (2014). Planning the innovation agenda for sustainable development in resource regions: A central Queensland case study. *Resources Policy*, 39: 42–53.

Long, R. Y., H. Chen, H. J. Li, & et al. (2013). Selecting alternative industries for Chinese resource cities based on intra- and inter-regional comparative advantages. *Energy Policy*, 57: 82–88.

State Council of the People's Republic of China (2013). National planning on sustainable development of resource-based cities (2013–2020).

Su, S., J. Yu & J. Zhang (2010). Measurements study on sustainability of China's mining cities. *Expert systems with applications*, 37, 6028–6035.

Yu, J., S. Z. Yao, R. Q. Chen & et al. (2005). A quantitative integrated evaluation of sustainable development of mineral resources of a mining city: a case study of Huangshi, Eastern China. *Resources Policy*, 30, 7–19.

Yu, J., Z. J. Zhang & Y. F. Zhou (2008). The sustainability of China's major mining cities. *Resources Policy*, 33, 12–22.

Zhang, W. Z., J. H. Yu, D. Wang & L. Chen (2014). Study on sustainable development of resource-based cities in China. Beijing: Science Press.

Civil, Architecture and Environmental Engineering – Kao & Sung (Eds)
© 2017 Taylor & Francis Group, ISBN 978-1-138-02985-9

PM$_{2.5}$ open-source component spectrum analysis in the harbor area

Yu Lin
Tianjin Research Institute for Water Transport Engineering, M.O.T, Tianjin, China

ABSTRACT: PM$_{2.5}$ is a major pollutant in the harbor area. The adoption of chemical mass balance model for PM$_{2.5}$ source profiles is an important way to reduce particulate matter pollution and research on the composition of particulates is the basis of the chemical mass balance model. This study collected fugitive dust (yard dust, resuspended dust, road dust, soil dust, construction dust) samples in the Tianjin port area. First, PM$_{2.5}$ samples were collected with the resuspension sampler. Then, the PM$_{2.5}$ source profiles of fugitive dust were established, based on which analysis and research on the characteristics of the composition spectrum were done.

1 INTRODUCTION

Atmospheric particulate matter is the general term for all solid and liquid granular materials except gases in the atmosphere. Among them, particles with aerodynamic diameter less than 2.5 μm are known as fine Particulate Matter (PM$_{2.5}$). The surface of PM$_{2.5}$ will easily adsorb toxic heavy metals (Pb, Zn, Cu, Cd, As, etc.), harmful organic compounds, bacteria, and viruses. So it has an impact on ambient air quality, human health status, and climate change (Almeida et al., 2006, Almeida et al., 2005, Hwang and Hopke, 2007, Harrison and Yin, 2000). The research on PM$_{2.5}$ pollution sources can be divided into four types: open source, fixed coal source, motor vehicle exhaust, and biomass dust. The biggest difference between open-source dust and fixed source dust is that the former does not pass through the exhaust funnels, causing an irregular, unorganized discharge.

The pollution situation of PM$_{2.5}$ in the harbor area has its own characteristic, which is the diffusion of dust in the atmospheric environment as one of the pollution components in the port area. During the loading and unloading operation, materials falling on the road are repeatedly raised by traffic, and piled coals, metal ores, building materials and other materials in the bulk cargo storage yard are likely to produce more dust due to the flat ground and strong wind. Since Tianjin port is a typical harbor area, we first collected fugitive dust samples in the Tianjin port area. We then used the resuspension sampler to collect PM$_{2.5}$ samples, establishing PM$_{2.5}$ source profiles of fugitive dust and eventually discussing the environmental risk and health risk assessment of heavy metals in the fugitive dust.

2 COLLECTION AND ANALYSIS OF SAMPLES

Tianjin port, the largest comprehensive port in the Bohai Sea, as well as the largest comprehensive foreign trade port and an important logistics center in northern China functions as the main hub of the whole country. It is located in the Binhai New Area. Binhai New Area is a sub-provincial district of Tianjin City under the jurisdiction of the national district and national comprehensive reform pilot area with a resident population of 2.97 million. Therefore, Tianjin port area was chosen as a typical harbor area for the research in this paper.

The PM$_{2.5}$ open sources in the Tianjin port area mainly include yard dust, resuspended dust, road dust, soil dust, and construction dust. Due to the large open-source emission surface, low intensity, and strong interference from the surrounding environment, field sampling is often difficult to obtain. In order to match the particle size in the atmospheric environment and to simulate the real process of particles going into the ambient air, resuspension sampler was put into use while sampling. During the simulation, a sampler for resuspended particulates in the environment and pollution sources cut the powder samples into the size of PM$_{2.5}$. The selection of the volume sampler was in strict accordance with national standards and the environmental standard. Teflon filter membrane and quartz filter membrane were the two types that were adopted according to the need for chemical analysis after sampling and the membranes' own characteristics in the sampling filters for collection. The Teflon filter membrane prepares for analysis of elements while the quartz filter membrane for ion and carbon analysis.

3 ANALYSIS OF OPEN-SOURCE COMPONENT SPECTRUM

3.1 Composition of component spectrum

In this study, we collected the samples of coal yard dust; ore yard dust, resuspended dust, road dust, soil dust, and construction dust in Tianjin port area and established those component spectrums (Figs 1–6).

In order to facilitate comparison, the chemical composition of the component spectrums was divided into crustal elements, trace elements, ions, Organic Carbon (OC) and Elemental Carbon (EC). Crustal elements include Ti, Al, Si, Mn, Mg, Ca, Fe, K, and Na. Trace elements include V, Cr, Co, Ni, Cu, Zn, As, Cd, and Pb. Ions include NO_3^-, SO_4^{2-}, Cl^-, K^+, Na^+, Ca^{2+}, Mg^{2+}, and NH_4^+ (Fig. 7).

It can be seen from the figures that the composition of storage yard dust is highly relevant with coal yard dust materials. EC and OC content are high with the total carbon content reaching 36%. The Fe element content of dust in the ore yard is the highest reaching 37%. Construction dust with components such as cement and other building materials have Ca, Si, Fe, and other content higher. The crustal elements and carbon content in urban dust are higher, reaching 29% and 35% because the resuspended dust in the port area is influenced by the adjacent coal yard. Crustal elements are the most

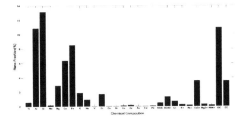

Figure 4. Chemical composition of resuspended dust.

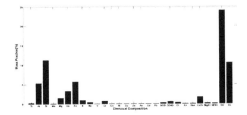

Figure 5. Chemical composition of road dust.

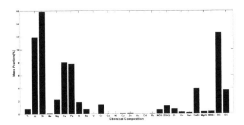

Figure 6. Chemical composition of soil dust.

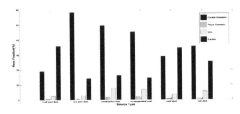

Figure 7. Chemical composition of fugitive dust.

abundant species in road dust and soil dust, which accounts for 45% and 49% of the chemical composition. The content of trace elements is the highest in road dust accounting for 2.11% of the total content followed by re-suspended dust accounting for 1.73% of the total content, which shows that road dust and resuspended dust are more susceptible to anthropogenic sources of pollution.

3.2 Reconstruction of crustal material composition

Open-source particle emission of inorganic chemistry groups divided into the spectrum of unknown components are mainly crustal elements existing in

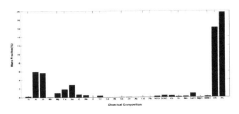

Figure 1. Chemical composition of coal yard dust.

Figure 2. Chemical composition of ore yard dust.

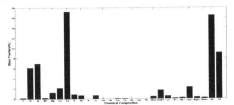

Figure 3. Chemical composition of construction dust.

the form of oxides, so Ho (Ho et al., 2003) rebuilt soil dust and road dust in the crust by Al, Si, Ca, Ti, and Fe mass percentage content and compared it with the total component content. The formula for calculating the percentage content of crustal materials is as follows:

$$\text{Crustal material} = 1.16\,(1.90\ Al + 2.15\ Si + 1.41\ Ca + 1.67\ Ti + 2.09\ Fe) \quad (1)$$

The 1.16 is to compensate for the mass of MgO, Na_2O, K_2O, and H_2O in the composition of the earth's crust that cannot be calculated. Derived from Formula 1, and expressed as the percentage content in $PM_{2.5}$ of urban dust, road dust, and soil dust was 76.47%, 59.67%, and 88.67%, respectively. The results show that the soil crustal elements occupy the highest proportion in the total component content of $PM_{2.5}$ in the dust area of Tianjin port indicating that soil dust is mainly composed of the crust. The source of urban dust and road dust is relatively complex, prone to be affected by vehicle exhaust dust, cement dust. and coal dust.

4 ENVIRONMENTAL RISK ASSESSMENT OF HEAVY METAL ELEMENTS

Heavy metal pollution of open-source dust is due to persistence, biodegradation, bioaccumulation and toxicity characteristics. Long-term heavy metals pose a great threat to the environment, so the evaluation of the extent of heavy metal pollution in open-source dust and the potential ecological risk is necessary. At present, there is no uniform standard for judging the pollution degree of heavy metals in China and abroad, such as the enrichment factor analysis method, the accumulation index method, the potential ecological risk assessment, and so on.

The Enrichment Factor (EF) is an important method to identify and evaluate elements of natural sources and human sources and is an important indicator to evaluate the degree of element enrichment. The method was first proposed to study the atmospheric particulate matter in the Antarctic and is now widely used to study the sources of heavy metals in atmospheric particulate matter (Banerjee, 2003). The calculation formula is as follows:

$$EF = \frac{C_n/C_{ref}}{B_n/B_{ref}}$$

where C_n measures the value of element n in open-source samples, C_{ref} is the reference element value in open-source samples, B_n is the value of element n in the reference system, and B_{ref} is the reference element value in the reference system.

The selection of a reference element is considered necessary in the analysis of enrichment factors. Different reference elements lead to different calculation results of the enrichment factors, thus the information of pollutants coming from the analysis will vary. Due to the above-mentioned statement, when using enrichment factors to evaluate the elementary pollution status in soil dust, it is crucial to choose variable reference elements according to the actual situation. Those with more stability in the soil, less mobility, and manufactured pollution will make comparatively better options. In China, Cr and Mn are typical pollution elements in the soil. Ca, whose chemical character is active, is the characteristic element in the construction dust, receiving more impact from urban construction and the concrete industry. Construction dust contributes considerably to the rising rate of particles. Highly influenced by manufactured sources, the geo-accumulation index (I_{geo}) of Na tends to be greater in particular areas. This research has picked Al which is that of a steady chemical property, the highest analysis with respect to accuracy and the most widespread use as the reference element.

The background values of heavy metal elements in Tianjin are shown in Table 2.

To calculate the concentration factors of heavy metals in the urban source, soil source, and road source in Tianjin port area refer to Figs 8–10.

It can be seen from Figures 8–10 that the enrichment factors of Mn, V, Co, Ni, Zn, As, and Pb are all less than 1, showing no sign of enrichment. The factor of Cu is between 1 and 2 indicating slight enrichment. The factor of Cd is between 2 and 5 indicating moderate enrichment. The factor of Cr is between 5 and 20, a noteworthy enrichment.

Table 1. Classification criteria of EF values.

Pollution level	EF value	Degree of enrichment
I	$EF \leq 1$	No enrichment
	$1 \leq EF \leq 2$	Slight enrichment
II	$2 \leq EF \leq 5$	Moderate concentration
III	$5 \leq EF \leq 20$	Significant enrichment
IV	$20 \leq EF \leq 40$	Strong enrichment
V	$EF \geq 40$	Very strong enrichment

Table 2. Soil background values of heavy metals in Tianjin (mg.kg⁻¹).

Heavy metals	Cr	Mn	V	Co	Ni
Background value	84.2	686	85.2	13.6	33.3
Heavy metals	Cu	Zn	As	Cd	Pb
Background value	28.8	79.3	9.6	0.09	21

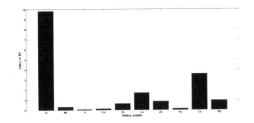

Figure 8. Enrichment factors of heavy metals in re-suspended dust of the Tianjin port area.

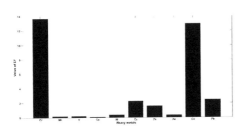

Figure 9. Enrichment factors of heavy metals in the soil dust of Tianjin.

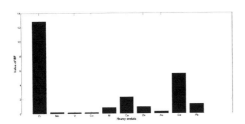

Figure 10. Enrichment factors of heavy metals in the road dust of Tianjin.

In soil dust, the enrichment factors of Mn, V, Co, Ni, and As are all less than 1 suggesting that there is no enrichment. Zn has an enrichment factor of a value between 1 and 2 indicating slight enrichment. A moderate enrichment taking place, Cu and Pb have factors valuing between 2 and 5, while Cr and Cd factors are between 5 and 20, appearing to be a significant enrichment phenomenon.

In road dust, the enrichment factors of Mn, V, Co, Ni, Zn, and As are all less than 1 suggesting that there is no enrichment. Pb has an enrichment factor of a value between 1 and 2 indicating slight enrichment. Moderate enrichment takes place with Cu having factors valuing between 2 and 5, while Cr and Cd factors are between 5 and 20 appearing to be a conspicuous enrichment phenomenon.

The element that has the highest amount of enrichment factor in these three types of dust is Cr, indicating a more than significant enrichment. Cu, Pb, and Cr manifest moderate enrichment

while receiving the impact from manufactured sources and natural sources as well. The rest of the elements due to their enrichment factors are all less than 2, mainly coming from crust materials.

5 CONCLUSIONS

The conclusion derived from the research on chemical characteristics of the open-source area in Tianjin port is that the yard dust components are related to storage materials, building materials, and components of cement and other relative construction dust, which is acquired. The earth's crust elements and carbon content in the urban dust are higher because the urban dust in the port area is affected by the adjacent coal yard. Crustal elements account for the highest proportion of road dust and soil dust and the highest percentage of trace elements in road dust followed by urban dust show that road and urban dust were more susceptible to anthropogenic sources of pollution. Through the reconstruction of the crustal materials, crustal elements in the total composition of soil dust $PM_{2.5}$ have the highest proportion indicating that soil dust is mainly composed of the crust. Urban dust and road dust source are complex, prone to be influenced by vehicle exhaust dust, cement dust, and coal dust.

ACKNOWLEDGEMENTS

This work was supported by the Central Level of Scientific Research Institutes of Public Welfare Funds for Special Projects (tks150104).

REFERENCES

Almeida, S.M., Pio, C.A., Freitas, M.C., Reis, M.A. & Trancoso, M.A. 2005. Source apportionment of fine and coarse particulate matter in a sub-urban area at the Western European Coast. Atmospheric Environment, 39, 3127–3138.

Almeida, S.M., Pio, C.A., Freitas, M.C., Reis, M.A. & Trancoso, M.A. 2006. Approaching $PM_{2.5}$ and $PM_{2.5-10}$ source apportionment by mass balance analysis, principal component analysis and particle size distribution. Science of the Total Environment, 368, 663–674.

Harrison, R.M. & Yin, J. 2000. Particulate matter in the atmosphere: which particle properties are important for its effects on health? Science of the Total Environment, 249, 85–101.

Ho, K.F., Lee, S.C., Chow, J.C. & Watson, J.G. 2003. Characterization of PM_{10} and $PM_{2.5}$ source profiles for fugitive dust in Hong Kong. Atmospheric Environment, 37, 1023–1032.

Hwang, I.J. & Hopke, P.K. 2007. Estimation of source apportionment and potential source locations of $PM_{2.5}$ at a west coastal Improve site. Atmospheric Environment, 41, 506–518.

Civil, Architecture and Environmental Engineering – Kao & Sung (Eds)
© 2017 Taylor & Francis Group, ISBN 978-1-138-02985-9

Distribution of oil spill risk on offshore facilities in the Bohai area based on ETA

Yunbin Li
School of Navigation, Wuhan University of Technology, Wuhan, China

Jingxian Liu
Hubei Key Laboratory of Inland Shipping Technology and School of Navigation, Wuhan University of Technology, Wuhan, China

Weihuang Wu
School of Navigation, Wuhan University of Technology, Wuhan, China

ABSTRACT: This paper proposes to quantitatively analyze the oil spill risk of the offshore facilities in the Bohai area. In particular, the oil spill risk probability of different offshore facilities was calculated by using the Event Tree Analysis (ETA) method. This method was presented based on the comprehensive analysis of the distribution of offshore oil facilities in the Bohai area and the oil spill accidents. To calculate the spatially variant oil spill risk probability, the Bohai area was divided into different grids and the oil spill risk probability was calculated based on the geometrical information related to the grids. Finally, the distribution of oil spill risk in the Bohai area was obtained to provide theoretical support for the oil spill warning in the Bohai area's offshore facilities.

1 INTRODUCTION

Bohai is China's inland sea surrounded on three sides by land, and only connects to the Yellow sea through the Strait of Bohai in the East and has a poor self-purification capacity. Bohai is a pioneer in operating oil and gas developments in the sea. Fixed production platforms account for similar platforms of more than 90% of the total, which include the earliest exploited work and the largest number of drills. The risk of marine oil spill

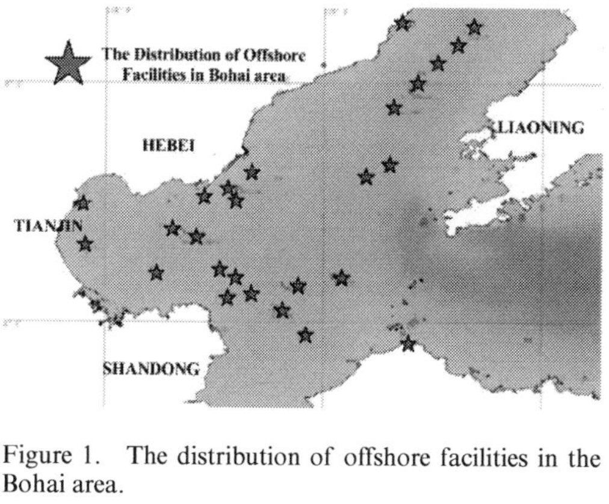

Figure 1. The distribution of offshore facilities in the Bohai area.

pollution is increasing because of offshore oil and gas operations, storage, and transportation. At the same time, along with the Bohai rim region and its hinterland's rapid economic growth, maritime traffic density is increasing with the occurrence of traffic accidents among ships, offshore facilities are increasing in the Bohai area and frequency of collision accidents are high. Therefore, the research of distribution of oil spill risk has the significance of oil spill warning and control for the Bohai area's offshore oil facilities.

2 STATISTICAL ANALYSIS OF OIL SPILL ACCIDENTS IN THE BOHAI AREA

From 2002 to 2016, there are 27 oil spill accidents from offshore facilities that have occurred in the Bohai area, which include 10 oil spills of submarine pipeline accidents and 16 crude oil leak accidents. The highest frequency is between 2011 and 2012, which is 5 times each year. In the 27 oil spill accidents, the cumulative booms (gas and solid) and suction drag bar used were 11,380 meters and 28,700 meters, respectively. Oil spill source for submarine pipeline accidents accounted for 37% and crude oil spill accounted for 59%.

3 THE OIL SPILL RISK OF OFFSHORE FACILITIES IN BOHAI

3.1 *Risk source analysis of offshore facilities*

The oil spill sources of offshore facilities in the Bohai area mainly include:

1. The oil spill of offshore facilities
Offshore oil facilities refer to the facilities engaged in offshore oil and gas exploration, development and production operations (including drilling ship) of the mobile offshore drilling rig, fixed offshore platform, single point mooring floating production, storage devices, submarine pipeline, etc. At present, there are 128 offshore oil platforms, 6 Floating Production Storage and Offloading (FPSO) vessels, 177 oil and gas pipelines (1653 km) and 32 oil tanks in the Bohai area. According to development planning, 205 oil platforms will be built in the Bohai area from 2013 to 2030. The average annual number of drilling wells is about 70 and the total export drilling adjustment wells will be 423. At that time, oil production will exceed 40 million squares every year and the new laying oil and gas pipeline will be 1890 km.

The damage to offshore facilities, poor seal, fire and explosion, operational errors, blowout, and so on will lead to oil spill accidents. Offshore facilities are the main risk sources of oil spill accidents.

2. The operations of offshore facilities
Offshore facilities operations include oil exploration, mining, oil transportation, oil refining, and other related products. The oil operation spill refers to the spill that occurs in the process of oil operation and the main reasons include the company's poor environmental awareness, careless thinking, nonstandard operations, human factors, etc.

3.2 *The analysis method of oil spill risk*

Event tree analysis (referred to as ETA) originated in the decision tree analysis, which is a kind of possible consequence derived from the initial event by time sequence, meaning it is a logical deduction method. When an early event was provided, analysis of events may lead to the result of the sequence of events, and the characteristics of qualitative and quantitative evaluation system, and help the analyst to arrive at the right decision. It is often used as a security system for accident analysis and system reliability analysis.

As the sequence of events is a graphical representation and looks like a fan, so it is named as event tree. Event tree analysis method has the advantage of being strong and flexible. The accident probability model based on ETA can significantly improve the scientific and effective probability calculation of oil spill accidents.

The steps of ETA are as follows:

1. Determining or looking for reasons which may lead to serious consequences of events and classifying them. For those which may lead to the same event tree for different events can be divided into the same class.
2. When we build the event tree, the function event tree is built first followed by the system event tree.
3. Simplify the event tree.
4. Make a quantitative calculation of events.

A branch of the event tree is shown in Figure 2.

From the process shown in the event tree, the probability of event F2 can be expressed as:

$$\begin{aligned} P(F_2) &= P(IS_1F_2) + P(IF_1F_2)P(IS_1F_2) \\ &= P(I) \cdot P(S_1) \cdot P(F_2) \\ P(IF_1F_2) &= P(I) \cdot P(F_1) \cdot P(F_2) \end{aligned} \tag{1}$$

As the reason of oil spill accidents occurring in offshore oil platforms is relatively simple, it is suitable to use the ETA) to analyze the oil spill risk as shown in Figure 3. The paper will calculate the probability of oil spill accident on the offshore oil platform by constructing the event tree.

According to the planned number of drilling developments in each area of our country, the average annual number of drilling wells is about 70 and the total export drilling adjustment wells will be 423 in the Bohai area from 2016 to 2030. However, the average annual number of drilling wells and adjustment wells is 70 and 90, respectively in the eastern and western part of the South China Sea from 2016 to 2018.

Thus, the average annual total of oil drilling platforms built is far higher in the Bohai area than other areas of our country, and the oil spill risk is more prominent. The annual number of drilling planned in other areas only accounts for 8% to 18% of the Bohai area, and as such the oil spill risk probability is reduced accordingly.

The oil spill probability of wells was calculated based on the ETA method in the Bohai area, which

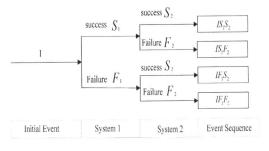

Figure 2.   Bifurcation diagram of ETA.

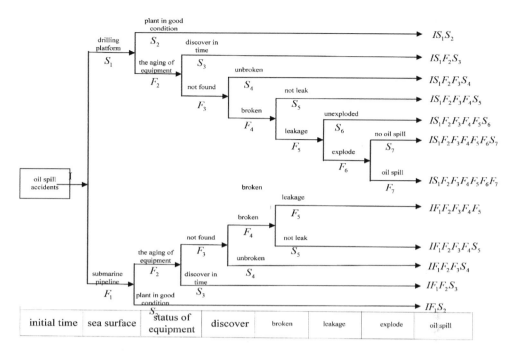

Figure 3. ETA of oil spill from a maritime structure.

Table 1. The oil spill probability of wells in the Bohai area.

Risk	1 ton	10 tons	100 tons	1000 tons
Oil spill probability of exploited well	0.014	0.007	0.004	0.002
Oil spill probability of development well	0.176	0.088	0.044	0.022
Oil spill probability of oil-well blowing	0.152	0.076	0.038	0.019

Table 2. The oil spill probability of submarine pipelines in the Bohai area.

Risk	1 ton	10 tons	100 tons
Oil spill probability of submarine pipelines	0.031	0.011	0.003

is shown in Table 1. The oil spill probability of submarine pipelines in the Bohai area is shown in Table 2.

According to the construction plan of submarine pipelines in the Bohai area, by 2030, the newly laid submarine pipelines will amount to 1890 km and accumulated massive oil spill probability will reach more than 0.001. The oil spill probability of FPSOs in the Bohai area is shown in Table 3.

Table 3. The oil spill probability of FPSOs in the Bohai area.

Risk	Oil spill probability			
	1 ton	10 tons	100 tons	1000 tons
Oil spill probability of FPSOs	0.050	0.015	0.005	0.001

* The large-scale oil spill risk probability of FPSOs in the Bohai area is 0.001.

4 THE DISTRIBUTION OF OIL SPILL RISK ON OFFSHORE FACILITIES IN THE BOHAI AREA

4.1 Risk superposition of oil spill on offshore facilities

According to the analysis of offshore facilities' characteristics and operation mode, the risk superposition of oil spill should be considered when calculating the probability of oil spill. The following are situations where the risk superposition should be considered:

1. Offshore oil facilities are combined together;
2. More than one submarine pipeline has the same route;
3. Production platform is combined with exploitation platform by trestle bridge;

Table 4. Oil spill probability of eight offshore facilities in the Bohai area.

Position of oil spill	BZ25-1	BZ34-2/4	CFD11	JZ9-3	LD	PL19-3	QHD32-6	YingBei
Rate of oil spill accidents	0.461	0.406	0.296	0.339	0.435	0.496	0.376	0.298

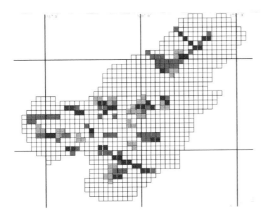

Figure 4. The distribution of oil spill risk grades.

Table 5. Corresponding relations between oil spill risk level and colors.

Colors	Red	Pink	Blue	Green	White
Oil spill risk level	High	Less high	Medium	Less low	Low

4. Production operation and exploitation operation are operated, respectively;
5. The design output of newly built offshore oil production facilities are more than 500 tons (3750 barrels) of crude oil within 10 nautical miles radius of target area in Bohai;
6. The newly built submarine pipeline's length is more than 40 km with a diameter of 800 mm length of target area in Bohai;

According to the above method, eight offshore facilities, Tianjin port, Dalian port, and onshore terminal in the Bohai area were dispersed and all the risks of oil spill sources (blowout, production oil spill, FPSO, etc.) was superimposed. After the calculation, the oil spill probability of dispersed areas in Bohai was obtained, which is as shown in Table 4.

4.2 The distribution of oil spill risk

Bohai area was divided into 905 6′ × 6′ squares according to the latitude and longitude. Considering the characteristics of oil spill accidents in the Bohai area, combined with the statistical data of oil spill accidents of platforms (drilling spill, development wells spill, oil blowout and production spill), submarine pipelines' spill, FPSOs' spill and terminals' spill, the risk level of different areas was provided based on assessment data of risk probability. Finally, different risk levels were plotted on the electronic chart with different colors. It was included in the oil spill risk grade map in the Bohai area, which is as shown in Figure 4. Corresponding relations between oil spill risk level and colors is shown in Table 5.

5 CONCLUSION

In this paper, the oil spill risk of offshore facilities in the Bohai area was identified by using the ETA method, which concluded the oil spill risk probability based on the statistical analysis of oil spill accidents over the years. According to the distribution characteristics of the Bohai offshore oil facilities, The Bohai area was dispersed into different grids for oil spill risk source and the distribution of oil spill risk grade was provided at last, which can lay the foundation for oil spill risk early warning, prevention, and control in the Bohai area.

ACKNOWLEDGEMENTS

This research was supported by the Chinese National Foundation under Grant No. 51179147 and the Ministry of Transport Construction Projects of Science and Technology (2015318 J34090).

REFERENCES

Ge Wang, John Spencer, Yongjun Chen. 2002. Assessment of a ship's performance in accidents. Marine Structures, 15:313–333.

IMO. Guidelines for Formal Safety Assessment (FSA) for use in the IMO rule-making process [S]. London: Author, 2002.

National Academy of Science (NAS), Risk assessment in federal government: managing the Process, National academy Press, Washington DC, 1983.

Tony Rosqvist, Risto Tuominen. 2004. Qualification of Formal Safety Assessment: an exploratory study, Safety Science, 42:99–120.

Civil, Architecture and Environmental Engineering – Kao & Sung (Eds)
© *2017 Taylor & Francis Group, ISBN 978-1-138-02985-9*

Developing a low-carbon economy and constructing a low-carbon city—a Xuzhou perspective

W.Y. Zhang & C.L. Zhou
School of Environment Science and Spatial Informatics, CUMT, Xuzhou, Jiangsu, China

F. Chen & J. Ma
Low Carbon and Energy Institute, CUMT, Xuzhou, Jiangsu, China

ABSTRACT: Developing a low-carbon economy and constructing a low-carbon city has become the trend in the world. China started a national low-carbon city pilot project in 2010. Xuzhou city is a mineral resource city that has a long history of coal mining. Its heavy machinery industry is also well-developed. However, Xuzhou faces several problems such as coal resource exhaustion and mining subsidence. Due to extensive energy consumption, high emissions, as well as other problems of traditional pillar industries, the low-carbon city construction of Xuzhou was hindered. Considering the development situation of the current economy and society and the restrictive factors of low-carbon economic development, this paper provides some suggestions for low-carbon city construction planning: (1) perfecting the law and policy system for low carbon economy; (2) transforming coal mining abandoned land and subsidence land, as well as developing characteristic tourism; (3) accelerating technological innovation and promoting industrial structure adjustment.

1 INTRODUCTION

Economic recovery and population explosion have gradually sharpened the contradiction between environment and development since the Second World War. Scholars began to think about the relationship between humans and nature at the same time. Ecological issues also became more and more important. As the global climate problem draws more international attention, the problem of carbon dioxide emission has become an important issue (Yang 2011). Since the United Nations put forward the concept of "Sustainable Development" in 1992, various countries have started paying more attention to the ecological environment and climate change.

Climate change originated from the interaction between nature and humans. Human beings have seriously affected the natural process through endless developmental activities, and thus the climate and natural environment have changed. In a word, climate change issue is a continuously developing problem. Therefore, if we want to improve the climate, we should change the way how humans survive and develop.

Several scholars have proposed theories that combined environmental protection with economic development. Based on the above-mentioned knowledge, we can understand these theories. In the 1960s,

American economist Kenneth Balding proposed "circular economy" on the basis of his "spaceship economy" theory. He also put forward the concept of ecological economy for the first time in 1966. The UK has proposed "energy white paper" in 2003 and put forward the concept of "low-carbon economy" (Xue 2011). These three concepts came into being in the same era, but had different emphasis as follows: the core of ecological economy was the sustainable development of economy and nature; circular economy emphasis on resource recycling and low carbon economy underlined low carbon emissions and improved the climate.

In July 2010, NDRC (National Development and Reform Commission) released "The notice on carrying out the pilot work of low carbon provinces and low carbon cities". China started the low carbon city pilot project. According to the local declaration situation, working foundation, and pilot representative information, the government has decided to carry out the pilot work in Guangdong, Liaoning, Hubei, Shanxi, Yunnan provinces and Tianjin, Chongqing, Shenzhen, Xiamen, Hangzhou, Nanchang, Guiyang, and Baoding cities first. Xuzhou won the national first batch of "the most competitive low carbon city" awards in 2010 which was unique in Jiangsu. All these conditions showed that constructing a low carbon city in Xuzhou was feasible and beneficial.

2 LOW CARBON ECONOMY AND LOW CARBON CITY

2.1 *The definition of low carbon economy and low carbon city*

Under the effect of the concept of sustainable development, low-carbon economy is becoming a new economic forum with characteristics of "low emissions, low energy consumption, low pollution" on energy consumption. Technical innovation, institutional innovation, industrial transformation, new energy development, and low carbon economy are designed to realize the human economic development and ecological environment protection (Xu 2009).

The low-carbon city is a city which was used as a unit to implement low-carbon economy into practice. Its purpose was to realize the unification of ecological benefit, economic benefit, and social benefit within the nature-human-social economy composite system. Also, gradually reducing carbon source and increasing carbon sinks in the city's economic development process is one of the goals. Meanwhile, city industries have transformed from the traditional high carbon emission industry to low carbon emission industry, as well as the urban consumption transforming from high carbon consumption to low carbon (Fang 2009).

2.2 *Low carbon economy development—present situation at home and abroad*

In 2003, the British government issued a report: "Our Energy Future: Creating a Low Carbon economy", which proposed the concept of a low carbon economy for the first time. Once put forward, the concept has received wide attention.

China released the "National Assessment Report on Climate Change" in 2006 and "China's National Program to Address Climate Change" in 2007. The USA released "The Low Carbon Economy Act". Then the UN climate change conference formulated the "Bali roadmap". These events marked the efforts of low carbon economy development from all over the world. In December 2009, the World Climate Conference convened in Copenhagen, which became the symbol of world comprehensive transformation into the low carbon era. Countries such as France, Japan, and Canada have taken corresponding policies and measures. Since January 2005, using renewable energy equipment has gotten tax breaks from 40% to 50% in France. With support from the Japanese government, a lot of low carbon technologies in Japan have led the world, such as green residential technology which could slash energy consumption by comprehensive utilization of solar energy and heat insulation material, wastewater treatment technology and plastic recycling technology (Xu 2009). The transition to low carbon economy has become the major trend of world economic development.

2.3 *Low-carbon city development in China*

China has been making positive efforts in the construction of a low carbon city.

1. Our country has been gradually formulating and maintaining laws and policies for energy conservation and emission reduction. China has formulated a series of laws such as "Law of Cleaning Product Development", "Water Pollution Prevention and Control Law of China", etc. These laws and regulations have played an important role in guidance and protection for various industries in the process of low carbon economic development. Some scholars have established an energy conservation and emissions reduction database which includes 1195 policies issued from 1978 to 2013. (Zhang 2014).
2. China is actively developing new energy and renewable energy actively. In recent years, China has positively developed hydroelectric, solar energy, nuclear energy, wind energy, and other sources of clean energy. In 2012, this clean energy accounted for 8.3% of the total energy consumption. In terms of the scope of application and technical level, there is still a large gap between China and the developed countries, but the momentum of application of new energy is good.
3. The country has been actively conducting forest plantation actively to increase carbon sinks.

In addition, the Chinese Society for Urban Studies discussed the implementation of China's low-carbon eco-city development steps in 2009, distributed in 44 years from 2007 to 2050 into three stages. China would realize the "low-carbon city" dream step-by-step with this implementation.

3 OPPORTUNITIES AND CHALLENGES OF LOW-CARBON CITY DEVELOPMENT IN XUZHOU

3.1 *Xuzhou overview*

Xuzhou, named as "Pengcheng" in ancient times is located in the northwest of Jiangsu province where the Longhai and Jinghu railways meet. The Beijing-Hangzhou Grand Canal runs through Xuzhou. Xuzhou is also the center of Huaihai economic zone, which has the reputation of being called "the city of engineering and machinery". The terrain is dominated by plains. It has warm temperate zone with a half moist monsoon climatic region with four distinct seasons.

Xuzhou has rich resources and the combination condition is superior. It has large coal reserves, multiple layers, thick coal seam and high quality. It is also an important coal-producing area and the power base of east China (Qu 2010). According to GDP size sorting, the Chinese Academy of Sciences Sustainable Development Strategy Research Group has selected 50 representative cities and classified them based on the differences of urban functions. Xuzhou was categorized as an industrial city (Liu 2013). After years of efforts on industry structure adjustment, the third industry growth rate was more than the rate of the second industry in the year 2014, whereas the proportion of output value was almost unchanged (Hao 2014). In 2015, the GDP of Xuzhou region had achieved 532 billion RMB. The third industrial production structure adjusted to 9.5:44.3, 46.2, and exceeded the secondary industry for the first time.

3.2 *Restraint of low-carbon economy development*

The low carbon city development in Xuzhou was restricted by some factors which were as shown below:

1. The industrial structure biased towards heavy industry and evolved slowly (Liu 2013). Since 1980, the industrial structure of Xuzhou has been adjusted twice. The primary industry has changed from first to second and then dropped to the third position. Up to 2014, the second industry was still in the leading position. The number of heavy industry enterprises was dominant and its comprehensive energy consumption level was high (Li 2011a). In 2015, Xuzhou's tertiary industry exceeded the second industry for the first time and low carbon city construction achieved initial success.
2. The coal industry has strong negative externality and high exit cost. Mining activities have caused significant damage to the ecological and geological environments. Also, strong public negative externality and inter-generational negative externality have been provoked. Xuzhou had more than 100 years of mining history and gradually appeared to show the case of shallow coal resources as reserves fell sharply, and deep mining was too difficult. As a resource-based city, Xuzhou has also established a large number of supporting facilities and mature industry chains in the process of coal development. Once the coal industry development was forced to stop, the corresponding industry chain would be interrupted and atrophied, resulting in a large number of workers being unemployed and equipment getting shut down, which would affect local economic development seriously.
3. Xuzhou was backward in science and technology and economic development emerged in apparent conflict. Carbon productivity was determined by the technical level, so technology was one of the decisive factors for success in being a low-carbon economy city. Compared with the south of Jiangsu, Xuzhou's science and technology and economic development level was backward, which hindered the low carbon city construction and development. Under the guidance of the economic development goal, stimulating economic growth and promoting consumption were necessary. However, low carbon economy required the reduction of continuous resource consumption and reducing expenditures. There was a significant contradiction between stimulating material consumption and development of "low-carbon economy". If Xuzhou wanted to achieve these two goals at the same time, the government should reach a higher level of economic development at a relatively low energy consumption level. This also meant that Xuzhou needed to assume the economic development requirements of the third-tier city and environmental protection requirements of the first-tier city at the same time. Challenge is huge.

3.3 *Present measures of existing low carbon city development in Xuzhou*

Xuzhou has implemented a series of low carbon measures. The first one was promoting the low-carbon concept actively, such as publishing the low carbon policy, slogans in newspapers, Web sites, billboards, etc. The second was constructing a low-carbon living environment. The traditional architectural crafts were encouraged to be low carbonization, such as reclaimed water cycling, charging low-emission car owners for lower parking fees, increasing community green rate, etc. The third was to promote industrial low carbon transformation. Ecological agriculture, modern service industry, and the energetic photovoltaic industry were developed, while the development of emerging industries was promoted and the industrial structure was adjusted simultaneously. On the other hand, traditional enterprises were rebuilt and transferred. The main measures included: (1) high energy consumption and high pollution enterprises were rectified or banned; (2) government would give some preferential policies and funding support to low carbon transformation of traditional enterprises; (3) high technology was developed to support the transition of traditional pillar industries. At present, there have been three new industrial parks of solar photovoltaic, new energy equipment manufacturing and clean production in the Xuzhou economic development zone, which is remarkable.

4 XUZHOU LOW-CARBON CITY CONSTRUCTION PLANNING

Even with hard restrictions of the above factors, development of a low-carbon city in Xuzhou still had feasibilities because the low carbon transition of Xuzhou was not only a challenge but also an opportunity. The 100-year-old coal-mining city has been experiencing the bottleneck in development because of the gradual exhaustion of coal resources. But the government's support for low carbon city constructions has also brought rising high-tech industries and the construction of livable environment, which led to correct consumption patterns in the people and economic recovery. Xuzhou's tertiary industry output value exceeded the achievement of the second industry in 2015 for the first time, which inspired people and let them see the possibility of a low carbon city being built.

The low carbon city construction planning of Xuzhou was not perfect, and at present, there were still some problems worthy of thinking deeply in Xuzhou's future low carbon road. Many scholars have studied the low carbon economy development of Xuzhou and provided many suggestions (Yang 2010, Tao 2011, Chang 2012, Li 2012, Zhang 2011). Combined with scholars mentioned above, this paper proposed our own advice as follows:

1. Developing and completing the law and policy for low carbon economy. For the construction of a low-carbon city, China has established laws and policy systems preliminarily. The government of Xuzhou city should adjust the measures according to local conditions and set policies according to local practice for perfection. In the implementation process, the government should pay more attention to quality and avoid getting hot-headed. In this way, good results might be obtained.
2. Transforming mining abandoned land and developing characteristic tourism. Reforming abandoned mines and mining subsidence land was necessary. Safe and well-preserved abandoned mines might be transformed into museums and mining subsidence land can be converted into ecological wetland parks after field investigation and feasibility analysis. There is already a built case named "Pan'an Lake wetland park".
3. Speeding up scientific and technological innovation and promoting industrial structure adjustment. As mentioned above, Xuzhou is an industrial city. Heavy industries had a high proportion consuming intensive energy and producing heavy pollution with high exit cost. On one hand, the government should strengthen scientific and technological innovation, support the transformation and upgrading of traditional pillar industries with science and technology, and perfect pollution treatment. On the other hand, the government should develop new energy sources and train talent for new energy development comprehensively. The coal industry chain and services established during the development of the old coal industry should be applied for the second time so that the coal industry exit cost might be reduced.

5 CONCLUSION

Economic recovery and population explosion have stimulated problems between social situation and environment since the Second World War. Various governments, institutions, and scholars have gradually dabbled in the theory and practice research of the "low-carbon economy" and "low-carbon city". Xuzhou has been making efforts.

According to the social and economic development situation of Xuzhou, the restriction factors of low carbon city construction have been summarized. There were three main factors: (1) the emphasis is biased towards heavy industry and the industrial structure evolution advanced slowly. (2) Coal industry had strong negative externality and high cost to quit. (3)The science and technology level of Xuzhou was backward and economic development showed apparent conflict.

Combined with other research, the development countermeasures have been proposed in this paper: (1) Developing and completing the law and policy for low carbon economy. (2) Transforming mining abandoned the land and developing the characteristic of the tourism business. (3) Speeding up scientific and technological innovation and promoting adjustment of industrial structure.

REFERENCES

Chang, J., Z. Jin Cui & L. Ya Bo (2012). Research on low-Carbon industry development of Xuzhou economic developing area. *J. Modern Urban Research.*4, 87–91.

Fang, D.C. & Z. Ming Xing (2011). The theoretical foundation and economic value of the low carbon economy. *J. China Population, Resources and Environment.* 7, 91–95.

Hao, J.Y. & S. Jing Zhi (2014). Thinking of Xuzhou's industry transformation under the Low carbon economy background. *J. Journal of Xuzhou Institute of Technology (Social Science Edition)*.5, 49–54.

Li, P. (2011). How far away from the low-carbon city of Xuzhou. *J. Commercial Research*.5, 210–216.

Li, Y.B. & C. Jiang (2012). A Study on Low-carbon Industrial Space Layout Optimization of Xuzhou Economic and Technological Development Zone. *J. Northern Economy and Trade*.4, 32–33.

Liu, N.N., S. Yan Dong & J. Xiao Bin (2013). Low carbon economy development promotes industry transformation and upgrading of Xuzhou. *J. Territory & Natural Resources Study*.1, 7–9.

Qu, J.F., Z. Shao Liang & L. Gang (2010). Strategy of rejuvenating Xuzhou's old industrial base Based on the perspective of low carbon economy. *J. HuaiHai Digest*.19, 15–18.

Tao, F. & W. Xiu Lin (2011). Low-carbon industrial structure path optimization of the Xuzhou economic and technological development zone based on industrial cluster. *J. Science and Technology Management Research*.19, 76–79.

Xu, D.Q. (2009). The practice and referential experience of developed countries to develop low carbon economy. *J. World Economics and Politics*.6, 112–116.

Xue, W.Z. (2011). The analysis of the relationship between low carbon economy, ecological economy, the circular economy and green economy. *J. Significant Attention*.2, 50–52.

Yang, W.L., N. Fu Tang & Z. Yuan (2010). Developing Low carbon economy to achieve the leap development of Xuzhou. *J. Guide of Sci-tech Magazine*.29, 28–29.

Yang, Y.X. (2011). Discrimination of ecological economy, circular economy, green economy and low-carbon economy. *J. Forward Position*.8, 94–97.

Zhang, G.X., G. XiuLin & W. Ying Luo (2014). Management Review, The measure, synergy and evolution of energy conservation and emissions reduction policy in China—based on data study of policies in 1978–2013. *J. China Population, Resources and Environment*.12, 62–73.

Zhang, L.N. & W. Xiu Lin (2011). Low carbon industrial structure optimization path selection of Xuzhou economic development zone. *J. Science and Management*.4, 77–80.

Civil, Architecture and Environmental Engineering – Kao & Sung (Eds)
© 2017 Taylor & Francis Group, ISBN 978-1-138-02985-9

Geophysical and geochemical characteristics and mineralization potential in the Taonatu area, Inner Mongolia

Q. Liu & R.M. Peng
School of Earth Sciences and Resources, China University of Geosciences, Beijing, China

ABSTRACT: The Taonatu area located in the western part of the north margin of the North China Craton in Inner Mongolia has good ore-forming potential. In this paper, a comprehensive prospecting model was applied based on geological, geophysical and geochemical characteristics. By using the 1:50,000 high-precision magnetic survey, we found that the magnetic anomaly is associated with the Agulugou formation and Zenglongchang formation in the Taonatu area, especially with the Agulugou formation. By using the 1:50,000 stream sediment geochemical survey in this area, anomalies of Fe, Zn, Au, Ag, and other polymetallic elements are extracted. Based on the geological study, this paper successfully applied a high-precision magnetic prospecting and stream sediment geochemical survey to explore the mineralization potential in the Taonatu area. It is believed that this area is a potential area for finding polymetallic deposits.

1 INTRODUCTION

With the increasing difficulty of prospecting, it is becoming more and more important to study different kinds of comprehensive prospecting models and interpretation methods for geophysical and geochemical exploration (Nabighian, et al. 2005 & Guan, et al. 2002). The high-precision magnetic survey has been used widely to find hidden magnetic metal deposits and also used in the regional geological survey, and oil and gas exploration. It has got a lot of valuable results. In this paper, combining geological exploration information extracted from the stratum, magmatic rocks, and various mineralized signatures and the relationship between geochemical anomaly and mineralization we reduced the prospecting target area so as to achieve the purpose of mineralization potential analysis in this area.

The Taonatu area is located at the boundary between Alashanzuoqi and Wulatehouqi in Inner Mongolia. It belongs to the western part of the north margin of the North China Platform, which has good metallogenic potential. In the eastern part of the study area, some large and medium-sized copper polymetallic deposits (for example Dongshengmiao, Huogeqi, Tanyaokou) have been found. There are favorable ore-forming background and ore-forming conditions in this area. Therefore, it has good prospecting potential. In this paper, the high-precision magnetic survey and stream sediment geochemical survey are combined to evaluate the mineralization potential in the Taonatu area, Inner Mongolia.

2 GEOLOGICAL BACKGROUND

The Taonatu area is located in the western section of the north of the North China Craton (Peng et al. 2007). Owing to the multi-stage tectonic compression, reworking, and magma intrusion, the strata outcrop incompletely. The exposed stratum are mainly Paleoproterozoic Baoyintu group, Mesoproterozoic Zhaertaishan group, Cretaceous and Quaternary (Figure 1).

Since the late Proterozoic, multi-stage tectonic movements occurred in the Taonatu area, therefore, the intrusive rocks were distributed widely, and the variety of types were complicated from ultrabasic to acidic rocks. The lithologies are mainly granite, diorite, gabbro, etc. The most common rock type is acidic intrusion rock.

3 GEOPHYSICAL CHARACTERISTICS

The instrument used for collecting data is the GSM-19T proton magnetometer which was made in Canada. It has high sensitivity and measurement accuracy. The data collection process includes instrument detection, diurnal variation correction, data observation, data preprocessing and so on, which was done according to "High-precision ground magnetic survey technology regulations". The magnetic survey of 1:50,000 was carried out with space in both 500 m survey lines and 100 m survey points. If data anomaly is obvious, high survey density measurement is needed.

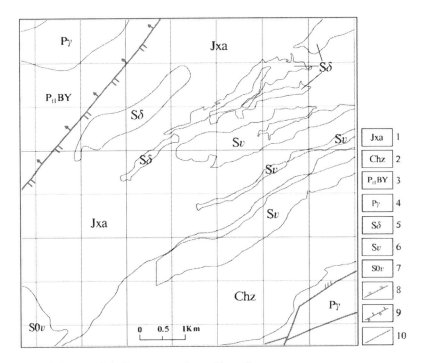

Figure 1.   Geological diagram of the Taonatu area, Inner Mongolia.
1-Agulugou group, 2-Zenglongchang group, 3-Baoyintu group, 4-granite, 5-Quartz diorite, 6-Gabbro, 7-Variable horn-blende gabbro, 8-Compressional torsion fracture, 9-Measured reverse fault, 10-Unknown fault.

Figure 2.   Contour map of magnetic anomaly in the Taonatu area, Inner Mongolia.

The contour map is as shown in Figure 2.

From Figure 2, the overall area has high magnetic anomaly but the peak was lower. The magnetic anomalies are distributed in the northeastern direction, which is consistent with the distribution of stratum. The distribution of positive and negative magnetic anomalies is consistent with the main geological formation. The positive magnetic anomaly is located mainly in the Agulugou formation and Zenglongchang formation, especially in the Agulugou formation. In addition, there are some anomalies in the basic rock. Negative anomaly distributes mainly in the area of granite. The upward continuation was done and the results show that the rock mass causing anomaly is a whole with a common magma chamber in depth. The locally superimposed anomalies are caused on the one hand by mineralization and alteration near the surface and on the other hand by the hidden magnetic body in depth.

Agulugou formation with magnetic anomaly is a favorable metallogenic area in the Taonatu area. It has good mineralization potential and is beneficial to find iron deposits. It will provide important information for the next geological prospecting work.

4 GEOCHEMICAL CHARACTERISTICS

The sampling medium is stream sediments. The sampling density is 4 samples per square kilometer in the 1:50,000 stream sediment geochemical survey.

Geochemical data processing is the basis for the extraction of anomaly information (Allegre & Lewin 1995). It is a basic problem for the determination of geochemical anomaly threshold in exploration geochemistry (Ren & Wu 1998). The traditional method to determine geochemical

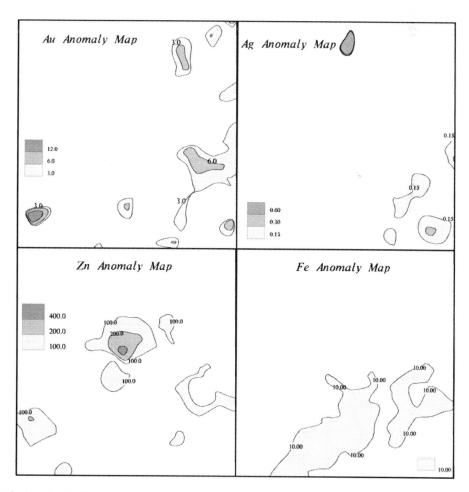

Figure 3.   Ag, Au, Fe, Zn element anomaly maps in the Taonatu area.

anomaly threshold is based on an assumption that the data obey the normal or lognormal distribution. The mean value (C) and the standard deviation (σ) are calculated for the data obeying the normal or lognormal distribution, then the anomalous threshold can be calculated by $C+k\sigma$ (k is a multiple of 2).

After determining the geochemical anomaly threshold, we could draw single element geochemical maps. In this paper, GeoExpl software was used to complete geochemistry maps (Figure 3). Based on the comparison of various gridding methods and the spatial structure of geochemical elements in the study area, exponential weighting method was chosen to do interpolation for the geochemical data and establish geochemical anomaly maps. Some obvious anomalies were evaluated and verified preliminarily, which would provide help for the evaluation and analysis of geochemical anomaly.

It has been found that this area has the multi-element combination characteristics of a geochemical anomaly, which is favorable for large-scale metallogenic conditions but there were no deposits in the anomaly area. The metallogenic elements have high anomaly intensity, obvious concentration zones, and centers (Figure 3).

The main geochemical anomaly contour maps of Fe, Zn, Au, and Ag are demonstrated in Figure 3. From Figure 3 we can see:

1. The anomalies of Fe, Zn, Au, and Ag in this area are located mainly in the Agulugou formation and Zaolongchang formation. The anomaly direction is in accordance with the approximate formation and distribution.
2. The Au, Fe, and Zn anomalies which are distributed mainly in the Agulugou formation and Zenglongchang formation have large areas of banding distribution. Concentration centers of Zn and Au are concentrated and the content gradients are obvious.
3. Ag anomalous area is smaller but the anomaly concentration is concentrated and the content gradient is obvious. Ag anomaly needs to be verified in the next work.

5 CONCLUSION

Based on the study of geological, geophysical, and geochemical characteristics, it is considered that the magnetic anomaly distributed in the Agulugou formation have favorable metallogenic potential in the Taonatu area. Those areas have good metallogenic potential and are favorable locations to find iron deposits. It has provided important information for the next mineral prospecting.

The geochemical anomaly characteristics have multiple elements in superimposition. Metallogenic elements show high anomalous intensity, obvious zone structure, and concentration centers in this area. There are good metallogenic conditions in this area but not any deposits, so we think that this area has good potential to find polymetallic deposits.

This paper applied high-precision magnetic survey and stream sediment geochemical survey to geological prospecting successfully in the Taonatu area, which achieved the aim of metallogenic prediction with comprehensive methods.

ACKNOWLEDGMENTS

I thank Professor Peng RM for his help in amending the manuscript. Thanks also to Chen Junlin for helping me.

REFERENCES

Allegre, C.J. & Lewin, E. 1995. Scaling laws and geochemical distributions. *Earth and Planetary Letters.* 132:1–13.

Guan, Z.N. Hao, T.Y. & Yao, C.L. 2002. Prospect of Gravity and Magnetic Exploration in the 21st Century. *Progress in Geophysics,* 17(2):237–244.

Nabighian, M.N. Grauch, V.J.S. Hansen, R.O. et al. 2005. 75th Anniversary-The historical development of the magnetic method in exploration. *Geophysics,* 70(6):31–61.

Peng, R.M. Zhai, Y.S. Han, X.F. et al. 2007. Mineralization response to the structural evolution in the Langshan orogenic belt, Inner Mongolia. *Acta Petrologica Sinica,* 23(3):679–688.

Ren, T.X. & Wu, Z.H. 1998. Method and Technique of Abnormality Screening and Verifying in Regional Geochemical Exploration, Beijing: Geological Publishing House.

Ye, T.Z. 2004. Solid mineral prediction and evaluation techniques. Beijing: China Land Publishing House.

Civil, Architecture and Environmental Engineering – Kao & Sung (Eds)
© 2017 Taylor & Francis Group, ISBN 978-1-138-02985-9

Effects of genistein supplementation on exhaustive exercise-induced oxidative damage in mice

Qiongli Hu, Hui Huang & Xiaoling Huang
Department of Physical Education, Central South University, Changsha, Hunan, P.R. China

ABSTRACT: The aim of this study was to evaluate the protective effects of genistein on exercise-induced oxidative damage in mice. Male mice were randomly divided into four groups: a control or one of three genistein treated groups (7, 15, and 30 mg/kg). Genistein was orally administered to mice once a day for 4 weeks, while the control group received physiological saline. After 4 weeks, mice were required to run until exhaustion on the treadmill. Their Creatine Kinase (CK), lactate dehydrogenase (LDH), aspartate aminotransferase (AST) and alanine aminotransferase (ALT) in the serum, as well as superoxide dismutase (SOD), Glutathione Peroxidase (GPx) and malondialdehyde (MDA) in the muscle were determined. As a result, when compared to the control group, genistein supplementation decreased CK, LDH, AST, and ALT levels in the serum, as well as MDA levels in the muscle, and increased SOD and GPx levels in the muscle. The findings suggest that genistein has protective effects against exhaustive exercise-induced oxidative damage.

1 INTRODUCTION

It is well established that regular physical exercise has many beneficial effects to health. It improves cardiovascular function, vascular tone, muscular strength, and endurance (Hamurcu et al. 2010). However, intense or excessive exercise could cause oxidative damage in various kinds of cells and tissues due to excessive free radicals and Reactive Oxygen Species (ROS) produced from the increase in muscle oxygen consumption during exercise (Huang et al. 2013). The main sources of ROS during exercise are the mitochondrial respiratory chain, xanthine oxidase-catalyzed reaction, and neutrophil activation (Sureda et al. 2009). It was also suggested that exercise-induced oxidative damage may be prevented by optimizing nutrition, particularly by increasing the dietary content of nutritional antioxidants (Belviranl et al. 2012).

Flavonoids are found ubiquitously in the plant kingdom. Genistein, a flavonoid in legumes and some herbal medicines (Su et al. 2016) has various pharmacological activities such as antitumor, anti-inflammatory, neuroprotective, antiapoptotic, antidiabetic, and estrogenic activities, as well as protective effects against bone loss and cardiovascular diseases (Wang et al. 2013; Turner et al. 2013). This compound also had a significant antioxidant activity (Wei et al. 1995) and our previous study has shown that genistein had anti-fatigue effects (Hu 2016) suggesting that it may be useful in preventing exercise-induced oxidative damage. To test this hypothesis, the present study was designed to evaluate the protective effects of genistein on excessive exercise-induced oxidative damage in mice.

2 MATERIALS AND METHODS

2.1 Materials and chemicals

Genistein (98.5% purity) was purchased from Zhenzhun Biological Technology Co., Ltd (Shanghai, China). The commercial diagnostic kit for Creatine Kinase (CK) was purchased from Hunan Yonghe-Yangguang Science and Technology Co., Ltd (Changsha, China). The commercial diagnostic kits for lactate dehydrogenase (LDH), Super-Oxide Dismutase (SOD), Glutathione Peroxidase (GPx) and malondialdehyde (MDA) were purchased from Jiancheng Bioengineering Institute (Nanjing, China). The commercial diagnostic kits for aspartate aminotransferase (AST) and alanine aminotransferase (ALT) were purchased from Kangmei Biotech Co., Ltd (Shenzhen, China). All the other chemicals used were of the highest grade available.

2.2 Experimental animals

Male Kunming mice, weighing 20 ± 2 g, were supplied by the Hunan Biological Supplier (Changsha, China). The animals were bred in a temperature-controlled ($(23 \pm 2)°C$), ($(55 \pm 5)\%$) humidity room with a 12-hour light and 12-hour dark cycle. Standard mouse diet and water were available ad libitum. All animal handling procedures were performed in strict accordance with the principles of Laboratory Animal Care published by the National Institutes of Health and were approved by the Animal Ethics Committee of Central South University (Changsha, China).

2.3 Experimental design

After adaptation for 1 week, mice were divided into four groups, and each group consisted of eight mice.

i. Control (C) group: mice were allowed free access to a standard mouse diet and were treated with 1.0 mL of physiological saline.
ii. low-dose genistein treated (GT-L) group: mice were allowed free access to a standard mouse diet and were treated with 7 mg/kg bw of genistein.
iii. intermediate-dose genistein treated (GT-I) group: mice were allowed free access to a standard mouse diet and were treated with 15 mg/kg bw of genistein.
iv. high-dose genistein treated (GT-H) group: mice were allowed free access to a standard mouse diet and were treated with 30 mg/kg bw of genistein.

Genistein was dissolved in 1.0 mL of physiological saline and treatments were administered orally by gavage daily for 4 weeks. During the fourth week, mice were introduced to treadmill running with 15 min exercise bouts at 20 m/min and a slope of 5% for a week to accustom them to running.

After final treatment with genistein or physiological saline, mice were required to run to exhaustion on the treadmill at a final speed of 30 m/min, 10% gradient, and approximately 70–75% maximal oxygen consumption (VO_2 max) (Li et al. 2014; Xiao 2015). Exhaustion was defined as the mouse being unable to turn upright by itself when placed on its back (Huang et al. 2009), and the mean time to exhaustion was 84.72 min (range: 62.34–95.48 min).

2.4 Analysis of biochemical parameters

Immediately after exhaustive exercise, mice were anesthetized using sodium pentobarbital and sacrificed by decapitation. Blood samples were collected in the test tube and serum was isolated by centrifugation for the CK, LDH, AST, and ALT analyses. Next, the gastrocnemius muscles were carefully removed, washed with physiological saline, and frozen in liquid nitrogen for storage at −80°C until required for the SOD, GPx, and MDA analyses. Measurements were performed according to the recommended procedures provided by the commercial diagnostic kits.

2.5 Statistical analysis

Data were expressed as mean ± SD. Statistical analysis was performed using one-way Analysis Of Variance (ANOVA) followed by Tukey's *post hoc* multiple comparison test using SPSS version 16.0 statistical software. Compare value less than 0.05 was considered significant.

3 RESULTS AND DISCUSSION

3.1 Effects of genistein on CK and LDH levels in the serum of mice

CK and LDH are highly inducible enzymes that are released into the blood as a result of the destruction of the cell membrane by oxidative stress or tissue damage, as well as the direct destruction of the cell wall and tissue necrosis (Kim & Chae 2006). Several recent studies have reported that exhaustive exercise elevates CK and LDH levels in serum (Sugino et al. 2007; Huang et al. 2009). Therefore, serum CK and LDH can be used as indexes of muscle damage. Effects of genistein on CK and LDH levels in the serum of mice are shown in Figure 1.

As shown in Figure 1, the CK levels in the GT-L, GT-I and GT-H groups, as well as the LDH levels in the GT-I and GT-H groups were significantly lower compared with that in the C group (p < 0.05). The results indicated that genistein could attenuate muscle damage following exhaustive exercise.

3.2 Effects of genistein on AST and ALT levels in the serum of mice

Liver enzymes are normally found within the liver cells, but these enzymes are released into the blood when the liver is damaged. The most critical and most widely used liver enzymes are aminotransferases such as AST and ALT (Salahshoor et al. 2014), which catalyze the transfer of amino

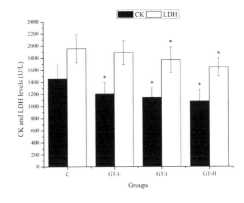

Figure 1. Effects of genistein on CK and LDH levels in the serum of mice. Data are expressed as mean ± SD. *, p < 0.05 compared with the C group.

groups from aspartate and alanine to ketoglutaric acid to generate oxaloacetic and pyruvic acids that are involved in energy production in the presence of oxygen in mitochondria (Bürger-Mendonça et al. 2008). Accumulating evidence indicates that exhaustive exercise temporally increases the activities of AST and ALT, and elevated AST and ALT levels in the absence of other evidence of liver disease should lead to the consideration of muscle damage, which is confirmed by the observed elevation of CK and LDH levels (Hammouda et al. 2012). Effects of genistein on AST and ALT levels in the serum of mice are shown in Figure 2.

As shown in Figure 2, the AST and ALT levels in the GT-I and GT-H groups were significantly lower compared with that in the C group ($p < 0.05$). The results provide further evidence that genistein could protect muscle damage after exhaustive exercise.

3.3 Effects of genistein on SOD and GPx levels in the muscles of mice

It has been well demonstrated that exhaustive exercise is associated with a dramatic increase in oxygen uptake by the whole body and particularly by the skeletal muscle and therefore enhances the production of ROS, such as superoxide anion radical, hydroxyl radical, or hydrogen peroxide (H_2O_2) (Cooper et al. 2002). As a result, accumulated excessive ROS can attack the vital biomolecules, such as plasma membrane lipids and proteins, and therefore deteriorate normal cellular functions (Yu et al. 2012). Antioxidant enzymes including SOD and GPx were regarded as the first line of defense against ROS generated during exhaustive exercise (Huang et al. 2008). SOD is a superoxide radical scavenging factor converting superoxide radicals to H_2O_2. GPx reduces H_2O_2 or hydroperoxides to H_2O and alcohol (Wu et al. 2013). The effects of genistein on SOD and GPx levels in the muscles of mice are shown in Figure 3.

As shown in Figure 3, the SOD levels in the GT-I and GT-H groups, as well as the GPx levels in the GT-L, GT-I, and GT-H groups were significantly higher compared with that in the C group ($p < 0.05$). The results indicated that genistein could effectively enhance antioxidant enzyme activities to protect exercise-induced oxidative damage.

3.4 Effects of genistein on MDA levels in the muscles of mice

Increasing evidence indicates that ROS generated by exhaustive exercise could attack polyunsaturated fatty acids, which will lead to lipid peroxidation (Qi et al. 2014). Lipid peroxidation is a complex phenomenon involving the generation of many products. MDA, one of the most important end-products of lipid peroxidation is a popular index of the first condition on oxidative damage (Lu et al. 2006). Effects of genistein on MDA levels in the muscles of mice are shown in Figure 4.

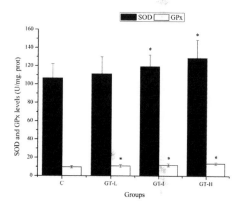

Figure 3. Effects of genistein on SOD and GPx levels in the muscles of mice. Data are expressed as mean ± SD. *, $p < 0.05$ compared with the C group.

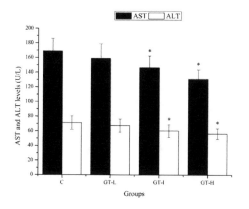

Figure 2. Effects of genistein on AST and ALT levels in the serum of mice. Data are expressed as mean ± SD. *, $p < 0.05$ compared with the C group.

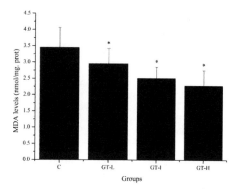

Figure 4. Effects of genistein on MDA levels in the muscles of mice. Data are expressed as mean ± SD. *, $p < 0.05$ compared with the C group.

As shown in Figure 4, the MDA levels in the GT-L, GT-I, and GT-H groups were significantly lower compared with that in the C group ($p < 0.05$). The results indicated that genistein could effectively reduce lipid peroxidation, and indirectly reflected genistein could protect oxidative damage after exhaustive exercise.

4 CONCLUSION

In conclusion, the results of the present study indicated that genistein supplementation afforded significant protection against exhaustive exercise-induced oxidative damage in mice. The protective effect is mediated by the lowering of CK, LDH, AST, and ALT levels in the serum, as well as MDA levels in the muscle, and elevating of the SOD and GPx levels in the muscle. However, further studies are needed to investigate the underlying mechanisms for the protective effects of genistein.

REFERENCES

Belviranl, M., Gökbel, H., Okudan, N. & Başaralı, K. 2012. Effects of grape seed extract supplementation on exercise-induced oxidative stress in rats. British Journal of Nutrition 108(2): 249–256.

Bürger-Mendonça, M., Bielavsky, M. & Barbosa, F.C. 2008. Liver overload in Brazilian triathletes after half-ironman competition is related muscle fatigue. Annals of Hepatology 7(3): 245–248.

Cooper, C.E., Vollaard, N.B., Choueiri, T. & Wilson, M.T. 2002. Exercise, free radicals and oxidative stress. Biochemical Society Transactions 30: 280–285.

Hammouda, O., Chtourou, H., Chaouachi, A., Chahed, H., Ferchichi, S., Kallel, C., Chamari, K. & Souissi, N. 2012. Effect of short-term maximal exercise on biochemical markers of muscle damage, total antioxidant status, and homocysteine levels in football players. Asian Journal of Sports Medicine 3(4): 239–246.

Hamurcu, Z., Saritas, N., Baskol, G. & Akpinar, N. 2010. Effect of wrestling exercise on oxidative DNA damage, nitric oxide level and paraoxonase activity in adolescent boys. Pediatric Exercise Science 22(1): 60–68.

Hu, Q.L. 2016. Anti-fatigue Effects of the Genistein Supplementation in mice subjected to forced swimming test. International Conference on Biomedical and Biological Engineering, Advances in Biological Sciences Research: 137–142.

Huang, C.C., Huang, W.C., Yang, S.C., Chan, C.C. & Lin, W.T. 2013. Ganoderma tsugae hepatoprotection against exhaustive exercise-induced liver injury in rats. Molecules 18(2): 1741–1754.

Huang, C.C., Lin, T.J., Lu, Y.F., Chen, C.C., Huang, C.Y. & Lin, W.T. 2009. Protective effects of L-arginine supplementation against exhaustive exercise-induced oxidative stress in young rat tissues. Chinese Journal of Physiology 52(5): 306–315.

Huang, C.C., Tsai, S.C. & Lin, W.T. 2008. Potential ergogenic effects of L-arginine against oxidative and inflammatory stress induced by acute exercise in aging rats. Experimental Gerontology 43(6): 571–577.

Kim, H.T. & Chae, C.H. 2006. Effect of exercise and α-lipoic acid supplementation on oxidative stress in rats. Biology of Sport 23(2): 143–145.

Li, M., Luo, L. & Yu, R. 2014. Effect of flavonoids from tartary buckwheat on the exhaustive exercise-induced oxidative stress of mice. Journal of Food, Agriculture & Environment 12(2): 128–131.

Lu, H.K., Hsieh, C.C., Hsu, J.J., Yang, Y.K. & Chou, H.N. 2006. Preventive effects of Spirulina platensis on skeletal muscle damage under exercise-induced oxidative stress. European Journal of Applied Physiology 98(2): 220–226.

Qi, B., Liu, L., Zhang, H., Zhou, G.X., Wang, S., Duan, X.Z., Bai, X.Y., Wang, S.M. & Zhao, D.Q. 2014. Anti-fatigue effects of proteins isolated from Panax quinquefolium. Journal of Ethnopharmacology 153(2): 430–434.

Salahshoor, T., Farzanegi, P. & Habibian, M. 2014. Synergistic Effects of omega 3 supplementation and exercise on markers of liver (ALP, AST, and ALT) and muscle (LDH and CK) damage in male karate athlete. Journal of Applied Science and Agriculture 9(1): 245–249.

Su, P., Zhang, J., Wang, S., Aschner, M., Cao, Z., Zhao, F., Wang, D., Chen, J. & Luo, W. 2016. Genistein alleviates lead-induced neurotoxicity in vitro and in vivo: Involvement of multiple signaling pathways. Neurotoxicology 53: 153–164.

Sugino, T., Aoyagi, S., Shirai, T., Kajimoto, Y. & Kajimoto, O. 2007. Effects of citric acid and l-carnitine on physical fatigue. Journal of Clinical Biochemistry and Nutrition 41(3): 224–230.

Sureda, A., Ferrer, M.D., Tauler, P., Romaguera, D., Drobnic, F., Pujol, P., Tur, J.A. & Pons, A. 2009. Effects of exercise intensity on lymphocyte H2O2 production and antioxidant defences in soccer players. British Journal of Sports Medicine 43(3): 186–190.

Turner, R.T., Iwaniec, U.T., Andrade, J.E., Branscum, A.J., Neese, S.L., Olson, D.A., Wagner, L., Wang, V.C., Schantz, S.L. & Helferich, W.G. 2013. Genistein administered as a once-daily oral supplement had no beneficial effect on the tibia in rat models for post-menopausal bone loss. Menopause 20: 677–686.

Wang, R., Tu, J., Zhang, Q., Zhang, X., Zhu, Y., Ma, W., Cheng, C., Brann, D.W. & Yang, F. 2013. Genistein attenuates ischemic oxidative damage and behavioral deficits via eNOS/Nrf2/HO-1 signaling. Hippocampus 23: 634–647.

Wei, H. Bowen, R. Cai, Q. Barnes, S. & Wang, Y. 1995. Antioxidant and antipromotional effects of the soybean isoflavone genistein. Proceedings of the Society for Experimental Biology and Medicine 208: 124–130.

Wu, C., Chen, R., Wang, X.S., Shen, B., Yue, W. & Wu, Q. 2013. Antioxidant and anti-fatigue activities of phenolic extract from the seed coat of Euryale ferox Salisb. and identification of three phenolic compounds by LC-ESI-MS/MS. Molecules 18(9): 11003–11021.

Xiao, N.N. 2015. Effects of resveratrol supplementation on oxidative damage and lipid peroxidation induced by strenuous exercise in rats. Biomolecules & Therapeutics 23(4): 374–378.

Yu, S.H., Huang, H.Y., Korivi, M., Hsu, M.F., Huang, C.Y., Hou, C.W., Chen, C.Y., Kao, C.L., Lee, R.P., Lee, S.D. & Kuo, C.H. 2012. Oral Rg1 supplementation strengthens antioxidant defense system against exercise-induced oxidative stress in rat skeletal muscles. Journal of the International Society of Sports Nutrition 9(1):23.

Civil, Architecture and Environmental Engineering – Kao & Sung (Eds)
© 2017 Taylor & Francis Group, ISBN 978-1-138-02985-9

Modeling technologies for evaluation system of Jilin Province environmental protection

Junyan Dong
Harbin Institute of Technology, Harbin, China
Changchun Institute of Technology, Changchun, China
Bangkok Thonburi University, Bangkok, Thailand

Wen Cheng
Harbin Institute of Technology, Harbin, China

ABSTRACT: Based on confirmation of the target of environmental protection planning text semantics, this study listed formal analysis and content analysis methods. Two kinds of modeling technology for planning evaluation are put forward: the modeling technology based on the driving force of the model and the modeling technology based on interest—driven model. The future research directions related to semantic modeling technology are suggested: 1) promoting computer information search and statistical techniques in the field of text form analysis. 2) paying attention to the stakeholders' claim, anxiety, dispute and the corresponding solutions during the evaluation process.

1 THE GOAL OF THE SEMANTIC ANALYSIS OF PLANNING TEXTS

In general, both planning of public sectors and the overall planning of private sectors are essentially a collection of 5 aspects: analysis of the present situation, prediction of future trends, goals in different stages, tasks to implement, and manpower, financial power and other institutional safeguard measures. The five-year environmental protection plan in Jilin Province, as an important part of the planning system of Jilin Province, covers contents of the above-mentioned 5 aspects.

The implementation of the five-year environmental protection plan of Jilin Province is evaluated, and the focus of the evaluation is placed on 3 aspects, that is, the attainment of staged goals, the progress of implementing specific tasks, and the implementation of manpower, financial and other institutional safeguard measures. Therefore, in the semantic analysis of planning texts, work should be done from 3 aspects to figure out which goals to attain in different stages, which tasks to complete, and who assumes responsibility of implementing the tasks and which sources provide funds for supporting the task implementation. To make clear the 3 questions, planning texts are usually analyzed through 2 analytical methods: form analysis and content analysis. The form analytical method is adopted to form key points and detailed rules for evaluation outlined by planning goals and tasks (Junyan Dong, 2014). On the basis of a form analysis, a content analysis is made of planning texts, to determine requirements for the progress of implementing the goals and tasks, and identify manpower, financial and other institutional safeguard measures, thus eventually forming detailed items for the evaluation of plans.

2 FORM ANALYTICAL METHODS OF PLANNING TEXTS

The planning text refers to an information structure which is made by specific people or organizations and has special symbols. Semantics of planning texts would inevitably reveal the specific standpoint, views, values, and interests of compilers (or principals). These specific standpoints, views, values and interests can be reflected by the form of texts, such as feature words, word frequency, word number, number and so on (Junyan Dong, 2011). Therefore, through a form analysis of planning texts, intention and motivation of planning compilers (or principals) can be roughly inferred, to provide direction and guidance for the content analysis of planning texts, and simplify and standardize contents of evaluating planning texts.

2.1 Selection of feature words

The selection of feature items is the basic work in the text analysis. A scientific and reasonable selection of feature items can have a multiplier effect on the text analysis and deconstruction. In accordance with the requirements of planning evaluation, feature words in the environmental protection plan of Jilin Province can be generally classified into 3 categories: a set of feature words about goals (such as goal, objective, main indictor, etc.), a set of feature words about tasks and projects (such as task, project, facility, supervision, etc.), a set of feature words about policies and measures (such as policy, measure, system, mechanism, law, standard, etc) (Junyan Dong, 2015). A complex collection of feature words can be further divided into sub-sets.

2.2 Statistics of the frequency of feature words

The word frequency of a given word refers to the number of occurrences of the given word in a text. The importance of a word increases along with the increase in the number of its occurrence in the text. Statistics is made of the word frequency of feature words selected from the environmental protection plan of Jilin Province (Junyan Dong, 2015). Through statistics of the word frequency, we can not only validate, select and identify feature words for the evaluation of this planning text, but also make a preliminary evaluation of the importance of evaluation indicators for the feature words.

2.3 Clustering analysis of feature words

In a given text, feature words often frequently appear in paragraphs with greater relevance in meaning, but rarely appear in paragraphs with smaller relevance in meaning. Through a clustering analysis of the position of feature words in a planning text, we can match the feature words with corresponding text modules, to facilitate the subsequent analysis of the word number, the number, other forms and contents of the text.

2.4 Statistics of the word count

Through a clustering analysis of feature words, paragraphs with greater relevance in the meaning of feature words are classified into the same part. Moreover, the clustering analysis can help to gain a more intuitive understanding of the importance of relevant evaluating contents. An analysis of text contents is combined to give further suggestions for specific perspectives and evaluation technology used to evaluate feature words.

3 CONTENT ANALYTICAL METHODS OF PLANNING TEXTS

Compared with the form analysis of texts, in the content analysis of texts, evaluators are required to have a more comprehensive structure of evaluation knowledge, make more detailed and complicated analyses and take a more subjective viewpoint. After a form analysis of planning texts, a sentence or several sentences containing a specific meaning group will be clearly separated. According to semantic contents of the text, text contents are further analyzed to develop basic items for the evaluation and analysis.

In general, basic elements (concepts) in the text content analysis include: event, action, role, subject/actor, object/receptor, thing, matter, etc. The purpose of the text content analysis is to identify the meaning of a sentence, paragraph or article. To put it simply, the text content analysis is to find answers to 4 W1H (Who, What, When, Where and How). In the "Environmental Protection Plan in Jilin Province", sentences have 4 kinds of semantics, including the statement of the past and present situation, prediction of related factors, guiding requirements and constraints of specific targeted issues. Specific methods for the text content analysis are as follows.

3.1 Requirements for developing evaluation items of declarative sentences

Generally speaking, in the "Five-year Environmental Protection Plan in Jilin Province", declarative sentences are mainly used to declare past achievements and existing problems. In the content analysis, existing problems should be listed in detail, in order to facilitate preliminary judgments of the problem-oriented logic for the subsequent implementation.

3.2 Methods for developing evaluation items of predictive sentences

In the five-year environmental protection plan of Jilin Province, predictive sentences are mainly used to predict the future social and economic development. In the content analysis, trend prediction should be quantified, and the impact on specific issues and goals in the planning text should be built, to facilitate planning evaluation.

3.3 Methods for developing evaluation items of restrictive sentences

In the five-year environmental protection plan of Jilin Province, restrictive sentences are mainly used to put forward requirements for goals and specific requirements for tasks and some policy measures. The plan-implementing situation is evaluated mainly based on the semantics of these sentences. In the process of developing items for restrictive sentences, focus should be placed on finding out the personal liable for the plan implementation (who), time of implementation (when), place of implementation (where), the goal of implementation (what) and the way of implementation (how), to facilitate subsequent work in planning evaluation.

Table 1. Constraint statement entry production demonstration table.

Who	When	Where	What	Ways to achieve
The Government of the city seat	To 2016	City and county	All to build sewage treatment facilities	Government investment and other
The Government of the city seat	To 2016	City and county	Sewage treatment rate is not less than 70%	Government investment and other
The Government of the city where the city is located	To 2016	Whole province	Sewage treatment capacity reached 10 thousand tons/day	Government investment and other

For example, in the "11th five-year environmental protection plan of Jilin Province", "in 2010, all cities should build sewage treatment facilities. The urban sewage treatment rate should not be lower than 70%, and the national urban sewage treatment capacity should reach 100 million tons/day." These sentences are restrictive ones. 3 items of evaluation are developed by analyzing contents of restrictive sentences. Specific analytical results are shown in Table 1.

From the perspective of form, many restrictive sentences contain numerals. However, some restrictive sentences do not contain numerals, but use the logic word "whether (this work is done" and introduce specific achievements. For example, "make total analysis and monitoring of the centralized drinking water quality at least once a year." "Launch comprehensive collection of fees on urban sewage, garbage, hazardous waste, medical waste disposal and radioactive waste collection and storage." These tasks have well-defined requirements, obvious achievements and great significance (see the plan-compiling specification and expert consultation, etc.). Items should be strictly developed for restrictive sentences in the text, to facilitate the follow-up planning evaluation.

The definition of the responsibility of restrictive sentences can be omitted, because the responsibility is usually allocated in advance, or the labor division is made in the jurisdiction competition, for completing goals, tasks or projects planned. The part with no well-defined liable person can be temporarily in the charge of the department of overall coordination.

4 SEMANTIC MODELING TECHNOLOGY FOR PLANNING EVALUATION

When the environmental protection plan of Jilin province is evaluated, the semantic model building is the end point for the constructive understanding of evaluation standards, as well as the starting point for the evaluation system building, so the semantic model building is a key link between the preceding and the following. The internal logical self-consistency and completeness of the semantic model design directly determine the veracity and applicability of the evaluation results.

Through a form analysis and content analysis of the environmental protection planning text of Jilin province, evaluation items which have clear semantics and basically fixed categories are developed. When analyzing simply-structured planning texts (such as texts which just contain goals and specific issues to implement), we can achieve semantic logical integration of evaluation items simply by observing the relationship of feature word sets, to establish a general evaluation model framework. However, when planning contexts whose evaluation items have causality, inclusion, overlapping and other complex relationships, correspondence, transformation, exhaustion, exclusion, deduction, induction and other logical analytical methods should be used to deeply analyze semantic logical relationships between the items. Meanwhile, the hierarchy process, logical framework analysis and PSR model analysis should be combined to build a semantic logical model for the planning texts. In general, semantic modeling technology for planning evaluation falls into two categories. One is the driving force-based model-building technology which has been quite mature, while the other is the interests-driven model-building technology which is emerging.

4.1 Driving force-based model-building technology

The "implementation of environmental protection plan in Jilin Province" as a comprehensive planning evaluation, in essence, is to evaluate multiple processes and issues in the environmental protection plan implementation. The driving force (mechanism) of these issues is a natural basis for building an evaluation semantic model. The driving force-based model-building method usually adopts the input-output, pressure-state-response or DPSIR (Driving force-Pressure-State-Impact-Response) model. This model-building technology is relatively mature and widely-applied, so unnecessary details would not be discussed in this paper.

4.2 Benefit-driven model-building technology

The stakeholder-based model building technology stresses that various stakeholders are involved in evaluation and the stakeholders have different opinions, run into disputes or even show anxiety in evaluation, so evaluators should focus on detecting the difference in stakeholders' demands for interests, coordinate or ease the difference according to the principle of equal rights and obligation, and eventually reach evaluation results that all stakeholders accept.

This evaluation model-building method should satisfy the following 2 conditions. On the one hand, evaluation is made up of opinions, anxiety and disputes of stakeholders. On the other hand, the constructivist paradigmatic methodology (that is, the methodology of hermeneutics: a common construction is established based on the same situation through continuous exposition, analysis, re-exposition and re-analysis of stakeholders. Evaluation is seen as a form of intents and actions which occur simultaneously and interact with each other.

1. Identify all stakeholders at risk in the planning evaluation;
2. Analyze each stakeholder group to figure out their construction of the evaluation object and desired opinions, anxiety and disputes related to the evaluation object;
3. Follow the information consistency, logical consistency, value consistency, rights-and-liability consistency and other principles of social consensus, so that different constructs, opinions, anxiety and disputes can be understood, criticized and considered;
4. Gain as much consensus as possible from the understanding, criticism and consideration;
5. Design one or more reports, so that stakeholders have convenience to communicate with each other about their constructs, opinions, anxiety and disputes;
6. Repeat the evaluation for those unresolved constructs and opinions, anxiety and disputes attached to the constructs. Results obtained in the last round of evaluation can be used for the next round of evaluation. Therefore, evaluation will never end, but can only be suspended.

5 CONCLUSION

It is recommended that the following evaluation technologies should be studied, in the process of semantic model building.

1. Popularize the application of computer information search and statistical technology in the analysis of text form, to improve the operation efficiency of the text form analytical technology.

2. The semantic modeling technology for evaluation should not just limited to get evaluation results based on the traditional cognition and modeling paradigm, but focus more on opinions, anxiety and disputes of stakeholders and their solutions and put forward more feasible recommendations for planning evaluation.

This paper introduced form analytical methods and content analytical methods of planning texts, presented 2 modeling technologies for planning evaluation, that is, driving force-based model-building technology and interests-driven model-building technology, and gave the following 2 recommendations for the future development of related semantic modeling technology. First of all, popularize the application of computer information search and statistical technology in the analysis of text form. Moreover, focus should be placed on opinions, anxiety and disputes of stakeholders and explore corresponding negotiated solutions.

ACKNOWLEDGMENTS

This work was financially supported by Science and Technology Project Foundation of Ministry of Housing and Urban-Rural Construction of the People's Republic of China (No. 2016R2005), Jilin Provincial Social Science Foundation (No. 2015BS83), Youth Foundation of Changchun Institute of Technology (No. 320140006) and Teaching reform project of Changchun Institute of Technology.

REFERENCES

Junyan Dong, Hong Jin, Jian Kang, Xi Chen. A pilot study of the acoustic environment in residential areas in Harbin, towards the questionnaire design. Journal of Harbin Institute of Technology. 2011, 18(2), pp. 319–322.

Junyan Dong, Hong Jin. The design strategy of green rural housing of Tibetan areas in Yunnan, China. Renewable Energy, Vol. 49, pp. 63–67.

Junyan Dong, Wen Cheng. Based on the Characteristics of Respondents and the Voice of the Urban Neighborhood Public Space Business Facilities Noise Environment Evaluation Research. Journal of Harbin Institute of Technology. 2014, 20(4), pp. 103–109.

Junyan Dong, Wen Cheng. Research on Optimized Construction of Sustainable Human Living Environment in Regions where People of a Certain Ethnic Group Live in Compact Communities in China. World Renewable Energy Congress XIV.2015.6.

Junyan Dong. The Acoustic Environment Research of Construction Exterior Based on the Ecology Idea. Materials Engineering and Environmental Science. 2015.9.

Civil, Architecture and Environmental Engineering – Kao & Sung (Eds)
© 2017 Taylor & Francis Group, ISBN 978-1-138-02985-9

Analysis and evaluation of the influence factors of waste packaging recycling in a circular economy

Ning Jiang, Kai Liu & Junjie Zhang
Department of Business Administration, Qingdao Vocational and Technical College of Hotel Management, Qingdao, China

ABSTRACT: Waste packaging is a kind of resource recycling. Recovery and recycling of waste packaging is an important part of developing a recycling economy and promoting energy savings and environmental protection. Now, the overall recovery rate of waste packaging is low in China. How to recycle and reuse the packaging material efficiently has become a subject, which everyone must face and participate in. This paper summarizes the related factors that affect the packaging recovery efficiency, adopts SPSS statistical software for factor analysis, finds out the main influence and corresponding evaluation index, and uses a practical example to verify its practicality and effectiveness.

1 INTRODUCTION

The circular economy is the mode of economic growth, which under the guidance of sustainable development achieves waste reduction, efficient use of resources, and recycling. The packaging industry is a resource consumption industry. Every year, the waste packaging material accounts for about one-third of urban domestic waste. Although some of the packaging has been recycled, such as cartons, bottles, and so on, the rate of most waste packaging is low. Therefore, putting the packaging industry into the orbit of the circular economy has become an urgent problem, and it is very much necessary to analyze and evaluate the factors that affect the recycling of waste packaging.

2 FACTORS AFFECTING THE RECOVERY OF PACKAGING

At present, the recycling of waste packaging, such as packaging collection, classification, transportation and other sectors are responsible for the third-part recycling enterprises. The packaging manufacturer is not directly involved in the recovery. It is only responsible for reprocessing of the "product" (Wu 2011). The main pattern of operation is shown in Figure 1.

Considering each node in the recycling network is referring to the previous research results of many scholars, we know that the recycling of waste packaging is affected by many factors such as manufacturing enterprises, recycling enterprises, consumers, and the government (Li 2013,

Liu 2008, Wang 2012). The influencing factors are shown in Figure 2.

Because there is a certain correlation between each factor, the information reflected will overlap to a certain extent, which causes information redundancy and reduces the actual operability. For this reason, the factor analysis method was used to screen and simplify the factor index, and to find and design one or more comprehensive factors.

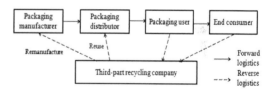

Figure 1. Operation pattern diagram of waste packaging recovery.

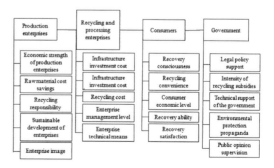

Figure 2. Influencing factors of waste packaging recycling.

3 BASIC PRINCIPLE OF FACTOR ANALYSIS

Factor analysis is an important branch of multivariate statistical analysis and its purpose is to condense data. The main information of the original variables is expressed by the correlation study.

Assuming there are n samples for the multi-criteria problem $X = (X_1, X_2,... X_k)$, and the reasons for the problem are various, in which the common reason is called the common factor, using F_j to express it, among them are mutually orthogonal. Each component of X_i has its special reason, called special factor, using e_i to express it and that they are not related to each other, and only have an effect on the corresponding X_i. F and E are independent of each other. So the mathematical model of factor analysis can be expressed as:

$$\begin{cases} X_1 = a_{11}F_1 + a_{12}F_2 + ...+ a_{1m}F_m + e_1 \\ ... \\ X_k = a_{k1}F_1 + a_{k2}F_2 + ...+ a_{km}F_m + e_k \end{cases} \quad m < k; a_{ij} \text{ is}$$

the load of a common factor, which is the j factor of i variable. Matrix $A = \begin{bmatrix} a_{11}a_{12}...a_{1m} \\ a_{21}a_{22}...a_{2m} \\ ... \\ a_{k1}a_{k2}...a_{km} \end{bmatrix}$ is a component matrix of the common factor. Finding out the common factor is done using some method to find out the factor component matrix A. For each factor, it is observed that there is a load on the variable that is larger and a load on the variable that is smaller, then according to the variable content of heavy load, the meaning of factor is explained. Factor analysis model can also be expressed as $X = AF + e$ (Wang 2002).

4 EXAMPLE ILLUSTRATION

In order to study the main reasons for the low recovery rate of waste packaging, SD province conducted a special investigation, including the above 20 influential factors. Through visiting inquiring, on-the-spot investigation, the results of the investigation were finally obtained in 10 regions. Because the content of the investigation contained many qualitative indicators, so the first thing to do was to quantify the qualitative indicators, and then conduct a comprehensive evaluation and scoring by experts, and finally obtain the impact factor table, as shown in Table 1.

By observing the data in Table 1, statistical analysis software SPSS would be used to carry out the factor analysis, where the factors involved will be analyzed respectively. Here take the recycling processing enterprise as an example to explain the process of factor analysis.

Using the factor analysis method to deal with the factors of enterprise recovery processing, we can get the eigenvalue and contribution of correlation matrix R (as shown in Table 2 total variance explained), and can draw the scree plot of the common factor and the eigenvalue as shown in Figure 3 (horizontal coordinate is the component number and the vertical coordinate is the eigenvalue).

Table 1. The influence factors on the reverse recovery of waste packaging in different regions.

District	Factors	1	2	3	4	5	6	7	8	9	10
Production enterprises	Economic strength of production enterprises	0.85	0.83	0.68	0.82	0.78	0.72	0.85	0.81	0.77	0.83
	Raw material cost savings	0.55	0.64	0.70	0.75	0.63	0.72	0.58	0.68	0.68	0.66
	Recycling responsibility	0.80	0.68	0.72	0.77	0.70	0.77	0.80	0.84	0.72	0.76
	Sustainable development of enterprises	0.66	0.71	0.62	0.80	0.58	0.82	0.64	0.88	0.77	0.60
	Enterprise image	0.79	0.64	0.62	0.75	0.70	0.74	0.76	0.68	0.62	0.80
Recycling and processing enterprises	Infrastructure investment cost	1910	1798	1673	1482	1490	1286	1394	1520	1865	1920
	Recycling channel construction	0.88	0.74	0.64	0.83	0.67	0.60	0.67	0.64	0.82	0.85
	Recycling cost	0.68	0.78	0.72	0.73	0.65	0.76	0.58	0.77	0.70	0.75
	Enterprise management level	0.62	0.76	0.83	0.80	0.77	0.80	0.83	0.78	0.64	0.65
	Enterprise technical means	0.78	0.83	0.75	0.76	0.80	0.82	0.73	0.85	0.68	0.73
Consumers	Recovery consciousness	0.80	0.82	0.71	0.72	0.81	0.77	0.65	0.85	0.80	0.75
	Recycling convenience	0.85	0.75	0.68	0.81	0.72	0.67	0.75	0.73	0.74	0.76
	Consumer economic level	0.57	0.60	0.61	0.64	0.66	0.71	0.74	0.70	0.73	0.72
	Recovery ability	0.78	0.80	0.71	0.72	0.67	0.72	0.75	0.63	0.58	0.67
	Recovery satisfaction	0.62	0.75	0.70	0.74	0.76	0.68	0.62	0.80	0.68	0.72
The government	Legal policy support	0.81	0.71	0.65	0.75	0.73	0.83	0.68	0.73	0.75	0.76
	Intensity of recycling subsidies	0.72	0.73	0.57	0.63	0.81	0.77	0.67	0.75	0.60	0.77
	Technical support of the government	0.78	0.72	0.58	0.68	0.68	0.72	0.66	0.76	0.70	0.72
	Environmental protection propaganda	0.73	0.85	0.68	0.73	0.76	0.75	0.83	0.78	0.84	0.83
	Public opinion supervision	0.58	0.77	0.67	0.75	0.60	0.81	0.75	0.78	0.72	0.71

Table 2. Total variance explained.

Component	Initial eigenvalues			Rotation sums of squared loadings		
	Total	% of variance	Cumulative %	Total	% of variance	Cumulative %
1	2.824	56.476	56.476	2.759	55.174	55.174
2	1.393	27.864	84.340	1.458	29.166	84.340
3	0.386	7.730	92.070			
4	0.240	4.795	96.865			
5	0.157	3.135	100.000			

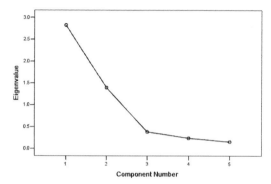

Figure 3. Scree plot.

Table 3. Rotated component matrix.

	Component	
	1	2
Infrastructure investment cost	0.930	0.049
Recycling channel construction	0.898	−0.169
Recycling cost	0.194	0.908
Enterprise management level	−0.926	0.030
Enterprise technical means	−0.439	0.775

Table 4. Rotated component matrix.

	Component	
	1	2
Economic strength of production enterprises	0.794	−0.253
Raw material cost savings	−0.513	0.719
Recycling responsibility	0.778	0.483
Sustainable development of enterprises	0.057	0.927
Enterprise image	0.827	−0.109

Table 5. Rotated component matrix.

	Component	
	1	2
Recovery consciousness	0.090	0.891
Recycling convenience	0.680	−0.083
Consumer economic level	−0.884	−0.230
Recovery ability	0.788	−0.351
Recovery satisfaction	−0.202	0.843

Table 6. Rotated component matrix.

	Component	
	1	2
Legal policy support	0.857	−0.112
Intensity of recycling subsidies	0.790	0.082
Technical support of the government	0.908	0.175
Environmental protection propaganda	0.184	0.835
Public opinion supervision	−0.080	0.812

From Table 2, if the common factor is extracted according to the standard of eigenvalue more than 1, then taking 2 public factors containing the amount of information, it can reach 84.340%. From Figure 3, the first 2 public factors change the largest, indicating that these 2 public factors can provide the original 5 indicators of sufficient information. At the same time, it is difficult to find the actual significance factor if the initial factor component matrix does not satisfy "simple structure principle", so we adopt a method of rotating component matrix to achieve the purpose of simplifying the structure (Wang 2002) as shown in Table 3.

According to the above steps, we can get the results of factor analysis of manufacturing enterprises, consumers, and the government as shown in Tables 4–6.

5 MODEL RESULT ANALYSIS

From the data in Tables 3–6, it can be analyzed:

For recycling and processing enterprises, factor 1 mainly explains infrastructure investment cost, recycling channel construction, enterprise management level, and indicates that the basic construction investment of recycling enterprises has a great influence on the recovery of waste packaging.

Factor 2 mainly reflects the impact of recycling cost, enterprise technical means, and indicates that the enterprise's recycling ability is also an important influence factor.

For production enterprises, factor 1 mainly explains the impact of the economic strength of production enterprises, the economic strength of production enterprises, sustainable development of enterprises.

Factor 2 mainly representative raw material cost savings, Enterprise image.

For consumers, factor 1 has a greater contribution on recycling convenience, Consumer economic level, recovery ability.

Factor 2 mainly reflects the impact of recovery consciousness, recovery satisfaction.

From the perspective of government, factor 1 has a greater contribution to legal policy support, the intensity of recycling subsidies, technical support of the government.

Factor 2 mainly reflects the impact of environmental protection propaganda, public opinion supervision.

Therefore, we can interpret above factors as recycling enterprises infrastructure investment, recycling enterprises processing capacity, production enterprise investment, the economic efficiency of production enterprises, residents' recycling literacy, residents' recycling environment, policy support of the government, social propaganda and supervision (Zhou 2012).

6 CONCLUSION

Factor analysis is a common method of multivariate statistical analysis to deal with high dimensional data. It can guarantee the minimum loss of data information and can quickly extract important information from the original database, which makes it possible for people to make a full and comprehensive understanding of the problem.

In this paper, we selected the relevant data of 10 regions in SD province and studied the influencing factors of the reverse recycling of packaging waste. Taking into account the desirability of data, the paper selected 20 indicators to build an index system, and avoid the one-sidedness of the single index. Through factor analysis, 8 common factors were extracted and the evaluation results were obtained.

However, due to limitations of the index selection, some indicators are excluded due to data sources, and the factor analysis may affect the comprehensive evaluation of the final.

REFERENCES

Li Xia. 2013. The Efficiency of Energy Utilization Evaluation Index System and Applied Research of China. *China University of Geosciences, Doctor Dissertation.* Wuhan.

Liu Xiao pei. 2008. Analysis on the Affecting Factors of End-of-life Vehicles and Design for System, *Chongqing University, Master degree thesis.* Chongqing.

Wang Jun jie. 2012. Packaging Waste Recycling Logistics System Research and Application, *Shanghai Jiao Tong University, Master degree thesis.* Shanghai.

Wang Zeng min. 2002. The application of factor analysis in the comprehensive analysis and evaluation of enterprise economic benefit, *Journal of applied statistics and management*: 10–13.

Wu Yu ping. 2011. Research on How to Select Packaging Waste Take-back Model Based on ERP, *Chongqing University of Technology, Master degree thesis.* Chongqing.

Zhou San yuan. 2012. Study on Influence Factors in Waste Appliance Recycling Based on PCA, *Logistics Technology*: 151–153.

Civil, Architecture and Environmental Engineering – Kao & Sung (Eds)
© 2017 Taylor & Francis Group, ISBN 978-1-138-02985-9

Construction and application of a water recycling system in crab–crayfish polyculture purification

Xiaoshuai Hang & Fei He
Nanjing Institute of Environmental Science, Ministry of Environmental Protection of China, Nanjing, China

Fangqun Gan
College of Environment and Ecology, Jiangsu Open University, Nanjing, China

ABSTRACT: Aquatic contamination caused by aquaculture has become a major concern in recent years, and the main pollutants are COD_{Mn}, Suspended Substance (SS), ammonia nitrogen (NH_3-N), Total Nitrogen (TN), and Total Phosphorus (TP). A water recycling system of attapulgite and artificial wetland was constructed to purify polyculture wastewater from crab (*Eriocheir sinensis* H. Milne-Edwards) and crayfish (*Procambarus clarkii*) pond breeding, and then the operation effect was also evaluated. Additional constructions were also designed in this system, including Aquaculture Pond (AP), Precipitation System (PS), Ecological Ditch (ED), Adsorption System (AS), Artificial Wetland (AW), water quality testing system (WS), and External Reservoir (ER). Attapulgite clay in AS had an excellent adsorption capacity of TP and NH_3-N in synthetic wastewater due to its structure and physicochemical properties. NH_3-N and TN concentrations in water decreased significantly in AW. Dissolved Oxygen (DO) in water increased in ED and AW, and the ED has good effects on SS and COD_{Mn} removal. The results of the recycling system indicate that it has an excellent performance of relieving aquatic pollution in crab–crayfish polyculture, without any wastewater discharge.

1 INTRODUCTION

The environment problem caused by the aquaculture industry has been increasingly serious in recent years and the major issue is water contamination and sediment pollution. The excreta of aquatic products, residual breeding baits, and medicines used in inland aquaculture farms could result in increasing accumulation of Nitrogen (N), Phosphorus (P) and COD_{Mn} into both water and sediments in downstream rivers and estuaries, which becomes an important source of eutrophication and other environmental problems (Boxall et al. 2004).

China is one of the countries with a long history of aquaculture in the word. According to the China fishery statistics yearbook 2016, there is 66,996,500 tons of total output aquatic product in China in 2015, and the amount of aquaculture is 49,379,000 tons. The national per capita share of aquatic products is 48.74 kg in China. The aquaculture area is 84,650 km² in China in the year 2015, mariculture and freshwater aquaculture area are 2317.6 km² and 61,472.4 km², respectively. The proportion of freshwater aquaculture is 72.6%, including ponds, reservoirs, lakes, ditches, and so on. Ponds form the largest area in freshwater aquaculture, which is 27,012.2 km². A great majority

of aquaculture is produced by inland-based freshwater or brackish water ponds, which rely on periodic effluent discharges into surrounding water bodies (Bostock et al. 2010).

Interactions between aquaculture and its surrounding environment are diverse and complex (Edwards 2015). The rapid expansion and intensification of pond aquaculture in China over the last two decades have attracted a lot of attention in the country, although there have been few studies on water quality used in aquaculture, the relationship between water quality and farm activity, and the impact of pond effluents on receiving waters. (Cao et al. 2007, Li et al. 2011). The Ministry of Agriculture in China has launched a nationwide initiative since 2006. The Action Plan for Promoting Healthy Aquaculture Development is to improve the efficiency and decrease the adverse environmental impact of pond aquaculture through improved land use and pond water recirculation with fewer effluents that cause pollution (Li et al. 2011).

There are many control measures of aquaculture tail water, such as physical methods, chemical methods, biological methods, and ecological methods. While most methods are expensive and also difficult to operate to meet the need of water recycling use, the artificial wetland is designed with the concept of ecological regeneration and it becomes more and more popular in aquaculture wastewater treatment. In addition, the aquaculture tail water treated by artificial wetland can be used to recycle water. It has been reported that attapulgite clay can remove P and N from polluted water (Gan et al. 2010, Shi et al. 2013, Yin et al. 2016). Based on the excellent performance of both attapulgite and artificial wetland, there is a hypothesis that combining attapulgite and artificial wetland could better purify pond feeding tail water and implement tail water recycling. So this paper's aims are: (1) to evaluate pollution status of crab–crayfish polyculture water of an aquaculture farm in Xuyi County of Jiangsu Province; (2) to assess the removal capacity and performance of attapulgite from Xuyi County; (3) to construct and evaluate the recycling system with attapulgite and artificial wetland for crab–crayfish polyculture water in the aquaculture farm in Xuyi county.

2 MATERIALS AND METHODS

2.1 The study area

Xuyi is an important county in central Jiangsu Province in east China with a total area of 2497 km^2 and a population of 0.80 million at the end of 2015. The region is located at 32°43'~33°13'N and 118°11'~118°54'E. It has a warm and humid subtropical climate with an annual temperature of 14.7°C, rainfall of 1005.4 mm. Xuyi County is located downstream of the Huaihe River, and south bank of Hongze Lake. The study aquaculture farm is located at western Xuyi County, and the area of the aquaculture pond is 135 ha. Its main breeds are crab and crayfish.

2.2 Sorption capacity of attapulgite from water

The natural attapulgite clay samples were collected from the study aquaculture farm at Xuyi County (China). The adsorbent sample used was manually ground and selected for particles <100 mesh.

Artificial phosphate and ammonia nitrogen solutions were used throughout the sorption tests. Initially, a stock solution of 1000 mg/L in phosphorus and ammonia nitrogen was prepared by dissolving a certain amount of chemically pure $K_2HPO_4•H_2O$ and NH_4Cl in distilled water, respectively. An aliquot of the stock solution was mixed with a certain volume of water so that phosphate and ammonia nitrogen solution were prepared at the desired experimental concentration.

To evaluate the phosphate and ammonia nitrogen sorption capacity of natural attapulgite, natural samples were used. The 25 mL of phosphate stock solution with phosphate concentrations of 0.4 mg/L and 1.0 mg/L was added to 0.25 g adsorbent. The mixture was stirred at 200 rpm in a thermostatic shaker for 24 h at 25°C to ensure approximate equilibrium. After phosphate sorption, the solution was filtered through a 0.22 mm membrane filter and then analyzed for P. The quantity of adsorbed phosphate (sorption capacity) was calculated from the decrease in phosphate concentration in solution.

The 25 mL of ammonia nitrogen stock solution with concentrations of 0.5 mg/L and 5.0 mg/L was added to 0.1 g adsorbent. The mixture was stirred at 200 rpm in a thermostatic shaker for 24 h at 25°C to ensure approximate equilibrium. After ammonia nitrogen sorption, the solution was filtered through a 0.22 mm membrane filter and then analyzed for ammonia nitrogen. The quantity of ammonia nitrogen removal (sorption capacity) was calculated from the decrease in ammonia nitrogen concentration in solution.

2.3 Construction of recycling system for crab–crayfish polyculture water

The recycling system for crab–crayfish polyculture wastewater treatment consisted of aquaculture pond, precipitation system, ecological ditch, adsorption system, artificial wetland, water quality testing system, and external reservoir (Figure 1).

1. Aquaculture pond: Crab and crayfish were bred in a pond with mixed aquaculture. The crab stocking density is 12,000 heads per hectare on March every year and crayfishes have natural reproduction. There are many aquatic plants in the ponds such as *Elodea nuttallii, Vallisneria natans* (Lour.) Hara, and *Hydrilla verticillata*. Some snails and silver carps are also being bred in the ponds. Through the food chain cycle of producers, consumers, and producers, a three-dimensional aquatic ecological environment is constructed.

2. Precipitation system: Tail water from the aquaculture pond enters into the PS through a siphon device. There are many shrimps and shellfishes in the PS, which can help suspended substances settle down in the aquatic pattern. The suspended matter will be removed through a sand tank at the end of the PS.

3. Ecological ditch: The ED part mainly cultivates a lot of small indigenous grass, as well as a small amount of grass-eating fish.

4. Adsorption system: The AS is a ditch with its sediment mainly composed of attapulgite clay.

5. Artificial wetland: There are many aquatic economic plants and native landscape plants in the AW and an appropriate amount of grass-eating

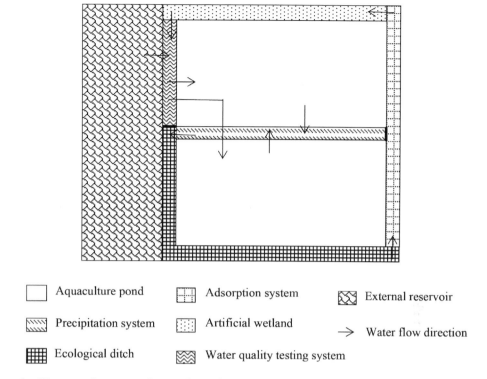

	Aquaculture pond		Adsorption system		External reservoir
	Precipitation system		Artificial wetland	→	Water flow direction
	Ecological ditch		Water quality testing system		

Figure 1. Water recycling system of aquaculture tail water.

fishes and aquatic animals are also being bred in the ponds in the AW.

6. Water quality testing system: The indices of water quality were analyzed after treatment, including physical and chemical index, and biological indicators. The water quality was also tested by certain aquaculture products. In order to improve the aquaculture water quality, water will be pumped from WS to AP after meeting the requirements of aquaculture.

7. External reservoir: In order to maintain a cyclic water balance, the ER is used to add fresh water to the WS.

2.4 Sampling and sample analysis

Since the crayfish is on sale from June 10th to September 30th and the crab comes into the market from September 20th to December 30th, the water quality is the poorest in September from crab and crayfish polyculture. To test the purification effect of pond tail water by the water recycling system, the water samples were collected in September 2013 and 2014, respectively. The pH, DO, SS, TP, TN, NH_3-N, and COD_{Mn} in water samples are determined by the portable water quality analyzer (Hach, DR2800).

3 RESULTS AND DISCUSSION

3.1 Pollution status of crab–crayfish polyculture water

The quality of crab–crayfish polyculture water in 3 ponds is presented in Table 1. According to the relative standard values of Water Quality Standard for Fisheries (GB11607-89), Environmental Quality Standards for Surface Water (GB 3838-2002), and Quality Standards for Surface Water Resources (SL63-94), the values of SS, COD_{Mn}, NH_3-N, and TN in pond water samples were seriously beyond their threshold. The TP values in water samples from pond 2 and pond 3 were above its threshold value. Meanwhile, DO values in water samples are also not achieving its standard. These data indicate that the crab–crayfish polyculture pond could lead to water pollution. The pollution sources are excreta of aquatic products and residual concentrations of aquaculture baits and medicines used in polyculture ponds.

3.2 Sorption ability of attapulgite in wastewater

The P and NH_3-N removal capacities of natural samples were analyzed from synthetic wastewater.

The results show that attapulgite adsorbent had a good phosphate removal capacity, which was 45.1% and 37.2% from initial synthetic wastewater with phosphate concentrations of 0.4 mg/L and 1.0 mg/L, respectively. Attapulgite adsorbent also had a good NH_3-N removal performance, which was 66.7% and 38.5% from initial synthetic wastewater with NH_3-N concentrations of 0.5 mg/L and 5.0 mg/L, respectively. Attapulgite is a vital non-metallic mineral resource at Xuyi and the total resource of attapulgite is 2.18 billion tons. Because of its structure and physicochemical properties, the attapulgite clay has excellent adsorption properties (Gan et al. 2009) and could be used as a potential adsorbent.

3.3 Application effect of recycling system in crab–crayfish polyculture water

The application effect of the water recycling system in September 2013 and 2014 were presented in Table 2 and Table 3, respectively. The results

Table 1. The water quality of crab and crayfish polyculture ponds.

Pond	Time	pH	SS	COD_{Mn}	NH_3-N	TN	DO	TP
Pond 1	Sep 2013	8.39	41	16	1.40	1.75	9.47	0.16
	Sep 2014	8.11	46	26	0.62	3.49	4.63	0.17
Pond 2	Sep 2013	8.27	42	54	1.51	2.20	7.19	0.19
	Sep 2014	7.77	52	69	2.49	5.09	4.47	0.44
Pond 3	Sep 2013	8.34	54	47	1.23	1.81	7.01	0.21
	Sep 2014	8.33	42	76	1.39	3.09	6.5	0.40
Limit value*		6.5~8.5	+10	–	–	–	5	–
III standard value**		6~9	30***	6	1.0	1.0	5	0.2

* Water quality standard for fisheries (GB11607-89).
** III standard value of environmental quality standards for surface water (GB3838-2002).
*** Three grade of quality standards for surface water resources (SL63-94).

Table 2. Operating results of the polyculture water recycling system in September 2013.

	pH	SS	COD_{Mn}	NH_3-N	TN	DO	TP
WS	8.01	25	15	0.53	0.91	6.3	0.03
Pond-1	8.39	41	16	1.40	1.75	9.47	0.16
Pond-2	8.27	42	54	1.51	2.20	7.19	0.19
Pond-3	8.34	54	47	1.23	1.81	7.01	0.21
ED	8.64	25	16	1.31	1.45	15.27	0.15
AS-1	6.67	28	15	1.23	1.23	8.64	0.13
AS-2	7.18	28	15	1.22	1.21	8.76	0.08
AW	8.55	17	7	0.92	0.96	21.9	0.03
ER	6.85	15	6	0.86	0.97	8.98	0.03
III standard value*	6–9	30	6	1.0	1.0	5	0.2

* III standard value of environmental quality standards for surface water (GB3838-2002).

Table 3. Operating results of the polyculture water recycling system in September 2014.

	pH	SS	COD_{Mn}	NH_3-N	TN	DO	TP
WS	8.26	34	18.2	0.32	2.93	10.9	0.08
Pond-1	8.11	46	26.5	0.62	3.49	4.6	0.17
Pond-2	7.77	52	69.6	2.49	5.09	4.5	0.44
Pond-3	8.33	42	76.3	1.39	5.09	6.5	0.40
ED	7.82	12	39.4	1.03	3.09	8.2	0.22
AS-1	8.65	28	18.3	0.69	1.93	12.1	0.07
AS-2	8.55	30	22.9	0.46	4.82	12.2	0.07
AW	8.68	8	12.7	0.13	1.53	11.7	0.02
ER	7.34	14	26	1.08	2.02	2.5	0.46
III standard value*	6–9	30	6	1.0	1.0	5	0.2

* III standard value of environmental quality standards for surface water (GB3838-2002).

suggest that the water recycling system had an excellent application effect. The ED has good effects on the SS and COD removal. The attapulgite in AS has a good adsorption capacity of TP in water. The NH_3-N and TN in water significantly decreased in AW. The DO in water was improved in ED and AW. The crab–crayfish polyculture water could wholly meet the III standard value of Environmental Quality Standards for Surface Water (GB3838-2002) after the water recycling system. Owing to external causes, the water quality in ER was poor in September 2014. To maintain the cycle of water balance, the recycling water flowed into WS on one hand, and fresh water from the ER also added to the WS on the other hand. Water recycling was implemented in the whole system, and there was no wastewater discharge.

4 CONCLUSIONS

The crab–crayfish polyculture could result in water pollution. Artificial wetlands designed with the ecological regeneration concept become increasingly popular in aquaculture wastewater purification. Attapulgite clays have a good adsorption capacity of P and NH_3-N in synthetic wastewater because of its structure and physicochemical properties. A combination of attapulgite and artificial wetland could be an enhanced method to purify polyculture tail water and the real application effect of the water recycling system indicates that it truly had an excellent performance on purifying the crab–crayfish polyculture water.

ACKNOWLEDGEMENTS

This work was financially supported by the science and technology project of Jiangsu Province (BN2012068) and the Major Science and Technology Program for Water Pollution Control and Treatment (2012ZX07506007).

REFERENCES

Bostock, J., B. McAndrew, R. Richards, K. Jauncey, T. Telfer, K. Laorenzen, D. Little, L. Ross, N. Handisyde, I. Gatward, & R. Corner (2010). Aquaculture: global status and trends. *Philos. T. R Soc. B.* 365, 2897–2912.

Boxall, A., L. Fogg, P.A. Blackwell, P. Blackwell, P. Kay, E. Pemberton, & A. Croxford (2004). Veterinary medicines in the environment. *Rev. Environ. Contam. T.* 180, 1–91.

Cao, L., W. Wang, Y. Yang, C. Yang, Z. Yuan, S. Xiong, & J. Diana (2007). Environmental impact of aquaculture and countermeasures to aquaculture pollution in China. *Environ. Sci. Pollut. R.* 14, 452–462.

Gan, F., J. Zhou, H. Wang, C. Du, & X. Chen (2009). Removal of phosphate from aqueous solution by thermally treated natural palygorskite. *Water Res.* 43, 2907–2915.

Li, X., J. Li, Y. Wang, L. Fu, B. Li, & B. Jiao (2011). Aquaculture industry in China: current state, challenges, and outlook. *Rev. Fish. Sci.* 19, 187–200.

Peter, E. (2015). Aquaculture environment interactions: Past, present and likely future trends. *Aquaculture* 447: 2–14.

Shi, W., Y. Duan, X. Yi, S. Wang, N. Sun, & C. Ma (2013). Biological removal of nitrogen by a membrane bioreactor-attapulgite clay system in treating polluted water. *Desalination* 317, 41–47.

Yin, H., M. Han, & W. Tang (2016). Phosphorus sorption and supply from eutrophic lake sediment amended with thermally-treated calcium-rich attapulgite and a safety evaluation. *Chem. Eng. J.* 285, 671–678.

Civil, Architecture and Environmental Engineering – Kao & Sung (Eds)
© 2017 Taylor & Francis Group, ISBN 978-1-138-02985-9

Sustainable energy industry coupled with social license to operate

Wan Shen & Ruifeng Li
Shenhua Science and Technology Research Institute, Beijing, China

ABSTRACT: From the present to the predictable future, coal remains the basic energy source around the world. Exploitation and utilization of coal resources plays a key role in sustainable development of world economy. Therefore, to ensure that coal resources can be exploited successfully, it's necessary to realize the sustainable development of the coal industry. The utilization of Social License to Operate in a scientific and reasonable way is closely associated with not only the successful implementation of an engineering project, but the healthy development of mining companies and social stability. This research conducts a data mining analysis on internet data in connection with coal exploitation in the past decade. A statistical review of factors affecting Social License to Operate of coal exploitation has been conducted on worldwide network media, search amount of internet users and report coverage of six representative print media. Through the above three channels, the order of importance regarding factors affecting Social License to Operate in connection with coal exploitation is obtained and verified. The results would provide strategic support and advice to world coal industry on sustainable mining on a global scale.

1 INTRODUCTION

With economic and social development, the requirement for energy is not limited to meet people's basic living standards. Comprehensive utilization of energy resources, reduction on damage to the ecological environment and concentration on the relationships with surrounding communities have become a new trend. With advancement of industrial technology, long term and significant development has been achieved for coal industry. The consumption of coal has occupied an important position in the energy mix. As one of main energies in the world, coal plays a key role in supplying energies required for economic growth and social development, as it has large reserve, relatively obtainable and highly reliable to satisfy energy needs in the world (Kraft John & Lee Chien-Chiang, 2008).

Coal is the most economic fossil fuel with large reserves and extensive distribution. The advantages also provide a guarantee for coal's leading role in the energy industry. However, the more the coal exploitation and consumption, the more ecological damage and environment pollution are and the more negative social image aroused. The coal industry will inevitably bring extensive, grave and even irreversible environmental problems while it guarantees economic development (Zhu Shuwen, 2010). The coal industry has caused large-scale damage to the land and vegetation, especially damage to the local landscape resulting from surface mining, which causes substantial imbalance between

regional environment and ecological systems. Coal gangue discharged during the production process causes severe pollution to lands, rivers and air. Industrial sewage discharged outside also cause severe pollution to underground water and rivers, lakes and seas. Meanwhile, a large amount of dust, CO_2, and SO_2 are produced in the process of coal production and consumption (Bian, H.I, 2010 & M.J. Chadwick, 1996). These problems have a direct effect on mining companies' development while negatively affect people's health physically and psychologically.

As the economy, politics and social culture are continuously developing, the relationships between coal companies, communities and society are also changing from time to time. Those conditions that coal companies depend on in their daily operations changes, thus coal companies must seek better approaches to deal with the relationships with local communities. Especially in recent years, as more and more people have kept their eyes on the social responsibilities of mining companies about environmental protection and sustainable development, the conflicts arising between coal companies and communities have become more furious. Coal mining offers job opportunities, builds industrial bases, and boosts taxation and incomes. On the other hand, coal mining has many negative effects on the society and environment, i.e. widening gap between the rich and poor, poor working conditions, corruptions, damage to environment and to physical and psychological health of employees. Thus coal industry has received more and more criticisms from government, non-government organizations and local communities. Due to coal companies' involvements in the energy sector, their social responsibilities will be given more attention from society.

Pursuit of harmonious development among resources, environment, economy and society has become the focal point. The major and important topic that needs to be resolved imperatively in the theoretical research and practical exploration of the current phase is how to find an emerging industrialized path with respect to harmonious advancement of coal mining and social development. The concept of Social License to Operate (SLO) indicates a direction for the development of the coal industry (R.G. Boutilier, 2014). At the United Nations Conference on Environment and Development held in 1992, companies' social and environmental responsibilities were raised on the agenda followed by establishment of the International Coal and Metal Commission, which is an authority organization for the resource industry. At the same time, the review commission required the mining industry to issue the Transparent Mining Initiative. Since then, this concept has been rapidly accepted by mining companies around the world. Enormous large international mining companies began to employ sociologists, anthropologists and even genealogists to offer advice on social and environmental problems encountered in coal industry. From then on, the Social License to Operate has progressively become a requirement for normal operation of coal companies (S. Bice, K, 2014 & D.M. Franks, 2010).

The acquisition of a Social License to Operate is necessary for the reduction of social conflict risks and to boost a company's reputation. If no Social License to Operate has been obtained, additional economic loss will occur. There will be difficulties in employing a labor force and all kinds of cost increases will incurred and consequently cause operating costs to rise, the company to stop its production and even cause the closure of the mine due to objections from communities. The social actions of the mining company will have a direct effect on the social evaluation and authentication from communities. Therefore, acquisition of the Social License to Operate from local communities would enable companies to avoid potential risks, conflicts and extra costs. Grant of the Social License to Operate means communities will benefit from the company's project and both parties will aim at the same objectives. In the whole life cycle of exploitation, the company will realize development toward a win-win situation.

As the mining companies also are part of the social system, not only their actions affect the surrounding environment, but also the surrounding environment and interested parties will affect mining companies in turn. And obtaining the Social License to Operate implies that this company acknowledges social responsibilities in the respect of environment and local communities, thus the local government would speed up to issue administrative approval and local residents will trust the company. Consequently, the company and the locals will form a positive relationship which is in favor of the development of the whole company. Once conflicts arise between the company and communities or any problem harmful to Social License to Operate caused by failure to maintain the relationships with people in the mining area, the company should find out factors affecting relationships and the influenccial extent, i.e. research factors affecting the Social License to Operate, then according to the extent of importance order to take measures. Only in this way most prominent social problems encountered by the company can be effectively identified and resolved in the priority order for specific purpose. In this case the risks incurred to the company arising from prominent problems will be avoided. Thus, the research on the Social License to Operate in connection with coal exploitation is of great significance for coal industries' sustainable development.

2 RESEARCH METHODS

This research applied data mining and analysis on mass information collected from the internet. First, this research carries out a statistical analysis on report coverage made by worldwide network media on factors affecting the Social License to Operate of coal exploitation. Based on the report coverage, the degree of attention to respective influencing factors by network media in the past decade is obtained. This degree of attention implies the significance of the order of respective factors affecting the Social License to Operate in connection with coal exploitation, thus we acquire the significance order of different influencing factors through the first channel. Next, we conduct a statistical analysis on the search amount on related influencing factors by worldwide internet users. According to the result, combined with the difference on search amount among respective factors, we obtain the degree of attention to respective influencing factors by worldwide internet users, where we acknowledge that the degree of attention to related factors by internet users is a decisive factor regarding significance of the Social License to Operate in connection with coal exploitation. Therefore, we acquire the significance order of different influencing factors through the second channel.

3 FACTORS AFFECTING THE SOCIAL LICENSE TO OPERATE

Social License to Operate (SLO) refers to the acceptance or support degree within local government, communities and media of the mining project developed by a mining company. SLO does not refer to a formal agreement or document, and does not have a formal or formalized assessment process or appropriate and institutionalized evaluation criteria. It depends on current creditability, reliability, and acceptance of mining companies and the projects. SLO mainly involves the relationships with stakeholders, especially with the local communities. The SLO is dynamic because stakeholders' perceptions can change over time and by region.

Firstly, the acquisition of the SLO has become a necessary process of mining companies to carry out mining operations. Similar to a business license, the SLO has become an indispensable part for mining companies to obtain mining rights. If they cannot obtain the SLO, it's hard to operate the mining operations in the local community due to difficulties in hiring employees, relocation or land occupancy.

Secondly, the acquisition and long-term maintenance of SLO is a necessary condition to reduce social conflicts. Company development is inseparable from the understanding and support of the community. This support is critical not only for dealing with "bilateral relations", but for their own long term development. Mining operations will inevitably bring to the local community problems, such as land occupancy, water pollution, etc. If the companies cannot get community support, then social conflicts may occur and result in delays or even coal mine shutdown.

Finally, the acquisition and maintenance of SLO can effectively reduce unnecessary expenditure and reduce financial costs. If a mining company cannot obtain the SLO, it means that the company has low social recognition, its credibility may be poor and investors will be more careful in investment. Thus, in order to attract business investors or social financing, the company will encounter big challenges to promote corporate image and earn business reputation. We must clearly understand that the cost to regain a SLO is far more than needed to maintain it.

In the above description, we have illustrated that coal companies cannot operate in isolation, and SLO has become a prerequisite for business success. Over the past decades, the coal industry has continued to make breakthroughs in technology, but the company profits increasingly depends on its ability to solve the "non-technical" issues, such as how to deal with the community, non-governmental organizations, government departments and other stakeholders. Technology is not the only tool leading to business success. For business operations, "social license" is also a prerequisite. For coal enterprises, consider how to get SLO the first problem to be solved is to identify factors which influence obtainment of SLO. By survey of key stakeholders and through a literature research, we identified that the main factors affecting coal mining operation can be divided into eight areas: environmental pollution, safety, land resources, water resources, human rights, community engagement, low-carbon technologies and greenhouse gases.

4 ANALYSIS

For the importance of factors, this paper carries out a statistical analysis on related data of eight factors relating to coal exploitation (environmental pollution, safety, land resources, water resources, human rights, community engagement, low-carbon technologies and greenhouse gases) through three methods, so as to obtain the significance order of the eight factors affecting SLO of coal exploitation, and provide guidance and advice for factors needed to be given priority. The three methods are report coverage of global Internet media, search amount of global internet users and report coverage of global print media.

Analysis on the significance of influencing factors through report coverage of global internet media is conducted from the perspective of internet media in order to analyze the degree of attention on related factors by network media. As the fastest news communication media, the degree of attention on related factors by internet media reflects to a certain extent the significance degree of such factors. Analysis on the significance of influencing factors through search amount of internet users is conducted from the perspective of social publics in order to analyze the degree of attention to related factors by publics. As the direct stakeholder for development of coal companies, the degree of attention to related factors by social publics directly determines the significance degree of such factors.

Analysis on the significance of influencing factors through report coverage of print media is conducted from the perspective of print media in order to analyze the degree of attention to influencing factors by the society. Articles published in print media are usually relative specific, objective and even authoritative, as in terms of social event, print media will publish more detailed reports. Commentary articles reflect the degree of attention on such events by specialists and scholars.

Thus to study the factors affecting the SLO of coal exploitation, the degree of attention also embodies to a certain extent the significance degree of related factors.

After the significance orders are obtained through above three methods, we conduct comparative analysis on each of the three orders so as to summarize and verify their consistencies and conduct analysis on their differences, thus determining the significance order of factors affecting the SLO of the coal industry. This order result will provide a basis for the World Coal Industry to identify priority areas for future strategies, and also advices on the priority of influencing factors for companies to obtain the SLO.

In the results of the statistical analysis regarding the report coverage from global internet media on factors affecting the SLO of coal mining, our data mainly come from search results by applying the Google search engine for "coal + 'influencing factor'" from September 2003 to October 2013. Statistics result is shown below:

From Figure 2, we can clearly observe, as time goes by, the report numbers of the Internet media on factors affecting the SLO of coal mining is constantly increasing, and with the simultaneous growth in this background, the report numbers of network media on each factor has apparent

Figure 1. Three methods used to research on factors affecting the SLO of coal mining.

difference and the report numbers from less to more are community engagement, low-carbon technologies, human rights, greenhouse gases, pollution, safety, land resources and water resources.

The Google trend platform is used for conducting statistical analysis on quantity of internet users' search on factors which affect coal mining and social license to operate. The statistics on search quantity of internet users still adopts the eight keywords in the above-mentioned research. By conducting statistics on search quantity of relevant factors between October 2003 and October 2013, we can obtain the search quantity collecting figure. By analysis on statistic results and information represented by the figure, the search quantity of influence factors and attention paid on influence factors are obtained in line with the sequence. The following figure shows the statistical result.

Figure 3 takes the maximum search numbers of the eight factors between October 2003 and October 2013 as 100, and other search values are respectively displayed in 0–100 compared to the maximum value. It can be clearly observed from the figure that the search quantity of every factor shows certain volatility in the past ten years, but there are obvious differences in search quantities on different factors. The search quantity can be ordered as community engagement, human rights, low-carbon technology, greenhouse gases, land resources, safety, and environmental pollution and water resources.

Whether there is pressure of social opinion or enterprise social responsibility, both make print media comply with strict malfunction system and regulations in the course of publishing information and auditing contributions, ensuring preciseness, authenticity and even authority, especially for some print media with long history and large social influence. The report quantity of print media in allusion to influence factors objectively embodies attention degree of relevant experts, scholars and even government departments on influence factors. The difference in the attention also determines importance degree of influence factors in

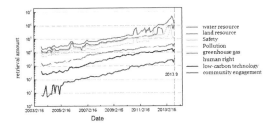

Figure 2. The reports numbers of global internet media on factors affecting the SLO of coal mining in the past decade.

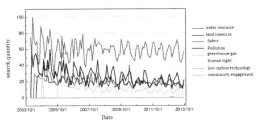

Figure 3. Search quantity of internet users on factors influencing coal mining and social license to operate in last ten years.

the view of experts, scholars and even government departments.

In allusion to the impact of internet media, print media also improves its operation mode to avoid being eliminated. Netzens can read and inquire periodicals published respectively via the website of print media. This research adopts the retrieval mode of using "coal + influence factors" in official websites of relevant print media, to obtain report quantity of news reports or commentary articles of a certain media on influence factors related to social license to operate of coal mining from January 2010 to October 2013, then gathering retrieved results of all media and drawing out the figure, to analyze importance degree of influence factors and conduct the significance sequence.

This research selects 6 print media including the New York Times, The Economist, The Times, Financial Times, Wall Street Journal and TIME magazine. The following information briefly introduces the six print media and the reason for selecting them as statistical objects. The six media have a long history with wide circulation, authority in news reporting and commentary articles and high influence on the global scale. The six media all meet the requirements of this research in quantity and quality, so they are selected as statistical objects.

In the following statistical results, a curve chart is used for analyzing statistical results of different media. The horizontal axis is time and the vertical axis is report quantity. The statistical results of New York Times as an example are listed in the Figure 4.

Following the importance ordering of the factors for the statistical results obtained through the above three channels, we could divided these factors into 3 tiers. The summary of the three tiers is as follows.

Through the above summary table we can see that the three channels have totally the same idea

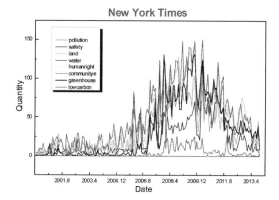

Figure 4. Quantity of report of New York Times on relevant factors in the past 13 years.

Table 1. Summary of the importance ordering of the factors obtained from three research channels.

	Extremely important	Important	Less important
Report amount by the global Internet media	Water	Land Safety Pollution	Human rights Greenhouse gas Low-carbon Community engagement
Search amount by the global Internet users	Water	Pollution Safety Land	Greenhouse gas Human rights Low-carbon Community engagement
Report amount by the print media	Water	Pollution Safety Land	Human rights Greenhouse gas Low-carbon Community engagement

on the importance of "water", which is in the "extremely important" position in the three orderings. Therefore, we consider "water" as the most important factor affecting the social license to operate for coal mining. Meanwhile, the three channels also have a high degree of agreement on the importance of the three factors "land resources", "safety problem" and "environment problem". Though a slight difference in the degree of importance exists in the interior of the level, it does not affect our view about the general importance degree. Therefore, due to the consistency on the results, we treat the "land resources", "safety problem" and "environment problem" as the "important" level. The three channels have certain differences on the importance degree for the four factors of "human rights", "greenhouse gas", and "low-carbon" and "community engagement" with the main reason that the report quantity and retrieval quantity for these four factors are relatively less; thus, we combine these four factors into one hierarchy, the "less important" level.

5 CONCLUSIONS

SLO has become necessary and mandatory for coal enterprises to remain in operation; for coal enterprises to obtain an SLO is not only required for them to launch a coal mining operation, but also an effective guarantee to save their costs and improve their social image and reputation. Therefore, the coal industry should evaluate the importance of SLO and promote it among its member enterprises and organizations. Meanwhile, by collecting and using the data on the Internet in the abovementioned research, we also obtained the important sequence of the factors which have influence on the SLO for coal mining, distribution charts of the global attention rate on each influencing factor and several well-known media's attention rates on relevant influencing factors. By making the research, we observed that the factor which has the most influence on the SLO for coal mining in current phase is water resources problems caused by coal mining, followed by land resources, environment and safety problems, and finally human rights, greenhouse gas, low-carbon technology and participation problems.

Based on the results, coal industry can take Public Relation (PR) strategies accordingly, providing specific countries or cities with help in respect to resources to improve the image of coal industry, promoting coal enterprises to obtain a wider SLO, winning understanding and support from the whole society. Simultaneously, we also research on influencing factors by different media, and we observed that different media pay different attention to different influencing factors, which is important for coal industry to select PR media and information releasing platforms.

REFERENCES

Bian, H., I. Inyang, J.L. Daniels, F. Otto, Environmental issues from coal mining and their solutions, Mining Science and Technology, 20 (2) (2010), pp. 215–223.

Bice, S., K. Moffat, Social licence to operate and impact assessment, Impact Assess. Proj. Apprais., 32 (2014), pp. 257–262.

Boutilier, R.G. Frequently asked questions about the social license to operate, Impact Assess. Proj. Apprais., 32 (4) (2014), pp. 263–272.

Chadwick, M.J., N.H. Highton, N. Lindman, Environmental Impacts of Coal Mining and Utilization, Pergamon Books Inc, Elmsford, NY (1996).

Franks, D.M., D. Brereton, C.J. Moran, Managing the cumulative impacts of coal mining on regional communities and environments in Australia, Impact Assess. Proj. Apprais., 284 (2010), pp. 299–312.

Kraft John, Arthur, Kraft, On the Relationship between Energy and GNP. J, Energy Development, 3 (1980), pp. 401–403.

Lee Chien-Chiang, Chang Chun-Ping, Energy Consumption and Economic Growth in Asian Economies, A more Comprehensive Analysis Using Panel Data. J, Resource and Energy Economics, 2008; (30): 50–65.

Zhu Shuwen, Ding Yongxia, et al. The Analysis of the Relationship among Energy consumption, Economic Growth and Carbon Emissions,. J, China Soft Science, 5 (2010), pp. 12–19.

Civil, Architecture and Environmental Engineering – Kao & Sung (Eds)
© 2017 Taylor & Francis Group, ISBN 978-1-138-02985-9

Lab-scale treatment of biologically pretreated landfill leachate by the electro-Fenton process in a continuous flow reactor

Shaopeng Yuan, Xia Qin, Li Zhang & Mengnan Zhou
College of Environmental and Energy Engineering, Beijing University of Technology, Beijing, P.R. China

ABSTRACT: The lab-scale treatment of biologically pretreated landfill leachate by electro-Fenton (E-Fenton) was carried out in a continuous flow reactor using $Ti/RuO_2\text{-}IrO_2$ mesh anodes and cathodes. The effects of important parameters including voltage, hydraulic retention time, H_2O_2/Fe^{2+} molar ratio, and H_2O_2/COD_0 molar ratio on COD removal were investigated, and the optimum conditions for this advanced oxidation process were found. There was an optimal H_2O_2/Fe^{2+} and H_2O_2/COD_0 molar ratio so that the highest COD removal rate was achieved. The Dissolved Organic Matters (DOM) in the leachate were analyzed using ultraviolet spectrum and Excitation Emission Matrix (EEM) spectrofluorimetry. About 84.58% COD was removed and colority was completely removed after the E-Fenton process. The results indicate that the E-Fenton process is an effective technology for the treatment of biologically pretreated landfill leachate.

1 INTRODUCTION

Landfill leachate is strongly polluted wastewater that contains complex pollutants including organic compounds, ammonia, heavy metals, inorganic salts, etc. (Wu, 2011). Conventional biological treatment is an economical way to remove biodegradable organic compounds in the leachate. However, mature landfill leachate contains significant amounts of recalcitrant organic compounds such as Humic Acid (HA), Fulvic Acid (FA), and hydrophilic fractions, which cannot be effectively removed by conventional biological treatment (Atmaca, 2009). Therefore, the effluent of biologically treated mature leachate usually contains considerable amounts of refractory organic pollutants, which may pose hazards to the environment. The biologically treated effluents then have to be properly treated before they can be discharged into the environment.

In the past two decades, Advanced Oxidation Processes (AOPs) have received great attention as alternative methods for the treatment of various wastewaters and they have advantages over other conventional treatment techniques such as high removal efficiency, minimum treatment time, and less sludge production (Antonopoulou, 2014; Oturan, 2014). Moreover, AOPs are capable of transforming nonbiodegradable pollutants into nontoxic biodegradable substances (Wang, 2011). Since the year 2000, the electro-Fenton process has been used for the treatment of mature landfill leachate (Lin, 2000; Zhang, 2007; Orkun,

2012). Compared with the classic Fenton process, E-Fenton oxidation could offer significant advantages such as the continuous regeneration of ferrous ion at the cathode and consequently minimization of sludge production (Brillas, 2009), and the synergistic effect between the electrochemical process and the Fenton process (Lin, 2000 & Zhang, 2007). The reaction mechanism of E-Fenton technology is complicated, but its main involved reactions can be described by the following equations (Brillas, 2009, Rosales, 2012; Wang, 2013).

$$O_2 + 2H^+ + 2e^- \rightarrow H_2O_2 \qquad (1)$$

$$H_2O_2 + Fe^{2+} + H^+ \rightarrow Fe^{3+} + \bullet OH + H_2O \qquad (2)$$

$$Fe^{3+} + e^- \rightarrow Fe^{2+} \qquad (3)$$

Generally, few studies have reported a lab-scale use of electro-Fenton technology in continuous mode for the treatment of landfill leachate. For this reason, a continuous E-Fenton reactor was developed in order to validate this technique in landfill leachate treatment at lab scale. The effects of various operating factors on E-Fenton performance were evaluated to determine the optimal reaction conditions, such as voltage, Hydraulic Retention Time (HRT), H_2O_2/Fe^{2+} molar ratio, and $H_2O_2/$ initial COD (COD_0) molar ratio. The results indicated that E-Fenton treatment was an effective way to degrade the refractory organics in biologically pretreated landfill leachate. The average removal efficiencies of COD and colority up to 85.58% and 100% were achieved with 40 min of the

E-Fenton treatment. In order to gain insight into the dissolved organic matter in the landfill leachate before and after treatment, leachate composition in the influent and effluent of the E-Fenton reactor was analyzed by ultraviolet spectrum and Excitation Emission Matrix (EEM) spectrofluorimetry.

2 MATERIALS AND METHODS

2.1 Leachate collection and characterization

The leachate samples used in this study were sampled from a municipal landfill site (Beijing, China). In the landfill plant, the leachate had been treated via a combined process of UASB–aerobic–oxic (A-O) and submerged membrane bioreactor treatment (MBR). Its characteristics on average are shown in Table 1.

2.2 Chemical reagents

Hydrogen peroxide (30%, w/w), ferrous sulfate ($FeSO_4$) and all other chemicals used were in analytical grade unless noted otherwise. All solutions were prepared with deionized water.

2.3 Experimental apparatus and treatment procedure

The experiments were carried out in a rectangular electrolytic reactor with 12 L of working volume. E-Fenton experiments were conducted with Direct Current (DC) power supply. Seven 25 cm × 17 cm mesh anodes (Ti/RuO_2–IrO_2) and seven same dimension anodes were positioned alternately, parallel to each other with the selected inter-electrode gap.

Sampling 5 L leachate, the initial pH value of the leachate was adjusted to 3 with diluted sulfuric acid and sodium hydroxide. When the DC power supply was initiated, the leachate was pumped into the reactor by a peristaltic pump. Then the hydrogen peroxide solution and ferrous iron solution were fed into the reactor separately. The treated waste water was discharged out of the reactor and

Table 1. The characteristics of landfill leachate.

Parameter	Value
COD (mg/L)	661.2~1016.74
Ammonia nitrogen (mg/L)	81.04~132.56
pH	7.98~8.06
Colority	850~1500
Color	Yellow

Table 2. Factor levels.

	Factors			
Level	Voltage/V	HRT/min	H_2O_2/COD_0	H_2O_2/Fe^{2+}
1	4.0	20	1	1
2	4.5	30	2	2
3	5.0	40	3	3

quenched the reaction by increasing pH to 8.5 by sodium hydroxide and diluted sulfuric acid.

2.4 Analytical methods

In order to determine the importance of various factors in treating biologically pretreated landfill leachate, an orthogonal experiment was designed to optimize the experiment condition. Factor levels are shown in Table 2.

After processing, the COD was measured by the potassium dichromate method. Dissolved Organic Matters (DOM) were measured with fluorescence and ultraviolet absorption spectrum.

3 RESULTS AND DISCUSSION

3.1 Analysis of the results

Orthogonal experiment design and results are shown in Table 3, and variance analysis shown in Table 4.

The range R_i is employed to evaluate the important order of the factors on the indicator. The factors with large range become the main factors while the ones with smaller values become secondary factors. From Table 3, the important order of these geometric factors on the indicator 'the COD removal rate' is as follows: HRT > Voltage> H_2O_2/Fe^{2+} molar ratio > H_2O_2/COD_0 molar ratio.

The ANOVA results for calculated models are shown in Table 4. The ANOVA indicates that HRT plays an important role in the COD removal rate, whereas, in the selected range, H_2O_2/Fe^{2+} and H_2O_2/COD_0 do not have significant effects on the removal of COD. Therefore, the process of practical application should focus on regulating HRT to enhance the COD removal rate.

Above all, the optimized conditions for the lab-scale continuous flow treatment of biologically pretreated landfill leachate by the E-Fenton process is a voltage of 5.0 V, HRT of 40 min, $H_2O_2/Fe^{2+}=1$, and $H_2O_2/COD_0=1$. A further experiment was performed under proposed conditions and the removal rate of COD is 84.58%, the COD value of effluent water is 101.96 mg/L, the colority is almost zero.

Table 3. L9 (3^4) orthogonal experiment design scheme and its results with the analysis of the indicator.

No.	Voltage (v)	HRT (min)	H_2O_2/COD_0	H_2O_2/Fe^{2+}	COD removal rate
		Test (COD_0 = 661.20 mg/L)			
1	4.0	20	1	1	45.26%
2	4.0	30	2	2	44.35%
3	4.0	40	3	3	57.68%
4	4.5	20	2	3	36.49%
5	4.5	30	3	1	65.05%
6	4.5	40	1	2	71.57%
7	5.0	20	3	2	45.08%
8	5.0	30	1	3	72.09%
9	5.0	40	2	1	84.49%
$\bar{K}_1$	49.10	42.28	62.97	64.93	
$\bar{K}_2$	57.70	60.50	55.11	53.67	
$\bar{K}_3$	67.22	71.25	55.94	55.42	
R_i	18.12	28.97	7.86	11.27	

Table 4. ANOVA of the COD removal rate.

Source of variance	Variance	F	F_α	Significant level
Voltage	246.5485	2.966	$F_{0.05} = 4.46$	
HRT	643.3960	7.739		*
H_2O_2/COD_0	56.0150	0.674		
H_2O_2/Fe^{2+}	110.2575	1.326		
Pooled error	83.1375			

Note: F = variance (factor)/variance (error).
F_α is obtained from F value table for analysis of variance.
* stands for playing an important role.

3.2 Ultraviolet spectrum analysis to the DOM before and after processing

$SUVA_{254}$ represents the aromatic constituents in DOM and has been widely accepted as an index of the aromatic structure during composting (Wang K, 2013). The $SUVA_{254}$ values decreased indicating that the degree of aromaticity becomes lower in the liquid phase of the leachate (Caricasole, 2010). E-Fenton can also oxidize macromolecular organic matter to small molecules compounds containing the conjugated system and the double bond so that the molecular weight become lower.

Using the E-Fenton process to deal with biologically pretreated landfill leachate, the water quality parameters before and after processing are as shown in Table 5.

It can be seen from Table 5, $SUVA_{254}$ has greatly reduced after processing, indicating that the E-Fenton technology has a good removal effect on the degree of aromaticity in the leachate and has a high conversion degree of macromolecular organic matter to small molecule compounds in the leachate.

3.3 Fluorescence analysis to the DOM before and after processing

Excitation-Emission Matrix (EEM) fluorescence spectroscopy has been widely used to characterize and monitor fluorescent Dissolved Organic Matter (DOM) in marine, freshwater, soil, and wastewater samples. EEM fluorescence spectroscopy is a highly sensitive, potent, and useful tool for characterizing DOM in wastewater (Maqbool, 2016). This technique has the advantages of being non-destructive, highly sensitive, rapid, and relatively inexpensive (Fellman, 2010). Fluorescence regional integration has been widely used for many environmental settings to characterize DOM, including landfill leachates (He, 2011, Wu, 2012). Filtered water samples are typically irradiated with excitation wavelengths from 200 to 450 nm and emission wavelengths recorded from 250 to 550 nm.

Generally, five regions in an EEM were operationally defined using consistent excitation (EX) and emission (EM) wavelength boundaries based on the fluorescence of model compounds and DOM fractions (Leenheer, 2003). The five defined regions are shown in Table 6 and Figure 1.

In Table 6, humic acid-like was a kind of degradation-resistant hydrophilic polymer based on diverse quinone and polyphenol as the aromatic core, which was formed by C, H, O, N, S, P, etc. elements. On the aromatic core, there has groups of carboxyl, phenolic groups, carbonyl, peptide, etc. In addition, there are a large number of benzene

Table 5. Water quality parameters before and after processing.

Water quality parameters	COD (mg/L)	$SUVA_{254}$
Before processing	661.2	4.736
After processing	101.96	1.805

Table 6. The five defined regions of EEM fluorescence spectrum.

Regions	Ex (nm)	Em (nm)	Description
I	200–250	280–330	aromatic protein I
II	200–250	330–380	aromatic protein II
III	200–250	380–550	fulvic acid-like
IV	250–400	280–380	soluble microbial byproduct-like
V	250–400	380–550	humic acid-like

rings and functional groups such as -OH, -COOH, >C = O, $-PO_3H_2$, $-NH_2$, $-CH_3$, $-SO_3H$, $-OCH_3$, etc. Under the action of the oxidant, the class humic acid can be oxidatively decomposed.

The EEM fluorescence spectrums before and after processing are shown in Figure 2a and 2b.

As can be seen from Figure 2a, only humic acid-like is in the fluorescence of DOM, Figure 2b shows that after processing, the intensity of region V has greatly reduced which is about 20. Most of the humic acid absolutely mineralizes into CO_2 and H_2O with little changes into small molecule intermediates like aromatic protein and fulvic acid-like.

Fluorescence index $f_{450/500}$ refers to the intensity of fluorescence emission spectrum in the ratio of 450 nm to 500 nm, when the excitation wavelength was 370 nm. The $f_{450/500}$ and the aromaticity of humic acid showed a negative correlation relationship, the greater the $f_{450/500}$ value, the weaker the aromaticity of humic acid and lesser the benzene ring structures.

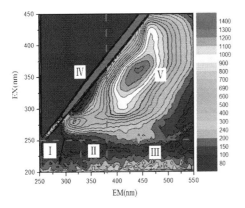

Figure 1.   The five defined regions of EEM fluorescence spectrum.

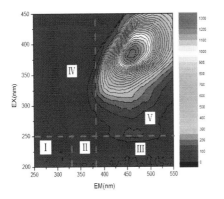

Figure 2a.   The EEM fluorescence spectrum before processing.

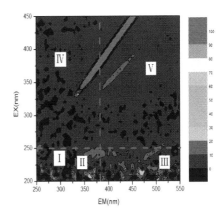

Figure 2b.   The EEM fluorescence spectrum after processing.

Table 7.   Index $f_{450/500}$ before and after processing.

Fluorescence intensity ratio	$f_{450/500}$
Before processing (a)	1.6
After processing (b)	3.1

Fluorescence spectral characteristics of DOM in the leachate samples before and after processing are shown in Table 7.

After processing, the $f_{450/500}$ value increased, indicating that after oxidation treatment, the aromaticity of humic acid attenuated in the leachate, so the benzene ring structure decreased, and this is consistent with the change of $SUVA_{254}$.

4   CONCLUSIONS

The results of the orthogonal experiment show the importance of the factors in treating biologically pretreated landfill leachate: HRT > voltage > H_2O_2/Fe^{2+} > H_2O_2/COD_0. Optimum conditions of dynamic continuous flow running mode for E-Fenton were found to be voltage 5V, HRT 40 min, H_2O_2/Fe^{2+} = 1, and H_2O_2/COD_0 = 1. Three parallel experiments were done under the conditions above, the average removal of COD and colority achieved were 84.58% and 100%, respectively. Hence, E-Fenton was recommended as a powerful technique for the degradation and decolorization of landfill leachate.

UV spectrum analysis results show that the E-Fenton process has a good removal effect on aromatic compounds. EEM spectrofluorimetry results show that E-Fenton oxidation technology has a good removal of macromolecular humic acid in the biologically pretreated landfill leachate.

Most of the organic matter can be decomposed into CO_2 and H_2O.

REFERENCES

Antonopoulou, M., et al (2014). A review on advanced oxidation processes for the removal of taste and odor compounds from aqueous media, J. Water Res. 53, 215.

Brillas, E.I., & M.A, Sires (2009). Oturan, Electro-Fenton process and related electrochemical technologies based on Fenton's reaction chemistry, J. Chem. Rev. 109, 6570–6631.

Caricasole, P (2010). Chemical characteristics of dissolved organic matter during composting of different organic wastes assessed by 13C CPMAS NMR spectroscopy, Bioresour. Technol. 101, 8232–8236.

Fellman, J.B., et al (2010). Fluorescence spectroscopy opens new windows into dissolved organic matter dynamics in freshwater ecosystems: A review. Limnol. Oceanogr. 55, 2452–2462.

He, X.S., et al (2011). Fluorescence excitation-emission matrix spectroscopy with regional integration analysis for characterizing composition and transformation of dissolved organic matter in landfill leachates, J. Hazard. Mater. 190, 293–299.

Leenheer J.A., et al (2003). Characterization and copper binding of humic and nonhumic organic matter isolated from the South Platte River: evidence for the presence of nitrogenous binding site, J.Environ. Sci. Technol. 37(1), 328–336.

Lin, S.H., & C.C. Chang (2000). Treatment of landfill leachate by combined electro-Fenton oxidation and sequencing batch reactor method, J. Water Res. 34, 4243–4249.

Maqbool Tahir, et al (2016). Characterizing fluorescent dissolved organic matter in a membrane bioreactor via excitation–emission matrix combined with parallel factor analysis, J. Bioresource Technology. 209, 31–39.

Orkun, M.O. & A. Kuleyin (2012). Treatment performance evaluation of chemical oxygen demand from landfill leachate by electro-coagulation and electro-Fenton technique, Environ. Prog. Sust. Energy. 31, 59–67.

Oturan, M.A. & J.J. Aaron (2014). Advanced oxidation processes in water/wastewater treatment: Principles and applications. A Review, Crit. Rev. Environ. Sci. Technol. 44, 2577.

Rosales, E., et al (2012). Advances in the electro-Fenton process for re-mediation of recalcitrant organic compounds, J. Chem. Eng. Technol. 35, 609.

Wang, K., et al (2013). Spectral study of dissolved organic matter in biosolid during the composting process using inorganic bulking agent: UV–vis, GPC, FTIR and EEM, J. Int. Biodeterior. Biodegrad. 85, 617–623.

Wang, M.Z., et al (2011). Optimization of Fenton process for decolouration and COD removal in tobacco wastewater and toxicological evaluation of the effluent, J. Water Sci. Technol. 63, 2471–2477.

Wang, Y (2013). Three-dimensional homogeneous ferrite-carbon aerogel: One pot fabrication and enhanced electro-Fenton reactivity, ACS Appl. Mater. Inter. 5, 842.

Wu, H.Y., et al (2012). Fluorescence-based rapid assessment of the biological stability of landfilled municipal solid waste, J. Bioresour. Technol. 110, 174–183.

Wu, P.X., et al (2011). Effect of dissolved organic matter from Guangzhou landfill leachate on sorption of phenanthrene by Montmorillonite, J. Colloid Interface Sci. 361, 618–627.

Zhang, H (2007). Treatment of landfill leachate by electro-Fenton process, J. Fresenius Environ. Bull. 16, 1216–1619.

Civil, Architecture and Environmental Engineering – Kao & Sung (Eds)
© *2017 Taylor & Francis Group, ISBN 978-1-138-02985-9*

The water and sediment characteristics numerical simulation of dig-in basin in strong tidal estuary

Gong-jin Zhang, Chuan-teng Lu, Xiao-feng Luo & Yu-fang Han
Nanjing Hydraulic Research Institute, Nanjing, China

ABSTRACT: Taking Dandong Port as the representative of dig-in basin in strong tidal estuary, the author intends to make a research on the water and sediment characteristics of dig-in basin. By means of numerical simulation method, the study suggests that the hydrodynamic and sediment concentration decrease gradually from the harbor entrance to terminus in Donggou 1# basin. The differences of hydrodynamic and sediment concentration in harbor entrance are larger than that of harbor terminus. The harbor sedimentation intensity of spring tide is higher than that of neap tide, and the harbor sedimentation intensity in harbor entrance is higher than that of harbor terminus.

1 INTRODUCTION

Dug-in basin has been widely used in domestic port because of the less occupied or not occupied deep water coastline and the convenience of operation with a good mooring condition in the basin. The study about the excavated-in basin may be classified into four kinds on the basis of the research content:

The layout:

The main goal of the layout of dug-in basin was to determine the water area, the shape of entrance, and the trend of the channel entering the port. The size of water area of the basin would take an impact on the project investment and dock operation. After analysing the data of domestic excavated-in basin, Han Shi Lin (2005) suggests that it is necessary to increase the basin width of the criterion for more than one berth in the port side. Chen Jie (2013) argues that the shape of the entrance in the port will affect the mooring condition and the amount of back silting in the basin. Moreover, Pan Bao Xiong (1994) has highlighted the waterway axis direction into port and the effect on sediment volume and stability conditions.

The hydro-dynamics and berthing stability:

The study of dynamics in port always focuses on the backflow. In most recent years, it has been analysed by physical model or numerical simulation. Shen Xiao Xiong (2003) has compared the difference of the plane flow of port in variable angle between water flow and the dock; Wang Jia Hui (2009) has researched the three-dimensional flow characteristics in dug-in port; Ge Jian Zhong (2013) has studied the effect on the surrounding waters by excavating the pool of Hengsha East

Shoal. The calculation of surge element by wave mathematical model has been the subject of many classic studies in the mooring stability condition of excavated-in harbor.

Back silt:

The sediment in dig-in port is always caused by the reflux and the density flow (Xu, 2008). It appears the condition that the slit near entrance is more than that inside (Yu, 2011). Several methods about deposition reduction have been proposed. For instance: Xu Ying (2008) has studied the effect of entrance direction and stream in dock bottom to back silt, while She Xiao Jian (2014) has discussed the influence of bulwark at outlet.

Exchange of water:

The capability of water exchange is associated with the water area, the size of entrance and the local hydro-dynamics condition. Liu Pei (2014) has investigated the influence of the width of the gate on the water exchange capacity of the water body of dig-in basin in Lvsi harbor of JiangSu province. He Jie (2007) has studied on the water exchange ability for the excavated-in harbor basin impacted by the linked river's position and dredged way.

This paper researches on the hydrodynamic and back-silting characteristics of excavated-in harbor in strong tidal stream estuary by numerical model, detailing the understanding of dig-in port.

Dandong harbor is located at southeast of the Liaodong peninsula, the west bank of Yalu River estuary, north to the Yellow Sea, adjacent to Dalian, facing Korean Peninsula to the east across the Yalu River. It is the convenient and appropriate sea passage of domestic trade and international trade in east of the Northeast China and the easternmost global commerce port in mainland

Figure 1. Sketch map of DanDong Port Dadong harbor district.

coastline. There are three main ports in Dandong: the East Port, the LangTou Port and the HaiYang-Hong port. The East port (Figure 1) is located in the south of Donggang City, the west shore of the Yalu River mouth.

2 MATHEMATICAL EQUATIONS AND COMPUTATIONAL METHODS

The water flow and sediment transport equations can be written as:

$$\frac{\partial z}{\partial t} + \frac{\partial (Hu)}{\partial x} + \frac{\partial (Hv)}{\partial y} = 0$$

$$\frac{\partial u}{\partial t} + u\frac{\partial u}{\partial x} + v\frac{\partial u}{\partial y} + g\frac{\partial z}{\partial x} - fv + g\frac{u\sqrt{u^2+v^2}}{C^2 h}$$
$$= N_x\frac{\partial^2 u}{\partial^2 x} + N_y\frac{\partial^2 u}{\partial^2 y}$$

$$\frac{\partial v}{\partial t} + u\frac{\partial v}{\partial x} + v\frac{\partial v}{\partial y} + g\frac{\partial z}{\partial y} + fu + g\frac{v\sqrt{u^2+v^2}}{C^2 h}$$
$$= N_x\frac{\partial^2 v}{\partial^2 x} + N\frac{\partial^2 v}{\partial^2 y}$$

$$\frac{\partial HS}{\partial t} + u\frac{\partial HS}{\partial x} + v\frac{\partial HS}{\partial y} - \frac{\partial}{\partial x}\left(D_x\frac{\partial HS}{\partial x}\right)$$
$$- \frac{\partial}{\partial y}\left(D_x\frac{\partial HS}{\partial y}\right) = F_s$$

Among them:
H-total depth (m); z-water level (m); u, v the velocity component (m/s) along the direction of x and y; t-time (s); f-Coriolis coefficient ($f = 2w\sin\varphi$, w is the angular velocity of the earth rotation, φ is the area dimension); g- acceleration of gravity (m/s²); C-Chezy coefficient (m$^{1/2}$/s); N_x, N_y The

turbulence viscosity coefficient in the direction of x, y (m²/s); S-sediment concentration (kg/m³); D_x, D_y suspended sediment diffusion coefficient in the direction of x, y (m²/s). F_s- sediment source function or sediment flushing silt function (kg/(m² s)); M_{ox}, M_{oy} the direction at the bottom of the river bed elevation changes in the direction of x, y; M_{fx}, M_{fy} the direction of the bottom friction term in the direction of x, y.

Using triangular element mesh, the finite volume method is used to analyse the flow equation (1), with reference to the literature (Lu, 2013).

3 ESTABLISHMENT OF MATHEMATICAL MODEL

3.1 Model scope and calculation parameters

The upper boundary of the model is Dandong port, the offshore boundary is 30 m depth isobath, the west side of the boundary is the ocean red, and the east boundary is to south Xuanchuan of North Korea. The total length of model is 90 km, and the width is 76 km. The grid of local sea area of 1#

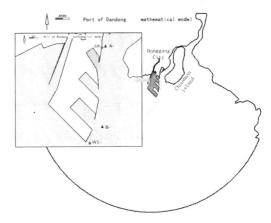

Figure 2. Sketch map of the range of mathematical model.

Table 1. Mathematical model calculation parameters.

Name	Parameter
Total number of units	109487
Minimum grid length	15 m
Time step	6 s
Roughness	0.013+0.013/h
The turbulence viscosity coefficient	1.5
Moving boundary	0.02 m
Sediment diffusion coefficient	10
Incipient velocity of sediment	Dou guoren formula
Sediment-carrying capacity	$S_* = \alpha U^2/(hw)$

harbor pool of DaDong port was compacted, as shown in Figure 2. The terrain of mathematical model is the latest data of Donggang District in 2013, and the calculation of parameters can be seen in Table 1.

3.2 Mathematical model validation

3.2.1 Water level verification
The validation data from the synchronous observation of large, middle and neap tide level included velocity and sediment concentration on May 8th to May 16th, 2014. The location of the station can be seen in Figure 1.

Tidal results are shown in Figure 3, the trend of validation in Figure 4, and the sediment concentration in Figure 5. We can see from the results that in addition to individual points, the calculated data is similar to the measured data, and the relative error was less than 10%. Due to the situation that it can't simulate the difference of different waters and sediment particle size, bulk density, particle shape in the process of calculation, the verification bias of sediment is relatively large, but overall, the verify of sediment concentration can reflect the variation characteristics of sediment station.

3.2.2 Verification of the deposition of the harbor basin
In May 2013, Donggou 1# harbor has been excavated to the design of the bottom elevation, according to the measured data analysis of harbor twice on November 18, 2013 and April 14, 2014. Figure 6 (left). The sedimentation in harbor entrance is large, about 0.3 m. The sedimentation is more than 0.5 m in some area of harbor entrance

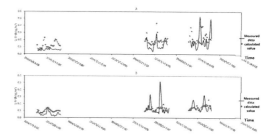

Figure 5. Verification of sediment concentration.

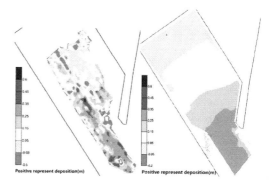

Figure 6. Verification of change of erosion and deposition of Donggou 1# (The left is the site map, The right is the mathematical model verification diagram).

inside. The more into the bottom of the basin, the smaller the sedimentation rate will be.

Figure 6 (right) is the mathematical model verification diagram. It can be seen from the chart that in Donggou 1#, the maximum sedimentation thickness is near the harbor entrance, and the sedimentation thickness is small in the internal harbour and smaller at the bottom end of the basin. The erosion and deposition population distribution of mathematical model is consistent with the actual measurement in general.

4 THE STUDY IN THE HYDRODYNAMIC FORCE AND SEDIMENT CHARACTERS FOR ESTUARY OF DIG-IN BASIN IN A STRONG TIDE AND WEAK FLOW.

The Yalu River estuary is a typical estuary of strong tide and weak flow in China, whose annual average tidal range is 4.51 m. The biggest rate of fluctuation is no more than 1.5 m/s during the spring tide. Then, we will study the hydrodynamic force and sediment characters of dredged harbor basins in the strong tide and weak flow estuary by using the Dandong Port, Dadong harbour district, 1# harbor basins as a representation.

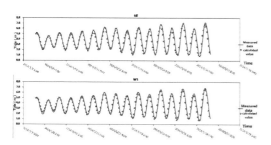

Figure 3. Validation of the tide.

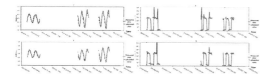

Figure 4. Validation of power flow.

4.1 Hydrodynamic characters of harbor basins

Figure 7 is the distribution of the biggest rate of fluctuation around the Dandong Port, Dadong harbour district, 1# harbor basins during a spring tide. We can see from the figure that at the entrance, the water power is relatively stronger. From the entrance to the basin interior, the water power weakens gradually. The water power of the central basin is stronger than that of both sides of the basin, which is related to the arrangement of entrance.

Figure 8 and Figure 9 are the comparison of hydrodynamic force entre spring tide and neap tide, and the sampling points are shown in Figure 7. From these figures, the hydrodynamic force of spring tide is obviously stronger than that of neap tide, and from the entrance to the internal part, hydrodynamic force will decrease. At the entrance, the biggest rate of fluctuation of spring tide is smaller than that of neap tide. Concerning the biggest rate of flood, spring tide is approximately 1 m/s, and neap tide is 0.5 m/s. Concerning the biggest rate of ebb tides, spring tide is approximately 0.8 m/s, and neap tide is 0.4 m/s. At the bottom of basin, the biggest rate of fluctuation of spring tide is larger than that of neap tide. The biggest rate of flood of spring tide is 0.04 m/s more than that of neap tide. However, the two biggest rates of ebb tides are approximate.

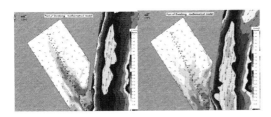

Figure 7. The biggest rate of fluctuation near the harbour.

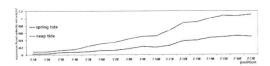

Figure 8. The flood maximum velocity distribution.

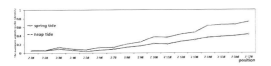

Figure 9. The ebb maximum velocity distribution.

4.2 Basin sediment concentration distribution characters

Figure 10 shows the basin average sediment concentration of spring tide and maximum sediment concentration distribution. The main basin sediment concentration comes from offshore into interior of basin at the entrance. From the entrance to the bottom of the basin, with the gradual decrease of hydrodynamic force, a decrease trend of sediment concentration is also found.

Figure 11 and 12 show the basin sediment concentration distribution. These figures tell us that the sediment concentration gradually decreases from entrance to the interior of basin. While it is a spring tide, the hydrodynamic force is fort, and the sediment concentration is obviously higher than that of a neap tide. From the entrance to the bottom of the basin, the difference of sediment concentration entre spring tide and neap tide weakens gradually. At the entrance, the average and the maximum sediment concentration for spring tide are 0.20 kg/m³ and 0.32 kg/m³, and for neap tide are 0.14 kg/m³ and 0.11 kg/m³. At the bottom, the difference entre spring tide and neap tide is much smaller.

4.3 Basin back-silting characters

Figure 13 shows us the back-silting intensity distribution of spring tide and neap tide. Due to the water sediment concentration of spring tide being significantly greater than that of the neap tide, the back-silting intensity of spring tide is also greater than that of the neap tide. While it is a spring tide, from the entrance to the bottom of the basin, back-silting intensity decreases at a fast

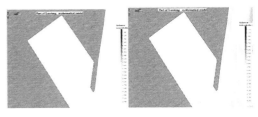

Figure 10. The average sediment concentration (left) and maximum sediment concentration (right) when spring tide in harbour.

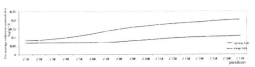

Figure 11. The average sediment concentrationwhen spring tide in harbour.

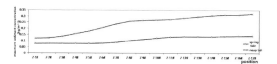

Figure 12. Maximum sediment concentration when spring tide in harbour.

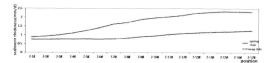

Figure 13. Characteristic distribution of the back and back in harbour.

rate. At the entrance, the back-silting intensity of Z-14#~Z-17# is similar, which is about 2.3 mm/d for spring tide and 1.2 mm/d for neap tide. From the entrance to the bottom, the back-silting intensity decreases. At the bottom, the back-silting intensity of spring tide and neap tide is close to each other, about 0.8 mm/d.

5 CONCLUSION

1. Based on the Dandong Port, the Dadong harbour district, and the 1# harbor basins, we can construct a tidal sediment mathematical model for estuary dredged harbor basins with a strong tide. The model hydrodynamic sediment and back-silting verify well.
2. For the Donggou 1# harbor basin, from the entrance to the bottom, the sediment concentration and hydrodynamic force decrease gradually. At the entrance, the difference of sediment concentration and hydrodynamic force entre spring tide and neap tide is bigger, and on the contrary, the difference is smaller at the bottom.
3. The back-silting intensity of spring tide is obviously greater than that of the neap tide, and at the same time, it is greater than that of the interior part at the entrance.

REFERENCES

Chen Jie, Zhang jun-wen. Definition and analysis of Certain wharf dug-in basin Entrance scale. China Water Transport, 2013, 24–25.

Ge Jian-zhong, Guo Wen-yun, etc. Analysis of effect of the Yangtze River estuary Hengsha shallow dug-in basin to flow field II: Impact on the surrounding flow field. Journal of East China Normal University (Natural Science), 2013, (4): 91–105.

Han Shi-lin, He Hui. River port dug-in basin scale Discussion [J]. Hunan Communication Science and Technology, 2005, 31(4): 141–143.

He Jie, Xin Wen-jie. Research on capacity of the dig-in basin water exchange. Journal of Hydraulic Engineering, 2007, S1: 330–333.

Liu Pei, Wang Hua, etc. Research on simulation dig-in basin water exchange. Renmin Zhujiang, 2014, (1): 23–28.

Lu Chuan-Teng, Chen Zhi-chang, etc. 2D and 3D nested tidal numerical simulation of the Yangtze River Estuary based on unstructured meshes. Hydro-Science and Engineering, 2013, (4): 18–23.

Pan Bao-xiong, Lin Rong-wang, etc. The determination of River port dug-in basin and entrance channel plane scale. China Water Transport, 1994, (4): 18–22.

She Xiao-jian, Zhang Lei, etc. Study on tidal sediment physical model for Caofeidian dig-in basin N.5 channel and breakwater project. China Harbour Engineering, 2014, (1): 32–38.

Shen Xiao-xiong, Han Shi-lin, etc. Study on experiment of inland river, dug-into basin reflux range. Journal of Changsha Communications University, 2003, 19(2): 49–54.

Wang Jia-hui, Li Yan, etc. Dug-in basin numerical simulation of flow characteristics. Yellow River, 2009, 31(6): 36–41.

Xu Ying, Liu Guo-long, etc. Dig—in Basins Sedimentation and experimental study of sedimentation reduction measures. China Harbour Engineering, 2008, (1): 31–33.

Yu Zhen, Zhang Wei, etc. Studies on the Yangtze River Estuary tidal part back-silting in dug-in basins. Journal of Waterway and Harbour, 2011, (1): 43–47.

Civil, Architecture and Environmental Engineering – Kao & Sung (Eds)
© 2017 Taylor & Francis Group, ISBN 978-1-138-02985-9

A method for detecting soil conditions of trenchless projects

D.B. Fu & C. Xia
Fuzhou Planning Design and Research Institute, Fuzhou, China

Y.F. Guo
Fujian Key Laboratory of Geohazard Prevention, Fuzhou, China

ABSTRACT: Trenchless technologies have been steadily growing in underground projects in urban areas. It is essential to obtain soil data at trenchless zones in order to successfully complete construction of underground projects. Consequently, a method for detecting soil conditions has been developed in this paper by the integrating Horizontal Directional Drilling (HDD) technique and cross-hole seismic tomography. Firstly, several small-diameter holes along trenchless projects are drilled by using the HDD technique, and Polyethylene (PE) pipes are installed along the designed path followed by placing detecting instrumentations in PE pipes. Travel times of each source–receiver pair are collected to reconstruct the velocity distributions between the boreholes. Finally, the soil conditions and anomalous zones are determined. To verify the feasibility of the proposed method, a main waterpipe constructed by the pipe jacking technique is taken as an example to detect obstacles distributed beneath bridges. The field results show that the proposed method is capable of detecting soil conditions for trenchless projects.

1 INTRODUCTION

Trenchless technologies such as pipe jacking and microtunneling have been used progressively in the installation and renovation of underground pipelines in urban areas owing to little disruption to traffic and nearby businesses (Struzziery et al. 1998). Compared to open-cut methods, the soil data is seen indirectly for trenchless methods, which may bring high risks during construction. Therefore, it is fundamental to develop a favorable method for detecting the subsurface conditions at a site.

Several geophysical exploration methods have been used extensively to detect underground information, which was primitively developed for natural resources exploration (Gupta & Roy 2007). Sakai & Takatsuka (1999) listed major geophysical exploration methods and their survey areas. Unlike borehole method, geophysical exploration methods assess ground conditions by measuring indirect data (such as velocity, attenuation, and wavelength of sound), which is controlled by ground properties. Among these existing geophysical exploration methods, the cross-hole seismic tomography is the most efficient exploration tool for detecting ground conditions. So far, a large number of applications and research about this method have been reported in the literature. Sakai & Takatsuk (1999) developed a new electric wave exploration system for detecting shallow strata boundaries in

urban areas and gave three experiments to verify its effectiveness. Bichkar et al. (1998) presented a genetic algorithm approach for detecting subsurface voids in cross-hole seismic tomography. Trivino & Mohanty (2015) assessed the crack initiation and propagation in rock from explosion-induced stress waves and gas expansion by cross-hole seismometry and FEM-DEM method. However, with the investigation and application increasing, problems gradually emerged in the cross-hole seismic method applied to detection of ground conditions, which involves:

1. Cross-hole configuration. Boreholes are perpendicular to the ground. Because underground pipelines in urban areas are usually linear underground structures, the borehole is simply able to show the results at the sections and the limit number along the planned route give insufficient information.
2. Image resolution and accuracy. The distances of boreholes directly determine the image resolution and accuracy. In the field surveys, the cross-hole configuration scheme needs to consider the distance of boreholes and the cost.
3. Complex surface conditions. Municipal pipelines are commonly seen beneath the road and constructed using the trenchless method when roads appear. Drilling the ground inevitably affects the traffic. In addition, difficulties exist in the detection of soil conditions beneath

existing structures when a subway crosses the existing buildings.

On account of the above issues, a geographical exploration technique for detecting obstacles of trenchless engineering (PN: ZL201310305447.5) is proposed (Xia et al. 2013). By integrating the HDD technology and cross-hole seismic tomography, the soil conditions in trenchless zones are perceived. The main water pipe constructed by pipe jacking technique is taken as an example to verify the feasibility of the proposed method.

2 A BRIEF DESCRIPTION OF EXISTING TECHNIQUES

2.1 Horizontal directional drilling technique

The HDD technology dates back to 1891 when the first patent was granted for equipment to place a horizontal hole from a vertical well (Carpenter 2002). The use of this technology has increased dramatically since the mid-1980s because of its environmentally friendly character.

The HDD process consists of three stages: pilot bore, back reaming, and pipe installation (Gokhale et al. 1999). The pilot bore stage consists of drilling a borehole along the designed path by using a small-diameter drill. Once the pilot hole is completed, the drill head is switched to a reamer. During the back-reaming stage, the reamer is pulled back to enlarge the bore path. On the final pipe installation stage, the pipe installation can either be performed in conjunction with the reaming stage or after the reaming is completed.

2.2 Cross-hole seismic tomography

The cross-hole seismic tomography involves the measurement of travel times of the ray path between two or more boreholes aiming to derive an image of seismic velocity in the intervening ground (Jackson et al. 2001).

The cross-hole seismic tomographic processing consists of measuring the travel times of all rays and restructuring the velocity distribution of target exploration zones. Firstly, the seismic sources and seismic receivers are placed in two boreholes. The travel times of each source–receiver pair are collected. Then the velocity distributions between the boreholes are reconstructed with the aid of mathematical inversion approaches such as the back-projection technique (Gilbert 1972), the algebraic reconstruction technique (Gordon 1974), and the genetic algorithm-based technique (Bichkar 1998) and so on. Because each soil property is corresponding to a velocity value, the anomaly

of the resulting velocity image can be regarded as obstacles.

3 A METHOD FOR DETECTING SOIL CONDITIONS OF TRENCHLESS PROJECTS

In this paper, a method based on HDD technique and cross-hole seismic tomography is developed to detect the soil condition in tunneling zones of trenchless projects, especially linear underground engineering (see Figure 1).

The proposed method includes five steps and is shown in Figure 2.

Step 1: Exploration scheme design
According to the field condition and exploration accuracy, the test scheme is designed, including determining the start point, end point, pipe number, and so forth. Three bore path schemes are delivered in Figure 3.

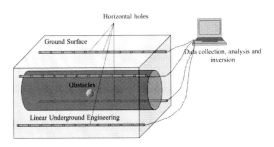

Figure 1. A method for detecting soil conditions of trenchless engineering.

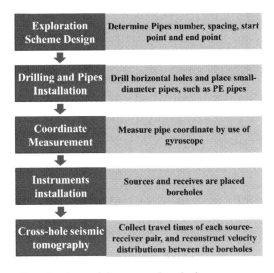

Figure 2. Steps of the proposed method.

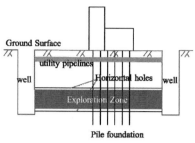

(a) Scheme 1

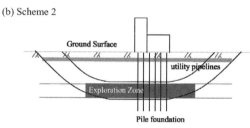

(b) Scheme 2

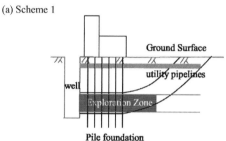

(c) Scheme 3

Figure 3.   Design path schemes.

Step 2: Drilling and pipe installation
Several horizontal holes along the tunneling direction are drilled by utilizing HDD technique. These holes are close to the tunneling zones. Small-diameter PE pipes (usually 90 mm) are placed in holes due to its flexibility, abrasion resistance, and toughness.
Step 3: Coordinate measurement
The coordinate of boreholes is measured by the use of a gyroscope to minimize the error of velocity image.
Step 4: Instrument installation
Sources and receivers are placed in boreholes. The borehole layouts are dependent on the tunneling area and exploration. Figure 4 illustrates three borehole layouts.
Step 5: Cross-hole seismic tomography
The velocity image is termed a tomogram by using measurements, which has taken a large number of rays with wide angular ray coverage (as illustrated in Figure 5). Consequently, the soil conditions and anomalous velocity zones are able to be determined.

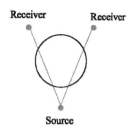

(a) Scheme 1

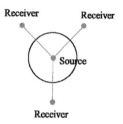

(b) Scheme 2

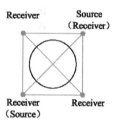

(c) Scheme 3

Figure 4.   Schematic diagrams of drill layouts.

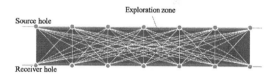

Figure 5.   Cross-hole configuration.

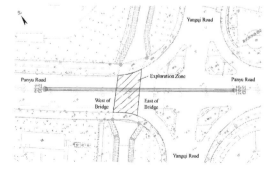

Figure 6.   Exploration zone.

1103

4 EXAMPLE

The proposed technique is applied to explore the tunneling zone of the main water pipe beneath the existing bridge. The pipe is constructed by the pipe jacking technique. The diameter and depth of

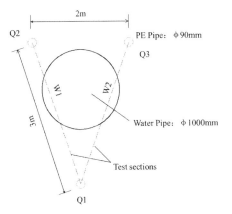

Figure 7. Source–receiver geometry and test sections.

Table 1. Anomalous zones.

Section	Number	Stake	Depth
W1	1	40.5 ~ 44	−10.5 ~ −11.5
	2	59.5 ~ 61	−10 ~ −11.5
	3	78.5 ~ 81.5	−9 ~ −9.5
	4	86.5 ~ 89	−10 ~ −11
W2	1	42.5 ~ 46	−8.5 ~ −9
	2	61 ~ 64	−10 ~ −11
	3	71 ~ 78	−9 ~ −10.5
	4	86 ~ 91	−8.5 ~ −10

water pipe are 1000 mm and 10.5 m. The bridge width is approximately 30 meters. The field condition is shown in Figure 6.

The obstacles such as the piers of the bridge, ripraps, and root may appear in the tunneling zones. Therefore, it is necessary to explore the distribution of obstacles for successfully constructing the water pipe.

4.1 Exploration scheme

The test zone is taken as the width of bridge plus 14 meters on both sides. The borehole path of HDD technique is designed as shown in Figure 3 (c). The entry angle and exit angle of PE pipes are 20°. The maximum pullback force and torque of the device for horizontal drilling are 380 kN and 12 kN·m. A total of three PE pipes defined as Q1, Q2, and Q3 are placed. The diameter of these pipes is 90 mm. The sparker source is placed in Q1, while the hydrophone receivers are placed in Q2 and Q3. The distance between the source and receiver are 3 meters. A total of two sections defined as W1 and W2 are tested. The source–receiver geometry and test sections are shown in Figure 7.

4.2 Results and discussion

Figures 8 and 9 show the result images of W1 and W2. It seems that the average velocity of test zones is 1800 m/s, and some anomalous zones (as listed in Table 1) appear in the exploration zone. These anomalous zones are beneath bridges, which may be stones. The following boreholes verify these results. It shows that the proposed method is able to detect the soil conditions and identify obstacles.

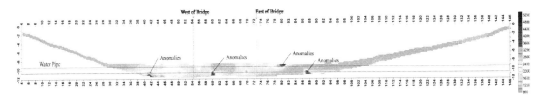

Figure 8. Inversion imaging results of W1.

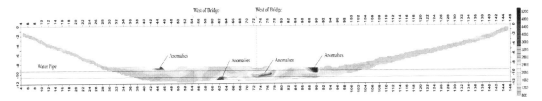

Figure 9. Inversion imaging results of W2.

5 CONCLUSIONS

A new method based on HDD technique and cross-hole seismic tomography is developed to detect the soil conditions in the tunneling zones of trenchless projects. A water pipe project constructed by pipe jacking technique is taken as an example to verify the feasibility of the proposed method. The test results demonstrate that the proposed method is capable of identifying the anomalous zones and has little impact on field conditions. Yet, this method has been applied to several projects and produced a good effect. Authors believe that this method would be able to expand on another similar engineering, such as subway, electric power tunnel, and utility tunneling, and so on.

REFERENCES

Bichkar, R.S., Singh, S.K. & Ray, A.K. 1998. Genetic algorithmic approach to the detection of subsurface voids in cross-hole seismic tomography. *Pattern recognition letters* 19(5): 527–536.

Carpenter, R. 2002. Embattled HDD industry sees change as necessary, ultimately beneficial: industry maturity, poor economy forcing massive market overhaul. *Underground Construction* 57(6): 30–32.

Gilbert, P. 1972. Iterative methods for the reconstruction of three-dimensional objects from projections. *Journal of theoretical biology* 36(105): 117–127.

Gokhale, S., Hamm, R. & Sterling, R. 1999. A comprehensive survey on the state of horizontal directional drilling in the North America provides an inside look at this increasingly growing industry, *Directional Drilling* 7: 20–23.

Gordon, R. 1974. A tutorial on ART (algebraic reconstruction techniques). *Nuclear Science, IEEE Transactions on* 21(3): 78–93.

Gupta, H.K. & Roy, Sukanta. 2007. Geothermal energy: an alternative resource for the 21st century. Elsevier: Holland.

Jackson, P.D., Gunn, D.A. & Flint, R.C., et al. 2001. Cross-hole seismic measurements for detection of disturbed ground beneath existing structures. *NDT & E International* 34(2): 155–162.

Sakai, S. & Takatsuka, T. 1999. Development of a geophysical exploration technique for detecting shallow strata boundaries. *Tunnelling and Underground Space Technology* 14: 21–29.

Struzziery, J.J., Spruch, A.A. & Blondin, C.A. 1998. Trenchless pipe repair for urban renewal project, *Journal of New England Water Environment Association* 32(2): 118–125.

Trivino, L.F. & Mohanty, B. 2015. Assessment of crack initiation and propagation in rock from explosion-induced stress waves and gas expansion by cross-hole seismometry and FEM–DEM method. *International Journal of Rock Mechanics and Mining Sciences* 77: 287–299.

Xia, C., Gao, X.L. & Chen, Y. A survey method for geotechnical engineering. China Patent: 201310305447.

Civil, Architecture and Environmental Engineering – Kao & Sung (Eds)
© 2017 Taylor & Francis Group, ISBN 978-1-138-02985-9

Comparison of thermal hydrolysis and wet air oxidation in sludge treatment in China

L. Peng & L.L. Hu
Sichuan College of Architecture Technology, Deyang, Sichuan, China

ABSTRACT: Wet Air Oxidation (WAO) and Thermal Hydrolysis (TH) are both associated with high temperature and high pressure. Both processes have been receiving increasing attention in sludge treatment over the last five years, especially TH as a pretreatment for Anaerobic Digestion (AD) has become a research hotspot in the world, including China. Sludge source and characteristics, energy consumption, and running cost technology are important factors. Both WAO and TH can reuse energy and heat by steam recycling, heat exchange, and system design. Typical commercial applications of WAO and TH in China are illustrated. WAO has been applied to sludge treatment to reduce moisture to 45%-48% from 80% with a running cost of 150–200 RMB per ton, while TH is used as an independent technology to achieve 50% of water content from 80% with 120–150 RMB per ton.

1 INTRODUCTION

As China is in the course of urbanization, the number and capacity of wastewater treatment plants are growing rapidly, and more and more attention is paid to sludge treatment and disposal. Sewage sludge commonly contains large amounts of water (including free water, interstitial water, and bound water), microorganisms, and mineral components (Qi and Thapa 2011, Vaxelaire and Cézac 2004). Chemical sludge from various factories usually includes toxic pollutants such as heavy metals, refractory organic matter, hormones, and colored substances.

Some current disposal and reuse practices are prohibited or questioned. Since land can no more be filled constantly, the landfill is not the recommended technology in China. Destruction of organic matter of biosolids is achieved by incineration, but secondary air pollutants and the high cost of building and running the facility makes it hard to be widely applied. Recycled sludge as an adsorbent, microbial consortium, plant, bulking agent (wheat husk) and nutrients are researched in the study of toxic materials in sludge (Nanekar and et al 2015). Compared to incineration, Wet Air Oxidation (WAO), which is a well-established process for toxic and hazardous organic wastewater treatment (Kanhaiya and Anurag 2015), generates minimal air pollution problems, maximum sludge solid reduction, and low environmental impact with a closed-loop water system. But industrial facilities used in the 1990s in the world were closed or capacity reduced because of high energy consumption with a temperature of 240–300°C. Anaerobic Digestion (AD), which is mostly applied in China (25 sludge treatment plants of 400 wastewater plants by the year of 2008) (Wu and et al 2008) is suggested by Chinese technology policy, but high cost in building and low efficiency of gas production raise barriers to its development. Thermal Hydrolysis (TH) is a kind of pretreatment usually used before AD to improve dewaterability of sludge by destroying microbial cells and increasing biogas production (Neyens and Baeyens 2003). To reduce the costs of sludge management and handling, sludge dewatering is crucial, no matter which technology is adopted.

WAO and TH are considered as optimal solutions to the sludge problem and have been wildly researched. However, no work has been attempted to compare TH and WAO in research studies and industrial applications in sludge treatment, especially in China. This paper summarized and compared the research studies of TH and WAO in research trends of sludge treatment and industrial factors combined with the actual situation in China. Finally, typical industrial facilities of WAO and TH in China and their conditions are illustrated.

2 RESEARCH STUDIES

To compare the research trends of two technologies in the world and China, date collection and analysis were accomplished in a database of Web of Science and China National Knowledge

Infrastructure, which reflect the research studies in the world and China, respectively.

In a brief search in Web of Science, hits were obtained for the term "wet air oxidation *sludge", compared with the term "thermal hydrolysis* sludge" (Figure 1).

WAO and TH both have a long history of sludge treatment. Thermal hydrolysis was mainly used to improve the sludge dewatering performance in the 1930s, with an increase in the anaerobic digestion performance sludge in the late 1970s and carbon source of denitrification in the 1990s (Zhi jun et al 2003). While research on Thermal Hydrolysis (TH) in sludge treatment have been booming over the past 10 years, it has contributed to the rapid development of Anaerobic Digestion (AD) technology, taking TH as a favorable pretreatment, as 80% of the TH studies have been related to AD. WAO was developed in 1967, and various techniques of WAO in sludge treatment were developed between 1990 and 1995 and applied for industrial use in Europe (Debellefontaine and Foussard 2000). For high temperature and pressure, WAO was regarded to be an energy-extensive solution to sludge, as well as resource wasting technology as it oxidizes the organic substance instead of reusing them.

Same keywords in Chinese were searched in China National Knowledge Infrastructure (CNKI), and results were as shown in Figure 2.

Research on WAO in China mostly concentrates in the introduction, parameter adjusting, and optimal conditions before the year 2000. In the recent 15 years, catalysis, dewatering character of sludge, and organic recomposing process were investigated (Baroutian et al 2013). TH technology focused on the transformation of pollutants in the process and influence followed in AD with different parameters (Feng et al 2014,) or effects of mixing with different waste (Lombardi and et al 2015).

Study trends of WAO and TH in China both show a good correlation, no matter the technology,

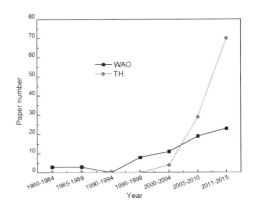

Figure 2. Published work number (data search in CNKI).

parameter, or material. The research on WAO in China has at least a 10 years lag than abroad (1980 compared with 1970), which simply reported the technology compared with the effect study with experiments, while the delay time of TH is shortened to 5 years (1998 compared with 2003) with the wide spread of Internet and awareness of the importance of dissolving problems of sludge disposal in China.

3 INDUSTRIAL APPLICATIONS IN CHINA

3.1 Sludge types

According to the source of sludge, it can be briefly divided into sewage sludge or similar sludge and industrial sludge, but in commercial applications, composition and characteristics of sludge are the key factors in selecting the treating technology.

In the past decades, a lot of research focused on the technology of treatment of the sludge from the sewage plant such as agriculture and soil reclamation, ocean disposal, sanitary landfill, burning method, and other recycling technology. (Yu and Tian 2014, Wang 2013). There are plenty of studies on the dehydration properties of sludge and the conclusion indicated that the pretreatment of TH has a positive impact on sludge dehydration (enhances gas production and accelerates reaction) (Xun and Wang 2009). Characteristics of sludge and property change in the process, especially water content and organic composition were researched widely. Based on sludge research, the application of WAO (Bengao and You 2014) or catalytic WAO in sludge treatment has been studied.

3.2 Energy consumption

In many cases, energy requirement impacts the commercial application of TH and WAO processes

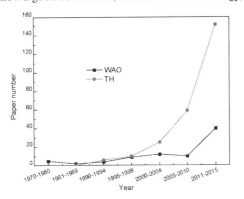

Figure 1. Published work number (data search in Web of Science).

in sludge treatment systems, especially in a developing country like China. TH and anaerobic digestion are usually discussed and evaluated together. It was found that the feed concentration of sludge must be increased (from 3%-7% in total solids) in order to become energetically self-sufficient in the TH system (Pérez-Elvira et al 2008). An energy integration scheme which considered heat recovery from the flash vapor outlet of the reactor can lower the required feed concentration.

To maintain the reaction temperatures, TH processes generally require more input of energy than WAO, which becomes auto-thermal at higher temperatures due to exothermic oxidation. As an independent process in sludge treatment, TH and WAO both need sludge heating while the energy in those processes is mostly spent. Implementing an energy integration scheme of the system to recover excess heat will be necessary for decreasing the heat energy consumed in TH and WAO process. There is another option that changes the energy source. The most widely used heat source now in China is steam, but electric heating is occasionally employed, as well as solar energy, microwave radiation, and heat pump.

Steam production needs a generator and it has been mostly used when a power plant or central steam supply is available around. Microwave heating has been widely used in system design as the system can achieve rapid heating both internal and external (Jones and Lelyveld 2002). As a microwave generator, developing and applying it in the industry can be a sound heat source for it can reduce reacting time of TH from 5 to 15 min and from 30 to 60 min (Qiao et al 2008). In sewage sludge treatment, microwaves are researched and used in sludge drying (Idris and Khalid 2004), stabilizing, and quality adjusting. Solar energy and heat pumps are usually applied to the drying process of sludge for energy reservation in a small temperature scale.

Despite consumption, energy recovery from waste treatment process is a trend in the world. Also, recycling and utilization of energy and resources of sludge are encouraged in the technology policy of China (Ministry of Environmental Protection of the People's Republic of China 2009).

3.3 Final disposal paths

For sludge, the final disposal of TH and WAO products from the industry can be divided into three phases, solid, liquid, and gas. The largest volume of sludge is in liquid form, but it is usually flowed back to wastewater treatment plants and decomposed by microorganisms, so that biodegradability and toxicity of liquids are vital parameters in evaluating. Gas contains organic and inorganic gases, which can be adsorbed by adsorbent or microorganisms, washed out or burned. Solid is most refractory for its high moisture and perishable organics, which makes it hard to transport, landfill, incinerate, and reuse. The content of cake dry solids is required to be greater than 40% (w/w) according to some China Standards (GB16889-2008, GB/T 24602-2009) if the solids take the landfills or incineration as the final disposal.

The final disposal path of solid products as an important part of sludge treatment is determined by the process adopted. According to requirements of the urban sewage treatment plant, sludge treatment and pollution control technology policy of China (2011), the principle of "safety and environmental protection, recycling, saving energy and reducing consumption, adjustment measures to local conditions, steady and reliable" are all raised in the sludge treatment and disposal process. The landfill, incineration, and compost are commonly chosen now in China, but TH combined with AD, even TH alone, and WAO receive more attention in industry applications for its green final disposal in liquid, solid, and gaseous state at the same time.

4 TYPICAL INDUSTRIAL FACILITIES IN CHINA

4.1 Yancang sludge disposal center

The Athos process by Veolia Water is one of the main WAO sludge treatment technologies currently provided commercially, which operates at temperatures between 250–300°C, using air or pure oxygen as the oxidant and claims to produce mineral products, clean gas emissions, and biodegradable liquids. As far as observed, no typical Athos process is employed in China.

A Chinese patent about Partial Wet Air Oxidation (PWAO) in sludge treatment was granted in the year 2013. The biggest feature of PWAO are lower reacting temperature and pressure (160–220°C, 1.5–3.0 MPa), with its export of solid moisture below 50%, and the solid disposal can be used as a plant food or matrix for improving soil quantity. Partial oxidation means oxidizing 30%–50% of organic matter that is unstable and perishable by macromolecules.

The first commercial PWAO in a sludge treatment plant began operating in 2014 in Yancang (Zhejiang province of China), to treat 100 tons per day (moisture 80%) of mixed sludge with 70% leather processing sludge and 30% municipal sludge from a wastewater treatment plant in the industrial zone. This system consists of three parts as shown in Figure 3.

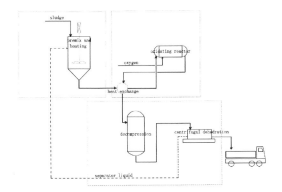

Figure 3.   Process flow diagrams of PWAO in Yancang.

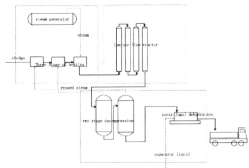

Figure 4.   Process flow diagrams of TH in Mianzhu.

The sludge treatment plant was located in the wastewater plant right beside the workshop of sludge dewatering. Mixed sludge and liquid flowed back forming slurry with 86%–92% moisture and was heated in the premix heating part. At the oxidation reactor, the oxidation of organic ingredients starts when temperatures rise between 160 and 180°C. The oxidized liquor in high temperature is piped back to the premixed part after centrifugal dehydration. Energy and heat recovery and exchange sections are designed to lower the energy consumption. The solid products of the line are used as a matrix of soil improvers or to be landfilled with the moisture between 45%–48%.

The heat source of the plant is electricity, which is flexible and plenty in many places. The global treatment cost is reported to be only 150–200 RMB for one ton of sludge with the moisture of 80%. If steam can be used as a heat source or the product has been sold out, the cost can be further decreased.

4.2 *Mianzhu municipal sludge treatment equipment*

To meet the standard which retained the moisture below 60%, dewater property can be enhanced by TH alone. Despite that it is not followed by AD, the product can be landfilled within a short reacting time (30–90 min) with a small-scale equipment instead of a large anaerobic digestion tank.

A commercial sludge treatment plant in Mianzhu (Sichuan province of China) was operating for over one year since 2013 with its capacity of 50 tons a day (80% moisture). The diagram of the system is as shown in Figure 4.

Sludge from the wastewater treatment plant is piped into three stage heating with reused steam and steam from the generator. Then the sludge is passed through the laminar flow reactor in order to maintain enough time at a given temperature (150–180°C). Two-stage decompressions make the

sludge easy to be dewatered by centrifugal dehydration and recycling the steam heat. The separated liquid is delivered back to the wastewater treatment plant or treated by a supporting device.

Sound and stable effect has been achieved. The line exported solid maintains 50% moisture, but additional section needs to be set if the solid has to be reused because of smell that comes from the unoxidized organic matter in solid form. The cost for the line is 120–150 RMB for one ton of sludge with the moisture of 80% since steam is applied as a heat source.

5   CONCLUSION

Wet Air Oxidation (WAO) and Thermal Hydrolysis (TH) were both paid increasing attention in sludge treatment in the last 5 years, especially TH as a pretreatment of Anaerobic Digestion (AD) is hot worldwide, as well as in China. Sludge source and characteristics, energy consumption and running cost of sludge treatment technology are important in selecting the method. Both WAO and TH can reuse energy and heat by steam recycling, heat exchange, and system designing. Typical commercial applications of WAO and TH in China are illustrated. Partial Wet Air Oxidation (PWAO) is developed from WAO and has been applied to sludge treatment, which can reduce moisture to 45%–48% with running cost of 150–200 RMB per ton (80% moisture). TH used as an independent technology to treat sewage sludge and 50% of water content can be achieved with 120–150 RMB per ton (80% moisture).

ACKNOWLEDGMENT

This work was supported by the Sichuan Education Department (project 16ZA0429) and the Deyang Science and Technology Branch Intellectual Property Office (project 2015ZZ038).

REFERENCES

Baroutian, S. Smit, A.M. Gapes, D. (2013). Relative influence of process variables during non-catalytic wet oxidation of municipal sludge. Bioresour. Technol. 148, 605–610.

Bengao, L. You, S. (2014), Mechanism biochemical excess sludge reduction by wet air oxidation", Pet. Process. Petroche.45(9), 85–89.

Debellefontaine, H. Foussard, J.N. (2000). Wet air oxidation for the treatment of industrial wastes. Chemical aspects, reactor design and industrial applications in Europe. Waste Manage. 20, 5–25.

Disposal of sludge from municipal wastewater treatment plant-Quality of sludge used in separate incineration, GB/T 24602-2009, Beijing, China.

Feng, G.H. Tan, W. Zhong, N. Liu, L.Y. (2014), Effects of thermal treatment on physical and expression dewatering characteristics of municipal sludge. Chem. Eng. J. 247, 223–230.

Gan Q. (2000), A case study of microwave processing of metal hydroxide sediment sludge from printed circuit board manufacturing wash water. Waste Manage. 20, 695–701.

Idris, A. Khalid, K. (2004), Drying of si li ca sludge using microwave heating. Appl. Therm. Eng. 24, 905–918.

Jones, D.A. Lelyveld, T.P. (2002), Microwave heating applications in environmental engineering-a review. Resour.Conserv. Recy. 34, 5–90.

Kanhaiya, L., Anurag, G. (2015), Catalytic wet oxidation of phenol under mild operating conditions: development of reaction pathway and sludge characterization, Clean Technologies and Environmental Policy, 17 (1), 199–210.

Lombardi, L. Carnevale, E.A. Corti, A. (2015), Comparison of different biological treatment scenarios for the organic fraction of municipal solid waste, International Journal of Environmental Science and Technology, 12 (1), 1–14.

Nanekar, S. Dhote, M. Kashyap, S. Singh, S.K. Juwarkar, A.A. (2015), Microbe assisted phytoremediation of oil sludge and role of amendments: a mesocosm study, International Journal of Environmental Science and Technology, 12 (1), 193–202.

Neyens, E. Baeyens, J. (2003), A review of thermal sludge pre-treatment processes to improve de-waterability, J. Hazard. Mater. 98, 51–67.

Pérez-Elvira, S.I., Fernández-Polanco, F., Fernández-Polanco, e.al, (2008), Hydrothermal multivariable approach: full-scale feasibility study. Electron. J. Biotechnol. 11 (4), 7–8.

Ploos van Amstel, J.J.A. Rietema, K. (1973), Wet-air oxidation of sewage sludge Part II:the oxidation of real sludges.Chem.Ing.Tech. 45, 1205–1211.

Qi, Y. Thapa, K.B. Hoadley, A.F.A. (2011), Application of filtration aids for improving sludge dewatering properties—a review, Chem. Eng. J. 171, 373–384.

Qiao, W. Wang, L. Pan, L. Xun, R. (2008), Sewage Sludge Microwave thermal hydrolysis process, Environ. Sci. 29(1), 152–157.

Shufan, Y. (2008), Research on the treatment of the sludge from urban sewage disposal plants by CWAO, M.Sc. thesis, Dept Resources. Civil Eng., Univ. Dongbei, Heping district culture road 3, China.

Standard of pollution control on the landfill site of municipal solid waste, GB 16889-2008.

Standard of pollution control on the landfill site of municipal solid waste, GB 16889-2008, The people's Republic of China Ministry of Environmental protection& General administration of Quantity supervision, inspection and Quarantine, 2008, Beijing, China.

Strong, P. Mcdonald, B. Gapes, D. (2011), Combined thermochemical and fermentative destruction of municipal biosolids: a comparison between thermal hydrolysis and wet oxidative pre-treatment. Bioresour. Technol. 102, 5520–5527.

The people's Republic of China Ministry of housing and urban-rural development of the people's Republic of China National Development and Reform Commission, Urban sewage treatment plant sludge treatment and pollution control technology policy(Trial).

Urban sewage treatment plant sludge treatment and pollution control technology policy (Trial), Ministry of Environmental protection of the people's Republic of China, Ministry of housing and urban-rural development of the people's Republic of China & Ministry of science and technology of the people's Republic of China, 2009, Beijing, China.

Vaxelaire, J. Cézac, P. (2004), Moisture distribution in activated sludges: a review, Wat Res., 38, 2215–2230.

Wang, G. (2013). Present status of treatment and disposal techniques of sludge at home and abroad. Enviro. Eng.31, 530–533.

Wu, J, Jiang, J, Zhou, H.M. Lei, B.I. (2008), current Operation Status of Sludge Anaerobic Digestion System in Municipal Wastewater Treatment Plants in China, China water & Wast water, 24(22), 21–24.

Xun, R. Wang, W. (2009), Water distribution and dewatering performance of the hydrothermal conditioned sludge. Environmen. Sci.30(3), 851–856.

Yang, Q.W. Gao, D. et al, (2012). Review of odor control technologies and materials for sewage sludge treatment disposal. ChinaWater & Wastewater 28(23), 145–148.

Yu, J. Tian, N.N. (2014). Analysis and discussion of sludge disposal and treatment of sewage treatment plants in China, J. Environ.Eng-China.1(1), 82–86.

Yu, J. Tian, N.N. (2014). Analysis and discussion of sludge disposal and treatment of sewage treatment plants in China, J. Environ.Eng-China.1(1), 82–86.

Zhi-jun Wang, Wei Wang, Fenfang Li, Development and application of thermal hydrolysis of sludge, China water & wastewater, 2003, 19(10):25–27.

Civil, Architecture and Environmental Engineering – Kao & Sung (Eds)
© 2017 Taylor & Francis Group, ISBN 978-1-138-02985-9

Enhancement of aerobic granulation with real domestic wastewater by powdered activated carbon addition

Changwen Wang, Bao Li, Shaoyan Gong & Ying Zhang
College of Resource and Environment, Linyi University, Linyi, China

ABSTRACT: Aerobic granulation with real domestic wastewater is difficult to be achieved and usually takes a long granulation time. This study investigated the feasibility of adding powered activated carbon to enhance the granulation process. It was found that only 37 days were needed to form granular-dominant sludge in the reactor, which could be comparable with the granulation process with artificial wastewater. After 55 days operation, COD and ammonia removal efficiency were 84.63% and 93.86% respectively. Effluent nitrite accumulation was observed in the reactor indicating partial nitrification could be achieved. This study presents a novel operational strategy to cultivate aerobic granular sludge with low COD level real domestic wastewater.

1 INTRODUCTION

After more than a score years' exploration, aerobic granular sludges have been successfully cultivated in well-controlled lab-scale reactors with high- or middle-strength synthetic wastewaters (Beun et al. 1999, Peng et al.1999, Moy et al. 2002, Arrojo et al. 2004, Su & Yu 2005). However, there are only a few studies that reported the aerobic granulation with real domestic wastewater which is usually characterized as a low-strength wastewater (Liu et al. 2011, Liu et al. 2010). How to cultivate active and compact aerobic granulars with the low-strength municipal wastewater is crucial for its full application. Moreover, the rapid aerobic granulation in pilot-scale or full-scale reactor is still the bottleneck that restricts its full application.

Fortunately, some forerunner explored the unknown area. de Kreuk (de Kreuk 2006) tried to cultivate aerobic granules by using pre-settled sewage as influent. After 20 days' operation at a high Chemical Oxygen Demand (COD) loading, heterogeneous aerobic granular structures were observed, with a Sludge Volume Index (SVI) after 10 min settling of 38 ml/g and an average diameter of 1.1 mm. This led to a pilot-scale plant setup at a WasteWater Treatment Plant (WWTP) in Ede, The Netherlands. Two bubble columns with a diameter of 0.6 m and height of 6 m were used. After about one year cultivation, dry weight concentration was 10 g/L, of which 80% consisted of granular sludge with a diameter larger than 0.212 mm. The influent concentration to the SBR in their work was in a range of 270–400 mg COD/L. These results provided valuable information for potential application of aerobic granules for the treatment of domestic sewage and municipal wastewater.

Although the long-term stability of aerobic granular sludge in pilot-scale reactor with real wastewater has obtained validation (Liu et al. 2010), the granulation process is still time consuming. 6–13 months were still needed to form granular dominant sludge in pilot-scale reactors with real wastewater (Liu et al. 2010, Ni et al. 2009), which is much longer than that required in the lab-scale reactors such as 2–4 weeks. Liu et al. reported that only 17 days were needed to form granule-dominant sludge and COD and NH_4^+-N removal efficiencies could reach above 80% and 90% respectively after 50-day operation, however, COD of the wastewater used in their experiment was about 800 mg/L on average (Liu et al. 2011). In general, the COD concentration of municipal wastewater in China is typically lower than 200 mg/L. Such a low COD level would severely impact the granulation process and could be a great barrier for the wide application of this novel technology. Moreover, adding external organic carbon source to raise the COD concentration is not an applicable engineering strategy. Successful granulation of activated sludge with such a low-strength municipal wastewater has not been reported yet. Therefore, exploring the possible operational strategies for rapid aerobic granulation with low-strength wastewater has very important scientific and engineering value. This work aims to enhance the aerobic granulation process with real domestic wastewater by adding Powdered Activated Carbon (PAC), which would be very useful for the development and application of aerobic granular sludge process.

2 MATERIALS AND METHODS

2.1 *Reactor setup and operation*

A bubble column with a diameter of 20 cm, a height of 60 cm and working volume of 18.8 L was used for the cultivation of aerobic granulars. The sand disk diffuser with the same cross section area was installed at the reactor bottom for air supply and mixing. The Dissolved Oxygen (DO) concentration in the aeration stage was in the range of 2.5–6.5 mg/L. Effluent was discharged from the middle port of the reactor and the exchange ratio was 60% with a 10 h hydraulic retention time. A 12 h operation cycle was implemented in each reactor operation, comprising 5-min feeding, 700-min aeration, 3-min settling (shortened from 30-min to 3-min gradually), 5-min decanting and 7-min idling.

2.2 *Seed sludge, wastewater and powered activated carbon*

The seed sludge, obtained from Gaobeidian wastewater treatment plant, Beijing, China, was seeded directly into the reactor without any acclimation. The initial Mixed Liquor Suspended Solid (MLSS) was 4200 mg/L. The wastewater used in this study was collected from the sewers of campus residential area. The characteristics of the wastewater are summarized in Table 1. Apparently, the feeding waster used here was typically low COD level.

Fine PAC particles were used to enhance aerobic granulation process with the real domestic wastewater. The PAC had a mean size of 50 µm having a specific surface area of 1130 m^2/g and an bulk density of 1.125 g/m^3. A PAC concentration of 4000 mg/L was added in the reactor. The whole experiment was carried out under room temperature, and the influent wastewater temperature was 20–23°C.

2.3 *Analytical procedures*

COD, NH_4^+-N, NO_2^--N, NO_3^--N, Sludge Volume Index (SVI) and MLVSS were measured according to standard methods (APHA, 1995). DO and pH were continuously monitored by pH/oxi340i meter with DO and pH probes (WTW Multi-340i, Germany). Microscopic examination was performed using an Olympus-BX51 (Olympus, Japan).

Table 1. The characteristics of feeding wastewater.

COD (mg/L)	221.80 ± 23.83	TP (mg/L)	5.52 ± 0.73
NH_4^+-N (mg/L)	77.17 ± 5.72	pH	7.0~7.8
TN (mg/L)	93.35 ± 6.82	Temperature (°C)	20~25

3 RESULTS AND DISCUSSIONS

Formation of aerobic granular sludge with powdered activated carbon addition.

The evolution of sludge morphologies were shown in Figure 1. After sludge was seeded into the reactor 4000 mg/L PAC was added following. Mixed by mixing function of aeration, PAC was dispersed in the reactor and contacted with activated sludge flocs. From Figure 1A, we could see that PAC is adhered onto the activated sludge flocs, so the settleability of the flocs could be improved thus more flocs could be maitained in the reactor. After 20 days' operation, from Figure 1B, granular sludges with compact structure and clear boundary could be observed, but flocculent sludge was still dominant. On day 37, granular sludges were obviously much more than before and the particle size grew bigger (Figure 1C). The domination of granular sludge in the reactor and the distinct spherical shape could state clearly that the aerobic granular sludge was cultivated successfully. The granulation process using only 37 days was as short as the granulation processes with artificial wastewater (Qin et al. 2004, Liu et al. 2005, McSwain et al. 2004). The results indicated the addition of powered activated carbon could enhance the granulation process with real domestic wastewater. The cultivated PAC-aerobic granular sludge appeared compact and small particle size (Figure 1D).

Shortening settling time to form selection pressure was employed to form aerobic granular sludge (Qin et al. 2004). The pattern of shortening settling time and variations of sludge characteristics (MLSS, MLVSS/MLSS and SV_{30}) were shown in Figure 2. After the inoculation of activated sludge, the settling time was set st 30 min. During the following operation, the settling time was

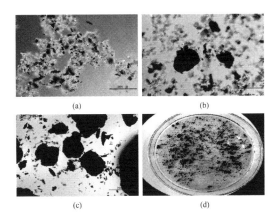

(a)　　　　　　　　(b)

(c)　　　　　　　　(d)

Figure 1. Images analysis of aerobic granulation process on different operation day 0 (a), 20 (b), 37 (c) (both by microscope) and 55 (d) (by digital camera).

shortened according to the sludge settling ability till to 3 min. Under the short settling time, the sludge flocs with poor settling ability were washed out, which directly led to the decrease of MLSS from 8000 mg/L (including 4000 mg/L PAC) to 720 mg/L. The corresponding SV_{30} were 32 and 7 respectively, which were improved by the addition of PAC. In the aerobic granulation process with no auxiliary substances addition, the MLVSS/MLSS ratio usually has a significant decrease due to the accumulation of inorganic substances in aerobic granulars, which is of great importance to aerobic granulation (Qin et al. 2004). In this study, because of the addition of PAC, the MLVSS/MLSS ratio was very low at the beginning, moreover, only had a slight decline from 0.69 to 0.5.

3.1 COD and Nitrogen removal

The wastewater influent substrate concentration of real domestic wastewater are varying compared with the artificial wastewater. As shown in Figure 3, the influent COD concentration varied from 45.15 mg/L to 173.10 mg/L. Although the

MLSS decreased during the operation period, the COD removal efficiency kept an increase trend. From day 35 to day 55, the MLSS was below 1000 mg/L and almost only 50% was the MLVSS, but the COD removal efficiency was kept at 85%. The results indicating that the addition of PAC could maintain the function bacteria. From day 0 to day 55, the average influent COD concentration was 113.15 mg/L, which was a low organic substances level. The wild fluctuation and low level of influent COD concentration of real domestic wastewater may lead to the slow granulation process. But it took only 37 days to form granular sludge in this study, which mainly due to the addition of PAC.

The variations of NH_4^+-N removal and effluent NO_2^--N, NO_3^--N were shown in Figure 4. The effluent NH_4^+-N was below 10 mg/L during the 55 days' operation except for several days at the process of granular sludge formation. The NH_4^+-N removal efficiency did not have an obvious fluctuation and average NH_4^+-N removal efficiency was 87.19%. Along with the shortening of settling time, the NH_4^+-N removal efficiency decreased slightly due to the wash out of biomass. But after granular sludge was formed from day 37, the NH_4^+-N removal efficiency kept a steady increase trend to about 95%.

3.2 Partial nitrification

The effluent NO_2^--N and NO_3^--N had a great change in the granulation process. From day 1 to day 23, NH_4^+-N was nitrified completely, so NO_3^--N was the mainly nitrogen in the effluent and there was no NO_2^--N. After that, the effluent NO_2^--N increased while the effluent NO_3^--N declined indicating a partial nitrification in the reactor.

The wash out of the activated sludge during granulation process will result in the incomplete nitrification, which is equivalent to stop aeration

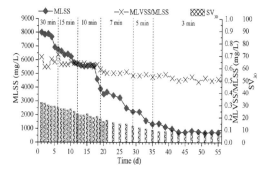

Figure 2. Profiles of reactor sludge MLSS, MLVSS/MLSS SV_{30}, during the whole operation period.

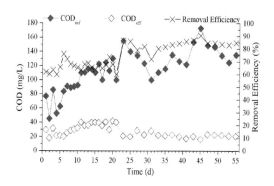

Figure 3. Performance of reactor with respect to COD removal over operation time.

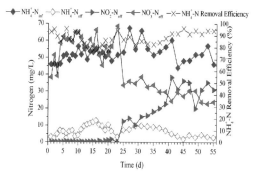

Figure 4. Performance of reactor with respect to nitrogen removal over operation time.

before "ammonia valley" (Guo et al. 2007, Yang et al. 2007). Moreover, the granular structure was suitable to maitain the slow growing Ammonia Oxidizing Bacteria (AOB) (Vázquez-Padín et al. 2010). Thus, the cultivation process of aerobic granular sludge is naturally to favor the formation of partial nitrification. From day 42, the effluent NO_2^--N was higher than NO_3^--N and kept an increasing trend. The nitrifying granular sludge was proposed as an alternative to obtain partial nitrification, but the previous reports mainly used the artificial wastewater. The results of this study indicated that stable partial nitrification with real domestic waster could be achieved by aerobic granular sludge.

4 CONCLUSIONS

Aerobic granulation process with real domestic wastewater was enhanced by the addition of PAC. It took only 37 days to form distinct and compact granular sludge, which could be comparable with those with artificial wastewater. MLSS was very low due to the short settling time, but wastewater treatment efficiency could be ensured. The COD and NH_4^+-N removal efficiency were 84.63% and 93.86% respectively after 55 days operation. Effluent NO_2^--N was accumulated after granular sludge formed indicating partial nitrification with real domestic wastewater could be achieved.

ACKNOWLEDGMENTS

The authors thanks for the financial supporting by the Beijing Municipal Education Commission general program (KM2012-10005028) and National Natural Science Funds of China (Grant No. 21207058).

REFERENCES

APHA. 1995. Standard Methods for the Examination of Water and Wastewater, 19th ed. *American Public Health Association/American Water Works Association/Water Envirionment Federation*, Washington, DC, USA.

Arrojo, B., Mosquera-Corral, A., Garrido, J.M., Mendez, R. 2004. Aerobic granulation with industrial wastewater in sequencing batch reactors. *Water Res.* 38(14–15): 3389–3399.

Beun, J.J., van Loosdrecht, M.C.M., Heijnen, J.J. 1999. Aerobic granulation in a sequencing batch reactor. *Water Res.* 36(3): 2283–2290.

de Kreuk, M.K. 2006. Aerobic granular sludge scaling up a new technology. *Ph.D. thesis*, Delft University of Technology, Delft, The Netherland.

Guo, J.H., Yang, Q., Peng, Y.Z. 2007. Biological nitrogen removal with real time control using step feed SBR technology. *Enzyme Microb. Technol.* 40(6): 1564–1569.

Liu, Y., Wang, Z.W., Qin, L., Liu, Y.Q., Tay, J.H. 2005. Selection pressure-driven aerobic granulation in a sequencing batch reactor. *Appl. Microbiol. Biotechnol.* 67(1): 26–32.

Liu, Y.Q, Moy, B.Y.P., Kong, Y.H., Tay, J.H. 2010. Formation, physical characteristics and microbial community structure of aerobic granules in a pilot-scale sequencing batch reactor for real wastewater treatment. *Enzyme Microb. Technol.* 46(6): 520–525.

Liu, Y.Q., Kong, Y.H., Tay, J.H., Zhu, J.R. 2011. Enhancement of start-up of pilot-scale granular SBR fed with real wastewater. *Sep. Purif. Technol.* 82: 190–196.

McSwain, B.S., Irvine, R.L., Wilderer, P.A. 2004. The influence of settling time on the formation of aerobic granules. *Water Sci. Technol.* 50(10): 195–202.

Moy, B.Y.P., Tay, J.H., Toh, S.K., Liu, Y., Tay, S.T.L. 2002. High organic loading influences the physical characteristics of aerobic sludge granules. *Lett. Appl. Microbiol.* 34(6): 407–412.

Ni, B.J., Xie, W.M., Liu, S.G., Yu, H.Q., Wang, Y.Z., Wang, G., Dai, X.L. 2009. Granulation of activated sludge in a pilot-scale sequencing batch reactor for the treatment of low-strength municipal wastewater, *Water Res.* 43(3): 751–761.

Peng, D., Bernet, N., Delgenes, J.P., Moletta, R. 1999. Aerobic granular sludge-a case study. *Water Res.* 33(3): 890–893.

Qin, L., Liu, Y., Tay, J.H. 2004. Effect of settling time on aerobic granulation in sequencing batch reactor. *Biochem. Eng. J.* 21(1): 47–52.

Su, K.Z. & Yu, H.Q. 2005. Formation and characterization of aerobic granules in a sequencing batch reactor treating soybean-processing wastewater. *Environ. Sci. Technol.* 39(8): 2818–2827.

Vázquez-Padín, J.R., Figueroa, M., Campos, J.L., Mosquera-Corral, A., Méndez, A. 2010. Nitrifying granular systems: a suitable technology to obtain stable partial nitrification at room temperature. *Sep. Purif. Technol.* 74(2): 178–186.

Yang, Q., Peng, Y.Z., Liu, X., Zeng, W., Mino, T., Satoh, H. 2007. Nitrogen removal via nitrite from municipal wastewater at low temperatures using real-time control to optimize nitrifying communities. *Environ. Sci. Technol.* 41(23): 8159–8164.

Civil, Architecture and Environmental Engineering – Kao & Sung (Eds)
© 2017 Taylor & Francis Group, ISBN 978-1-138-02985-9

Optimal water strategy for the United States

Hao-yu Zheng
Sheng Li College, China University of Petroleum, Donging, China

Jia-chen Lu & Mu Hu
School of Oil and Gas Engineering, Southwest Petroleum University, China

ABSTRACT: The main objective of this paper is to address the various uses of water to an ideal weight ratio, solve the storage and movement, form saline alkali, and protect the water during the process, covering the economic, physical, and environmental benefits. The Analytic Hierarchy Process (AHP) model is adopted to filter all the influencing factors and extract the final 10 criteria to determine the use of water in these three parts. After computing their weights, the abstract problem is successfully transferred to a mathematical model. The second step that we refer to is the grey correlation model to further determine the priority order of different states in the water using the calculated weight ratio. Due to the geographical location and climate change in the United States, in particular, we confirm the temperature coefficient, in this study, in every aspect of the water demand that can be in some densely populated areas and the lack of water in the area. By using these two models, through which we will model the multi-objective decision analysis model and genetic algorithm, and combined with all the water supply in each country's three regional branches of the classification, the final optimal solution can be obtained.

1   MATHEMATICAL MODEL

1.1   *AHP model*

The evaluation criteria for the distribution of rivers or lakes are involved in many aspects such as economy, water resources, and environment. In order to divide the United States according to the conditions of the natural environment, we need to figure out the social economic condition, development, and utilization degree of water resources and state of the ecological environment as the three swordsmen.

Social economic condition

- GDP.
- Modulus of agriculture output.
- Hydropower production.
- Urbanization level.

Development and utilization degree of water resources

- Proportion of the number of people by water resources (self-sustaining).
- Average temperature.
- Precipitation.
- Percentage of brine.

State of ecological environment

- Modulus of soil and water loss.
- Industrial sewage emission.
- Ratio of underground water in water use.

1.2   *Comparison matrix*

Starting from the rule hierarchy (the second criteria of the hierarchy figure above), comparison matrix is structured by the Comparison Method of the number 1–9 (each represents a different degree of importance). Table 1 shows the weights of the social economic condition, development, and utilization degree of water resources and state of the ecological environment.

From the four tables presented above, we have confirmed that the entire consistency ratio is correct. The combination weight vector of criterion with respect to the evaluation method of distribution of American river basins is shown in Table 5.

1.3   *Grey correlation model*

The steps in grey correlation analysis are as follows:

STEP 1: Note reference data series as the ideal comparison criteria. In general, reference data series consists of the optimal value of every criterion. We can also choose other reference data by different evaluation destination. The relationship is shown as follows:

$$A_0 = (a_0(1), a_0(2), \cdots, a_0(m))$$

In which, we choose every optimal value as reference data series, which is $A_0 = (1, 1, \ldots, 1)$.

Table 1. Pairwise comparison matrix of hierarchy.

Comprehensive impact	Social economic condition	Development and utilization degree of water resources	State of ecological environment	Weight
Social economic condition	1	1/3	5	0.2746
Development and utilization degree of water resources	3	1	8	0.6571
State of ecological environment	1/5	1/8	1	0.0683

By computing the weights of the three main factors, we get the final maximum eigenvalue $\lambda = 3.0444$, *consistency ratio = 0.0383*.

Table 2. Pairwise comparison matrix of hierarchy II–III.

Comprehensive impact	GDP	Modulus of agriculture output	Hydropower production	Urbanization level	Weight
GDP	1	6	4	2	0.5125
Modulus of agriculture output	1/6	1	1/2	1/4	0.0743
Hydropower production	1/4	2	1	1/2	0.1377
Urbanization level	1/2	4	2	1	0.2755

Based on the calculation of the weight of the four factors of the social economic condition, the maximum eigenvalue $\lambda = 4.0104$, *consistency ratio = 0.0039*.

Table 3. Pairwise comparison matrix of hierarchy II–III.

Comprehensive impact	Proportion of the number of people by water resources	Average temperature	Precipitation	Percentage of brine	Weight
Proportion of the number of people by water resources	1	1/2	3	5	0.2963
Average temperature	2	1	5	8	0.5351
Precipitation	1/3	1/5	1	2	0.1085
Percentage of brine	1/5	1/8	1/2	1	0.0601

After ciphering the weight of the three elements for the development and utilization degree of water resources, the maximum eigenvalue $\lambda = 4.0104$, *consistency ratio = 0.0039*.

Table 4. Pairwise comparison matrix of hierarchy II–III.

Comprehensive impact	Modulus of soil and water loss	Industrial sewage emission	Ratio of underground water in water use	Weight
Modulus of soil and water loss	1	1/2	5	0.3258
Industrial sewage emission	2	1	8	0.6039
Ratio of underground water in water use	1/5	1/8	1	0.0703

When it comes to the final three branches of the state of the ecological environment, the maximum eigenvalue $\lambda = 3.0055$, *consistency ratio = 0.0048*.

STEP 2: Calculate the absolute difference between comparison series and reference data series of every evaluated object. That is, $\Delta_i = (j) = |a_i(j) - a_0(j)|$. In which, I represents the sequence of states, j represents the sequence of criterion, $a_0(j)$ represents the reference data of the jth evaluation criterion.

STEP 3: Determine the value of p and q.

$$p = \min_{1 \le i \le n} \min_{1 \le i \le m} \{\Delta_i(j)\}$$

$$q = \max_{1 \le i \le n} \max_{1 \le i \le m} \{\Delta_i(j)\}$$

STEP 4: According to the equation, we can calculate the correlation coefficient with respect to every comparison series and reference data series.

$$y_i(j) = \frac{(p + q\beta)}{(\Delta_i(j) + q\beta)} \quad j = 1, 2, \cdots, m$$

STEP 5: To find the relationship between every evaluated object and reference data series, we calculate the average value of correlation coefficient of every evaluation criterion and reference data series. We define it as correlation, which is shown as follows:

$$r_j = \frac{1}{n} \sum_{i=1}^{n} y_i(j)$$

STEP 6: Calculate the weight of each evaluation criterion.

Table 5. Criterion's weights to overall objects.

Criterion	Weight
GDP	0.1407
Modulus of agriculture output	0.0204
Hydropower production	0.0378
Urbanization level	0.0757
Proportion of the number of people by water resources	0.1858
Average temperature	0.3516
Precipitation	0.0713
Percentage of brine	0.0395
Modulus of soil and water loss	0.0223
Industrial sewage emission	0.0412
Ratio of underground water in water use	0.0048

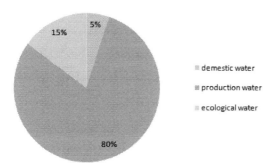

Figure 1. Final water allocation scheme concerning three ways of using water.

$$r_j' = \frac{r_j}{r_1 + r_2 + \cdots r_m} \quad j = 1, 2, \cdots m$$

STEP 7: Construct a general evaluation model.

$$Z_i = r_1' a_i(1) + r_2' a_i(2) + \cdots r_m' a_i(m) \quad i = 1, 2, \cdots, n$$

2 ERROR ANALYSIS

2.1 Error from statistical data

Data from Web sites and encyclopedia are not absolutely correct and precise. It has systematic errors and random errors. Since the water allocation plan is totally dependent on those data, the errors from the data we collected will bring about errors at the conclusion.

2.2 Error from the model

The weights of the evaluation criterion that we use in the AHP model have strong subjectivity and randomness; therefore, the weight of each criterion is not that credible. In the grey correlation model, we divided America into three parts solely according to the regional distribution, this division method will cause some errors in the conclusion because the river we calculate might not flow through the state while our MATLAB program deems it would. In other words, we did not consider the topography and geography. In the multi-objective decision analysis model, the use of trend prediction is not that precise because rivers have a periodical nature and also a constellation of whatever ecological and environmental-related affair does not fit our trend model; therefore, it also leads to error in the conclusion.

2.3 Error from the algorithm

The Genetic Algorithm has high computational requirements and cannot identify the global optimum. Because we have a two-object function, the ratio of priority of function is determined by intuition and common sense, neither of which is a scientific resort to obtain this key factor in computing the optimum solution.

3 CONCLUSIONS

We have come up with a water allocation scheme for 2013–2025 in America, in the industrial age, in which massive production is the main theme. In this sense, we should allocate most of the water to production. The massive production brings about massive pollution, in that we should give second priority to ecological water use. As the society develops, production will become more effective and environment-friendly and the water use in production will have a descending trend, and as the water saving concept is accepted by the people, the ratio of domestic water use will have a descending trend too. As we all know, during the last two centuries after the industrial revolution, the environment issue was intentionally or unintentionally overlooked for a long time, so the environmental and ecological condition was altered severely. In order to surrogate this, we should give more attention to environment conditioning. In this sense, environmental water supply should have an ascending trend.

REFERENCES

Guaimei Cheng. Research of regional water sustainable use based on multi-objective analysis [D]. Yunnan: Kunming University of Science and Technology, 2005.

Hongbo Zhang, Allocation model of Yellow River and method discussions [J]. Yellow River, Vol 28, No 1, Jan 2006.

Lei Li, Game analysis of allocation of water resources [J]. Commercial Research Vol 390, Dec.

Zhengfa Wang, Water resources distribution models on the river [J]. Water Resources and Hydropower Engineering, Vol 31 No 9, Sep 2000.

Civil, Architecture and Environmental Engineering – Kao & Sung (Eds)
© 2017 Taylor & Francis Group, ISBN 978-1-138-02985-9

A test study on the improved filler from blast furnace slag

Xiaoyi Liu & Guoyuan Xu
School of Civil Engineering and Transportation, South China University of Technology, Kwangtung, China

Dawei Huang
Engineering Research Center of Railway Environmental Vibration and Noise, Ministry of Education, East China Jiaotong University, Nanchang, China

ABSTRACT: Large direct shear test and compression test for the improved filler from blast furnace slag by the large direct shear apparatus were conducted, and relevant mechanical parameters of the improved filler were obtained. Test results show that water content has little influence on the mechanical index as fine particle cannot fill the interspace of coarse particles, so the filler has good water stability. There is little occlusion cohesion of the mixture filler when the vertical stress is less than 150 kPa and large occlusion cohesion if it is more than 150 kPa. The less the vertical stress, the more the shear shrinkage at the initial stage and the less the shear dilation at a later stage. When the vertical stress is more than 150 kPa, the shear dilation reduces to a greater degree than the shear zone formed by the fragmentation and breakage of coarse particles. Compression at the stage of loading is mainly concerned with the smashed edges and corners and long flat shape particles breaking off. The settlement calculation of the foundation should refer to the deformation modulus at the stage of reloading when the foundation is constructed by dynamic consolidation.

1 INTRODUCTION

Substantial amounts of solid wastes have been produced during the rapid development of the iron and steel industry and blast furnace slag is a major waste. The fact that there has been plenty of research on the application of blast furnace slag has not changed the situation of little consumption with a high cost. To reduce the land covered by slag, how to utilize and consume it remains the most urgent problem in the process of steel and iron production. Research and analysis indicate that the application of blast furnace slag in filling foundations or roadbeds is one of the major methods to substantially consume it (Jin Xia, 2005 & Wang Haifeng, 2007). Normally dynamic consolidation with heavy hammers are adopted when blast furnace slag are utilized as fillers for high-filled foundations or roadbeds to lower the possibility of settlement and improve the grain size of fillers (Yang Youhai, 2011; Gao Zhengguo, 2013; Yang Youhai, 2012; Huang Dawei, 2012;).Improved fillers with relatively large grain size are considered coarse particles. The shear strength and compression modulus of fillers are usually tested indoor to calculate the stability and high-filled foundation or roadbed as well as the possible settlement after construction (Zhu Jungao, 2011; Chen Xiaobin, 2007; Ding Zhouxiang, 2007; Huang Dawei, 2012).

In this article, a large-size direct shear apparatus was involved here to conduct improved filler direct shear test and compression test and we tried to analyze the fillers' direct shear property and compression property. We have analyzed the shear dilation of the blast furnace slag under different normal stressors by means of vertical displacement of the vertical loading plate and provided suggestions on compression modulus while calculating the settlement according to test results.

2 FILLERS' GRADATION PROPERTY

As solid wastes are produced during the process of steel and iron production, blast furnace slags feature irregular shapes, large grain size, and a certain amount of pores inside the grains with the average internal porosity being 10.5% and the solid density of blast furnace slag (break off the large-size slags before testing to ensure no closed pores are included and adopt the measuring flask method) being 2.535 g/cm³. The Curve 1 in Figure 1 shows the gradation curve of blast furnace slags according to the particle analysis while Table 1 shows composition parameters of slag particles. It can be seen from the two graphs that blast furnace slags are majorly composed of large size particles. Suppose the particles with a diameter larger than 5 mm are referred to as coarse particles represented by P_5 (proportion of slags with d larger than 5 mm) (Huang Dawei, 2012), it can be seen from Table 1 that P_5 is 98.99. Based on the calculation, the non-homogeneous coefficient and curvature coefficient of blast furnace slag are $C_U = 2.38$ and $C_C = 0.95$, respectively. According to the Rules of Geotechnical Testing (SL237-1999), some blast furnace slags are defined with bad gradation and filler category falls into the group of mixture fillers with giant particles.

Having doped with fly ashes, ground layer-by-layer and conducted using dynamic consolidation with heavy hammers, we again layer the blast furnace slag and get samples. Results show that the particle composition of each layer resembles a lot. The Curve 2 in Figure 1 is the gradation curve when all fillers are equally mixed. Also see particle composition in Table 1 where P_5 is 87.56, non-homogeneous coefficient $C_u = 6.58$, and curvature coefficient $C_C = 1.29$. Mixture fillers feature great gradation and filler category falls into gravel soil. From this, we can know that the mixture, compaction, and dynamic consolidation clearly increase the number of fine particles and gradation of

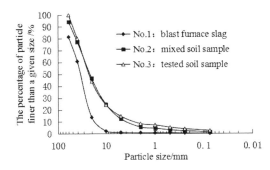

Figure 1. Grading curves of the blast furnace slag and the filler sample.

blast furnace slag as fillers are greatly improved. Through the back calculation of fine particle content, we get to know that the proportion of actual fly ashes mixed in total fillers on-site is 2%.

According to the Rules, the largest particle diameter of sample fillers for large-scale direct shear and compression testing should not be larger than 60 mm. In order to achieve this, we run mixture fillers through a soil analysis sieve of 60 mm pore diameter and obtain filler samples for compression testing, its gradation curve being Curve 3 in Figure 1 and particle composition being seen in Table 1. It can be read from the parameters of Figure 1 and Table 1 that the composition of fillers for compression test is similar to that of the mixture filler. The test results can be considered as compression performance parameters for on-site filling.

3 TESTING APPARATUS AND TESTING PROGRAM

3.1 Introduction to the apparatus and ways to prepare filler samples

The direct shear test that is to say large-scale shear test using large structural surface shear apparatus (shear apparatus for short) is suitable to be applied to the shear test when coarse fillers are involved. The large-scale shear apparatus involved here is a stress and strain double-controlled one, as shown in Figure 2. In this test, we adopt a strain-controlled apparatus whose direct shear rate is 2 mm/min. The direct shear box is in a rectangle shape, size being height*width*length = 200*400* 600 (the bottom shear box height is 100 mm, the upper shear box height 110 mm and 10 mm is left for the normal loading plate when packing). This kind of direct shear apparatus is only applicable when the largest diameter is no larger than 60 mm. There is a vertical displacement sensor on the

Table 1. Index of particle composition for the blast furnace slag and filler sample.

Sample	Fine granule content (%)	Coarse granule content (%)	Coefficient of uniformity C_u	Coefficient of curvature C_c	Grading	Soil classification
Blast furnace slag	0.15	98.99	2.38	0.95	Poor	Soil mixtures with giant grains
Mixed soil sample	1.27	87.56	6.58	1.29	Well	Gravelly soil
Tested soil sample	2.48	85.34	9.66	1.86	Well	Gravelly soil

Note: There is good particle gradation ($C_u \geq 5$, and $C_c = 1\sim3$) and bad gradation ($C_u < 5$, and $C_c \neq 1\sim3$), among which non-homogeneous coefficient $C_u = d60/d10$, and curvature coefficient $C_c = d302/(d10 \times d60)$, d10, d30, and d60 are in accordance with 10%, 30%, and 60% particle diameter in the particle gradation curve. Coarse granule means the soil of which particle size is more than 5 mm and fine granule less than 0.075 mm.

Figure 2. Large direct shear apparatus.

large direct shear apparatus, which imposes constant normal loading by means of liquid pressure. Also by collecting vertical displacement data from the vertical loading plate, we have obtained shear shrinkage and shear dilation property exposed in the direct shear test.

The large compression apparatus should be adopted in the case of coarse fillers for they are mostly composed of large-size particles. According to *Rules of Geotechnical Testing* (SL237-1999) (Chen Zhongyi, 1992), the ratio between the solidification container diameter (D) and its height (H) is 2~2.5 and generally, people use large gathering ring type compression apparatus. However, being different from large direct shear apparatuses, the large compression apparatuses are not widely utilized by science and research institutions. Considering this, the large shear apparatus is innovatively adopted in this test.

When a large direct shear apparatus is adopted in the direct shear test and compression test, there is difficulty in obtaining originally-shaped fillers and samples are normally obtained through disturbance. We learn from Figure 1 and Table 1 that in this test fillers majorly comprise of large size particles, which will probably lead to segregation between coarse and fine particles when preparing samples and the segregation is more serious than an on-site one. To avoid such things from happening, actions should be taken to separate coarse fillers and fine fillers before making samples (setting 10 mm as a benchmark). While preparing samples, dump one-third of the coarse filler into the direct shear box, smoothen it and then pour one-third of the fine filler. Again smoothen the surface and utilize an iron shovel and an iron bar to harden it until the coarse and fine fillers are evenly spread. Repeat the same procedure three times and make sure the thickness of each layer resembles one another. By doing all this, we make sure that the shear plane is even and the mechanical index obtained indoor is close to that of an on-site test.

Table 2. Experimental scheme.

Test number	Moisture content (%)	Weight of the dry sample (kg)	Dry density (g/cm³)	Initial void ratio
1	0	75	1.56	0.4602
2	0	79.4	1.65	0.3801
3	8	86.3	1.80	0.2687

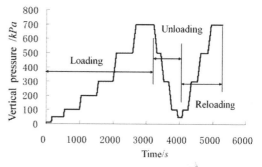

Figure 3. The curve of vertical compressive stress and time.

3.2 Test programs

The main purpose of this test is to find out the impact of shear strength and filler water content on basic mechanical parameters after the mixture, compaction, and dynamic consolidation of the blast furnace slag is completed. During the test, we have separately conducted a direct shear test on dry fillers (air drying) and wet fillers adopting the same filling method and compacting method. As for the dry filler, conduct the test while it remains dry; as for the wet filler, the water content is 8%. The dry density for both the dry filler and the wet filler is 1.78 g/cm³ during the test.

The main purpose of this large compression test is to find out the impact of compression parameters as well as filler water content on compression parameters after the mixture, compaction, and dynamic consolidation of the improved blast furnace slag is completed. We adopt the same filling and compacting procedure during the test. Table 2 reveals the filler parameters in three compression tests, in which dry fillers are air-dried indoor and water content of the wet filler is 8% in terms of fillers with a diameter less than 5 mm. Graded loading, graded unloading, and graded reloading are adopted in the compression test. The stability time for every loading (unloading) is determined by the vertical displacement of the normal loading plate. Usually, if no vertical displacement occurs in 2 min, another graded loading (unloading) will be triggered. Figure 3 is the relationship curve of "vertical compressive stress–time" in one compression test.

4 TEST RESULT ANALYSIS

4.1 *Shear strength parameter property analysis*

The direct shear test results of dry filler and wet filler under different vertical pressures are shown in Table 3 and Figure 4 and Figure 5. While analyzing, draw all dots on the coordinates together with a smooth curve and try to fit it into the linear tendency line. We have obtained shear strength parameters of the two fillers from the fitted linear tendency line and rendered them in Table 4. It can be learned from shear strength parameters of dry fillers and wet fillers that water content produces little impact on the fillers' shear strength parameters and fillers have great water stability. It can also be learned from particle analysis that fillers comprise majorly of large-sized particles (P5 is about 85.34); water mainly influences fine fillers' strength and since blast furnace slag boasts of great water stability, water exerts little influence on large-size particles. That is because when fine particles are not enough to fill up pores between coarse particles, fine particles will be free between pores while coarse particles make up as filler skeleton. Therefore, water content has limited influence on shear strength parameters.

Table 3. Direct shear scheme and results.

Moisture content: 0		Moisture content: 8%	
Vertical pressure (kPa)	Maximum shearing force (kN)	Vertical pressure (kPa)	Maximum shearing force (kN)
50	103.33	50	101.67
100	236.25	75	184.58
150	343.33	150	345.00
275	501.25	275	502.92
400	647.92	500	764.17
500	769.17		

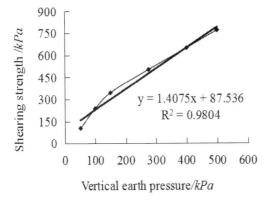

Figure 4. The relationship between shearing strength and vertical pressure of a dry sample.

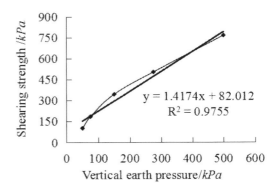

Figure 5. The relationship between shearing strength and vertical pressure of a wet sample.

It can be seen from shear strength parameters that despite the lack of cohesion between filler particles, filler cohesion can be seen from a direct shear test that is occlusion cohesion.

It can be seen from Figure 4 and Figure 5 that a smoothed curve is in good accordance with the linear tendency line, that is to say both corresponding dots of 150 kPa and 275 kPa normal stress are over the tendency line; the slope of circular curve tends to reduce; all coordinates with normal stress larger than 150 kPa are seen to lie in one line.

The destructive plane (shear plane) in the direct shear test is decided by workers. During the test, both strain and stress are non-homogeneous and quite complicated. Different coordinates have different stress and stress paths. The main stress varies surrounding the shear plane in a spiral direction. Here is the Coulomb equation:

$$\tau_f = c + \sigma \tan \varphi \qquad (1)$$

where

τ_f is the fillers' shear strength (kPa);

c is the fillers' occlusion cohesion (kPa);

σ is the normal stress imposed on the shear plane (kPa);

φ is the fillers' internal friction angle (°);

For non-cohesive fillers, the occlusion cohesion is $c = 0$ when shear strength is in direct proportion to the normal stress imposed on the shear plane, which indicates that the τ_f of non-cohesive fillers is determined by the frictional resistance between particles. However, the frictional resistance between particles is made up by sliding friction and additional resistance provided by mutual occlusion, and the value of frictional resistance is determined by the size, gradation, density, and coarseness of the particles. Hence, it can be understood that conditional on small vertical stress, blast furnace slag shows no occlusion cohesion and when vertical stress is relatively large (150 kPa for example), shear strength shows occlusion cohesion.

4.2 Shear dilation analysis

During the direct shear test, as soon as shear displacement reaches 20 mm, both upper and bottom shear boxes begin to gradually open up as shown in Figure 6. The analysis shows that shear dilation is responsible for the opening up of shear boxes and upper box fillers' side friction leads the upper box to translate. We have measured the vertical displacement of the vertical stress loading plate as a way to analyze the shear dilation in a large direct shear test, as shown in Figure 7. It can be seen from Figure 7 that downward displacement of the vertical stress loading plate can be observed at an early stage of the shear test and the largest downward displacement is about 1 mm. The larger the vertical stress is, the larger the vertical displacement will be at an early stage. That is to say, as the shear compression continues to increase with continuing increase of shear displacement, and the vertical stress loading plate starts to move upward. It can be seen from Figure 7 that larger the vertical stress is, larger is the shear compression at an early stage and smaller the shear dilation at a later stage.

Shear compression occurs when fillers' volume reduces because of mutual embedding and breaking off of long and flat particles during the shear deformation process. Hence, conditional on the same consolidation, the larger the normal stress is, the larger the shear compression will be. Shear dilation occurs when fillers' volume increases owing to particle re-arraying because of the shear moving and tumbling. However, conditional on larger normal stress, when the force that breaks a particle is smaller than the force that causes shear moving and tumbling, particles will break off. Conditional on the same consolidation, the smaller the normal stress is, the larger the shear dilation will be.

The destructive plane (shear plane) in the shear test is not a whole plane but a complex zone where coarse fillers impose force on each other. Under these complex mutual interactions, embedding, crushing, shear moving, tumbling, and breaking off are supposed to take place at the same time; where embedding and crushing can be regarded as a decrease of the volume; particle shear moving and tumbling are regarded as an increase in the volume; and particles breaking off are regarded as remaining same of the volume. During the direct shear test, particles work differently when under different normal stress or at different stages of shear displacement and particles breaking off mostly take place conditional on large vertical stress and large shear displacement. It can be seen from Figure 10 that when normal stress is larger than 150 kPa, the vertical displacement of the vertical loading plate is much smaller than the displacement when normal stress is 150 kPa and it is close to the normal stress at 50 kPa. Therefore, it can be deducted that when the fillers' normal stress is larger than 150 kPa, the breaking off of particles is mainly responsible for the shear plane. This is the reason for the obvious decrease in shear dilation. The fillers also show large occlusion cohesion at the same time. We learn from these test results that there are more breaking off particles when normal stress is over 150 kPa. See Figure 8.

Figure 6.   Opened shear box.

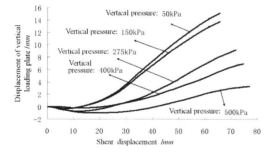

Figure 7.   The relationship between the vertical displacement of the upload plate and shear displacement for a dry sample.

Figure 8.   Broken particles by shear.

4.3 The compression property of improved blast furnace slag

Figure 9 shows the curve of settlement of the loading plate and time, where it can be seen that the longest time is needed for a stable vertical displacement of the uploading plate at each level stage, longer time for the reloading stage, and least time for the downloading stage.

Figure 10 shows the curve of "settlement-vertical stress" when the loading plate displacement is stable at different loading/downloading levels during the three compression tests. It can be seen that among the three compression tests, there is a slight difference in loading plate settlement between the prior three levels of the first loading and for other loading/downloading at different levels the settlement is pretty close. According to the definition of fillers' confining deformation modulus E_s from *Soil Mechanics*, if we randomly choose the slope of a small secant line as the confining deformation modulus in relevant range, then the equation will be:

$$E_s = \frac{\Delta \sigma'}{\Delta \varepsilon} \tag{2}$$

where

$\Delta \sigma'$ is effective stress increase

$\Delta \varepsilon$ is compression strain increase

Confining deformation modulus Es obtained from different stages of uploading/downloading stages according to Equation (2) are listed in Table 4, where it can be seen that the compression parameters of dry and wet improved blast furnace slag are pretty close indicating blast furnace slag particles have good water stability. Few fine particles are seen and they are barely involved in the formation of filler skeleton and that is why water content exerts little influence on the fillers' compression property.

It can be seen from Table 3 that when stress is growing from 100 kPa to 200 kPa, the E_s remains to be 15 MPa, and according to the reference value of filler compression from *Soil Mechanics* (when stress is growing from 100 kPa to 200 kPa, the confining compression modulus lower than 4 MPa will be regarded as high compression fillers, 4~20 MPa medium compression fillers, and larger than 20 MPa low compression fillers), the blast furnace slag adopted in this test falls in the group of medium compression fillers.

Draw curves of E_s from different uploading/downloading stages in Figure 11 (the abscissa of coordinates come from the average of vertical stress of both starts and ends of all levels of uploading and downloading). It can be seen from Figure 11 that E_s are small the first time when loading and the values are all smaller than 25 MPa; E_s (supposed to rebound modulus) will be generally larger when downloading, and with the increase of downloading it gradually decreases; when reloading, it will first increase and then decrease; it will be larger when reloading than the E_s at the first

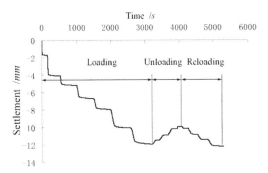

Figure 9. The curve of settlement of the loading plate and time.

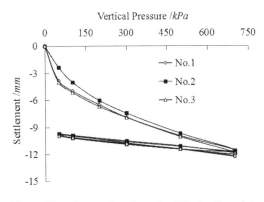

Figure 10. Curves of settlement of the loading plate and vertical compressive stress.

Table 4. Confining deformation modulus Es for different stages of loading and unloading (MPa).

Stages (kPa)		No. 1	No. 2	No. 3
Loading	0→50	2.59	4.22	2.47
	50→100	8.95	6.05	8.95
	100→200	13.11	9.93	12.85
	200→300	14.27	14.18	16.33
	300→500	19.42	17.75	18.75
	500→700	23.69	20.81	21.49
Unloading	700→500	81.50	85.40	85.11
	500→300	70.18	75.24	72.94
	300→100	55.94	62.02	57.28
	100→50	33.33	38.21	37.27
Reloading	50→100	48.50	54.56	54.05
	100→300	55.10	64.34	58.39
	300→500	62.00	68.76	65.76
	500→700	48.00	50.53	51.73

loading mainly owing to great deformation that has taken place during the first loading.

Improved blast furnace slag fillers mainly comprise of broken slags whose edges and corners are quite obvious. Samples are obtained through indoor disturbance. During the compression test, all edges and corners are compressed and therefore the stress is highly focused on breaking off particles; besides, the lack of fine particle content makes it impossible for fine particles to fill up space between coarse particles, which leads to the occurrence of tensile stress between large-size and flat particles as shown in Figure 12. Hence, the improved blast furnace slag adopted in indoor compression test belongs to the medium compressive filler. In order to make sure that fine particles will fill up the space between coarse fillers, a number of particles with a diameter smaller than 5 mm should not be less than 30% of total fillers.

Dynamic consolidation with heavy hammers is adopted while alternating foundation with improved blast furnace slag high fillers. During

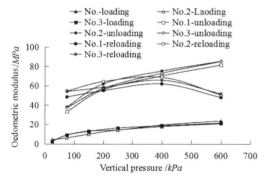

Figure 11. Curves of confining deformation modulus and vertical compressive stress.

Figure 12. Edges and corners smashed and long flat shape particles breaking off.

dynamic consolidation, a majority part of edges and corners of particles under concentrated stress will be broken off and large tensile stress takes place. Large size particles and flat particles are also broken off. When the heavy hammer falls on the ground, it produces stress much larger than the one that is produced by the later upper filler. The dynamic consolidation foundation is similar to preloading, however, filler samples prepared by indoor filler disturbance is difficult to match the actual contact between fillers and particles after dynamic consolidation. Hence, the compression modulus of reloading is highly recommended in light of the analysis of high fill foundation compression property of blast furnace slag after an on-site dynamic consolidation with a heavy hammer.

5 CONCLUSIONS

1. The results of the large direct shear test and compression test of the improved filler from blast furnace slag indicate that owing to the lack of fine particles (P_5 is about 85.34), they hardly involve in the formation of the fillers' skeleton and also blast furnace slag particles feature great water stability. Water content has little influence on the fillers' mechanical properties. Fillers adopted in this test have good water stability.
2. The direct shear test of improved blast furnace slag fillers also shows that conditional on the same compaction, lower normal stress leads to small shear compression at an early stage but a large shear dilation at a later stage and larger normal stress leads to a large shear compression at an early stage, but a small shear dilation at a later stage.
3. Conditional on normal stress less than 150 kPa, the shear plane formation of blast furnace slag mainly results from the particle shear moving and tumbling in the shear zone; conditional on normal stress larger than 150 kPa, the shear plane formation mainly results from the shear cutting and breaking off of particles in the shear zone.
4. The improved blast furnace slag mainly comes from the breaking of large size slag through compaction and dynamic consolidation with a heavy hammer. Sample fillers are mainly comprised of the skeleton formed by particles with obvious edges and corners in this compression test. The compression of the first loading in the indoor compression test is caused by the crushing of edges and corners of particles as well as the breaking off of long and flat large size particles. Hence, despite the fact that blast furnace slag improved fillers are coarse particles, they

still are medium compressive fillers at the first loading in the indoor large compression test.

5. Disturbed soil is adopted as filler samples in the improved blast furnace slag indoor compression test, which is difficult to match the practical contact between filler compaction and particles after dynamic consolidation. Considering the characteristics of dynamic consolidation, the compression modulus of reloading is recommended in light of the compressive analysis of improved blast furnace slag high fill foundation with dynamic consolidation.

REFERENCES

Chen Xiaobin, Zhang Jiasheng, Feng Zhipeng. Experimental study on rheol logical Engineering properties of coarsely granular red sandstone soil [J]. Chinese Journal of Rock Mechanics and Engineering, 2007, 26(3): 601–607. (in Chinese)

Chen Zhongyi, Zhou Jingxing, Wang Hongjin. Soil Mechanics [M]. Beijing: Tsinghua University Press, 1992. (in Chinese)

Cheng Zhanlin, Ding Hongshun, Wu Liangping. Experimental study on mechanical behaviour of granular material [J]. Chinese Journal of Geotechnical Engineering, 2007, 29(8): 1151–1158. (in Chinese)

Ding Zhouxiang, Zhu Hehua, Gong Xiaonan, et al. Large deformation description method for constitutive relation based on compression test [J]. Chinese Journal of Rock Mechanics and Engineering, 2007, 26(7): 1356–1364. (in Chinese)

Gao Zhengguo, Du Yulong, Huang Xiaobo, et al. Reinforcement mechanism and construction technology of broken stone fills by dynamic consolidation, 2013, 32(2): 377–384. (in Chinese)

Guo Qingguo. Coarse grain soil engineering properties and applications [M]. Zhengzhou. The Yellow River water Conservancy Press. (in Chinese)

Housing and urban-rural development of the People's Republic of China. GB 50007–2011 Code for design of building foundation [S]. Beijing: China Building Industry Press, 2011. (in Chinese)

Huang Dawei, Yang Youhai, Huang Jiqiang, et al. Study on the assessment indexes of compaction quality for railway Subgrade constructed by gobi coarse-grained soil filler [J]. China Railway Science, 2012, 33(2): 21–27. (in Chinese)

Huang Dawei, Yang Youhai, Lai Guoquan, et al. Analysis of dynamic deformation modulus for high-speed railway subgrade [J]. Rock and Soil Mechanics, 2012, 33(5): 1402–1408. (in Chinese)

Jin Xia, Li Liaosha, Dong Yuanchi. Development and prospect of recycling technology of blast furnace slag at home and abroad [J]. China Resources Comprehensive Utilization, 2005 (9): 4–7. (in Chinese)

Ministry of Water Resources of the People's Republic of China. SL 237–1999 Specification of soil test [S]. Beijing, China WaterPower Press, 1999. (in Chinese)

Wang Haifeng, Zhang Chunxia, Qi Yuanhong, et al. Present situation and development trend of blast furnace slag treatment [J]. Iron and steer, 42(6): 83–87. (in Chinese)

Wang Haifeng, Zhang Chunxia, Qi Yuanhong. Status and development trend of treatment and heat recovery technology for blast furnace slag [J]. China Metallurgy, 2007, 17(6): 53–58. (in Chinese)

Yang Youhai, Huang Dawei, Lai Guoquan, et al. Analysis of ground coefficient and modulus of deformation of gobi area filler in high-speed railway subgrade [J]. Rock and Soil Mechanics, 2012, 33(5): 1402–1408. (in Chinese)

Yang Youhai, Lai Guoquan. Research on Filling Test of High-speed Railway Subgrade in the Gobi Desert [J]. Journal of the China Railway Society, 2011, 33(3): 77–83. (in Chinese)

Zhu Jungao, Wang Yuanlong, Jia Hua, et al. Experimental study on resilience behaviour of coarse grained soils [J]. Chinese Journal of Geotechnical Engineering, 2011, 33(6): 950–954. (in Chinese)

Civil, Architecture and Environmental Engineering – Kao & Sung (Eds)
© 2017 Taylor & Francis Group, ISBN 978-1-138-02985-9

Environmental impact assessment of shipping construction based on the FCE method

C. Xu & D.M. Yu
Central South University, Changsha, Hunan, China

ABSTRACT: In comparison with highway and railway construction, shipping construction's concepts such as ecological and green construction and environmental protection are still in their infancy. In this paper, shipping construction' environmental impact assessment index system is presented. By introducing the Analytic Hierarchy Process (AHP) to determine the weight, this paper utilizes the Fuzzy Comprehensive Evaluation (FCE) method to conduct the comprehensive evaluation, with the 1st-phase of Xiangxi Autonomous prefecture shipping construction project in Hunan Province as the case study. The case results demonstrate that the impact of shipping construction is slightly negative on the overall environment. The research results can support theoretical guidance for the reasonable design, green construction, and environmental protection of shipping construction. In addition, they can provide auxiliary information and scientific basis for the development of the project of environmental management and economic development planning along the route.

1 INTRODUCTION

The purpose of shipping construction is to eliminate shoal, jet flow, rapids and other shipping obstruction phenomena, which are created by the rivers under natural conditions, and then to improve the channel dimension, the shipping conditions, and to expand the passing capacity. The measures taken for shipping engineering mainly include damming, dredge, bank revetment, reef explosion, and channelization. Shipping construction is a developmental activity that has a far-reaching impact on the society and the economy, and this construction and operations have an impact on the natural and social environment. Therefore, we need to efficiently handle the relationship between the project implementation and the environmental protection.

Vast research has been conducted on the environmental impact of shipping construction in China and foreign countries. Technological and industrial developmental project activities disturb the natural ecosystem and have multiple direct and indirect environmental consequences (Gupta, R. et al. 2003). Environmental Impact Assessment (EIA) is an anticipatory environmental management tool, which is of great significance to the sustainable development of the society (Jay, S. et al. 2007, Liu, K. F. R. et al. 2013). Information related to different environmental impacts produced by the execution of shipping construction activities and projects has been limited, described by semantic variables, and is affected by a high degree of inaccuracy and uncertainty (Peche, R. & Rodriguez, E. 2009). Environmental impact caused

by construction processes are reflected in three aspects: the impact on public, the impact on natural resources, and the impact on the ecological system (Zolfagharian, S. et al. 2012). The ecological environmental impact phase of highway construction project is divided into two periods: construction and operation period (Long, X. 2010). Desert railways, a kind of railways in special regions, have many particularities with respect to the climate, geography, and ecological environment. Due to these particularities, seven factors are taken into consideration to set up a factor set, which includes the water environment, soil environment, sand disaster, vegetation cover, wild flora and fauna, climate environment, and noise environment (Shao, L. et al. 2009). The environmental impact of inland waterway engineering is manifested in mainly six aspects: ecological environment, air environment, surface water environment, sound environment, solid waste, water loss, and soil erosion (Chen, G. J. 2011) From the perspective of "people-oriented", the discussion about human health cannot be ignored in terms of EIA (Bhatia, R. & Wernham, A. 2008, Harpet, C. & Le Gall, A.R. 2013). However, studies such as the current domestic one, aiming at assessing the degree of effect of shipping construction on the overall regional environment are few.

There are various kinds of evaluation methods, which include a complex method system. Foreign researchers use the Fuzzy Analytical Hierarchy Process (FAHP) to assess the impact of a mineral project on groundwater environment (Aryafar, A. et al. 2013) and evaluate the teaching performance

(Chen, J.F. et al. 2015). Online static security assessment and hazard assessment of organic chemicals are based on artificial neural networks (Sunitha, R. et al. 2013, Germashev, I. V. et al. 2009). A fuzzy multi-criteria approach is used to assess the ex-ante impact of food safety policies (Mario, M. et al. 2013). Chinese researchers use Fuzzy Comprehensive Evaluation (FCE) method to evaluate the landscape ecology system in highway and groundwater quality (Zhang, X. B. et al. 2012, Kou, W. J. 2013). Zhou Hong (2016) introduced the ecological footprint method and the energy analysis method of ecological economics, which can be applied to large engineering construction projects and transform comprehensive evaluation into ecological evaluation. Based on the principle of the Analytic Hierarchy Process (AHP) and the characteristics of evaluation index system about tourism, an improved AHP-TOPSIS model has been established. It aims at solving the problem of quantitative assessment of World Expo influencing the Shanghai tourism industry (Yang, J. K. et al. 2012). Through the comparative analysis of various methods and combining the characteristics of shipping construction, this paper uses the FCE method to conduct the assessment of shipping construction's EIA.

2 EVALUTION INDEX SYSTEM

Choosing evaluation index should follow certain principles, which mainly include hierarchical principle, system integrity principle, relative independence principle, intuition principle, and feasibility principle. To make the index system operational and comprehensive, after referring to the "Environmental impact assessment specification for inland waterway project" (JTJ 227-2001) and "Technical guidelines for environmental impact assessment" (HJ/T 2.1~2.3-93), literature searching, experts consulting, engineering analyzing, and other means, the evaluation index set is divided into three layers through analysis and selection. The first, target layer u; the second, criteria layer $\{u_1, u_2, u_3\}$ the third, index layer
$\{u_{11}, u_{12}, u_{13}, u_{14}, u_{15}\}, \{u_{21}, u_{22}, u_{23}, u_{24}\}$, and
$\{u_{31}, u_{32}, u_{33}, u_{34}\}$ In this paper, the structure and the specific meaning of each index are shown in Table 1.

3 ENVIRONMENTAL IMPACT COMPREHENSIVE EVALUATION MODEL OF SHIPPING CONSTRUCTION

3.1 *Evaluation set*

An evaluation index set is a kind of language description of every evaluation index level, which

Table 1. The index system of environmental impact assessment of shipping construction.

Target layer	Criteria layer	Index layer
Environmental impact assessment of shipping construction u	Natural environment u_1	Sound environment u_{11}
		Water environment u_{12}
		Atmospheric environment u_{13}
		Soil u_{14}
		Natural landscape u_{15}
	Ecological environment u_2	Terrestrial animal u_{21}
		Land resources u_{22}
		Land vegetation u_{23}
		Aquatic ecosystem u_{24}
	Social environment u_3	Land requisition and demolishing u_{31}
		Immigrant resettlement u_{32}
		Cultural facilities u_{33}
		Human health u_{34}

is a collection of comments given by an evaluator on evaluation indicators. The impacts of shipping construction on EIA are not the same; they can be either positive or negative and have different degrees. Degree set $v = \{v_1, v_2, v_3, v_4, v_5\}$ where $v_1 \sim v_5$ respectively denotes that shipping construction has a negative and serious impact; a negative and slight impact; a negligible impact; a positive and slight impact; and a positive and major impact on the environment. In order to facilitate the evaluation, this paper uses the score to show the evaluation criteria for each level. Scores of 50, 40, 30, 20, 10 denote the level of evaluation criteria v_1, v_2, v_3, v_4, v_5 respectively.

3.2 *The determination of the weight*

When conducting the FCE, the weight has an effect on the final evaluation results. Different weights can sometimes derive a completely different conclusion.

Therefore, the choice of weight being appropriate or inappropriate is directly related to the success or failure of the model (Du, D. & Pang, Q. H. 2006.). A number of different kinds of methods can be used to determine the weight, such as the expert estimation method, the AHP (Li, Z.Y. 2004), etc. The complexity of the system and the practical need are the basis of making choices. This paper uses the AHP to determine the weight.

1. Construct the judgment matrices

First, the experts construct the judgment matrix $S = (\mathcal{U}_{ij})_{n \times n}$ using Saaty scale (Saaty, T. L. 2008), and calculate the largest eigenvalue $\lambda_{\max}$ and its eigenvector A of judgment matrix using MATLAB. This feature vector is the importance ranking of each evaluation factor.

2. Check consistency

Certain errors arise between the judgment matrix constructed by the expert's scoring method and the theoretical judgment matrix. In order to minimize the error, we need to do a consistency check (Zhu, Y. et al. 2015). The standard for inspection is CI, and the calculation method is shown below:

$$CI = \frac{\lambda_{\max} - n}{n - 1} \quad (1)$$

where $\lambda_{\max}$ is the largest eigenvalue of the judgment matrix, and n is dimension of the judgment matrix.

When $CI = 0$, judgment matrix has complete consistency. The greater the value of CI, the poorer the consistency of the judgment matrix.

3. Find the corresponding mean random consistency index RI

Table 2 shows the mean random consistency index of $n \times n$ positive reciprocal matrix (n = 1, 2,... 8). The results of the mean random consistency indexes are calculated after 1000 times.

4. Calculation of the consistency ratio

Using the random consistency ratio to judge the result of AHP ranking, the formula is as follows:

$$CR = \frac{CI}{RI} \quad (2)$$

When $CR < 0.10$, the AHP has a satisfactory consistency. If the AHP does not have a satisfactory consistency, we would need to adjust the judgment matrix element value and re-assign the value of the weights.

3.3 *Single factor fuzzy judgment*

Single factor fuzzy judgment matrix, which makes assessments based on a single factor, determines the degree of membership that evaluation object has on evaluation set V. Suppose the degree

Table 2. Mean random consistency index.

n	1	2	3	4	5	6	7	8
RI	0	0	0.52	0.89	1.12	1.26	1.36	1.41

of membership that u_i ($i = 1,2,...,m$) has on v_j ($j = 1,2,...,n$) is r_{ij}, where v_j ($j = 1,2,...,n$) is the jth element of the evaluation set V. So we get the single gle factor fuzzy judgment matrix as follows:

$$R = \begin{bmatrix} R_1 \\ R_2 \\ \vdots \\ R_m \end{bmatrix} = \begin{bmatrix} r_{11} & r_{12} \cdots r_{1n} \\ r_{21} & r_{22} \cdots r_{2n} \\ \vdots & \vdots \quad \vdots \\ r_{m1} & r_{m2} \cdots r_{mn} \end{bmatrix} \quad (3)$$

Drop half trapezoid model (Song, H. J. 2014.) is used to determine the membership function. Suppose V_j and V_{j+1} are adjacent of two grading standards. Clearly, $V_j > V_{j+1}$; thus, the factors for membership function on V_j are defined as follows:

$$r(X) = \begin{cases} \dfrac{X - V_{j+1}}{V_j - V_{j+1}} & V_{j+1} \le X \le V_j \\ 0 & X < V_{j+1} 或 X > V_j \end{cases} \quad (4)$$

The membership function of v_{j+1} is defined as follows:

$$r(X) = \begin{cases} \dfrac{V_j - X}{V_j - V_{j+1}} & V_{j+1} \le X \le V_j \\ 0 & X < V_{j+1} 或 X > V_j \end{cases} \quad (5)$$

3.4 *Fuzzy comprehensive evaluation*

Using the appropriate mathematical operations to combine weight vector A with fuzzy judgment matrix R of evaluated objects, we get the result vector B of FCE of evaluated objects; b_j is the degree of membership that evaluation object has on evaluation set from an overall perspective. Thus, the equation is

$$B = A \cdot R = (a_1 \ a_2 \cdots a_m) \cdot \begin{bmatrix} r_{11} & r_{12} & \cdots & r_{1n} \\ r_{21} & r_{22} & \cdots & r_{2n} \\ \vdots & \vdots & \cdots & \vdots \\ r_{m1} & r_{m2} & \cdots & r_{mn} \end{bmatrix} = (b_1 \ b_2 \ \cdots \ b_n) \quad (6)$$

4 CASE ANALYSIS

4.1 *Use of the AHP to determine the index weight*

Through the project team interview experts and analysis of similar engineering, the environment impact report of the 1st-phase of XAP shipping construction project in Hunan Province (2009) is combined, which is compiled by the Hunan Province Research Institute of Environmental Protection. Those questionnaires are combined behind the report, after which, according to the

AHP, the judgment matrix of the second layer is established as follows:

$$S = \begin{bmatrix} 1 & 4 & 3 \\ 1/4 & 1 & 1/3 \\ 1/3 & 3 & 1 \end{bmatrix}$$

$$CR = \frac{CI}{RI} = \frac{0.037}{0.52} = 0.07 < 0.10$$

$$CI = \frac{\lambda_{max} - n}{n - 1} = \frac{3.074 - 3}{3 - 1} = 0.037$$

Use MATLAB to calculate the judgment matrix's largest eigenvalue $\lambda_{max} = 3.074$. Then, check Table 2, when n is 3, $RI = 0.52$ and the random consistency ratio:

Matrix S meets the consistency requirement, and the distribution of weights is reasonable. As a normalized processing, the criteria layer's weight vector is $A = (0.615, 0.117, 0.268)$. Similarly, judgment matrix of the third layer is established as follows:

$$S_1 = \begin{bmatrix} 1 & 1 & 2 & 3 & 5 \\ 1 & 1 & 2 & 3 & 4 \\ 1/2 & 1/2 & 1 & 2 & 3 \\ 1/3 & 1/3 & 1/2 & 1 & 3 \\ 1/5 & 1/4 & 1/3 & 1/3 & 1 \end{bmatrix}$$

$$S_2 = \begin{bmatrix} 1 & 5 & 3 & 1 \\ 1/5 & 1 & 2 & 1/5 \\ 1/3 & 1/2 & 1 & 1/3 \\ 1 & 5 & 3 & 1 \end{bmatrix}$$

$$S_3 = \begin{bmatrix} 1 & 1 & 5 & 1/2 \\ 1 & 1 & 5 & 1/2 \\ 1/5 & 1/5 & 1 & 1/5 \\ 2 & 2 & 5 & 1 \end{bmatrix}$$

Using MATLAB to calculate judgment matrix's largest eigenvalue and checking the consistency, the weights of natural environment, ecological environment, and social environment are as shown below:

$$A_1 = (0.326, 0.314, 0.181, 0.119, 0.060)$$

$$A_2 = (0.394, 0.112, 0.100, 0.394)$$

$$A_3 = (0.254, 0.254, 0.061, 0.431)$$

4.2 Single factor fuzzy judgment

Based on the actual situation, the scores of sound environment, water environment, atmospheric environment, soil, and natural landscape are (38, 42, 32, 34, 25); the scores of terrestrial animal, land resources, land vegetation, and aquatic ecosystem are (31, 17, 32, 36); the scores of land requisition and demolishing, immigrant resettlement, cultural facilities, and human health are (35, 36, 17, 31). According to the formulae (4) and (5), the

corresponding single factor evaluation matrix is thus the following:

$$R_1 = \begin{bmatrix} r_1 \\ r_2 \\ r_3 \\ r_4 \\ r_5 \end{bmatrix} = \begin{bmatrix} 0 & 0.8 & 0.2 & 0 & 0 \\ 0.2 & 0.8 & 0 & 0 & 0 \\ 0 & 0.3 & 0.7 & 0 & 0 \\ 0 & 0.4 & 0.6 & 0 & 0 \\ 0 & 0 & 0.5 & 0.5 & 0 \end{bmatrix}$$

$$R_2 = \begin{bmatrix} r_1 \\ r_2 \\ r_3 \\ r_4 \end{bmatrix} = \begin{bmatrix} 0 & 0.1 & 0.9 & 0 & 0 \\ 0 & 0 & 0 & 0.7 & 0.3 \\ 0 & 0.2 & 0.8 & 0 & 0 \\ 0 & 0.6 & 0.4 & 0 & 0 \end{bmatrix}$$

$$R_3 = \begin{bmatrix} r_1 \\ r_2 \\ r_3 \\ r_4 \end{bmatrix} = \begin{bmatrix} 0 & 0.5 & 0.5 & 0 & 0 \\ 0 & 0.6 & 0.4 & 0 & 0 \\ 0 & 0 & 0 & 0.7 & 0.3 \\ 0 & 0.1 & 0.9 & 0 & 0 \end{bmatrix}$$

4.3 Fuzzy comprehensive evaluation

Putting weight vector and single factor evaluation matrix into formula (6), the first level results of FCE are shown below:

$$B_1 = (0.063, 0.614, 0.293, 0.030, 0)$$

$$B_2 = (0, 0.296, 0.592, 0.078, 0.034)$$

$$B_3 = (0, 0.322, 0.617, 0.043, 0.018)$$

The second level results of FCE are shown below:

$$B = A \cdot R = (0.615, 0.117, 0.268) \cdot \begin{bmatrix} 0.063 & 0.614 & 0.293 & 0.030 & 0 \\ 0 & 0.296 & 0.592 & 0.078 & 0.034 \\ 0 & 0.322 & 0.617 & 0.043 & 0.018 \end{bmatrix}$$
$$= (0.039, 0.498, 0.415, 0.039, 0.009)$$

To sum up, based on the principle of maximum degree of membership, the impact degree of the 1st-phase of XAP shipping construction project in Hunan Province is negative and slight in the aspect of natural environment. The impact is negligible on ecological environment and social environment. The impact is negative and slight on the overall environment.

5 CONCLUSIONS

1. Compared with highway construction and railway construction's EIA, the contents of shipping construction's EIA have not formed a complete index system yet. Shipping

construction's environmental impact index system established in this paper can better reflect the impact of shipping construction on the environment.

2. Currently, the preparation of environmental assessment report is based on the technical guidance and evaluation specification, including the evaluation of the scope, contents, key points, and so on; however, it does not propose a specific comprehensive evaluation method. The model established in this paper combines the quantitative analysis with the qualitative analysis, which is feasible, and provides a scientific, reasonable, and simple method for the comprehensive evaluation of impact of shipping construction on the environment.

3. Evaluation results obtained from the comprehensive evaluation can provide decision-making basis for scheme comparison and selection during the shipping construction planning stage; provide auxiliary information and scientific basis for the development of the project of environmental management and economic development planning along the route; and promote the sustainable development of the economy and the environment in the areas along the route.

ACKNOWLEDGEMENTS

The research presented in this paper was funded by the Department of Transportation of Hunan Province (the impact research of 1st-phase of Xiangxi Autonomous Prefecture shipping construction project in Hunan Province on the tourism industry of Xiangxi, contract no. 201545).

REFERENCES

Aryafar, A. Yousefi, S. & Ardejani, F.D. 2013. The weight of interaction of mining activities: groundwater in environmental impact assessment using fuzzy analytical hierarchy process. *Environmental Earth Sciences* 68(8):2313–2324.

Bhatia, R. & Wernham, A. 2008. Integrating human health into environmental impact assessment: An unrealized opportunity for environmental health and justice. *Environmental Health Perspectives* 116(8): 991–1000.

Chen, G.J. 2011. The impact analysis of inland waterway project on the environment and related environmental protection measures. *Pearl River Water Transport* 14:74–77.

Chen, J.F. Hsieh, H.N. & Do, Q.H. 2015. Evaluating teaching performance based on fuzzy AHP and comprehensive evaluation approach. *Applied Soft Computing* 28:100–108.

Du, D. & Pang, Q.H. 2006. *Modern comprehensive evaluation method and cases*. Beijing: Tsinghua University Press.

Gmashev, I.V. Derbisher, E.V. Aleksandrina, A.Y. & Derbisher, V.E.2009. Hazard assessment of organic chemicals with the use of artificial neural networks. *Theoretical Foundations of Chemical Engineering* 43(2):212–217.

Gupta, R. Kewalramani, M.A. & Ralegaonkar, R.V. 2003. Environmental impact analysis using fuzzy relation for landfill siting. *Journal of Urban Planning and Development* 129(3):121–139.

Harpet, C. & Le Gall, A.R. 2013. Trends in the consideration of environmental and health factors in urban planning: Need for an integrated approach. *Environmental Risques & Sante* 12(3):231–241.

HJ/T 2.1~2.3–93, *Technical guidelines for environmental impact assessment*.

Hunan Province Research Institute of Environmental Protection. 2009. *The environment impact report of the 1st-phase of Xiangxi Autonomous Prefecture shipping construction project in Hunan province*. Hunan: Hunan Province Research Institute of Environmental Protection.

Jay, S. Jones, C. Slinn, P. & Wood, C. 2007. Environmental impact assessment: Retrospect and prospect. *Environmental Impact Assessment Review* 27(4):287–300.

JTJ 227-2001, *Environmental impact assessment specification for inland waterway project*.

KOU, W.J. 2013. Application of modified fuzzy comprehensive evaluation method in the evaluation of groundwater quality, *South-to-North Water Transfers and Water Science & Technology* 02:71–75.

Li, Z.Y. 2004. *Environment quality evaluation principles and methods*. Beijing: Chemical Industry Press.

Liu, K.F.R. Ko, C.Y. Fan, C.H. & Chen, C.W. 2013. Incorporating the LCIA concept into fuzzy risk assessment as a tool for environmental impact assessment. *Stochastic Environmental Research and Risk Assessment* 27(4):849–866.

Long, X. 2010. Research on the eco-environmental impact assessment of highway construction project and case study. Guangzhou: Jinan University.

Mario, M. Ragona, M. & Zanoli, A. 2013. A fuzzy multicriteria approach for the ex-ante impact assessment of food safety policies. *Food Policy*38:177–189.

Peche, R. & Rodriguez, E. 2009. Environmental impact assessment procedure: A new approach based on fuzzy logic. *Environmental Impact Assessment Review* 29(5):275–283.

Saaty, T.L. 2008. Decision making with the analytic hierarchy process. *International Journal of Services Science* 1(1):83–98.

Shao, L. & Lin, B.L. 2009. Environment impact assessment of desert railway based on fuzzy comprehensive evaluation. *Journal of the china railway society* 31(5).

Song, H.J. 2014. Research of grey fuzzy evaluation method for risk analysis of earth-rock dam. Zhengzhou: Zhengzhou university.

Sunitha, R. Kumar, S.K. & Mathew, A.T.2013. Online static security assessment module using artificial neural networks. *IEEE Transactions on Power Systems* 28(4):4328–4335.

Yang, J.K. Dong, W.G. Du, C. & Ma,C. 2012. The Quantitative assessment of World Expo influencing the Shanghai tourism industry. *Mathematics in Practice and Theory* 16:13–20.

Zhang, X.B. Xiong, X.B. Dai, H.F. Fan, Z.Q. & Chen, K. 2012. Study on fuzzy AHP comprehensive evaluation of the landscape ecology system in highway. In IEEE(eds), *2012 International Conference on Fuzzy Theory and Its Applications*:379–383. New York: IEEE.

Zhou, H. 2016. *The theory and methods of engineering project ecological evaluation*. Beijing: China Architecture & Building Press.

Zhu, Y. Yang, R. Li,D.S. & Liu, Y.S. 2015. Comprehensive assessment of regional environmental impact of Beijing-shanghai high speed railway construction project. *Journal of the china railway society* 11:117–121.

Zolfaghaian, S. Nourbakhsh, M. Irizarry, J. Ressang, A. & Gheisari, M. 2012. Environmental impacts assessment on construction sites. *2012 Construction Research Congress*: West Lafayette, 21–23 May 2012. Reston: American Society of Civil Engineers.

Civil, Architecture and Environmental Engineering – Kao & Sung (Eds)
© *2017 Taylor & Francis Group, ISBN 978-1-138-02985-9*

Analysis and prediction of energy sustainability in the developed regions of China

Ping Wang
Science Experimental Research Center, Harbin Sport University, Harbin, China

Xinjun Wang
Department of Environmental Science and Engineering, Fudan University, Shanghai, China

Zhongliang Wang
Harbin Normal University, Harbin, China

Yibin Ren
Heilongjiang Research Academy of Environmental Sciences, Harbin, China

ABSTRACT: Energy, which accounts for two-thirds of today's greenhouse gas emissions, is the key toward reducing greenhouse gas emissions and decelerating global warming. In this paper, the IPCC-recommended reference approach and scenario analysis are applied to evaluate the dynamic change in the energy supply and energy-related carbon dioxide emissions during the period of 1995–2010 in the Yangtze River Delta Region (YRDR). The results show that energy importing reliance reached 85% in 2010, and that the energy structure has become more diversified in the YRDR. In addition, the per capita CO_2 emission is significantly higher whereas the carbon intensity is lower than that of the national average. Under the Low Carbon (LC) scenario, CO_2 emissions will begin to fall for the first time in 2017. Hence, if energy-saving and emission-reduction strategy and regional planning for the YRDR are implemented thoroughly, the YRDR will achieve the national emission reduction targets by 2020 and will have a great CO_2 mitigation potential in the future.

1 INTRODUCTION

For a century, both carbon intensity and global temperature have been on the rise. Furthermore, at the same time, human society has gradually become aware of the consequences of climate change due to the "greenhouse effect", especially of carbon dioxide emissions, making it one of the crucial international political issues at present. Energy, which accounts for two-thirds of today's greenhouse gas emissions, is at the heart of the problem and therefore must form the core of the solution (IEA 2015). In recent years, the famous international institutions such as Intergovernmental Panel on Climate Change (IPCC), International Energy Agency (IEA), World Bank (WB) and the Energy Department of Germany, England and America have issued energy reports periodically in order to mitigate greenhouse gas emissions and climate change. The detailed inventory of energy and future projections in light of the recent economic and policy developments come from these institutions, and other institutions and countries are increasingly influencing other countries. However, the regional-scale studies of energy and CO_2 emissions are comparatively scarce (Moriarty 2009, WB 2014).

The Yangtze River Delta Region (YRDR), China's most developed region (GDP was 22.1% of the nation's total, NBSC 2015), is now in the middle and later periods of industrialization, possessing energy-intensive and capital-intensive features. The rapid urbanization and industrialization are still escalating YRDR's rigid demand of energies. At the moment, enhancing energy security, maintaining economic growth and reducing greenhouse gas emission are becoming the major issues that challenge the future of the region. Therefore, we estimated the past and the present energy supply and CO_2 emissions in the YRDR by applying the IEA and IPCC-recommended approaches and projected the future trends of energy demand and CO_2 emissions based on China's Medium and Long Term Energy Strategy and 2011–2020 regional planning for the YRDR. According to the results, we have put forward a sustainable development project for the YRDR's future energy planning and policy making, which may provide some general insights into the effectiveness of regional-level energy conservation and GHG reduction for other regions as well.

2 METHODOLOGY AND DATA

2.1 Estimation of energy supply

Total Primary Energy Supply (TPES) is made up of production, imports, exports, international marine bunkers, international aviation bunkers and stock changes. Beginning with the 2009 IEA edition, international aviation bunkers are subtracted from the supply in the same way as international marine bunkers (IEA 2010). According to energy balance table data of the YRDR, energy supply can be calculated using the following expression:

$$E = \sum_i (P_i + R_i + l_i \pm S_i) \qquad (1)$$

where E is the total primary energy supply (Mtce, million tons of coal equivalent); the subscript i represents various fuels; and P_i, R_i, I_i, E_i and S_i denote primary production, recoveries, imports, exports and stock changes of fuel i, respectively.

For the analysis of the final energy consumption in the YRDR, the total and sectoral energy data for the period of 1995–2010 were obtained from the energy balance tables of the Shanghai Statistical Yearbook by Shanghai Statistical Bureau; Jiangsu Statistical Yearbook by Jiangsu Statistical Bureau; Zhejiang Statistical Yearbook by Zhejiang Statistical Bureau; and Yearbook database. The with final energy consumption sector is divided into four parts: primary industry (such as agriculture, forestry, animal husbandry and fishery), secondary industry (such as industry and construction), tertiary industry (such as transport, storage, post and communications, wholesale, retail sales, catering trade and others) and household consumption. This study considers mainly three types of energy: primary energy (such as coal, oil, natural gas), secondary energy (washed coal, other coal washing, mold coal, coke, coke oven gas, other gases, gasoline, kerosene, diesel oil, fuel oil, liquefied petroleum gas, refinery gas, other petroleum products, other coking products, heat and electricity) and other energy in the energy balance tables of the YRDR.

2.2 Estimation of CO_2 emissions

Based on the energy data, the following expression provides the detailed calculation to estimate carbon dioxide (CO_2) emissions from fossil fuel combustion from the YRDR using the IPCC-recommended reference approach (1996).

$$E_{CO_2} = \sum_i \left(A_i e_i c_i \times 10^{-3} - S_i \right) o_i M \qquad (2)$$

where E_{CO_2} is the total carbon dioxide emitted from fossil fuel combustion in kilograms (kt CO_2); A_i is the apparent consumption of fuel i (kt or 100 million m³ for natural gas); e_i is the net calorific value of fuel i (TJ/kt); c_i is the carbon emission factor of fuel i (t C/TJ), $S_i = A_{si} e_i c_i s_i$, is the carbon in products from non-energy uses of fossil fuels (kt C), o_i is the carbon oxidation rate of fuel i, and M is the molecular-to-atomic weight ratio of CO_2 to carbon (44/12). Calorific values and emission factors of fuels were gathered from "the Revised 1996 IPCC Guidelines for National Greenhouse Gas Inventories (The IEA is still using the 1996 IPCC Guidelines)".

2.3 Scenario analysis

Scenarios are self-consistent storylines of how a future energy system might evolve over time under a particular set of conditions (Kadian 2007). In order to analyze possible effects of series policies for energy saving and emission mitigation, two scenarios were set up: the Business As Usual (BAU) scenario and the Low Carbon (LC) scenario. The BAU scenario assumes that the indicators of energy-saving and emission reduction will be the same as before. Considering the development of new and renewable energy, the rates of energy-saving and emission reduction during 2011–2020 under the LC scenario will be increased by 0.5–1% compared with BAU. The calculation of energy consumption per unit of output value (CSUS 2009) and the major assumptions for the BAU and LC scenarios are as follows (Table 1):

Table 1. The basic assumptions for key variables in the YRDR.

Key variables	2010[a]	2015	2020
Population[b] (millions)	149.91	157.71	165.92
Population growth rate[c] (%)	1.02	1.02	1.02
GDP per capita[d] (yuan)	53434	82000	110000
GDP[e] (billions of yuan)	8010	12932	18251
GDP growth rate (%)	14[f]	10	7

(Continued)

Table 1. (Continued)

Key variables		2010[a]	2015	2020
Saving rate of	BAU	4.0	4.0	4.0
energy intensity[g] (%)	LC	4.5	5.0	5.0
Reduction rate of	BAU	6.5	6.5	6.5
carbon intensity[h] (%)	LC	7.0	7.5	7.5

*[a]The key variables are from the YRDR Statistical Year-book, 2011.
[b]The population used in this paper is the resident population in the YRDR.
[c]1.02% was the average annual population growth rate during 1995–2010.
[d]GDP per capita for different time periods are assigned based on the 2011–2020 regional planning for the YRDR.
[e]GDP = Population × GDP per capita.
[f]14% was the average annual GDP growth rate during 1995–2010.
[g]Saving rates of energy consumption per unit of GDP for different time period are assigned based on the 11th Five-Year Plan for National Economic and Social Development in Shanghai, Jiangsu, Zhejiang.
[h]Reduction rates of CO_2 emissions per unit of GDP for different time period are assigned based on the average annual reduction rates of energy consumption per unit of GDP during 1995–2010 in the YRDR and Medium and Long-Term Development Plan for Renewable Energy in China.

$$\frac{E_0}{G_0}(1-m)^t = \frac{E_t}{G_t} = \frac{E_t}{G_0(1+n)^t} \quad (3)$$

$$E_t = E_0\left[(1-m)^t \times (1+n)^t\right] \quad (4)$$

$$\frac{C_0}{G_0}(1-m')^t = \frac{C_x}{G_x} = \frac{C_x}{G_0(1+n)^t} \quad (5)$$

3 RESULTS AND DISCUSSION

3.1 The energy status in the YRDR

3.1.1 Energy supply in the YRDR
Based on energy balance table, the YRDR's energy demand rose rapidly with the continuing growth in population and economic. During 1995–2010, the energy supply increased from 170.51 to 436.55 Mtce, representing an annual rate of increase of 8.11%. However, primary energy production was only 40.45 Mtce in 2010, an increase of 60.32% over 1995 levels; however, it was only 9.26% of the YRDR's total supply, compared to 14.79% in 1995. An average of nearly 85% of energy supply for the period of 1995–2010 must have been met by imports from overseas and other domestic regions. Oil imports were 79.73 Mtce in

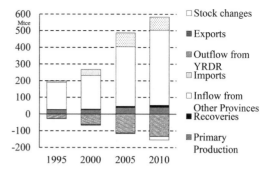

Figure 1. Energy supply by source in the YRDR.

2010, which was 10 times as large as in 1995, as shown in Figure 1.

3.1.2 The final energy consumption in the YRDR
The analysis showed that the final energy consumption in the YRDR rose from 169.32 Mtce in 1995 to 399.73 Mtce in 2010, accounting for 16.7% of total final energy consumption of China (2538.61 Mtce in 2010). In recent years, the energy structure has become more diversified in the YRDR: the coal share has decreased from 74% in 1995 to 63% in 2010, and the share of crude oil, natural gas and outdoor electricity has increased by 1%, 2% and 6% in 2010, respectively (Figure 2). From 1995 to 2010, the energy consumption from industry contributed the largest shares, followed by household consumption and transport, as shown in Figure 3. In 2010, the total final energy consumption of the industry sector in the YRDR reached 274.87 Mtce and accounted for 69% of the total. This was a substantial decline of 8 percentage points from 77% in 1995. Yet, construction sector was the smallest energy consumer in the energy consumption. The construction sector consumed 6.82 Mtce in 2010, accounting for just 2% of total final energy consumption. In contrast to its small share, however, the construction sector's energy consumption grew at a leading rate of 16% a year on an average during the period of 1995–2010.

3.1.3 CO_2 emissions in the YRDR
CO_2 emissions from fuel combustion in the YRDR increased from 361.29 million tons in 1995 to 901.86 million tons in 2010, with an average annual growth rate of 7.88%. The emission of CO_2 per capita in the year 2010 was 6.3 tons (based on resident population), which is higher than the world average and the national average in China. In contrast to per capita emissions, carbon intensity, measured as emissions per unit of GDP, was 37% lower than that of China's average and 25% higher than that of the world average level in the

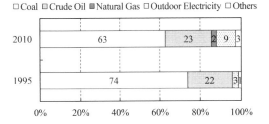

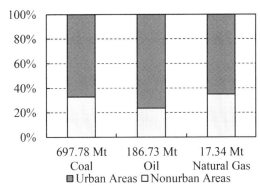

Figure 2. Shares of various fuel types in final energy consumption.

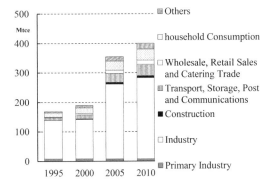

Figure 3. Final energy consumption by sector in the YRDR.

Table 2. Comparisons of CO_2 emission indicators, 2010.

	World	China	India	USA	YRDR
Population (million)	6614	1327	1123	306	144
Share of world population (%)	100	20	17	5	11
GDP ($2011 trillion, PPP)	67.2	7.6	3.1	14.1	1.8
Share of world GDP (%)	100	11	5	21	22
Share of world CO_2 emissions (%)	100	21	5	20	15
CO_2 emissions per capita (t)	4.4	4.6	1.2	18.7	6.3
CO_2 intensity (kg/$2011, ppp)	0.4	0.8	0.4	0.4	0.5

*a,b,c Share of population, GDP and CO_2 emissions in China, respectively.
Sources: IEA analysis (2012).

same period (Table 2). On the other hand, due to the rapid urbanization and economic development of urban areas in the YRDR, CO_2 emissions from urban areas were 623.36 million tons in 2010, which was 69% of the total (Figure 4).

Figure 4. Urban carbon dioxide emissions as percentage of the total.

3.2 The energy future in the YRDR

3.2.1 Energy demand
Based on the assumptions of socio-economic development in the YRDR and the various parameters for scenarios in the calculation, the total energy demand under the BAU scenario will increase to 814.88 Mtce in 2020, which is twice as large as in 2010 (Figure 5). Under the LC scenario, it will reach to 722.46 Mtce in 2020, suppressing the growth of total energy demand by 0.98% annually compared to BAU. The energy intensity was 0.76 tce/10,000 yuan in 2010, and it would be 0.44 and 0.39 tce/10,000 yuan in 2020 for the BAU and LC scenarios, respectively. While the energy intensity has improved in the YRDR, the improvements have only been able to decrease the demand growth to some extent, but have not halted it, owing to rapid economic and population growth.

3.2.2 CO_2 emissions
Two important turning points occurred in 2017 (Figure 6): under the BAU scenario, CO_2 emissions will maintain a 0.03% lower average annual growth rate during the period of 2017–2020 compared to 3.54% during 2010–2017. Due to the series of energy-saving and emission-reduction policies and measures, for the first time in 2017, CO_2 emissions under the LC scenario will begin to fall. CO_2 emissions will reach 1055.70 million tons in 2020, suppressing the growth of total CO_2 emissions by 0.95% annually compared to BAU. In per capita terms, this means 7.20 and 6.36 tons in 2020 under the BAU and LC scenarios, respectively, compared to 6.30 tons in 2010. The CO_2 intensities of the economy for the BAU and LC scenarios will be reduced by 58% and 63% in 2020 compared with that in 2010, respectively. The results show that the YRDR will achieve the national emission reduction targets (a 40–45% reduction target, Uwasu 2010) only when it continues to implement

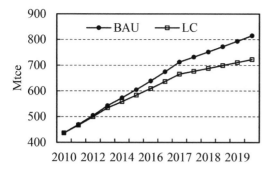

Figure 5.   Trends of energy demand for two scenarios.

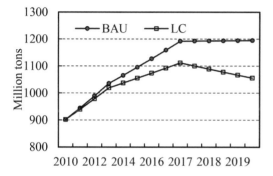

Figure 6.   Trends of CO_2 emissions for two scenarios.

its past emission reduction targets and regional plans related to energy saving and environmental protection.

4   CONCLUSIONS

In this study, the IPCC-recommended reference approach and scenario analysis were applied to the YRDR to analyze the dynamic change in the energy supply and energy-related CO_2 emissions for the period of 1995–2010 and project the energy demand and CO_2 emissions up to 2020 under the BAU and BP scenarios. The main conclusions drawn from the present study may be summarized as follows: 1) The GDP and energy consumption of the YRDR occupied nearly 22.1% and 16.7% of our country in 2010, respectively. However, it was one of the serious short-energy regions, and energy importing reliance reached 85% during the period 1995–2010; this affected the development of the local socio-economy and made the regional energy security situation grimed. 2) At present, despite the energy structure becoming more diversified in the YRDR, the energy system is heavily dominated by coal. In addition, industry

has long been the largest final user of energy, reaching 69% of total final consumption in 2010. The construction sector was the smallest, at 2%; however, its energy consumption grew at a leading rate of 16% a year on an average over the period 1995–2010. 3) CO_2 emissions from fuel combustion in the YRDR reached 901.86 million tons in 2010, and annual CO_2 emission growth rate was 1/2 of that of GDP. In the YRDR, the per capita emission was significantly higher, whereas carbon intensity was lower than that of the national average in China. In addition, CO_2 emissions from urban areas reached 69% of the total in 2010. 4) Combining low carbon development with the 2011–2020 regional planning for the YRDR, scenario analysis results show that total energy demand will remain relatively high and continue to grow. However, under the LC scenario, CO_2 emissions will begin to fall for the first time in 2017, and the growth rates of total energy demand and CO_2 emissions will be smaller, 0.98% and 0.95% separately, than the BAU scenario in the YRDR from 2011 to 2020. Thus, if these energy-saving and emission reduction plans are implemented thoroughly, the YRDR will achieve the national emission reduction targets in 2020.

Based on the above research results, the following strategies should be undertaken in order to increase geographic and fuel-supply diversity, to curb the growth in energy demand and to mitigate greenhouse gas emissions: 1) In the YRDR, energy security has emerged as a central policy issue and is increasingly affecting regional economic and development policy. The government should ensure regional supply by diversifying the geographic sources of oil and physical supply route, strengthening energy infrastructure construction and increasing domestic production of conventional fuels. 2) We must continue implementing recent energy-saving and ejection-decreasing policies to promote the adjustment and upgrading of industrial structure, as prescribed in China's Medium and Long Term Energy Conservation Plan. Development and deployment of clean energies should be carried out in a more cost-effective manner to phase in the fuel switch from coal dominance to more shares of clean energy types such as natural gas and electricity (Hannah 2016). 3) Despite a rapid decline in carbon intensity in the YRDR, there is still much potential for carbon intensity to decline further, especially in some urban areas. Therefore, pursuing closer cooperation between cities and building integrated regional management platform are becoming the major tasks, which will further accelerate the spread of the information technique and ensure that energy policy challenges facing the region are addressed in a consistent manner.

REFERENCES

Chinese Society for Urban Studies. 2009. China's Low-carbon Eco-city Development Strategy. Beijing: China City Press.

Hannah, J.W. & Hari M.O. 2016. Regional Energy Governance and U.S. Carbon Emissions. *Ecology Law Quarterly* 43(1): 22–24.

International Energy Agency. 2010. CO_2 Emissions from Fuel Combustion Highlights. http://www.iea.org/publications/free_all.asp.

International Energy Agency. 2015. World Energy Outlook 2014 Edition. http://www.iea.org/publications/free_all.asp.

IPCC. 1997. Revised 1996 IPCC Guidelines for National Greenhouse Gas Inventories. Paris: IPCC/UNEP/OECD/IEA.

Kadian R., R.P. Dahiya, & H.P. Garg. 2007. Energy-related emissions and mitigation opportunities from the household sector in Delhi. *Energy Policy* 35(12): 6195–6211.

Moriarty P., & D. Honnery. 2009. What energy levels can the Earth sustain? *Energy Policy* 37(7): 2469–2474.

National Bureau of Statistics of China. 2015. China Statistical Yearbook 2014. Beijing: China Statistics Press.

Uwasu M., Y. Jiang, & T. Saijo. 2010. On the Chinese carbon reduction target. *Sustainability* 2(6): 1553–1557.

World Bank. 2014. World Development Report 2015: Development and Climate Change. http://www.worldbank.org/INFOSHOP1/Resources.pdf.

Civil, Architecture and Environmental Engineering – Kao & Sung (Eds)
© *2017 Taylor & Francis Group, ISBN 978-1-138-02985-9*

The interaction between outflow dynamics and removal of NH_4^+-N in a vertical flow constructed wetlands treating septage

Y.Y. Tan, F.E. Tang, A. Saptoro & E.H. Khor
Faculty of Engineering and Science, Curtin University Sarawak Malaysia, Miri, Sarawak, Malaysia

ABSTRACT: This study investigated the influence of outflow dynamics on the treatment efficiency of NH_4^+-N in a pilot-scale Vertical Flow Constructed Wetland (VFCW) designed for septage treatment. Continuous samplings had been carried out to measure the temporal effluent flux and the associated concentration of nitrogen compounds. The effluent shows a dependency on the sludge thickness at the wetland surface. The proposed system demonstrated a promising treatment for Total Nitrogen (TN), where the average removal reaches 69.21 ± 16.75%. Nevertheless, it was observed that the overall removal of NH_4^+-N is still below 50% and the concentration of NO_3^--N is high in the effluent. Comparisons between the peak of effluent flux and removal rate and linear regression analysis of the outflow and NH_4^+-N dynamics revealed that the removal of NH_4^+-N was greatly affected by the hydraulic behavior. As the nitrification is the main process in removing NH_4^+-N, the percolating rate through the wetland bed determines the contact time with the attached-growth biofilm which eventually govern the treatment performance.

1 INTRODUCTION

Septage is a sludge-based wastewater removed from the septic tanks (Crites and Tchobanoglous, 1998). Vertical Flow Constructed Wetland (VFCW) has become an attractive system in septage treatment due to its reasonable cost, low energy consumption and simple operation (Tan et al., 2015). The raw septage is loaded on the wetland surface with respect to the specified hydraulic or solid load. Then, it percolates through the wetland bed and then is freely drained from the bottom of the bed. In general, the VFCW system is fed after the previous batch of raw septage has been completely drained to allow the restoration of oxygen during the draining period. Studies have indicated that this type of septage treatment system showed a promising treatment for solids content and organic matter, where the removal of these parameters generally reach 80% in the existing system (Lienard and Payrastre, 1996; Koottatep et al., 2001; Paing and Voisin, 2005; Kengne et al., 2009; Troesch et al., 2009; Jong and Tang, 2014). Nevertheless, the removal of nitrogen is less consistent, where the removal efficiency of ammonium-nitrogen (NH_4^+-N) is below 60% in some systems (Lienard and Payrastre, 1996; Troesch et al., 2009) and the concentration of nitrate-nitrogen (NO_3^--N) is generally increased after the treatment (Lienard and Payrastre, 1996; Koottatep et al., 2001; Paing and Voisin, 2005; Kengne et al., 2009; Troesch et al., 2009; Jong and Tang, 2014).

Nitrogen content in the raw septage, which mostly exist in the form of organic nitrogen (Org-N) and NH_4^+-N, are removed by means of physical and biochemical processes (Vymazal, 2007). The particulate Org-N is physically filtered at the wetland surface with other solids content to form a layer of sludge deposit. Meanwhile, the removal of dissolved Org-N and NH_4^+-N mainly rely on sequential ammonification-nitrification. Nevertheless, the absence of denitrification due to the dominant aerobic environment results in excessive NO_3^--N (Vymazal, 2010). In VFCW system, the microorganisms involved in the biochemical process develop as a biofilm attached on the surface of porous medium (Kadlec and Wallace, 2008), and thus the treatment performance is governed by the flow rate through the wetland bed.

The influence of the system configuration (e.g. design of wetland bed, vegetation) and operational regimes (e.g. hydraulic and solids loading rate) to the overall nitrogen removal in VFCW-based septage treatment system have been well studied (Koottatep et al., 2001; Jong and Tang, 2014). Nevertheless, the study relevant to the relation between the outflow dynamics and effluent nitrogen content within a feed-drain cycle is very limited in the literature. Therefore, this research work aims to carry out a preliminary study to enhance the nitrogen treatment in the first stage treatment of a pilot-scale, two-staged VFCW-based septage treatment plant by controlling the hydraulic behaviour in the system. The hydraulic efficiency of the system including the recovery of water and peak flux in the effluent was evaluated in respect to the hydraulic load, solids load and sludge thickness. Then, the outflow dynamics were compared to the concentrations of NH_4^+-N and NO_3^--N concentration to identify their effect on the efficiency of NH_4^+-N removal.

2 METHODOLOGY

Experimental data were collected from the first stage treatment of a pilot-scale, two-staged VFCW-based septage treatment system located at Curtin University Sarawak Malaysia. Two wetland beds were constructed in 1 m^3 polyethylene tanks, which have a surface area of 1.1 m^2 as shown in Figure 1. A 20 cm drainage layer was built using large gravel ($\varnothing > 5$ cm) at the bottom of the bed. The main layer consists of two different sizes of crushed gravel, where a 27 cm thick medium-sized gravel layer ($\varnothing$ 2.5–3.75 cm) was topped by a 27 cm thick small-sized gravel layer ($\varnothing$ 0.95–1.25 cm). Two vertical ventilation pipes were installed along the wetland bed to enhance the reaeration. Nine clumps of common reeds (*Phraugmites Karka*) were transplanted to each wetland bed. Prior to the data collection, the wetland bed was acclimatized with diluted raw septage for 2 months to stimulate the growth of vegetation and biofilm.

Raw septage was fed into the pilot-scale wetland bed through batch loading process in between November 2015 to February 2016. The raw septage was collected from the household area and was stored in two 400-gallon polyethylene tanks. During the operation, the raw septage was discharged to a 250-gallon feeding tank and was manually homogenised with a PVC stirrer. The hydraulic loads were varied at five different values of 50 l, 75 l, 100 l, 125 l and 150 l. Volume of the effluent was sampled and measured with respect to the time after loading. The percentage of water recovery was calculated as a fraction of the volume of effluent and influent:

$$Water\ Recovery\ (\%) = \frac{\sum_{i=0}^{N} V_{e_i}}{V_0} \times 100\% \quad (1)$$

where N is the number of sample collections (-), V_e is the volume of effluent (l), and V_0 is the hydraulic load (l). The effluent flow rate (l min^{-1}) was converted to per unit area (cm min^{-1}) in the analysis.

Concentrations of Dissolved Oxygen (DO), NH_4^+-N and NO_3^--N were measured using a HACH HQ40d portable multi-parameter meter with specific probes. The concentration Total Nitrogen (TN) in the raw septage and effluent were determined using Test'N Tube method with HACH DR2800 spectrophotometer. As the water loss was mainly attributed to the incomplete infiltration of influent, the Mass Removal Rate (MRR%) is estimated using a conservative equation to minimize overestimation of treatment performance:

$$MRR(\%) = \frac{C_0 \sum_{i=0}^{N} V_{e_i} - \sum_{i=0}^{N} C_{e_i} V_{e_i}}{C_0 \sum_{i=0}^{N} V_{e_i}} \times 100\% \quad (2)$$

where C_0 is the influent concentration (mg l^{-1}) and C_e is the effluent concentration (mg l^{-1}).

3 RESULTS

Eighteen sets of data were collected from the experimental work. Each case of hydraulic load was carried out for four runs except for 150 l, which was only carried out for two runs. Table 1 summarizes the operating conditions, comprising the hydraulic loads, thickness of surface deposit and solids content, and the experimental results of hydraulic performance, including the water recovery and peak effluent flux.

Due to the continuous operation and regular maintenance, the thickness of sludge deposit varied from 3 cm to 12 cm. Moreover, different batches of raw septage collection and varying hydraulic load led to the fluctuating solids load throughout the experiment, ranging from 105 g to 2325 g. The water recovery shifts between 9.72% and 90%. In a similar way, the peak effluent flux ranged from 0.004 to 0.073 cm min^{-1}. The mean water recovery and peak flux are 58.28 ± 23.54% and 0.027 ± 0.018 cm min^{-1}, respectively.

The water recovery and peak effluent flux are highly dynamic with respect to hydraulic load and solids load. On the other hand, both parameters appears to be lower when the thickness of sludge

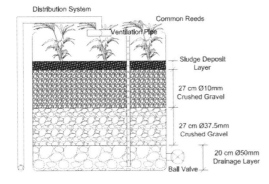

Figure 1.   The configuration of proposed wetland bed.

Table 1. Hydraulic performance of the pilot-scale VFCW system with varying factors.

Case	Sludge thickness (cm)	Solids load (g)	Water recovery (%)	Peak flux (cm min⁻¹)
50A	3	390	62.48%	0.034
50B	4	445	67.43%	0.016
50C	10	1200	24.49%	0.004
50D	6	735	56.43%	0.032
75A	5	255	85.08%	0.026
75B	6	360	71.92%	0.015
75C	7	105	85.87%	0.026
75D	5	105	68.14%	0.065
100A	6	1280	73.41%	0.043
100B	5	460	70.10%	0.019
100C	8	1260	46.94%	0.017
100D	3	1950	47.59%	0.012
125A	12	1188	14.57%	0.004
125B	9	525	9.72%	0.007
125C	3	275	90.00%	0.031
125D	5	2325	56.43%	0.032
150A	3	480	81.37%	0.073
150B	4	225	35.67%	0.022
Mean	5.78	753.50	58.28%	0.027
SD*	2.51	624.35	23.54%	0.018

*Standard deviation.

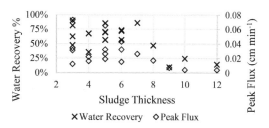

Figure 2. Influence of sludge thickness upon water recovery and peak flux.

Table 2. Treatment efficiency of nitrogen compounds and R^2 between outflow dynamics and concentration of NH_4^+-N.

Case	TN	NH_4^+-N	R^2
50A	42.72%	33.95%	0.81
50B	67.25%	67.55%	0.53
50C	77.87%	57.73%	0.79
50D	76.77%	9.53%	0.60
75A	41.70%	50.51%	0.85
75B	73.41%	49.70%	0.85
75C	71.59%	67.36%	0.80
75D	79.55%	23.70%	0.83
100A	77.35%	19.56%	0.81
100B	96.37%	47.08%	0.86
100C	77.17%	38.40%	0.91
100D	89.96%	57.73%	0.71
125A	76.57%	96.51%	0.30
125B	69.06%	76.94%	0.87
125C	31.99%	36.86%	0.75
125D	84.56%	35.51%	0.69
150A	68.73%	24.09%	0.81
150B	55.94%	55.11%	0.90
Mean	69.21%	47.10%	0.76
SD	16.75%	21.35%	0.15

deposit layer reaches 8 cm as described in Figure 2. This observation highlights the importance of sludge thickness to the hydraulic performance. Nevertheless, neither high water recovery nor rapid effluent flux was solely influenced by a thin sludge deposit layer, where these hydraulic performances are still affected by varying hydraulic and solid load.

The DO concentration is extremely low in the raw septage (0.38 ± 0.32 mg l⁻¹). The average concentration of influent TN is 170.78 ± 104.96 mg l⁻¹, while the average influent concentration of NH_4^+-N and NO_3^--N are 53.11 ± 12.49 mg l⁻¹ and 8.90 ± 4.63 mg l⁻¹, respectively. The high SD indicates the highly variable quality of septage throughout the experiment. As the concentration of nitrite in raw septage was very low and can be assumed negligible, the concentration of Org-N was estimated from the nitrogen balance. It was found that an average value of Org-N is 126.90 ± 96.44 mg l⁻¹. Table 2 displays the removal performance of nitrogen compounds in each case. In general, the treatment efficiency of TN was between 60% and 80%, which gave an average efficiency of 69.21 ± 16.75%. The treatment of NH_4^+-N is less effective since the average removal rate is less than 50%. The influent was well aerated and the concentration of NO_3^--N increased throughout the treatment due to the favourable nitrification. Figure 3 illustrates that the removal rate of NH_4^+-N tends to be proportional to the peak effluent flux, where a lower peak effluent flux most likely resulted in a better removal rate.

Figure 4 shows an example of the comparisons between outflow dynamics and the concentration of NH_4^+-N and NO_3^--N in effluent of case 75A. The effluent flux typically reached maximum rate at the early stage of the feed-drain cycle. Then, the effluent flux gradually decelerated and stabilized at the late stage. The dynamics of NH_4^+-N suggests a similar trend with the outflow dynamics, where the concentration is proportional to the effluent flux. Meanwhile, the NO_3^--N appears to show a progressive rise. Linear regression analysis was employed to establish the relation between

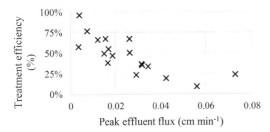

Figure 3. Relation between peak effluent flux and treatment efficiency of NH_4^+-N.

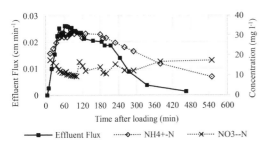

Figure 4. Comparison between outflow dynamics and concentrations of NH_4^+-N and NO_3^--N (Case 75A).

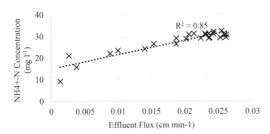

Figure 5. Linear regression analysis of outflow dynamics and concentrations of NH_4^+-N (Case 75A).

the outflow dynamics and NH_4^+-N as shown in Figure 5. The results of coefficient of determination (R^2) are presented in Table 2. There are only three out of eighteen cases having a R^2 value below 0.7, and the average value of R^2 is 0.76 ± 0.15.

4 DISCUSSION

The batch loading in the pilot-scale VFCW-based septage treatment system is described using a feed-drain cycle. Such a cycle is crucial in generating a variably-saturated condition to aerate the wetland bed during the dry period. Under this condition, the hydraulic behavior in the substrate

can be highly fluctuating, especially related to the alteration of hydraulic load and solids load Theoretically, a higher hydraulic load creates a larger head difference at the wetland surface that accelerates the infiltration, and thus eventually results in a rapid effluent flux (Richards, 1931). Nevertheless, the solids content in the raw septage is retained and forms a sludge deposit layer instantaneously at the wetland surface during infiltration and subsequently influences the hydraulic behavior in the wetland bed.

The water recovery and peak effluent flux are found to be governed by the thickness of this low permeable layer, which is similar to the finding by Molle (2014). However, the sludge thickness is greatly affected by the operating strategies and influent quality.

These parameters is unable to be altered easily and directly. As a result, although the hydraulic and solids load did not show any direct impact on the water recovery and the peak effluent flux, these parameters are still crucial in governing the accumulation rate of sludge deposit. Xu et al. (2013) highlighted that higher hydraulic and solids load tend to generate a rapid sludge accumulation and ultimately lead to the clogging of wetland bed.

Org-N and NH_4^+-N are found to be the major forms of nitrogen compounds in the raw septage. Meanwhile, the concentration of NO_3^--N is typically low. The Org-N exists in particulate and dissolved form. The former is mainly filtered at the wetland surface and the latter is hydrolyzed to NH_4^+-N by ammonification. The favorable aerobic condition in a VFCW system promotes nitrification to oxidized NH_4^+-N to NO_3^--N. It should be noted that the nitrification is a transformative process. It is not an ultimate removal mechanism of nitrogen, where the changes of TN may be insignificant from this process. Therefore, the major removal of TN in the system is attributed to the straining of particulate Org-N in the wetland bed. This also explains the relatively consistent removal rate of TN compared to NH_4^+-N.

Adsorption is another important NH_4^+-N removal pathway in the VFCW system. The adsorbed NH_4^+-N is nitrified during resting and is washed out in the following feeding, so-called a two-step process (Woźniak et al., 2007). The flush out of NO_3^--N is observed at the early phase of effluent, which is illustrated in Figure 5. The limitation of anaerobic condition in the VFCW system inhibits the occurrence of denitrification (Vymazal, 2007), and thus the concentration of NO_3^--N generally remained high in the effluent throughout the experiment. The plant uptake is a potential mechanism for NH_4^+-N and NO_3^--N treatment, however, its contribution to the overall nitrogen treatment is insignificant (Vymazal, 2007). In summary, the current system still requires further improvements in delivering a complete nitrogen removal, especially for NO_3^--N. A secondary treatment is required to enhance the removal of NH_4^+-N and to eliminate excessive NO_3^--N in the effluent.

As the nitrification involves biochemical mechanisms, the contact time between influent, attached-growth biofilm and oxygen content ultimately drive the treatment efficiency (Torrens et al., 2009). Under the operating regime of feed-and-drain, the Hydraulic Retention Time (HRT) in the VFCW system can be described as the percolating rate throughout the wetland bed. Figure 4 indicates that the overall removal of NH_4^+-N shows a dependency on the peak effluent flux. The outflow and concentration of NH_4^+-N are linearly correlated as their linear regression has high values of average R^2. Accordingly, the treatment of NH_4^+-N can be improved by controlling the percolating rate in the wetland bed.

The poor treatment efficiency of NH_4^+-N in this study can be attributed to the rapid percolating rate throughout the wetland bed, where the HRT is relatively short to enhance nitrification. The experimental results indicate that the peak effluent flux is required to be control below 0.02 cm min^{-1} to achieve 50% of NH_4^+-N removal.

According to Table 1, neither the loading regime with hydraulic load nor solid load effectively controls the HRT, as both factors show less effect on the effluent flux. Low HRT can be addressed by accumulating a thicker sludge deposit layer to decelerate the effluent flux. Nevertheless, the extended infiltration period is impractical from the view of productivity. Therefore, a further study is required to optimize the removal of NH_4^+-N with a reasonable HRT. In addition to the optimization of operation, the identification of the correlation between outflow and nitrogen dynamics in a VFCW system treating septage will gains useful insight towards its numerical modelling. Langergraber and Šimůnek (2005) highlighted the importance of hydraulic simulation in modelling the VFCW system. Linear relationship proposed in this study will be able to simplify the calibration procedure in the modelling of nitrogen dynamics.

5 CONCLUSION

The first stage treatment of a pilot-scale VFCW system was tested and it was found that it demonstrates a promising treatment for TN. Nevertheless, the overall removal of NH_4^+-N and NO_3^--N were less efficient. Linear regression analysis was employed to obtain a correlation between the effluent flux and the associated NH_4^+-N concentration. The high average R^2 (0.76 ± 0.15) indicates that the NH_4^+-N dynamics is greatly affected by outflow dynamics. In a VFCW system, the percolating flow rate determines the contact time between the contaminants and attached-growth biofilm. This parameter subsequently governs the nitrification rate and ultimately affects the overall treatment efficiency. A thicker sludge thickness is capable of extending the retention time of influent in the bed, however, it may reduce the productivity of the system at the same time. Therefore, a further study is required to optimize the treatment of nitrogen compounds with a reasonable HRT. These outcomes gain useful insights towards the optimization of operation and numerical modelling.

ACKNOWLEDGEMENT

This study was supported by the Faculty of Engineering and Science, Curtin University Sarawak Malaysia. Authors would like to express grateful acknowledgements to Mr. Jason Bui Jie Xiang and Mr. Brian Chin Wei Yang for their assistance in experimental works.

REFERENCES

Crites R and Tchobanoglous G. (1998) *Small and decentralized wastewater management systems*, New York: McGraw-Hill Book Company.

Jong VSW and Tang FE. (2014) Organic matter and nitrogen removal at planted wetlands treating domestic septage with varying operational strategies. *Water Science & Technology* 70: 9.

Kadlec RH and Wallace S. (2008) *Treatment Wetlands, Second Edition*: Taylor & Francis.

Kengne IM, Dodane PH, Akoa A, et al. (2009) Vertical-flow constructed wetlands as sustainable sanitation approach for faecal sludge dewatering in developing countries. *Desalination* 248: 291–297.

Koottatep T, Polprasert C, Oanh NTK, et al. (2001) Septage dewatering in vertical-flow constructed wetlands located in the tropics. *Water Science & Technology* 44: 8.

Langergraber G and Šimůnek J. (2005) Modeling Variably Saturated Water Flow and Multicomponent Reactive Transport in Constructed Wetlands. *Vadose Zone J.* 4: 924–938.

Lienard A and Payrastre F. (1996) Treatment of sludge from septic Tanks in a reed-bed filters pilot plant. *5th international conference on wetland systems for water pollution control*. Vienna: Austria.

Molle P. (2014) French vertical flow constructed wetlands: a need of a better understanding of the role of the deposit layer. *Water Science & Technology* 69.

Paing J and Voisin J. (2005) Vertical flow constructed wetlands for municipal wastewater and septage treatment in French rural area. *Water Science & Technology;* 51: 11.

Richards LA. (1931) Capillary conduction of liquids through porous mediums. *Journal of Applied Physics* 1: 318–333.

Tan YY, Tang FE, Saptoro A, et al. (2015) Septage Treatment Using Vertical-Flow Engineered Wetland: A Critical Review. *Chemical Engineering Transaction* 45.

Torrens A, Molle P, Boutin C, et al. (2009) Impact of design and operation variables on the performance of vertical-flow constructed wetlands and intermittent sand filters treating pond effluent. *Water Research* 43: 1851–1858.

Troesch S, A. Liénard, P. Molle, et al. (2009) Treatment of septage in sludge drying reed beds: a case study on pilot-scale beds. *Water Science & Technology* 60: 11.

Vymazal J. (2007) Removal of nutrients in various types of constructed wetlands. *Science of The Total Environment* 380: 48–65.

Vymazal J. (2010) Constructed Wetlands for Wastewater Treatment: Five Decades of Experience†. *Environmental Science & Technology* 45: 61–69.

Woźniak R, Dittmer U and Welker A. (2007) Interaction of oxygen concentration and retention of pollutants in vertical flow constructed wetlands for CSO treatment. *Water Science and Technology* 56: 31–38.

Xu Q, Cui L, Zhang L, et al. (2013) The Effect of Two Factor Combination of Three Kinds of Loading on the Soil Clogging in Vertical Flow Constructed Wetland. *Frontier of Environmental Science* 2.

Water use efficiency of the Lancang-Mekong River basin region in "the Belt and Road Initiative"

Chunyan Xie
Business School, Hohai University, Nanjing Jiangsu, China

Jingchun Feng & Ke Zhang
Business School, Hohai University, Nanjing Jiangsu, China
Institute of Project Management, Hohai University, Nanjing Jiangsu, China
Jiangsu Provincial Collaborative Innovation Center of World Water Valley and Water Ecological Civilization,
Nanjing Jiangsu, China

ABSTRACT: Water is an important resource in "the Belt and Road Initiative". As the most important transboundary river in Asia, the Lancang-Mekong River plays a significant role in the development of aquatic ecosystems in "the Belt and Road Initiative". For this reason, in this paper, non-radial direction distance function is used to analyze the water use efficiency of the Lancang-Mekong River Basin region from two perspectives (total-factor and water-factor indicators). Then, the influencing factors of water use efficiency in the Lancang-Mekong River are discussed using gray correlation analysis. The result shows that water use efficiency of Thailand, Cambodia, Yunnan and Qinghai is at a relatively optimal level, whereas for Vietnam, Tibet, Laos and Myanmar, it is relatively low. Economic development level, technological progress, industrial structure and trade and foreign investment have an impact on water use efficiency. Among the influencing factors, economic development level has the greatest impact.

1 INTRODUCTION

In September 2013, Chinese President Jinping Xi put forward "the Silk Road Economic Belt and the 21st-Century Maritime Silk Road" (referred to as "the Belt and Road") during his visit to the Central Asian and Southeast Asian countries. The Silk Road Economic Belt region is ecologically fragile, comprising arid inland areas, where clean water is limited and per capita water resources are deficient. Serious water issue has become one of the most important resource factors that constraints "the Belt and Road Initiative".

In March 2016, the Lancang-Mekong cooperation mechanism was established around the implementation of sustainable development and cooperation in the Lancang-Mekong River in order to promote regional integration process that includes water resource management, disaster response, etc. In fact, in 1992, the Asian Development Bank (ADB) had initiated the Greater Mekong Sub-regional Economic Cooperation Program (GMS) to strengthen the economic ties among the basin countries.

With the progress of "the Belt and Road", many scholars studied the problem of water resources of Lancang-Mekong River. In China, the scholars are concerned with issues such as sustainable water utilization (Guo et al., 2013) and co-management among Mekong river basin countries (Piao et al., 2013). Scholars from other countries pay more attention to China's water exploitation on upper Mekong River from the perspective of international relations and its political effect (Menniken, 2007; Hensengerth, 2009).

The importance of Lancang-Mekong River has concerned scholars, but existing research mostly analyzes water utilization from a macro level. However, due to the challenge of increasing scarce water resources, improving water use efficiency seems to be the effective solution to improve water imbalance between supply and demand and for sustainable use of water (Allan, 1999; Deason et al., 2001). Former EPA Administrator William Reilly points out that water use efficiency is not high in the world, and typically 50% of the water is wasted. The UK government ensures improvement of water use efficiency and reduction of water pollution to achieve both environmental protection and sustainable utilization of water resources. China's *Water Law* clearly indicates the improvement in water use efficiency through reasonable water resource allocation of agricultural water, industrial water and domestic water.

In this view, we studied water utilization of Lancang-Mekong River Basin regions from the perspective of water use efficiency using quantitative analysis. Because of the difference of geographical conditions, economic development level and the dependence on water resources, water use efficiency is different. Then, we explored the factors that influence water use efficiency on the basis of the estimated water use efficiency. Finally, some suggestions have been put forward to increase the water use efficiency in order to achieve sustainable development of the water resources of the Lancang-Mekong River Basin region.

2 METHODOLOGY

2.1 Non-radial directional distance functions

In the current research (Zhang et al., 2014), non-radial directional distance function is established to estimate water use efficiency from two perspectives: total-factor indicator and water-factor indicator.

Suppose there are N regions, and each region uses capital (K), labor (L), and water (W) as inputs, and gross domestic product (Y) as output. Then, the production technology can be described as follows:

$$T = \left\{ \begin{array}{l} (K, L, W, Y): \sum_{n=1}^{N} \lambda_n K_n \leq K, \sum_{n=1}^{N} \lambda_n L_n \leq L, \\ \sum_{n=1}^{N} \lambda_n W_n \leq W, \sum_{n=1}^{N} \lambda_n Y_n \geq Y, \lambda_n \geq 0, n = 1, 2 \cdots N \end{array} \right\} \tag{1}$$

The non-radial directional distance functions of water use efficiency can be defined as follows:

$$\begin{aligned} &\vec{D}(K, L, W, Y; g) \\ &= \sup \left\{ w^T \beta :: ((K, L, W, Y) + g \times diag(\beta)) \in T \right\} \end{aligned} \tag{2}$$

where, $w = (w_K, w_L, w_W, w_Y)^T$ denotes the normalized weight vector relevant to the numbers of inputs and outputs; $\beta = (\beta_K, \beta_L, \beta_W, \beta_Y)^T \geq 0$ denotes a vector of scaling factors representing individual inefficiency measures for each input/output; and $g = (-g_K, -g_L, -g_W, g_Y)$ is the explicit directional vector representing the expectation of each input/output to reduce/increase.

Considering water resource investment and other inputs, the total-factor non-radial direction distance function can be calculated using the following DEA-type model:

$$\vec{D}(K, L, W, Y; g) = \max w_K \beta_K + w_L \beta_L + w_W \beta_W + w_Y \beta_Y$$
$$s.t. \sum_{n=1}^{N} \lambda_n K_n \leq K - \beta_K g_K, \sum_{n=1}^{N} \lambda_n L_n \leq L - \beta_L g_L, \tag{3}$$
$$\sum_{n=1}^{N} \lambda_n W_n \leq W - \beta_W g_W, \sum_{n=1}^{N} \lambda_n Y_n \geq Y + \beta_Y g_Y,$$
$$\lambda_n \geq 0, n = 1, 2, 3, \ldots, N, \beta_K, \beta_L, \beta_W, \beta_Y \geq 0$$

If $\vec{D}(K, L, W, Y; g) = 0$, then the region to be evaluated is located along the best-practice frontier in the g direction. Assume that the input and output are equally important, both to be given weight 1/2. Then, the average is given the weight 1/2 to three inputs. Therefore, we set the weight vector as $w^T = (1/6, 1/6, 1/6, 1/2)$ and the directional vectors as $g = (-K, -L, -W, Y)$.

Suppose that $\beta_K^*, \beta_L^*, \beta_W^*, \beta_Y^*$ represent the optimal solution to Eqs. (3). We define the average efficiency of each factor as Total-Factor Efficiency Index (TEI). Then, the TEI can be formulated as follows:

$$\begin{aligned} TEI &= \frac{1}{3} \left[\frac{(1 - \beta_K^*) + (1 - \beta_L^*) + (1 - \beta_W^*)}{1 + \beta_Y^*} \right] \\ &= \frac{1 - \frac{1}{3}(\beta_K^* + \beta_L^* + \beta_W^*)}{1 + \beta_Y^*} \end{aligned} \tag{4}$$

To measure the pure water use efficiency, it is better to fix non-water inputs. Assuming that capital and labor are fixed, the maximum reduction ratio of water and the maximum increase ratio of output are evaluated. We set the weight vector as $w^T = (0, 0, 1/2, 1/2)$ and the directional vectors as $g = (0, 0, -g_W, g_Y)$. Then, the Water-Factor Efficiency Index (WEI) is expressed as follows:

$$WEI = (1 - \beta_W^*)/(1 + \beta_Y^*) \tag{5}$$

The TEI and the WEI both lie between zero and unity. If the TEI/WEI is equal to unity, it means that the region is located along the frontier of best practice.

2.2 Gray correlation analysis

Gray correlation analysis model can be formulated as follows:

Let reference sequence $X_0 = (x_0(1), x_0(2), \ldots, x_0(n))$ and compare sequence $X_1 = (x_1(1), x_1(2), \ldots, x_1(n))$, $X_2 = (x_2(1), x_2(2), \ldots, x_2(n)), \ldots, X_m = (x_m(1), x_m(2), \ldots, x_m(n))$.

For $\rho \in (0,1)$, let

$$\gamma(x_0(k), x_i(k)) = \frac{\min_i \min_k |x_0(k) - x_i(k)| + \rho \max_i \max_k = |x_0(k) - x_i(k)|}{|x_0(k) - x_i(k)| + \rho \max_i \max_k |x_0(k) - x_i(k)|} \tag{6}$$

$$\gamma(X_0, X_i) = \frac{1}{n} \sum_{k=1}^{n} \gamma(x_0(k), x_i(k)) \tag{7}$$

where, ρ denotes the distinguishing coefficient, usually with a value of 0.5. The value does not affect the results of the sort, only changes the size of the relative value; $\gamma(X_0, X_i)$ denotes gray correlation degree between X_0 and X_i.

2.3 *Data*

In comparison with other five countries, China has a vast territory and complicated geographical features, and the economic scale and population also vary greatly. In order to increase the comparability, we selected Qinghai, Tibet and Yunnan, which are the regions the Mekong flows through in China, as the research object was to analyze water use efficiency of the Mekong River Basin regions.

Due to the availability of data, we selected the statistical data of eight regions of the Mekong River Basin in 2013. The data was respectively collected from the *World Bank WDI database, IMF database, WTO database, UNCTAD FDI database* and *China Statistical Yearbook*.

We measured the output (Y) of each region by GDP and the capital input (K) by gross fixed capital formation. The labor input (L) was measured by the number of employees in each region. Due to different statistical standards, the water input of Laos, Myanmar, Thailand, Cambodia and Vietnam was measured by annual freshwater withdrawals, whereas the water input of Qinghai, Tibet and Yunnan was measured by the annual total water use.

The following factors influencing water use efficiency were considered. (1) Water use efficiency is affected by regional economic and social development. Economically developed regions are more likely to show greater water efficiency (Zhang et al., 2014). Economic development level was measured by the per capita GDP of each region. (2) Upgrading water-saving technology is the direct path to improve water use efficiency. We measured technological progress by the ratio of capital to labor. (3) The difference of water use efficiency stems from different economic structures (Deng et al., 2016). Agriculture is the largest sector of water consumption. The ratio of added value of agricultural sector to GDP was measured to influence water use efficiency. (4) Imports of virtual water can increase the supply of water in the importing country and reduce their water pressure. Therefore, trade inevitably influences water consumption and use efficiency. Therefore, import trade (import/GDP) and export trade (export/GDP) are assumed to affect water use efficiency. (5) Foreign investment helps improve water use efficiency through technology spillover. We measured foreign investment in terms of foreign direct investment.

3 RESULTS AND DISCUSSION

3.1 *Estimation results of water use efficiency*

Water use efficiency is calculated and showed in Figure 1. From Figure 1, it can be observed that the average TEI of the Mekong River Basin region is 0.8239. Thailand, Cambodia, Yunnan and Qinghai show the highest TEI values (unity), indicating that water use efficiency of these regions is at a relatively optimum level. The TEIs of Vietnam, Tibet and Laos are 0.6995, 0.6702 and 0.5962, respectively. The TEI in Myanmar is the lowest, with the value of 0.4492. The average WEI of the Mekong River Basin region is 0.7082. Thailand, Cambodia, Yunnan and Qinghai still show the highest WEI values (unity). In these regions, the level of water usage and output are in optimal frontier. They have less improved space. The WEI in Vietnam, Tibet, Laos and Myanmar are all less than 0.5, with the value of 0.4698, 0.3469, 0.3042 and 0.2528, respectively. The average TEI is higher than the average WEI, indicating that the Mekong River Basin regions show better performance in TEI than WEI.

TEI and WEI both indicate that in the process of economic development, the capital, labor and water accomplish optimum allocation in Thailand, Cambodia, Yunnan and Qinghai of water resources. In Vietnam, Tibet, Laos and Myanmar, the capital, labor and water are largely the inputs, but the output efficiency is low. This means that water allocation deviates from the optimal configuration because of capital, labor and water input redundancy or output deficiency. Extensive mode of development and low water recycling rate may be the causes of low water use efficiency in these regions. There is huge potential in these areas in terms of saving water resources; improving water use efficiency; and increasing the output.

3.2 *Factor analysis of water use efficiency*

Gray correlation degree between TEI/WEI and its factors are shown in Table 1. It can be observed

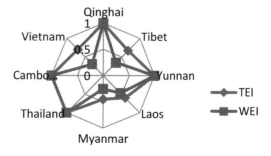

Figure 1. Comparison of TEI and WEI for Mekong River basin regions.

Table 1. Gray correlation degree between water use efficiency and its factors.

Efficiency index		TEI	WEI
Economic development level		0.8166	0.8579
Technical progress		0.7469	0.8037
Industrial structure		0.6031	0.6244
Trade	Import	0.7693	0.7976
	Export	0.7266	0.7607
Foreign investment		0.7003	0.7265

that the correlation degree between economic development level and TEI is the highest, with values of 0.8166; the next factor is import trade, and correlation degree is 0.7693; the correlation degree of technological advances and TEI, export trade and TEI, foreign investment and TEI are 0.7469, 0.7266 and 0.7003, respectively; the correlation degree between industrial structure and TEI is the lowest, with the value of 0.6031. Compared with correlation degree between TEI and its factors, the order of correlation degree between WEI and its factors is slightly different. Specifically, the correlation degree between economic development level and WEI is still the highest, with the value of 0.8575, whereas the correlation degree between industrial structure and WEI is still the lowest, with the value of 0.6244; the correlation degree between technological progress and WEI comes in the second place, with the value of 0.8037; and the correlation degree of import trade, export trade and foreign investment are 0.7976, 0.7607 and 0.7265, respectively.

Overall, the result shows the following. (1) Economic development level has a great impact on the water use efficiency. Higher level of economic development provides support for the water use efficiency and stabilizes the water infrastructure construction. (2) Technological progress improves the water-saving technology and equipment, thereby improving water use efficiency directly. (3) The effect of import trade is greater than export trade. Using import and export commodity, considering local water use efficiency and water use by using comparative advantage theory, the region can reduce water consumption and improve water use efficiency. (4) Local enterprises can get access to advanced international knowledge and production management experience through technology spillover from foreign-invested enterprises. It is conducive to improve water use efficiency. (5) The higher the proportion of agriculture in the economic structure, the more the agricultural water consumption. Low efficiency of agricultural water consumption reduces overall water use efficiency.

4 CONCLUSIONS AND SUGGESTIONS

Using non-radial direction distance function, this paper analyzes water use efficiency of the Lancang-Mekong River Basin regions from the perspective of total-factor and water-factor indicators. Then, the influencing factors of water use efficiency are discussed using gray correlation analysis. The result shows that water use efficiency of Thailand, Cambodia, Yunnan and Qinghai are at a relatively optimal level, whereas Vietnam, Tibet, Laos and Myanmar are relatively low. Economic development level, technological progress, industrial structure and trade and foreign investment have an impact on water use efficiency. Among the influencing factors, economic development level has the greatest impact, followed by technological progress and trade, and the effect of industrial structure is relatively low.

On this basis, we propose the following suggestions. (1) The modernization level of water conservancy and paddy field irrigation can be improved, water-saving agricultural technology can be spread, and methods for low water consumption in agriculture can be developed. China already has certain mature water-saving irrigation techniques such as sprinkling irrigation and drip irrigation. Through technical output and cooperation with the Mekong River Basin regions, regions with low water use efficiency can reduce water waste and improve water use efficiency. (2) Regions with low water use efficiency can import water-intensive commodity, while using inadequate water for economic activities can produce higher economic returns. (3) The agricultural structure can be adjusted. The Lower Mekong countries are agricultural countries, and rice is the agricultural pillar industry. However, the water consumption of rice is large. To alleviate water shortage, the regions can reduce rice cultivation while enriching cash crops species. (4) Institutionally, as one of the economic instruments that optimizes water resource allocation and water shortage relieving, establishing water price rationally can encourage farmers to increase crop production and save irrigation water. For crops with high yield per unit area and low water consumption, low water prices can be levied; for crops with low yield per unit area and high water consumption, high water prices can be levied. (5) At the national level, the basin countries play a complete role in "the Belt and Road Initiative" and Lancang-Mekong cooperation mechanisms. On the one hand, they enhance environmental protection and increase the total amount of water resources, and on the other hand, they strengthen communication and coordination to achieve reasonable allocation of water resources and benefit sharing.

ACKNOWLEDGMENT

We are grateful to the financial support provided by the National Social Science Foundation of China (No.12 AZD108) and the National Science Foundation of China (No.71401052).

REFERENCES

Allan T. 1999. Productive efficiency and allocative efficiency: why better water management may not solve the problem [J]. Agricultural Water Management, (40): 71–75.

Deason J.P., Schad T.M., Sherk G.W. 2001. Water policy in the United States: a perspective [J]. Water Policy, 3(3): 175–192.

Deng G., Li L. 2016. Song Y. Provincial water use efficiency measurement and factor analysis in China: Based on SBM-DEA model [J]. Ecological Indicators, 69: 12–18.

Guo Y.J., Ren N. 2013. Water resources development and environmental protection in lower Mekong basin: policy options and basin governance [J]. World Economics and Politics, (07): 136–154.

Hensengerth O. 2009. Transboundary River Cooperation and the Regional Public Good: The Case of the Mekong River [J]. Contemporary Southeast Asia, 31(2): 326–349.

Menniken T. 2007. China's Performance in International Resource Politics: Lesson from the Mekong [J]. Contemporary Southeast Asia, 29(1): 97–120.

Piao J.Y., Li Z.F. 2013. Water cooperation governance: new Issues of the regional relationship construction in the Langcang-Mekong river basin [J]. Southeast Asian Studies, (05): 27–35.

Zhang N., Kong F., Choi Y., et al. 2014. The effect of size-control policy on unified energy and carbon efficiency for Chinese fossil fuel power plants [J]. Energy Policy, 70: 193–200.

Civil, Architecture and Environmental Engineering – Kao & Sung (Eds)
© 2017 Taylor & Francis Group, ISBN 978-1-138-02985-9

An accessibility study of elevator Braille signage system in Da Nang City, Vietnam

C.Y. Hsia & C.M. Huang
Program in Civil and Hydraulic Engineering, Feng Chia University, Taichung, Taiwan

Liang Tseng
Department of Architecture, Feng Chia University, Taichung, Taiwan

ABSTRACT: This paper is about elevator equipment in Da Nang City, Vietnam. The research project is based on the accessibility norms of the Braille system of barrier-free elevator equipment set up, and the differences are analyzed. The research purposes are as follows: first, Da Nang City elevator Braille system's analysis and explanation of the meaning of its contents; second, according to the Vietnamese TCXDVN 264-202, houses and buildings, basic rules of accessible design and construction are applied for people with disabilities, where 64.3% of Braille panel used English Braille and 21.4% used Korean Braille. Elevator equipment should comply with the provisions of the concept of universal design.

1 INTRODUCTION

The subject of this study is accessible elevator Braille systems in Da Nang City, Vietnam. The research background and motivation and objectives of this study are as follows.

1.1 Background and motivation

This study investigates the limitations of inaccessible elevators with Braille System and the literal meanings of Braille symbols, locations, and floor setups in elevators in public transport systems, departmental stores, office buildings, and hospitals in Da Nang City with reference to Vietnamese Construction Standards TCXDVN 264:2002. The study compares the elevator specifications from three aspects: design, use, and construction.

1.1.1 Design aspect

According to section 5.8.3 of the Vietnamese Construction Standards TCXDVN 264:2002 specifications, Braille systems must be provided in accessible elevators in public buildings, venues, and public transport systems for the aid of the disabled, as shown in Fig. 1.

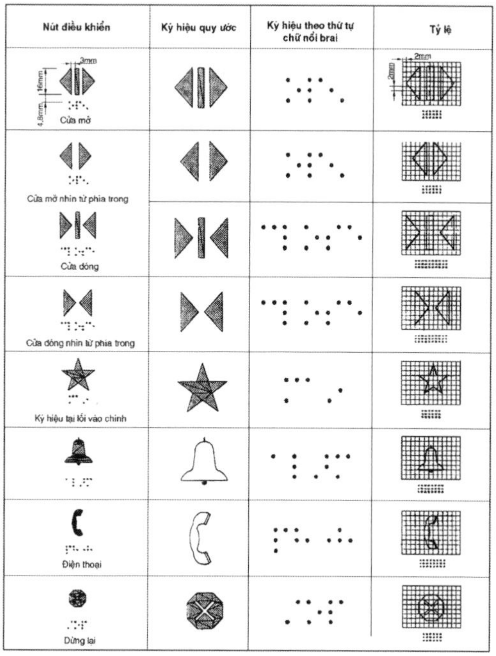

1.1.2 Use aspect

During the research that examined the accuracy of Braille symbol placement in elevators, field studies were conducted and photographs were captured. Braille chips are typically placed in two locations: (a) at the lower part of the

Figure 1. TCXDVN 264:2002 specifications.

control buttons, and (b) at the right-hand side of the control buttons. The subjects of this study include one airport, two office buildings, nine hotels, and two department stores in Da Nang City, Vietnam. The details of the analysis of the

Table 1. Da Nang City, to investigate the object tables.

Code	Name of building	Intended use	Languages	Up meanings	Down meanings	Open meanings	Close meanings	Floor meanings	Basement meanings	Alarm meanings
AP01	Da Nang Airport	Airport	English	×	×	o p n	sh u t	G	×	p h "one"
OB01	MB Bank	Office	English	u p	d n	o p n	c l s	G	×	t e l
OB02	AAC Auditing Accounting	Office	Korean	상 sang	하 ha	개 ke	폐 pye	G	×	Korean a l a r m
HT01	Da Nang Port Hotel	Hotel	Korean	상 sang	하 ha	개 ke	폐 pye	G	×	Korean a l a r m
HT02	Fansipan Hotel	Hotel	Korean	상 sang	하 ha	개 ke	폐 pye	G	×	Korean a l a r m
HT03	HAGL Plaza Hotel	Hotel	Chinese	"sh"ang" h "ying"	h"ya" h"ying"	k"ai" m"en"	g "wan" m"en"	G	B	d "yan" "wa" g"ying" l"ying"
HT04	HAGL Plaza Hotel Service	Hotel	Chinese	u p	dow n	k"ai" m"en"	g "wan" m"en"	G	B	d "yan" "wa" g"ying" l"ying"
HT05	Novotel Premier Han River	Hotel	English	×	×	o p "en"	"sh" u t	1	×	p h "one"
HT06	Orchid Hotel	Hotel	English	×	×	o p n	c l s	1	×	t e l
HT07	Van Son Hotel	Hotel	English	u p	d o n	o p n	"sh" u t	1	×	p h "one"
HT08	Olalani Restort	Hotel	English	u p	d "ow" n	o p e n	shut	1	B	a l m
HT09	Azura da-nang	Hotel	English			o p n	"sh" u t	G	B	a l "ar" m
DS01	CO.OP Mart	department store	English	u p	down	open	close	G	B	a l "ar" m
DS02	Lotte Mart	department store	English	u p	d "ow" n	o p n	"sh" u t	1	×	a l m

current situation of accessible elevator facilities are listed below.

Representations of these buildings are presented in Table 1. According to the Vietnamese construction law regulations, Braille symbols must be placed "at the lower part of control buttons." Therefore, the cases collected from the field studies are compared against the current regulations in Vietnam for an accurate analysis.

1.1.3 Construction aspect

Since construction workers may not have sufficient knowledge of Braille, they may not accurately place the symbols, which could lead to incorrect arrangement of Braille symbols. This study will review the contents of Braille, construction, location, and whether or not they are in the correct proportions.

1.2 Research objectives

The refinement of elevator Braille signage systems that better suit elevator deployment scenarios is subsequently based on the survey results in areas as follows: 1. Braille systems in elevator facilities; 2. Functional meanings and the positioning of control buttons; 3. Braille systems in use. It is expected that the findings of this study can be used as a reference for accessible elevator Braille chip system design and construction in Taiwan, in addition to as a reference for a universal design complying with international standards.

2 RESEARCH SUBJECT AND PROCEDURE

The subjects of this study include one airport, two office buildings, nine hotels, and two department stores of Da Nang City, Vietnam. The details of the analysis of the current situation of accessible elevator facilities are listed below.

Representations of these buildings have been shown in Table 1.

2.1 Procedures of research and analysis

2.1.1 Determining the subject
Four types of organizations were selected as the research subjects: airports, office buildings, hotels, and departmental stores.

2.1.2 Aspect analysis
This analysis included the design aspect, user aspect, and construction aspect.

2.1.3 Problem definition
A. Based on the design aspect, problems were divided into two categories: Braille systems and literal meanings.
B. Based on the user aspect, problems were divided into two categories: chip positioning and chip recognition.
C. Based on the construction aspect: the problem definition was from the point of view of whether the construction workers misplace the Braille signs.

2.1.4 Functional divisions
Functions were divided into internal and external functions.

2.1.5 Individual buttons
Up, down; open, close; floor; and alarm.

2.1.6 Solutions
Solutions for problem resolution are provided using the following analysis.

2.1.7 Conclusion
A Braille symbol table is developed for design referencing.

3 CASE ANALYSES OF VARIOUS ASPECTS

In this study, accessible elevator facilities in airports, office buildings, hotels, and departmental stores in Da Nang City were examined. Various aspects of the data were gathered from a survey and processed with statistical procedures for the analysis.

Analysis from the design aspect: (Tables preceding with letter A): two codes (A1 and A2) are used for elevator Braille panel design and their meanings.

A. Braille System (i.e., English Braille, numeric, and Korean letters). B. Meanings of words.

3.1 A1–1: Analytical comparisons on the landing call buttons of the Braille signage system

A. The number of cases adopting A1 Braille system in elevators with English Braille systems and Braille signage systems with another language were statistically analyzed.
B. Functional division: A1–1, [up 、 down]; the up and down buttons of the elevator outside the control panel buttons.
C. Statistical comparison: A1–1 [up 、 down]; function buttons using Braille signage system proportion. 50.00% were with English Braille; 21.43% were with Korean Braille; 7.14% were with Braille Chinese Roman pronunciation, and 21.43% systems did not have Braille signs.

The Vietnam construction standard TCXDVN 264:2002 does not regulate up 、 down Braille.

3.2 A2–2: Analytical comparisons of the literal meanings of the open and close buttons

A. Problem definition: Statistical analysis of the variance of A2 word meanings against the collected cases.
B. Functional division: A2–1, open and close section: opening or closing the door by controlling the open and close buttons on the internal control panel.
C. Statistical comparisons: Analysis of the Braille system open buttons, function buttons content word meaning [opn] with 42.86%, [open] with 14.29%, [op"en"] with 7.14%, Korean Braille [개 ke] with 21.42%, Chinese Braille [k "ai" m "en"] with 14.29%. Analysis Braille close buttons, Function buttons literal content [cls] with 14.29%, ["sh"ut] with 35.71%, [close] with 7.14%, [shut] with 7.14%, Korean Braille [폐 pye] with 21.43%, Chinese Braille [g "wan" m "en"] with 14.29%.

3.3 B1–2: Analytical comparisons between chip positions in open and close buttons

A. Problem definition: Statistical analysis of the case for the position of the elevator Braille signage with the buttons' positioning.
B. Functional division: B1–2, open and close section: the elevator control panel of the button inside the elevator door button.
C. Statistical comparisons: open and close buttons. The positional relationship between the proportion of function buttons with Braille signage. Analysis Braille position [open] buttons, Braille

elevator buttons located at the lower part of the button comprise 71.43%, 21.43% are located on the left-hand side, and 7.14% are located on the left-hand side of the button. Analysis of position of Braille [close] buttons: 71.43% are located in the lower part of the button; 7.14% are located on the left-hand side of the button; 14.29% are located on the right-hand side of the button; and 7.14% are located on the left-hand side of the button.

4 PROBLEMS AND COUNTERMEASURES

Fourteen public buildings of Da Nang City, Vietnam were selected for the analysis from various aspects, and the problems and countermeasures are as follows.

4.1 *Design aspect*

A. Problems: Inconsistent abbreviations of English usage in the Braille system.

a. Braille systems with English characters comprised 64.29%, other 21.43% used Korean Braille, and 14.29% used Chinese Braille (Roman alphabet).
b. For example, among the close buttons, up to 35.7% used the English word "sh u t", which is different from the word "close" as recommended by Vietnamese standard.
c. The up and down buttons used the English words "up" and "down", with three elevators adopting the Korean Braille system. Two elevators adopted the Chinese Braille system.
c. Several English words for the alarm buttons were used, including "alm" and "alarm". The "alm" did not conform to the Vietnamese standard, but was used in 14.28% of the cases.

4.2 *The use aspect*

A. Problems: Braille symbols should accommodate the tactile-based reading behaviors of people with visual impairments. Therefore, Braille symbols should be prominently positioned at the bottom of the control buttons for easy recognition. Currently, ground floors are represented by multiple variations of the English Braille symbols, including 1 and G, causing confusion.
B. Countermeasures: In accordance with Vietnamese standards, the English words representing the floor landing call buttons have not been standardized. Users with visual impairments spend a lot of time confirming the control buttons.

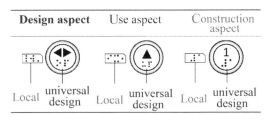

Figure 2. Braille clip site-position of U.D.

4.3 *The construction aspect*

A. Problems: For Braille signs located on the button below, whether a tactile behavior will affect the visually impaired, the Vietnamese construction standards TCXDVN 264:2002 specification left.
B. Countermeasures: For the Braille tactile position, the behavior of the visually impaired must also be considered. The use of left button keyboard as keys to increase the area. Therefore, Braille located beneath the bump button is easy to discern with clarity, and using the function keys helps in reducing the time for confirmation by the visually impaired people.

This is in accordance with the Vietnamese construction standards TCXDVN 264:2002 specification left.

5 CONCLUSION

In Da Nang, Vietnam, 14 public buildings' (barrier-free elevator) elevator Braille systems were compared and analyzed, and the following resulting conclusions and solution proposals were made:

A. According to the Vietnamese construction standards TCXDVN 264:2002 standard, the proportion abbreviation system is as follows: [close] 7.14%, [open] 14.29%, [alarm] 21.44%, and Braille [phone] 14.28%. The efficiency of the implementation details need to be discussed.
B. In Vietnam, usually, floor 1 (ground floor) begins with number [1] and G Floor (Ground Floor), proposal on the ground floor label with the number [1] F (most used) representative, while still underground floors with B1, B2, B3 (the current situation most used).
C. The universal design is divided into the two parts: first, with the universal design, Braille is located in the lower part of the button using English Braille; second, the local language Braille System is located at the left-hand side of the button. The English Braille and the sign should be printed into a single button. The local

language Braille should be placed on the left-hand side separately (as shown in Fig. 2).

D. The universal design should be installed in all the accessible elevators in public transportation facilities where frequent international travelling occurs, such as in airports or mass rapid transition, e.g. the subway stations.

REFERENCES

Li, M.Q., C.C. Tang, L., Lin, C.M., & J.X. Wang, (2011). Braille system of research facilities elevator accessibility East Regional Hospital. Architectural Institute of Taiwan—23nd second annual meeting building research results publication to meeting collection (p. 119) Taipei, Taiwan: National Taipei University of technology.

Su, M.B. (2010). "Research the public works non-barrier elevator braille system application—take Hong Kong and Taiwan as the example," Feng-Chia University master the paper.

Tseng, L. (2015). An Accessibility Study of Elevator Braille System in Chiang Mai Province, Thailand, Architectural Science, 11, 29–43.

Tseng, L., C.C. Tang, & C.J. Sun, (2013). A Study on the Braille Elevator Signage System in Public Buildings: The QFD Perspective., Original Research Article Science Direct (Procedia-Social and Behavioral Sciences), 85p152–163.

Civil, Architecture and Environmental Engineering – Kao & Sung (Eds)
© *2017 Taylor & Francis Group, ISBN 978-1-138-02985-9*

Investigation and analysis of thermal comfort and IAQ in naturally ventilated primary school classrooms

Fusheng Ma & Changhong Zhan

School of Architecture, Heilongjiang Cold Climate Architecture Science Key Laboratory, Harbin Institute of Technology, Harbin, China
Institute of Architecture Design and Research, Shenyang Jianzhu University, Shenyang, China

Xiaoyang Xu & Yu Tang

School of Architecture and Urban Planning, Shenyang Jianzhu University, Shenyang, China

ABSTRACT: This research studied the variables of the thermal environment and Indoor Air Quality (IAQ) of classrooms in a naturally ventilated primary school in Shenyang during winter. Some features of thermal sensation and IAQ satisfaction of pupils and teachers in the classrooms were revealed. Field measurements were carried out in 6 classrooms occupied with 32 teachers and 197 pupils in total from November to December 2015, while the heating system was used. The results indicated that classrooms' IAQ was bad. Thermal comfort and IAQ satisfaction of pupils and teachers in the classroom were significantly different. Most of the pupils could not make accurate judgments of the indoor air quality. It was demonstrated that the existing ventilation mode could not meet the needs of thermal comfort and IAQ of classroom in winter.

1 INTRODUCTION

Thermal environment and IAQ are two essential components to evaluation on quality of indoor environment in severe cold region. Researchers started to evaluate the indoor environment of schools from the students' perspective on thermal comfort in the 1970s (Table 1).

By studying children' thermal comfort and preference, Humphreys found that most children under the age of seven, usually less sensitive to thermal environment than adults, can understand and use simple expressions to describe their thermal sensations (Humphreys 1977). Sander ter Mors, a Dutch scholar, found that the comfortable temperatures of pupils were lower than previous theoretical predictions from thermal comfort models and they prefer a cooler environment (Mors et al. 2011). Teli, a British scholar, found that students had different thermal sensations compared to adults in the classrooms (Teli et al. 2012).

Different regional groups usually have different thermal demand due to the theory of thermal comfort of the human body. Although the researches on the thermal comfort of the children were gradually enriched recently, there were few researches on thermal comfort of minors under natural ventilation in winter. In winter of Shenyang China, the main type of ventilation in classrooms is natural ventilation. However, teachers and pupils control

Table 1. Thermal comfort field studies in school classrooms.

Country	Reference	Ventilation type	Age group
UK	Humphreys (1977)	NV*	7–11
Singapore	Wong & Khoo (2003)	NV	13–17
Japan	Kwok & Chun (2003)	NV + AC*	13–17
Italy	Corgnati, Filippi, & Viazzo (2007)	NV	12–23
Kuwait	Al-Rashidi, Loveday, & Al-Mutawa (2009)	MM*	11–17
Taiwan	Hwang, Lin, Chen, & Kuo (2009)	NV	11–17
Netherlands	Mors, Hensen, Loomans, & Boerstra(2011)	NV	9–11
Taiwan	Liang, Lin, & Hwang (2012)	NV	12–17
Italy	De Giuli, Da Pos, De Carli (2012)	NV	9–11
UK	Teli, James, & Jentsch (2012)	NV	7–11
Iran	Haddad, Osmond, King, & Heidari (2014)	MM	10–12
Chile	Trebilcock, Soto, & Figueroa (2014)	NV	9–10
China	Xu, Liu, & Liu (2014)	AC	≤13
Turkey	Emir (2016)	NV+AC	13–15

*NV = natural ventilation; *AC = air-conditioning; *MM = mixed-mode ventilation.

the ventilation into the classrooms by thermal comfort rather than air quality. The indoor air quality was poor due to the lack of ventilation to keep the indoor air temperature. The above suggests that there was a need for research on the thermal perception of children in Shenyang's school classrooms to obtain a deeper understanding of pupil thermal preferences. The aim of this study was to extend the knowledge base of pupils' response towards their classrooms' thermal environment. The main objectives were as follows; to record the current situation of the environment, to analyze the influence of gender on indoor thermal sensation, to compare the difference of thermal sensation between pupils and teachers, to test whether pupils have the ability to evaluate the quality of the classroom.

2 THERMAL COMFORT FIELD STUDY

2.1 *The school for case study*

The object of this study was located in Shenyang, that northeast of China (Fig. 1).In the Chinese building climate district, Shenyang is belong to the severe cold region. The average annual temperature of Shenyang was 6.2 to 9.7 degrees, and the average temperature in January was less than or equal to −10°C. Shenyang had 5 months in winter, so from the beginning of November to March of next years, the central heating be used.

This school had two buildings. The case study part was denoted as B on Figure 1. The case study building had 6 floors. It was a typical frame structure without thermal insulation structure. The case study building used hot water radiation. The main ventilation type was natural ventilation.

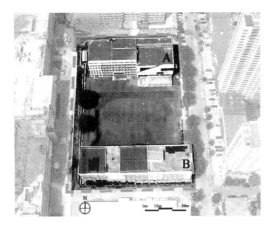

Figure 1. Main plan of school building and school ground.

Occasionally the teachers opened doors and windows for ventilation.

The case study building had 34 classrooms. These classrooms were arranged in two sides along the corridor. In this study, we selected 6 classrooms in the fourth floor. There were 4 classrooms in south orientation and 2 classrooms in north orientation. The classroom size was 6600 mm × 8400 mm. There were two doors, three windows and one internal window in each classroom. The door size was 2100 mm × 900 mm. the window size was 2100 mm × 1500 mm, and the internal window size was 600 mm × 1500 mm. Inside windows and outside windows were sliding windows. Ventilation was achieved through manually opening the classroom windows, which was usually undertaken by the teachers or students near the windows.

There were 197 pupils in 6 classrooms aged 9~12, in grades 5~6. There were 97 boys and 100 girls. The information of classroom was shown in Table 2.

2.2 *Measurement instruments*

Indoor temperature and relative humidity were selected to denote the comfort level of indoor thermal environment. The CO_2 concentration was measured as an important indicator of IAQ. The test were lasted from November 24th to December 24th in 2015, using RR002 temperature/RH logger and handhold Telaire 7001 CO_2 tester (Fig. 3). Figure 4 showed outdoor temperature of Shenyang during test period.

The above parameters were measured at a height of 1.1 m except for the middle measured point. The middle measured point in the classrooms was measured at a height of 0.75 m. The measuring performance specification was shown in Table 3.

2.3 *Survey questionnaire*

The questionnaire was divided into teachers' questionnaire and students' questionnaire. The questionnaire includes; 1) gender, age 2) the thermal sensation, humidity sensation 3) the satisfaction evaluation of thermal sensation 4) the satisfaction degree of the air quality evaluation.

The questionnaires were issued from November to December in 2015. 197 questionnaires were distributed to the students in this study, and 197 valid questionnaires were recovered. 32 questionnaires were distributed to the teachers, and 32 valid questionnaires were recovered. The thermal sensation.

(TSV) scales and satisfaction scales used in the questionnaire were shown in Table 4.

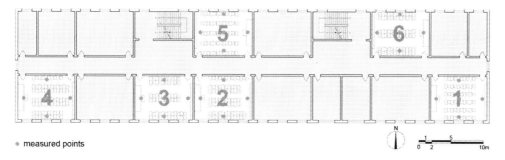

measured points

Figure 2.   The fourth plan of case study building.

Table 2.   The information of classrooms.

No.	Room	Grade	Orientation	Number of boys	Number of girls	Total	Age group
1	401	5	South	18	12	30	10–11
2	405	5	South	16	16	32	10–12
3	406	5	South	15	23	38	9–11
4	409	6	South	20	11	31	11–12
5	412	5	North	16	14	30	10–12
6	414	5	North	12	24	36	9–12

 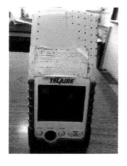

Figure 3.   RR002 temperature/RH logger and hand-hold AZ7752 temperature/CO$_2$ tester.

Table 3.   The measuring range and accuracy of instruments.

Environmental parameter	Range	Accuracy
Air temperature	−40°C~+85°C	±0.5°C (−10°~+85°C)
Relative humidity	0~100% RH	±3% RH (10~95% RH, +25°C)
CO$_2$ concentration	0~10000 ppm	±5% (0~10000 ppm)

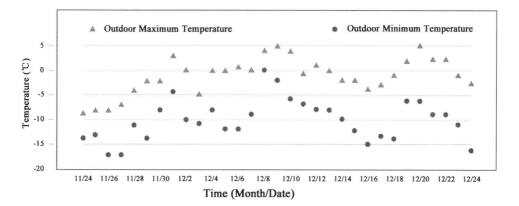

Figure 4.   Outdoor temperature during test period from Shenyang Meteorological Administration.

1161

Table 4. Thermal sensation (TSV) scales and satisfaction scales used in the questionnaire.

TSV scale	Hot (5)	Warm (4)	OK (3)	Cool (2)	Cold (1)
Temperature satisfaction	Very satisfied (5)	Satisfied (4)	OK (3)	Dissatisfied (2)	Very dissatisfied (1)
Humidity satisfaction	Very satisfied (5)	Satisfied (4)	OK (3)	Dissatisfied (2)	Very dissatisfied (1)
IAQ satisfaction	Very satisfied (5)	Satisfied (4)	OK (3)	Dissatisfied (2)	Very dissatisfied (1)

Table 5. The values of the main environmental parameters.

Classroom	1	2	3	4	5	6
Indoor temperature (°C)						
Mean	19.4	17.5	17.6	18.6	19.6	19.5
SD	1.2	1.4	1.2	1.2	0.7	0.9
Maximum	21.2	19.4	19.7	20.4	21.3	21.3
Minimum	15.5	13.6	13.7	14.8	17.6	17.0
Relative humidity (%)						
Mean	68.9	45.7	67.6	48.8	52.9	60.6
SD	4.7	8.0	5.5	7.1	7.0	9.5
Maximum	79.0	62.0	76.0	63.0	67.0	73.0
Minimum	57.0	30.0	51.0	31.0	35.0	30.0
CO_2 (ppm)						
Mean	2356.0	3134.4	2783.8	2908.4	3493.9	3914.6
SD	914.8	1736.6	1505.0	1682.3	1672.8	1615.3
Maximum	4334.0	7070.0	6254.0	6883.0	6642.0	6987.0
Minimum	722.0	640.0	828.0	758.0	794.0	1295.0

3 RESULTS AND DISCUSSION

3.1 Indoor thermal environment and air quality in winter

Table 5, which is organized by classroom, gives the mean, Standard Deviation (SD), maximum and minimum values of each environmental parameter measured during the surveys. It was found the 6 classrooms almost did not open windows in class during test period. The air speed was tested several times as 0m/s while windows kept closed. Hence, the study does not treat air speed as the environment parameter.

Table 5 showed that the temperatures of classroom No.1 to No.4 varied greatly. Indoor temperatures of classroom No.2 and No.3 were lower than the standard value (18°C). That because windows of classroom No.1 to No.4 were opened randomly by teachers or students during the class interval. Times of opening windows in classroom No.2 and No.3 were more than other classrooms. Classroom No.5 and No.6 almost don't open the window. So the temperatures of them were higher than the other classroom. Thus, winter ventilation by opening windows is the main cause of indoor heat loss in severe cold region.

In the aspect of indoor relative humidity, the relative humidity of the six classrooms was significantly different. The relative humidity of classroom No.1 and No.3 was slightly higher than the standard value (40%-60%).

The concentration of carbon dioxide in the six classrooms was significantly higher than that of the national standard (1500ppm). Because of insufficient ventilation rate. Ventilation for a short time could not solve the problem of indoor air quality.

3.2 User evaluation of thermal environment and air quality

Pupils' evaluation of indoor air quality was illustrated in figure 5. Generally speaking, students' evaluation of indoor air quality is satisfactory. Table 5 shows that the six classrooms IAQ were poor in the actual measurement, but pupils had a good evaluation of indoor air quality. There was a significant difference between the subjective evaluation of pupils and the objective measurement of IAQ.

The main reason was divided into the following points; 1) Pupils' knowledge of IAQ is not enough. Most pupils know little about IAQ. 2) It might

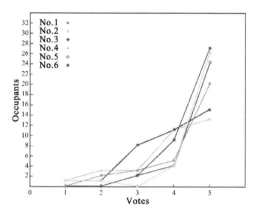

Figure 5. Pupils' evaluation of indoor air quality.

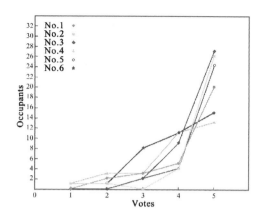

Figure 7. Pupils' evaluation of relative humidity.

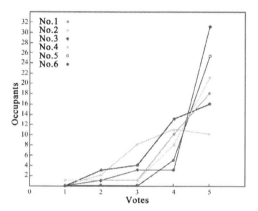

Figure 6. Pupils' evaluation of indoor temperature.

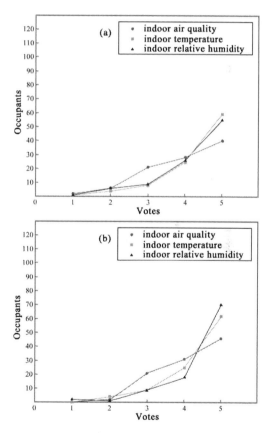

Figure 8. Influence of gender on evaluation (a) The satisfaction votes of boys, (b) The satisfaction votes of girls.

has a certain adaptability to the indoor air quality, because pupils had to remain staying in the classrooms for a relatively long time.

Study shows that pupils have a certain ability to evaluate the indoor air quality, but the gap between the evaluation and the actual measurement results was huge.

Figure 6 was shown that the pupils' evaluation of indoor temperature was very good. Among them, pupils in the classroom 4 had the highest degree of satisfaction with the temperature. Although there were two classroom temperatures slightly lower than 18°C and the excessive temperature differentials.

Figure 7 describes the degree of satisfaction with the relative humidity of the pupils. In general, pupils were satisfied with the relative humidity of the classrooms. Most students know little about relative humidity. But under the guidance of teachers, pupils were able to make the evaluation of indoor relative humidity.

According to the comparison figure 8 (a) and (b), there was no significant difference between boys and girls of the evaluations of IAQ, indoor temperature and indoor relative humidity. It means there was no influence of gender on evaluation of IAQ, indoor temperature and indoor relative humidity.

Figure 9 was shown that pupils' and teachers' evaluation in the classroom of IAQ, temperature, and relative humidity was very different. Pupils' evaluations to the classroom air quality, temperature, relative humidity were very high, but teachers' evaluations to the classroom air quality, temperature, relative humidity were not so good. The reasons that lead to the difference between teachers and students were; differences in educational background, attention of environment and evaluation criteria.

4 CONCLUSION

The study shows the existing ventilation mode couldn't meet the needs of IAQ in the classroom during heating seasons. The indoor temperature varies greatly due to the ventilation by randomly opening window. Ventilation by open window would make a great waste of heat. Therefore, the appropriate ventilation mode should be used in the classroom during winter. The objective test showed that the air quality in the six classrooms was quite poor, because the concentration of carbon dioxide in all of the six classrooms exceeded the level for body health. There were obvious differences between pupils and teachers in the subjective evaluation results. Pupils could not make accurate evaluation of indoor air quality. Appropriate monitoring equipment should be used in school classroom.

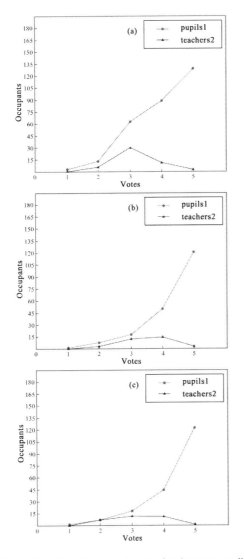

Figure 9. The difference of evaluation between pupils and teachers (a) The evaluation of IAQ, (b) The evaluation of indoor temperature, (c) The evaluation of relative humidity.

REFERENCES

Ahmed, A. S., Ali, K. A., Ahmed. H & S. Ookawara. 2006. An analysis of thermal comfort and energy consumption within public primary schools in Egypt. *IAFOR Journal of Sustainability Energy and the Environment*. 3(1), 51–64.

Bai, LJ., Yang, L., Li, S. & Song, B. 2015. Study on the thermal environment of middle and primary school classrooms in spring in Xi'an. *J. Xi'an Univ. of Arch. & Tech. (Natural Science Edition)*. 47(3), 408–411.

Despoina, T., Patrick, A. B. & Mark, F. J. 2013. Thermal comfort in naturally ventilated primary school classrooms. *BUILDING RESEARCH & INFORMATION*. 41(3), 301–316.

Griffiths, M. & M, Eftekhari. 2008. Control of CO_2 in a naturally ventilated classroom. *Energy and Buildings*. 40, 556–560.

Humphreys, M. A. 1977. A study of the thermal comfort of primary school children in summer. *Building and Environment*. 12, 231–239.

Li, B., Zhu, MS., Zhan, CH. & Cai, WH. Field study on IAQ and thermal comfort of university classrooms in winter of Harbin. *ENERGY CONSERVATION TECHNOLOGY*. 28(162), 336–373.

Li, BZ., Liu, J. & Yao, RM. 2007. Investigation and analysis on classroom thermal environment in winter in Chongqing. *HA & AC*. 37(5), 115–117.

Maureen, T., Jaime, S. & Rodrigo, F. 2014. Thermal comfort in primary schools: a field study in Chile. Proceedings of 8th Windsor Conference: Counting the Comfort in a changing world Cumberland Lodge, Windsor, UK, London: Network for Comfort and Energy Use in Buildings. 10–13.

Mors, S. T., Hensen, J. L. M., Loomans, M. G. L. C. & Boerstra, A. C. 2011. Adaptive thermal comfort in primary school classrooms: creating and validating PMV-based comfort charts. *Building and Environment*. 46, 2454–2461.

Silay, E. 2016. The Evaluation of Thermal Comfort on Primary Schools in Hot-Humid Climates: A Case Study for Antalya. *European Journal of Sustainable Development*. 5(1), 53–62.

Teli, D., Jentsch, M. F. & James, P. A. B. 2012. Naturally ventilated classrooms: An assessment of existing comfort models for predicting the thermal sensation and preference of primary school children. *Energy and Buildings*. 53, 166–182.

Valeria, D. G., Osvaldo, D. P. & Michele, D. C. 2012. In door environmental quality and pupil perception in Italian primary schools. *Building and Environment*. 56, 335–345.

Xu, H. & Ou, D. Y. 2016. Investigation and evaluation of classroom acoustical environment in primary and secondary schools in Fuqing city. *Building Science*. 32(4), 77–86.

Xu, J., Liu, J. P. & Liu, D. L. 2014. Testing and evaluation of indoor thermal environment of classrooms in rural primary schools in Guanzhong area in winter. *Building Science*. 30(2), 47–50.

Zahra, S. Z., Mohammad, T. & Mohammadreza, H. 2016. Thermal comfort in educational buildings: A review article. *Renewable and Sustainable Energy Reviews*. 59, 895–906.

Civil, Architecture and Environmental Engineering – Kao & Sung (Eds)
© 2017 Taylor & Francis Group, ISBN 978-1-138-02985-9

Improvement of the ensemble forecast of typhoon track in the Northwestern Pacific

J.Y. Yuan
State Key Laboratory of Hydrology-Water Resources and Hydraulic Engineering, Hohai University, Nanjing, China

Y. Pan & Y.P. Chen
College of Harbor, Coastal and Offshore Engineering, Hohai University, Nanjing, China

ABSTRACT: An accurate forecast of typhoon track is essential for the effective mitigation of typhoon-induced loss in the coastal areas. Based on the dynamical analysis of forecast errors of typhoon track at four weather forecast centers, i.e. China Meteorological Administration (CMA), Japan Meteorological Agency (JMA), Joint Typhoon Warning Center (JTWC) of USA, and Taiwan Meteorological Center (TMC), in this paper, the ensemble forecast method is proposed and further improved by using the running training scheme during the entire typhoon process. The performance of the method is examined by the forecasting of two typical typhoons, i.e. "Damrey" (No. 1210) and "Fitow" (No. 1323), in the region of Northwestern Pacific. The results show that better accuracy is achieved using the ensemble forecast method compared to the results from the four individual forecast centers and the existing method for both the typhoons.

1 INTRODUCTION

The region of northwestern Pacific is often attacked by typhoons. During a typhoon, the sea water level rises rapidly, and storm surge disasters appear under the effect of strong wind and low pressure, which may cause serious economic losses and causalities. In order to reduce the typhoon-induced damages, it is essential to provide a high-precision forecast service of storm surges in coastal areas, which needs a reliable forecasting of typhoon tracks. In recent years, many weather forecast centers have been using numerical models for the typhoon forecast, which, however, cause some errors to arise from the inaccurate initial boundary conditions, physical parameters, numerical schemes, etc. In order to reduce such kinds of errors, an ensemble of forecast methods based on the initial disturbance or the mode perturbation are usually employed (Zhang and Krishnamurti, 1997; Zhu and Dai, 2000; Zhou et al, 2003; Wang and Liang, 2007). These methods can generate multiple groups of forecast results through the combination of different initial conditions, physical parameters, or even different meteorological models (e.g. Duan and Wang, 2004; Rao and Srinivas, 2014). On the basis of statistical analysis, the methods have been proved to be able

to present more reliable typhoon track forecast. However, the computational time cost by using these methods turns out to be very high. In order to tackle this problem, Krishnamurti et al. (1999, 2000) proposed a multi-mode super-ensemble forecasting method. It contains more than one independent modal forecasting result, which can rapidly achieve multi-sample ensemble forecast, and effectively improve the results of wind field forecast. Due to its low cost of computational time, simplicity, and practicability, this method has been widely used in meteorological and hydrological forecasting in recent years (Zhi et al., 2009; Cane et al., 2013; Chen et al., 2014). The relevant research shows that ensemble forecast is a method that makes full use of the results of central modal forecast to reduce systemic deviation. It gives a significant direction to the current numerical forecast technological development. This paper focuses on the application of super ensemble method in the forecast of typhoon tracks. An improved ensemble method is proposed and applied in the forecast of two typical typhoons that occurred in the region of Northwestern Pacific. The accuracy of the method is examined in comparison to the individual forecast results from four different forecast centers. The details are presented in the following sections.

2 METHODOLOGY

2.1 General introduction

This super ensemble method is proposed based on the weighted average of forecast results from different weather centers. Without any loss of representativeness, the 24-hour forecast data from four operational weather forecast centers—China Meteorological Administration (CMA), Japan Meteorological Agency (JMA), Joint Typhoon Warning Center (JTWC) of USA, and Taiwan Meteorological Center (TMC)—is used for the study. The forecast procedure is divided into two periods: training period and forecasting period. The weighted coefficients for each center are determined by the center's performance of typhoon forecast in the training period. A higher weighted coefficient will be assigned to the center that performed more accurately in the training period.

2.2 Existing method

The existing method proposed by Ding et al. (2015) uses the dynamical training scheme to calculate the weight coefficients along with the changes of forecasting typhoons in more than one typhoon forecast. The forecasting period is for a single typhoon, whereas the training period is for the previous k typhoons adjacent to the forecasting one, as illustrated in Figure 1. As the weight coefficients are adjusted according to the changes of the training samples, the forecasting result is related to the value of k.

With a series of numerical experimental tests, the scheme of k = 40 without "double typhoon" samples is recommended. Compared with static training method, this method has following merits: (1) the most recent typhoons can be taken into account in the training period; (2) the overall forecast errors become quite stable when k ≥ 40. However, this method also has some limitations: (1) it requires a large amount of data; (2) the weight coefficients for the different weather forecast centers remain unchanged during the entire forecasting period; (3) the size of the training sample is 40, but the specificity, type, and landing area of the 40

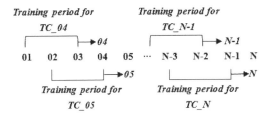

Figure 1. Illustration of a moving training method.

typhoons are not completely equivalent. Therefore, there is a need for this method to be improved.

2.3 Improved method

In order to tackle the limitations of the existing method, an improved method is proposed, which is based on the dynamic analysis of forecast errors at the four weather forecast centers during the entire typhoon process. By studying the typhoon track graph, we found that the forecast tracks are consistent with the actual typhoon track in the general trend. Based on this finding, we used forecast and the corresponding measured values in the first two moments as raw data. Then, the raw data was extended linearly to obtain the 24-hour forecast value and the corresponding measured value of the certain typhoon in every start time. The revised deviation between the linearized forecast value and the real forecast value was regarded as the revised error in ensemble forecast, and the distance of forecast deviation of each model was gained. The smaller the deviation is, the bigger the weighted coefficient (α_i) will be. It can be described as

$$\alpha_i = \frac{E_i}{\sum_{i=1}^{N} E_i} \qquad (1)$$

in which α_i is the weighted coefficient for the i-th forecast center; and E_i is the multiplicative inverse of the mean forecast errors of i-th forecast center in the training period.

Once the weighted coefficient is determined, the forecast values can be calculated by using the bias-removed ensemble mean (WEM) method. As the typhoon track cannot be absolutely smooth, a distance correction parameter is introduced. The parameter value is 0.3 according to experience, which is described as

$$\begin{cases} F_{WEM} = \sum_{i=1}^{N} \alpha_i F_i \\ F_i = f_i + D_i * 0.3 \end{cases} \qquad (2)$$

in which F_{WEM} is the forecast value; F_i is the forecast value of the i-th forecast center; N is the number of the forecast centers, with N = 4 in this study; f_i is the real forecast value of the i-th center; and D_i is the forecast deviation of the i-th center in the training period.

3 ENSEMBLE FORECAST ANALYSIS

To examine the performance of the modified method on the typhoon track forecast, two typical typhoons that occurred in the region of Northwestern Pacific, i.e. "Damrey" (No. 1210) and "Fitow" (No. 1323), were taken as examples for the case studies. The results are presented as follows.

Typhoon Damrey is the strongest typhoon that landed in the north of the Yangtze River since 1949. It strengthened slowly in the early stage, and its track direction was changeable. Besides, it moved slowly and strengthened stably offshore. The forecast tracks and the actual track are shown in Figure 2, and the average track forecast error is shown in Figure 3.

It can be observed from Figure 2 that with the advancement of typhoon to the coasts, the forecast results tend to be closer to the actual track. The corresponding forecast errors decrease significantly. This is because at the beginning, the movement of the typhoon center has large uncertainties, and there is no obvious regularity of the typhoon track; therefore, there is fluctuation of the forecasting results. With the ensemble method, the forecast tracks are generally closer to the actual ones compared to the results from single forecast centers because of the real-time revision of the forecast errors; this shows that revised errors have a great influence on the accuracy of typhoon track forecast. As it can be seen from Figure 3, the track error of ensemble forecasting method decreased evidently compared with JTWC, JMA, CMA, and TMC by 20.14%, 21.26%, 19.45%, and 11.69%, respectively. Compared to the existing method, the proposed ensemble method decreased by 6.82%. The existing method also revised the weight coefficients and track errors in the forecasting of new typhoon; however, their coefficients remained unchanged during one single typhoon, and a large amount of typhoon data were required. Furthermore, there might be major deviance when using typhoons from the history. This is why the dynamical training method uses the same typhoon for training and forecasting period, which can improve the forecast precision.

Based on the track error analysis of "Damrey", this paper chose 1323 "Fitow" for ensemble forecast to further demonstrate the improvement in the method created on the forecast precision of a single typhoon.

A severe tropical typhoon, Fitow (1323) was the strongest typhoon landing that was encountered in China's mainland after October. It caused economic losses that amounted to 62.33 billion RMB to China and ranked only second to Typhoon Herb (9609). It formed on September 30th in 2013 in the eastern ocean of the Philippines and slowly developed in the early stage. It intensified into a strong typhoon on October 4th and landed at Fuding in Fujian Province of China at 01:15 on October 7th. The typhoon was weakened into a tropical storm at 5 a.m. on the same day and disappeared near the coast of Jianou in Fujian Province of China at 9 a.m. on the same day. The similar results for the typhoon "Fitow" are shown in Figures 4 and 5.

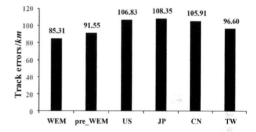

Figure 3. Damrey's average track forecast errors.

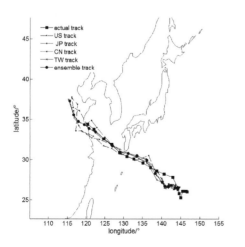

Figure 2. Damrey's forecast tracks and the actual track.

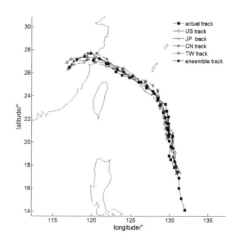

Figure 4. Fitow's forecast tracks and the actual track.

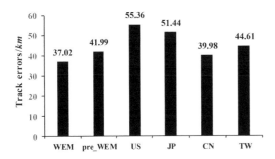

Figure 5.   Fitow's average track forecast errors.

As it can be observed from the figures, in the process of typhoon track forecast of "Fitow", the forecasting track error obtained by revised ensemble forecast diminished by 9.31% compared with the best station (TMC). Compared to the existing method, the track forecast error diminished by 11.83%. Through these case studies, we found that the modified training method evidently diminished the track error and greatly improved the forecast accuracy.

4   CONCLUSION

Based on the forecast data obtained from CMA, JMA, JTWC of USA, and TMC, the ensemble forecasting of 24-hour typhoon tracks was conducted for two typical typhoons that occurred over the Northwestern Pacific. We used the forecast and the corresponding measured values at the first two moments as the raw data, and they were extended linearly to calculate the weighted coefficients and bias for each weather forecast center. The revised deviation between the linearized forecast value and the real forecast value is regarded as the revised error in ensemble forecast. The results are listed as follows:

a. The idea of estimation proposed in this paper is feasible. According to the statistical results, the revised deviation has a great impact on the ensemble forecast value. The distance correction parameter is introduced when calculating the ensemble forecast, and 0.3 is reasonable as the correction parameter. The modified method could improve the accuracy of track forecast.
b. According to the analysis of some typical typhoons like "Damrey" and "Fitow", we found that the track error of ensemble forecasting method is less than that of the single stations. In addition, the precision of track forecast is better compared with the existing method. Although the existing method also revised the weight coefficients and track errors in the forecasting of new

typhoon, their coefficients remained unchanged during a single typhoon. Moreover, there might be major deviance when using typhoons from the history. This is why the dynamical training method uses the same typhoon for training and forecasting period, which can improve the forecast precision.
c. The modified method does not need to use historical data, and it builds a strong operability system in forecasting typhoon track. Besides, the modified method provides forecast with high accuracy and realizes the real-time correction of the weight coefficient and forecast deviation.

REFERENCES

Cane, D., Baebarino, S., Renier, L.A., et al. (2013). Regional climate models downscaling in the Alpine area with multimodel superensemble, J. Hydrology and Earth System Sciences 17, 2017–2028.

Chen, Y.P., Gu, X. & Ding, X.L. (2014). Fast Generation of Ensemble Typhoon Wind Field and Its Application. *Proceedings of the 24th International Ocean and Polar Engineering Conference*, Busan, Korea, v3, 438–442.

Ding, X.L., Chen, Y.P., Yuan, J.Y. & Pan, Y. (2015). Super-ensemble forecasting of typhoon tracks over the Northwestern Pacific. *3rd IMA International Conference on Flood Risk*, Swansea, UK.

Duan, M.K. & Wang, P.X. (2004). Advances in researches and applications of ensemble prediction, J. Journal of Nanjing Institute of Meteorology 27, 279–288.

Krishnamurti, T.N., Kishtawal, C.M., Larow, T.E., et al. (1999). Improved weather and seasonal climate forecasts from multimodel superensemble, J. Science 285, 1548–1550.

Krishnamurti, T.N., Kishtawal, C.M., Shin, D.W., et al. (2000). Improving tropical precipitation forecasts from a multianalysis superensemble, J. Journal of climate 13, 4217–4227.

Rao, D.V.B. & Srinivas, D. (2014). Multi-Physics ensemble prediction of tropical cyclone movement over Bay of Bengal, J. Natural Hazards 70, 883–902.

Wang, C.X. & Liang, X.D. (2007). Ensemble Prediction Experiments of Tropical Cyclone Track, J. Journal of Applied Meteorological Science 18, 586–593.

Zhang, Z. & Krishnamurti, T.N. (1997). Ensemble forecasting of hurricane tracks, J. Bull. Amer. Meteor. Soc. 78, 2785–2795.

Zhi, X.F., Lin, C.Z., Bai, Y.Q., et al. (2009). Superensemble forecasts of the surface temperature in Northern Hemisphere middle latitudes, J. Sciences Meteorological Sinica 32, 569–574.

Zhou, X.Q., Duan, Y.H. & Zhu Y.T. (2003). The ensemble forecasting of tropical cyclone motion I: using a primitive equation barotropic model, J. Journal of Tropical Meteorology 19, 1–8.

Zhu, Y.T. & Chen, D.H. (2000). An ensemble forecasting scheme of the tropical cyclone track based on dynamic interpretation method, J. Scientia Meteorologica Sinica 20, 229–238.

Civil, Architecture and Environmental Engineering – Kao & Sung (Eds)

An investigation into the system of feasibility study for the revitalizing of idle public facilities through outsourcing of venue management

Jyh-Harng Shyng

Department of Environment and Property Management, Jin-wen University of Technology, Taipei, Taiwan

ABSTRACT: This study found that any proposed revitalization strategy through outsourcing of venue management must be subject to proper feasibility studies and negotiations. Public departments shall refer to the "profitability" and "economic benefits" of the public facilities to formulate various tendering criteria. Private entities, on the other hand, must determine whether these tendering criteria would be compliant to their investment expectations and "financial benefits". Under reasonable "financial benefits", feasibility studies and negotiations shall then be carried out in order to determine whether the revitalization strategy of the public facility is capable of complying with the features of the public facilities and the objectives of providing socio-economic and public benefits.

1 INTRODUCTION

1.1 Research background

The Executive Yuan of Taiwan has initiated a "Program for Promoting the Revitalization of Idle Public Facilities" and established a cross-departmental taskforce responsible for promoting revitalization work. A number of taskforce evaluation meetings have been held. In 2016, the Public Construction Commission of the Executive Yuan continued to compile a list of idle public facility construction projects that were completed but failed to be used according to its planned objectives or were underutilized, as well as public facility construction projects that had been suspended for extended periods of time but may offer social economic potential. This list will help an "evaluation to be made of the feasibility of revitalizing idle public facilities" and initiate a diverse selection of revitalization strategies that include "improvement of facility functions", "transformation and reutilization", and "outsourcing".

1.2 Experimental objectives and scope

The objectives of this study would be to introduce novel forms of management for public assets by private entities. Such public assets and facilities have the advantages of added support from public departments and the dynamism and innovative management approaches of private entities, resulting in a mutually beneficial opportunity for both public and private organizations involved.

The scope of this research focused on the negotiation process between public departments and private entities to establish revitalization strategies and decision making mechanisms needed for feasibility studies that take place for the "outsourcing of venue management".

2 LITERATURE REVIEW

2.1 Studies of outsourcing processes of idle public facilities in other countries

According to past experiences in other countries and current trends of privatization and diversification, the "outsourcing of venue management" model was adopted to introduce the dynamism and management experiences of private enterprises. This would be a key strategy for the revitalization of public construction and facilities. The following lists the experiences of outsourced management by the governments of the UK, the US, Canada, and Japan (Chang and Chang, 2006):

1. Outsourcing experiences in the UK: Bevir, Mark & David O'Brien (2001), Woods and Rober (2000), Clarke and John (2000), and Falconer, Peter K. & Kathleen McLaughlix (2000) provided the following reference: Step-by-step process of "activation, search for potential partners, collaboration, and project investigation" for public-private partnerships.
2. Outsourcing experiences of the US: Raffe, Jeffrey A., Debar A. Auger, & Kathryn G. Denhardt (1997), Savas, E. S. (2000) and DeHoog & Ruth Hoogland (1990) provided the following reference: Using "clear" and

"flexible" principles to construct a contract-based management system.

3. Outsourcing experiences of the federal and provincial government of Canada: "Partnership contracts" were used to establish a positive partnership between government and private organizations in order to review the necessity of government tasks.

4. In the 21st century, Japan utilized the concept of "transformation" in order to promote effective utilization of urban areas.

2.2 *Outsourcing theories and their suitability for public facilities*

According to Li Tsung-Hsun (2002), the theoretical basis for the outsourcing of government processes adopted the general concept of "best value", which would be used to introduce various other theoretical perspectives.

Savas, E.S. (2000) proposed that certain degrees of categorization should be employed for the nature of various public facilities. There is a number of steps that the public department must exercise control over during the overall outsourcing process. These steps include: leading feasibility evaluations, creating a competitive environment, establishing benefits or qualifications criteria, and preparing a tendering manual.

2.3 *Key aspects that influence the outsourcing process*

Experiences of outsourced public construction projects and literature review found that "financial issues" were the key determinant of the success of the outsourcing process (Huang, 2003). Outsourcing was also defined as a "revolutionary revitalization strategy for management models".

The mutual influences included within this model could be divided into 4 categories, including: (1) "Differences between the roles of the

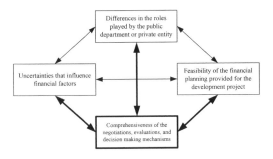

Figure 1. Aspects that may influence the negotiations and evaluation system for the feasibility of outsourced projects.

public departments and private entities". Public departments must comply with specifications and guidelines defined by law or policy. (2) "Uncertainties that influence financial factors". (3) "Feasibility of the financial planning provided for the development project". (4) "Comprehensiveness of the negotiations, evaluations, and decision-making mechanisms" as shown in Figure 1.

3 RESULTS OF INVESTIGATIONS CARRIED OUT INTO IDLE CASES

This study referred to other studies and compiled a total of 6 feasibility study indicators for the outsourced management of public facilities. Relevant studies as well as the context and results of evaluations conducted by the Public Construction Commission were also referenced to generate 6 major feasibility categories with 37 primary variables. The categories were "Policy and legal systems"; "economic environment"; "market environment"; "project planning"; "project finances"; and "project negotiation system". The targets reviewed in this study included 153 public facilities that had been listed by the Executive Yuan Public Construction Commission. A total of 119 effective results were recovered for a recovery rate of 77.8%.

4 ANALYSIS OF THE INDICATORS AND DIMENSIONS FOR THE FEASIBILITY STUDY OF THE OUTSOURCING PROCESS

This study analyzed the reliability values of 3 dimensions. Factor 1 was the dimension of "concession criteria and factors"; factor 2 was the "external criteria and packages"; and factor 3 was "influences of the economic environment". The reliability scores for each of the 3 dimensions exceeded 0.9. Cumulative variance of the 3 dimensions was 70.765%, showing that these dimensions provided adequate representativeness. The reliability of the dimensions was very close and consistent with the overall dimension. Reliability for each of the factor dimensions exceeded 0.9, which was acceptable.

Several dimensions were extracted after the aforementioned factor analysis. These were: "discussion of concessions, external criteria and packages", and "influences of the economic environment". This also helped to clarify the interpretation and meaning of the "feasibility study of the outsourcing of public facilities" in order to construct indicators and dimensions of the feasibility study (Figure 2).

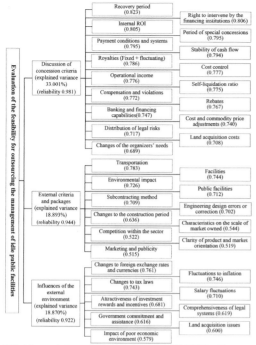

* Total cumulative explained variance was 70.765% with an eigenvalue of 21.682.

Figure 2. Indicators and dimensions for the feasibility study of the outsourcing of idle public facilities.

5 FEASIBILITY STUDY OF THE OUTSOURCING

Revitalization of public facilities through outsourcing would be mainly carried out through public-private partnerships. Under this arrangement, both parties must arrive at a consensus and establish a partnership model.

5.1 Using "management profitability" to categorize factors that may influence the feasibility study

The term "Differences between the roles of the public departments and private entities" mainly refers to the legal or policy standards guidelines with which public guidelines must comply. These would be the prerequisites for establishing public facilities.

The nature and type of the public facilities shall be divided into 2 categories of "profitable public facilities" and "non-profit public facilities". The 3 dimensions listed above would therefore each include 2 hidden factors listed in the following:

1. "Financial influence" could be regarded as "uncertainties that influence the financial factors" and "feasibility of the financial planning of the development project".
2. "Planning environment" could be regarded as "comprehensiveness of the negotiations, evaluations, and decision-making system" which would be similar to the factor of influences of the economic environment.

These two categories of factors could therefore be compiled as contents of the "financial dimension of the overall development project" used to formulate the "tendering criteria". These criteria would be key during negotiations between public departments and private entities needed to establish partnerships between the two.

5.2 Evaluations and decision-making processes when employing outsourcing as a revitalization strategy

The Executive Yuan raised a total of 3 revitalization strategies: "improving facility functions"; "transformation and reutilization"; and "outsourcing". The profitability of public facilities would be used to conduct feasibility studies and negotiations for the revitalization process. This study specifically analyzed the feasibility study procedure for the third revitalization strategy of "outsourcing" (Figure 3).

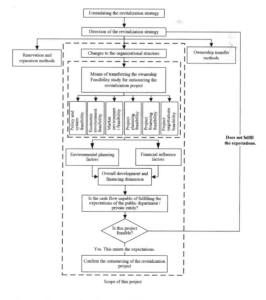

Figure 3. Evaluation and decision-making processes when using outsourcing as a revitalization strategy for idle public facilities.

Public and private departments and entities must comply with the principles of partnerships. Where planning was led by public departments, these requirements would be known as the "tendering criteria". If private entities submit development applications to the public departments pursuant to the articles of the "Act for Promotion of Private Participation in Infrastructure Projects", these requirements would be regarded as "application criteria".

Regardless of the type of criteria used, public-private partnerships must include factors that influence the planning process, details that may influence the finances, and maintain a certain degree of flexibility for the adjustment process. Negotiation systems and flexibility must be provided for. The decision-making procedure and factors influencing the development project will differ according to the tendering mode along with factors that influence project finances. Hence, the negotiation system for the revitalization of public facilities through outsourcing would also generate different revitalization strategies.

6 CONCLUSION AND RECOMMENDATIONS

Decision making processes for the revitalization strategies of idle public facilities must undergo feasibility study or a relevant system. The first step must involve the characterization of the "management profitability" of the specified public facility. A suitable feasibility study must then be selected to carry out feasibility studies and evaluations by both the public department and private entities. Various performance indicators and factors under the categories of "environmental planning" and "financial criteria" must then undergo negotiations and sensitivity analysis. It is only by performing these previous steps that planners will be able to conduct accurate feasibility studies and generate proper details of the revitalization strategy.

Public departments must consider the economic benefits (employment opportunities and encouragement of peripheral industries) of the public facilities. Private entities must refer to the tendering criteria of the public departments to determine whether these projects are able to meet their investment expectations and "financial benefits". During the phase of the feasibility study, "tendering criteria" proposed by the public departments and financial reviews by the interested private entities

will generate a reasonable "financial benefits plan" to review the project and determine whether the revitalization strategy selected is capable of meeting the expected socio-economic benefits as well as public policy objectives for developing public facilities.

REFERENCES

Bevir, Mark & David O'Brien 2001. "New Labour and the Public Sector in Britain," Public Administration Review. 61 (5): pp. 535–547.

Chang Chiung-Ling & Chang Li-Ya 2006. "Outsourcing of government processes and its operations—A case study of the Taipei City Government Department of Social Welfare". *Hua-gang Journal of Social Sciences:* 19, pp. 31–60.

Clarke, John 2000. "Leisure: Managerialism and Public Space," in John Clark, Sharon Gewirtz & Eugene McLaughlin (eds.), New Managerialism New Welfare? London, UK.: SAGE Publications, pp. 186–201.

DeHoog, Ruth Hoogland. 1990. "Competition, Negotiation, or Cooperation: Three Models for Service Contraction, "Administration & Society. 22(3): pp. 317–340.

Falconer, Peter K. & Kathleen McLaughlin. 2000. "Public-Private Partnerships and the 'New Labour' Government in Britain," in Stephen P. Osborne(ed.), Public-Private Partnerships: Theory and Practice in International Perspective. London, UK.: Routledge, pp. 120–133.

Huang Shih-Chieh. 2003. Financial issues would be the key factor in determining the success of BOT projects. *Journal of Engineering Biology*, 76, Issue 2, pp. 102–106.

Li Tsung-Hsun. 2002. Government outsourcing: theories and practice. *BEST-WISE PUBLISHING CO., LTD.*, p. 26.

Polhill, R.M. 1982. *Crotalaria in Africa and Madagascar.* Rotterdam: A.A. Balkema., pp. 1–31.

Raffe, Jeffrey A., Debar A. Auger, & Kathryn G. Denhardt 1997. Competition and Privatization Options: Enhancing Efficiency and Effectiveness in State Government. New York, DE: Institute for Public Administration, University of Delaware, p. 56.

Savas, E.S. 2000. Privatization and Public-Private Partnerships. New York: Chatham House. Ibid., pp. 174–210.

Savas, E.S. 2000. "Privatization and Choice in New York City Social Services," Public Administration Review, 62(1): pp. 82–91.

Woods, Rober. 2000. "Social Housing Managing Multiple Pressures," in John Clarke, Sharon Gewirtz & Eugene McLaughlin(eds.), New Managerialism New Welfare? London, UK.: SAGE Publications, pp. 137–151.

Civil, Architecture and Environmental Engineering – Kao & Sung (Eds)
© 2017 Taylor & Francis Group, ISBN 978-1-138-02985-9

Distribution characteristics of ammonia emission from the livestock farming industry in Hunan Province

Xing Rong, Chunhao Dai, Ying Deng & Pufeng Qin
Resources and Environmental Sciences, Hunan Agricultural University, Changsha, Hunan Province, China

ABSTRACT: The distribution characteristics of livestock ammonia emissions in Hunan Province were investigated by applying the foreign average emission factor as the calculating factor. The total ammonia emissions of poultry feeding industry in Hunan province are 39.84×10^4 t in 2014, among which the maximum city emission was in Hengyang City as much as 5.50×10^4 t·a^{-1}. Moreover, the intensity of ammonia emissions in 2014 from livestock in Hunan Province averaged 1.88 t/km^{-2}, and the Xiangtan City emission intensity went up to 3.87 t/km^{-2} as the maximum. The results showed the reduction of ammonia emissions was mainly attributed to the gradual expansion of exit and prohibited area. Furthermore, pig breeding was found to be the largest source of ammonia emission, as the proportion to 54.82%.

1 INTRODUCTION

Atmospheric fine particulate matters have posed a serious threat to human health and ecological environment, attracting much public attention in recent years. The formation mechanism of fine particles in the atmosphere is complex, including primary particles that are directly formed and secondary particles that form indirectly. Primary particles are mainly composed of dust particles and generated by combustion of carbon black (organic carbon) fuel particles. Besides, secondary particles are mainly water-soluble aerosol particles (sulfates, nitrates and ammonium salts), which account for about 57% of mass concentration of the atmospheric fine particles.

Ammonia (NH_3) is one of the most important trace gases in the atmosphere involved in the nitrogen cycle. As the most important alkaline substances in the atmosphere, ammonia plays an important role in cushioning the underlying atmosphere acidification, acid deposition and the forming of secondary particles. Meanwhile, the NH_3 emissions into the atmosphere would produce the greenhouse effect. Back in 1997, ammonia was discovered as an important precursor of fine particulate matter ($PM_{2.5}$). SO_4^{2-}, NO_3^-, and NH_4^+ were the major components of $PM_{2.5}$ while NO^{3-}, NH^{4+} and ammonia were closely related. Recently, due to the association between ammonia emission and formation of the secondary particles, which posed a considerable threat to human health, a number of researches on ammonia emissions were carried out. Ammonia and nitrate-forming nitrate aerosols will significantly increase the total particulate matter; for example, much of the ammonium nitrate aerosol could be found in $PM_{2.5}$ particles.

Ammonia emissions include anthropogenic and natural sources. Ammonia emissions from anthropogenic sources were increasingly serious, among which the ammonia emissions from livestock were the largest source and caused the most negative impact increasingly. Hunan province is a big producer of livestock and poultry breeding in China because of its breeding scale. It is in the forefront of China with regard to livestock breeding, which leads to more serious ammonia emission pollution. Obviously the characteristics of ammonia distribution influences the prevention and control of atmospheric pollution, provided the basic data on the fine particulate matter source spectra is available. Therefore, the main goal of the present study was to investigate the distribution of ammonia emission from livestock and poultry breeding in Hunan province in 2004–2014.

2 RESEARCH METHOD

Research areas include 13 prefecture-level cities as follows: Changsha, Zhuzhou, Xiangtan, Shaoyang, Huaihua, Changde, Yiyang, Yueyang, Hengyang, Yongzhou, Chenzhou, Loudi, Zhangjiajie City and Xiangxi Tujia and Miao autonomous prefecture, which were studied from 2004 to 2014.

2.1 Calculation method

Up to now, the emission factor method was adopted in the domestic and international research on the source of ammonia emission inventory. Emission factor multiplied by the corresponding number

Table 1. Ammonia emission factors in the livestock farming industry/kg·a^{-1}.

Species	References [10]	[11]	[12]	[13]	[14]	[15]	[16]	[17]	[18]	[19]	[20]	[21]	This study
Pig		2.25	6.4	5.92	9.2	4.8	4.8		4.8				5.4
Dairy cattle	13.1–55		28	31.0	39.7	17.4	19.4–24.8	40	28.5	29	27		29.4
Beef				18.6	39.7	10.0	9.5–9.9	28	14.3	14	6.8		16.7
Laying hen		0.1	0.1	0.41	0.60	1.24	0.32		0.38		0.45	0.22	0.43
Broiler chicken		0.1	0.1		0.17	0.24	0.18		0.27	0.28	0.23	0.22	0.20
Goat		1.1–3.0	1.34			1.2	1.2	1.9					1.30

(from the Hunan Provincial Environmental Protection Office 2004–2014 annual environmental statistics) was applied to calculate ammonia emissions from farming. Ammonia emissions divided by each region area were equal to the emission intensity.

2.2 Determination of the emission factor

Due to the complexity of many factors affecting the ammonia emission factors, it is difficult to determine the emission factor. Meaningful reports and review articles on these studies have appeared and have been published abroad. Therefore, much literature on ammonia emissions of livestock and poultry breeding industry almost refer to the foreign emission factors. However, foreign emission factors largely depend on the specific situation, which caused some differences between China and worldwide and pose a remarkable negative impact on the estimation accuracy of ammonia emission. Thus, referring to some data on foreign emission factors, the average value was used as the calculation values in this study. These values are shown in the Table 1.

3 RESULTS AND DISCUSSION

According to the calculation, all kinds of livestock and poultry ammonia emissions in Hunan province during 2004–2014 are shown in Table 2.

3.1 Spatial distribution of ammonia emission from the livestock and poultry breeding industry in Hunan Province

Fig.1(a) showed the spatial distribution of livestock and poultry breeding in Hunan province in 2014. The total amount of ammonia emission from livestock and poultry breeding industry in Hunan province was 39.84 × 10^4 tons, within which the district-level emission per unit was 2.85 × 10^4 tons on average. One of the biggest emissions was Hengyang, whose emission was as much as 5.50 × 10^4 tons, followed by Changde city, Shaoyang city and Changsha city. Comparatively, Zhangjiajie was the minimal emission source, at the level of 0.48 × 10^4 tons and the key reason was that Zhangjiajie was the tourist area and therefore covered less amount of livestock and poultry breeding than Hunan Hengyang.

3.2 Spatial distribution of livestock and poultry breeding in Hunan province ammonia emission intensity

The spatial distribution of ammonia emission from livestock and poultry breeding in Hunan province in 2014 is shown in Figure 1(b). The Xiangtan city was the emitter with the strongest intensity, with a value of 3.87 t·km^{-2} in 2014. It was followed by Hengyang and Changsha city, which emitted significantly more than 3 t·km^{-2}. Meanwhile, there was a large proportion of the population and industrial emitter in Changsha and Xiangtan, which could make it easier to transform the pollutants into secondary aerosols. Therefore, Changsha and Xiangtan city were chosen for the management and treatment of livestock and poultry breeding ammonia emissions. Apart from these results, the lowest emission was 0.37 tons per square kilometers in western Hunan noticeably.

According to some literature and reports on emission intensity of ammonia due to the livestock and poultry breeding, Beijing, Tianjin, Hubei province, Jiangxi province, Jiangsu and Guizhou were 2.95 2.54, 1.27, 0.59, 2.74, 1.09 tons per square kilometers, respectively. Compared to these areas, the average emission intensity in Hunan province was 1.88 t·km^{-2}, which belongs to the medium level among the above-mentioned areas.

3.3 Temporal distribution of ammonia emissions from livestock and poultry breeding in Hunan province

Temporal distribution of livestock and poultry breeding in Hunan province in terms of ammonia

Table 2. Ammonia emission in the livestock farming industry/10^4t·a^{-1}.

Year	Pig	Dairy cattle	Beef	Laying hen	Broiler chicken	Goat	Total
2004	22.2	5.7	6.2	2.5	4.5	0.8	41.8
2005	23.5	6.0	6.4	0.3	1.3	0.9	38.4
2006	23.9	6.1	6.5	0.3	1.5	0.9	39.3
2007	23.7	6.0	6.4	0.4	2.1	0.9	39.6
2008	20.4	4.2	4.5	1.9	5.6	0.7	37.3
2009	21.1	0.9	10.5	3.0	6.1	0.7	42.4
2010	21.8	2.2	7.8	3.6	5.7	0.7	41.7
2011	21.8	4.4	4.7	3.6	4.7	0.7	39.8
2012	21.1	5.3	5.8	3.0	3.8	0.6	39.6
2013	19.5	5.2	5.7	0.3	1.8	0.5	33.0
2014	19.4	5.3	5.5	0.5	1.1	0.5	32.2

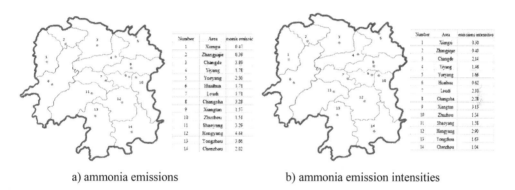

a) ammonia emissions   b) ammonia emission intensities

Figure 1. Spatial distribution of ammonia emissions from the livestock farming industry in 2014 in Hunan/10^4t.

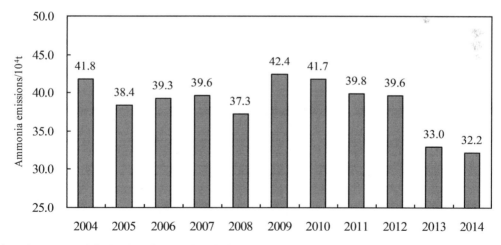

Figure 2. Temporal distribution of ammonia emissions from the livestock farming industry in 2014 in Hunan.

emissions in 2014 is illustrated in Figure 2. It could be seen that the total annual ammonia emission fluctuated and showed a downward trend since 2001. However, there was a sharp maximum in 2009 and the total emission reached 42.4×104t·a^{-1} in the Hunan province. The primary cause was the significantly increased pig-breeding quantity.

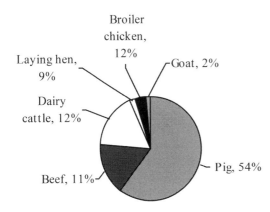

Figure 3. Ammonia emissions of different livestock in 2014 in Hunan.

3.4 Distribution of livestock and poultry ammonia emissions in Hunan province

The ammonia emissions of different livestock in 2014 in Hunan are showed in Figure 3. In proportion to 54.82%, the pig-breeding was the largest ammonia emitter. Although its emission factor is not high, nearly 0.22 million slaughters and huge pig-producing quantities led to the strong emissions. Differing from the situation in Pearl River Delta of China and Beijing, the pig-breeding was further manifested as the main ammonia emitter in Hunan province.

4 CONCLUSIONS

Ammonia emissions from livestock and poultry farming in Hunan province ranged from 32.2 to $42.4 \times 10^4 t\cdot a^{-1}$ when foreign emission factor is applied in the calculation. In 2014, the strongest emitter was Hengyang city, with an emission of $5.50 \times 10^4 t\cdot a^{-1}$. In contrast, Zhangjiajie is the least polluter, with an emission of $0.48 \times 10^4 t\cdot a^{-1}$. Moreover, the total annual emission fluctuated and showed a downward trend since 2001 due to the expanded area of banning and retirement. Furthermore, the average emission intensity in Hunan province was 1.88 t·km^{-2}, among which the biggest emissions was Hengyang, with a value of 5.50×10^4 tons and the lowest emission was 0.37 tons per square kilometers in western Hunan. The results also show that pig-breeding contributed most (54.82%) to ammonia emission.

REFERENCES

Asman Willem AH. Ammonia emission in Europe: Updated Emission and Emission Variations. Bilthoven, the Netherlands: National Institute of Public Health and Environmental Protection, 1992.

Barnard W R. Development of a National Emission Inventory to Support Revision of the Particulate National Ambient Air Quality Standard. Toronto, Canada: 1997.

Battye William, Aneja Viney P, Roelle Paul A. Evaluation and improvement of ammonia emissions inventories. *Atmospheric Environment*, 2003, 37(27): 3873–3883.

Behera S N, Sharma M. Investigating the potential role of ammonia in ion chemistry of fine particulate matter formation for an urban environment Science of the Total Environment, 2010, 408:3569–3575.

Bouwman A F, Lee D S, Asman WAH, et al. A global high-resolution emission inventory for ammonia. *Global Biogeochemical Cycles*, 1997, 11(4): 561–587.

Bouwman A F, Van Der Hoek K W. Scenarios of animal waste production and fertilizer use and associated ammonia emission for the developing countries. *Atmospheric Environment*, 1997, 31(24): 4095–4102.

Chitjian Mark, Mansell Gerard. An Improved Ammonia Inventory for the WRAP Domain[R]. WRAP missions Forum, 2003.

Dong wenxuan, Xing jia, Wang shuxiao. Temporal and Spatial Distribution of Anthropogenic Ammonia China:1994–2006. *Environment Science*, 2010, 31(7): 1457–1463.

Ecetoc J. Ammonia Emissions to Air in Western Europe. European Centre for Ecotoxicology and Toxicology of Chemicals-ECETOC, 1994.

Fu Joshua, Kim Yunhee, Davis Wayne, et al. Quality Improvement for Ammonia Emission Inventory. Washington: United States Environmental Protection Agency, 2005.

Kang C M, Lee H S, Kang B W, et al. Chemical characteristics of acidic gas pollutants and PM2.5 species during hazy episodes in Seoul, South Korea, Atmospheric Environment, 2004, 38: 4749–4760.

Klimont Zbigniew. Current and Future Emissions of Ammonia in China. Laxenburg, Austria: International Institute for Applied Systems Analysis (IIASA), 2001.

Lin Y C, Cheng M T. Evaluation of formation rates of NO_2 to gaseous and particulate nitrate in the urban atmosphere. Atmospheric Environment, 2007, 41:1903–1910.

McInnes Gordon. Atmospheric Emission Inventory Guidebook: Volume 1. Copenhagen, Denmark: European Environment Agency, 1996.

McNaughton Daniel J, Vet Robert J. Eulerian Model Evaluation Field Study (EMEFS): a summary of surface network measurements and data quality. Atmospheric Environment, 1996, 30(2): 227–238.

Misselbrook T H, Van Der Weerden T J, Pain B F, et al. Ammonia emission factors for UK agriculture. *Atmospheric Environment*, 2000, 34(6):871–880.

Van Der Hoek K W. Estimating ammonia emission factors in Europe: summary of the work of the UNECE ammonia expert panel. *Atmospheric Environment*, 1998, 32(3): 315–316.

Pinder Robert W, Strader Ross, Davidson Cliff, et al. Ammonia Emissions from Dairy Farms: Development of a Farm Model and Estimation of Emissions from the United States[R]. Washington: United States Environmental Protection Agency, 2003.

Wang wenxing, Lu xiaofeng, Pang yanbo, et al. Geographical distribution of NH_3 Emission Intensi-time in China. *Acta Scientiaecirum Stantiae*, 1997, 17(1): 3–8.

Wei Yu-xiang, Yang Wei-fen, Yin Yan, et al. Pollution Characteristics of Nanjing Water-soluble Ions in Air Fine Particles under Haze Days. Environmental Science and Technology, 2009, 32(11): 66–71.

Yang zhipeng. Estimation of Ammonia Emission from Livestock in china based on Mass-flow Method and Regional Comparison. Beijing: Beijing University, 2008.

Yang F, Tan J, Zhao Q, et al. Characteristics of $PM_{2.5}$ speciation in representative megacities and across China. Atmospheric Chemistry and Physics, 2011:5207–5219.

Zhang Q, Jimenez J L, Canagaratna M R, et al. Ubiquity and dominance of oxygenated species in organic aerosols in anthropogenically-influenced Northern Hemisphere midlatitudes. Geophysical Research Letters. VOL.34,L13801,doi:10.1029/2007GL029979, 2007.

Civil, Architecture and Environmental Engineering – Kao & Sung (Eds)
© 2017 Taylor & Francis Group, ISBN 978-1-138-02985-9

Application of aerobic bioremediation to cleanup octachlorinated dibenzofuran polluted soils

J.L. Lin, Y.T. Sheu & C.M. Kao
Institute of Environmental Engineering, National Sun Yat-Sen University, Kaohsiung, Taiwan

W.P. Sung & T.Y. Chen
Department of Landscape Architecture, National Chin-Yi University of Technology, Taichung, Taiwan

ABSTRACT: In this study, effectiveness of using *Pseudomonas mendocina* NSYSU (*P. mendocina* NSYSU) on the bioremediation of octachlorinated dibenzofuran (OCDF)-polluted soils was evaluated through batch and bioreactor experiments under aerobic conditions. The goal of the research were to assess the feasibility of biodegradation of OCDF by indigenous soil bacteria and isolated bacterial strain (*P. mendocina* NSYSU) from OCDF-polluted soils, and Results show that *P. mendocina* NSYSU was able to degrade OCDF through the aerobic cometabolic mechanisms with the addition of carbon substrates. Up to 62% of OCDF was removed after a 50-day operation with carbon substrate supplement. Results indicate that primary substrate supplement is required for the enhancement of aerobic biodegradation of OCDF, and OCDF could not been used as the sole carbon source for the growth of *P. mendocina* NSYSU. Results reveal that an aerobic bioremediation system using *P. mendocina* NSYSU as the inocula would be a cost-effective system to remediate furan-polluted soils.

1 INTRODUCTION

Polychlorinated dibenzofuran (PCDF) isomers, which are usually produced thermal processes, have been classified as the mutagens and carcinogens (Coutinho et al. 2015; Squadrone et al. 2015). Incinerators and boilers have been considered as the major causes of PCDFs production after waste burnings (Wittsiepe et al. 2015; Pongpiachan et al. 2016). Production of PCDFs cause the ecosystem and environmental media contamination (Klees et al. 2015; Kruse et al. 2014; Shin et al. 2016). PCDFs have strong environmental persistent characteristics and they are also subjected to biomagnification as well as bioaccumulation effects in many living organisms (Hanano et al. 2014; Wu et al. 2014; Yang et al. 2015). Different furan isomers including octachlorinated dibenzofuran (OCDF, a highly chlorinated furan), have been observed in different environmental media (e.g., sediments, soils) in many industrialized areas (Govindan and Moon 2015; Urban et al. 2014).

As a result of their hydrophobic and xenobiotic nature, OCDFs usually are very persistent in ecosystems (Liu et al. 2014; Zhao et al. 2015). Due to their highly adsorptive, less biodegradable, and highly toxic natures, remediation of PCDF-polluted media (e.g., soils, sediments) can be a necessity but costly (Anasonye et al. 2014; Zhao et al. 2015). Compared to physical and chemical remedial methods, biological method can reduce the cleanup cost for the PCDF-polluted sites if significant amounts of media need to be remediated (Tue et al. 2016; Vallejo et al. 2015). The microbial species involved in bioremediation technologies include aerobic and anaerobic processes depending on the nature of contaminant and microorganisms (Megharaj et al. 2014). Compared to anaerobic process, aerobic bioremediation of PCDF-polluted media can be more efficient, and thus, remediation time can be reduced (Chen et al. 2016; Hanano et al. 2014).

One promising remedial method for the remediation of chlorinated compounds polluted soils is aerobic bioremediation by aerobic or facultative bacterial species (Futagami et al. 2008; Lai and Becker 2013). Some aerobic microorganisms have specific metabolic processes, which enable them to biodegrade less-biodegradable and persistent chlorinated compounds.

Supplement of carbon sources for the use of primary substrates is necessary to enhance the aerobic cometabolic mechanisms (Kruse et al. 2014; Liu et al. 2014; Zhen et al. 2014). Currently, the information related to the biodegradation of higher chlorinated PCDFs under aerobic conditions is limited (Kuokka et al. 2014; Liu et al. 2014; Tu et al. 2014). A pentachlorophenol (PCP)-biodegrading bacterial strain, *Pseudomonas mendocina* NSYSU (*P. mendocina* NSYSU), was isolated from PCP, dioxin and furan-contaminated soils (Kao et al. 2005). The site was also polluted by OCDF with concentrations up to 10.8 mg/kg (NSC 2012). In this study, a biodegradation study was performed to assess if *P. mendocina* NSYSU could bioremediate OCDF-polluted soils in an aerobic system. The major tasks of this study were as follows: (1) evaluation of the feasibility of improving OCDF biodegradation by *P. mendocina* NSYSU under aerobic conditions, and (2) assessment of the existence of functional genes for the anaerobic OCDF biodegradation.

2 MATERIALS AND METHODS

2.1 Medium and growth of P. mendocina NSYSU

P. mendocina NSYSU culture was incubated in nutrient medium. The components of the media included the following: The mineral medium contained the following components (mg/L): $Mg_2SO_4.7H_2O$, 98.6; H_2PO_4, 326.4; NH_4Cl, 10.7; $CaCl_2.2H_2O$, 44.1; Na_2HPO_4, 1263.8; and 3.35 mg of trace elements ($CuCl_2.2H_2O$, 0.25; $Na_2B_4O_7.10H_2O$, 0.25; $MnSO_4.4H_2O$, 1; $FeSO_4.7H_2O$, 1; $CoCl_2.6H_2O$, 0.25; $ZnCl_2$, 0.25; NH_4VO_3, 0.1; $(NH_4)_6Mo_7O_{24}.4H_2O$, 0.25). The pH of this buffer solution was 7.5. The medium solution was autoclaved before use. The *P. mendocina* NSYSU solution was cultured at 200 rpm for 48 h in a 50 mL flask (sealed with butyl rubber stopper) at 20°C under aerobic conditions. Density of the bacteria was analyzed by the spectrophotometer (Hach Co., USA).

2.2 OCDF biodegradation experiment

The biodegradation of OCDF under aerobic conditions was investigated in the batch experiment. In this study, *P. mendocina* NSYSU and soils collected from the OCDF-contaminated site (OCDF concentration = 10.8 mg/kg) were used as the inocula. Glucose, which was used as the primary substrate, was supplied in the microcosms. Approximately 500 mg of glucose was added in the batch bottle to serve as the primary substrate for microorganisms and *P. mendocina* NSYSU. Each batch bottle contained 5 mL of glucose solution (100 g/L) (or 5 mL of mineral medium solution), 15 g of site soils, 5 mL of *P. mendocina* NSYSU solution as inocula (or 5 mL of mineral solution), and 35 mL of mineral medium (autoclaved before use) in a 70-mL serum bottle, which was sealed with Teflon-lined rubber septa.

Table 1 lists the constituents of different microcosms. Batch A was kill control batch containing 500 mg/L NaN_3 and 250 mg/L $HgCl_2$, and the soils were autoclaved before use. Batch B was live control batch containing OCDF-contaminated soils and *P. mendocina* NSYSU, but no glucose addition. Batch C was also live control batch containing OCDF-contaminated soils and glucose, but no *P. mendocina* NSYSU addition. Dead (Batch A) and live controls (Batches B and C) were prepared to assess the effects of substrate (glucose) and inocula addition on OCDF removal.

Batch D bottles contained sterilized soils, glucose, and *P. mendocina* NSYSU, and Batch E bottles contained unsterilized soils, glucose, and *P. mendocina* NSYSU. *P. mendocina* NSYSU was incubated aerobically in nutrient and glucose medium solution and the microcosms were operated at room temperature (20°C). Duplicate samples were analyzed for OCDF concentrations for each sampling event. The degradation efficiency of OCDF was calculated as a percentage of the concentration on day 0. The procedures for *P. mendocina* NSYSU incubation, soil extraction procedures, and OCDF analytical methods were described in Tu et al. (2014).

Table 1. Components of five batches of biodegradation experiments.

Microcosm	Inocula	Components
A (Kill control)	Sterilized soils	Sterilized OCDF-contaminated soils + nutrient medium solution + glucose + 250 mg/L $HgCl_2$ + 500 mg/L NaN_3
B (Control-no glucose)	Soils + *P. mendocina* NSYSU	OCDF-contaminated soils + nutrient medium solution + *P. mendocina* NSYSU
C (Control-no strain)	Soils	OCDF-contaminated soils + nutrient medium solution + glucose
D	Sterilized soils + *P. mendocina* NSYSU	Sterilized OCDF-contaminated soils + nutrient medium solution + *P. mendocina* NSYSU + glucose
E	Soils + *P. mendocina* NSYSU	OCDF-contaminated soils + nutrient medium solution + *P. mendocina* NSYSU + glucose

3 RESULTS AND DISCUSSION

Figure 1 presents the remained OCDF in A to E batches during the 50 days of operation. Slight OCDF removal [approximately 3% (Batch A) to 8% (Batch C) removal] was observed in kill-control batch (Batch A), control-no glucose batch (Batch B), and control-no NSYSU strain (Batch C). Results demonstrate that soil bacteria could not biodegrade OCDF effectively under aerobic conditions. This might be due to the fact that the furan-degrading bacteria were not the dominant bacteria in site soils. Results also reveal that supplement of primary substrate is a necessity to enhance the anaerobic dechlorination. Results demonstrate that OCDF could not be used as the carbon source by *P. mendocina* NSYSU or soil bacteria. Slight decrease in OCDF concentration in Batch B (control-no glucose) bottles was because natural organic carbon was consumed by *P. mendocina* NSYSU for primary substrate.

In Batch D bottles (sterilized soils with *P. mendocina* NSYSU addition), significant drop of OCDF was observed and about 62% of OCDF was degraded after 50 days. Results demonstrate that efficient OCDF biodegradation could be obtained aerobically by *P. mendocina* NSYSU with the supplement of glucose. The glucose could be used as the carbon sources by *P. mendocina* NSYSU. Thus, OCDF could be degraded through aerobic cometabolic mechanisms using glucose as the carbon sources. In Batch E bottles (non-sterilized soils with *P. mendocina* NSYSU addition), relatively lower efficiency OCDF degradation (27%) was detected. Results might be due to the fact that indigenous bacteria competed the supplied carbon sources with *P. mendocina* NSYSU resulting in the decreased efficiency of OCDF removal.

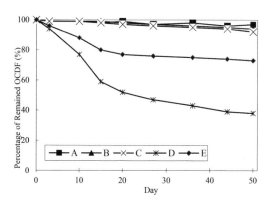

Figure 1. The remained OCDF in microcosms during the 50-day incubation period.

4 CONCLUSIONS

This study was conducted to evaluate the capability of *P. mendocina* NSYSU on the aerobic biodegradation of OCDF. Under aerobic cometabolic conditions, *P. mendocina* NSYSU had the capability to biodegrade OCDF. Glucose could be used as the primary substrate resulting in increased aerobic biodegradation of OCDF. The removal efficiencies for OCDF reached 65% in microcosms with the supplement of glucose under anaerobic conditions, respectively. Results demonstrate that OCDF could not be used as the carbon source for *P. mendocina* NSYSU and indigenous soil bacteria under aerobic conditions. Therefore, addition of an appropriate substrate was required to enhance the OCDF biodegradation. Results reveal that an on-site bioreactor or in situ bioremediation using *P. mendocina* NSYSU as the inocula would be a cost-effective and acceptable remedial system to remediate furan-polluted soils.

REFERENCES

Anasonye, F., E. Winquist, T.B. Kluczek, M. Räsänen, K. Salonen, K.T. Steffen, & M. Tuomela. (2014). Fungal enzyme production and biodegradation of polychlorinated dibenzo-*p*-dioxins and dibenzofurans in contaminated sawmill soil. *Chemosphere 110*, 85–90.

Chen, W.Y., J.H. Wu, S.C. Lin, & J.E. Chang. (2016). Bioremediation of polychlorinated-*p*-dioxins/dibenzofurans contaminated soil using simulated compost-amended landfill reactors under hypoxic conditions. *J. Hazard. Mater. 312*, 159–168.

Coutinho, M., M. Albuquerque, A. P. Silva, J. Rodrigues, & C. Borrego. (2015). Long-time monitoring of polychlorinated dibenzo-*p*-dioxins and dibenzofurans over a decade in the ambient air of Porto, Portugal. *Chemosphere 137*, 207–213.

Futagami, T., M. Goto, & K. Furukawa. (2008). Biochemical and genetic bases of dehalorespiration. *Chem. Rec. 8 (1)*, 1–12.

Govindan, M., & I.S. Moon. (2015). Expeditious removal of PCDD/Fs from industrial waste incinerator fly ash using electrogenerated homogeneous Ag(II) ions. *Chem. Eng. J. 272*, 145–150.

Hanano, A., I. Almousally, & M. Shaban. (2014). Phytotoxicity effects and biological responses of Arabidopsis thaliana to 2,3,7,8-tetrachlorinated dibenzo-*p*-dioxin exposure. *Chemosphere 104*, 76–84.

Kao, C.M., J.K. Liu, Y.L. Chen, C.T. Chai, & S.C. Chen. (2005). Factors affecting the biodegradation of PCP by *Pseudomonas mendocina* NSYSU. *J. Hazard. Mater. 124*, 68–73.

Klees, M., P. Hiester, P. Bruckmann, K. Molt, & T.C. Schmidt. (2015). Polychlorinated biphenyls, polychlorinated dibenzo-*p*-dioxins and dibenzofurans in street dust of North Rhine-Westphalia, Germany. *Sci. Total Environ. 511*, 72–81.

Kruse, N.A., J. Bowman, D. Lopez, E. Migliore, & G.P. Jackson. (2014). Characterization and fate of polychlorinated biphenyls, polychlorinated dibenzo-*p*-dioxins and polychlorinated dibenzofurans in soils and sediments at the Portsmouth Gaseous Diffusion Plant, Ohio. *Chemosphere 114*, 93–100.

Kuokka, S., A.L. Rantalainen, M. Romantschuk, & M.M. Häggblom. (2014). Effect of temperature on the reductive dechlorination of 1,2,3,4-tetrachlorodibenzofuran in anaerobic PCDD/F-contaminated sediments. *J. Hazard. Mater. 274*, 72–78.

Lai, Y.J., & J.G. Becker. (2013). Compounded Effects of Chlorinated Ethene Inhibition on Ecological Interactions and Population Abundance in a *Dehalococcoides-Dehalobacter* Coculture. *Environ. Sci. Technol. 47 (3)*, 1518–1525.

Liu, H., J.W. Park, & M.M. Häggblom. (2014). Enriching for microbial reductive dechlorination of polychlorinated dibenzo-*p*-dioxins and dibenzofurans. *Environ. Pollut. 184*, 222–230.

Megharaj, M., K. Venkateswarlu, & R. Naidu. (2014). Bioremediation A2-Wexler, Philip. In *Encyclopedia of Toxicology (Third Edition). Reference Module in Biomedical Sciences 1*, 485–489.

NSC. (2012). Development of Treatment Technologies to Remediate Toxic Chemical Contaminated Sites. *National Science Council, Taipei, Taiwan Report No. 101–2622-E-006–001-C-C1.*

Pongpiachan, S., T. Wiriwutikorn, C. Rungruang, K. Yodden, N. Duangdee, A. Sbrilli, M. Gobbi, & C. Centeno. (2016). Impacts of micro-emulsion system on polychlorinated dibenzo-*p*-dioxins (PCDDs) and polychlorinated dibenzofurans (PCDFs) reduction from industrial boilers. *Fuel 172*, 58–64.

Shin, E.S., J.C. Kim, S.D. Choi, Y.W. Kang, & Y.S. Chang. (2016). Estimated dietary intake and risk assessment of polychlorinated dibenzo-*p*-dioxins and dibenzofurans and dioxin-like polychlorinated biphenyls from fish consumption in the Korean general population. *Chemosphere 146*, 419–425.

Squadrone, S., P. Brizio, R. Nespoli, C. Stella, & M.C. Abete. (2015). Human dietary exposure and levels of polychlorinated dibenzo-*p*-dioxins (PCDDs), polychlorinated dibenzofurans (PCDFs), Dioxin-Like Polychlorinated Biphenyls (DL-PCBs) and Non-Dioxin-Like Polychlorinated Biphenyls (NDL-PCBs) in free-range eggs close to a secondary aluminum smelter, Northern Italy. *Environ. Pollut. 206*, 429–436.

Tu, Y.T., J.K. Liu, W.C. Lin, J.L. Lin, & C.M. Kao. (2014). Enhanced anaerobic biodegradation of OCDD-contaminated soils by *Pseudomonas mendocina* NSYSU: Microcosm, pilot-scale, and gene studies. *J. Hazard. Mater.* 278, 433–443.

Tue, N.M., A.Goto, S. Takahashi, T. Itai, K.A. Asante, T. Kunisue, & S. Tanabe. (2016). Release of chlorinated, brominated and mixed halogenated dioxin-related compounds to soils from open burning of e-waste in Agbogbloshie (Accra, Ghana). *J. Hazard. Mater. 302*, 151–157.

Urban, J.D., D., Wikoff, A.T.G. Bunch, M.A. Harris, & L.C. Haws. (2014). A review of background dioxin concentrations in urban/suburban and rural soils across the United States: Implications for site assessments and the establishment of soil cleanup levels. *Sci. Total Environ. 466–467*, 586–597.

Vallejo, M., M. Fresnedo San Román, I. Ortiz, & A. Irabien. (2015). Overview of the PCDD/Fs degradation potential and formation risk in the application of advanced oxidation processes (AOPs) to wastewater treatment. *Chemosphere 118*, 44–56.

Wittsiepe, J., J.N. Fobil, H. Till, G.D. Burchard, M. Wilhelm, & T. Feldt. (2015). Levels of polychlorinated dibenzo-*p*-dioxins, dibenzofurans (PCDD/Fs) and biphenyls (PCBs) in blood of informal e-waste recycling workers from Agbogbloshie, Ghana, and controls. *Environ. Int. 79*, 65–73.

Wu, T.W., J.W. Lee, H.Y. Liu, et al. (2014). Accumulation & elimination of polychlorinated dibenzo-*p*-dioxins and dibenzofurans in mule ducks. *Sci. Total Environ. 497–498*, 260–266.

Yang, C.Y., S. L. Chiou, J.D. Wang, & Y. L. Guo. (2015). Health related quality of life and polychlorinated biphenyls and dibenzofurans exposure: 30 years follow-up of Yucheng cohort. *Environ. Res. 137*, 59–64.

Zhao, L., H. Hou, T. Zhu, F. Li, A. Terada, & M. Hosomi. (2015). Successive self-propagating sintering process using carbonaceous materials: A novel low-cost remediation approach for dioxin-contaminated solids. *J. Hazard. Mater. 299*, 231–240.

Zhen, H., S. Du, L.A. Rodenburg, G. Mainelis, & D.E. Fennell. (2014). Reductive dechlorination of 1,2,3,7,8-pentachlorodibenzo-*p*-dioxin and Aroclor 1260, 1254 and 1242 by a mixed culture containing *Dehalococcoides mccartyi* strain 195. *Water Res. 52*, 51–62.

Civil, Architecture and Environmental Engineering – Kao & Sung (Eds)
© 2017 Taylor & Francis Group, ISBN 978-1-138-02985-9

Experimental study on water-soil interaction influence for environmental change of marine soft soil

Han-min Liu, Dong Zhou, Heng Wu, Wen-can Jiao & Ye-tian Wang
College of Civil Engineering and Architecture, Guangxi University, Nanning, China
Key Laboratory of Disaster Prevention and Structural Safety of the Ministry of Education, Guangxi University, Nanning, China

ABSTRACT: In this paper, the research object is marine soft soil in Linhai Park of Qinzhou Port and Qisha Park of Fangchenggang Port. The marine soft soil environment factors, additional load stress of coastal reclamation layer and calcium ions of pore water chemical composition, were simulated by using a soil soaking-load linkage. The soil soaking-load linkage is mechanism simulation device. Impact on water-soil interactions of marine soft soil was analyzed for additional load stress of coastal reclamation layer and calcium ions of pore water chemical composition. The conclusion is as follows. Firstly, from the point of geotechnical engineering, there are mainly hydraulic connection, water chemical field and additional load stress of coastal reclamation layer for the soil and water environmental change of marine soft soil in coastal reclamation district. Secondly, the soil and water environmental change of marine soft soil in coastal reclamation district is simulated by using a soil soaking-load linkage. Thirdly, under the same circumstances of the calcium ions concentration of soak solution, the consolidation deformation of artificial soil samples come up under the action of additional load stress, consequently, artificial soil samples' porosity decreases. Artificial soil samples' porosity and osmotic coefficient decreases along with the increase of additional load stress of coastal reclamation layer. Artificial soil samples' calcium ions content decreases along with the increase of additional load stress of coastal reclamation layer. Fourthly, under the same circumstances of the additional load stress of coastal reclamation layer, artificial soil samples' calcium ions content increases along with the increase of calcium ions concentration of soak solution. Fifthly, artificial soil samples' porosity and osmotic coefficient decrease slightly along with the increase of the calcium ions concentration of soak solution under the same circumstances of the additional load stress of coastal reclamation layer with the exception of 180 kPa in Linhai Park of Qinzhou Port.

1 INTRODUCTION

Since the 21st century, large-scale coastal reclamation engineering was created in our country (Ge et al. 2014, Sun et al. 2012, Dong et al. 2014). Much of the coastal reclamation engineering is industry and urban construction land such as Linhai Park of Qinzhou Port and Qisha Park of Fangchenggang Port. There is a mass of soft soil in coastal reclamation district (Liu et al. 2014, Yi et al. 2015, Ou et al. 2015). The kind of project such as the coastal reclamation engineering has the following characteristics. First, the area of coastal reclamation land is large, according to the overall planning of Qinzhou City during 2008 to 2025, the total coastal reclamation area will reach about 79 km². Second, the schedule of coastal reclamation land is fast, the time is shorter that the coastal reclamation land use for industrial and urban construction land from construction to use. The coastal reclamation engineering of Linhai Park of Qinzhou Port and Qisha Park of Fangchenggang Port completed by hydraulic filled sand, and applied directly to the engineering facilities construction after foundation treatment was fulfilled (Liu et al. 2012, Hu 2009). Third, the load of coastal reclamation layer to underlying layer strata is great. Coastal reclamation district extends gradually from tidal flats to shallow sea, along with the construction technology of Coastal reclamation become mature and demand for land use area increase. Thickness of coastal reclamation layer increase sharply, therefore the load of coastal reclamation layer to underlying layer strata also increase sharply. The depth of railway feeder hydraulic filled sand project from Dalanping to Qinzhou Free Trade Area is as deep as decade meters. Disturbance of environment of marine soft soil is extremely severe because of coastal reclamation compare with the formation of the natural evolution process of marine soft soil. Marine soft soil and its formation of the natural geological environment are a relatively stable state of dynamic balance before coastal reclamation. The state of marine soft soil deposition would change rapidly after a violent disturbance of coastal reclamation, and its environment would achieve a new dynamic equilibrium.

2 BASIC PHYSICAL PROPERTIES OF SOIL SAMPLES IN COASTAL RECLAMATION DISTRICT

Test results of basic physical properties of soil samples in coastal reclamation district are shown in Table 1.

Particle size distribution of marine soft soil was measured by sieving method and densimeter method in Linhai Park of Qinzhou Port and Qisha Park of Fangchenggang Port. Test samples of marine soft soil mixed sodium hexametaphosphate that its concentration is 4%. Test analysis results of particle size distribution are as follows.

Content of fine grained soil which the particle size is less than 0.075 mm reach 80.55%, content of fine grained soil which the particle size is less than 0.005 mm reach 43.69% in Linhai Park of Qinzhou Port.

Content of fine grained soil which the particle size is less than 0.075 mm reach 82.2%, content of fine grained soil which the particle size is less than 0.005 mm reach 33.3% in Qisha Park of Fangchenggang Port.

3 ANALYSIS OF THE SOIL AND WATER ENVIRONMENTAL CHANGE OF MARINE SOFT SOIL FOR COASTAL RECLAMATION

That coastal reclamation impact on environment of marine soft soil is multifaceted. From the point of geotechnical engineering, in order to coastal reclamation engineering of Linhai Park of Qinzhou Port and Qisha Park of Fangchenggang Port as the research object, water-soil environmental change of marine soft soil was analyzed as follows.

First, environmental change of the hydraulic connection was analyzed. Hydraulic connection environment of marine soft soil will be significant changes before and after coastal reclamation engineering. Marine soft soil contact with seawater

directly before coastal reclamation engineering, hydraulic connection is strong, marine soft soil was separated by coastal reclamation layer of hydraulic filled sand, hydraulic connection is weaken. Marine soft soil contact with seawater by means of pore water seepage of coastal reclamation layer of hydraulic filled sand. The surface of coastal reclamation layer of hydraulic filled sand is ground surface which is exposed to air layer, the bottom contact with marine soft soil, one side of coastal reclamation layer of hydraulic filled sand contact with seawater, the other side contact with land directly. The nature of the changes of the original boundary conditions have taken place. Pore water source of coastal reclamation layer will be controlled by its boundary conditions.

Second, environmental change of water chemical field was analyzed. With marine soft soil environmental change of the hydraulic connection, discharge and supply conditions of marine soft soil's pore water will change, water chemical field and its chemical components of marine soft soil's pore water will change before and after coastal reclamation engineering. Material sources of coastal reclamation layer's pore water chemical composition divided into three parts. Section one of pore water chemical composition is derived from seawater lied in the side of the sea, section two of pore water chemical composition is derived from the groundwater seepage lied in the side of the land, section three of pore water chemical composition is derived from the surface seepage. Soil-water interactions of marine soft soil will be new changes, because the material composition of water field change.

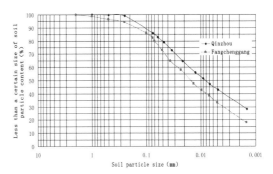

Figure 1. The cumulative curve of grain size in Qinzhou and Fangchenggang.

Table 1. The physical properties of marine soft soil.

	Natural moisture content (%)	The proportion of soil particles	Plastic limit (%)	Liquid limit (%)	Plasticity index	Liquidity index
Qinzhou	52.57	2.70	20	42.5	22.5	1.45
Fangchenggang	65.07	2.69	25.3	49.2	23.9	1.67

Third, additional load stress of marine soft soil layer was analyzed due to coastal reclamation layer. Coastal reclamation layer consists of hydraulic filled sand. Coastal reclamation layer of hydraulic filled sand exert new additional load stress to marine soft soil layer. Additional load stress will produce corresponding geotechnical engineering geology effect to marine soft soil.

4 MECHANISM ANALYSIS OF SOIL-WATER INTERACTIONS

Water—soil—electrolyte system consists of marine soft soil, pore water and pore water chemistry composition. Ion composition exchange will happen between pore water of soft soil and soak liquid in the process of soaking, pore water ionic concentration and its components will change. The original dynamic equilibrium of Soil-water interactions is broken between soft soil clay mineral particles and pore water, until a new dynamic equilibrium is reached.

First, ion exchange adsorption occur between soft soil clay mineral particle and pore water solution when the pore water ionic concentration and its components change. The surface state of soft soil clay mineral particles will change because of ion exchange adsorption, and mechanical properties and engineering properties of soft soil soil will change. The characteristics of ion exchange adsorption is that dissociation ions of surface of soft soil clay mineral particles substitute for disappearing ions in the pore water solution. Ion exchange adsorption is the result of the interaction between clay mineral particles and pore water solution, the strength of the ion exchange adsorption depends on two aspects that are clay mineral particles and pore water solution.

Second, settling—dissolution chemical action of ion chemical compositions of pore water solution come up when the ion composition concentration of the pore water solution change. Calcium ions concentration was increased in in the process of the test, calcium ions precipitated, the generated solid cement filled in the pores of the soil particles, so that the pores of the soil particles would decrease.

5 EXPERIMENTAL STUDY ON WATER-SOIL INTERACTION INFLUENCE FOR ENVIRONMENTAL CHANGE OF MARINE SOFT SOIL

Water-soil environment of marine soft soil was simulated by using a soil soaking-load linkage (Wu et al. 2013) when calcium ions concentration of

soft soil pore water changes. Water-soil interaction taken place between soak liquid calcium ions and soft soil. And then meso-structures of soft soil changed, consequently, physical and mechanical properties of soft soil transformed.

Environmental change of marine soft soil was simulated under the action of additional load stress of coastal reclamation layer and soak liquid of calcium ions in Linhai Park of Qinzhou Port. Combination plans of working condition are a total of nine combinations are shown in Table 2.

Environmental change of marine soft soil was simulated under the action of additional load stress of coastal reclamation layer and soak liquid of calcium ions in Qisha Park of Fangchenggang Port. Combination plans of working condition are a total of nine combinations are shown in Table 3.

5.1 Calcium ions of soft soil

Ion components of soft soil are measured by using the instruments that is model number 5300DV

Table 2. The working condition simulation combination plans of marine soft soil in Qinzhou.

Soak solution	Additional load stress of coastal reclamation layer (kPa)		
	80 kPa	130 kPa	180 kPa
Seawater in Qinzhou (calcium ions concentration is 0.025%)	No. 1	No. 2	No. 3
calcium ions concentration is 0.25%	No. 4	No. 5	No. 6
calcium ions concentration is 2.5%	No. 7	No. 8	No. 9

Table 3. The working condition simulation combination plans of marine soft soil in Fangchenggang.

Soak solution	Additional load stress of coastal reclamation layer (kPa)		
	80 kPa	130 kPa	180 kPa
Seawater in Fangchenggang (calcium ions concentration is 0.044%)	No. 10	No. 11	No. 12
calcium ions concentration is 0.44%	No. 13	No. 14	No. 15
calcium ions concentration is 4.4%	No. 16	No. 17	No. 18

Table 4. Calcium ions content in Linhai Park of Qinzhou Port (mg/kg).

Soak solution	Additional load stress of coastal reclamation layer (kPa)			Before soaked
	80 kPa	130 kPa	180 kPa	
Seawater in Qinzhou (calcium ions concentration is 0.025%)	299.6	271.1	256.0	328.8
calcium ions concentration is 0.25%	440.9	418.3	385.9	
calcium ions concentration is 2.5%	1927	1758	1589	

Table 5. Calcium ions content in Qisha Park of Fangchenggang Port (mg/kg).

Soak solution	Additional load stress of coastal reclamation layer (kPa)			Before soaked
	80 kPa	130 kPa	180 kPa	
Seawater in Fangchenggang (calcium ions concentration is 0.044%)	145.0	129.2	117.8	146.1
calcium ions concentration is 0.44%	241.3	207.9	190.4	
calcium ions concentration is 4.4%	1615.1	1494.3	1403.0	

Inductively Coupled Plasma Atomic Emission Spectrometer made in Optima Company.

Calcium ions content of disturbed soft soil samples is initial value before disturbed soft soil samples were soaked in Linhai Park of Qinzhou Port and Qisha Park of Fangchenggang Port. Calcium ions content of the center of the artificial soil samples were measured after the completion of the soft soil samples soaked. Test method is that extractive content of soft soil samples' calcium ions in deionized water is evaluation standard.

The determination results of soft soil samples' calcium ions content are shown in Table 4 and Table 5. The unit, mg/kg, shows calcium ions content in per kilogram soil samples.

Change laws of calcium ions content are shown in Figure 2 and Figure 3 along with the additional load stress of coastal reclamation layer.

Figure 2 and Figure 3 show that calcium ions content of artificial soil samples decreases along with the increase of the additional load stress of coastal reclamation layer under the same circumstances of calcium ions concentration of soak solution. With the combination of Figure 6, Figure 7, Figure 10 and Figure 11, we can infer that calcium ions inlet will decrease from soak solution to artificial soil samples along with the decrease of porosity and osmotic coefficient in the process of soaking.

Change laws of calcium ions content are shown in Figure 4 and Figure 5 along with the calcium ions concentration of soak solution.

Figure 4 and Figure 5 show that calcium ions content of artificial soil samples increases along with the increase of the calcium ions concentration of soak solution under the same circumstances of the additional load stress of coastal reclamation layer. The reason is that calcium ions inlet will increase from soak solution to artificial soil samples along with the increase of the calcium ions concentration of soak solution.

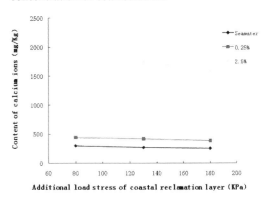

Figure 2. The change chart of calcium ions content along with the additional load stress of coastal reclamation layer in Linhai Park of Qinzhou Port.

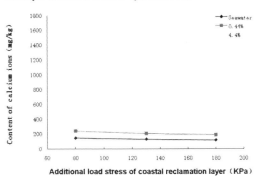

Figure 3. The change chart of calcium ions content along with the additional load stress of coastal reclamation layer in Qisha Park of Fangchenggang Port.

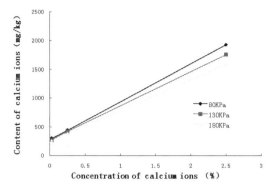

Figure 4. The change chart of calcium ions content along with the calcium ions concentration of soak solution in Linhai Park of Qinzhou Port.

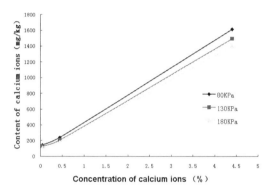

Figure 5. The change chart of calcium ions content along with the calcium ions concentration of soak solution in Qisha Park of Fangchenggang Port.

5.2 Porosity

Porosity of artificial soil samples are shown in Table 6 and Table 7 under different conditions in Linhai Park of Qinzhou Port and Qisha Park of Fangchenggang Port.

Change laws of artificial soil samples' porosity are shown in Figure 6 and Figure 7 along with additional load stress of coastal reclamation layer.

Figure 6 and Figure 7 show that artificial soil samples' porosity decreases along with the increase of additional load stress of coastal reclamation layer under the same circumstances of the calcium ions concentration of soak solution. The reason is that the consolidation deformation of soil samples come up under the action of additional load stress, consequently, soil samples' porosity decreases.

Change laws of artificial soil samples' porosity are shown in Figure 8 and Figure 9 along with calcium ions concentration of soak solution.

Figure 8 and Figure 9 show that artificial soil samples' porosity decreases slightly along with

Table 6. Porosity of artificial soil samples in Linhai Park of Qinzhou Port.

Soak solution	Additional load stress of coastal reclamation layer (kPa)		
	80 kPa	130 kPa	180 kPa
Seawater in Qinzhou (calcium ions concentration is 0.025%)	1.10	1.00	0.85
calcium ions concentration is 0.25%	0.96	0.95	0.94
calcium ions concentration is 2.5%	0.90	0.83	0.77

Table 7. Porosity of artificial soil samples in Qisha Park of Fangchenggang Port.

Soak solution	Additional load stress of coastal reclamation layer (kPa)		
	80 kPa	130 kPa	180 kPa
Seawater in Fangchenggang (calcium ions concentration is 0.044%)	1.47	1.38	1.31
calcium ions concentration is 0.44%	1.43	1.36	1.29
calcium ions concentration is 4.4%	1.33	1.24	1.19

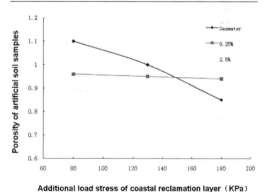

Figure 6. The change chart of artificial soil samples' porosity along with additional load stress of coastal reclamation layer in Linhai Park of Qinzhou Port.

the increase of the calcium ions concentration of soak solution under the same circumstances of the additional load stress of coastal reclamation layer with the exception of 180 kPa in Linhai

1189

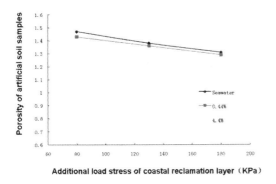

Figure 7. The change chart of artificial soil samples' porosity along with additional load stress of coastal reclamation layer in Qisha Park of Fangchenggang Port.

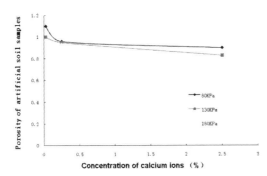

Figure 8. The change chart of artificial soil samples' porosity along with calcium ions concentration of soak solution in Linhai Park of Qinzhou Port.

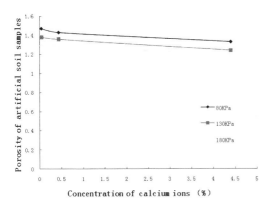

Figure 9. The change chart of artificial soil samples' porosity along with calcium ions concentration of soak solution in Qisha Park of Fangchenggang Port.

Park of Qinzhou Port. With the combination of Figure 4 and Figure 5, we can infer that artificial soil samples' porosity decreases slightly along with the increase of the calcium ions concentration of

Table 8. Osmotic coefficient k of artificial soil samples in Linhai Park of Qinzhou Port (10^{-8} cm/s).

Soak solution	Additional load stress of coastal reclamation layer (kPa)		
	80 kPa	130 kPa	180 kPa
Seawater in Qinzhou (calcium ions concentration is 0.025%)	14.1	11.2	6.09
calcium ions concentration is 0.25%	12.0	9.65	7.92
calcium ions concentration is 2.5%	10.3	7.65	4.94

Table 9. Osmotic coefficient k of artificial soil samples in Qisha Park of Fangchenggang Port (10^{-8} cm/s).

Soak solution	Additional load stress of coastal reclamation layer (kPa)		
	80 kPa	130 kPa	180 kPa
Seawater in Fangchenggang (calcium ions concentration is 0.044%)	15.8	12.5	8.62
calcium ions concentration is 0.44%	14.1	10.6	8.02
calcium ions concentration is 4.4%	11.0	9.35	7.49

soak solution because part of the pores was filled by cementing material precipitation of calcium ions. The reason of the exception of 180 kPa in Linhai Park of Qinzhou Port may be that impact on artificial soil samples' porosity is little for the increase of soak solution's calcium ions concentration. Variable quantity of artificial soil samples' porosity was concealed by error in the process of experiment.

5.3 Osmotic coefficient

Osmotic coefficient of artificial soil samples are shown in Table 8 and Table 9 under different conditions in Linhai Park of Qinzhou Port and Qisha Park of Fangchenggang Port.

Change laws of artificial soil samples' osmotic coefficient are shown in Figure 10 and Figure 11 along with additional load stress of coastal reclamation layer.

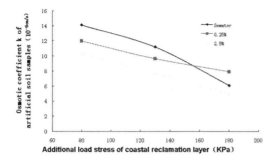

Figure 10. The change chart of artificial soil samples' osmotic coefficient along with additional load stress of coastal reclamation layer in Linhai Park of Qinzhou Port.

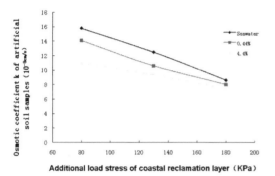

Figure 11. The change chart of artificial soil samples' osmotic coefficient along with additional load stress of coastal reclamation layer in Qisha Park of Fangchenggang Port.

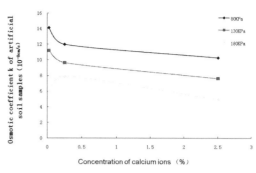

Figure 12. The change chart of artificial soil samples' osmotic coefficient along with calcium ions concentration of soak solution in Linhai Park of Qinzhou Port.

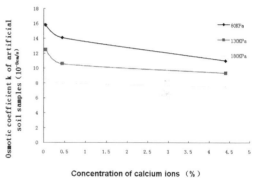

Figure 13. The change chart of artificial soil samples' osmotic coefficient along with calcium ions concentration of soak solution in Qisha Park of Fangchenggang Port.

Figure 10 and Figure 11 show that artificial soil samples' osmotic coefficient decreases along with the increase of the additional load stress of coastal reclamation layer under the same circumstances of calcium ions concentration of soak solution. The change law of artificial soil samples' osmotic coefficient is consistent with change law of artificial soil samples' porosity.

Change laws of artificial soil samples' osmotic coefficient are shown in Figure 12 and Figure 13 along with calcium ions concentration of soak solution.

Figure 12 and Figure 13 show that artificial soil samples' osmotic coefficient decreases slightly along with the increase of the calcium ions concentration of soak solution under the same circumstances of the additional load stress of coastal reclamation layer with the exception of 180 kPa in Linhai Park of Qinzhou Port. The change law of artificial soil samples' osmotic coefficient is consistent with change law of artificial soil samples' porosity.

6 CONCLUSION

In this paper, the research object is marine soft soil in Linhai Park of Qinzhou Port and Qisha Park of Fangchenggang Port. The marine soft soil environment factors, additional load stress of coastal reclamation layer and calcium ions of pore water chemical composition, were simulated by using a soil soaking-load linkage. The soil soaking-load linkage is mechanism simulation device. Impact on water-soil interactions of marine soft soil was analyzed for additional load stress of coastal reclamation layer and calcium ions of pore water chemical composition. The conclusion is as follows.

1. From the point of geotechnical engineering, there are mainly hydraulic connection, water chemical field and additional load stress of coastal reclamation layer for the soil and water environmental change of marine soft soil in coastal reclamation district.

2. The soil and water environmental change of marine soft soil in coastal reclamation district is simulated by using a soil soaking-load linkage.
3. Under the same circumstances of the calcium ions concentration of soak solution, the consolidation deformation of artificial soil samples come up under the action of additional load stress, consequently, artificial soil samples' porosity decreases. Artificial soil samples' porosity and osmotic coefficient decreases along with the increase of additional load stress of coastal reclamation layer. Artificial soil samples' calcium ions content decreases along with the increase of additional load stress of coastal reclamation layer.
4. Under the same circumstances of the additional load stress of coastal reclamation layer, artificial soil samples' calcium ions content increases along with the increase of calcium ions concentration of soak solution.
5. Artificial soil samples' porosity and osmotic coefficient decrease slightly along with the increase of the calcium ions concentration of soak solution under the same circumstances of the additional load stress of coastal reclamation layer with the exception of 180 kPa in Linhai Park of Qinzhou Port. The reason of the exception of 180 kPa in Linhai Park of Qinzhou Port may be that impact on artificial soil samples' porosity is little for the increase of soak solution's calcium ions concentration. Variable quantity of artificial soil samples' porosity was concealed by error in the process of experiment.

REFERENCES

Dong, D.X., Li, Y.C., Chen, X.Y., et al. (2014). The influence of large-scale reclamation on the hydrodynamic environment of Qinzhou Bay. Guangxi Sciences, 21(4), 357–364, 369.

Ge, Z.P., Dai, Z.J., Xie, H.L., et al. (2014). The northern Gulf coastline of the temporal and spatial variation characteristics. Shanghai land and resources, 35(2), 49–53.

Hu, C.L. (2009). The Guangxi Beibu Gulf reclamation engineering foundation and construction method. Exploration Engineering, 36(6), 51–54.

Liu, H.M., Wu, H., Zhou, D. (2012). Mechanism analysis and application of dynamic consolidation method in treatment of blown sand foundation. Construction Technology, 41(S1), 58–61.

Liu, H.M., Zhou, D., Yuan, H.B., Wu, H. (2014). Effect of variation of water chemical field in reclamation area on fine structure of fine grained soil. Journal of Civil, Architectrual & Environment Engineering, 36(S1), -4, 8.

Ou, X.D., Pan, X., Yin, X.T., et al. (2015). Guangxi beibu gulf artificial reclaimed soil for land consolidation test research [J]. Rock and Soil Mechanics, 36(1), 28–33.

Sun, Y.G., Gao, J.G., Zhu, X.M. (2012). Qinzhou bonded port of reclamation reclamation project on marine environment. Marine Sciences, 36(12), 84–89.

Wu, H., Liu, H.M., Zhou, D., et al. (2013). The device of soil soaking-load linkage device, Chinese patent, CN103454154A.

Yi, Y.L., Gu, L.Y., Liu, S.Y. (2015). Microstructural and mechanical properties of marine soft clay stabilized by lime-activated ground granulated blastfurnace slag. Applied clay science, 103(1), 71–76.

Civil, Architecture and Environmental Engineering – Kao & Sung (Eds)
© 2017 Taylor & Francis Group, ISBN 978-1-138-02985-9

Energy use and CO_2 emissions in China's pulp and paper industry: Supply chain approach

Cheng Chen & Rongzu Qiu
School of Transportation and Civil Engineering, Fujian Agriculture and Forestry University, Fuzhou, Fujian, China

Jianbang Gan
Department of Ecosystem Science, Texas and Management A&M University, College Station, TX, USA

ABSTRACT: This study is to assess greenhouse gases and energy use of China's PPI using an accounting framework that makes up the entire supply chain. A cradle-to-grave tracking is applied in the calculation. We consider three groups of raw materials—wood, waste paper, and non-wood fiber—yet focus on wood and waste paper, which are further divided into two subclasses according to the sources: domestic and imported. The different impacts in the major exporting countries to China's PPI are analyzed by considering the land use change. In addition, sensitivities to changes in major factors in the supply chain are examined. The results of this study are useful in framework designing and decision making of regional strategy department of PPI.

1 INTRODUCTION

Greenhouse Gas (GHG) emissions have become a focus of international attention as the impacts of global climate change increase. Burning of fossil fuels is a major source of GHG emissions, and industries are major users of fossil energy. The Pulp and Paper Industry (PPI) is the fourth largest industrial energy user in the world; electricity and heat are needed for chemical pulping, paper making, black liquor evaporation, and other operations (Chen et al., 2012). Consequently, the PPI is one of the biggest emitters among all the industrial sectors (Nilsson et al., 1995; Gasbarro et al., 2012).

Since 2008, China has overtaken the US as No.1 emitter of GHGs in the world, accounting for 23.33% of global emissions in that year. Furthermore, carbon leakage may exist between Organization for Economic Co-operation and Development (OECD) countries and the newly emerging economies like China as polices in the OECD countries aim at the reduction of domestically produced GHGs, which may result in the relocation of emission-intensive industries to countries with less stringent emission standards. China is also ranked as the largest carbon exporter (Homma et al., 2012; Bruckner et al., 2010).

According to China Paper Manufacturers Association (2010), there were more than 3700 mills in China's PPI, annually producing 73.18 million tonnes of pulp and 92.70 million tons of paper and paper board, respectively. The gross output value of China's PPI was about USD 95 billion (585 billion Yuan), which accounts for 3.64% of the total GDP (gross domestic product) in China's manufacturing sector. The total consumption of paper and paper board in 2010 was 91.73 million tons, a 7.05% increase from the previous year; per capita consumption of paper and paper board in China was 68 kg, about 23% of the average per capita consumption in developed countries (300 kg). Obviously, there is a great growth space for China's PPI. Thus, energy consumption and GHG emissions in China's PPI will increase inevitably in conjunction with the growing market.

In addition, China's PPI is facing great shortage of domestic raw material supply (Manda et al., 2012). In 2010, China imported 35.43 million m^3 of industrial roundwood, an increase of 26.6% from the previous year, which accounts for 34.6% of its total domestic roundwood production (FAO, 2012). China is also a net importer of pulp; 35% of total pulp consumption during 2006–2010 was imported. Moreover, China is the world's biggest buyer of recovered paper (BIR, 2011). Over 62% of China's pulping materials in 2010 was recycled paper, of which 24.35 million tonnes was imported (CTAPI, 2011).

The huge market size of China's PPI highlights the need to investigate the energy use and GHG emissions. Although it has been proved that the composition structure of raw materials has a great relationship with the energy use and GHG emissions of PPI (Hammett et al., 2001; Wang et al., 2012), imported materials along the supply chain also have a great impact on that because of different forest types and management practices across exporting countries (James, 2012). The perspective of whole supply chain is essential to assess the energy use and GHG emissions of PPI properly.

This paper will look into the energy use and GHG emissions of China's PPI using the supply chain approach and life cycle assessment. We will measure and compare energy use and GHG emissions of paper products made from different raw materials originating from domestic and foreign sources.

2 METHODOLOGY AND DATA

2.1 *Assumptions*

To achieve study goal, necessary assumptions are made as follows:

1. For raw materials, we included wood fiber, recovered fiber and non-wood fiber. All types of fiber are further divided into domestic and imported fiber except for non-wood fiber that is predominantly domestic grain stoves.
2. Imported wood fiber refers to chips exclusively under the consideration of transportation convenience and industry and employment opportunities protection.
3. Wood pulp is roughly divided into to two categories: chemical pulp and mechanical pulp, to achieve a balance between calculation simplicity and obvious differences of energy consumption that exist in different types of wood pulp.
4. This study focuses on fiber only; therefore, other substances (such as fillers, chemicals) are excluded.
5. Carbon sink resulting from domestic forest cultivation to supply wood fiber to PPI is included in the total CO_2 emissions, while the carbon stock in the wooden product used as a raw material in the supply chain is considered as carbon emission.
6. Activities occurring in the final product utilization phase are not incorporated.
7. All used paper is assumed to be incinerated to generate electricity if not recovered, and energy contained in recovered paper is considered to be equivalent to the same amount contained in wood fiber.
8. Being a territorial accounting, transportation activities that occur outside of the study area are excluded consequently.
9. All measurements in this study are based on the production and consumption of 1 ton of paper, and energy is expressed on the Lower Heating Value (LHV) basis.

2.2 *System boundary*

The system considered in this study consists of wood harvesting, transportation, pulping, paper making, and waste paper disposing as well as importing of raw materials, intermediate products and final products. The system boundary is shown in Figure 1.

First, carbon sequestration in forest cultivation is calculated as emissions mitigation while those associated with wood and waste paper incineration, are ignored since sustainable forestry is assumed. Second, wood cutting and chipping operations are supposed to be carried out in-field before transportation, which is widely accepted as a rational option (Emer et al., 2011; Kallio et al., 2011; Acuna et al., 2012). Third, details of non-wood fiber production are not considered. Finally, energy contained in biomass or/and recycled paper is calculated as energy input, while waste paper discarded is assumed to be incinerated with municipal solid waste, which is considered as energy generation.

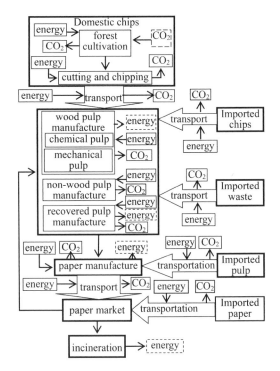

Figure 1.   Details of the system boundary.

2.3 Calculation method

2.3.1 Energy consumption calculation

Energy consumption along the supply chain is calculated by subtracting the amount of total energy generation from the amount of total energy input (energy use).

Energy use in wood cutting and chipping operations are included in energy input with cultivation activities. Besides wood, imported pulp and waste paper put into the supply chain are converted into the wood equivalent to calculate biomass energy use. Energy use of non-wood pulp making is assumed to be 28.7 GJ/Adt, including raw material (mainly wheat straw) production, harvesting, chipping, transportation and pulping (Kissinger et al., 2007). The average transportation distance (one way) of domestic goods is assumed to be 150 km and 500 km is assumed for imported goods (Manda et al., 2012).

By-products in pulping processes can be used for energy generation. Generally, there are two sources of energy generation extracted from by-products: black liquor and rejects.

End-products also contain energy. Two end processes are assumed for waste paper: recycled and incineration. The recycle rate of waste paper in China in 2010 is 0.4 (Chen Q.W., 2010).

2.3.2 CO2 emissions calculation

CO_2 emissions are calculated by multiplying total consumption of each type of energy with the carbon content of the energy source, as follows:

$$
\begin{aligned}
CO_2 emission \\
= \left(\sum_i E_{el_i} - \sum_j E_{elg_j} \right) * \alpha + \left(\sum_i E_{he_i} - \sum_j E_{heg_j} \right) * \beta \\
+ (Energy_{input-cultivation} + Energy_{input-nonwood}) * \gamma \\
+ Energy_{input-transport} * \theta - D_w * \tau
\end{aligned} \tag{1}
$$

where α (kg CO_2/kwh) is the emission factor of electricity; β (kg CO_2/GJ) is the emission factor of steam, relying on main fuel used and convert efficiency in objective region; γ is the emission factor of energy used in wood cultivation and non-wood pulp acquisition; θ is the emission factor of energy used in transportation; and τ (kg CO_2/m³) is unit CO_2 emissions mitigation of forest cultivation.

The emissions factors used are the CO_2 emissions factors for fuel combustion as recommended by the IPCC (1996). Emission factors of national primary energy are used for the calculation of carbon emissions in the cultivation and non-wood pulp acquisition, which is accurate enough because the amount of energy use in these activities is minimal.

Table 1. Data of energy consumption and generation.

	Steam (GJ/Adt)	Electricity (kwh/Adt)
	(Energy consumption/generation)	
Chemical pulp	22.2/22.2[b]	700[a]/2463[c]
Mechanical pulp	–/5.4[d]	2200[b]/–
Recovered pulp	0.4[c]/0.42[e]	390[c]/–
Paper making	5.6[c]/–	608[c]/–
paper incineration		–/1200[e]
Wood incineration		–/1750[f]

a. Nilsson et al. (1995); b. Gullichsen & Fogelholm (2000); c. Farahani et al. (2004); d. Holmberg & Gustavsson (2007); e. Laurijssen et al. (2010); f. Dornburg et al. (2006).

2.4 Inputdata

Current state-of-the-art technologies are supposed in estimating the parameters used in this study.

Table 1 presents the amount of energy use and/or generation in different processes. Emissions with energy used in forest cultivation are assumed to be equivalent to coal combustion in this study, because the primary energy used in China is coal (IEA, 2012). Energy used in transportation is assumed to be diesel. The average energy conversion efficiency set in this study is 35% for electricity and 90% for heat (Graus et al., 2007).

3 COMPUTATION RESULTS

3.1 Energy input

The total amount of six parts of energy input are 2914E6 GJ, 955.57E6 GJ, 537.2E6 GJ, 455.44E6 GJ, 372.24E6 GJ, 1.1E6 GJ for biomass, paper making, transportation, pulping, non-wood pulp production and cultivation, respectively, and the total energy input is 5235.55E6 GJ (Figure 2).

Energy use in cultivation is so minimal that it cannot be seen in the diagram. Energy use in pulping is not as much as it used in paper making in China's PPI that corresponds to the large amount of imported pulp as well as large amount of energy use in transportation. Energy contained in all materials put into the supply chain is accounted for, which can explain for the largest energy use among all parts.

3.2 Energy generation

Energy generation in pulping activities is 347.56E6 GJ. Energy generated from waste paper incineration is 679.34E6 GJ, and 1221.88E6 GJ is the energy contained in the recovered paper.

In 2010, there is a small amount of net exported paper and paper board whose energy contained is calculated as energy generation with conversion factor between paper and wood equivalent. In all, the total energy generation in China's PPI in 2010 is 2281.08E6 GJ.

3.3 Energy consumption

Based on energy use and energy generation, the total energy consumption in China's PPI can be calculated and the result is 2954.47E6 GJ. The average unit energy use per pulp consumption is 32.21 GJ/t paper. This figure is of the same magnitude with unit energy use in Dutch PPI (Laurijssen et al., 2010), which also versifies the accounting framework used in this study, Even though the value is higher than that in Netherland due to higher recycling rate in Netherland and differences in those two accounting frameworks.

3.4 CO₂ emissions

Total CO_2 emission from China's PPI in 2010 is 147.69E6 t CO_2 and average carbon emission per tonne of paper is 1610 kg CO_2. Energy use and CO_2 emission in some countries are listed in Table 2.

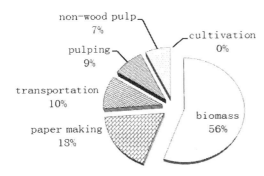

Figure 2. Energy input composition in China's PPI.

Table 2. Energy use and CO_2 emission of producing every 1 t paper in different countries.

Country	Energy use (GJ/t)	CO_2 emission (kg/t)
China	32.2	1610
Canada	18.2[a]	260[a]
USA	32.5[b]	787[c]
Netherland	14[d]	200[d]
Sweden	29.3[b]	204[b]

a Adès J. et al. (2012), Only natural gas and oil products emit CO_2; electricity is not considered to emit CO_2; b Farahani S. et al. (2004); c Heath L. S. et al. (2010); d Laurijssen J. et al. (2010).

Energy use and carbon emissions depend strongly on energy efficiency, primary energy used and accounting framework. The high value of carbon emissions in this study is partly due to the energy contained in imported materials that is included, but the corresponding carbon sink in forest cultivation activities is excluded. In fact, a net importer of wood product should be responsible for carbon storage in imported wood materials (Heath, 2010).

3.5 Comparison with production-based accounting

Within production-based or territorial accounting calculation method (Bruckner et al., 2010), the total energy consumption in China's PPI in 2010 is 1974.0E6 GJ (21.3 GJ/t paper) and the total carbon emission is 118.26E6 t CO_2 (1275.7 kg CO_2/t paper). These two figures both smaller than the ones within incorporating impacts of imported goods, which may indicate that China is a carbon importer in terms of PPI.

3.6 Sensitivity Analysis

3.6.1 Importing

1. Change in quantity of imports

We investigate the sensitivity to the change in quantities of imported raw materials and final product of the supply chain (see Figure 3). To keep comparativeness, when the amount of imported chips changes, the amount of domestic chips changes accordingly and all other parameters are assumed to be unchanged in order to satisfy the paper market's demand. As the amount of imported chips increasing, unit energy use and CO_2 emission both increase (see Figure 3a).

Imported chips consume more energy in transportation and less energy in cultivation when compared to domestic chips; however, the energy use saved in cultivation activities cannot offset the extra energy used in transportation due to longer distance transportation of imported chip. With decreasing demand of domestic chips, carbon sink accompanied with cultivation activities decreases as well, which contributes to the increasing in carbon emissions. The marginal unit energy use is 0.03 GJ/t paper and the marginal unit carbon emission is 15.03 kg CO_2/t paper.

Figure 3b shows the sensitivity of unit energy use and unit carbon dioxide emission to quantity of imported pulp, which can give us insight of the influence of the amount of imported pulp. The quantity of domestic pulp changes with quantity of imported pulp changes to keep total paper production and consumption unchanged and the proportion of all kinds of raw materials to produce

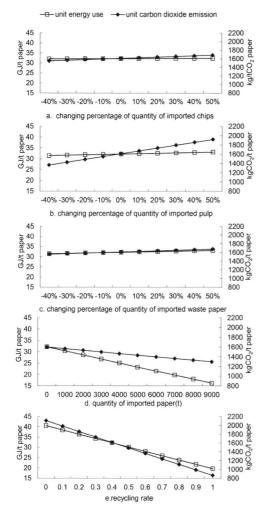

Figure 3. Sensitivity to the change in quantities of imported raw materials and final product of China's PPI.

domestic pulp is fixed invariably. With the amount of imported pulp increasing, unit energy use and unit carbon emissions go up. If the amount of imported pulp increases by 10%, unit energy use and unit carbon emission will have an increment of 0.18 GJ/t paper and a jump of 61.35 kg CO_2/t paper respectively. With conjunction of energy use and generation, the net energy use in pulping process is negative as well as carbon emissions.

A similar phenomenon is also found in sensitivity analysis for imported waste paper (see Figure 3c). The marginal unit energy use is 0.18 GJ/t paper and the marginal unit carbon emissions are 12.95 kg CO_2/t paper.

Figure 3d illustrates unit energy use and carbon emissions from one ton of paper products

consumed in China at different quantity of imported paper. With more paper imported, less energy consumption is required and less carbon is released. If all paper consumption in China is satisfied by import, the unit energy use is 15.89 GJ/t paper and the unit carbon emissions are 1291.97 kg CO_2/t paper.

Although importing paper can result in less energy use and carbon emissions, replacement of domestic paper production by imported paper may cause a decline of PPI in the country and severe issues of unemployment.

2. Different import sources

In our accounting framework, carbon emissions are calculated from the perspective of a country, however, impacts on carbon emissions associated with export countries should be taken into account.

Based on FAO data, China imports chips mainly from Vietnam, Thailand, Australia, Indonesia, New Zealand, Russian Federation, Cambodia, USA, and Malaysia, while imports pulp mainly from Canada, Brazil, USA, Indonesia, Chile, Russian Federation, Uruguay, Finland, Japan, New Zealand, Sweden, South Africa and, paper and paper board mainly from USA, Indonesia, Japan, Republic of Korea, Sweden, Russian Federation, Canada, Finland, Australia, Brazil, Germany, New Zealand, Thailand and Belgium. We convert the different products into Wood Raw Material Equivalent (WRME) underbark based on the research of James K. (2012).

Although carbon contained in wood fiber display slightly differences in different countries, the gap in carbon emissions among countries by considering land use change is significant as conditions of land use in different countries vary greatly. Don et al. (2011) identified absolute losses of 10.2–15 t C ha^{-1} where primary forest is converted to secondary forest (including plantation), while Munoz et al. (2007) found that soil carbon under 50% lower than primary forest. The main exporters to China's PPI are divided into several groups, according to their changes of primary forest and planted forest, see Figure 4.

Based on the facts described above, countries in the top right corner in the diagram with increasing cover both in primary forest and plantation are perfect exporters within PPI in terms of carbon emissions. Then are the countries in the middle part from top to bottom, and countries in the top left corner should decrease their exports. This analysis illustrates that importing forest woody materials from Indonesia may cause worst carbon impacts among the countries under scrutiny.

Using the quantifying method proposed in James (2012), change in carbon stores associated with removing 1 m^3 of WRME underbark of each country can be measured based on FAO data, which is listed in Table 3. With the first group concerned, forest extends with carbon sink increasing. So the change in carbon stock with removing 1 of WRME can be considered as carbon mitigation in forestation. The calculation results illustrate that Russia is the best exporter in group 1. However, in group 2, with data available, it can be judged that as forest product exporter South Africa is better than Canada, and Thailand is better than Malaysia. Also, New Zealand is better than Cambodia in group 4, and in group 5, Australia may be the worst exporter since decreasing carbon stock and increasing forest area, which indicates the degradation of forest.

3.6.2 China's domestic recycle rate

Figure 3e shows trends of unit energy use and unit carbon dioxide emissions as the paper recycling rate in China increases from zero to one with anything else unchanged. The greater the domestic recycling rate is, the less the carbon emissions and energy use become. When no waste paper is recycled (recycling rate = 0), the unit energy use is 40.59 GJ/t paper and the unit carbon emissions are 2107.15 kg CO_2/t paper. The value of unit energy use will decrease by 2.1 GJ/t paper and the value of unit carbon emissions will increase by 124.29 kg CO_2/t paper with a 10% increase of recycling rate. A higher recycling rate indicates a reduction in waste paper incineration. Because paper incineration causes more GHG emissions than recycling, carbon emissions decline.

If ignore different uses of recycled waste paper, that is to say, all recycled waste paper is reused

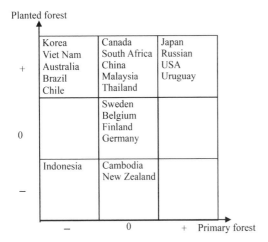

Figure 4. Dividing export countries into different groups according to change of primary forest and planted forest.

Table 3. Changes in carbon stock associated with the removal of 1 m³ WRME underbark, 2005–2010.

Group	Country	CO_2 eq (t/m3)	Group	Country	CO_2 eq (t/m3)
1	Japan	58.04	3	Sweden	N/A
	Russian	80.24		Belgium	6.47
	USA	20.72		Finland	N/A
	Uruguay	N/A		Germany	N/A
2	Canada	N/A	5	Korea	−60.34
	South Africa	N/A		Viet Nam	4.52
	Malaysia	7.87		Australia	N/A
	Thailand	7.42		Brazil	3.50
	China	1.83		Chile	1.94
4	Cambodia	2.30	6	Indonesia	14.59
	New Zealand	−7.10			

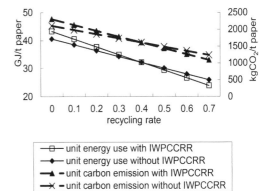

Figure 5. Comparison of two kinds of assumption of recycled paper on energy use and carbon emission.

in PPI, the amount of imported waste paper will decrease. Figure 5 shows the comparison of results between the two kinds of assumption on recycled paper. It is noted that when recycling rate reaches 0.7, domestic recycled waste paper can match the total amount of waste paper used in China's PPI supply chain and 0.7 is just the recycling rate in developed countries.

As depicted in Figure 5, the slope of the two curves with the assumption that Imported Waste Paper Change Corresponding to Recycling Rate (IWPCCRR) are both steeper than the corresponding curve without IWPCCRR, which strengthens the argument that recycling is beneficial with regard to emission and energy use.

4 CONCLUSION

In response to climate change, balancing energy use, GHG emissions, and resource scarcity in China's PPI is an important task in forest resource planning and allocation and national sustainability development policy making. This study investigates the environmental impacts of PPI in China using energy use and CO_2 emissions as indicators and presents a new accounting framework from the viewpoint of supply chains, which can compensate for the deficiency of LCA widely used in research. With the proposed accounting framework, total energy use and total CO_2 emissions in China's PPI in 2010 are 2954.47 E6 GJ and 147.69 E6 t CO_2, respectively.

The results of this study are useful for decision making of the strategy departments of PPI. GHG emissions accounting framework designed from the viewpoint of supply chain can provide a new account sight to assess energy use and GHG emission of certain industry in a country or a region and an insight into the strategy such as resource allocation and sourcing decisions.

Issues that need further research are (i) combination of economic indicators and environment indicators in assessing different industries of a country or a region; (ii) environment impacts of allocation of limited raw materials among different industries.

ACKNOWLEDGEMENT

This work was funded by the University Development Foundation of Fujian Agriculture and Forestry University (No. 113-612014018). The authors are indebted to the anonymous referees for their thorough comments and suggestions.

REFERENCES

Acuna, M., Mirowski, L., Ghaffariyan, M. R., Brown, M. 2012. Optimising transport efficiency and costs in Australian wood chipping operation, biomass and bioenergy. Biomass and Bioenergy 46, 291–300.

Adès, J., Bernard, J. T., González P. 2012. Energy use and GHG emissions in the Quebec pulp and paper industry, 1990–2006. Canadian Public Policy 38, 71–90.

Bruckner, M., Polzin, C., Giljum, S. 2010. Counting CO_2 Emissions in a Globalised World. German Development Institute.

Bureau of International Recycling (BIR). 2011. World Markets for Recovered and Recycled Commodities. Brussels, Belgium.

Chen, H. W., Hsu C.H., Hong G. B. 2012. The case study of energy flow analysis and strategy in pulp and paper industry. Energy Policy 43, 448–455.

Chen, Q.W., 2010. The future prospect of paper industry with recycled fibers. China pulp & paper industry 31, 28–38. (in Chinese)

China Technical Association of Paper Industry, 2011. Almanac of China paper industry 2011. China light industry press, Beijing.

Don, A., Schumacher, J., Freibauer, A. 2011. Impact of tropical land-use change on soil organic carbonstocks—a meta-analysis. Global Change Biology 17, 1658–1670.

Dornburg, V., Faaij, A., Meulemam, B. 2006. Optimising waste treatment systems. Part A. Methodology and technological data for optimising energy production and economic performance. Resources, Conservation and Recycling 49, 68–88.

Emer, B., Grigolato, S., Lubello, D., Cavalli, R. 2011. Comparison of biomass feedstock supply and demand in Northeast Italy. Biomass and Bioenergy 35, 3309–3317.

FAOSTAT. Food and Agricultural Organization of the United Nations. FAOSTAT database on website. Figures from 2010 [latest access on 28.12.2012].

Farahani, S., Worrell, E., Bryntse, G. 2004. CO_2 free paper? Resources, Conservation and Recycling 42,317–36.

Gasbarro, F., Rizzi, F., Frey, M. 2012. The mutual influence of Environmental Management Systems and the EU ETS: Findings for the Italian pulp and paper industry, European Management Journal http://dx.doi.org/10.1016/j.emj.2012. 10.003.

Graus, W., Voogt, M., Worrell, E. 2007. International comparison of energy efficiency of power generation. Energy Policy 35, 3936–3951.

Gullichsen, J., Fogelholm, C-J, editors, 2000. Chemical pulping. Papermaking Science and technology, Book 6 A. Finland: Fapet Oy. Jyväskyä.

Hammett, A. L., Youngs, R. L., Sun, X.F., Chandra, M. 2001. Non-wood fiber as an alternative to wood fiber in China's pulp and paper industry. Holzforschung 55, 219–224.

Heath, L. S., Maltby, V., Miner, R., Skog, K. E., Smith J. E., Unwin J., Upton B. 2010. Greenhouse gas and carbon profile of the U.S. forest products industry value chain. Environmental Science & Technology 44, 3999–4005.

Holmberg, J. M. & Gustavsson, L. 2007. Biomass use in chemical and mechanical pulping with biomass-based energy supply. Resources, Conservation and Recycling 52,331–350.

Homma, T., Akimoto, K., Tomoda, T. 2012. Quantitative evaluation of time-series GHG emissions by sector and region using consumption-based accounting. Energy Policy 51,816–827.

Intergovernmental Panel on Climate Change (IPCC), 1996. Greenhouse gas inventory book—1996 IPCC guidelines for national greenhouse gas inventories. Geneva, Switzerland: IPCC.

International Energy Agency (IEA), 2012. IEA data services. http://www.data.iea.org. Paris, France: IEA.

James, K. 2012. An investigation of the relationship between recycling paper and card and greenhouse gas emission from land use change. Resources, Conservation and Recycling 67, 44–55.

Kallio, A. M. I., Anttila, P., McCormick M. 2011. Are the Finnish targets for the energy use of forest chips realistic-Assessment with a spatial market model. Journal of Forest Economics 17, 110–126.

Kissinger, M., Fix, J., Rees, W.E. 2007. Wood and non-wood pulp production: Comparative ecological footprinting on the Canadian prairies. Ecological economics 62,552–558.

Laurijssen, J., Marcsidi, M., Westenbroek, A., Worrell E., Faaij, A. 2010. Paper and biomass for energy? The impact of paper recycling on energy and CO_2 emissions. Resources, Conservation and Recycling 54,1208–1218.

Manda, B.M. K., Blok, K., Patel, M. K. 2012. Innovations in papermaking: An LCA of printing and writing paper from conventional and high yield pulp. Science of the Total Environment 439, 307–320.

Munoz, C., Ovalle, C., Zagal, E. 2007. Distribution of soil organic carbon stock in Alfisol profile in Mediterranean Chilean Ecosystems. Journal of Soil Science and Nutrients 7:15–27.

Nilsson, L.J., Larson, E.D., Gilbreath, K.R., Gupta, A. 1995. Energy efficiency and the pulp and paper industry. Washington, DC, USA: American Council for an Energy Efficient Economy.

Wang, L., Templer, R., Murphy, R. J. 2012. A Life Cycle Assessment (LCA) comparison of three management options for waste papers: bioethanol production, recycling and incineration with energy recovery. Bioresource Technology 120, 89–98.

Civil, Architecture and Environmental Engineering – Kao & Sung (Eds)
© 2017 Taylor & Francis Group, ISBN 978-1-138-02985-9

Research on the limit equilibrium circle theory of entry support

Qing-Xiang Huang, Yan-Peng He & Zhong-Qing Shi
School of Energy Engineering, Xi'an University of Science and Technology, Xi'an, China

Jie Chen
School of Material Engineering, Xi'an University of Science and Technology, Xi'an, China

ABSTRACT: In order to discover the effect of floor and rib failure on the whole entry stability in the three-soft coal seam, through site measurement and numerical simulation, the failure circle around the entry is obtained, and its mechanism is explored. Based on Protodyakonov's arch theory, the limit equilibrium arch function of entry roof is set up by considering the effect of floor and rib failure. It is found that the height of roof equilibrium arch is proportional to the equivalent width of the entry, and is inversely proportional to the friction coefficient of the rib and the roof. When the rib fail, the equivalent width of entry increases, and the increment of roof equilibrium arch height is proportional to the depth of rib failure zone. When floor fails, it causes the equivalent height of entry to increase, and results in the equivalent width of entry increasing and the height of roof equilibrium arch increasing. The increment of roof equilibrium arch height is proportional to the floor failure depth. Thus, the theory of equilibrium circle of entry is established based, and it is pointed that strengthening the ribs and floor is important to improve the stability of the whole entry.

1 INTRODUCTION

Usually, the roof is paid more attention than the ribs, and the floor is always ignored in the entry support. The main reason is unclear in the interactions of "floor–ribs–roof". There are a lot of research on soft rock entry support theory, but less study on the influence of the floor and ribs on the roof. Practice shows the soft floor has a direct effect on the rib deformation and influences the stability of the roof. Currently, the main theories on the entry and roadway support including Protodyakonov's arch theory (Li 1986), broken rock zone theory (Deng et al. 1994), key circle theory of entry (Kang 1997), self-stable arch theory (Huang 2014), and roof stability (Gou et al. 2006, Tan 2003, Xu et al. 2013).

There is some research dealing with the shape of the roof arch by considering the initial horizontal stress (Zhao 1978), and the surrounding rock stability of the roadway (Miao 1997). Some researches found that strengthening the floor and the two ribs had important impact on the whole stability of entry (Bai et al. 2011, Guo et al. 2013). But there is less study on the interaction mechanism of "floor–ribs–roof".

This research analysis involves the interaction mechanism of "floor–ribs–roof", and provides a new theory of entry support in the soft coal seam

with soft roof and floor, which is so called the three-soft coal seam.

2 LIMIT EQUILIBRIUM CIRCLE OF ENTRY

2.1 *Failure circle measured around entry*

The study entry is in the Gaocheng coal mine in Henan Province of China. The thickness of the No. 21 coal seam is 4.8 m and the average depth is about 450 m, the dip angle of coal seam is 8°–13°. The roof is in sliding structural zone, which is extremely broken, and the floor is soft sandy mudstone. It belongs to typical three-soft coal seam. The entry size is width 6.0 m and height 3.4 m. The mechanical parameters of coal and rock are shown in Table 1. The compressive strengths are 0.92–17.0 MPa, water softening coefficients are 0.24–0.36, and the elastic modulus are 33.0–413.8 MPa. It is clear that the surrounding rock is typical soft rock.

The original support of the entry is U section steel in arch-shaped yielding supports (Fig. 1).

The convergence between roof and floor is 1.5 ~ 2.0 m, and between ribs is 1.5 m. It belongs to large deformation soft entry. According to the radar scan and borehole observation in the No. 21051 roadway, the failure zone around the

Table 1. The physical and mechanical parameters of roof and floor in the No. 21 coal seam.

| Stratum | Lithology | Uniaxial compressive strength (MPa) | Thickness (m) | Water softening coefficient | Tensile strength (MPa) | Shear strength | | Elastic modulus (MPa) | Poisson's ratio | Unit weight (t/m³) |
						C(MPa)	Φ(°)			
Coal seam	No.21 coal	0.92	4.86	0.32	0.10	0.70	20	33.0	0.28	1.35
Roof	Sandstone	16.35	8.00	0.24	2.97	5.12	35	351.2	0.20	2.65
Floor	Mudstone	15.44	15.00	0.36	3.32	0.86	25	413.8	0.20	2.44

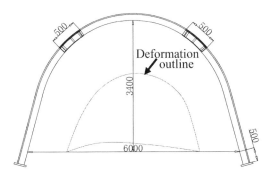

Figure 1. Deformation of entry section.

 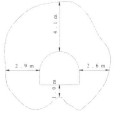

(a) Entry without repaired   (b) Entry repaired once

Figure 2. Entry failure circle by site measurement.

entry is circle shape (Fig. 2). The height of roof arch is 3.8 ~ 4.1 m, and the thickness of failure zone in ribs is 2.6 ~ 3.1 m, and the deepness in floor is 1.0 ~ 1.5 m.

In general, the surrounding rock self-stable equilibrium circle refers to the circular boundary of the surrounding rock in the limit state of self-stability. In the entry roof, it is approximate to tensile stress boundary. And in ribs and floor, it is approximate to plastic zone and tensile failure zone boundary. Generally, the failure circle zone around entry refers the limit equilibrium self-stable circle.

2.2 Numerical simulation of failure circle

In order to explore the interaction and influence of "floor-rib-roof", simulation models are set up by Flac3D and the result is shown in Fig. 3.

When the roof is weak (c = 0.6 MPa, σ_t = 0.4 MPa) and the floor is stronger (c = 1.8 MPa, σ_t = 1.6 MPa), the strength of ribs reduce from c = 1.4 MPa to 0.5 MPa, and σ_t = 1.2 MPa to 0.3 MPa, the height of roof arch increases 15%. Based on that, when the strength of floor is reduced from c = 1.8 MPa to 0.4 MPa and σ_t = 1.4 MPa to 0.2 MPa, the thickness of rib failure zone increases by 20%, and the height of roof arch increases by 25%. The total increment of roof arch height due to the influence of soft and soft floor is 43%. So, it is obvious that the soft floor and soft ribs have a significant impact on the development of roof failure arch of the three-soft entry.

2.3 Natural limit equilibrium arch of entry roof

If only the influence of soft roof is considered, the natural equilibrium arch in the entry roof can be determined by Protodyakonov's arch theory (Wang 1996). Actually, the arch is in limited equilibrium state, and the arch calculation model is shown in Fig. 4.

Assume that there is only axial compression force, and no shear force and bending moment on any cross-section of the limited equilibrium arch. Choose any point $M(x,y)$ in the arch curve, the bending moment of all the external force to the point $M(x,y)$ is

$$\Sigma M = Ty - qx^2/2 = 0 \qquad (1)$$

where q is vertical uniformly distributed load on the arch, MPa; T is horizontal force at arch section, MN.

According to static equilibrium equation, $T = T'$

Based on Protodyakonov's arch theory, the horizontal force T' at the arch feet must meet the condition:

$$T' \leq qaf \qquad (2)$$

where T' is the horizontal force at arch feet, MN; f is internal friction coefficient of roof. Put $T = qaf$ into Eqs. (1), and there is

$$y = x^2/2af \qquad (3)$$

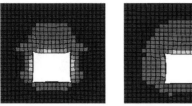
A. soft roof

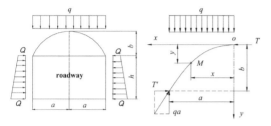
B. soft roof and soft ribs

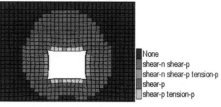
C. three-soft

Figure 3. The failure circle development in three-soft entry.

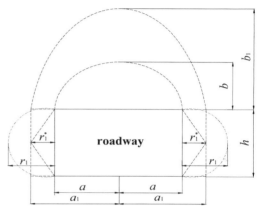

Figure 4. Calculation model of the limit equilibrium arch.

The height of the nature equilibrium arch is:

$$b = a/2f \qquad (4)$$

where b is the height of the arch with stable ribs, m; a is the half width of entry, m; f is the internal friction coefficient of the roof.

2.4 *Height of the limit equilibrium arch with rib failure*

When there is a failure in ribs, the maximum failure depth in rib can be calculated as:

$$r_1 = h/2f_b \qquad (5)$$

where r_1 is the maximum depth of rib failure, m; h is the entry height, m; f_b is the internal friction coefficient of rib.

Rib failure is equivalent to the entry width increase, and leads to the development of roof equilibrium arch (Fig. 5). The equivalent half-width increment is given by $r_1^* = kr_1$, where r_1^* is equivalent width increment of rib, m; k is equivalent coefficient.

According to fracture mechanics, the equivalent increment of crack length is equal to half-width of the plastic zone. So, the equivalent width coefficient $k = 0.5$, and the equivalent half width of entry is

$$a_1 = a + r_1^* = a + 0.5r_1 \qquad (6)$$

Figure 5. Limit equilibrium arch of entry with rib failure.

where a_1 is the equivalent half entry width with ribs failure, m.

Put $x = a_1$, $y = b_1$ into the Eqs. (3), the height of the limit equilibrium arch is:

$$b_1 = a_1/2f = (a + 0.5r_1)/2f = b + h/8ff_b \qquad (7)$$

where b_1 is the height of equilibrium arch with ribs failure, m.

3 LIMIT EQUILIBRIUM CIRCLE THEORY

3.1 *Limit equilibrium circle theory of entry support*

In the three-soft entry, floor heave will cause ribs damage and result in the roof arch increment (Fig. 6).

Assume equivalent floor failure depth h_d is half floor failure depth, there is:

$$h^* = h + h_d \qquad (8)$$

where h^* is equivalent height with floor failure, m; h_d is equivalent depth of floor failure, m.

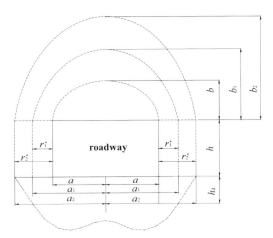

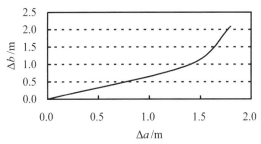

Figure 7. Increment of arch height vs. half width.

Figure 6. Equilibrium circle of entry with floor and rib failure.

If the internal friction angles of ribs and floor are same, the width of floor influence by Eqs. (5) is:

$$r_2 = r_1 + h_d / 2f_b \qquad (9)$$

where r_2 is the maximum depth of the rib failure zone influencing floor failure, m.

In the same way, the equivalent width of entry is:

$$a_2 = a + 0.5r_2 = a_1 + h_d / 4f_b \qquad (10)$$

where a_2 is equivalent half width of floor failure.

So, the height of the limit equilibrium arch by considering the effects of floor and ribs failure is:

$$b_2 = a_2 / 2f = b + (h + h_d) / 8ff_b \qquad (11)$$

where b_2 is the height of the limit equilibrium arch with floor and rib failure, m;

The increment of arch height by floor failure is:

$$\Delta b_2 = b_2 - b_1 = h_d / 8ff_b \qquad (12)$$

3.2 Principles of entry support

Assume, $a = 2.5$ m, $h = 4.0$ m, $h_d = 1.0$ m, and $f = f_b = f_d = \tan 35° = 0.7$, the roof arch height increment Δb with the width increment Δa is in Fig. 7.

Based on above analysis, the entry support principle is put forward:

1. Entry support design must put "roof–ribs– floor" as an integral whole. Rib strengthening is

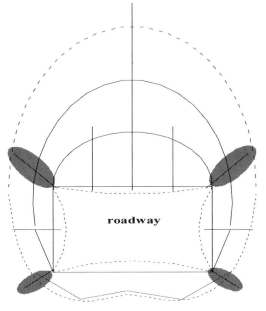

Figure 8. Entry equilibrium circle evolution.

important to maintain the stability of the roof. Floor strengthening is helpful to rib stability.
2. Integration of "rib–floor" and "roof–rib" (shadow zone in Fig. 8) are key areas in the entry.
3. When the height of roof equilibrium arch is less than the general bolt length, bolt length needs to be higher than the arch height.
4. For larger equilibrium arch, cables should be used and cable length should be longer than the equilibrium arch height.

4 CONCLUSIONS

In the three-soft coal seam, the failure zone in the surrounding rock of entry is observed as a circle type. The failure circle can be simply expressed in the limit equilibrium state. The size of the circle is influenced by "floor–ribs–roof" interaction.

The height of the roof equilibrium arch is proportional to the width of the entry. Entry rib failure equals to increase the equivalent width of the entry. Strengthening ribs is helpful to decrease the development of the limit equilibrium arch, and improve the stability of the roof.

Entry floor failure will cause the rib failure and equivalent to increase the entry width. The equivalent width of the entry is proportional to the floor failure depth.

The purpose of entry support is to control the stability of rock mass in the limit equilibrium circle. The length of cable bolt can be determined by the height of the limit equilibrium arch of roof. Entry support should pay attention to the floor and rib control. The entry support design must take roof-rib-roof as a whole support system.

ACKNOWLEDGEMENT

This research is financially supported by National Nature Science Foundation of China (granted by No. 51174278 and No.51674190).

REFERENCES

Bai, J.B. & Li, W.F., et al.(2011). Mechanism of floor heave and control technology of entry induced by mining. Journal of Safety Science and Technology, 28(3), 1–5.

Dong, F.T. & Song, H.W., et al. (1994). Entry support theory based on broken rock zone. Journal of China Coal Society, 19(1), 22–32.

Gou, P.F. & Hu, Y.G.(2006). Effect of faults on movement of roof rock strata in gateway. Journal of Mining & safety Engineering, 23(3), 285–288.

Gou, P.F. & Xin, Y.J, et al.(2013). Stability analysis and mechanism of two side anchorage body in deep mine entry. Journal of Safety Science and Technology, 30 (1), 7–13.

Huang, Q.X. & Liu, Y.W.(2014). Ultimate self-stable arch theory in roadway support. Journal of Mining & safety Engineering, 31(3), 355–358.

Kang, H.P.(1997). The key circle theory of roadway surround rock. Mechanics and Practice, 19(1): 34–36.

Li, S.P. (1986). Concise course of rock mechanics. Xu Zhou: China University of Mining and Technology Press, 139–150.

Miao, X.X.(1990). Natural equilibrium arch and stability of surrounding rock of roadway. Journal of Mining & safety Engineering, 7 (2), 55–57.

Tan, Y.L. Wang, C.Q. & Gu, S.T.(2003). Anchor reinforcement of entry roof stability potential incentive analysis. Journal of Rock Mechanics and Engineering, 22(1), 2211–2213.

Wang, W.G.(1996). Advanced rock mechanics theory. Beijing: Metallurgical Industry Press, 75–82.

Xu, J.K. Zhu, Y.F. & Song, D.Z.(2013). Study on the effect of blasting on the stability of entry disturbance based on the theory of shear beam. Journal of Safety Science and Technology, 9(7), 26–31.

Yan, H. & Zhang, J.X.(2014). Simulation study on key influencing factors to the roof abscission of the entry with extra-thick coal seam. Journal of Mining & safety Engineering, 31(5), 682–686.

Zhao, P.N.(1978). The shape of natural equilibrium arch of the roadway roof by considering the initial horizontal stress. Journal of China University of Mining & Technology, 7(1), 51–58.

Civil, Architecture and Environmental Engineering – Kao & Sung (Eds)
© *2017 Taylor & Francis Group, ISBN 978-1-138-02985-9*

Evolution characteristics of rainfall erosivity area based on the frequency analysis method in Zhejiang Province

Jinjuan Zhang, Gang Li & Fangchun Lu
Zhejiang Institute of Hydraulics and Estuary, Hangzhou, Zhejiang, China

ABSTRACT: Based on the daily rainfall data of 84 meteorological stations from 1980 to 2009 in Zhejiang Province, the power function model was used to calculate rainfall erosivity. The frequency analysis method, the Mann–Kendall test method and the cumulative deviation method were used to mine data changes. And the spatial variation characteristic was analyzed by the radial basis function interpolation method. The results showed that the relationship of distribution area in different periods was R_l (low rainfall erosivity) > R_e (extreme rainfall erosivity) > R_m (medium rainfall erosivity) > R_h (high rainfall erosivity). The area mutation of the R_l and R_m were detected in the early 21st century. And in 1980s, the area under R_h and R_e also mutated. The variation process of rainfall erosivity in the 30 years can be divided into three successive stages. In spatial distribution, central lower grade rainfall erosivity gradually spread into marginal higher grade rainfall erosivity.

1 INTRODUCTION

The potential ability of rainfall to infuse soil erosion is called rainfall erosivity. It is an objective evaluation index for soil separation and transport caused by rainfall. But it is not the concept of "force" in physics, and the rainfall erosivity R factor is calculated by the measured rainfall and rainfall intensity. The R factor is a statistical index to characterize the impact of rainfall on soil erosion and it can be used to evaluate the potential impact of regional climate on soil erosion (Fan et al. 2003). Thus, it becomes an important factor in soil erosion prediction model (Wischmeier and Smith 1978). At present, in the study on the large scale, the daily rainfall data are obtained from conventional meteorological stations are usually used to estimate the rainfall erosivity (Richardson et al. 1983; Zhang et al. 2002). Domestic scholars have made a lot of studies on the temporal and spatial evolution characteristics of rainfall erosivity in different regions of China. For instance, Wang (2013) studied the distribution law of rainfall erosion in the upper Yangtze River. Wu (2011) analyzed the change characteristics of rainfall erosivity in Three Gorges Reservoir area. Mu (2010) discussed the spatial-temporal characteristics of rainfall erosivity on the Loess plateau. Liu (2012) explored the rainfall erosivity in the southwest mountain area. However, many studies were aimed at measuring rainfall erosivity on a certain scale. How to grade the rainfall erosivity and the temporal-spatial variation characteristics of different grades rainfall erosivity were less studied. Therefore, this study used the hydrological frequency method to grade rainfall erosivity. The coverage area changes of different rainfall erosivity were analyzed. It is aimed to provide the basis for comprehensive management of soil and water conservation and ecological environment construction.

2 DATA AND METHODS

Considering the integrity and unity of data sequence, it was ultimately determined to use the daily rainfall data of 84 meteorological stations from 1980 to 2009. The power function model widely used in China was selected to calculate the rainfall erosivity (Zhang et al. 2003). The average annual rainfall erosivity was graded by the hydrological frequency analysis method (Guo et al. 2010). At the same time, the mutation points of the time series of rainfall erosivity were determined by the non-parameter Mann–Kendall test method (Li et al. 2011). The different rainfall erosivity phased variation characteristics were explored by the accumulation of the measured data distance away from average (Wang et al. 2007), abbreviated as ADDA. In the spatial analysis, the spatial interpolation of rainfall erosion force is analyzed by using the radial basis function interpolation method in ARCGIS10.2 software.

3 RESULTS AND ANALYSIS

3.1 The gradation standard of rainfall erosivity

In the frequency analysis of hydrology in China, the Pearson III type curve has been used as the hydrology frequency curve. The parameters of hydrological frequency curve usually were obtained by curve-fitting method. In other words, based on the empirical frequency point and under certain curve-fitting conditions, parameters were solved. The best estimation method that can meet the requirement of hydrologic frequency analysis was determined. Figure 1 showed three types of Pearson III type curve with different parameters. The three curves had good fitness, but by comparison, the curve that features skewness coefficient equal to two times the variation coefficient has a higher adaptability. Therefore, this paper adopted this curve to evaluate the frequency distribution of rainfall erosivity. According to the calculation of the data series and the parameters, the values of rainfall erosivity under different frequency were 8361 ($p = 0.5$), 9572 ($p = 0.2$) and 10249 ($p = 0.1$). Based on the calculated results at different frequencies, the gradation standard of rainfall erosivity was established (in Table 1).

3.2 The area of different rainfall erosivity

The area of different grades rainfall erosivity had an obvious difference (Figure 2). The relationship of different grades of rainfall erosivity in different periods were $R_l > R_e > R_m > R_h$. In 1990s, the area where the R_l and R_e occurred obviously change. In 1980s and the early 21th century, the area proportion of the R_l was 57.5% and 58.8%, respectively. But only 38.3% in 1990s. On the contrary, the area proportion of R_e in 1990s (33.5%) was obviously higher than that in 1980s (14.9%) and the early 21st century (20.9%). For medium rainfall erosivity, in 1980s and 1990s, the area accounted for nearly 17.8%, respectively. The proportion (13.9%) relatively slightly lower in the early 21st century. The area of the R_h was also relatively close in 1980s (9.9%) and 1990s (10.4%), and the proportion decreased to 6.4% in the early 21st century. Through rainfall erosivity gradation and area statistics, the area of the R_l was found to be the largest, and R_e next highest, which mutated in the 1990s and early 21st century. Their area proportion accounted for 76%. It followed that the distribution of different gradation rainfall erosivities appeared to be polarization. The area of R_l in each period broadly remained unchanged. The values of R_m and R_h decrease, but R_e increases.

3.3 Time variation trend of area

Figure 3 clearly shows the annual variation trend of areas with different rainfall erosivities. The average area of the R_l was 5.32×10^4 km², and the inter-annual variation coefficient less than 0.5 (Table 2). The time series showed a weak downward trend ($Z < 0$). The mutation test showed obviously increased mutation occurring in 2003 (5.74×10^4 km²) and 2006 (1.13×10^4 km²),

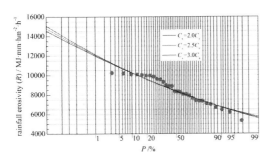

Figure 1. Pearson-III frequency curve of rainfall erosivity.

Table 1. Gradation standard of rainfall erosivity.

Grades	Range	Code
Low	$R < 8361$	R_l
Medium	$8361 \leq R\ 9572$	R_m
High	$9572 \leq R\ 10249$	R_h
Extreme	$R \geq 10249$	R_e

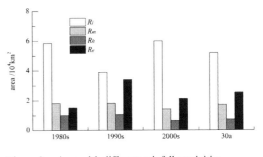

Figure 2. Area with different rainfall erosivities.

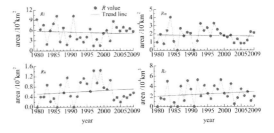

Figure 3. Time series characteristics of area.

1208

Table 2. Gradation standard of rainfall erosivity.

Code	Cv.	Z value	My.	Vr.
R_l	0.47	−0.357	2003	5.74↑
			2004	1.80↓
			2006	1.13↑
R_m	0.52	−0.500	2002	0.48↑
R_h	0.66	0.749	1986	0.34↓
			2005	0.18↓
R_e	0.68	0.785	1984	5.01↓
		0.785	1987	4.05↑

Note: *Cv.* stands for variation coefficient. *Z* value is the statistical eigenvalue of Mann–Kendall. *My.* stands for mutation year. *Vr.* stands for variation range (Unit is 10^4 km²).

and a significantly decreased mutation in 2004 $(1.80 \times 10^4$ km²).

The average area of R_m was 1.79×10^4 km² and the inter-annual variation coefficient 0.52. The annual variation of area also showed a downward trend. The mutation occurred in 2002 $(0.48 \times 10^4$ km²). The average area of R_h was 0.61×10^4 km², the inter-annual variation was highly variable. The area variation trend was increasing, and a significantly decreased mutation appeared in 1986 $(0.38 \times 10^4$ km²) and 2005 $(0.18 \times 10^4$ km²). In terms of extreme rainfall erosivity (R_e), the area was an average of 2.47×10^4 km², Inter-annual variation was highly variable. The time series increased, and the mutation occurred in 1984 $(5.01 \times 10^4$ km²) and 1987 $(4.05 \times 10^4$ km²).

The ADDA processes of different rainfall erosivity were different (Figure 4). The variation process of the R_l can be divided into three stages, which was in the continuing increasing stage (1980–1986), decreasing stage (1987–2002) and gradually increasing stage (2003–2009). The process of R_m was relatively complex, and its area was fluctuating between 1980 and 1992, which belonged to volatile stage. The ADDA curve showed a continuous increase trend (in 1993–2000). But in the nearly ten years, the area was decreasing. The distribution area of R_h also mainly presented three variation stages. The process showed a downward trend in 1980–1992, increasing continuously in 1993–2002, and decreasing in 2003–2009. There were also three great variation stages for the R_e. The area gradually shrank in 1980–1987. The continuous fluctuation in the increasing stage occurred in 1988–2002. But the process showed a slight downward trend in 2003–2009.

3.4 Spatial distribution characteristics

The distributions of rainfall erosivity in different periods were shown in Figure 5. It can be seen that

in 1980s the R_l widely distributed in the north and center region of Zhejiang. The R_m and R_h distributed along the boundary of R_l. The distribution of the Re showed three zones. By comparison, in 1990s the distribution of different rainfall erosivities showed a drastic change. The distribution area shrunk from three sides to the middle. The distribution shape of the Rm and Rh did not change, but the area underwent a change. The area of the Re expanded obviously, and was widely distributed in the west and the southeast coast of Zhejiang. By the beginning of 21st century, the distributions of different rainfall erosivities were further changed. The area of R_l expanded to the west of Zhejiang. The distribution belt of R_m in the west changed short and wide, and the area in the southeast changed narrow. The distribution belt of the R_h in the west disappeared, and the belt in the southeast changed narrow. The R_e was concentrated in the southeastern coast of Zhejiang.

In the distribution of average rainfall erosivity of thirty years, the whole north and center region

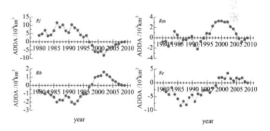

Figure 4. Time series of the ADDA.

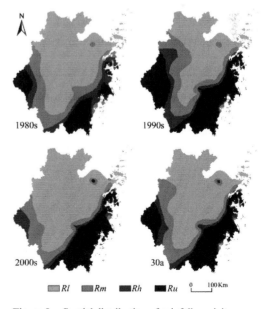

Figure 5. Spatial distribution of rainfall erosivity.

belonged to low rainfall erosivity range. The medium rainfall erosivity mainly showed a long belt distribution in the edge of low rainfall erosivity. The high rainfall erosivity distributed along the edge of medium rainfall erosivity, but the belt was narrower. Extreme rainfall erosivity concentrated in the west and the southeastern coast of Zhejiang.

4 CONCLUSIONS

The study concluded that the relationship of different rainfall erosivities in different periods was $R_l > R_e > R_m > R_h$. It showed that the composition of rainfall erosivity presented polarization. The average area of the low rainfall erosivity accounted for 51%, the extreme take up 25%, and the medium and high were responsible for 24%.

Through time series analysis of the area of different rainfall erosivities, the low and medium rainfall erosivity showed a moderate variation. The time series showed a decreasing trend, and the mutation occurred in the early 21st century. The high and extreme rainfall erosivity showed the highest variation, and the time series had an increasing trend. The mutation mainly occurred in 1980s. The variation process of different rainfall erosivities can be divided into three stages. The process of low rainfall erosivity showed the trend of increase-decrease-increase. The medium rainfall erosivity showed the trend of fluctuation-increase-decrease. The high and extreme rainfall erosivity showed a trend of decrease-increase-decrease.

The low rainfall erosivity was mainly distributed in the north and the center region of Zhejiang. The medium and high rainfall erosivity showed a belt distribution in the west and southeast. The extreme rainfall erosivity showed a concentrated distribution in the west and southeast. During thirty years evolution, the distribution of different rainfall erosivities had changed obviously.

ACKNOWLEDGEMENTS

This work was supported by the Science and Technology Planning Project of Zhejiang Province, China under Grant No. 2015F50060.

REFERENCES

Fan, S.X., Jiang, D.M., Alamusa, et al. 2003. Studies on throughfall model in forest area. Acta Ecologica Sinica, 23(7):1403–1407.

Guo, T.X., Liu H. 2010. The drawing of Pearson-III frequency curve based on excel. Journal of Water Resources Research, 31(1):39–40. (in Chinese)

Li, R., Guo, Q., Gao H.J., et al. 2011. Analysis of climate change features in Tai'an city from 1951 to 2008. Journal of Shang Dong Agricultural University (Natural Science), 42(1):95–101. (in Chinese)

Liu, B.T., Tao, H.P. & Song, C.F. 2012. Study on annual variation of rainfall erosivity in southwest China using gravity center model. Transactions of the Chinese Society of Agricultural Engineering, 28(21):113–120. (in Chinese)

Mu, X.M., Dai, H.L., Gao, P., et al. 2010. Spatial-temporal characteristics of rainfall erosivity in northern Shanxi region in the Loess Plateau. Journal of Arid Land Resources and Environment, 24(3):37–43. (in Chinese)

Richardson, C.W., Foster, G.R. & Wright D.A. 1983. Estimation of erosion index from daily rainfall amount. Transactions of the ASAE, 26(1):153–156.

Wang, A.J., Li Zhi G. & Liu F. 2013. Spatiotemporal distribution of rainfall erosivity for water erosion district in upper reaches of Yangtze River. Bulletin of Soil and Water Conservation, 33(1):8–11. (in Chinese)

Wang, F., Mu, X.M., Jiao, J.Y., et al. 2007. Impact of human activities on runoff and sediment change of Yan He River based on the periods divided by sediment concentration. Journal of Sediment Research, (4):8–13. (in Chinese)

Wischmeier, W.H. & Smith, D.D. 1978. Predicting rainfall erosion losses: a guide to conservation planning. Agriculture handbook, No. 537 USDA.

Wu, C.G., Lin D.S., Xiao, W.F., et al. 2011. Spatiotemporal distribution characteristics of rainfall erosivity in Three Gorges Reservoir Area. Chinese Journal of Applied Ecology, 22(1):151–158. (in Chinese)

Zhang, W.B. & Fu, J.S. 2003. Rainfall erosivity estimation under different rainfall amount. Resources Science, 25(1):35–41. (in Chinese)

Zhang W.B., Xie, Y. & Liu B.Y. 2002. Rainfall erosivity estimation using daily rainfall amounts, Geographical Research, 22(6):705–711. (in Chinese)

Civil, Architecture and Environmental Engineering – Kao & Sung (Eds)
© *2017 Taylor & Francis Group, ISBN 978-1-138-02985-9*

Research on the Norbulingka garden art characteristics under the background of Han-Tibetan cultural blending

Chang Liu & Min Wang
Huazhong Agricultural University, Wuhan, China

ABSTRACT: Norbulingka is one of the centers of politics, religion and culture in Tibet. Only in Norbulingka, a perfect combination of architectural art, religious art and garden art can be seen. Garden art is the detailed embodiment of people's life. This paper, through the analysis of garden art of Norbulingka, explores the world outlook, philosophy of life and values of the Tibetan. In the meanwhile, the research on the common features between Tibetan garden art and traditional Hans garden art reflects the two nationalities have many similarities when they are dealing with the relationship between man and nature.

1 INTRODUCTION

The Han and Tibetan people created a completely different living space in order to meet the needs of their own lives because of the different living environments. Especially the Tibetan people's living environment is more difficult than the Han people. But on the other hand, the Han and the Tibetan people living together formed a mutually dependent and harmonious ethnic relation. The culture of the two nationalities influenced each other.

Garden is a reflection of the higher level of people's spiritual and material life called "linka" in Tibet. "Linka" means a small forest, trees or gardens originally. And the same time, the meaning of garden is same with the Han Chinese garden. But there are great differences in the form and style of the garden between "linka" in Tibetan and garden in inland of China. In the Tibetan language, "linka" describes a beautiful place with lush flowers and trees and the environment is favorable. On the other hand, "linka" is a small piece of artificial forest surrounded by earth, stone or fences in order to serve the ruling class of the old Tibet to play and feast. This paper just focuses on the characteristics of garden art as a traditional garden now. "Linka" is an important form of China's classical gardens in Tibetan Gardens.

Norbulingka is the representative of Tibet Garden "linka". Its gardens, sculptures, murals and architecture have a very high artistic status. As far as the garden art of Norbulingka is concerned, some of its characteristics are related to the Han culture. Through studying the relationship between Han culture and Norbulingka, we can observe that the philosophy of life and values of Tibetan and Han are the same.

2 NORBULINGKA

2.1 Introduction

"Norbulingka" means the garden is precious as pearls in the Tibetan language. It covers an area of 36 hectares located in the southwest of Potala Palace spanning at least 2 kilometers. Norbulingka was listed as a world heritage site in 2001 by United Nations Educational, Scientific and Cultural Organization. It was founded in the middle of the eighteenth century to handle affairs, held ceremonies and religious activities, spend summer holiday for Darailama. The three main groups of palace are "gesang" palace, "gesang" palace and "dadanmijiu" palace. Norbulingka is the most distinctive of the large-scale garden in Tibet, which sets gardens, palaces and buildings as one. It shows the unique history, culture, nature and other features of Tibet with high value in history, science and art.

2.2 Origin

This area was originally a shrub in which the trees were lush and the birds and animals often appeared. People called this scenic place "lawacai". According to the records of Tibetan documents, there was a fountain said to be able to cure people. So the VII Dalai Lama "gesangjiacuo" usually went there to have rest. At the time of the Ambans, a simple structure for rest called "wuxiao palace" which means "pavilion palace" was designed for the emperor Qian Long in the Qing dynasty. This is the initial state of the Norbulingka.

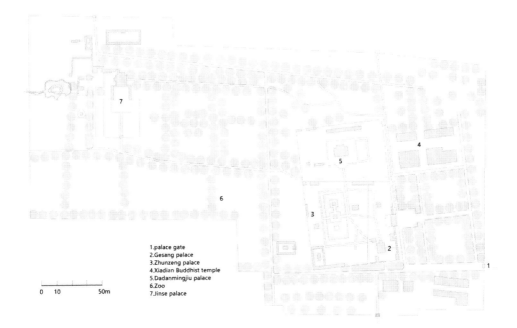

Figure 1. The plan of Norbulingka.

1.palace gate
2.Gesang palace
3.Zhunzeng palace
4.Xiadian Buddhist temple
5.Dadanmingjiu palace
6.Zoo
7.Jinse palace

0   10   50m

2.3 *Layout*

Norbulingka can be divided into two parts of East and West according to the distribution of buildings ordinarily. The western part is called "Jinselinka" and the eastern is called "Norbulingka" by the local people. But according to the overall landscape effect, Norbulingka can be divided into the palace area, before the palace area and forest area. Each area has its different landscape characteristics. The palace area had palaces built in different periods. The area before the palace was built was used to hold meetings. The forest area is the back garden for Dalai Lama. But the three areas connected perfectly. The overall layout of the space is very sparse with distinct levels, and the layout of the form is very diverse.

3 LANDSCAPE FEATURES

3.1 *Terrain*

Basically the terrain is flat without artificial construction rockery because Tibetan people are very devout to the mountains. So Tibetan people respect the original landform extraordinarily when they created the gardens. On the other hand, the vast ground is very rare in the plateau region. At the meanwhile, this type of site selection was done to simulate the "Elysian Fields" depicted

Figure 2. The color of the building is very bright.

in Buddhist scriptures because the ground of the "Elysian Fields" is flat like a palm.

3.2 *Planting*

The description for the Elysium in Buddhist scriptures is "Seven roads to form a road network, seven lines of trees by the roadside". The cedar and poplar are planted by column and occasionally planting peach and willow to create a fresh literary scene. The way of planting mainly used cypress to memory Shakya Muni feed by cedar seed.

3.3 *Decoration*

The altitude is high and the air is thin in Tibet plateau. Because of the air with high transparency, the various colors of objects are constantly changing under sunlight. So the bright colors are used to decorate buildings. The interior decoration of buildings is affected by the Hans techniques such as wood windows and screens.

4 SIMILARITIES

The Han and the Tibetan people have formed special national characteristics because of different ways of life. But on the other hand, the two nationalities never stop blending. The traditional culture of Han and Tibetan people communicates from past to present. So there are many similarities in their way of daily life. In garden art aspect, the characteristics of Norbulingka have many resemblances with classical garden for Hans.

4.1 *A lake with three hills*

There is a pool in the west of the Gesang palace. Three islands are arranged in a straight line. One of the islands is called "Huxingong", and another one is "Longwangdian". A lot of trees planted on the last island at the south. "A Lake With Three Hills" is a classical model of terrain design of Chinese classical gardens first appeared in "Shanglin yuan" in the Qin Dynasty. This composition simulated the three mountains called "Penglai" "Fangzhang" "Yingzhou," which is a paradise living many gods in Chinese fairy tales.

4.2 *The combination of palace and garden*

The palace and the garden for imperial families, which is in organic combination, is an important template of the later imperial gardens planning represented by the "Jianzhanggong" from Han dynasty. The palace is in the north and the garden is in the south in Jinse palace. This pattern is very similar to the Imperial Palace and Jingshan in Beijing.

Figure 3. Huxingong is the middle one of the three islands in water.

4.3 *The function for animal farm*

The early Chinese gardens have the function of animal farm as early as in the Shang Dynasty. This function is handed down in the Norbulingka. There is a zoo in Norbulingka covering an area of 4000 square meters and breeding a variety of rare animals.

The palace and the garden for imperial families, which is in organic combination, is an important template of the later imperial gardens planning represented by the "Jianzhanggong" from the Han dynasty. The palace is in the north and the garden is in the south in Jinse palace. This pattern is very similar to the Imperial Palace and Jingshan in Beijing.

4.4 *Planting*

The planting way in Norbulingka mostly is similar to the Temple of Heaven in Beijing. It is called determinant planting, which is the form of an extension of the building to set the great and solemn mood for the building. On the other hand, this way forms a sense of sequence of space creating a holy and sacred tour experience.

4.5 *Garden in the garden*

Garden in the Garden is one of the structural layouts of the Royal classical gardens. The small garden is the basic units for the whole garden such as the "Xiequyuan" in the Summer Palace. We can see this pattern in Norbulingka that the boundaries between each atom are distinct. But this layout in the Norbulingka is formed because each dominator builds a park. So there is no organic link between the respective small gardens. At this point, the Tibetans and Hans are still very different.

4.6 *Scenery-borrowing*

The scenery-borrowing is an important artistic technique in the Constitution. Its core is the universe view of "harmony between man and nature".

Figure 4. Trees are planted in rows.

Figure 5. The garden is divided into different small parts by the plants.

Figure 6. At the end of the road before the palace, we can see the other palace and the mountains in the distance.

The layout in Norbulingka particularly emphasizes the placement technique to create a discrete echo around the landscape, which sets off each other.

5 CONCLUSIONS

As part of a culture, classical garden is not only closely related to literature, painting and other art forms, but also to the detailed embodiment of the people's way of life. Though the Han and the Tibetan people created a completely different national culture due to different living environments, they had influenced each other over the past two thousand years. Classical garden is typical of the blending of the two cultures.

Classical garden is the combination of traditional culture, both of Han and Tibetan people. They are the embodiment of pursuit of the material and spiritual life. In terms of the development degree, linka has not reached the level of garden art. But classical garden is typical. Linka not only has been affected by the culture of Han in many ways but also absorbed it. The styles and methods of gardening and some ideas and life styles spread to Tibet. People build gardens and feel the beauty. People can experience both the culture of Han and Tibet in one garden, which shows the combination of different culture.

REFERENCES

Chenchen L, D. (2010). Chinese Feng Shui and environmental art. Xi'an Academy of Fine Arts.

Chuanli D.D. (2005). The analysis of the Tibetan traditional garden. Southwest Jiao Tong University.

Huajun Y. & Changwen C, J. (2006). A comparative study of Tibet "Linka" and Chinese traditional garden. 79–80.

Qiusi D.D. (2004). The cultural connection of Chinese traditional garden. Chongqing University.

Shuiche Y.M. (2012). The Tibetan art. Jilin Publishing Group.

Weiquan Z, M. (2008). History of Chinese Classical Gardens. China Building Industry Press. 754–756.

Xuezhi J, M. (2006). Chinese Garden Aesthetics. China Building Industry Press. 132–134.

Zhong J, J. (2002). Discussion on landscape design in Tibet. 11.

Civil, Architecture and Environmental Engineering – Kao & Sung (Eds)
© 2017 Taylor & Francis Group, ISBN 978-1-138-02985-9

Releasing flux and characteristic of nitrous oxide from wastewater land treatment systems

X.X. Zhang, Y.H. Li, Y. Wang & H.Y. Weng
School of Resourses and Civil Engineering, Northeastern University, Shenyang, Liaoning, China

ABSTRACT: Wastewater land treatment system can remove pollutants deeply through a combination process of physical, chemical and biological functions. During the complex process of biological metabolism, greenhouse gases, such as nitrous oxide (N_2O), are released from the substrate bed, causing the block of land treatment system and the destruction of atmospheric environment. This paper reviews the production mechanism of N_2O released from wastewater land treatment system. Furthermore, the influence of impact factors (types of system, plant and temperature) on N_2O release characteristic was summarized in constructed wetland system, slow rate land treatment system, rapid infiltration system, overland flow system and subsurface wastewater infiltration system. The method of accounting N_2O releasing flux was introduced. Finally, the viewpoint was pointed out that temperature, matrix composition and operation conditions affect N_2O releasing flux considerably.

1 INTRODUCTION

Wastewater land treatment system utilizes self-regulation mechanism and comprehensive purification function of soil–microorganism–plant to make sewage reclamation harmless, including rapid infiltration system, slow rate land treatment system, overland flow system, wetland system, and subsurface wastewater infiltration system (Li et al. 2015). Compared with traditional processing for wastewater such as activated sludge process, wastewater land treatment system gains the advantages of high decontamination efficiency, low operating costs and ecological services function (Yang et al. 2007, Zachrit & Fuller 1993, Zheng et al. 2012). However, wastewater land treatment system still has some shortcomings in the design and practical application. The greenhouse gas releasing from this kind of system is mainly nitrous oxide (N_2O). Besides, the decomposition product of N_2O is the main source of NO, which is the key substance in the chemical chain reaction of ozone layer destruction (Lin et al. 2008). Wastewater land treatment system is a useful and necessary complementary form of centralized sewage treatment methods. It has been applied widely in rural, suburban, and other areas without uniform sewage networks. So the releasing flux and characteristics of N_2O has aroused worldwide concern.

2 MECHANISM OF N_2O EMISSIONS

At present, domestic and foreign scholars have basically reached a consensus on the release mechanism of N_2O in land treatment systems. According to literature, N_2O is generated by the following three kinds of biochemistry processes (Capra et al. 2007, Lin et al. 2008, Liao et al. 2010): ① Nitrification process: In the process of N converted into NO_2^{-1} and NO_3^{-1} as electronic receptor, N_2O releases as side-product (Fig. 1) (Qu et al. 2008). This process is controlled by aerobic nitrifier and Nitrosobacteria; ② Denitrification process: in the process of $NO_3^{-1} - N$ converted into N_2, N_2O is produced as an intermediate product (Fig. 2) (Qu et al. 2008). This process is controlled by anaerobic denitrifying bacteria; ③ Non-biological nitrogen (N) converting processes: NO_3^{-1} and NO_2^{-1} react with organic or inorganic matter to N_2O or N_2. Most of the reports indicate that processes of nitrification and denitrification are the main process of N_2O produced.

According to the release mechanism, the factors affecting process of nitrification and denitrification in the substrate layer is also the main factor influencing N_2O emissions. The influencing factors are as follows.

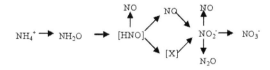

Figure 1. N_2O release from the nitrification process.

$$NO_3^- \longrightarrow NO_2^- \longrightarrow NO \longrightarrow N_2O \longrightarrow N_2$$

Figure 2. N_2O release from the denitrification process.

3 INFLUENCING FACTORS OF N_2O EMISSION

3.1 *Effect of system types*

Due to the differences in building forms, process, and operation parameters between the different types of land treatment systems, there are big differences in the process of N_2O release. Research (Wu et al. 2009) shows that both Subsurface Flow (SF) and Free Water Surface (FWS) constructed wetland system to be emission source of N_2O, and the SF process promotes N_2O emissions. The mean N_2O fluxes are 0.2965 mg·m^{-2}·h^{-1} and 0.0282 mg·m^{-2}·h^{-1} respectively in SF and FWS constructed wetland systems. This is mostly because of the wetland types and substrate water content. In FWS constructed wetland, sewage on soil layer reduces the permeability of the soil. The anaerobic conditions promote denitrification process, and hinder N_2O diffusing to the atmosphere, thus increasing the rate of N_2O converted to N_2. So FWS process can reduce N_2O emissions.

3.2 *Effect of plant types*

Plants are important part of the soil–microorganism–plant system. Its type and coverage can impact soil nitrogen load, Dissolved Oxygen (DO) etc. and then affect N_2O release. Reeds, cattail and sedge are common aquatic plants in wetland. Wetland plants work in two aspects: ① they can absorb some of the nutrients; ② their roots provide places for the metabolism of microbes and the degradation of nutrients (Li et al. 2007). Plants remove nitrogen through absorption. More plant biomass can absorb more N, which means fewer N_2O emissions. Plant types, environment, nutrition and plant adaptation can affect the ability of N_2O removal (Wang et al. 2013). Research (Li et al. 2010) indicates that N_2O fluxes are quite different in vegetation cover and vegetation-free sediment systems. Compared with vegetation-free sediment systems, reeds in vegetation sediment systems show improvement on N_2O emissions, and during the growing season, the effect is more pronounced, increasing the emissions from 0.71 µg·m^{-2}·h^{-1} to 566.28 µg·m^{-2}·h^{-1}. Under the typical dry-wet alternate condition of the high tidal flat, plant physiological activities $NO_3^- - N$, $NO_4^+ - N$, temperature and water content affect the nitrification–denitrification process.

3.3 *Effect of substrate types*

Substrate is the living place of microbes. Its composition, structure, and physico-chemical properties greatly affect N_2O emissions by affecting microbiological activity. For example, higher porosity substrate means higher DO level. High DO concentration hinders N_2O release. Microbiological denitrification is a strictly anaerobic process. When DO concentration is above 0.2 mg/L, denitrification is difficult to happen (Lim et al. 2001). High soil pH is not conducive to N_2O release. Generally, the optimum pH for nitrification is 7.5–8.6, while 7–8 for denitrification (Al-Omari et al. 2003). When pH is lower than 6.0, denitrification is rapid as pH decreases. Sun et al. (2007) show that humus marsh soil and meadow marsh soil have similar change ranges of N_2O emission rate. But the latter has a much higher N_2O emission rate; the change ranges of their denitrification rate are not similar. Meadow marsh soil has a higher denitrification rate. Nitrification plays an important part in N_2O release and nitrogen removal in meadow marsh soil, while denitrification is important in humus marsh soil.

3.4 *Effect of temperature*

Nitrogen cycle in the soil mainly depends on catabolic metabolism of microbes and absorption of plants. Temperature greatly impacts the growth and metabolism of plants and microbes. Microorganisms will be inactive in low temperature, so the influence of biochemical process can be ignored. Above a certain temperature, the rate of biochemical reaction increases rapidly following Arrhenius formula (Rodriguez-Aragón et al. 2005). The optimal temperature of nitrification is 25°C–35°C. The nitrification will be restricted if the temperature is over 40°C or under 4°C. Temperature for denitrification is wide with a range of –4°C to 70°C, and the optimal temperature is 10°C–25°C (Lang et al. 2012). Researches show that N_2O fluxes and water temperature have a significant positive correlation ($p < 0.05$) in constructed wetland (Wu et al. 2009).

Grunfeld (1999) believes that solar radiation can accelerate gas exchange of plants. Radiation increases the temperature. On the one hand, molecular diffusion is faster in high temperature, conducive to the spread of N_2O. On the other hand, high temperature increases microorganism activities, conducive to N_2O generation. N_2O is the intermediate product of nitrogen transformation. Under stable environmental conditions, soil nitrogen remains relatively stable forms. Environment changes cause the changes in nitrogen forms. Therefore, N_2O emissions not only depend on the environmental conditions, but also the intensity and frequency of environment changes. Wang et al. (2007) analyzed N_2O emission characteristics in freezing–thawing cycles. The result shows that freezing–thawing cycles impact the process and strength of nitrogen transformation, increasing

N$_2$O emissions. On the one hand, freezing–thawing cycles change the aqueous phase in soil, breaking the stability of the soil aggregates. Thus, N$_2$O is released. On the other hand, the decomposing of microbial residues releases many nutriments. Nitrification and denitrification use these nutriments to produce N$_2$O.

3.5 *Effect of hydrography*

Soil water content affects wetlands N$_2$O generation primarily through impacting microbial activity and DO content (Wang et al. 2013). Low soil moisture and long sustained flooding are not conducive to the growth of nitrifying and denitrifying bacteria. Pore water saturation of soil has been an important controlling factor of N$_2$O release for a long time. It affects the degradation of organic matter, ammonification rate, denitrification rate and the path of N$_2$O emissions. So pore water saturation reflects the humidity conditions of sediment better (Lang et al. 2012, Zheng et al. 2014). Studies (Arriaga et al. 2010) show that when the soil pore water saturation is 30%–60%, N$_2$O emissions are mainly from nitrification, and when 60%~90%, mainly from denitrification. Soil pore water saturation too high or too low are detrimental to N$_2$O emissions. Research (Wolf et al. 2000) indicates that nitrification and denitrification take place concurrently. Water content determines the main reaction. When soil moisture is unsaturated, nitrification is the main reaction, and NO production was significantly higher than that of N$_2$O production. When soil moisture is saturated, denitrification plays a leading role, but nitrification cannot be ignored, and N$_2$O production is higher in unsaturated moisture conditions.

3.6 *Effect of Oxidation–Reduction Potential (ORP)*

Oxidation–Reduction Potential (ORP) is an important indicator of soil permeability and oxygen condition. It can affect the production and emission of greenhouse gases. Kong et al. (2002) found that there is a significant positive correlation between ORP and N$_2$O emission. When ORP remains above +200 mV with good ventilation, N$_2$O production is half of that in no ventilation. To ensure the efficiency of the system, Subsurface Wastewater Infiltration System (SWIS) will continuously run from aerobic to anaerobic environments. This provides different ORP conditions, so nitrification and denitrification can go on at the same time. In denitrification process, oxygen will inhibit N$_2$O from transforming to N$_2$, making N$_2$O final metabolite of denitrification. So, high ORP conditions can increase N$_2$O emission.

4 ACCOUNTING FOR N$_2$O RELEASING FLUX

At present, there is no accurate and unified quantitative method for N$_2$O emissions accounting in wastewater land treatment system. Scholars mostly reference the estimating methods and models of agro-ecological system to calculate. Intergovernmental Panel on Climate Change (IPCC) guideline recommends a method (Liu et al. 2013). It introduces the calculation process of N$_2$O emission in detail. Coefficients in the emission process are important. Calculation of N$_2$O emission can be divided into two parts: ① N$_2$O releases after sewage or treated sewage is discharged into receiving waters, and the emission factor is 5 g N$_2$O–N/kg N; ② N$_2$O releases from the process of nitrification and denitrification in the sewage treatment plant, and the emission factor is 3.2 g N$_2$O–N/(person· a) (Wang et al. 2010). However, it is just a rough calculation based on per capita emissions, and it is based on amount of nitrogen input and emission factors. Therefore, estimation results are uncertain. Based on the IPCC methods, Zheng (2008) and others have developed a regional nitrogen cycle model, IAP-N. IAP-N model fully considers every process of nitrogen cycle in ecosystems, so the results are more scientific than the IPCC method. Li et al. (2014) used IAP-N model to estimate N$_2$O emissions in a system similar to constructed wetland ecosystem. IAP-N models have been used for estimating N$_2$O emissions in large-scale ecosystems. In addition, to simulate the complex process of organic nitrogen cycle in ecosystems, some models such as DNDC, CASA, CENTURY and EXPERT-N are expected to be used for N$_2$O emissions accounting in sewage land treatment system.

5 SUMMARY AND PROSPECTS

Wastewater land treatment system has been widely used, especially in underdeveloped towns and scattered scenic spots lacking uniform drainage systems. Emission mechanism of N$_2$O was generally accepted. Among the dominant factors, temperature, matrix composition and operation conditions (such as influent loading, hydraulic retention time and distribution depth) affect N$_2$O releasing flux considerably. However, the research on the releasing characteristic of N$_2$O is inadequate. Biological-physical-chemical interactions and the migration and transformation of nitrogen need further research.

ACKNOWLEDGEMENT

This work was supported by the National Natural Science Foundation of China under Grant

[41571455, 51578115]; Basic Science Research Fund in Northeastern University under Grant [N140105003].

REFERENCES

Al-Omari, A. & Fayyad, M. et al. 2003. Treatment of domestic wastewater by subsurface flow constructed wetlands in Jordan. *Desalination* 155(1): 27–39.

Arriaga, H., Salcedo, G. & Calsamiglia, S. et al. 2010. Effect of diet manipulation in dairy cow N balance and nitrogen oxides emissions from grasslands in northern Spain. *Agriculture, Ecosystems and Environment* 135: 132–139.

Capra. & Seicolone et al. 2007. Recycling of poor quality urban wastewater by drip irrigation systems. *Journal of Cleaner Production* 15: 1529–1534.

Grunfeld, S. & Brix, H. 1999. Methanogenesis and CH_4 emissions: effects of water table, substrate type and presence of Phragmites australis. *Aquatic Botany* 64(1): 63–75.

Kong, H. N. & Kimochi, Y. 2002. Study of the characteristics of CH_4 and N_2O emission and methods of controlling their emission in the soil-trench wastewater treatment process. *The Science of the Total Environment* 290: 59–67.

Lang, M., Li, P. & Zhang, X. et al. 2012. Effects of land use type and incubation temperature on soil nitrogen transformation and greenhouse gas emission. *Chinese Journal of Applied Ecology* 18(10): 2361–2366.

Liao, Q. & Yan, X. et al. 2010. Models of N_2O Emission from Agricultural Fields: A Review. *Journal of Agro-Environment Science* 29(5): 817–825.

Li, H., Xia, H. & Xiong, Y. et al. 2007. Research progress of factors affecting soil formation and emissions of greenhouse gases. *Ecology and Environment* 16(6): 1781–1788.

Li, J., Wang, Y. & Wang, L. et al. 2015. Evaluation indices of greenhouse gas mitigation technologies in cropland ecosystem. *Chinese Journal of Applied Ecology* 26(1): 297–303.

Lim, P. E., Wong, T. F. & Lim, D. V. et al. 2001. Oxygen demand, nitrogen and cop-per removal by free-water-surface and subsurface flow constructed wetlands under tropical conditions. *Environment International* 26: 425–431.

Lin, K., Feng, J. & Zhu, N. et al. 2008. Discussion on Disadvantages and Solutions in Wastewater Land Treatment Technology. *Environmental Science and Management* 33(11): 76–82.

Liu, H., Li, X. & Luo, L. et al. 2013. Constructed rapid infiltration wastewater treatment system of GHG emission calculation and comparison. *Proceedings of the 2013 annual academic meeting of the Chinese society for environmental sciences* 868–873.

Li, Y., Liu, M. & Lu, M. et al. 2010. Phragmites australis effects on N_2O emission in the Chongming eastern tidal flat. *Acta Scientiae Circumstantiae* 30(12): 2526–2534.

Li, Y., Wang, Y. & Wang, C. et al. 2014. Analysis of N_2O emissions from the agro-ecosystem in Fujian Province. *Chinese Journal of Eco-Agriculture* 22(2): 225–233.

Qu, S., Gao, B. & Wang, J. et al. 2008. Study on generation mechanism and influencing factors of agricultural soils greenhouse gas. *Ecology and Environment* 17(6): 2488–2493.

Rodriguez-Aragón, L. J. & López-Fidalgo, J. et al. 2005. Optimal designs for the Arrhenius equation. *Chemometrics and Intelligent Laboratory Systems* 77(1): 131–138.

Sun, Z., Liu, J. & Yang, J. et al. 2007. N itrification-denitrification and N_2O emission of typical Calamagrostis angustifolia wetland soils in Sanjiang Plain. *Chinese Journal of Applied Ecology* 18(1): 185–192.

Wang, H. 2010. Reducing carbon emissions of Municipal wastewater treatment field. *Water & Wastewater Engineering* 36(12): 1–4.

Wang, H., Jia, X. & Gao, B. et al. 2013. Soil GHG emissions responding to carbon and nitrogen added in different soil types. *Acta Pedologiga Sinica* 50(6): 1172–1182.

Wang, L., Cai, Y. & Xie, H. et al. 2007. Relationships of soil physical and microbial properties with nitrous oxide emission under effects of freezing-thawing cycles. *Chinese Journal of Applied Ecology* 18(10): 2361–2366.

Wolf, I. & Russow, R. 2000. Different pathway of formation of N_2O, N_2 and NO in black earth soil. *Soil Biology and Biochemistry* 32(2): 229–239.

Wu, J., Zhang, J. & Jia, W. et al. 2009. Nitrous Oxide Fluxes of Constructed Wetlands to Treat Sewage Wastewater. *Environmental Science* 30(11): 3146–3151.

Yang, W., Liu, C. & Wen, H. 2007. Discussion on wastewater land treatment system. *Chinese Journal of Soil Science* 38(2): 394–398.

Zachrit, W. H. & Wayne, F. J. 1993. Performance of an artificial wetlands fiber treating facultative lagoon effluent at Carville, Louisiana. *Water Environmental Research* (65): 46.

Zheng, J., Pan, G. & Cheng, K. et al. 2014. A discussion on quantification of greenhouse gas emissions from wetlands based on "2013 Supplement to the 2006 IPCC Guidelines for National Greenhouse Gas Inventories: Wetlands". *Advances in Earth Science* 29(10): 1120–1125.

Zheng, J., Zhang, Y. & Chen, L. et al. 2012. Nitrous oxide emissions affected by tillage measures in winter wheat under a rice-wheat rotation system. *Acta Ecologica Sinica* 32(19): 6138–6146.

Zheng, X. H., Liu, C. Y. & Han, S. H. et al. 2008. Description and application of a model for simulating regional nitrogen cycling and calculating nitrogen flux. *Advances in Atmospheric Sciences* 25(2): 181–201.

Civil, Architecture and Environmental Engineering – Kao & Sung (Eds)
© *2017 Taylor & Francis Group, ISBN 978-1-138-02985-9*

A study of development planning for conservation areas in Taiwan

Tung-Ming Lee
China Institute of Technology, Taipei, Taiwan

ABSTRACT: The conservation of ancient monuments has special meaning in its times. As time goes by, the area preserves the original style of the area. More and more conservation areas are formed step by step in the procedure of the changes and development. And they are still the units in the construction of a developing city. Nowadays, we are still paying much attention to the hardware in the case of which the conservation areas of ancient monument and conservation areas are delimited in few cases. The great majority of ancient monuments are still persevered without integrity conservation concepts.

1   INTRODUCTION

The protection of cultural heritage assets by the United Nations Educational, Scientific, and Cultural Organization (UNESCO) focuses mostly on the authenticity and integrity of such assets. In addition to buildings being adequately preserved, proximal environments should be protected to the same level, and attention should be paid to the historical and cultural pathways and spaces to maintain the integrity and authenticity of monuments or historical buildings. In the twenty-first century, the concept of cultural heritage protection focuses on an integrity conservation approach, in which the authenticity and integrity of a conservation area should be examined. The famous Irish poet Oscar Wilde said: "The one charm of the past is that it is the past." In light of this statement, protecting conservation areas is the most onerous and arduous task among initiatives for protecting historical and cultural heritage. A monument and its conservation area are a historical and cultural phenomenon marking the history of a specific era. Preserving the memory of each period, protecting the continuity of history, and retaining the context of human civilization are essential for the future development of human civilization. The development and context of initiatives for protecting the historical and cultural heritage internationally involve (1) protecting not only architectural and artistic treasures (e.g., palaces, churches, official residences, and temples) but also historic buildings that reflect the lifestyles of regular people (e.g., houses and shops) and (2) protecting all the relevant historical structures, whether an individual building, an environment, or a district.

The current development in cultural asset conservation area is not very effective; the problems are the major issues. It lacks in integrity conservation concept. Legal regulations are rudimentary.

Administrative coordination is difficult. Data integration was not employed.

Although Articles 33–36 of the Cultural Heritage Conservation Act stipulate the definition of a conservation area and regulate its development, the topic of conservation areas is complicated, and thus current legal regulations continue not to satisfy the demands of developing conservation areas. Therefore, relevant examples of conservation areas in Taiwan are scant, and no concrete achievements serve as a reference for communities in the Taiwanese society. Neither are there legal regulations that can be followed in developing a conservation area, thereby creating a predicament where the management of conservation areas focuses on tangible but not intangible assets. Future development of conservation areas should first focus on preserving intangible assets and on elevating the Quality Of Life (QOL) of local residents. In addition to preserving tangible and intangible cultural assets, the following should be incorporated into future development projects: (1) integration of the techniques required for meeting demands of daily living and (2) safety and disaster prevention projects that are critical to residents' lives and property safety. Establishing the parameters of cultural tourism for providing local cultural tourism resources can further the application of cultural resources and the development of cultural tourism, thus enabling culture to serve as a basis for elevating the cultural and economic value and local economic competitiveness of the conservation areas.

2   CURRENT SITUATION AND PROBLEMS OF CONSERVATION AREAS IN TAIWAN

Currently, most research reports and survey results on monuments in Taiwan have discussed how conservation areas should be designated, but such designations have not progressed beyond the reporting stage. However, the required management methods and measures remain at the recommendation stage. Although a few monument conservation areas have been officially designated "conservation areas for urban planning" by urban planning departments, the integrity of designated conservation areas and the conservation of monuments remains a problem; hence, Taiwan's Directions for the Operation of Urban Design Review or Regulations for Urban Land Use and Control cannot be effectively implemented in conservation areas, and designated conservation areas for urban planning cannot be adequately managed and controlled. Accordingly, conservation areas for urban planning exist for just name sake; this problem extends to matters related to coordinating relevant organizations and preserving intangible assets.

The aforementioned problems can be divided into five major issues:

A. Unclear Concept of Integrity Conservation.
B. Legal Regulations on Tangible Asset Conservation.
C. Ambiguity in Future Benefits and Relevant Measures.
D. Difficulty of Administrative Coordination between Urban Planning and Cultural Authorities.
E. Difficulty of Data Integration.

3 THEORY OF CONSERVATION AREA DEVELOPMENT IN THE TWENTY-FIRST CENTURY

3.1 Household, neighborhood, and community units in designated conservation areas

In earlier times, a characteristic of historical spaces was that they generally included areas that were within a walking distance. Household and neighborhood relationships were established through interpersonal and social organization, blood lineage, and geopolitical relations. Such relationships enabled establishing friendly, harmonious, and stable living domains with specific characteristics, including temples, churches, plazas, public spaces, and units that were within walking distance of other units. The current significance attached to areas designated for preserving monuments involves the shared consciousness of the residents living near the areas; in other words, the significance of monument conservation areas is associated with the household, neighborhood, and community units within the areas.

3.2 Design and maintenance of urban public spaces in designated conservation areas

From traditional cities in Europe to Taiwan, public buildings (e.g., churches, temples, and private residential buildings) are often surrounded by plazas,

temple courtyards, roads, or parks and green spaces, thus forming systems of public spaces in urban cities. Accordingly, the maintenance of public spaces must be considered in maintaining and preserving monuments and their surrounding areas with the aim of enhancing the quality of environments designated for monument conservation.

3.3 Fair resource redistribution and rational benefit distribution in designated conservation areas

Because designating conservation areas for monuments inevitably increases or decreases the rights and interests of certain section of people, most countries offer government subsidization (e.g., subsidies, tax reductions, reward systems, or transfers of development rights) to compensate these people. Therefore, fairness in redistributing resources must be considered when designating conservation areas for monuments, and the rationality of benefit compensation must be maintained after the redistribution of resources.

3.4 Integrated coordination in designated conservation areas

Designating conservation areas for monuments entails managing and coordinating landscapes and public facilities (e.g., ornamental plants, landmarks, architectural styles, and water and electricity infrastructure) through urban planning and urban design measures. For example, expanding the designation of a conservation area through urban renewal measures must be coordinated with the regulations on the transfer of development rights. Accordingly, conservation areas should be designated according to a set of integrated methods.

3.5 Expansion of designated conservation areas

Designating conservation areas for monuments should include improving undesirable surrounding environments, using resources effectively, and maintaining and enhancing the area's socioeconomic activities. The designation should expand in coordination with overall development projects for cities and towns to ensure that the designated conservation areas are embedded with specific significance.

3.6 Revitalization of designated conservation areas

A lack of management and maintenance at monuments and their surrounding environments can cause these valuable assets to be lost someday. Accordingly, after conservation areas are designated, governments should provide subsidies for the purpose of preserving and maintaining the environments of these areas, thereby improving and revitalizing the QOL of local residents.

4 NEW ORIENTATIONS FOR CONSERVATION AREA DEVELOPMENT IN THE TWENTY-FIRST CENTURY

Humans experienced two world wars in the twentieth century that led to major social, economic, and cultural problems such as ecological damage, social disintegration, and environmental pollution. In the twenty-first century, humans have reflected on the various catastrophes and conflicts caused previously and started to construct new values, cultures, and lifestyles. Developing conservation areas provides an opportunity for people to preserve and examine the various cultural heritage sites left by their predecessors over the last century as well as to learn about universal values and formulate personal goals. Recently, theories regarding the subsequent development of heritage assets have facilitated cultivating potential developmental orientations and goals.

4.1 *Conservation of the outstanding universal value of monuments: a central initiative and goal of conservation area development*

Previously, monuments were mostly preserved or protected because of personal affection and historical remembrance. Since the emergence of the "world heritage" concept, the conservation and protection of monuments has focused on sustaining and furthering the authenticity and integrity of monuments with universal value. Accordingly, the surrounding environments related to the monuments must be incorporated into the areas designated for conservation and protection. Therefore, the concept of conservation area development does not merely focus on monuments but involves the historical fabric and context, aiming to fully and prominently reveal the universal value of such monuments. Therefore, the central goal of developing conservation areas in the future should focus on promoting the universal value of monuments.

4.2 *Both prominent monuments and common ancient buildings must be preserved to realize the practical significance of conservation area development*

Previously, efforts aimed at preserving monuments have focused primarily on prominent monuments while neglecting the ancient buildings belonging to ordinary people, which are mostly located on the periphery of prominent monuments but are essential for establishing the overall universal value of monuments. However, the lack of an appropriate concept of conservation area development has frequently caused common ancient buildings (e.g. those of ordinary people) to be demolished and razed in protecting the prominent monuments, thus destroying the integrity of the monument areas, an element essential for establishing their universal value. Accordingly, an appropriate concept of conservation area development should not involve designating conservation areas according to the social status of historical buildings. Instead, designating conservation areas according to whether the historical buildings are essential to the integrity and universal value of the preserved monuments is more adequate.

4.3 *Monument conservation is no longer limited to preserving monumental cultural heritage sites: industrial heritage is extensively emphasized in recent years and is critical to conservation area development*

The scale of the conservation areas for industrial heritage is large; thus, independently preserving each monument located within an industrial heritage area diminishes the authenticity and integrity of the universal value of the overall heritage area. Hence, the concept of conservation area development must be similarly applied to industrial heritage conservation, in which all the relevant historical buildings located within an industrial heritage area should be incorporated to highlight its unique historical value and industrial characteristics. An appropriate concept for developing a conservation area in the future will involve extensively incorporating various types of historical buildings and reasonably expanding the scope of the conservation area to ensure that initiatives for preserving a variety of heritage assets can be successfully implemented.

4.4 *Reducing the impact of contemporary construction in monument complexes: crucial to concretizing and expanding conservation areas, protecting the urban fabric, and structuring urban development programs*

The emergence of contemporary buildings within monument complexes frequently degenerates or destroys the integrity and universal value of such a complex; nevertheless, contemporary construction is inevitable in urban cities and thus may lead to a dilemma. Applying the concept of conservation area development can combine the measures and goals of urban development programs and conservation area development schemes to minimize the impact caused by newly constructed buildings within monument complexes, to create functional spaces benefiting the complexes, and to generate development benefits by equally emphasizing the fair use and conservation of relevant areas and sites (e.g., urban disaster prevention and favorable urban life functions). An appropriate concept of conservation area development in the future can be formulated from the perspective of the existing urban fabric, thus enabling the urban fabric to become a part of the overall urban development program and ensuring the successful implementation of development initiatives for conservation areas.

4.5 Conservation of a cultural landscape that emphasizes and reflects human life and culture and land use: a new direction of and the key to successful conservation area development

In recent years, much attention has been focused on cultural landscape, which is a cultural asset that can reflect the relationships between and the universal value of human life, culture, and land use. A cultural landscape generally does not entail a building or a construction complex; nevertheless, it is associated with universal value and thus is a critical cultural asset to preserve. Because the scope of a cultural landscape is generally large, the concept of conservation area development must be applied to preserving cultural landscapes. The conservation of cultural landscapes is a new domain that is key to developing conservation areas and critical to formulating new directions and topics for conservation area development, which are critical to implementing initiatives for preserving cultural landscapes and successfully developing conservation areas.

5 CONCLUSION

In the 20 years since the Council for Cultural Affairs was established in Taiwan, the concept and consciousness of conservation has become deeply rooted in the Council, which has developed relevant conservation techniques and accumulated considerable achievements. In the twenty-first century, preserving a comprehensive heritage area that simultaneously expresses cultural significance and historical value will be the focus of future conservation schemes.

Previously, the conservation and maintenance of monuments neglected the surrounding environments and areas. However, subtle changes in the conservation and maintenance schemes of international cultural heritage areas have occurred in recent decades. The focus of monument conservation and maintenance has expanded to include relevant building complexes (not merely individual historical buildings), the essence or pathways of historical developments in cities and towns, and even the associated intangible values, operations management, and education projects in preserved areas. In recent years, attention has been focused on maintaining and protecting heritage areas; hence, more appropriate concepts have been developed to comprehensively preserve the internal and external environments of a heritage. This approach should be the key to developing theories on conservation area development in the twenty-first century.

Because of differences in the social, economic, and educational conditions of countries committed to preserving cultural heritage areas, the time devoted to and the intensity, methods, and goals of developing conservation areas may differ. Nevertheless, fundamental requirements for environmental landscapes, transportation, conservation, health, and safety should be consistent. The changes in residential areas and environments around preserved area result in interrelated issues of economic benefits, convenience, and safety.

Consequently, how to adequately formulate the purposes, goals, legal regulations, strategies, and instruments for developing designated areas of cultural heritage conservation is not only an essential topic for enhancing the conservation of cultural heritage assets but also a major task for preserving, maintaining, and developing Taiwan's monuments. Theory on developing conservation areas for cultural heritage must be constructed from the perspectives of social equality and justice, environmental ecology, and economic activity to achieve sustainable development.

REFERENCES

Tung-Ming Lee, Jun Hatano. (1998). The History of Arcade (Din-A-Ka) in Di-Hwa Street, Taipei City. The 2nd International Symposium on Architectural Interchange in Asia. Biol. 147, 195–197.

Tung-Ming Lee, Jun Hatano. (2001). The Design and Formative Background of Town-Houses. Journal of Architecture and Planning (Transactions of AIJ) No. 547, p. 237~p. 242.

Tung-Ming Lee. (2004). Study for Generally Conservation and Sustainable Development of Traditional Town Area in Taiwan. Research Project Report of Architecture and Building Research Institute, Ministry of the Interior, Taiwan, R.O.C.

Tung-Ming Lee. (2005). Study of Spatial and Facade Character of Town-House in Traditional Town Area, Taiwan. Research Project Report of Architecture and Building Research Institute, Ministry of the Interior, Taiwan, R.O.C.

Tung-Ming Lee, Alex Yaning Yen. (2007). A Preliminary Study of Historical Conservation Areas in Taiwan. The 10th Session of the Conference on the Science of Conserving and Reusing Cultural Properties (Ancient Remains, Historic Buildings, Settlements, and Cultural Landscapes).

Civil, Architecture and Environmental Engineering – Kao & Sung (Eds)
© 2017 Taylor & Francis Group, ISBN 978-1-138-02985-9

The study on the control optimization and strategy of indoor visual comfortable environment system

Dan Wang, Hongwei Yin, Pu Dong & Yifei Chen
China Agricultural University, Beijing, China

ABSTRACT: The primary goal of the daylighting and shading system is to reduce the consumption of light and meet occupants' visual comfort. Thinking for many related control systems, they were designed that based on different algorithms, architectural models, simulation software and evaluation systems to meet the requirement. For these considerations, there is a need to propose optimization programs and develop suitable control algorithms based on the control constraints, the control strategies and experimental models. This paper summarizes several control systems of the indoor visual comfortable environment and finally gives some recommendations for the future research. Presented current indoor visual environment control methods and the characteristics were analyzed in this paper, and it can be found that the automatic control strategy has great improved space for optimization by studying on an intelligent control method as well as its simulation.

1 INTRODUCTION

The issued Directive 2010/31/EU of the European Parliament and Council of 19 May 2010 on the energy performance of buildings requires all new buildings to be "Nearly Zero Energy Buildings" (NZEB) by 2020 (EPBD, 2010), thus the energy efficiency of building environment has been the top priority (Nielsen 2011). According to recent studies, lighting energy consumption represents between 40% and 70% of total energy consumption within the tertiary sector, and moreover, it is expected that this consumption will increase in the future (Alcalde 2012). At this regard, energy efficiency in lighting is one of the key elements of nearly zero-energy buildings.

With the continuous improvement of the life quality, creating a comfortable visual environment with low energy consumption has been the most basic requirements of the occupants. In order to achieve both energy-efficient lighting and visual comfort, people have made great efforts in the relevant field of study. Advanced lighting control is the most effective way to improve energy efficiency, and it can meet the requirements of the occupants by way of regulation based on human perception. The settings of lighting and shading devices for example lighting blocking louver blinds are the key factors in regulating the indoor visual environment. One of the methods is to exploit the daylight coming into the indoor areas more effectively. If the lighting can be automatically turned off according to the level of daylight flowing indoors, the amount of electrical energy consumed for artificial lighting can be thereby reduced. The application of the dynamic operation of daylighting and shading systems can make the available resources, such as solar radiation and natural light, to get a better exploitation, so that the lighting energy saving 30%–77% (Yang 2010, Tzempelikos 2007), the energy efficiency and visual comfort can be improved at the same time (Lee 1998). According to the paper findings, the dynamic operation of daylighting and shading systems has many advantages in comparison to traditional control, for example, it can use the natural light efficiently, maximize energy efficiency and improve the quality of indoor lighting. However, there are some limits on the dynamic operation of daylighting and shading systems, such as the stability and versatility of the system, and the optimization of visual comfort control systems in energy-efficient buildings will be the focus of the future researches.

2 CONTROLLING ELEMENTS

2.1 Visual comfort

There are many factors influencing in the control of indoor visual comfortable environment, generally including architectural factors and non-construction factors, such as illumination intensity, solar radiation, climate, lamps as well as their power, windows' orientation and size, area ratio between the window and wall, indoor reflectance, sun-shading measures and so on (Chen 2004). In accordance with international standard UNE EN-12665 (E. 2011) visual comfort can be defined as a subjective condition of visual well-being induced by the visual environment. Rodrıguez et al. (2015) took visual comfort as a function of appropriate levels of both illuminance and glare. Some indexes for these parameters are defined, and some methodologies are applied to calculate the indexes. For instance, Table 1 can be observed the appropriate levels of indoor illuminance, glare rating and color rendering for common tasks which are developed inside offices.

More specifically, visual comfortable condition can be defined as function of the luminance distribution, illuminance and its distribution, glare, color of light, color rendering, flicker rate and amount of daylight (Castilla 2014).

2.2 Peripheral equipment

In fact, solar energy can be used in daylight and solar heating gain, benefiting both energy performance and psychophysical comfort, providing the use of that dynamic facade elements and integrated multi-criteria decision methods. It is very helpful for the optical data exchange technology (Castilla 2008) to have a wide range of lighting and light control techniques in the construction of the external envelope.

Shading devices mainly includes shutter, discount blinds, vertical blinds and venetian blinds which are classified by the static or dynamic shading device, manual or automatic control. The structure and material of shading devices largely affect the quality of indoor light and the indoor temperature. Decker M. (2013) found that the shape of movement triggered with the application of electric charge blocks which directly gains solar heat or allows daylight into the building. According to the literature reviewed on Phase Change Materials (PCM) by Silva et al. (2016) PCMs increase thermal comfort conditions in buildings and therefore occupants' satisfaction. Besides the conventional

Table 1. Recommendation for indoor lighting of offices.

Activity	Maintained Illuminance[lux]	Limiting glare rating
Filing, copying, etc.	300	19
Writing, typing, reading, et.	500	19
Technical drawing	750	16
CAD work stations	500	19
Conference and meeting-rooms	500	19
Archives	200	25

components, there are dynamic technologies under development, which are based on new, "smart" materials. Lelieveld C. (2007) and Thompson R. (2014) put forward that new smart materials are being developed. Martirano et al. (2016) noted that the lighting energy demand is sensitive to lighting type and control technique—in particular, two types of lighting (fluorescent versus LED) and two lighting control techniques (dimming versus switching)—its sensitivity affects the overall energy demand.

2.3 Occupants' behavior

Occupants' satisfaction and comfort are invaluable and relate to their well-being and working-performance (Heschong 2002, Veitch 1996). Occupants' satisfaction is influenced by their visual and thermal comfort as well as their ability to control the conditions of their working environment. Their response might be the cause for discrepancies between calculated and measured data due to the fact that the occupants' behavior is not often well represented in simulation models (Roetzel 2010). Occupants' response to dynamic daylighting and shading systems relates to their comfort and also affects buildings energy performance. Some advanced controls add more control variables, for example, user preferences may be integrated into the controller by automatically learning user preferences (Guillemin 2002).

3 CONTROL SYSTEM

3.1 Control strategy

In the existing literatures, the most common control objects are venetian blinds and roller shades. From the traditional control strategies (Colaco & Kurian 2008) to more advanced control systems, which have demonstrated their positive impact on energy performance and comfort. Rodrıguez et al. (2015) focused on the development of a preliminary fuzzy logic control system which allows maintaining visual comfortable conditions and its validation inside a meeting-room. More specifically, the control system has been tested through real tests inside a meeting-room of a bioclimatic building. These preliminary results have shown that the proposed control architecture has been able to maintain the specified visual comfortable conditions with the available means while decreasing the use of artificial light. Another example of advanced control uses an adaptable hierarchical fuzzy blind controller optimized by genetic algorithms that consider various state variables (Daum 2010). Similarly, there are many control algorithms have been developed in the control systems. Li (2008) focus on the perspective of intelligent control way, based on the physical model of indoor shading system and aluminum blinds with electric shade, to control the blind turning angle by fuzzy neural network to adjust the light in the room and make ensure that the indoor illumination meets the visual comfortable requirements. Lang (2008) proposed a computer indoor lighting control system based on two-dimensional pure fuzzy control algorithm. Referring to the fuzzy rules, applying the hardware interface can control the angle of blinds and the fill light angle effectively. Chen (2010) proposed an adaptive fuzzy control method to control the rotation angle of blinds adaptively based on the relationship analysis between the illumination and interior shading blinds rotation angle.

3.2 Energy assessment

In order to evaluate the energy-saving and visual comfort of control system, Shen et al. (2014) compared the energy and visual comfort performance of seven independent and integrated lighting and daylight control strategies. For independent closed-loop control and closed-loop dimming control window, blinds are always fully deployed and controlled by the solar position and incident light angle. During daytime, the blind slat angle rotates to the cut-off angle to block solar incident light. Lighting is controlled to maintain a set point illuminance of 500 lx on the work plane surface. It can be also controlled by occupancy-based switching. For the fully integrated lighting and daylight controlling with blind tilt angle, the goal is to maintain task illuminance close to the required lighting levels and minimize energy consumption for electric lighting. Blinds are always fully deployed and operated in the energy-saving and the comfort mode. Blinds operate as described previously in the energy-saving mode. In the comfort mode, when there is insufficient light, the integrated controller opens the blind slats to admit daylight to ensure no glare. In case of the target illuminance is not reached, the electric lights are dimmed on for insufficient daylight. Respectively, when the lighting levels are above the set-point, the electric lights are slowly dimmed.

The article focused on the quantitative analysis of performance of the seven lighting and daylight control strategies. The co-simulation platform includes EnergyPlus, BCVTB and Matlab, which is developed to estimate the energy and visual comfort performance in a typical medium office building. The results showed that the fully integrated lighting and daylight control with blind tilt angle (Figure 1) performed well in most cases and reached the lowest level of power consumption.

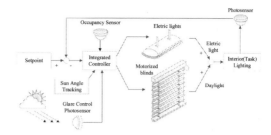

Figure 1. Integrated lighting and daylighting control.

Similarly, Shen and Hong (2009) developed an integrated daylight and lighting control, evaluated the energy savings in an office model using Energy-Plus simulations, and found lighting energy savings of 64–84%, and heating, mechanical ventilation and airconditioning (HVAC) savings from 3% to 43% depending on location, compared to a base case with clear windows and all lighting fully on.

In general, lighting control plays a key role in NZEB project. The strategies used concern: (1) reducing wasted hours of lighting unoccupied spaces; (2) automatically adjusting electric light levels in synchrony with available daylight via the dynamic façade (Cziker 2007, Cziker 2008). According to the actual situation and the review of the literatures, the current situation of the dynamic operation of daylighting and shading systems can be summarized as follows conclusions:

1. Daylighting varies according to climatic conditions, building orientation and area ratio of the window and wall. But the systematic studies of literatures are in a particular environment, which is based on the set conditions for commissioning.
2. There is a limit to system energy estimation.
3. The acceptance level by users in the control system used to be a great impact on the effective implementation of the control system. The current developed control strategies developed are too complex or too simple for users, which cause difficulties of operation or their unwariness to the problem.

4 AN INTELLIGENT CONTROL METHOD

This article presents RBF neural network control approach that efficiently ensures visual comfort and energy saving. This paper analyzes the impact factors and the control rules of the indoor visual environment, introduces the learning algorithm of RBF neural network and builds control system of indoor visual comfortable environment based on the RBF neural network control (Figure 2). And finally the simulating experiments are carried out under the environment of MATLAB. By comparing the differences of the control performance between the BP neural network and the RBF neural network as shown in Table 2, the indoor visual environment control system based on RBF neural network has higher control accuracy in rotation angle of venetian blinds and lighting system. Experimental results (Figure 3) show that this control method not only achieves the purpose of saving energy, but also improves the

control quality of the indoor visual comfortable environment.

The innovation of this project is the use of artificial neural network control algorithm, which can fully approximate to any complex nonlinear relationship. And the self-learning function of controller increases the precision of the control system. On the other hand, this paper constructs a multi-input multi-output system which is suitable for different weather conditions and thus matches practical condition for the control system.

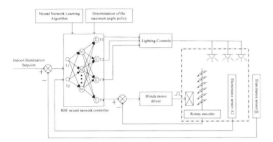

Figure 2. The system structure of indoor visual environment based on neural network control. *Illuminance sensor B is used to measure the outdoor illumination, Illuminance sensor A2 is used to measure the indoor illumination.

Table 2. Error comparison.

Output	RBF average error	BP average error	RBF error variance	BP error variance
Blinds angle of rotation	0.3428	0.6629	0.7077	40.7511
Light A2	0.0094	0.2029	0.0010	0.0638

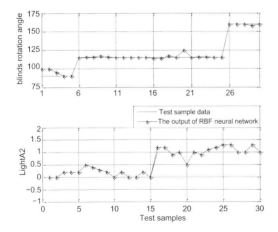

Figure 3. The fit of the RBF neural network output curve.

5 CONCLUSIONS

According to the previous discussion, the following recommendations are meant to guide future researches in the optimization of the systems:

1. It is efficient for the control system to improve the control performance by using more advanced control algorithms special in sun shading control.
2. In future studies, the control system optimization criteria may include personal preferences in order to realize more advanced intelligent control, and embedded into Smart Home System.
3. Future research will be considered using Radiance for daylight imulation and EnergyPlus for energy simulation to realize more accurate simulation results, and using more realistic and advanced simulation strategies.
4. At present, few researches deal with global optimization of energy demand and users' comfort, see e.g. (Ippolito 2014, Tagliabue 2012). Future Studies should be committed to the application of advanced smart materials, which are the key to reach energy saving and visual comfort.
5. With aim to obtain significant results, solar radiation, electric lighting, HVAC and shading systems cannot be considered separately but have to be handled as a whole, because solar radiation affects both lighting and HVAC and it impacts both energy performance and comfort, introducing sunlight and overheating possibly uncomfortable for occupants. Therefore, reasoning one controlled integrated goal is worth to be researched.
6. This system is one of special mixed system, and studying one valid control method to this mixed system is also next work.

REFERENCES

Alcalde, E. (2012). Potencial de ahorro y eficiencia energetica en instalaciones de iluminacion (in Spanish). CIRCE, Tech. Rep.

Castilla, M., J. Alvarez & F. Rodrıguez (2014). Comfort Control in Buildings. Springer. Advances Series in Industrial Control. ISBN: 978-1-4471-6346-6.

Chen hongbing, Li deying & Shao zongyi (2004). Study on the Effect of Window on Building Energy Consumption. Journal of Beijing Institute of Civil Engineering and Architecture. 20(4):9–12.

Chen xueliang & Chen yifei (2010). Research of self-adaptive fuzzy control on sun blind.

Colaco S & Kurian (2008). Prospective techniques of effective daylight harvesting in commercial buildings by employing window glazing, dynamic shading devices and dimming control—a literature review. Build Simul 2008;1(4):279–89.

Cziker, A., M. Chindris & A. Miron (2007). Implementation of Fuzzy Logic in Daylighting Control. 2007. 195–200.

Cziker, A., M. Chindris & A. Miron (2008). Fuzzy controller for a shaded daylighting system. 2008. 203–208.

Daum D & Morel N (2010). Identifying important state variables for a blind controller. Build Environ 2010;45(4):887e900.

Decker M (2013). Distributed sensing and actuation in building skins. In: Proceedings of energy forum 2013 conference. Bressanone, Italy; December 5–6.

E. 12665 (2011). Light and lighting—Basic terms and criteria for specifying lighting requirements. Brussels: European Committee for Standardization.

Guillemin A & Molteni S (2002). An energy-efficient controller for shading devices self-adapting to the user wishes. Build Environ 2002;37(11):1091e7.

Heschong L. (2002). Daylight and human performance. ASHRAE J 2002;44:65–67.

Ippolito, M.G., E. Riva Sanseverino & G. Zizzo (2014). Impact of building automation control systems and technical building management systems on the energy performance class of residential buildings: An Italian case study. Energy and Buildings. 69(0):33–40.

Lang caien & Chen yifei (2008). Lighting Fuzzy Controller Based on Visual Comfort Indoors. China Instrumentation. 2008;2:57–59.

Lee, E.S., D.L. DiBartolomeo, & S.E. Selkowitz (1998). Thermal and daylighting performance of an automated venetian blind and lighting system in a full-scale private office. Energy and Buildings. 29(1):47–63.

Lelieveld C, Voorbij A & Poelman W (2007). Adaptable architecture. Building stock activation. Tokyo: TAIHEI Printing Co; 2007. 245–252.

Li huai & Chen yifei (2008). Research on indoor illuminance control system based on fuzzy neural network. Building Electricity. 2008;7(27):27–30.

Luigi Martirano, Matteo Manganelli, Luigi Parise & Danilo A. Sbordone (2016). Design of a fuzzy-based control system for energy saving and users comfort. Renewable and Sustainable Energy Reviews 2016;16:268–283.

Nielsen, M.V., S. Svendsen, & L.B. Jensen (2011). Quantifying the potential of automated dynamic solar shading in office buildings through integrated simulations of energy and daylight. Solar Energy. 85(5):757–768.

Rodrıguez, J.M., M. Castilla, & J.D. Alvarez (2015). A fuzzy controller for visual comfort inside a meeting-room. IEEE.

Roetzel A, Tsangrassoulis A & Dietrich U (2010). A review of occupant control on natural ventilation. Renew Sustain Energy Rev 2010;14:1001–1013. http://dx.doi.org/10.1016/j.rser.2009.11.005.

Shen E & Hong T (2009). Simulation-based assessment of the energy savings benefits of integrated control in office buildings. Build Simul e Springer; 2009:239e51.

Shen E, Hu J & Patel M (2014). Energy and visual comfort analysis of lighting and daylight control strategies. Build Environ 2014;78:155–170.

Silva T, Vicente R & Rodrigues F (2016). Literature review on the use of phase change materials in glazing and shading solutions. Renew Sustain Energy Rev 2016;53:515–535.

Tagliabue, L.C., M. Buzzetti, & B. Arosio (2012). Energy Saving Through the Sun: Analysis of Visual Comfort and Energy Consumption in Office Space. Energy Procedia. 30(0):693–703.

Thompson R & Yan Ling EN (2014). The next generation of materials and design [Chapter 14]. In: Karana E, Pedgley O, Rognoli V, editors. Materials experience. Boston: Butterworth-Heinemann; 2014. 199–208.

Tzempelikos A & Athienitis AK (2007). The impact of shading design and control on building cooling and lighting demand. Sol Energy 2007;81(3):369e82.

Urbano R & Andersen M (2008). A new perspective for searching and selecting light control technologies as a designer. Proceedings of the 25th passive and low energy architecture conference. PLEA2008-25th passive and low energy architecture conference. Dublin, Ireland: October 22–24.

Veitch JA & Gifford R (1996). Assessing beliefs about lighting effects on health, performance, mood, and social behavior. Environ Behav 1996;28:446–470. http://dx.doi.org/10.1177/0013916596284002.

Yang I & Nam E (2010). Economic analysis of the daylight-linked lighting control system in office buildings. Sol Energy 2010;84(8):1513e25.

Civil, Architecture and Environmental Engineering – Kao & Sung (Eds)
© 2017 Taylor & Francis Group, ISBN 978-1-138-02985-9

A study of a low-carbon and intelligent city index system

Yu-Sheng Shen
Academia Sinica, Taipei City, Taiwan

Hsiao-Lan Liu
Department of Land Economics, National Chengchi University, Taipei City, Taiwan

ABSTRACT: The low-carbon and intelligent city is the city combines the technology, digital information, low-carbon, and sustainable development together. In order to evaluate the status of low-carbon and intelligent city effectively, city governments need to develop an index system. However, most city governments lack the suitable index system of low-carbon and intelligent city. Therefore, this paper seeks to establish the appropriate index system of low-carbon and intelligent city by correlation analysis and factor analysis rating method, and evaluate the status of low-carbon and intelligent cities within the Taiwanese domain. According to the empirical results, there are five evaluative aspects of the low-carbon and intelligent city: (1) economic and intelligent technology, (2) consumption of energy and resource, (3) health and environmental conservation, (4) local development, and (5) social status. Based on the comprehensive performance of the low-carbon and intelligent city index system, Taipei City shows the best performance.

1 INTRODUCTION

The low-carbon and intelligent city is the city combines the technology, digital information, low-carbon, and sustainable development together. The development of low-carbon and intelligent city not only can integrate the digital information and apply advanced technology, but also can solve the environmental problems and carbon emissions of urbanization. Effective evaluation of low-carbon and intelligent degree are the most important tasks to develop low-carbon and intelligent cities. The evaluation can reveal the current conditions of urban developments, and the gaps between those conditions and the prospective goal. Moreover, the evaluation can guide the cities to become low-carbon and intelligent cities, and examine the efficiency of cities setting policies in motions. The evaluation can also help to monitor the status of developing low-carbon and intelligent cities, and implement corrective action as needed.

In order to evaluate the status of low-carbon and intelligent cities effectively, city governments need to develop the index system. However, most city governments lack the index system of low-carbon and intelligent cities, while they strive towards the goal of becoming low-carbon and intelligent cities. Besides, most of the evaluation indicators tend to lean toward particular area and lacked comprehensive consideration, thus influencing the final results. Therefore, this paper seeks to establish the appropriate evaluation index system, based on virtues, ideas, and goals of low-carbon and intelligent cities. Furthermore, the paper utilizes the opportunity to evaluate the status of low-carbon and intelligent developments of cities within the Taiwanese domain, in order to provide reference for urban development in the future.

2 RESEARCH DESIGN

2.1 Analytical framework

Because the past studies lack operational framework, this study proposes the analytical framework, which includes seven steps (Figure 1). In addition to an analytical framework for this paper, it also can apply to other cities in different regions and countries due to its general rule.

2.2 Method

In order to avoid the subjectivity of selecting and determining index system, correlation analysis and factor analysis rating method are used to construct the index system in this paper. The factor analysis rating method decides the index system by factor analysis, factor analysis is a multivariate analysis that simplifies a lot of complicated indicators to the few meaningful aspects/component using the correlation matrix. Therefore, factor analysis is not

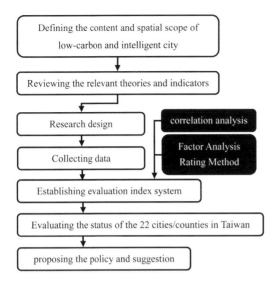

Figure 1. Analytical framework.

only used to simplify the numbers of indicators and select the meaningful indicators in this paper, but also determine the relative weight of each aspect and indicators. In the empirical analysis of factor analysis, principal-axis factoring analysis (PFA) is used as the way of factor extraction, and varimax is used as the way of rotation.

The weight of low-carbon and intelligent city evaluation index system established in this paper includes aspects weight and indicators weight. The former comes from the total variance of certain aspect divided by the total variance of selected aspects, as in Formula (1); the latter is obtained by multiplying the ratio of factor loadings in each aspect with aspects weight, as in Formula (2).

$$A_i = a_i / \sum_{i=1}^{n} a_i \tag{1}$$

$$B_{ij} = A_i \times (|b_{ij}| / \sum_{j=1}^{m} |b_{ij}|) \tag{2}$$

A_i: weight of aspects
a_i: total variance of aspects
B_{ij}: weight of indicators in certain aspects
b_{ij}: factor loading of indicators
i: aspects
j: indicators of certain aspects

2.3 *Defining indicators*

Low-carbon and intelligent city is the combined concept of low-carbon and intelligent systems, therefore, this paper reviews and organizes the indicators of low-carbon development

and intelligent systems respectively, in defining indicators. Moreover, quantifiable indicators are preferred in this paper to maintain the objectivity, verifiability and comparability of the indicator system and evaluation results. According to the research on low-carbon (Chen and Zhu 2013; Su et al. 2012; Price et al. 2013) and intelligent system (Deakin 2012; Deakin 2013; Deakin 2014), this paper generalizes the 48 quantifiable indicators for the evaluation of low-carbon and intelligent cities.

Due to the difference in data, type, unit and range of the indicators, formula (3) and (4) are used to convert the original data. If the indicator is maximizing form (the bigger the value, the better the performance), formula (3) is used; If the index is minimizing form (the smaller the value, the better the performance), formula (4) is chosen.

$$x_l^c = (x_l - x_{min})/(x_{max} - x_{min}) \tag{3}$$

$$x_k^c = (x_{max} - x_k)/(x_{max} - x_{min}) \tag{4}$$

x_l^c: indicator value after maximizing conversion
x_k^c: indicator value after minimizing conversion
x_l: original value of indicator l
x_k: original value of indicator k
x_{max}: maximum value in all (selected) indicators
x_{min}: minimum value in all (selected) indicators

3 DATA

Based on considering consistency in data and accuracy and coherence in the analysis, 22 cities in Taiwan were chosen for applying spatial scope of empirical analysis. As for data type, all indicators are quantitative, quoted from Urban and Regional Development Statistics, and statistical database of Directorate-General of Budget, Accounting and Statistics in Taiwan. Considering the limitations of the governmental data content, samples for factor analysis come from the data of the 22 cities between 2006 to 2010. Data of the Taiwan's 22 cities in 2010 is used for evaluating low-carbon and intelligent status of cities.

4 EMPIRICAL ANALYSIS AND DISCUSSION

4.1 *Establishing the evaluation indicator system for low-carbon and intelligent city*

4.1.1 *Model verification*
In the verifying suitability of factor analysis for data, the results passed the Bartlett's test of sphericity, and the Kaiser–Meyer–Olkin measure of sampling adequacy (KMO) is 0.817. Therefore, the data was suitable for factor analysis. In verifying the number of components by extracting,

Table 1. Results of eigenvalue and variance by factor analysis.

Aspect	I	II	III	IV	V
Eigenvalue	38.64	11.36	5.43	3.73	2.62
Total variance (%)	49.54	14.56	6.96	4.78	3.35
Cumulative variance (%)	49.54	64.11	71.08	75.87	79.23

the results achieved the three standards (such as Kaiser criterion, Cattell criterion, cumulative variance criterion). The six aspects were the best number of factor extraction as the cumulative variance reaches 81.76%, and all eigenvalue of each component was above 1 (Table 1).

4.1.2 Results analysis and discussion

After factor extraction and rotation through factor analysis, the result of evaluation indicator

Table 2. Evaluation index system for low-carbon and intelligent cities.

Evaluation indicator system	Content of indicator	Aspect weight	Indicated weight
F1: economy and intelligent technology		0.6	
Low-carbon and low pollution industries market value ratio	Ratio of market value in tertiary industries to total market value		0.162
Low-carbon and low pollution industries employment ratio	Ratio of employed population in tertiary industries to total employed population		0.156
Number of network users	Number of network users		0.144
Construction of network hardware	Number of construction of network hardware		0.138
F2: consumption of energy and resource		0.18	
Fossil energy consumption	Fossil energy consumed by living and production activities		0.0324
Electric power consumption	Electric power resource consumed by living and production activities		0.0306
Renewable energy consumption	Renewable energy consumed by living and production activities		0.0288
Water consumption	Water resource consumed by living and production activities		0.0306
F3: health and environmental conservation		0.09	
PSI exceeding 100 ratio	Ratio of days with PSI exceeding 100 to total days investigated		0.0189
Sewage emissions	Sewage emissions generated by living and production activities		0.0180
Waste generated	Waste generated by living and production activities		0.0180
Hospital beds	Number of hospital beds in medical institutions		0.0162
Medical employees	Number of medical employees in medical institutions		0.0189
F4: local development		0.06	
Green building	Number of green building		0.0096
Public facilities land area	Area of public facilities land		0.0070
Forest coverage area	Coverage area of forest		0.0091
Urban land density	Ratio of urban planning area to total land area in city		0.0078
F5: social status		0.04	
Unemployed population	Number of unemployed population		0.0064
Gini coefficient	Ratio of the area between Lorenz curve and line of equality to area below line of equality		0.0068
Resource recycling	Number of resource recycling in all organization and communities		0.0068
Car and motorcycle ownership	Number of automobiles and motorcycles ownership with license		0.0064

Note: PSI is Pollutant Standard Index that reports the concentration of five air pollutants, particulate matter, sulfur dioxide, carbon monoxide, ozone, nitrogen dioxide.

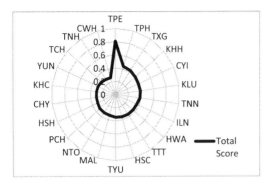

Figure 2. Results of evaluation of low-carbon and intelligent.

system, which were established for low-carbon and intelligent city included economic and intelligent technology, consumption of energy and resource, health and environmental conservation, local development, and social status (Table 2). The weight of aspects and indicators could be obtained through the formula (1) and (2). Based on Table 2, the aspect of economic and intelligent technology was the most important component for its variance reporting 49.54, and the indicators (such as low-carbon and low pollution industries market value ratio, low-carbon and low pollution industries employment ratio, number of network users, and construction of network hardware) were also the most important indicators. Therefore, the aspect of economic and intelligent technology served as a foundation for low-carbon and intelligent city. While consumption of energy and resource occupied the position as the secondary aspect, and various indicators included by this aspect for the development of low-carbon and intelligent city. The rest of indicators still were importance and effectiveness for low-carbon and intelligent city.

4.2 Evaluation and strategy for low-carbon and intelligent development of Taiwan's cities

The results of evaluating the low-carbon and intelligent status of Taiwan's 22 cities in 2010 were shown in Figure 2. Taipei city scored 0.806 in total score, showing the best performance among the 22 cities in Taiwan. Taipei city also does particularly well in aspects of economic and intelligent technology, consumption of energy and resource, and local development. Moreover, the future improvement of Taipei city (on low-carbon and intelligent status) should be centered on two aspects, including health and environmental conservation and social status. On the contrary, the worst performance was Changhua county, with the total score of 0.261. Therefore, in order to ameliorate the degree

of low-carbon and intelligent system, Changhua county should not only improve its economic and intelligent technology primarily, but also elevate its performance in local development and social status.

5 CONCLUSION AND SUGGESTION

Environmental issues regarding global warming and human habitability have received much emphasis recently, as a result of which low-carbon and intelligent systems gradually assumed prominence as the goal of urban development. Despite the fact that numerous cities in the world have set low-carbon and intelligent systems as developmental goal, a corresponding evaluation system is missing. This paper has attempted to construct an objective and quantifiable evaluation indicator system by factor analysis, and to evaluate the status of the 22 cities in Taiwan. The results of the above can be regarded as references for the urban development and policy in the future. The results of empirical analysis show that the index system should encompass five aspects and 32 indicators in total. The low-carbon and intelligent status of all cities in Taiwan, Taipei city shows the best performance, whereas Changhua county shows the worst performance. Application of evaluation indicator system for low-carbon and intelligent city depends greatly on the local features. The indicator system may call for specific adjustments to adapt to various cultures when applied to different regions. Moreover, the indicator system must be updated according to the latest changes and developments of indicators. If the database is sufficient, it is strongly recommended to include analysis and comparison of various regions at different times in the future researches.

REFERENCES

Chen, F., and Zhu, D. (2013). Theoretical research on low-carbon city and empirical study of Shanghai. *Habitat International*, 37: 33–42.

Deakin, M. (2012). *Creating Smarter Cities*. New York: Taylor & Francis.

Deakin, M. (2013). *Smart Cities: Governing, Modelling and Analysing the Transition*. New York: Taylor & Francis.

Deakin, M. (2014). *From Intelligent to Smart Cities*. New York: Taylor & Francis.

Price, L., Zhou, N., Fridley, D., Ohshita, S., Lu, H. Y., Zheng, N. N., and Fino-Chen, C. (2013). Development of a low-carbon indicator system for China. *Habitat International*, 37: 4–21.

Su, M. R., Chen, B., Xing, T., Chen, C., and Yang, Z. F. (2012). Development of low-carbon city in China: Where will it go? *Procedia Environmental Sciences*, 13: 1143–1148.

Civil, Architecture and Environmental Engineering – Kao & Sung (Eds)
© 2017 Taylor & Francis Group, ISBN 978-1-138-02985-9

Exploring the lean strategies to obtain the banking loan for energy service companies

Albert K.C. Mei
Department of Business Administration, China University of Technology, Taipei, Taiwan, R.O.C.

Y.W. Wu
Department of Architecture, China University of Technology, Taipei, Taiwan, R.O.C.

Anny Y.P. Wu
School of Business Administration, Hubei University of Economics, Wuhan, China

C.C. Liu
*Management Consulting Division, Ahead International Consulting Company,
Taipei, Taiwan, R.O.C.*

ABSTRACT: In recent years, Energy Service Companies (hereinafter called ESCO) emerged from concerns of the entire world on enterprises' energy saving and environmental protection issues. The success of financing from banks to ESCO would be no doubt the key factor influencing ESCO's development. In this study, we have interviewed some financing experts to realize what key factors will be evaluated when banks are facing this new business financing cases, and we tried to analyze with nonlinear quantified model, Fuzzy Dematel, to find out key principles and factors to ESCO's development when they are under limited resources, which could be their lean strategies in the future.

1 INTRODUCTION

In 1980s, increasing scientific evidence indicated the correlation between rising greenhouse gas emissions and global climate change. According to International Energy Agency (IEA, IEA Energy Technology Perspective 2010), 38% of global carbon emissions will be reduced from better energy efficiency (IEA Energy Technology Perspective 2010). It is inevitable for governments and private enterprises to develop energy conservation equipment and high-efficiency equipment, however, their high set-up costs for new energy-saving equipment could not meet the market requirement. From the 21st century, most governments have been promoting ESCO to speed up energy conservation and reduction of carbon emission.

Most studies of ESCO concentrated on operational models, development strategies and energy-saving efficiencies. In recent years, some scholars have studied and discussed on financing operation (Huang, 2007; Chen, 2011; Chen, 2014). Earlier studies mainly discussed ESCO's financing methods and models. According to Chen (2014), investigations indicated only 34% enterprises in Taiwan applied financing to banks and only 12.5% of

them succeeded. It is because of the market size of ESCO's that could not successfully get banks' attention. Bank crediting personnel did not get clear of ESCO's operational models, ESCO's were most small or medium enterprises, and most of them were newly established. Therefore, the purpose of the study is to explore the key factor to influence bank financing of ESCOs, when banks are under limited resources, and then analyze the causation and degree of influence between these factors to find out key strategy factors of bank financing. At last, the key strategy factors can be used by ESCO to adjustment of management and strategy.

2 LITERATURE REVIEWS

2.1 What is ESCO

The WTO has broadly defined "Energy Services Company (ESCO)" as the enterprises running for oil/electricity trade service, electricity generation/transportation/supply/distribution, water resources, energy-saving, and management of coal/electricity/gas/nuclear/power/oil/renewable energy. Ministry of Economic Affairs in Taiwan then defined it as the enterprises running for new clean energy, energy-saving, energy efficiency promotion or peak demand transfer equipment/system/ engineering/planning, feasibility study/design/ installation/construction/maintenance/detection/ thirdgeneration/hardware and software construction/ related technical service.

2.2 *Management models of ESCO*

World Bank has classified ESCO's business model into eight types (World Bank GEF Energy Efficiency Portfolio Review and Practitioners' Handbook, 2004), which provided initial direction of business model planning. In practice, however, some extended and change models have produced due to individual differences between energy users, ESCO and financial institutions. Chen (2011) found that almost all models have been tried in Taiwan and two types were mostly accepted by the market: "Energy Efficiency Guarantee Type" and "Energy Efficiency Share Type". Energy Efficiency Guarantee Type means energy users sign contracts with ESCO to install equipment to lower energy use and improve energy efficiency, which results in the capital demand of energy users. To energy users, though the guarantee contracts allow them to fulfill payment after energy-saving effects have been achieved, they still have to be responsible for the financing. (Energy Charter Secretariat, 2003)

To strengthen energy users to execute energy-saving projects, ESCO promoters have developed "Energy Efficiency Share Type". Energy Council (2008) has explained that ESCO should be the sponsor of bank financing and energy users could use the amount saved by energy efficiency guarantee project to pay ESCO. In that way, ESCO has to run the risk of project achievement and the repayment pressure after financing. Financial institutions still run the risk of repayment ability of debtors, while energy users only provide energy efficiency to ESCO in a certain period of time. (Energy Council, 2008)

In recent years, several scholars have studied ESCO financing issues. For example, Huang (2007) has probed ESCO's financial mechanism models in Taiwan through in-depth interviews and put forward three models, ESCO credit guarantee model, local ESCO development fund or concessional loan project model, and local ESCO firefly fund project model. They all fit the developing model of ESCO financial in Taiwan. Chen (2011) also put forward three ESCO financing models, direct financing plus credit guarantee fund, contract financing plus accounts receivable, and lease. After experts' confirmation, Chen then proceeded with in-depth interviews with bank institutions and ESCO to discuss local ESCO development and verify the feasibility of the three models.

3 RESEARCH METHODS

3.1 *Research variables and factors*

From document study and expert interviews, we have concluded five principles and 18 key factors in Table 1 as key for ESCO financing strategies of banks. Five principles are People, Purpose, Payment, Protection, and Perspective. People principle includes three factors of ESCO's situation: business performance, and equipment and technical skill. Purpose principle includes three factors of use of funds: energy-saving financial plan, and concentrated purpose of funds. Payment principle includes four factors of attainability of energy efficiency, quality of accounts receivable, project business finance ratio, and business dispute. Protection principle includes four factors of guaranteed capability of the ESCO's capital, responsible persons, collaterals, and external security. Perspective principle includes four factors of ESCO's patents, capacity of resource integration, and government policies.

3.2 *Fuzzy Dematel*

Fuzzy Dematel is usually used for social science and the study of disorganized phenomena (Tamura, Nagata & Akazawa, 2002). It is a research tool to explain complicated relationships (Yang & Tzeng, 2011). It is a systematic structure designed by using professional knowledge (Liou, Yen, & Tzeng, 2008), and through the specific characteristics of each factor we can construct mutually interacting relationship between factors and transform this causation into a systematic structure model, which is the "Causal Diagram". The main purpose of Causal Diagrams is to solve many problems in strategic decisions (Hu, 2003). Through combination and conclusion of questionnaires, we can use analyzing steps of Fuzzy Dematel to get a Total Relation Matrix (T), calculating prominence (D+R) and relation (D–R).

According to the Causal Diagram divided by average value of D+R and D–R, if D+R is bigger and D–R is above the average value, the

Table 1. Five principles and 18 factors of bank lending evaluation to ESCO.

Principles	Factors
People	Situation, Performance, Skills
Purpose	Funds, Plans, Concentrated
Payment	Attainability, Quality, Ratio, Dispute
Protection	Guaranteed, Responsibility, Collaterals, Security
Perspective	Others, Patents, Integration, Policies

coordinate locates in the first quadrant, which means the factor is closer to the core determining factors to problem resolution and should be taken into first priority. When D+R is smaller and D–R is above the average value, the coordinate locates in the second quadrant, which means this factor is highly independent and only has influence on certain characteristics. When D+R is small and D–R is below the average value, the coordinate locates in the third quadrant, which means this factor is highly independent and will be influenced by some certain characteristics. When D+R is bigger and D–R is below the average value, the coordinate locates in the fourth quadrant, which means this factor is a very significant core determining factors.

4 ANALYSIS AND RESULTS

4.1 *Background analysis*

Interviewees/respondents of the study are mainly ESCO financing personnel, who are mostly chiefs responsible for small/medium enterprise financing business, including Land Bank (8 persons), Taiwan Business Bank (14 persons) and Far Eastern International Bank (7 persons). There are totally 29 persons interviewed. Twenty-five of them have more than 6 years of experience in financing and 21 of them have 4 years' experience. Twenty-eight of them are bank chiefs. They are all appropriate interviewed objects.

4.2 *Data analysis*

The result of 29 questionnaires taken back and analyzed by FUZZY DEMATEL indicates the conclusion given in Table 2. And we transform these values into a Causal Diagram by Microsoft Excel in Figure 1.

From the causal diagram of Figure 1, the highest prominence principle (People; F1) indicates in general when banks deal with ESCO financing, key factors are ESCO's situation, business performance and equipment and technical skills. However, the

Table 2. The values of prominence and relation for five principles by Fuzzy Dematel analysis.

Principles	D	R	D+R	D–R
People (F1)	1.61	1.63	3.24	–0.02
Purpose (F2)	1.46	1.54	3.00	–0.08
Payment (F3)	1.37	1.37	2.74	0.01
Protection (F4)	1.38	1.35	2.73	0.04
Perspective (F5)	1.47	1.42	2.89	0.05

principle "People" is still influenced by principles "Perspective (F5)" and "Protection (F4)".

As a whole, on analysis of Figures 2 and 3, the most significant factors in the principle "Perspective" is "Government policies", because it can have great influence on ESCO's "Patents & Skills" and "Capacity of Resource Integration". The second significant factor in the principle "Protection" is "External Security".

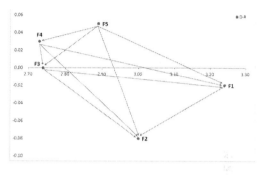

Figure 1. The causal diagram of 5 principles by Fuzzy DemateI analysis.

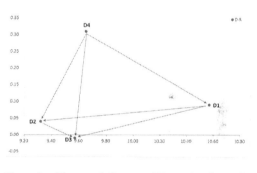

Figure 2. The causal diagram of Protection factors by Fuzzy DemateI analysis.

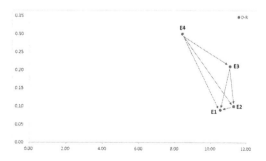

Figure 3. The causal diagram of Perspective factors by Fuzzy DemateI analysis.

1235

Table 3. The values of prominence and relation for 18 factors by Fuzzy Demetel analysis.

Principles	Factors	D	R	D+R	D–R
People (F1)	Situation (A1)	5.53	5.41	10.9	0.12
	Performance (A2)	6.17	5.98	12.2	0.19
	Skills (A3)	5.55	6.09	11.6	–0.50
Purpose (F2)	Funds (B1)	5.14	5.38	10.5	–0.20
	Plans (B2)	5.43	5.75	11.2	–0.30
	Concentrated (B3)	5.06	5.42	10.5	–0.40
Payment (F3)	Attainability (C1)	5.18	5.71	10.9	–0.50
	Quality (C2)	4.71	4.77	9.48	–0.10
	Ratio (C3)	4.97	4.86	9.84	0.11
	Dispute (C4)	4.72	4.24	8.96	0.49
Protection (F4)	Guaranteed (D1)	5.33	5.24	10.6	0.09
	Responsibility (D2)	4.68	4.64	9.32	0.05
	Collaterals (D3)	4.78	4.79	9.57	–0.01
	Security (D4)	4.98	4.67	9.64	0.31
Perspective (F5)	Others (E1)	5.32	5.23	10.5	0.09
	Patents (E2)	5.69	5.59	11.3	0.09
	Integration (E3)	5.64	5.43	11.1	0.22
	Policies (E4)	4.36	4.06	8.42	0.30

5 CONCLUSIONS AND SUGGESTIONS

Financial support has been playing an important role in the development of emerging industries. When global warming and climate change come into fact, energy conservation and reduction of carbon emission have become an obligation. This study used Fuzzy Demetel analysis to discuss what evaluation principles banks should take to face new type of business model, such as ESCO, to make sure that creditor's rights as well as actively assisting in ESCO's development needs are fulfilled. Meanwhile, we can use results of the analysis as for lean strategy and management key point of industrial development, so that to achieve the goal of energy-saving and reduce energy cost.

After discussions between experts from industries, government authorities and academies, this study has concluded banks' five evaluation principles and 18 specific factors to ESCO financing business, and has defined various kinds of factors. From the study, we know the highest principle of the prominence is "People", which indicates when banks deal with ESCO financing, key factors are ESCO's situation, business performance and equipment and technical skills. However, the principle "People" is still influenced by principles "Perspective", "Protection", and "Payment". Therefore, when banks are faced with ESCO financing, they should put emphasis on the principles "Perspective", "Protection", and "Payment". From the aspect of "Perspective", the most

important factor is "Government policies", which reflects actual support and industrial policy of the government is still the key factor to decide if the bank will accept financing application.

To ESCO, execution of energy efficiency projects is the way they gain profits. When banks do not understand relevant engineering and equipment techniques in the process of evaluation while ESCO owns only limited managing resources, ESCO should try to strengthen their "Patents & Skills" and "Capacity of Resource Integration" in principle "Prospective" as well as the "Guaranteed capability of the ESCO's capital" and "External Security" in principle "Protection". Thus financial personnel can improve their professional knowledge in energy-saving and then ESCO can get their capital successfully.

REFERENCES

Chen, H.J. (2011). ESCO industry development and trends, 2011 Cross Strait Forum of Climate Change and Energy Sustainability, Taipei.

Chen, S.H. (2014). An Analysis of Impacts on Energy Services Company (ESCO) Industry Using Economic Model and Its Internationalization, the Project for Bureau of Energy, Ministry of Economic Affairs, Taipei.

Chen, S.W. (2011). The Development and Financing Patterns of Energy-Saving Service Industry in Taiwan, Master Paper, Providence University.

Hu, S.C. (2002). The Thesis for the Degress of Complexity in Enterprise Problems and for Quantification in Enterprise Problems—by Using DEMATEL as an Analyzing Tool, Master Paper, Chung Yuan Christian University, Taiwan.

Huang, H.J. (2007). A Study of The Financing Mode for Technology Service Industry—The Case of Energy Service Companies (ESCO), Master Paper, National Sun Yat-Sen University, Taiwan.

IEA (2010). Energy Technology Perspective, International Energy Agency.

Liou, J.J.H., Yen, L. and Tzeng, G.H. (2008). 'Building an effective safety management system for airlines'. Journal of Air Transport Management, 14, 20–26.

Tamura, H. and Akazawa K. (2002). Structural modeling and systems analysis of uneasy factors for realizing safe, secure and reliable society. Journal of Telecommunications and Information Technolegy, 3, 64–72.

World Bank Environment Department (2004). World Bank GEF Energy Efficiency Portfolio Review and Practitioners' Handbook, World Bank Environment Department.

World Energy Council (2008). Energy Efficiency Policies around the World: Review and Evaluation, London: World Energy Council.

Yang, J.L. and Tzeng, G.H. (2011). 'An integrated MCDM technique combined with DEMATEL for a novel cluster-weighted with ANP method'. Expert Systems with Applications, 38(3), 1417–1424.

Civil, Architecture and Environmental Engineering – Kao & Sung (Eds)
© 2017 Taylor & Francis Group, ISBN 978-1-138-02985-9

In-plane lateral load resistance of cast in-situ mortar panels

R. Iskandar, S. Suhendi & C. Lesmana
Department of Civil Engineering, Universitas Kristen Maranatha, Bandung, Indonesia

ABSTRACT: Cast in-situ mortar panel was a model of mortar wall that was designed to achieve effectiveness in terms of time, worker, and construction waste by using the unused sand from sand-filter as a construction material for building the wall. Although the wall panel is a non-structural element, the performance in residential buildings under wind and seismic loading is a major concern. The research aims to conduct an experiment of the mortar panels under lateral loading. Two panel specimens of mortar wall using a moving formwork were tested. The mortar panels demonstrated that the wall systems are advantageous in construction of residential or industrial buildings. The moving formwork was used for cast in-situ mortar panels to obtain solution of saving time and workers. The performances were acceptable, which was shown in the obtained results that pointed to higher strength with acceptable displacement for non-structural elements. It should be pointed out that the connection between the wall panels and the footing should be designed carefully to obtain an efficient load transfer mechanism in the structure. The inter-connection that attained from the moving formwork should be further studied since the presence of minimum dowel bars can significantly improve the lateral behavior by providing more ductility and higher strength. The cast in-situ mortar panels as alternative solution for non-structural construction material are recommended for the applications in the residential or industrial buildings.

Keywords: cast in-situ, lateral load, wall, mortar, moving formwork

1 INTRODUCTION

A more sustainable construction practice is encouraged in the population expansion. The construction of infrastructure facilities produces an amount of waste and consumes natural resources. A wall material such as masonry is commonly practiced for construction. Many researchers are eager to find the new building materials to find innovative solutions for environmental concern, such as composite masonry constructed with recycled building demolition waste and cement stabilized rammed earth (Jayasinghe et al., 2016), fine recycled aggregates from ceramic waste in masonry mortar manufacturing (Jiménez et al., 2013), recycled glass-fly ash geopolymer as low-carbon masonry units (Arulrajah et al., 2016), and so on. The technology of wall and mortar building materials has been developed to seek a viable alternative to reduce demolition waste and sustainable development (Jiménez et al., 2013). However, the more visible solutions remain unsolved since it depends on time, place, and availability materials.

A mortar is a workable paste from a mixture of sand and cements that often used to bind building blocks. The important characteristics of hardened mortar are mechanical strength, modulus of elasticity, water permeability, adhesive strength and resistance to weathering, and those of fresh

mortar are workability (NenoI et al., 2013, ASTM C270, 2014). The mortar usually uses a common construction material for bond the bricks or blocks but lack of research can be found in a wall panel by mortar, especially mortar that makes use of unused sand from sand-filter from the construction waste.

As a non-structural element, wall system must be easy to build, durable, strong enough to support self-weight and cost effective. Some of the evidence shows that many masonry walls crack due to wind or lateral loads (Cumhur et al., 2016, Triantafillou, 1998, Taghdi et al., 2000). Those evidences demonstrate the need to have minimum lateral response although the wall is a non-structural element.

The construction techniques and the eliminating of intermediate processes in selecting and processing materials can effectively reduce material and transportation cost. A moving formwork has high degree automation and short construction period. It is suitable in the case of the remote area, fixed dimension and flat form of structure (Yang et al., 2015). A mortar panel using moving formwork can become a solution in wall material for the remote area.

In this paper, the utilization of construction waste such as unused sand from sand-filter for mortar panels can help to demolish the waste in order to create more sustainable construction practices is discussed. The experimental work was conducted to observe the effectiveness of moving formwork for mortar panels. Besides, the mortar panels were tested in compressive and lateral strength to ascertain the possibility of the material as an alternative solution of sustainable materials for residential or industrial buildings in remote area.

2 EXPERIMENTAL INVESTIGATION OF MORTAR PANELS

As part of the investigation, two full-scale cast in-situ mortar panels were prepared using unused sand from sand-filter from the construction waste. The variable considered in this investigation were the compressive strength and performance of the lateral resistance of the mortar panels due to in-plane loading.

2.1 Materials

The mortar consisted of cement (pc), fine aggregate (fa), and water. The fine aggregate was using unused sand from sand-filter from the construction waste. The proportion of the material was mixed based on volume ratio of cement and fine aggregate as 1 pc: 7 fa with water/cement ratio 0.7. The weight volume of fine aggregates was 2.63 (SSD) with the absorption 6.07%. The water content was 8.46%.

2.2 Test specimens

The experimental program for mechanical characterization of mortar panel included tests in compressive strength. The cube specimens, $150 \times 150 \times 150$ cm, were used to obtain the compressive strength in 7, 14, and 28 days. The total number of specimens was 9 as three specimens were prepared for each conservative day.

Two cast in-situ mortar panels are prepared using unused sand from sand-filter from the construction waste. The wall is constructed using a moving framework. The benefit of this type of wall are the surface of the wall is flat and relatively smooth, hence, the construction time is faster than traditional brick wall. The mortar is first made with the right proportion to obtain acceptable workability. The mortar is filled into moving formwork up to 80%, a direct compaction should be done and the formwork can move to upper section. The mortar needs to fill until the wall reaches the expected height. The dimension of the specimen was $1700 \times 1460 \times 100$ mm that can be seen in Figure 1.

The in-plane lateral load was applied the tested specimens to obtain the resistance of the wall. The tests are performed at the Structural Testing Laboratory of Civil Engineering Department of Universitas Kristen Maranatha, Indonesia.

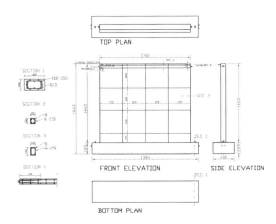

Figure 1. Plan elevation of the mortar panel.

Figure 2. Horizontal strengthened rebar in the middle of the wall for Specimen W2.

2.3 Experimental test program

The walls were subjected to in-plane monotonic load. The horizontal loading was applied until failure occurred. The walls were fixed to the floor and left free at the top. The horizontal displacement was measured at the top transfer beam shown in Fig. 1.

Two types of walls were prepared: W1 was the ordinary mortar wall and W2 was the mortar wall with the horizontal strengthened rebar for the connection joint.

3 DISCUSSION OF TEST RESULTS

3.1 Compressive strength

Table 1 shows the average compressive strength for the cube mortar specimens. The average compressive strength of the mortar (12,62 MPa) was higher than that of the brick material (5–7 MPa). The reused material from unused sand for replacing fine aggregate in mortar wall was recommended to use. The result showed acceptable compressive strength for non-structural element.

Table 1. Compression test results.

Type	Day [hari]	Area [mm²]	Weight [N]	Load [N]	Compressive strength [MPa]	Average compressive strength [MPa]
A-1	7	22837.50	58.96	139000	6.09	
A-2	7	21780.00	60.53	205000	9.41	
A-3	7	23179.06	64.65	151000	6.51	
B-1	14	22303.13	57.39	198000	8.88	
B-2	14	22425.00	61.51	242000	10.79	12.62
B-3	14	22459.00	61.80	205000	9.13	
C-1	28	22535.63	57.39	215000	9.54	
C-2	28	23293.13	64.45	345000	14.81	
C-3	28	22948.25	63.08	310000	13.51	

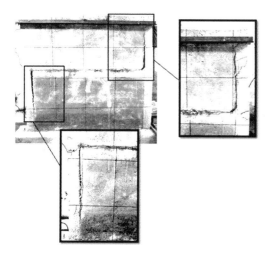

Figure 3. Crack pattern of Specimen W1.

3.2 Cracking

In the mortar wall specimen W1, the first major crack was observed at approximately in 4 mm of lateral displacement, which occurred in the middle of the wall. The horizontal crack widely spread from the middle and became wider and spread to create the vertical in the edge connection between mortar panel and frame as shown in Figure 3.

Different from specimen W1 that has relatively wider thickness crack about 30 mm, the specimen W2 has a crack of only about 5 mm thick. The horizontal strengthened rebar has shown good performance in strengthening the interconnection edge of the mortar wall due to the construction practice from moving formwork. The crack pattern in Figure 4 shows the mortar panel only has vertical crack.

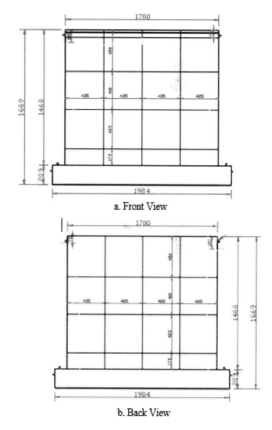

a. Front View

b. Back View

Figure 4. Crack pattern of Specimen W2.

3.3 Strength

The strength of the mortal specimen W1 was lower than specimen W2. This result indicated that the horizontal perpendicular rebar can strengthen the panels. Good lateral resistance can be observed from

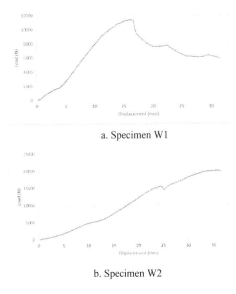

a. Specimen W1

b. Specimen W2

Figure 5. Load displacement of mortar wall.

Figure 5. The ultimate strength of specimen W1 was 11.40 kN in 15.95 mm horizontal displacement. Figure 5 illustrated that after the strength reached a maximum, the stiffness of the wall was decreased. The specimen W2 showed higher strength (20.2 kN) but the slip can be indicted in 15.60 kN.

4 CONCLUSIONS

The contribution of two different specimens to the in-plane lateral resistance of the mortar wall was experimentally investigated. The conclusions are summarized as follows:

1. The walls showed vertical cracks, with small thickness for the wall with strengthened and larger thickness for the wall without strengthening. The anchorage in the frame should be installed to have better load distribution.
2. High lateral resistance in the mortar wall of the specimen W2 is produced by the perpendicular strengthened horizontal rebar placed in the middle of the wall.
3. It is necessary to perform more tests to improve the models so the strength can be predicted.

4. The mortar wall that made from unused sand for replacing fine aggregate is recommended to use since it has acceptable compressive strength for non-structural elements and good lateral resistance.

ACKNOWLEDGEMENT

The authors acknowledge the support and financial assistance provided by the Universitas Kristen Maranatha in the form of research funds from Center of Research and Community Services.

REFERENCES

Arulrajah, A., Kua, T.-A., Horpibulsuk, S., Phetchuay, C., Suksiripattanapong, C. & Du, Y.-J. 2016. Strength and microstructure evaluation of recycled glass-fly ash geopolymer as low-carbon masonry units. *Construction and Building Materials,* 114, 400–406.

ASTM C270 2014. Standard Specification for Mortar for Unit Masonry. USA: ASTM.

Cumhur, A., Altundal, A., Kalkan, I. & Aykac, S. 2016. Behaviour of brick infill walls strengthened with expanded steel plates. *Bulletin of Earthquake Engineering,* 1–28.

Jayasinghe, C., Fonseka, W.M.C.D.J. & Abeygunawardhene, Y.M. 2016. Load bearing properties of composite masonry constructed with recycled building demolition waste and cement stabilized rammed earth. *Construction and Building Materials,* 102, Part 1, 471–477.

Jiménez, J.R., Ayuso, J., López, M., Fernández, J.M. & De Brito, J. 2013. Use of fine recycled aggregates from ceramic waste in masonry mortar manufacturing. *Construction and Building Materials,* 40, 679–690.

Nenoi, C., Britoi, J.D. & Veiga, R. 2013. Using fine recycled concrete aggregate for mortar production. *Material Research,* 17.

Taghdi, M., Bruneau, M. & Saatcioglu, M. 2000. Seismic Retrofitting of Low-Rise Masonry and Concrete Walls Using Steel Strips. *Journal of Structural Engineering,* 126, 1017–1025.

Triantafillou, T.C. 1998. Strengthening of Masonry Structures Using Epoxy-Bonded FRP Laminates. *Journal of Composites for Construction,* 2, 96–104.

Yang, J., Zhou, J., Li, X., Shan, A. & Kuang, Y. 2015. Numerical Analysis of the Moving Formwork Bracket Stress during Construction of a Curved Continuous Box Girder Bridge with Variable Width. *Modern Applied Science,* 9, 56–63.

Civil, Architecture and Environmental Engineering – Kao & Sung (Eds)
© 2017 Taylor & Francis Group, ISBN 978-1-138-02985-9

A discussion on how people perceive co-housing that facilitates intergenerational communication

Shu-Ying Tsai & Cheng-Huang Ou
Department of Architecture, National Taipei University of Technology, Taipei, Taiwan

ABSTRACT: Diverse co-living has been a priority in the agenda of most nations in the world. This study attempts to investigate how people perceive co-housing from the perspective of intergenerational communication. To this end, a questionnaire survey was conducted for people living in the Taipei Metro Area. Totally, 192 valid copies of a specially designed questionnaire were collected, and the survey results indicated that the two most significant factors are the respondents' age and housing type, and that in co-housing, the public facilities most demanded by the respondents are public vegetable gardens/flower gardens, group dining space, and parking lots.

1 INTRODUCTION

In Taiwan, rapid population growth and overexpansion of cities have led to problems related to aging and high population density in cities, which in turn has degraded the quality of life. According to the demographic analysis of 2015 conducted by the Department of Statistics, the Ministry of the Interior of Taiwan, R.O.C., by the end of 2015, there were 2.95 million people aged 65 or above in Taiwan, i.e., 12.51% of the total population. This percentage is close to the 14% threshold of aging society as defined by the WHO. In addition, the birthrate of people at 0–14 has seen a decrease from 19.34% to 13.57% in the last ten years. These facts suggest that Taiwan might face the impact of both aging society and sub-replacement fertility, making diverse co-living one of the most important issue during the transformation of social structure in Taiwan.

As an approach to improve intergenerational relationship and the quality of life, co-housing helps different age groups that are alien to each other to live together and get along with each other and provides various social facilities that facilitate the elders' interaction with other age groups and thereby support good quality of life. The study particularly aims at two objectives:

1. To discuss how residents from both the elderly and other age groups perceive intergenerational communication and co-housing; and
2. To suggest public function that co-housing is expected to have for facilitating intergenerational communication to consider future planning of communities.

2 LITERATURE REVIEW

Aging population are present worldwide, and there is a transformation of population profile and limited contact between school-age children and the elderly. Hence, for reducing this intergenerational distance, interaction and relationship between generations should be rebuilt through their living in some common facilities that make possible their living together, supporting each other and so on, but not through mere one-time activities. Lin (2006) studies how the elderly use rooms in senior citizens' housing and finds that most elderly are capable of taking care of themselves and housework and that they value interpersonal relationship. There are various forms of senior citizens' housing around the world for meeting the needs of the elderly with different lifestyles. It is proper to plan for senior citizens' welfare starting from housing because senior citizens' housing of good quality can function as a great platform for them to enjoy independence in the latter part of their lives where they are respected and are not in despair (Lo, 2015). In terms of the outdoor environment, public facilities are essential to quality life for the elderly to interact with their neighbors and the society. It is necessary to think on the basis of the experience of elderly in living and interacting with the environment about how living space facilitates interactivity and provides support to the elderly, and thereby helps the elderly adapt themselves to interpersonal relationship, increase self-worth, and maintain their physiological health. Moreover, intergenerational relationship, living arrangement, and aging society are major concerns for future social development. This study probes into issues related to communication between generations and co-housing.

2.1 Contents of intergenerational communication

With changes seen in the areas of society, economy, life expectancy, population, and family structure, people may have more time to share with the seniors from not only their own families but also other families, and this presents opportunities for both the youngsters and the elders to understand each other's demands for development (Liu, 2006). The trend of intergenerational integration is an aspect of an ideal model of community development that suits the aging society (Masato & Huang, 2011). Kaplan (2007) asserts that different generations realize the levels of close intergenerational interaction in accordance with how deeply they participate. The levels of intergenerational interaction are intergenerational relationship, learning, support, and solidarity.

2.1.1 Intergenerational relationship

The parties of intergenerational relationship are not limited to people kind to each other, but what is important is to foster good relationship between different generations (Chen, 2009). Given the changing social structure in Taiwan as well as demographical increase in the elderly and decrease in children, it is essential to strengthen the connection between generations by redefining intergenerational relationship, so as to build social cohesion in the future where different groups live together and help each other irrespective of whether they share blood relation.

2.1.2 Intergenerational support

Intergenerational support is important for the health of the elderly and has been proven as an effective way to improve elders' physical and cognitive abilities and to educate children (Bangerter et al., 2015). Xu (2007) suggests two types of intergenerational support: (1) emotional support: providing suggestion, listening, or soothing through conversation with others; and (2) instrumental support: helping the elders or people in need by performing housework or sharing things with them.

2.1.3 Intergenerational learning

By sharing experience with and gaining knowledge and skills from each other, people of different generations may change their perception of each other, as they become equal in terms of learning (Chen, 2014). Intergenerational learning refers to a process where different generations learn from and interact with each other through sharing and exchanging their experiences so that both the elderly and younger groups can grow and gain knowledge with such interaction (Huang, 2007; Li & Tang, 2008).

2.1.4 Intergenerational solidarity

With regards to the issue of aging, the United Nations has addressed the following tasks for promoting intergenerational equality and reciprocity and intergenerational solidarity (UN, 2002): (1) provide civic education to people to create awareness about population aging and its great importance in our society; (2) ensure that policies support intergenerational solidarity and advocate social integration; (3) implement systems that facilitate intergenerational communicate and make the elderly irreplaceable valuable social resources; (4) encourage the elderly to get together with the other age groups, thereby closing the gap between generations; (5) strengthen intergenerational link and mutual support so as to achieve successful social development; and (6) diversify the elderly lifestyle with consideration to family co-living in different cultures and backgrounds.

2.2 Development of co-housing

Co-housing originated in North Europe when a large number of housing units sprang up. The core thinking is that housing should be built and used with consideration of people living there and that it should support multi-generation community and friendly neighborhood that exhibit diverse cooperation (Wen et al., 2015; Sanguinetti, 2014). Following this trend, some autonomous organizations have been established to build housing that meets residents' needs, and their efforts have also been seen in planning indoor and outdoor facilities for public use. These facilities are designed to have various functions that cater to different user groups, with the common goal to improve the residents' mental and physical health and welfare. To the residents, such housing facilities are community resources that provide for their needs and change their life (Scharlach, 2012).

As to the benefits of co-housing, Chatterton (2013) states that by concentrating residences and enriching the surroundings with public facilities that facilitate lifestyle with high interactivity, critical mass can be created. The following are some of the effects of co-housing: (1) opportunities of engagement, (2) meeting of user needs through negotiation among residents, (3) a website as a platform for community interaction, (4) social welfare, (5) feeling of safety, (6) residents' care and support for the elders, (7) sharing private courtyard with other residents, (8) residents having the authority to manage all the information in the website, (9) reduced living costs thanks to agglomeration, (10) reduced number of vehicles as well as separation between pedestrian and vehicles, so as to create a zero-car environment that is safe and has less carbon emission, and (11) significant engagement and ownership.

The combination of intergenerational communication and co-housing satisfies people's internal feeling and external needs not only by reducing alienation among people in the urban scenario but also by meeting residents' needs in a better manner. These two inseparable aspects jointly form an optimal solution to urban coldness. During the implementation of co-housing, we face various problems that need to be addressed. This study primarily focuses on how to overcome these problems before accomplishing the vision of popularizing co-housing around the world.

3 SUBJECTS AND METHODS

3.1 Scope and subjects

The Taipei Metro Area is the most urbanized city in Taiwan and has a high population density that leads to urban coldness. This study thus focuses on the people living in the Taipei Metro Area.

3.2 Methods

This study investigates into how people perceive intergenerational communication and co-housing by reviewing relevant theories and articles, conducting a questionnaire survey, and analyzing in terms of people's perception, acceptance, and expectation of future promotion.

3.2.1 Questionnaire survey

A questionnaire was created on the basis of the understanding gained through literature review about how people perceive intergenerational communication and co-housing. A total of 215 copies of the questionnaire were issued; 192 collected copies were valid, while 23 were invalid.

3.2.2 Data analysis

The responses to the questionnaire were input to SPSS (Statistical Package for the Social Sciences), Chinese version 22.0, for data analysis.

4 ANALYSIS AND DISCUSSION

4.1 Analysis of individual subjects

This study summarizes the respondents' profiles as shown in Table 1. The results indicate that most respondents are female (69.3% of the total subjects); the majority occupation is housekeeping (24%); most respondents are in the education level of colleges and universities (63%); most respondents are satisfied with the status quo of their housing (81.8%); and most respondents are satisfied with the status quo of their neighborhood/communities (79.2%).

After the questionnaire analysis, the average of Cronbach's alpha is 0.706 for a total

Table 1. Respondents' profiles.

Item			Item		
Age	Count	Percentage	Housing type	Count	Percentage
15–24	31	16.1%	House	58	30.2%
25–44	34	17.7%	Townhouse	11	5.7%
45–64	62	32.3%	Apartments without elevators	80	41.7%
65+	65	33.9%	Mansions with elevators	39	20.3%
			Other	4	2.1%
Are you satisfied with your house?			Are you satisfied with your community?		
Yes	100	52.1%	Yes	157	81.8%
No	92	47.9%	No	35	18.2%

number of items of 55. Generally, Cronbach's alpha > 0.70 means the questionnaire used has good reliability.

4.1.1 People's perception of intergenerational communication

As many as 52.1% of the respondents are aware of co-housing and see intergenerational communication as helpful for better understanding groups outside their families (80.2%), gaining knowledge through interaction with each other (76%), reducing intergenerational distance (75%), improving belongingness and reciprocal relationship (72.9%), maintaining good relationship (72.4%), frequent communication (46.4%), concern and care for other groups (59.4%), and participation in activities held in the neighborhood/community (41.1%).

For identifying the factors that affect how people perceive intergenerational communication, this study considers the following assumptions:

Assumption I: respondents of different ages have different views on intergenerational communication; and

Assumption II: respondents who are satisfied and not satisfied with the status quo of their communities have different views on participation in activities held in the neighborhood/community.

4.1.2 Significance of questionnaire and cross analysis

Table 2 shows that the variable "age" displays the greatest significant difference, which is followed by occupation and housing type. Therefore, this study performs a cross analysis of significant items and summarizes the descriptive analysis according to the cross table, thereby clarifying the relationship between the items.

4.2 Discussion

4.2.1 Respondents' perceptional difference on intergenerational communication

According to the significant difference for the perceptions listed in Table 2, it is confirmed that "age" and "housing type" mostly affect the respondents' perception of communicate with people outside their families, as stated below:

1. Perceptional difference about communicate with people by age

The respondents aged 65 and above reported that they often communicate with people outside their families (55.4%). Far few respondents aged 15–24 often communicate with people outside their families (51.6%). Their responses to this question indicate that age is a major factor affecting one's perception of communicate with people outside their families.

2. Perceptional difference about communicate with people by housing type

The respondents living in houses show a higher frequency of communicate with people outside their families (55.2%). Here, 35.9% of the respondents living in mansions with elevators do not often communicate with people outside their families, showing that the housing type is a major factor affecting how often the respondents communicate with people outside their families.

4.2.2 Demands for public facilities of co-housing by age

Some demands common to the respondents of different age groups seen in Table 3 are public gardens, group dining space, and parking lots. This means provision of these facilities can be an effective way to encourage intergenerational communication, making them the priorities of co-housing.

Due to urbanization, mansions with elevators form the dominant housing type. However, the resulting community/neighborhood environment is unfavorable to the residents' social communication and in turn leads to different levels of alienation from people of not one's family among different age groups. This problem is highly related to the global population aging and sub-replacement fertility. Thus, we should contemplate how to improve communicate between groups by building residences as co-housing that provides more diverse public facilities that facilitate communicate between generations and encourage residents to share time with people outside their families.

Table 2. Significant difference among people's perception of intergenerational communication.

Item	Question		Gender	Age	Occupation	Education	Housing type
People's perception of intergenerational communication		Do you know what is co-housing for?	**0.023***	0.335	0.693	0.209	0.165
	Relationship	Maintaining good relationship	0.220	0.779	0.585	0.998	0.462
		Better understanding	0.158	0.502	0.224	0.970	0.354
	Learning	Learning knowledge from getting along with each other	0.240	0.736	0.750	0.303	0.272
		Idea communicate with other groups	0.981	**0.031***	0.252	0.676	**0.044***
	Support	Improved belongingness and reciprocal relationship	0.384	0.359	0.598	0.768	0.872
		Concern and care for other groups	0.654	**0.035***	0.089	0.860	0.541
	Solidarity	Reduced intergenerational distance	0.280	0.390	0.952	0.266	0.250
		Participation in activities held in the neighborhood/community	0.891	**0.000*****	**0.024***	0.626	0.984

Significance Levels: *P < 0.05 = Significant, **P < 0.01 = Very Significant, ***P < 0.001 = Extremely Significant.

Table 3. People's demands for public facilities of co-housing.

		Public facility					
		Public kitchen	Public reception room	Public garden	Playground	Group dining space	Parking lot
Age group	15–24	38.7%	38.7%	64.5%	54.8%	58.1%	54.8%
	25–44	32.4%	44.1%	76.5%	50.0%	55.9%	64.7%
	45–64	31.1%	36.1%	91.8%	49.2%	57.4%	60.7%
	65+	16.9%	30.5%	79.7%	40.7%	39.0%	37.3%

5 CONCLUSION AND RECOMMENDATION

5.1 Conclusion

Many cases are seen around the world in which co-housing successfully improves psychic isolation by transforming the urban high-density environment into a place where people can have commensalism, co-living, and sharing. This study finds that the respondents aged 15–24 and living in mansions with elevators form the group lacking communication with other people the most. To improve this, addition of public facilities such as public flower gardens/vegetable gardens, group dining space, and parking lots to their housing is a solution to encourage the residents to communicate with and support each other. In addition, some space currently occupied by buildings may be reconstructed into public facilities for public interest, so that in addition to parks and green lands, people can enjoy indoor and outdoor public facilities in their street blocks for community events, group dining, leisure, and so on. This idea may receive objections from residents at the initial stage, but it is believed that through environmental education and participatory design, people will see the benefits and turn to support co-housing.

5.2 Recommendation

1. Intergenerational communication is all about communicating with and supporting people outside one's own family and with no blood relationship in diverse ways.
2. It is recommended that events are held within and between communities and neighborhoods with common interests for residents to dine together and share things with each other, thereby fostering extensive interpersonal networks.
3. As a norm for co-housing design, subject to minimal requirements of private space, public facilities that serve the public interest should take the larger proportion.

4. Co-housing should be promoted as an important remedy for intergenerational barrier and as an effective approach to better quality of life and creation of an environment friendly atmosphere for the elderly. This is an issue we have to address in future.

ACKNOWLEDGMENTS

This study was funded by "The Research of Aging Collective Housing Using the Intergenerational Communication, MOST 105-2410-H-027-008-" research project [MOST 105-2410-H-027-008-] of the Ministry of Science and Technology. We are grateful for the assistance provided by relevant staff.

REFERENCES

Bangerter, L. R., Kim, K., Zarit, S. H., Birditt, K. S. & Fingerman, K. L. (2015). Perceptions of Giving Support and Depressive Symptoms in Late Life. *Gerontologist. 55(5)*, 770–779.

Chen, J. Y. (2009). Development Strategies of Intergenerational Education. *Intergenerational education Study Notes.* 3–18, Taichung.

Chatterton, P. (2013). Towards an Agenda for Post-carbon Cities: Lessons from Lilac, the UK's First Ecological, Affordable Cohousing Community. *International Journal of Urban and Regional Research. 5(37)*, 1654–1674.

Chen, Y. C. (2014). Learning Process and Effects of Incorporating Intergenerational Learning in a Social Gerontology Course. *Journal of research in education sciences. 59(3)*, 1–28.

Huang, G. C. (2007). Intergenerational Learning Ispire Elder Education. *Community Development Journal.* (118), 265–278.

Kaplan, M. (2007). Intergenerational Approaches for Community Education and Action. Contained in the National Institute for Environmental Education in Taichung University of Education (Ed.). *The new intersection of Adult Education and Environmental Education: Workshop Proceedings intergenerational programs.* 67–79, Taichung.

Liu, X. J. (2006). Children how to get along with the elders. –Respect for elders and caring family to start from the home education. *The pelican parenting monthly. May.*

Lin, S. C. (2006). A study on the behavior tendency in room of elderly housing—case studies of three elderly housings in Taipei area–. *Chung Yuan Christian University Dissertations.*

Li, Q. F. & Tang, X. M. (2008). The research of intergenerational learning in taiwan. *Population structure change and family Education.* 213–233.

Lo, Y. P. (2015). The research and prospects of luxury residences for the aged. *National Tsing Hua University Dissertations.*

Masato, S. & Huang, Q. H. (2011). *Life Plus.* 2.

Scharlach, A. (2012). Creating Aging-Friendly Communities in the United States. *Springer Science+Business Media.* 35, 25–38.

Sanguinetti, A. (2014). Transformational practices in cohousing: Enhancing residents' connection to community and nature. *Journal of Environmental Psychology.* 40, 86–96.

United Nations. (2002). Intergenerational solidarity. *Political Declaration and Madrid International Plan of Action on Ageing.* 5, 28.

Wen, F., Wang, Z. & Qiu, Z. (2015). Analysis of Multi—generational Living Mode Under Cohousing in Germany: Case Study on Dortmund City. *HUAZHONG ARCHITECTURE.* 5, 20–25.

Xu, C. D. (2007). The Intergenerational Support, Norms, Ambivalence and Living Arrangement in Later Life of The Mid-Generation. *National Taiwan Normal University Dissertations.*

Civil, Architecture and Environmental Engineering – Kao & Sung (Eds)
© 2017 Taylor & Francis Group, ISBN 978-1-138-02985-9

A note on the ABAQUS Concrete Damaged Plasticity (CDP) model

W.S. Li
Electric Power Research Institute of Guangdong Power Grid Co. Ltd., Guangzhou, China

J.Y. Wu
State Key Laboratory of Subtropical Building Science, South China University of Technology, Guangzhou, China

ABSTRACT: In this work, the ABAQUS Concrete Damaged Plasticity (CDP) model is investigated both theoretically and numerically. The interrelations between this model and its two precursors, i.e., Barcelona model (Lublinear et al. 1988) and Lee & Fenves (1998) model, are clarified. In particular, they are analyzed with respect to the issue of mesh size dependence. On the one hand, as damage evolution is not accounted for in the definition of fracture energy, Barcelona model (Lublinear et al. 1988) and Lee & Fenves (1998) model cannot suppress the issue of mesh-size dependence. On the other hand, an incorrect definition of the cracking displacement is used in the ABAQUS CDP, resulting in not only the issue of mesh size dependence, but also unit system sensitivity. Accordingly, the input data of stress and damage variable necessary for the ABAQUS CDP model can only be given in terms of the cracking strain, whereas the ones defined in terms of crack displacement should be used with great care unless this issue is removed in the future release.

1 INTRODUCTION

Being the most widely used building material in engineering structures and infrastructure, concrete exhibits complex behavior. With plenty of well-documented experimental tests (Kupfer et al. 1969) and research efforts, it is now generally accepted that the exhibited nonlinearities in concrete behavior are mainly attributed to two types of micro-structural mechanisms: (i) evolution of microcracks and microvoids, and (ii) plastic flows along some preferred crack lips. The aforesaid physical mechanisms have to be addressed appropriately in a constitutive model for concrete.

Among the different alternatives that are available, the plastic-damage model, combining the plasticity theory and Continuum Damage Mechanics (CDM), has been widely adopted in the constitutive modeling of concrete and applied to the nonlinear analysis of structures (see Krajcinovic (2003)). Owing to its great success in reproducing the nonlinear behavior of concrete, several commercial finite element packages, e.g., ABAQUS, LS-Dyna, etc., developed their own plastic-damage models or incorporated those proposed in the literature. Particularly, a Concrete Damaged Plasticity (CDP) model for concrete was developed in ABAQUS version 6.3 based on the well-known Barcelona model. This model is able to capture the typical nonlinear behavior of concrete rather well, greatly extending the capability of ABAQUS and pushing forward its application in civil engineering.

For concrete like quasi-brittle materials exhibiting strain softening, it is important to guarantee the objectivity of the numerical results when using a material constitutive model to evaluate the stresses at quadrature points. That is, the numerical results have to be independent of the size and alignment of the finite element mesh, such that the load capacity and deformations of structures can be evaluated correctly.

However, as will be shown later, the aforementioned mesh size independence cannot always be guaranteed for the ABAQUS CDP model. Even worse, the numerical results obtained using different SI units do not always coincide. Consequently, the numerical results may be misleading and sometimes erroneous. We address in this work the relevant issues and stress that the ABAQUS CDP model should be used with great care until the exhibited problems are eliminated in future releases.

This paper is organized as follows. After this introduction, the ABAQUS CDP model is discussed in Section 2. Particularly, the similarities and differences between the CDP model and its two precursors, i.e., Barcelona model and Lee & Fenves (1998) model, are clarified. Section 3 is devoted to the theoretical analysis of the CDP model regarding the issue of mesh size dependence. Numerical examples are presented in Section 4 to support the theoretical results. The most relevant conclusions are drawn in Section 5 to conclude this paper.

2 THE ABAQUS CDP MODEL AND ITS PRECURSORS

In all the models to be discussed, the simplest stress *versus* strain relation is expressed as

$$\sigma = (1-d)\bar{\sigma}, \quad \bar{\sigma} = \mathbb{E}_0 : (\varepsilon - \varepsilon^p) \tag{1}$$

where ε and ε^p denote the second-order strain tensor and its plastic component, both being infinitesimal; σ and $\bar{\sigma}$ signify the nominal and effective stress tensors, respectively; $\mathbb{E}_0$ represents the fourth-order elasticity tensor of the material.

Before we discuss the ABAQUS CDP model, let us first introduce its two precursors, i.e., Barcelona model and Lee & Fenves (1998) model.

2.1 Barcelona model

In the Barcelona model (Lubliner et al., 1988), the plastic strain tensor ε^p is determined by the classical plasticity theory, i.e.,

$$\begin{cases} \varepsilon^p = \dot{\lambda}\dfrac{\partial F_p(\sigma)}{\partial \sigma} \\ F(\sigma,q) \le 0, \quad \dot{\lambda} \ge 0, \quad \dot{\lambda}F(\sigma,q) \equiv 0 \end{cases} \tag{2}$$

where $F(\sigma,q)$ and $Fp(\sigma)$ represent the plastic yield and potential functions, respectively; $\dot{\lambda} \ge 0$ is the Lagrangian multiplier, with the non-vanishing value $\dot{\lambda} > 0$ upon plastic loading determined in terms of the consistency condition $\dot{F}(\sigma,q) = 0$.

One important contribution of the Barcelona model is that a simple plastic yield function $F(\sigma,q)$ is postulated for concrete under biaxial stress states

$$F(\sigma,q) = \frac{\alpha I_1 + \sqrt{3J_2} + \beta\langle\sigma_1\rangle}{1-\alpha} - q \le 0 \tag{3}$$

where $I_1 = \mathrm{tr}\,\sigma$ and σ_1 represent the first invariant and the major principal value of the stress tensor σ, respectively; $J_2 = \frac{1}{2}s:s$ denotes the second invariant of the deviatoric stress tensor $s := \sigma - (I_1/3)I$; the Macaulay brackets $\langle\cdot\rangle$ are defined as $\langle x \rangle = \max(x,0)$. The stress-like internal variable q signifies the cohesion; material parameters α and β are given by

$$\alpha = \frac{f_{bc}/f_c - 1}{2f_{bc}/f_c - 1}, \quad \beta = (1-\alpha)\frac{f_c}{f_t} - (1+\alpha) \tag{4}$$

in terms of the equi-biaxial compressive strength f_{bc}, the uniaxial compressive one f_c and the uniaxial tensile one f_t, respectively.

In order to postulate the plastic and damage evolution laws, an internal variable κ is introduced as

$$\dot{\kappa} = w\dot{\kappa}^+ + (1-w)\dot{\kappa}^- \tag{5}$$

where the weight function $w(\sigma) := \sum_i \langle\sigma_i\rangle / \sum_i |\sigma_i|$ discriminates the dominant tension and compression; $\kappa^\pm$ represent the normalized plastic work done in the uniaxial tension and compression, i.e.,

$$\kappa^\pm := \frac{1}{g_f^\pm}\int_0^{\varepsilon^{p\pm}} \sigma^\pm \, d\varepsilon^{p\pm}, \quad g_f^\pm = \int_0^\infty \sigma^\pm \, d\varepsilon^{p\pm} \tag{6}$$

with

$$\dot{\varepsilon}^{p+} = \mathbf{w}\dot{\varepsilon}^p_{max}, \quad \dot{\varepsilon}^{p-} = -(1-\mathbf{w})\dot{\varepsilon}^p_{min} \tag{7}$$

where $\dot{\varepsilon}^p_{max}$ and $\dot{\varepsilon}^p_{min}$ represent the maximum and minimal principal values of the plastic strain rate tensor $\dot{\varepsilon}^p$, respectively. Once the uniaxial tensile and compressive stresses, $\sigma^\pm(\varepsilon^{p\pm})$, or, equivalently, $\sigma^\pm(\kappa^\pm)$, are calibrated, the evolution law for κ can be determined.

Both evolution laws for the cohesive q and the damage d are expressed as the following rate forms:

$$\dot{q} = \dot{\sigma}^- = h_p \cdot \dot{\kappa}, \quad \dot{d} = h_d \cdot \dot{\kappa} \tag{8}$$

in terms of the strain-like internal variable κ, where h_p and h_d denote the plastic and damage hardening/softening moduli, both determined from experimental test data.

The Barcelona model can capture the nonlinear behavior of concrete under monotonic loading, but the predictions under cyclic one are not so satisfactory. More specifically, as only one single damage variable d is adopted, the stiffness recovery due to the Microcracks Closure-Reopening (MCR) effects cannot be described. Similarly, the yield function (3) with a single plastic internal variable q is incapable of modeling distinct strength softening under dominant tension and compression.

2.2 Lee & Fenves model

In order to extend the Barcelona model to cyclic loading, Lee & Fenves (1998) proposed an improved plastic-damage model. The stress *versus* strain relation is still given by Eq. (1). However, the damage variable d is related to another two independent scalar ones through

$$1 - d = (1 - wd^+)[1 - (1-w)d^-] \tag{9}$$

where d^+ and d^- represents the damage variables under dominant tension and compression, respectively, with distinct evolution laws.

When compared to the classical plastic evolution laws (2), the so-called effective stress plasticity (Ju, 1989) is considered, leading to

$$\begin{cases} \dot{\varepsilon}^p = \dot{\bar{\lambda}} \dfrac{\partial \bar{F}_p(\bar{\sigma})}{\partial \bar{\sigma}} \\ \bar{F}(\bar{\sigma},\bar{q}) \le 0, \quad \dot{\bar{\lambda}} \ge 0, \quad \dot{\bar{\lambda}}\bar{F}(\bar{\sigma},\bar{q}) \equiv 0 \end{cases} \tag{10}$$

where the yield function $\bar{F}(\bar{\sigma},\bar{q})$ and potential one $\bar{F}_p(\bar{\sigma})$ are both expressed in terms of the effective stress $\bar{\sigma}$, with $\bar{q}$ being the effective cohesion; $\dot{\bar{\lambda}} \ge 0$ is the effective Lagrangian multiplier, with its non-vanishing value $\dot{\bar{\lambda}} > 0$ upon plastic loading determined through the consistency condition $\bar{F}(\bar{\sigma},\bar{q}) = 0$.

In particular, the effective stress-based yield function $\bar{F}(\bar{\sigma},\bar{q})$ is given by the similar form as Eq. (3)

$$\bar{F}(\bar{\sigma},\bar{q}) = \frac{\alpha \bar{I}_1 + \sqrt{3\bar{J}_2} + \bar{\beta}\langle \bar{\sigma}_1 \rangle}{1 - \alpha} - \bar{q} \le 0 \tag{11}$$

where $\bar{I}_1 = \text{tr}\,\bar{\sigma}$ and $\bar{\sigma}_1$ represent the first invariant and the major principal value of the effective stress tensor $\bar{\sigma}$, respectively; $\bar{J}_2 = \frac{1}{2}\bar{s}:\bar{s}$ denotes the second invariant of the effective deviatoric stress tensor $\bar{s} := \bar{\sigma} - (\bar{I}_1/3)I$. The parameter α is coincident with Eq. (4), but $\bar{\beta}$ and $\bar{q}$ are both variables, expressed as

$$\begin{aligned} \bar{\beta}(\kappa) &= (1-\alpha)\frac{\bar{\sigma}^-(\kappa^-)}{\bar{\sigma}^+(\kappa^+)} - (1+\alpha) \\ \bar{q}(\kappa) &= \bar{\sigma}^-(\kappa^-) \end{aligned} \tag{12}$$

in terms of the strain-like plastic internal variables $\kappa := \{\kappa^+, \kappa^-\}$ defined in Eq. (6). The effective stresses $\bar{\sigma}^\pm$ under uniaxial tension and compression are determined by

$$\bar{\sigma}^\pm(\kappa^\pm) = \frac{\sigma^\pm(\kappa^\pm)}{1 - d^\pm(\kappa^\pm)} \tag{13}$$

with the stresses $\sigma^\pm(\kappa^\pm)$ and damage variables $d^\pm(\kappa^\pm)$ calibrated from test data.

In both the Barcelona and Lee & Fenves models, the internal variables $\kappa^\pm$ are defined as the normalized plastic work with respect to the total ones $g_f^\pm$ during the whole deformation process. Furthermore, it is assumed that they are the specific fracture energies $g_f^\pm := G_f^\pm / l_{ch}$, for the fracture energies $G_f^\pm$ under uniaxial tension and compression, respectively, and the characteristic length l_{ch}.

However, on the one hand, the monotonically increasing condition $\dot{\kappa}^\pm \ge 0$ cannot be guaranteed in the softening regimes. On the other hand, energy dissipation due to damage evolution is not accounted for. Consequently, it is not appropriate to assume the normalized plastic work as the specific fracture energy, nor select it as the internal variable.

2.3 ABAQUS CDP model

Perhaps noticing the aforesaid issues, the ABAQUS CDP model suggested adopting the cumulative (equivalent) plastic strains as the internal variables, i.e.,

$$\begin{cases} \kappa^\pm = \varepsilon^{p\pm} := \int \dot{\varepsilon}^{p\pm}\,dt \\ \dot{\kappa}^+ = w\dot{\varepsilon}^p_{max}, \dot{\kappa}^- = -(1-w)\dot{\varepsilon}^p_{min} \end{cases} \tag{14}$$

With the above re-definitions, the non-negative property $\dot{\kappa}^\pm \ge 0$ can always be guaranteed.

When using the ABAQUS CDP model, the stresses $\sigma^\pm$ and damage variables $d^\pm$ under uniaxial tension and compression can be given in terms of the cracking (or inelastic) strains $\varepsilon^{cr\pm}$ defined as

$$\varepsilon^{cr\pm} = \varepsilon^{p\pm} + \frac{d^\pm}{1-d^\pm} \cdot \frac{\sigma^\pm}{E_0} \tag{15}$$

Accordingly, the input data $\sigma^\pm(\varepsilon^{cr\pm})$ and $d^\pm(\varepsilon^{cr\pm})$ are then transformed into the required curves $\sigma^\pm(\varepsilon^{p\pm})$ and $d^\pm(\varepsilon^{p\pm})$ such that the effective uniaxial stresses $\bar{\sigma}^\pm$ and other quantities $(\bar{\beta},\bar{q})$ in the yield function (11) can be determined. It is very important that the plastic strains $\varepsilon^{p\pm}$, transformed from Eq. (15), be positive. Otherwise, the analysis would be terminated.

Alternatively, for plain concrete structures in tension, ABAQUS suggests using the stress σ^+ and damage variables d^+ in terms of the following cracking displacement u^{cr}:

$$u^{cr+} = u^{p+} + \frac{d^+}{1-d^+} \cdot \frac{\sigma^+}{E_0} \cdot l_0 \tag{16}$$

where the length scale l_0 is assumed as $l_0 = 1$, independent of the actual mesh size or adopted unit system. Similarly, the non-negative plastic strains $\varepsilon^{p+} := u^p / l_{ch}$, with l_{ch} being a characteristic length, are determined from the given data $\sigma^+(u^{cr+})$ and $d^+(u^{cr})$.

As will be shown, the definition (16) of cracking displacement is problematic. Consequently, the mesh-size dependence cannot be suppressed, and even worse, the numerical results are not objective regarding the adopted unit system.

1249

3 DISCUSSION OF THE ABAQUS CDP MODEL

When using the ABAQUS CDP model, the key issue is to estimate the evolution of uniaxial stresses and damage variables in terms of either the cracking strain or cracking displacement. In this section, the above issues are analyzed theoretically.

3.1 Input data

In order to suppress the issue of mesh size dependence accompanying strain softening materials, Hillerborg et al. (1976) proposed the fictitious crack model based on the concept of cohesive crack (zone). Specifically, the softening traction t (equal to the uniaxial tensile stress σ; in this section we omit the superscript '+' for uniaxial tension) versus separation relation u^{cr} is suggested, with the area surrounded by the $\sigma(u^{cr})$ curve being a constant as

$$G_f = \int_0^\infty \sigma \, du^{cr} \tag{17}$$

for the mode-I fracture energy G_f usually regarded as material property.

It is not convenient to use the above fictitious crack model in finite element analyses. To this end, the so-called characteristic length l_{ch} is introduced, leading to the following cracking strain ε^{cr}

$$\varepsilon^{cr} = \frac{1}{l_{ch}} u^{cr} = \varepsilon - \frac{\sigma}{E_0} = \varepsilon^p + \frac{d}{1-d} \frac{\sigma}{E_0}$$

or, equivalently,

$$u^{cr} = l_{ch} \varepsilon^{cr} = l_{ch} \cdot \varepsilon^p + \frac{d}{1-d} \frac{\sigma}{E_0} \cdot l_{ch} \tag{18}$$

Accordingly, the condition (17) becomes

$$\frac{G_f}{l_{ch}} = \int_0^\infty \sigma \, d\varepsilon^{cr} = \int_0^\infty \sigma \, d\varepsilon \tag{19}$$

This is the crack band theory proposed by Bažant & Oh (1983). It implies that, for a uniaxial tensile stress versus strain curve $\sigma(\varepsilon)$, the surrounding area is not a constant, but is inversely proportional to the characteristic length l_{ch}. With the above regularization of a softening material model, the mesh size dependence of numerical results can be largely suppressed.

Generally, the permanent displacement u^p is assumed to be a fraction of the crack one, i.e., $u^p = \delta u^{cr}$ and $\varepsilon^p = \delta\varepsilon^{cr}$. Accordingly, it follows from Eq. (18) that

$$d = \frac{(1-\delta)\varepsilon^{cr}}{(1-\delta)\varepsilon^{cr} + \sigma/E_0} = \frac{(1-\delta)u^{cr}}{(1-\delta)u^{cr} + \sigma l_{ch}/E_0} \tag{20}$$

For the particular case $\delta = 0$, no permanent displacement is present upon unloading and an elastic damage model is recovered. Eq. (20) can be used to produce the data of damage variable d necessary for the ABAQUS CDP model.

As can be seen, the cracking strain (15) in the ABAQUS CDP model coincides with the definition of (18a) suggested in the crack band theory. However, the cracking displacement (16) is different from Eq. (18b), unless the element characteristic length $l_{ch} = l_0 = 1$. That is, the two types of input data are inconsistent with each other. Consequently, numerical results independent of the mesh size cannot be guaranteed provided the stress σ and damage variable d are given in terms of the crack displacement (16). Even worse, the unit of length scale also affects the numerical results, violating the objectivity requirement.

3.2 Element tests

Let us consider a square plate of unit thickness, with edge length $L = 200$ mm. The plate is stretched by increasing displacements u^* with opposing directions at the left and right edges. At the initial stage, the material behaves linear elastic with Young's modulus $E_0 = 3.0 \times 10^4$ MPa. After the stress σ arrives at the tensile strength $f_t = 3.0$ MPa, the material enters softening regimes, with the fracture energy $G_f = 0.12$ N/mm.

A single irreducible quadrilateral element Q1 is used to discretize the plate, resulting in a characteristic length $l_{ch} = L$ 200 mm. For the sake of simplicity, a linear softening curve is considered, i.e.,

$$\sigma = \max(0, f_t + H \cdot u^{cr}) = \max(0, f_t + h \cdot \varepsilon^{cr}) \tag{21}$$

for the softening moduli $H = -37.5$ MPa/mm of the $\sigma - u^{cr}$ curve and $h = -7.5 \times 10^3$ MPa for the $\sigma - \varepsilon^{cr}$ one. The analytical solution of the reaction force F^* versus the imposed displacement u^* is depicted in Figure 1. Note that the result does not depend on the damage variable d since only the monotonic responses are concerned in this case.

For the given softening curve (21), the damage variable d can be generated with an assumed plastic fraction parameter $\delta \in [0,1]$. As expected, provided that the stress and damage are given in terms of the cracking strain, the numerical predictions, independent of the parameter δ, are coincident with the analytical result; see Figure 1.

However, if the stress and damage are given in terms of the cracking displacement, the numerical predictions are totally different. As shown in

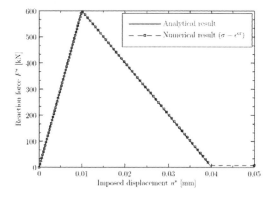

Figure 1. Numerical results using $\sigma - u^{cr}$ data.

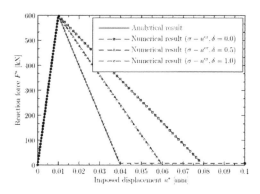

Figure 2. Numerical results using $\sigma - \varepsilon^{cr}$ data.

Figure 2, the numerical results not only are dependent on the parameter δ, but also are not identical to the analytical one except for the plastic model ($\delta = 1.0$) in which the value l_0 is irrelevant.

4 APPLICATION TO CONCRETE MODELING

In this section, a notched concrete beam under three-point bending (Petersson 1981) is considered. Figure 3 depicts the geometry of the specimen: a plain concrete beam of dimensions $2000 \times 200 \times 50$ mm³ with a notch of sizes $20 \times 100 \times 50$ mm³ at its bottom center. The beam is simply supported and an increasing displacement is enforced downward at its top center. The material properties are taken from Rots (1985), i.e., Young's modulus $E_0 = 3.0 \times 10^4$ MPa, Poisson's ratio $\nu_0 = 0.2$, tensile strength $f_t = 3.33$ MPa and fracture energy $G_f = 0.124$ N/mm. A linear softening curve (21) with the damage Eq. (20) is assumed, with two finite element meshes, the coarse and fine ones shown in Figure 4, are used in the numerical

simulations. Furthermore, two unit systems, i.e., SI-mm and SI-m, are considered.

The input data of stress and damage are first given in terms of the cracking strain. As can be seen from Figure 5(a), owing to the crack band theory, the numerical results are almost mesh size independent. Furthermore, the unit systems adopted in the simulations have no effects as expected.

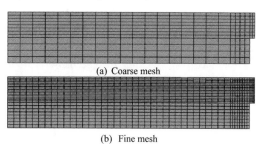

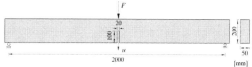

Figure 3. Geometry, boundary and load conditions.

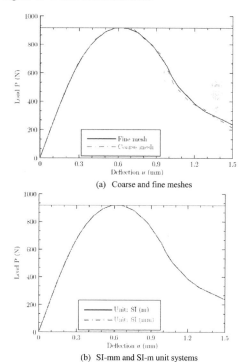

(a) Coarse mesh

(b) Fine mesh

Figure 4. Finite element meshes.

(a) Coarse and fine meshes

(b) SI-mm and SI-m unit systems

Figure 5. Numerical results using the cracking strain.

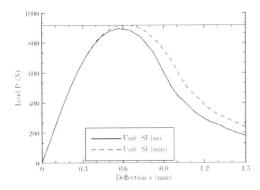

Figure 6. Different units using cracking displacement.

The input data of stress and damage are then given in terms of the cracking displacement. As can be seen from Figure 6, the numerical results are heavily dependent on the unit systems adopted in the simulations. This unphysical problem, caused by the incorrect cracking displacement (16), is completely unacceptable.

5 CONCLUSIONS

In this work, the Concrete Damaged Plasticity (CDP) model in ABAQUS is investigated both theoretically and numerically. The interrelations between this model and its two precursors, i.e., Barcelona model (Lublinear et al. 1988) and Lee & Fenves (1998) model, are clarified. In particular, they are analyzed with respect to the issue of mesh size dependence. Owing to the inappropriate strategies adopted, mesh size-independent numerical results cannot be guaranteed in the aforesaid models. On the one hand, the issue exhibited in the Barcelona model (Lublinear et al. 1988) and Lee & Fenves (1998) model is attributed to the fact that damage evolution is not accounted for in the definition of fracture energy. On the other hand, and incorrect cracking displacement is used in the ABAQUS CDP model, resulting in not only the issue of mesh size dependence, but also the loss of unit system objectivity. Therefore, the input data of stress and damage variable necessary for the ABAQUS CDP model can only be given in terms of the cracking strain, and the ones in terms of crack displacement should not be used unless the relevant issues are removed in the future release.

ACKNOWLEDGMENTS

This work was supported by the Science and Technology Project (GDKJ00000030) of China Southern Power Grid Co. The second author (J.Y. Wu) acknowledges support from the State Key Laboratory of Subtropical Building Science (2015ZB24, 2016 KB12) and the Fundamental Research Funds for the Central University (2015ZZ078).

REFERENCES

Bazant, Z.P. & Oh, B., 1983. Crack band theory for fracture of concrete. *RILEM Material of Structure*, 16: 155–77.

Hibbitt, Karlsson & Sorensen (HKS) inc., 2002. *Abaqus analysis user's manual*. Providence, Rhode Island, United States.

Hillerborg, A., Modeer, M. and Petersson, P.E., 1976. Analysis of crack formation and crack growth in concrete by means of fracture mechanics and finite elements. *Cement Concrete Res.*, 6: 773–782.

Ju, J.W., 1989. On energy-based coupled elastoplastic damage theories: Constitutive modeling and computational aspects. *International Journal of Solids and Structures*, 25(7): 803–833.

Krajcinovic, D., 2003. *Damage mechanics*. Elsevier B.V., Netherlands.

Kupfer, H., Hilsdorf, H.K. & Rush, H., 1969. Behavior of concrete under biaxial stress. *ACI Journal*, 66, 656–666.

Lee, J. & Fenves, G.L., 1998. Plastic-damage model for cyclic loading of concrete structures. *Journal of Engineering Mechanics*, ASCE, 124(8): 892–900.

Lubliner, J., Oliver, J., Oller, S. & Onate, E., 1989. A plastic-damage model for concrete. *International Journal of Solids and Structures*, 25(3): 299–326.

Ortiz, M., 1985. A constitutive theory for inelastic behaviour of concrete. *Mechanics of Materials*, 4, 67–93.

Petersson P.E., 1981 *Crack growth and development of fracture zones in plain concrete and similar materials*. Report No. TVBM-1006, Division of Building Materials, University of Lund, Lund, Sweden.

Rots, J.G., et al., 1985. Smeared crack approach and fracture localization in concrete. *Heron*, 30: 1–47.

Civil, Architecture and Environmental Engineering – Kao & Sung (Eds)
© 2017 Taylor & Francis Group, ISBN 978-1-138-02985-9

The cutting parameter model of energy consumption and its characteristics analysis

Shaohua Hu & Shilin Li
Chongqing Vocational Institute of Engineering, Chongqing, China

Jun Xie
Chongqing University of Technology, Chongqing, China

ABSTRACT: With the soar of the energy price and the increasingly serious environmental problems, the energy efficiency of manufacturing is becoming the focus of academia and industry gradually. It is able to provide theoretical basis and decision support for process planning and the energy-saving optimization of cutting parameters to master energy consumption characteristics. While the relationship between the energy consumption and the material removal rate has been conducted, the research on the influence characteristics of the cutting parameters on the energy consumption of machining process is still lacking. Therefore, this paper proposes a cutting parameter model of the energy consumption of the machine tools, and then discusses the influence characteristics of the cutting parameters on the energy consumption under a specific machining process based on the proposed model, and at last the influence characteristics were verified by experiments.

1 INTRODUCTION

With the soar of the energy price and the increasingly serious environmental problems, the energy efficiency of manufacturing is becoming the focus of academia and industry gradually (Jia, 2013). The machining systems that mainly consist of machine tools are numerous and widely used in industries, In China, for example, machining involves over 7 million machine tools, whose total power is greater than 70 million kilowatts. However, Many researches indicated that the energy efficiency of machining process is very low. For instance, the energy efficiency of a case described by Gutowski is only 14.8% (Gutowski, 2009). As a result, it has great potential for energy savings in machining processes.

In recent years, many famous universities, enterprises and international organizations have done a lot of work about the energy consumption of machine tools and machining system. Consortium on Green Design and Manufacturing (CGDM) of University of California Berkeley researched the impact of environment and the consumption of resource from mechanical design, processes planning, manufacture system modeling and environmentally conscious manufacturing (Munoz, 1995). Liu et al. established a multi-period energy model of electro-mechanical main driving system during the service process of machine tools that can

provide a theoretical basis for the follow-up study about the energy consumption of machining process (Liu, 2012; Wang, 2012).

In conclusion, some significant research on the energy consumption of machine tools has been performed. However, the research on the influence characteristics of the cutting parameters on the energy consumption of machining process is still lacking. This paper established the cutting parameter model of the energy consumption of the machine tools, which took the influence of the cutting parameters on the energy consumption of machining processes into consideration; then discussed the influence characteristics of the cutting parameters on the energy consumption based on the application of the model under a specific machining process; and the conclusions about influence characteristics were verified by experiments.

2 THE PARAMETER MODEL OF ENERGY CONSUMPTION OF THE MAIN DRIVING SYSTEM

2.1 *The power model of the main driving system*

The main driving system of machine tool consists of the motor driving system and the transmission system. Liu established a power equation of the main driving system in previous studies (Liu, 2012).

$$P_{in} = P_{Fe} + P_{Cu} + P_{ad} + P_{mco} + \frac{dE_{mm}}{dt} + \frac{dE_{ke}}{dt}$$

$$+ \sum_{j=1}^{r} P_{mcj} + \frac{d\sum_{j=1}^{r} E_{kj}}{dt} + P_c \tag{1}$$

P_{in} is the input power of the machine tool. P_{Fe} is the core loss of the motor. P_{Cu} is the winding loss of the motor. P_{ad} is the additional load loss of the machine tool. P_{mco} is the mechanical power loss of the rotor. E_{mm} is the magnetic field energy of the motor. E_{ke} is the kinetic energy of the motor rotor. P_{mcj} is the mechanical power loss of the mechanical transmission system. E_{kj} is the kinetic energy of the mechanical transmission system. P_c is the cutting power.

Some of the parameters in equation (1) are difficult to obtain, so we usually treat the power as the fixed power and named it as idle power P_u. When the machine tool operates in a stable state at a given speed, $\frac{dE_{mm}}{dt} = 0$. Therefore, when the machine tool runs steadily, the power equation of the machining process can be written briefly as follows:

$$P_{in} = P_u + P_{ad} + P_c \tag{2}$$

The idle power is the essential power which ensures the machine tool operation normally. The main driving system is of high inertia, when the machine tool runs at a specific spindle speed, the value of the idle power is a constant. For traditional machine tools, we can measure the idle power at every speed beforehand, and establish the idle power database as follows.

$$P_u(n) = \{P_u(1), P_u(2), \cdots, P_u(j)\} \ (n = 1, 2, \ldots, j) \tag{3}$$

For CNC machine tools whose main driving system have While for the step-less speed, we can measure the idle power at several selected speed and construct the idle power fitting function.

$$P_u(n) = g(n) \tag{4}$$

Due to the additional load loss power is difficult to be obtained by measuring, the additional load loss power P_{ad} can be fitted by a quadratic function of the cutting power (Hu, 2010).

$$P_{ad} = a_2 P_c^2 + a_1 P_c \tag{5}$$

The parameters a_1 and a_2 in the formula are the additional load loss coefficients of the main driving system which can be obtained by experimental method (Hu, 2010).

2.2 The cutting power model of the main driving system

The cutting power is the power required by the tool tip to remove the material of the workpiece, which is the output power of the main driving system. The cutting power is given by the following.

$$P_c = F_c \times \frac{v_c}{60} \tag{6}$$

The cutting force has strongly related to the cutting parameters, and the relationship between the cutting force and its impact factors is so complicated that there is still lack of an authoritative method to calculate the cutting force. As a result the empirical method is usually used to calculate the cutting force in practical application, the model is as follows.

$$F_c = C_{F_c} a_p{}^{x_{F_c}} f^{y_{F_c}} v_c{}^{n_{F_c}} K_{F_c} \tag{7}$$

In equation (7), a_p denotes the depth of cut (mm), f denotes the feed rate (mm/r), $v_c = \frac{\pi D_0 n}{1000}$ denotes the cutting speed (m/min), $C_{F_c}, x_{F_c}, y_{F_c}, n_{F_c}$ and K_{F_c} are the coefficients related to the machining process, the coefficients can be obtained from the look-up tables in the manual.

According to the equation (6) and (7), the cutting power is given by the following.

$$P_c = \frac{C_{F_c}}{60} a_p{}^{x_{F_c}} f^{y_{F_c}} v_c{}^{(n_{F_c}+1)} K_{F_c} \tag{8}$$

2.3 The model of energy consumption of the main driving system

According to the analysis described above, the parameter model of the input power can be obtained as follows.

$$P_{in} = \alpha_2 \left(\frac{C_{F_c}}{60} a_p{}^{x_{F_c}} f^{y_{F_c}} v_c{}^{(n_{F_c}+1)} K_{F_c} \right)^2$$
$$+ (\alpha_1 + 1) \frac{C_{F_c}}{60} a_p{}^{x_{F_c}} f^{y_{F_c}} v_c{}^{(n_{F_c}+1)} K_{F_c} + P_u \tag{9}$$

Furthermore, the energy consumption of the main driving system can be calculated by the integral of input power over the time of the machining process.

$$E_{in} = \int_{t_1}^{t_2} \left[\alpha_2 \left(\frac{C_{F_c}}{60} a_p{}^{x_{F_c}} f^{y_{F_c}} v_c{}^{(n_{F_c}+1)} K_{F_c} \right)^2 + (\alpha_1 + 1) \frac{C_{F_c}}{60} a_p{}^{x_{F_c}} f^{y_{F_c}} v_c{}^{(n_{F_c}+1)} K_{F_c} + P_u \right] dt \tag{10}$$

In most cases, the load of machining process is continuous and stable, when the cutting parameters keep constant, the input power can be treated as a constant. For the variable load cutting process, each tiny period can be regarded as a constant load cutting process. Hence, this paper analyzes the influence characteristics of the parameters on the energy consumption under the constant load conditions, equation (10) can be changed into equation (11)

$$E_{in} = \frac{60V\left[\alpha_2\left(\frac{C_{F_c}}{60}a_p^{x_{F_c}} f^{y_{F_c}} v_c^{(n_{F_c}+1)}K_{F_c}\right)^2 + (\alpha_1+1)\frac{C_{F_c}}{60}a_p^{x_{F_c}} f^{y_{F_c}} v_c^{(n_{F_c}+1)}K_{F_c} + P_u\right]}{a_p f v_c} \tag{11}$$

3 THE INFLUENCE CHARACTERISTICS OF THE CUTTING PARAMETERS ON ENERGY CONSUMPTION

3.1 Introduction of the machining process

The parameter model of the energy consumption mentioned above is a comprehensive model which can be used in different machining processes. For ease of understanding, taking a turning machining as an example, the influence characteristics of the cutting parameters on the energy consumption is discussed. The work-piece is shown as the Fig. 1. The machining process of CNC lathe (C2-6136HK/1) is chosen as a machining case. The specification of machine tool is shown in Table 1, the information of the cutter and blank material is listed in Table 2, and the cutting parameters are listed as Table 3, respectively.

According to the parameters listed in Table 1 and Table 2, the empirical parameters obtained from the look-up tables in the manual are listed in Table 4. According to the Fig. 3, we can draw the conclusion that with the increase of the cutting parameters, the input power increase too, and the input power of machine tool and the cutting parameters have positive correlation. On the other hand, the spindle speed has the strongest influence on the input power, and then is the depth of cut and the feed rate.

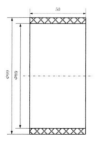

Figure 1.   The blank drawing of the work-piece.

Table 1.   Specification of machine tool.

Type: C2-6136HK/1
Rated power of the main motor (kW): 5.5
Revs area of low gear (rpm): 100–1000
Revs area of high gear (rpm): 300–2100

Table 2.   Information of the cutter and blank material.

Type: WNMG080404-MA UE6020
Material of the cutter: cemented carbide (coated)
Blank material: S45C Carbon steel
Diameter (mm): Φ99
Length (mm): 50

Table 3.   Parameters of processes.

No.	Parameters		
	$n(rpm)$	$f(mm/r)$	$a_p(mm)$
1	20–1000	0.2	2
2	500	0.05–0.3	2
3	500	0.2	0.1–4

Table 4.   Empirical parameters of cylindrical turning.

Material	Parameters				
	C_{F_c}	x_{F_c}	y_{F_c}	n_{F_c}	K_{F_c}
S45C Carbon steel	2795	1	0.75	–0.15	0.92

3.2 The influence characteristics of cutting parameters on energy consumption

3.2.1 Univariate influence characteristics on the energy consumption

The input power can only represent the transient state of the energy consumption of machine tools, and cannot reflect the characteristics of the energy consumption of the entire machining process. So the influence characteristics on the energy consumption will be studied following.

1255

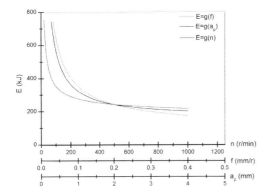

Figure 2. The univariate influence characteristics on the energy consumption.

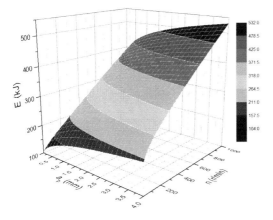

Figure 3. The multivariate influence characteristics on the energy consumption.

Substituting the parameters listed in Table 3 and Table 4 into the equation (11), and the univariate influence characteristics is shown in Fig. 2. According to the Fig. 2 we can see that the energy consumption of machining process is inversely proportional to the machining rate. As a result, in the actual process of rough machining, to reduce the energy consumption, we should choose large machining rate for processing.

3.2.2 Multivariate influence characteristics analysis on the energy consumption

The analyses mentioned above are all about the univariate influence characteristics of the input power and the energy consumption, and can not reflect the influence characteristics of the cutting parameters. In the actual machining process, the processing is usually conducted at the request of a specific processing rate. So this paper will analyze the multivariate influence characteristics at a specific processing rate.

Machining at the request of a specific processing rate means the material removal rate of the process is a constant. To simplify the analysis below, the material removal rate is set as 0.5 cm^3/s.

$$MRR = \frac{a_p f v_c}{60} = \frac{a_p f n \pi D}{60000} = 0.5 \qquad (12)$$

Substituting the parameters listed in Table 3 and Table 4 into the equation (11), the multivariate influence model of the energy consumption is as follows.

The Fig. 3 demonstrates that under the condition of a specific processing rate, the effect of the three parameters on the energy consumption is different. According to the Fig. 3, in the actual process of rough machining with the specific processing efficiency, to reduce the energy consumption, we should choose a large feed rate, a lower spindle speed and then choose a proper depth of cut according to the request machining efficiency.

4 EXPERIMENTAL VALIDATION

The equipment applied to measure the energy consumption is HIOKI 3390 power analyzer which can continuously record the input power of a machine tool at a 0.05 s sampling interval. The cutting parameters and the experimental results are listed in Table 5. According to the results of the experimental validation, we can draw the conclusions as follows: 1) we can see that when the cutting parameters (n, f, a_p) increase, the value of input power

Table 5. Parameters of process and experimental results.

	Parameters					
No.	n (rpm)	f (mm/r)	a_p (mm)	Power (W)	Time (s)	Energy (J)
1	100	0.2	1.5	1044	600	551250
2	250	0.2	1.5	1753	240	370860
3	400	0.2	1.5	2329	150	312637.5
4	550	0.2	1.5	3017	109.08	299725.57
5	700	0.2	1.5	3410	85.72	262731.8
6	400	0.05	1.5	1570	600	879000
7	400	0.1	1.5	1653	300	456675
8	400	0.2	1.5	2329	150	312637.5
9	400	0.3	1.5	2853	100	250225
10	400	0.4	1.5	3427	75	222956.25
11	400	0.2	0.5	1314	375	492750
12	400	0.2	1.0	1810	187.5	339375
13	400	0.2	1.5	2329	150	312637.5
14	400	0.2	2.0	2730	112.5	270375
15	400	0.2	2.5	3085	75	231375

increases too. The power consumption is positively associated with the cutting parameters; 2) we can see that the increment of the power consumption of spindle speed is the largest, then is the depth of cut, and feed rate is the smallest; 3) the energy consumption decreases with the increase of each cutting parameters, and the energy consumption is inversely proportional to the cutting parameters; 4) it can be concluded that to reduce the energy consumption, we should choose a large feed rate, a lower spindle speed and a proper depth of cut. The experimental result is consistent with the conclusions from the analysis of the model.

5 CONCLUSIONS

This paper established the parameter model of and the energy consumption of machining processes based on the previous studies, which take the influence of the cutting parameters on the energy consumption of machining processes into consideration respectively; then discussed the influence characteristics of the cutting parameters on the energy consumption based on the application of the model under a specific machining process.

1. With the increase of the cutting parameters (n, f, a_p), the power consumption increases. Furthermore, the spindle speed (n) has the strongest influence on the input power while the feed rate (n) is the smallest.

2. The energy consumption of machining process was inversely proportional to the machining rate. As a result, in the actual process of rough machining, we should choose large machining rate for processing, which can not only reduce the energy consumption, but also improve processing efficiency.

ACKNOWLEDGMENTS

The authors gratefully acknowledge the financial support from Chongqing Natural Science Foundation research project (cstc2013jcyjA70014).

REFERENCES

Editorial Board of Mechanical Engineering Manual Mechanical engineering handbook. (China Machine Press, Beijing, 1997).

Gutowski T, Branham M, Dahmus JB, Jones AJ, Thiriez A, Sekulic DP. Environmental Science & Technology 43, (2009).

Hu S, Liu F, He Y, Peng B. Journal of Advanced Mechanical Design, Systems, and Manufacturing 4,7 (2010).

Jia S, Tang R, Lv J. Computer Integrated Manufacturing Systems. 19, 5 (2013).

Liu F, Liu S. Journal of Mechanical Engineering 21, (2012).

Liu S, Liu F, Wang Q. Journal of Mechanical Engineering 48, 23 (2012).

Munoz A, Sheng P. Journal of Materials Processing Technology 53, (1995).

Civil, Architecture and Environmental Engineering – Kao & Sung (Eds)
© *2017 Taylor & Francis Group, ISBN 978-1-138-02985-9*

Tibetan question parsing integrated with phrase structure and event feature

Ning Ma, Hongzhi Yu, Fucheng Wan & Xiangzhen He
Key Laboratory of National Language Intelligent Processing, Gansu Province (Northwest University for Nationalities), Lanzhou, Gansu, China

ABSTRACT: We use syntactic parsing technology based on phrase structure and event feature for analyzing Tibetan question sentences in the public information service. The core approaches include Tibetan question sentence pretreatment, syntactic parsing, and question semantic representation. Through an experiment, we obtained very successful results on accuracy rate and recall rate, which proves that the introduction of the event feature in Tibetan question sentence analysis is very useful.

1 INTRODUCTION

Big data appears with the rapid development of computer networks and the information on the Internet becomes large and semi-structured. Now we find it more and more difficult to search information we are interested in as the researching results on the Internet can no longer satisfy us. Thus, we need some new ways to find useful information.

The automatic question answering system depends on it but is quite different from the retrieving technology. It can be questioned by sentences which are denoted in the natural language, and the returning answers will never be collections of only related web pages. Recently, we have not found any mature Tibetan automatic question answering system in literature. This paper researches the Tibetan automatic question answering system for public information service and focuses on the strategy of Tibetan question parsing.

2 RELATED RESEARCH

The question answering system is a typical application example of natural language processing and also the focus and difficulties for academic research. In foreign countries, the automatic question answering system of English, Japanese, and German languages has been applied in a restricted domain (Kayes, 2015). In China, there are some successful automatic question answering systems, for example, the campus navigation system of Tsinghua University, the frequently asked question answering system of Harbin Institute of Technology, and the bank question answering system of Beijing Institute of Technology, and so on.

At present, with further research on natural language processing technology, the lexical, syntactic, semantic, and pragmatic research achievements are beginning to be applied in the automatic question answering system. Wei Chuyuan et al. used the valid information in event modeling to enhance the accuracy of questions in semantic representation annotation in the Chinese question answering system (Nguyen, 2015). Kang Haiyan et al. set up a Web intelligent question answering system based on the questions in semantic representations. Tang Suqin et al. introduced dependency grammar in Chinese sentence parsing and extracted the semantics of questions and the limitation relationship among each component of the question so as to obtain the question understanding results aiming towards the domain ontology knowledge base.

Research of Tibetan information processing develops slowly, and the research on automatic question answering system still stays in the lexical analysis domain. The syntactical research is still in the initial stages and focuses on dependency grammar.

3 TIBETAN QUESTION PARSING FOR PUBLIC INFORMATION SERVICE

3.1 *Strategy of technology*

As shown in Figure 1, we divide the Tibetan automatic question answering system into four steps: Tibetan question input, Tibetan question parsing, the solution of Tibetan FAQ, and Tibetan difficult questions, and the last Tibetan answer output. This paper mainly focuses on the research of Tibetan question parsing as shown in the dashed box.

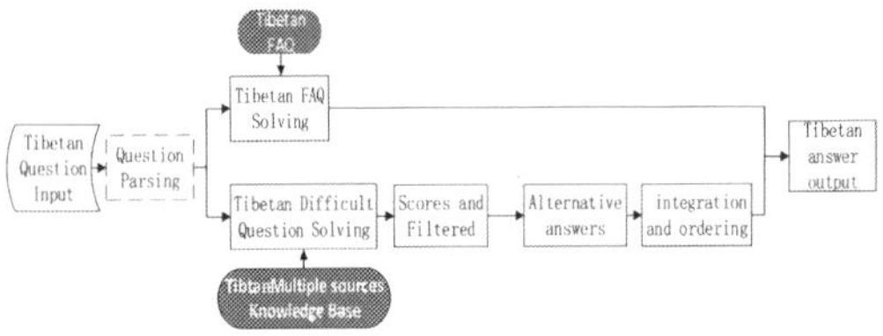

Figure 1. Technology strategy of the Tibetan automatic question answering system.

3.2 *Tibetan question pretreatment*

In the Tibetan automatic question answering system, there are many redundant components like stopping words, for example, "འདྲི་ཞུ་གནང་རོགས། (hello)", "ཁྱེད་ཀྱིས་ང་ལ་བཤད་རོགས། (would you please tell me)", "ཁྱེད་ལ་དོ་བསྣལ་རེད། (excuse me)", and "སུས་ང་ལ་བཤད་ཆོག (is there anybody who can tell me)". In order to analyze the Tibetan question parsing, at first we need to get rid of these redundant components and then continue to parse them effectively. We set up a Tibetan redundant words table to solve this problem.

After removing these redundant components, we use the Tibetan word segmentation system developed in our laboratory to segment the Tibetan words and get some line sequences of Tibetan words to prepare for the next steps.

3.3 *Tibetan phrase syntactic parsing integrated with functional semantic information*

We construct a small amount of Tibetan phrase syntactic treebank in early research and the treebank is mainly used for Tibetan event information. This paper is based on Tibetan phrase syntactic treebank and integrates with Tibetan functional semantic information, as shown in Table 1. The main argument roles are agent labeled as NP-SUB(ARG0) and patient labeled as NP-OBJ(ARG1), and the secondary argument roles are time labeled as ARGM-TEMP, location labeled as ARGM-LOC, manner labeled as AGRM-MNR, etc. so that Tibetan functional semantic information can be integrated into the Tibetan phrase syntactic tree.

Through the above labeling strategy, we successfully integrate Tibetan phrase syntactic tree with functional semantic information. Take the sentence "པེ་ཅིན་ དེ་ ཀྲུང་གོ་ འི་ རྒྱལ་ས་ ཡིན། (Beijing is the capital of China)" for example. Its syntactic tree is as shown in Figure 2.

3.4 *Tibetan question semantic representation*

Question semantic representation is a special representation method for question semantic parsing, which represents Tibetan questions as the most

Table 1. Tibetan functional semantic information labeling table.

Agent	ARG0-SUB	Patient	AGR1-OBJ	Time	AGRM-TMP
Location	AGRM-LOC		Manner	AGRM-MNR	

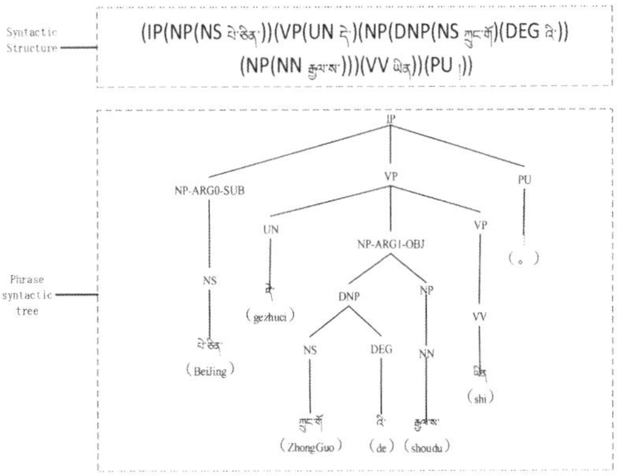

Figure 2. Structure of the Tibetan phrase syntactic tree.

intuitive questions in focus, with the main part and event information presented in an abstract labeling manner, so as to conduct semantic parsing for Tibetan questions. The Tibetan question semantic parsing representation labels Tibetan question semantic structure by a triple form as shown in the equation (1) below,

$$TQS(TQ) = \{TQf, TQt, TQe\} \tag{1}$$

In equation (1), TQS (Tibetan Question Semantic) is the Tibetan question semantic representation structure, TQf is Tibetan question focus blocks, TQt is Tibetan question topic blocks, and TQe is Tibetan question event blocks.

$$\begin{aligned} TQe &= < TA, TL, TT, TD > \\ &= < TQEa, TQEl, TQEt, TQEd > \end{aligned} \tag{2}$$

Formula 2 represents the semantic structure of Tibetan question event blocks, TQEa represents event trigger element of Tibetan question event blocks, TQEl represents space element of Tibetan question event, TQEt represents time element of Tibetan question event, and TQEd represents the extent and effects element of Tibetan question time blocks, while the main semantic element is denoted with a vector. For example, TQ: ཁྱེད་ལ་དོ་བསྣལ་རེད། འདྲི་ཞུ་གནང་རོགས།ང་ད་ལྟ་ལན་གྲུ་ཆུ་མིག་ལྟ་ཅན་རི་ནས་ལན་གྲུ་ནུབ་ཕྱོགས་ས་ཚུགས་སུ་འགྲོ་ན་སྤྱི་སྤྱོད་རླངས་འཁོར་ཨང་དུ་པར་བསྐྱོད་ན་འགྲིག། (Excuse me, which bus can I take best from Wuquanshan in Lanzhou to west station in Lanzhou?).

3.4.1 Recognition of Tibetan question focus blocks

Question focus blocks are the most related elements with question information, including the combinations of question words and phrases, for example, the combinations of question words with noun phrases, verb phrases, prepositional phrases, and so on. "སྤྱི་སྣུང་འཁོར་ཨང་དུ་བ (which bus)" is the combination of question words and noun phrases.

3.4.2 Recognition of Tibetan question topic blocks

Question topic blocks are the type and background of Tibetan questions. This paper researches the Tibetan automatic question answering system for public information service, which divides question types into traffic, education, science and technology, culture, health, and sports. Obviously, the topic type of the example sentence is འགྲིམ་འགྲུལ (traffic).

3.4.3 Recognition of Tibetan question event blocks

Question event represents the event information or some actions that take place at certain times and places. Tibetan question event blocks are composed of a tetrad including action, time, space, and effect. The above-mentioned example sentence involves the element of action "བླངས(take)", the element of time "ད་ལྟ(now)", the element of space "ལན་གྲུ་ཅུ་མེག་ལྷ་ཚན་རེ་ནས་ ལན་གྲུ་ནུབ་ཕྱོགས་ས་ཚིགས (from Wuquanshan in Lanzhou to west station in Lanzhou)", the element of effect "འགྲིག (best)". The final Tibetan question semantic representation is shown as below (3) and (4),

TQS(TQ) = {TQf (སྤྱི་སྣུང་འཁོར་ཨང་དུ་བ (which bus)),
  TQt (འགྲིམ་འགྲུལ (traffic)), TQe} (3)

TQe = <TA, TL, TT, TD> = <TQEa (བླངས (take))
  TQEl (ལན་གྲུ་ཅུ་མེག་ལྷ་ཚན་རེ་ནས་ ལན་གྲུ་ནུབ་ཕྱོགས་ས་ཚིགས (from Wuquanshan in Lanzhou to west station in Lanzhou)), TQEt (ད་ལྟ (now)),
  TQEd (འགྲིག (best))> (4)

The element of action is directly related to the trigger word of the Tibetan question. In general, the verb is the most important element of a Tibetan sentence, and the Tibetan verb is the trigger word of the event. If you can recognize Tibetan trigger words, you will be able to recognize the core verb in a Tibetan sentence. However, a lot of verbs are not included in the training corpus. For example, " དགེ་རྒན་ལ་བཀུར་བ (respect the teachers)". Assuming that, " བཀུར (respect)" is not

included in the training corpus as an event trigger word, we cannot recognize this event. But the meanings of " བཀུར (respect)" and "གུས་བཀུར (esteem)" are similar. In view of this, we carry out Tibetan word semantic similarity computation on the basis of the Tibetan verb lexicon and lexicon trigger words in order to automatically expand the trigger words and cover various types of event trigger words as much as possible. Then the acquired seed trigger word and its information type will form a tuple combination (trigger, type), for example, (chairman, Person/Respect). Regarding one trigger word corresponding to multiple categories in special circumstances, due to its small probability, we mostly choose the category with the largest probability of occurrence in the training corpus. Through this method, we construct a table of "Tibetan seed trigger words and event information category", in which the first column is event trigger words and the second column is information category.

In Table 2, the first column corresponds to seed trigger words, and the second column corresponds to the event information category of each trigger word. Each trigger word corresponds to only one event information category and the trigger words in the table cover all categories of event information.

"Time" is the time information element of the event in Tibetan question, "space" is the space information element of the event in Tibetan question, including location. To recognize the time and space, syntactic analysis features are used. The time element is labeled as NT, and the spatial element is labeled as NL. In syntactic parsing model, more features are introduced to train, in order to decode the test data, and finally identify the time and space elements.

"Effect" is the extent and effects element of the Tibetan question events. They are generally adverbs, which often exist in a Prepositional Phrase (PP) or adverb phrase (ADVP) in the whole sentence. We usually identify the effect element by combining block recognition with corresponding rules.

Table 2. "Tibetan seed trigger words and event information category".

Seed trigger words	Event information category
ཧེབ་ས (bankrupt)	Business/ Declare-Bankruptcy
འཁྲོག་རྩོད (contend)	Conflict/Attack
ཧོས་ཐོན (flee)	Movement/Transport
ཟར་སྟོན (demonstration)	Conflict/Demonstrate
...	...

4 EXPERIMENTS AND EVALUATION

4.1 Set up of experiment

For corpus selection, currently, there is no special corpus and standard evaluation criteria for the Tibetan language question answering system. In order to show our work effectively, we annotate 300 sentences for the corpus of Tibetan question answering system and include 70 sentences in the test data. The precision rate and recall rate are used for evaluation.

4.2 Experimental results

The results are shown in Table 4. The evaluation indexes are precision rate and recall rate. Since the Tibetan question-answering system research is still in its infancy stage, we do not compare the experiment results with a baseline system, but horizontally compare the triggering word recognition, syntactic parsing, and question parsing.

4.3 Result analysis

Table 4 shows that the accuracy of the Tibetan question parsing method in this paper is relatively high. The reasons are mainly as given below: (1) Early research has covered the removal of redundant words as well as word segmentation so that the accuracy rate of this part is high, but we have not considered the effect of functional words in Tibetan sentences. Take the example sentences in section 2.4. There are one position prepositions ষ and two-state prepositions - ন and ৰ in this sentence. In the next step, we will further research functional words in the Tibetan sentence; (2) In early laboratory research, we have built a number of Tibetan phrase syntactic treebanks, in which the accurate rate of syntax analysis for a sentence

composed of less than 15 words is relatively good. While in questions for public information service, the sentences are mostly composed of less than 15 words, and as such the accurate rate is also good in syntax analysis; (3) In Tibetan sentences with expression, the core verb is always located in the end of a predicative sentence, so it is easy to identify the triggered words; (4) The recognition of an event block is directly related with syntax parsing. Meanwhile, the labeling of linguistic features like time and space elements are conducive to event element recognition. This also proves that in the analysis of questions, the introduction of event features can help improve accuracy.

5 CONCLUSIONS AND FUTURE WORK

Tibetan question parsing model is based on phrase structure and ontology features, and on the basis of Tibetan syntactic parsing, it integrates with functional semantic information, time, and space information in the training step so as to provide support for Tibetan information extraction.

This paper provides a lexical, syntactic, and shallow semantic solution for the Tibetan question analysis, including Tibetan semantic information classification, Tibetan semantic information labeling, and Tibetan syntactic parsing.

The contribution for our work is to provide a question analysis scheme for other languages including Mongolia language, Uyghur language, and other minority languages.

Our work lies in the syntax and shallow semantic level. In the next step, we will enhance our research work at the lexical-syntactical level and research the Tibetan lexical syntax integration model, thus enabling the Tibetan information extraction model without word segmentation, i.e., a Tibetan parsing model based on syllable sequence so as to support the Tibetan question parsing.

Table 3. Set up of experimental corpus (sentence).

Corpus quantity / Corpus category	Training corpus	Test corpus
Total	300	70

Table 4. Analysis of Tibetan information extraction results.

Result / Evaluation	P	R
Phrase syntactic parsing	83.96%	81.47%
Trigger word recognition	93.20%	89.16%
Tibetan question parsing	91.76%	87.28%

ACKNOWLEDGMENT

This research was supported by the project zyp2015001 of the Key Laboratory of National Language Intelligent Processing, Gansu Province (Northwest University for Nationalities).

REFERENCES

Antol S, Agrawal A, Lu J, et al. VQA: "Visual Question Answering"[C]// 2015 IEEE International Conference on Computer Vision (ICCV). IEEE, 2015.

Hu D, Wang W, Liu S, et al. "Text Segmentation Model Based LDA and Ontology for Question Answering in Agriculture"[M]// Proceedings of 2013 World

Agricultural Outlook Conference. Springer Berlin Heidelberg, 2014:307–319.

Kaur J. *"Effective Question Answering Techniques and their Evaluation Metrics"*[J]. International Journal of Computer Applications, 2013, 65(12).

Kayes I, Kourtellis N, Quercia D, et al. *"The Social World of Content Abusers in Community Question Answering"*[C]// Proceedings of the 24th International Conference on World Wide Web. International World Wide Web Conferences Steering Committee, 2015.

Liu K, Zhao J, He S, et al. *"Question Answering over Knowledge Bases"*[J]. Intelligent Systems IEEE, 2015, 30(5):26–35.

Liu R, Nyberg E. *"A phased ranking model for question answering"*[C]// ACM International Conference on Conference on Information & Knowledge Management. 2013:79–88.

Ma K, Abraham A, Yang B, et al. *"Intelligent Web Data Management of Social Question Answering"*[M]// Intelligent Web Data Management: Software Architectures and Emerging Technologies. Springer International Publishing, 2016.

Nguyen D Q, Dai Q N, Pham S B. *"Ripple Down Rules for Question Answering"*[J]. Eprint Arxiv, 2015.

Olvera-Lobo M D, Gutiérrez-Artacho J. *"Question answering track evaluation in TREC, CLEF and NTCIR"*[J]. Advances in Intelligent Systems & Computing, 2015, 353:13–22.

Pavlić M, Han Z D, Jakupović A. *"Question answering with a conceptual framework for knowledge-based system development "Node of Knowledge""*[J]. Expert Systems with Applications, 2015, 42(12):5264–5286.

Perera R, Nand P. *"Interaction History Based Answer Formulation for Question Answering"*[J]. Communications in Computer & Information Science, 2014, 468:128–139.

Przybyła P. *"Question Analysis for Polish Question Answering"*[C]// Meeting of the Association for Computational Linguistics, Proceedings of the Student Research Workshop. 2013:96–102.

Ryu P M, Jang M G, Kim H K. *"Open domain question answering using Wikipedia-based knowledge model"*[J]. Information Processing & Management, 2014, 50(5):683–692.

Ryu P M, Jang M G, Kim H K. *"Open domain question answering using Wikipedia-based knowledge model"*[J]. Information Processing & Management, 2014, 50(5):683–692.

West R, Gabrilovich E, Murphy K, et al. *"Knowledge base completion via search-based question answering"*[J]. PROFESSION, 2014.

Yang B, Manandhar S. *"Tag-based expert recommendation in community question answering"*[C]// Advances in Social Networks Analysis and Mining (ASONAM), 2014 IEEE/ACM International Conference on. IEEE, 2014:960–963.

Civil, Architecture and Environmental Engineering – Kao & Sung (Eds)
© *2017 Taylor & Francis Group, ISBN 978-1-138-02985-9*

Study on energy planning and partner selection problems in strategic alliances management of new energy industry

Jinzhi Zhai
School of Business Administration, Northeastern University, Shenyang, China
Liaoning Vocational Technical College of Modern Service, Shenyang, China

Xinan Zhao
School of Business Administration, Northeastern University, Shenyang, China

ABSTRACT: The strategic alliances can respond rapidly to market changes and makes resource sharing more efficient among manufacturing partners. The partner selection problems in strategic alliances management extensively appear in social and management fields such as the selection of dynamic alliance strategies, assigning task problems, portfolio selection, agile production and supply chain management, etc. In this paper, a mathematical model, 0-1 programming model, is proposed on partner selection problems in strategic alliances management of new energy industry. The model is concerned with the former-latter relationship of the items and the dynamic receiving and paying of the capital flow. The established model is in conformity with the reality. It lays basis for further analyzing and solving the partner selection problem of new energy industry.

1 INTRODUCTION

Nowadays, more and more companies have established business strategic alliances to compete in a global changing environment and benefited greatly. The alliances meet the country's proposal of "building the technology innovation system with enterprises as mainstay, the market as guide, and the integration of production, education and research" (Heidl, 2014). Establishment of the alliances is based on enterprise requests of internal development and general common interests. Features of the alliances include joint development, complementary advantages, profit and risk sharing (Constantinos, 2016). Goals of the alliances are to enhance innovation abilities of industry and technology, to promote the optimization and upgrade of industrial architecture, and to advance the transformation of economic growth mode from extensive type to intensive type. The alliances will make efforts to realize economy sustainable development, improving comprehensive strength and core competition of the country (Donald, 1982).

However, according to management literature, about half of the alliances established turned out to be failures. One of the key factors for this failure is the choice of a wrong partner. The strategic alliances is a management or operation unit, which has the intention of profiting but without the concrete organization (specially, it doesn't organize

the concrete production activities and operation processes) (Ellram, 1990). The mode of the operation and management of such strategic alliances is that they contrive to gain projects or production tasks, decompose them, select the coordinate partners by call for tenders, and distribute the decomposed projects or production tasks to the selected coordinate partners to benefit from the process. It is evident that the amount of benefit relies on the selection of the coordinate partners. Under the condition of accomplishing the project or task, choosing the best coordinate partners to make enterprise benefit maximally is called the Partner Selection Problems (PSPs) in the strategic alliances management, which extensively appears in such social and management fields as the selection of dynamic alliance strategies, assigning task problems and the investment combination, etc.

The importance and meaning of the partner selection problems is clear, because it is not only the main parts of any strategic alliances management, but also the chief factor which affects upon the profits of any enterprise (Stevan, 2009). One of the focuses about strategic alliances management is to manage and coordinate the cooperation relationship between enterprises (Brouthers, 1995; Wu, 2010). However, the prerequisite of establishing good cooperative relationship needs to select the cooperative partner (Jin, 2008). At present, the correlative theories are mainly based on sharing total information and complete cooperation between the core enterprise and the supplier (Mehralian, 2012). However, enterprises separate from each other to a certain extent, and have business secrets of their own. In addition, the trust mechanism of our country has a big risk, and the application of information technology in enterprises is not perfect (Holjevac, 2008). Then it is impossible that the complete cooperation and sharing all information can take place between enterprises (Kokangul, 2009). So it is a competitive-cooperative relationship of different degrees between enterprises because of these characteristics. And basing on cooperation of various degrees, the supplier' s selecting standards vary greatly (Lee, 2009). So the enterprise should classify the partners at first, and then choose different kinds of cooperative partners. Taking the above factors into consideration, we propose the process of partner selection. The first stage is to define the type of the enterprises offering the different values, and the second stage is to select the cooperative partners of strategic alliances offering the same value on the basis of the first stage.

For recent years, more and more researchers have been interested in the PSPs, which are publicly considered as the hot problems in the agile production and strategic alliances management. this article puts forward a strategic alliance partner selection and evaluation model of new energy industry based on the operational research theory hoping to provide some useful references for the enterprise managers and relevant researchers. This paper establishes a mathematical model, 0-1 programming model, of the partner selection problems. The established model involves the former-latter relationship of the items and the dynamic receiving and paying of the capital flow. Therefore it is more conformable to reality and suitable for optimal algorithms. It is the theoretical basis of solving partner problem of new energy industry.

2 THE MATHEMATICAL MODEL OF PARTNER SELECTION PROBLEMS BASED ON OPERATIONAL RESEARCH

Many previous studies on partner selection and evaluation defined numerous evaluation criteria and selection frameworks for supplier selection. For example, Dickson surveyed buyers to identify factors they considered in awarding contracts. Out of the 23 factors considered, Dickson concluded that quality, delivery, and performance history are the three most important criteria. Another study by Weber et al. derived key factors thought to influence partner selection decisions. These factors were taken from 74 related articles that have

appeared since Dickson's well-known study. Based on a comprehensive review of vendor evaluation methods, they surmised that price was the highest-ranked factor, followed by delivery and quality. These empirical researches revealed that the relative importance of various selection criteria such as price, quality, and delivery performance is similar.

However, the establishment of a partner evaluation model from the above studies mostly did not focus on the needs of end consumers from the perspectives of the strategic alliances or the strategy of the enterprise itself. Their key factors were mostly obtained by questionnaires or professional interviews. Therefore, this study thus proposes a structured methodology for s partner selection and evaluation of new energy industry based on the integration architecture, to help leading enterprises establish a systematic approach to selecting and evaluating potential partners of new energy industry. The continued importance of fast paced innovation, market pre-emption, and risk management has led firms to stretch their organizational boundaries, accessing resources from an increasingly diverse set of partners. A range of studies have established that during the past decades, firms of new energy industry were no longer able to 'go it alone' in their creation of innovative products, processes, and services within a reasonable time to market and hence entered dyadic alliances. The literature also reveals that under specific conditions, dyadic alliances might not suffice and the inputs of multiple partners will be required.

Suppose that an enterprise of new energy industry gets a project with many branch items, but does not have the enough capability and resources to accomplish the entire project, so it has to invite public bidding for each of the branch items, or the enterprise itself is virtual and the main means of its gaining profit is to earn the difference benefit by winning a tenders and distributing it to contractor. Assume a project is composed of n items, indicated respectively by 1, 2, …, n, which have a fixed sequence being done, so form an activity networks (Jin, 2008). If item k can not begin until item i is completed, namely, item i is previous to item k (we call item i is a closely former operation of item k), we denote it by binomial relationship $(i, k) \in H$, where H is a set of all join-ship. For simplicity, suppose that $i < k$ holds for any $(i, k) \in H$, and the last term is denoted by n.

We have the following notations:

n: the number of the items of the project.
D: the project span-time stipulated by bidding contract.
e(t): the investment flow of the project, namely the investment extra at the moment t, $t = 1, 2, …, D$.

r(t): the interest rate of borrowing and loaning at the moment t.

$\beta(t)$: the rate of penalties, because of tardiness of t unit time than the project span-time stipulated by bidding contract and it is usually a piecewise increase-successively function.

mi: the number of the candidates of item i, i = 1, 2, ..., n.

xij: the project span-time of item i promised by candidate j (span-time of item i done by candidate j), j = 1, 2, ..., mi, i = 1, 2, ..., n.

fi(xij): the bidding of candidate j to the item i (or the expenses of item i done by candidate j), j = 1, 2, ..., mi.

C: the actual project span-time.

This paper only considers selecting partners in terms of the item expense and completing time of the project.

In addition, we suppose that the enterprise pays afi(xij) at the beginning (for the candidate j of item i) and pay the rest, (1-a)fi(xij), of the item expenses at the end of the item. In production practice the total predetermined fund of a large scale project is not appropriated once. It is usually paid by stages. When the appropriated fund is not enough to pay to the contractor, the enterprise has to loan from bank and pays interest to bank. If the project is not completed on time, the enterprise will be fined and suffers from some economical loss in a certain degree. Consequently, the total costs, Z, of the project are composed of the item expense, the interest of the bank and tardiness penalties, denoted by Z_1, Z_2, Z_3 respectively, therefore $Z = Z_1 + Z_2 + Z_3$.

For any i = 1, 2, ..., n, j = 1; 2, ..., mi, we introduce variable

$$w_{ij}(t)\begin{cases} 1, & \text{if item } i \text{ is contracted to candidate } j \\ & \text{and starts at the moment } t \\ 0, & \text{else} \end{cases}$$

Then Z1 is represented as

$$Z_1 = Z_1\left(w_{ij}(t)\right) = \sum_{i=1}^{n}\sum_{j=1}^{m_i} f_i\left(x_{ij}\right)\sum_{t=1}^{C} w_{ij}(t). \qquad (1)$$

The item expense having been paid by enterprise by the moment t is the sum of the expense of items being done and the expense of the items have been done, so the sum is

$$A = A\left(w_{ij}(t)\right) = \alpha\sum_{i=1}^{n}\sum_{j=1}^{m_i}\sum_{k=1}^{t} f_i\left(x_{ij}\right)w_{ij}(k)$$
$$+ (1-\alpha)\sum_{i=1}^{n}\sum_{j=1}^{m_i}\sum_{k=1}^{t} f_i\left(x_{ij}\right)w_{ij}\left(k-x_{ij}\right).$$

Since the amount of the investment received by enterprise at the moment t is $B = B(t) = \sum_{k=1}^{t} e(k)$. We introduce the notation

$$[A-B]^{+} = \begin{cases} A-B, & A > B, \\ 0, & A \le B. \end{cases}$$

Then $[A-B]^{+}$ is the amount of the money loaned from bank by the moment t. The amount of interest paid to bank is

$$Z_2 = Z_2\left(W_{ij}(t)\right) = \sum_{t=1}^{C} r(t)[A-B]^{+}$$
$$= \sum_{t=1}^{C} r(t)\Bigg[\alpha\sum_{i=1}^{n}\sum_{j=1}^{m_i}\sum_{k=1}^{t} f_i\left(x_{ij}\right)w_{ij}(k)$$
$$+ (1-\alpha)\sum_{i=1}^{n}\sum_{j=1}^{m_i}\sum_{k=1}^{t} f_i\left(x_{ij}\right)w_{ij}\left(k-x_{ij}\right)$$
$$- \sum_{k=1}^{t} e(t)\Bigg]^{+}. \qquad (2)$$

It is evident that the expression of the tardiness penalties is denoted as

$$Z_3 = \beta\left([C-D]^{+}\right)[C-D]^{+}.$$

Thus, the total expenses of the project, paid by the enterprise, can be represented as

$$Z\left(w_{ij}\right)(t) = Z_1\left(w_{ij}(t)\right) + Z_2\left(w_{ij}(t)\right) + Z_3\left(w_{ij}(t)\right)$$
$$+ \sum_{i=i}^{n}\sum_{j=1}^{m_i} f_i\left(x_{ij}\right)\sum_{t=1}^{C} w_{ij}(t)\sum_{t=1}^{C} r(t)\Bigg[\alpha\sum_{i=1}^{n}\sum_{j=1}^{m_i}\sum_{k=1}^{t} f_i\left(x_{ij}\right)w_{ij}(k)$$
$$+ (1-\alpha)\sum_{i=1}^{n}\sum_{j=1}^{m_i}\sum_{k=1}^{t} f_i\left(x_{ij}\right)w_{ij}\left(k-x_{ij}\right) - \sum_{k=1}^{t} e(t)\Bigg]^{+}$$
$$+ \beta\left([C-D]^{+}\right)[C-D]^{+}. \qquad (3)$$

The intention of selecting partners and determining the starting time of every items (in other words, $w_{ij}(t), i = 1, 2, ..., n, j = 1, 2, ..., mi, t = 1, 2, ..., C$) is to minimize the above total expenses $Z(w_{ij}(t))$, which is obviously the final purpose of PSP, namely the comprehensive object of PSP is

$$\min\left\{ Z\left(w_{ij}(t)\right) \mid w_{ij}(t) \in \{0,1\}\right\}. \qquad (4)$$

However, the above minimization must satisfy such constraints as

1. The 0-1 constraints

$$w_{ij}(t) \in \{0,1\}, \qquad \forall i, j, t.$$

1267

2. The non-conflict constraints

$$\sum_{j=1}^{m_i}\sum_{t=1}^{C} w_{ij}(t) = 1, \qquad i = 1, 2, \ldots, n.$$

Namely, each of the items must have and only have one contractor, and must start at the moment t.

3. The former-later relationship constraints of items

$$\sum_{j=1}^{m_i}\sum_{t=1}^{C}\left(t + x_{ij}\right)w_{ij}(t) \le \sum_{j=1}^{m_k}\sum_{t=1}^{C} tw_{kj}(t), \quad \forall (i, k) \in H,$$

i.e., if item i is a former operation of item k ($\forall$ (i, k) $\in$ H), then item i must have been completed when k begins.

4. The complete constraint

$$\sum_{j=1}^{m_n}\sum_{t=1}^{C}\left(t + x_{nj}\right)w_{ij}(t) = C,$$

i.e., the completing time of the n-th item is the finishing time of the entire project.

In summary, we can derive the following 0-1 programming model of PSPs

$$\min \quad Z\left(w_{ij}(t)\right) \tag{5}$$

s.t.

$$w_{ij}(t) \in \{0,1\}, \quad i = 1, 2, \ldots, n \quad j = 1, 2, \ldots, m_i,$$
$$t = 1, 2, \ldots, C, \tag{6}$$

$$\sum_{j=1}^{m_i}\sum_{t=1}^{C} w_{ij}(t) = 1, \quad i = 1, 2, \ldots, n, \tag{7}$$

$$\sum_{j=1}^{m_n}\sum_{t=1}^{C}\left(t + x_{ij}\right)w_{ij}(t) \le \sum_{j=1}^{m_k}\sum_{t=1}^{C} tw_{kj}(t), \quad \forall (i,k) \in H, \tag{8}$$

$$\sum_{j=1}^{m_n}\sum_{t=1}^{C}\left(t + x_{nj}\right)w_{nj}(t) = C, \tag{9}$$

where Z(wij(t)) is determined by (3).

The number of variables in the above 0–1 programming model is $L = C \times \sum_{i=1}^{n} m_i$, but the scale of the searching space is $L = C \times \prod_{i=1}^{n} m_i$. It is obvious that even for a small scale problem, the searching space of this model is very big. However, the biggest advantage of the model is that, except such non-linear terms as [A–B]+, all terms of the objective function are linear, so we can use the revised linear programming method to solve the model.

3 CONCLUSIONS

Enterprise strategy is a plan that develops through the process of an enterprise analyzing internal organizational strengths and weaknesses, as well as external environmental threats and opportunities. Enterprises use such plans to establish competitive models for promoting their competitiveness. In a dynamic competitive environment, if an enterprise wishes to maintain its competitive strengths, it must develop partner relationships at a strategic level rather than just focusing on products and prices. From the viewpoint of strategic alliances, an enterprise must also consider the purpose of establishing competitive strengths and developing long-term relationship while planning enterprise strategy. Consequently, enterprise competitive strategy and relationship strategy are both crucial influences on supplier selection and evaluation for strategic alliances. To straighten internal relationship of the alliance, to propel substantive and high level cooperation. The alliance should pay highly attention to R&D of key technologies based on national strategy, so as to improve its ability of original innovation, apply long-term and high level strategic cooperation. The partner selection is also an essential prerequisite for the alliance development.

In this paper, the 0-1 programming model is established on partner selection problem of new energy industry in strategic alliances management. The model is concerned with the former-latter relationship of the items and the dynamic receiving and paying of the capital flow. It is in conformity with the reality. The established model is the theoretical basis for further analyze and solve the partner selection problem of new energy industry. Solving the model has significant guidance for selecting a group of coordinate partners to collaboratively accomplish a project and to gain the best profit. Therefore this paper tries to analyze the principles in the selecting course of strategic alliances partner selection of new energy industry at first, provide a 0-1 programming model of partner selection and its realization methods next. Selecting different type of partners on the basis of defining the types of cooperative partner needs different evaluating models, so that it can make the course of partner selection of new energy industry more rational.

ACKNOWLEDGMENT

This work is Supported by National Natural Science Foundation of China (71271048).

REFERENCES

Brouthers KD, Brouthers LF, Wilkinson TJ. Strategic alliances: choose your partners. Long Range Planning. Vol. 28, No. 3, p. 28–25. (1995).

Chong Wu, David Barnes. Formulating partner selection criteria for agile supply chains: A Dempster? Shafer belief acceptability optimization approach. International Journal of Production Economics. Vol. 125, No. 2, p. 284–293. (2010).

Constantinos S. Lioukas, Jeffrey J. Reuer, Maurizio Zollo. Effects of Information Technology Capabilities on Strategic Alliances: Implications for the Resource-Based View. Journal of Management Studies. Vol. 53, No. 2, p. 161–183. (2016).

Donald R Lehmann, O' Shaughnessy. Decision criteria used in buying different categories of products. International Journal of Purchasing and Materials Management. Vol. 81, No. 1, p. 9–14. (1982).

Ellram, L. The supplier selection decision in strategic partnerships. Journal of Purchasing Material Management. Vol. 26, No. 4, p. 8–14. (1990).

Heidl, R., K. Steensma and C. Phelps. Divisive faultlines and the unplanned dissolutions of multipartner alliances. Organization Science, Vol. 25, No. 5, p. 1351–1371. (2014).

Holjevac, I.A. Business ethics in tourism—as a dimension of TQM. Total Quality Management and Business Excellence. Vol. 19, No. 10, p. 1029–1041. (2008).

Jin Wenfeng, Li Naicheng, Xu Zongben. Mathematical Model on Partner Selection Problems in Virtual Enterprise Management. Chinese Journal of Engineering Mathematics. Vol. 25, No. 4, p. 729–734. (2008).

Kokangul, A. and Z. Susuz. Integrated analytical hierarch process and mathematical programming to supplier selection problem with quantity discount. Applied Mathematical Modeling. Vol. 33, No. 3, p. 1417–1429. (2009).

Lee, A.H.I. A fuzzy supplier selection model with the consideration of benefits, opportunities, costs and risks. Expert Systems with Applications. Vol. 36, No. 2, p. 2879–2893. (2009).

Mehralian, G., A. Rajabzadeh Ghatari, H. Morakabzti, and H. Vatanpour. Developing a suitable model for supplier selection based on supply chain risks: an empirical study from Iranian pharmaceutical companies, Iranian Journal of Pharmaceutical Research. Vol. 11, No. 1, p. 209–219. (2012).

Stevan R Holmberg, Jeffrey L Cummings. Building success-ful strategic alliances: strategic process and analytical toolfor selecting partner industries and firms. Long Range Planning. 42(2):164–193. (2009).

Civil, Architecture and Environmental Engineering – Kao & Sung (Eds)
© 2017 Taylor & Francis Group, ISBN 978-1-138-02985-9

A fractal study on the surface topography of shearing marks

B.C. Wang
Shenzhen University, Shenzhen, China

C. Jing & L.T. Li
Guangdong Hengzheng Forensic Clinical Medicine Institute, Guangzhou, China

ABSTRACT: The surface topography of shearing marks is studied by using the fractal theory. The irregularity and random distribution of the surface topography shows the fractal characteristics. The data of the surface topography of shearing marks are collected digitally by using surface topography tools, and the fractal dimensions are calculated. The result shows that surface topography of shearing marks and fractal dimensions are different when metallic materials are sheared by different tools. Therefore, the fractal dimensions can be used as a characteristic index to describe the surface topography of shearing marks. It can help analyze the characteristics of surface topography of shearing marks effectively. It can also provide a reference for the judgment of shearing tools. It is a new method of quantitative inspection of shearing marks.

1 INTRODUCTION

On crime locations, various traces will be left behind by criminals using tools to destroy objects, the mark of a striation tool is one of them. It is usually composed of pliers and scissors and other tools. These tools will form a continuous striation seen visually or microscopically visible on every section in the process of cutting the objects. Traditionally, the tool marks are mostly inspected by visual comparison. It is to compare the striation tool marks collected from the crime scene with the test marks formed by suspected tools. The tool marks are determined based on striation comparison. This process is called "form inspection". Practically, the surface topography of the striation tool marks is very complex and irregular, and also random and disorderly. Therefore, the surface topography of the striation tool marks cannot be described by digital description tools for a long time. The inspection of digital analysis cannot be used.

Fractal geometry theory can be used to describe irregular and disorderly graphics. The fractal dimension is an important parameter which is used to quantitatively characterize the "singularity" degree of chaotic attractor. It is widely used to describe the digital characteristics of a nonlinear system of behavior. The surface topography of a material has been studied by using the fractal theory. The research data shows that fractal theory gets certain achievements of surface topography in characterizing. (Brown & Savary 1991, Gagnepain & Roques-Carmes 1990, Ganti & Bhushan 1995). In this paper, the characteristics of the surface topography of shear marks are studied by applying the fractal geometry theory and the literature on the application of fractal research (Mou & Yang 2010, Su et al. 2014, Li et al. 2014). The purpose of this study is to apply the fractal theory to the inspection of shearing marks. The modern theory and the application technology can be combined to deepen the theory of mark inspection. It introduces new inspection technologies into the inspection of shearing marks during the investigation process which involves shearing tools.

In order to improve the automation and reliability of inspection, it is necessary to use a certain number to express the complex striation tool marks to achieve digital inspection and identification. To achieve this purpose, this study adopts scissors, steel pliers, and wire-breaking pliers as shearing tools to make samples of shearing marks. The sample surface was marked by the software to get the profile curve which is perpendicular to the surface of marks. The fractal dimensions of the profile curve are calculated by applying general fractal dimension methods. The characteristic parameters of the dimension's energy spectrum are calculated according to the fractal dimension spectrum. The energy spectrum of different shearing marks is found according to the characteristic parameters of the dimension spectrum and then extracted and characteristics of different shearing marks recognized. It greatly improves the reliability of inspection and recognition by introducing the characteristics of the dimension energy spectrum.

2 EXPERIMENTAL CONDITIONS AND DATA ACQUISITION

Scissors and wire cutters are used as shearing tools to shear the lead wires. Several samples of shear marks are made. The samples are observed under the stereomicroscope and those samples which can reflect the stability characteristics of the shear marks are chosen to be collected digitally.

In this experiment, we used the Austria infinite focus auto-zoom three-dimensional surface topography measurement device with its objective magnification 10 times and the sampling resolution 1.1 μm. The surface of shearing marks is digitally collected and a three-dimension of shearing marks is stored in the computer. The stable part of the marks character is marked by the application software. The profile curve which is perpendicular to the marks surface is obtained. The profile curve which is perpendicular to the marks surface formed by scissors is as shown in Figure 1(a). The profile curve which is perpendicular to the marks surface formed by wire cutters is as shown in Figure 1(b). The profile curve which is perpendicular to the marks surface formed by wire breaking pliers is as shown in Figure 1(c).

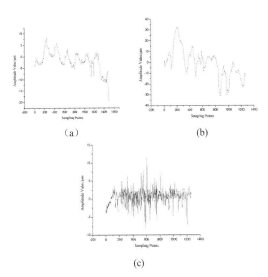

(a)   (b)

(c)

Figure 1.   Three profile curves.

3 GENERAL FRACTAL THEORY AND THE GENERAL FRACTAL DIMENSION SPECTRUM OF THE PROFILE CURVE

The fractal theory provides the theoretical basis for the description of a nonlinear behavior system. It is used to quantitatively characterize the singularity of chaotic attractor. The fractal dimension makes up for the deficiency of the traditional analysis model to a certain extent. In order to describe the nonlinear system behavior characteristics by fractal dimension, the various definitions and calculations methods are provided such as the box dimension method, the power spectrum method, the information dimension method, the correlation dimension method, and the multi-scaling dimension method, and so on (Wang et al. 2015). In this paper, the fractal dimension of the profile curves which are perpendicular to the marks surface formed by scissors, pliers, and wire pliers are calculated by the general dimension method.

The general fractal is used to describe the non-uniform random probability distribution of the fractal geometry on different layers; the general fractal dimension is an important parameter to describe the general fractal. This is the general definition of the fractal dimension. The expression is:

$$D_q = -\lim_{\varepsilon \to 0} \frac{I_q(\varepsilon)}{\ln \varepsilon} \qquad (1)$$

$I_q(\varepsilon)$ is the formation entropy of the Renyi definition for which the expression is:

$$I_q(\varepsilon) = \begin{cases} \dfrac{1}{1-q} \ln\left[\displaystyle\sum_{i=1}^{N} P_i(\varepsilon)^q \right] & q \neq 1 \\ \displaystyle\sum_{i=1}^{N} P_i(\varepsilon) \ln P_i(\varepsilon) & q = 1 \end{cases} \qquad (2)$$

$$q = -\infty, \ldots, -1, 0, \ldots, +$$

$P_i(\varepsilon)$ is the cover rate, Dq depends on parameter q. It is: When the locket with a side length of ε is used to cover fractal collect, $P_i(\varepsilon)$ is the rate of a dot in fractal collect falling into the locket. When q is different, it means the difference of the fractal dimension. For example, if q = 0, 1, 2, Dq is respectively equal to the box dimension, the information dimension, correlation dimension, and so on.

The most important problem in practical applications is how to calculate the general fractal dimension according to experimental results and observation data, and how to seek and find the multi-scaling characteristics according to the general fractal dimension spectrum q–D_q. Calculation measures of D_q spectral include: covering method, fixed radius method, fixed quality method, and so on.

Covering method can be used for all kinds of simple or complex fractal objects. It is one of the most common methods of the fractal study. For a multi-scaling fractal set, the object of study is covered by the box of scales ε, a probability distribution function of cover collection $P_i(\varepsilon)$ shall be first defined. The expression is:

$$P_i(\varepsilon) = \frac{A_i}{\sum\limits_{i=1}^{N} A_i} \qquad (3)$$

This formula is actually to be replaced with the approximate probability with frequency. Where A_i is covering "points" of the i box, $\Sigma_{i=1}^{N} A_i$ is the total "points" of the entire box covering. N is the total box quantity. So $P_i(\varepsilon)$ is the probability of points of fractal sets falling in the i box. The information entropy of the Renyi definition can be calculated according to the probability function $P_i(\varepsilon)$, scale ε and weighting factor q sequence.

The algorithm is as follows:

First, the analog signal or the continuous fractal contour curve $Z(x)$ is A/D converted conversion. It is called discretization. If the sampling length is L, the sampling interval is $\Delta\varepsilon$, then the number of sampling points is $N = L/\Delta\varepsilon$. The length of the signal or the curve is discretized into a one-dimensional array $Z = \{Z_1, Z_2, Z_3 \ldots Z_N\}$.

Second, the meshing size division is ε, $\varepsilon_i = 2i\Delta\varepsilon$ is the grid width, the number of rows and columns of the grid is $m = L/\varepsilon_i$. If the k row and the t columns of the grid are called kt grid, i is the mesh type, then kt grid coordinates are $\{(k\varepsilon_i, (t-1)\varepsilon_i), (k\varepsilon_i, t\varepsilon_i), ((k-1)\varepsilon_i, (t-1)\varepsilon_i), ((k-1)\varepsilon_i, t\varepsilon_i)\}$. The discerning data points of the grid is $B = \{Z/\varepsilon_i + 1\}$, while the kt grid covering the discrete data points is recorded as $A_{k,t}$, then kt grid covers the collection probability and can be as follows according to equation (3):

$$P_{k,t}(\varepsilon_i) = A_{k,t} \Big/ \sum_{k=1}^{m}\sum_{t=1}^{m} A_{k,t} = A_{k,t}/N \qquad (4)$$

A series of information entropy $I_q(\varepsilon)$ can be calculated according to this formula. A scale-free region is found in $\ln\varepsilon \sim \ln[I_q(\varepsilon)]$ figure fitting the slope of the segment by using the least squares method, the absolute value of this slope is D_q, a given value q.

The spectral values of the general fractal dimension are obtained with three kinds of profile curves according to the formula of the general fractal dimension. The dimension spectrum curve is shown in Figure 2.

From the general fractal dimension spectrum value, characteristic values are extracted such

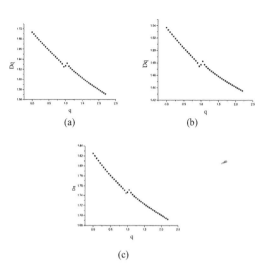

(a)   (b)

(c)

Figure 2.   Three kinds of dimension curve.

Table 1.   Characteristic values of the multi-fractal dimension spectrum.

Dq	Sample a	Sample b	Sample c
D_0	1.6932	1.5364	1.8251
$D_{0.95}$	1.6251	1.4746	1.7445
D_1	1.6270	1.4772	1.7462
$D_{1.05}$	1.6321	1.4826	1.7513
D_2	1.5713	1.4354	1.6925

as D_0, $D_{0.95}$, D_1, $D_{1.05}$, D_2, and they are as shown in Table 1.

To take the iteration order of q = 0.05 in the calculation of the general fractal dimension, a series of fractal dimensions are obtained as shown in Figure 2. From the fractal dimensions shown in Figure 2, it can be found that the dimensional numerical showed monotonicity and was decreasing with the continuous increase in the value of q, that is, the dimensional numerical was decreasing with the q value increasing, but the fractal dimension value in the q = 0.95 and q = 1.05 generates mutations. From the general dimension spectrum value of three kinds of profile curves shown, the dimension values at q = 0.95 and q = 1.05, it respectively obtains high mutation and low mutation. So the dimension at the two points and fractal dimension of D_0, D_1 and D_2 combine together to describe the profile curve feature, and then the fractal dimension group of identification rules is established. Even if the fractal dimension (D_0, $D_{0.95}$, D_1, $D_{1.05}$ and D_2) at q = 0, 0.95, 1, 1.05, 2 and five special points are taken as a group to distinguish the different profile curve, it can be found from the

Table 2. The value of general dimension energy spectrum of three kinds of profile curves.

Profile curve	Sample a	Sample b	Sample c
Energy spectrum	13.28774	10.97557	15.35509

results, that three kinds of profile curves are different, and dimension spectra are also different.

4 THE ANALYSIS OF THE PROFILE CURVE CHARACTERISTIC BASED ON ENERGY SPECTRUM OF GENERAL FRACTAL DIMENSION

The energy used and spatial distribution is different when using three different tools to cut the object according to the energy theory. The dimension values are different when the fractal dimension is used to describe the different profile curves. The energy of the shearing process can be expressed as follows:

$$E = \sum_{i=1}^{n} |x_i|^2 \qquad (5)$$

From the form of the energy equation (5) for the introduction of the general fractal energy spectrum to describe different profile curve characteristics, its expression is:

$$E_{Dq} = \sum |D_q|^2 \qquad (6)$$

$q = 0, 0.95, 1, 1.05, 2$

General fractal energy spectra of three kinds of profile curves can be calculated by the formula (6). The calculated results are shown in Table 2.

It can be found from the results that the dimension spectrum of three kinds of profile curves is significantly different and the distinction is big.

5 CONCLUSIONS

The general dimension spectrum of three kinds of profile curves are calculated respectively, and the profile curve and dimension spectrum are also different. The dimension value of the spectrum is analyzed in different q values. According to the characteristics of the spectrum, the profile curve characteristics are described by the characteristic

value of five dimensions (D_0, $D_{0.95}$, D_1, $D_{1.05}$ and D_2), that is to identify the profile curve by using the fractal dimension group, the recognition rate is more reliable than using a single dimension.

The dimension energy spectrum is obtained based on this calculation. As the dimension energy spectrum value is combining 5 characteristic values, the profile curve characteristics can be reflected more comprehensively and accurately. It greatly improves the reliability of recognition of the profile curves formed by different tools. The results show that the dimension energy spectrum of three kinds of profile curves is significantly different and the distinction is big. Therefore, the characteristic value of the dimension and dimension energy spectrum can be used as the characteristic parameters of different profile curves. It can be used as a criterion for identifying different tools. The study provides a basis for the inference of shearing tools. It also provides a new method for quantitative inspection of shearing marks.

ACKNOWLEDGMENTS

This work was supported by the Nature Science Fund of China (NSFC), No. 61571307.

REFERENCES

Brown, C.A., G. Savary (1991). Describing ground surface texture using contact profilometry and fractal analysis. Wear, 147:211–226.

Gagnepain, J.J., C. Roques-Carmes (1990). Fractal approach to characterizations of rubber wear surfaces as a function of load and velocity. Wear, 141: 73–84.

Ganti, S., B. Bhushan (1995). Generalized fractal analysis and its applications to engineering surfaces. Wear, 180: 17–34.

Li, Y.X., M.Z. Gao, K. Ma (2014). Developments In Calculation Theory of Fractal Dimension of Rough Surface, Advances in Mathematics, Vol. 41, No. 4, P. 397–408, (In Chinese).

Mou, L., M. Yang (2010). Research on Criminal Tool Wear based on Multi-scale Fractal Feature of the Texture Image. ICCASM 2010, pp: 685–688, ISBN: 978-1-4244-7236-9.

Su, Y.W, W. Chen, A.B. Zhu, et al. (2014). Contact and Wear Simulation Fractal Surfaces, Journal of Xi An Jiaotong University, Vol. 47, No. 7, P. 52–55, (In Chinese).

Wang, B.C., Z.Y. Wang, C. Jing (2015). Study on Examination Method of Striation Marks Based on Fractal Theory. Applied Mechanics and Materials, Vol. 740: 553–556.

Computer simulation & computer and electrical engineering

Civil, Architecture and Environmental Engineering – Kao & Sung (Eds)
© 2017 Taylor & Francis Group, ISBN 978-1-138-02985-9

Investigation of denoising methods of infrared imagery

Ming-kai Yue
Beijing Institute of Technology, Beijing, China
Shenyang Ligong University, Shenyang, Liaoning, China

Jia-hao Deng
Beijing Institute of Technology, Beijing, China

ABSTRACT: As for the infrared imaging system, the denoising effect of infrared imagery, a key to influencing the infrared imaging quality of the target, directly determines the precision of the target segmentation and identification. The noise characteristics of infrared imagery are discussed and an integrated filtering method was proposed based on the contrasting advantages and disadvantages of two classic methods, median filtering and mean filtering. The experimental results and quantitative analysis indicate that the proposed method has an excellent denoising effect for infrared imagery containing mixed noise.

1 INTRODUCTION

The infrared imaging system has been widely applied in various weapon equipment systems due to many advantages, such as good disguising, strong anti-interference, high angular resolution, the capability to work all day long, etc. Since there is interference in every step during the course of obtaining, transmitting, and displaying infrared imagery, the noise will pollute infrared imagery and reduce its quality. Consequently, the processing and usage of infrared imagery will inevitably involve the denoising problem. As for the infrared imaging system, the denoising effect of infrared imagery directly determines the precision of the target segmentation and identification, therefore influencing the performance of the infrared imaging system. Hence, directly or through denoising, obtaining the higher Signal to Noise Ratio (SNR) infrared imagery is the primary premise to increase the performance of the infrared imaging system.

2 NOISE CHARACTERISTICS OF INFRARED IMAGERY

As infrared imagery is the premise and basis of auto-identification of the infrared imaging system and its noise characteristics decide which method to process, it is necessary to firstly discuss the noise characteristics. The infrared imaging system has the function to convert infrared light into the visible one. The optical system converts the infrared emanation signal into the electrical one by the

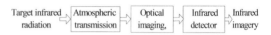

Figure 1. Schematic diagram of the infrared imaging system.

infrared detector and displays the video signal reflecting the target infrared emanation distribution on the monitor by processing electronically, which realizes the reconversion from electricity to light and obtains visible imagery reflecting the target thermal imagery. Figure 1 shows the schematic diagram of the infrared imaging system.

It can be seen from Figure 1 that infrared radiation of the target can convert into infrared imagery after a series of processing such as atmospheric transmission, optical imaging, infrared detector, etc. Hence, both the environment outside the infrared imaging system and the physical parameter inside it influence the infrared imagery. The infrared image reflects the space distribution of the infrared radiation between the target and the background and its radiance distribution is mainly decided by the temperature of the observed target and the emissivity of the infrared wave. Since noise is the biggest factor to influence the quality of infrared imagery, denoising is the key point in pre-processing of the infrared imagery.

The infrared detector is the main noise source of the infrared imaging system and the thermal and $1/f$ noise produced by it can be approximately presented by Gaussian distribution, i.e., Gaussian noise. At the same time, the uniformity of the infrared detector can produce some blind pixels and pulse noise, which is recognized as salt and pepper noise in infrared imagery. Hence, the main noise in infrared imagery can be approximately considered as a Gaussian one with some salt and pepper one, so the filtering in the pre-processing of infrared imagery mainly aims at the Gaussian noise and the salt and pepper. The infrared imagery is a little dark and has low definition and resolution. Also, the target and background with blur edges have low contrast. Consequently, the denoising must aim at the characteristics of infrared imagery, i.e., the factors disadvantageous to the post-processing of the infrared target identification system, such as the blur edges produced during denoising must be reduced as far as possible.

3 FILTERING OF INFRARED IMAGERY

The imagery filtering methods mainly include frequency domain and space. The former has heavy computation and bad real-time. The noise of the infrared imagery mainly consists of Gaussian noise and salt and pepper. The better processing method for the former is mean filtering while that for the latter is median.

3.1 Median filtering

The median filtering, one of the space domain filters belongs to nonlinear filtering technology and has light computation and is easy to realize.

The salt and pepper noise is also named as pulse one and its probability density function is

$$p(z) = \begin{cases} p_a & z = a \\ p_b & z = b \\ 0 & \text{else} \end{cases} \quad (1)$$

The noise pulse can either be a positive one or a negative one. When $a = 0$, $b = 255$ for the common eight-digit image, the positive pulse will be displayed as white points whilst the negative one as black points.

The fundamental of median filtering can be described as follows: the window W of odd pixels is firstly set up and every pixel lines up according to the gray scale size, and then the center gray scale $g(x, y)$ takes the place of the original $f(x, y)$ one, i.e.,

$$g(x, y) = median\{f(x - k, y - l), (k, l \in W)\} \quad (2)$$

The median filtering not only preferably protects the image edge information, but also has a better effect for many random noises, especially for the pulse noise because the darker or brighter points

among the pixel domain are forcedly converted into the median brightness for the $n \times n$ median filter and simultaneously wipes out the isolated pixels set, whose domain is less than $n^2/2$.

3.2 Mean filtering

The mean filtering can also be named as linear mean filtering and its fundamental is to mask the gray scale of all pixels by filtering to take the place of the gray scale of the pixels, i.e., using the gray

(a) Containing Gaussian noise

(b) After median filtering

(c) After mean filtering

Figure 2. Experimental effects using different filtering algorithms.

scale mean of deleted neighborhood S to take the place of every pixel gray scale:

$$g(x, y) = \frac{1}{M} \sum_{(i,j) \in S} f(i, j) \qquad (3)$$

where $f(i, j)$ is the original image, $g(x, y)$ is the smoothed image, M is the pixel number in the deleted neighborhood S.

Although the mean filtering has a noteworthy effect by processing Gaussian noise, it often weakens the gradient change of the image gray scale and causes the blurred edges, and even weakens the effective feature in identification. Therefore, the sharpening process must be conducted in order to deepen the image edges after the mean filtering.

The most transient noise of infrared imagery belongs to the Gaussian one. The probability density function of the Gaussian random variable is

$$p(z) = \frac{1}{\sqrt{2\pi}\sigma} e^{-(z-\mu)^2/2\sigma^2} \qquad (4)$$

where z is the gray scale, μ is the expectation of z and σ is the standard deviation of z.

The infrared tank imagery containing Gaussian noise was processed by two noise linear algorithms and the experimental results were shown as in Figure 2. Here, the radius of the filtering window was and σ was 10.

It can be seen from Figure 2 that median filtering has a better effect than the mean one in the aspect of restraining the Gaussian noise, but meanwhile, both methods delete the part detail and cause image edge information deficiency to some degree.

4 INTEGRATED FILTERING

The infrared imagery often contains both Gaussian noise and salt and pepper. It can be known from the above that the median and mean filtering have different effects when processing different noises, but the two methods have certain limits, i.e., weakening the image edge information and covering up some feature deficiencies of the target. In order to overcome their deficiency, i.e., protecting useful information (such as edge information), an integrated filtering method will be proposed to apply different denoising methods according to different noises.

The noise testing, a primary task of the integrated filter is the first and also key step and it lays the foundation for correct classification of the pixel points in the image and creates conditions for further denoising. The main principle of noise testing classification can be described as follows: the pixel points in the image can be classified into three types below according to the gray scale of salt and pepper noise staying in the extreme black (less than 10) or the extreme white (more than 245) scopes: the pixel value of more than 245 is extreme white point, that of less than 10 is extreme black point, and that between 10 and 245 is a common point.

The methods used to judge extreme black or white points is to pick up the masking template of 5×5, and transverse the whole image and $X(x, y)$ is the center pixel point of this window. When $X(x, y) < 10$ or $X(x, y) > 245$, calculate the number of common pixels in the sliding window. If the number is larger than the threshold value, then this point will be the one of salt and pepper noise. The salt and pepper noise points of the image will be conducted with median filtering of the masking of 3×3 while the other points will be done with mean filtering in the same way. The main advantages of this method are to restrain the noise when using a 5×5 sliding window to classify the noise while assuring not to weaken the image edge information when using the masking of 3×3 as far as possible.

5 RESULTS AND DISCUSSION

The Gaussian noise with variance 0.005 and the salt and pepper with variance 0.05 were added up to the infrared image to better test the effect of the above denoising algorithm. The experimental results are shown in Figure 3.

It can be seen from Figure 3 that although the mean filtering weakens the noise, the target edges become blurred and the salt and pepper noise is still obvious; the effect of median filtering when processing the salt and pepper noise is obviously better than that of mean filtering; the proposed integrated filtering not only wipes out both the Gaussian and salt and pepper noise but also assures not weakening the image edge information.

6 QUANTITATIVE ANALYSIS OF THE FILTERING EFFECTS

The imagery filtering effect can be analyzed from visual and quantitative views. The quantitative analysis measured by Mean Square Error (MSE) and SNR can be employed if it is difficult to analyze the filtering effect for the visual view. For a specified image, less the MSE and bigger the SNR after filtering, the better the denoising effect.

MSE can be defined as

$$MSE = \sqrt{\frac{1}{N} \sum_{i=1}^{N} \left(x(i) - \hat{x}(i)^2 \right)} \qquad (5)$$

(a) Adding up noise

(b) After median filtering

(c) After mean filtering

(d) After integrated filtering

Figure 3.   Denoising effects contrasting the different filtering methods of infrared imagery.

(a) Directly collecting using CCD

(b) Median filtering

(c) Mean filtering

(d) Integrated filtering

Figure 4.   Visual effects contrasting different denoising
of infrared imagery.

SNR can be defined as

$$SNR = 10\log\left[\frac{\sum_{i=1}^{N}\left(x(i)\right)^2}{\sum_{i=1}^{N}\left(x(i)-\hat{x}(i)\right)^2}\right] \quad (6)$$

where N is the point of data, $x(i)$ and $\hat{x}(i)$ are
original data and processed ones.

Figure 4 shows the visual effects contrasting of
different denoising infrared imagery under better meteorological conditions. The definition of
the four pictures has a unobvious difference from
visual view.

However, the describing filtering effect by SNR
and MSE has an obvious difference from quantitative analysis. The filtering data of the three methods are shown in Table 1.

It can be seen from Table 1 that the denoising
effect of the proposed integrated filtering is best
contrasting the median and mean filtering based
on the calculated results of SNR and MSE from
the quantitative view.

Table 1. Filtering data contrast of three methods shown in Figure 4.

Filtering methods	SNR/dB	MSE
Median filtering	37.788	12.55
Mean filtering	38.279	11.59
Integrated filtering	40.635	6.38

7 CONCLUSIONS

The infrared imagery noise can be approximately regarded as mainly consisting of Gaussian noise with the addition of some salt and pepper noise, so pre-processing filter mostly aims at these two noises. The noise characteristics of the infrared image were discussed, as well as the median and mean filtering, two classic methods. An integrated filtering method was proposed based on analyzing the advantages and disadvantages of the two methods to effectively filter the mixed noise of Gaussian and salt and pepper. The experimental results and quantitative analysis showed that the proposed method has an excellent restraining effect for mixed noise and the SNR of the target image gets a remarkable enhancement and the background noise level is also effectively restrained after denoising, which creates a strong premise for infrared target testing.

REFERENCES

Edwin H Land (1997). The Retinex Theory of Color Vision [J]. Scientific American, 237, 108–128.

Rafael C Gonzalez, Richard E Woods (2002). Digital Image Processing [M]. Biejing: Publishing House of Electronics Industry.

Xia Deshen, Fu Desheng (1997). Modern Picture Processing Technology and Application [M]. Nanjing: Southeast University Press.

Xu Jun (2003). Research on the Detection of Small and Dim Targets in Infrared Images [D]. Xi'an: Xidian University.

Xun Ganqing (1998). Infrared Physics and Technology [M]. Xi'an: Xidian University Press.

Zhang Xiufeng, Lou Shuli, Zhang Yanni (2006). Research of Infrared Image Preprocessing in Pattern Recognition [J]. Electro-optic Technology Application, 10, 58–60.

Civil, Architecture and Environmental Engineering – Kao & Sung (Eds)
© *2017 Taylor & Francis Group, ISBN 978-1-138-02985-9*

Feature optimization approach to improve performance for big data

Hua Zhu
College of Engineering, University of Michigan, Ann Arbor, MI, USA

ABSTRACT: The rapid growth of data has become a serious challenge and valuable opportunity for many industries. So, feature optimization approaches are important for large-scale complex data processing to maintain the original characteristics of the feature space. Feature selection, as a field of feature optimization, contains filter, wrapper, and embedded models, which can eliminate outliers and reduce the dimensionality. After feature selection, the data analysis takes less time and uses less computing resources. In this paper, different kinds of feature optimization methods are introduced.

1 INTRODUCTION

In big data, a large amount of data needs to be analyzed, which costs much time and computation. So some methods are used to reduce data size, simplify machine learning models, and improve performance. In this motivation, feature optimization approaches are widely adopted to reduce data dimension, which makes the model more transparent and more comprehensible and provide a better explanation of the system model.

Feature optimization approaches are divided into many kinds based on different angles containing high-dimensional reduction and feature selection. High-dimensional reduction refers to a mapping method that maps data points in the original high dimensional space to the low dimensional space. High-dimensional reduction approaches include Principal Component Analysis, Linear Discriminant Analysis, Isomap, Local Linear Embedding, Laplacian Eigenmaps, Local Preserving Projection, and Maximum Variance Unfolding (Liu et al. 2010). Meanwhile, feature selection does not change the original feature space, it only selects a part of the important features from the original high-dimensional feature space to a new low-dimensional space. According to the working order between feature selection algorithms with machine learning classifications, feature selection methods can be divided into three categories such as filter, wrapper, and embedded methods (Guyon and Elisseeff 2003).

A lot of research have proven that feature optimization approaches can effectively eliminate irrelevant and redundant features. The result of feature optimization approaches also enhances the understanding of results of the study. On the other hand, research results show that the performance of feature selection methods is better than that of high-dimensional reduction methods (Yu and Liu 2003).

Therefore, in this paper, we emphasize feature selection methods which can improve the efficiency of mining tasks and the performance of prediction.

2 FEATURE OPTIMIZATION APPROACHES

Feature selection is often viewed as a search problem in a space of feature subsets, which is divided into three categories: filter, wrapper, and embedded models. Filter model analyzes the general characteristics of features and calculates its relevance without involving any machine learning algorithm, such as Information Gain (IG) and Chi-squared Test (CHI) (Yang 1997). The wrapper model combines a predetermined learning algorithm and identifies the best features in the optimization process to fit the performance, such as Particle Swarm Optimization (PSO) (Duch et al. 2004). The embedded model incorporates feature selection as a step in the learning process, such as Decision Tree (DT), which splits training recording into successively purer subsets based on IG (Lin et al. 2008).

2.1 Chi-squared test

The calculation process of Chi-squared Test (CHI) is based on the theory of hypothesis testing in Statistics. At first, assuming features and categories are not related directly, then the deviation of observed value and expected value are calculated. If the deviation is bigger, the original hypothesis is denied (Forman 2003).

In the sample set, N is the size of samples, and A is the number of samples with feature x existing and belonging to category y_k. B is the number of samples with feature x existing but not belonging to category y_k. C is the number of samples with feature x existing but belonging to category y_k. D is the number of samples with feature x neither existing nor belonging to category y_k and $N = A + B + C + D$. A, B, C, and D are observed value, while E_A, E_B, E_C, and E_D are the corresponding expected value.

The function for CHI is:

$$E_A = \frac{(A+B)(A+C)}{n}, \quad E_B = \frac{(A+B)(B+D)}{n}$$
$$E_C = \frac{(C+D)(A+C)}{n}, \quad E_D = \frac{(C+D)(B+D)}{n}$$

$$\chi^2(x,y) = \sum \frac{(\text{observed} - \text{expected})^2}{\text{expected}}$$
$$= \frac{(A-E_A)^2}{E_A} + \frac{(B-E_B)^2}{E_B}$$
$$+ \frac{(C-E_C)^2}{E_C} + \frac{(D-E_D)^2}{E_D}$$
$$= \frac{N(A*D - C*B)}{(A+C)(B+D)(A+B)(C+D)}$$

According to the deviation between theoretical value and practical value sorted in descending order, we can get the best features which have a strong correlation with categories.

2.2 Particle Swarm Optimization

Particle Swarm Optimization (PSO) is a random optimization method based on population (called a swarm). These particles move around in space according to their own best position and the entire swarm's best-known position. When improved positions are being discovered, conversely, the optimization result can guide the movement of the swarm. The process is repeated until the recycling conditions are satisfied (Kennedy 2011, Wang et al. 2007). PSO is used to select the best features from all features by fitting high detection accuracy.

Let the number of particles be S, and each has a position $x_i \in X$ and a velocity v_i in the sample space. The search termination condition is set as iteration number. The process of the algorithm is as follows:

Initialization: The position of the particle is created with a uniform matrix. The value of every element is limited to the boundary of the feature space. The velocity of particles is defined by the likely position.

Building Fitness: Fitness is defined to value the current result of the particle. In this work, the detection accuracy of the training set is treated as fitness.

Finding optimal positions: Comparing fitness and optimal position history to find the optimal position p_i of every particle and global optimal position p_{gi} of all particles.

Updating position and velocity: According to the following functions:

$$v_i = w \cdot v_i + c_1 \cdot rand() \cdot (p_i - x_i)$$
$$+ c_2 \cdot rand() \cdot (p_{gi} - x_{gi})$$
$$x_i = x_i + p_i$$

where w is the inertia weight coefficient which balances global search ability and local search ability c_1 and c_2 are learning factors which control particle movement to global optimal positions and local optimal positions, usually $w \in [0.9, 1.2]$ and $c_1 = c_2 = 2$.

Optimization termination condition: Looping number is usually defined as termination condition of the particle finding optimal positions. The value can be taken as less than 200.

2.3 Decision Tree

Decision Tree (DT) classifiers use a tree structure for making predictions. The internal nodes of the tree represent a feature and the leaf nodes represent the ultimate decision result of the algorithm (Tan et al. 2006). In the decision tree, each internal node is labeled with an input feature. DT is considered based on a set of if-then decision rules. The deeper the tree, the more complex are the decision rules and fitter is the model.

A decision tree is constructed in a recursive partitioning by splitting the training records into successively purer subsets based on Information Gain (IG) to select the best features. When the nodes are assigned to the same value of the target variable or values are no longer added to the prediction, then the recursion is completed. Let D_t be the set of training features that are associated with note t and $y = \{y_1, y_2, ..., y_n\}$ be the class labels. The process of DT is as following:

Step 1: if all instances in the dataset D_t belong to the same class y_t, then t is a leaf node which is labeled as y_t.

Step 2: if D_t contains records belonging to less than one class, an attribute test condition is selected to partition the records into smaller subsets. Then the records in D_t are distributed to children based on outcomes. The method is then recursively used to each child node.

DT tends to over-fit on data with a large number of features. However, pruning algorithms are effective to solve this problem, which takes global optimum into consideration. Anyhow, DT is simple to understand and interpret. The result tree can be visualized.

3 EXPERIMENT AND DISCUSSION

Almost in all previous papers, Confusion Matrix and related evaluation metrics, such as accuracy and Area Under the Curve (AUC) are utilized to evaluate and guide the performance of machine learning classifiers. The larger the AUC score or accuracy, the better the classifier performs.

The testing environment is set as Intel(R) Xeon(R) CPU E5-2620v3@2.40GHz with 32.0 GB RAM and 64-bit Windows 7 operating system.

3.1 Sample dataset and features

The dataset is used to make Android malicious application detection, which contains 1260 benign applications and 1210 malicious applications. Benign applications are downloaded from Google Play Store, and malicious applications are collected from the authoritative malicious application Web sites.

After decompiling, the features are extracted from source code files concluding. smali files and AndroidManifest.xml files (Aafer et al. 2013). Totally, we combine 429 permissions and 27,650 APIs as the feature universal set and the number is 28,079. In the next part, we just take the total 28,079 combined permissions and APIs as input data of feature selection models. Then the machine learning models are used to continue analyzing the result data to prove the validity of the feature selection (Peiravian and Zhu 2013).

3.2 Comparison with feature selected methods

IG represents the filter methods' select best features from high-dimension features dataset according to the relevance between features, while PSO representing wrapper methods get the global optimum value to identify the best features according to fitness of the training accuracy of predetermined machine learning algorithm. The selected feature is a subset of all features and in order to enhance the comparability of methods, DT is used as the malware classifier.

Figure 1 shows the accuracy of DT classifier with 100 to 600 features by feature selection methods. The performance of PSO is better than others for classification results. According to the trend of curves, we find that if the number of selected features is big, the accuracy becomes high. Moreover, the the performance of 300 selected features increases faster than all other features. We think 300 features selected by feature selection methods are optimal for machine learning classifiers.

According to the rules of data operation, we know that less number of features will take less time and use less computing resources, and some-

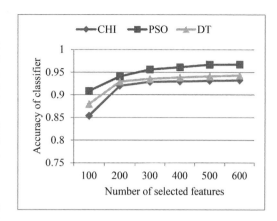

Figure 1. Comparison with filter and wrapper methods.

times we have to lose accuracy to exchange high performance. Just from the accuracy point, PSO is a good choice, but from the efficiency point, CHI outperforms. Therefore, in real-world applications efficiency is an issue. We should measure effectiveness and efficiency to choose feature selection methods.

3.3 Machine learning classification model comparison with feature selection

After feature selection, we can find 300 features selected just thought to be best features. Then we input these features to several machine learning classification models to explore how they affect performance.

Table 1 shows the performance for comparing seven machine learning methods based on feature selection. The difference between the comparing results originates from different processes of machine learning algorithm for data processing. kNN classifier relays on instances with the nearest neighbors which have no solution to outliers, so the performance is not good. While the final result of NB depends on the product of posterior probability from per features, the larger the sample dataset, the smaller is the detection rate. DT constructs the best tree based on the IG index to split the sample space and uses pruning algorithms to avoid the over-fitting problem. LR applies a logistic function to guarantee classification correctly in a training set while regularization is used to avoid over-fitting to ensure high accuracy in the testing set. SVM not only uses a kernel function to map low-dimension space to high-dimension space to improve accuracy but also uses slack variables and penalty factor to deal with noise data. So SVM has high accuracy but takes more time to process data. In order to detect correctly, Adaboost is adapted

Table 1. Machine learning classification model comparison with feature selection.

	kNN	NB	DT	LR	SVM	Adaboost	k-Means
Accuracy	0.9130	0.792	0.9356	0.9574	0.9434	0.9607	0.5104
AUC	0.9342	0.8436	0.9562	0.9698	0.9602	0.9815	0.5218
TPR	0.9429	0.8941	0.9715	0.9783	0.9773	0.9885	0.6510

to learn a series of weak classifiers to generate a boosted one. This process is so complex and loops many times and takes more time. In sum, DT, LR, SVM, and Adaboost all adopt appropriate strategies to solve the over-fitting problem so that test detection accuracies are higher while kNN and NB do nothing.

On the other hand, the performance of supervised learning methods is better than unsupervised learning methods like k-Means. It is mainly because class label used in supervised learning methods is helpful to classification while unsupervised learning methods do not pay attention to the class label. In fact, class label is a valuable information for classification. Therefore, supervised learning methods are recommended to use in the real world.

4 CONCLUSION

Feature selection methods containing filter, wrapper, and embedded models are used to eliminate outliers and reduce the amount of input data of machine learning models. Experiment results indicate that feature selection methods can improve efficiency and performance of machine learning classifiers. On the other hand, according to the rules of data operation, we know that less number of features will take less time and use less computing resources, but sometimes we have to lose accuracy to exchange high performance. Just from the accuracy point, PSO is a good choice, but from the efficiency point, CHI outperforms them. Therefore, in real-world applications efficiency is an issue. We should measure effectiveness and efficiency to choose feature selection methods.

REFERENCES

Aafer Y, Du W, Yin H. 2013. DroidAPIMiner: Mining API-level features for robust malware detection in Android [M]//Security and Privacy in Communication Networks. Springer International Publishing: 86–103.

Duch W, Wieczorek T, Biesiada J, et al. 2004. Comparison of feature ranking methods based on information entropy [C]//Neural Networks, 2004. Proceedings. 2004 IEEE International Joint Conference on. IEEE, 2: 1415–1419.

Forman G. 2003. An extensive empirical study of feature selection metrics for text classification [J]. Journal of machine learning research, 3(Mar): 1289–1305.

Guyon I, Elisseeff A. 2003. An introduction to variable and feature selection [J]. Journal of machine learning research, 3(Mar): 1157–1182.

Kennedy J. 2011. Particle swarm optimization [M]//Encyclopedia of machine learning. Springer US: 760–766.

Lin S W, Ying K C, Chen S C, et al. 2008. Particle swarm optimization for parameter determination and feature selection of support vector machines [J]. Expert systems with applications, 35(4): 1817–1824.

Liu H, Motoda H, Setiono R, et al. 2010. Feature Selection: An Ever-Evolving Frontier in Data Mining [J]. FSDM, 10: 4–13.

Peiravian N, Zhu X. 2013. Machine learning for Android malware detection using permission and api calls [C]// Tools with Artificial Intelligence (ICTAI), 2013 IEEE 25th International Conference on. IEEE: 300–305.

Tan P N, Steinbach M, Kumar V. 2006. Classification: basic concepts, decision trees and model evaluation [J]. Introduction to data mining, 1: 145–205.

Wang X, Yang J, Teng X, et al. 2007. Feature selection based on rough sets and particle swarm optimization [J]. Pattern Recognition Letters, 28(4): 459–471.

Yang Y, Pedersen J O. 1997. A comparative study on feature selection in text categorization [C]//ICML. 97: 412–420.

Yu L, Liu H. 2003. Feature selection for high-dimensional data: A fast correlation-based filter solution [C]//ICML. 3: 856–863.

Civil, Architecture and Environmental Engineering – Kao & Sung (Eds)
© 2017 Taylor & Francis Group, ISBN 978-1-138-02985-9

Influence research on file of software defective testing configuration

Yangxin Yu & Liuyang Wang
Faculty of Computer and Software Engineering, Huaiyin Institute of Technology, Huai'an, China

ABSTRACT: Software defect testing technology is important means to find and remove software defects. Software defect detection is a part of software inspection process. Software reading techniques could be used by software inspectors in detecting defect from object-oriented design documents. Around the software defect detection, this paper deeply analyzes the strengths and weaknesses of current popular software defect testing technology and the key problems. Traceability-based reading technique was one of the scenario-based reading techniques which had been empirically validated to help software inspectors detect defects more effectively than the widely used checklist based reading technique due to its readability. This study was to discover whether software inspectors with different background can perform defect detection differently using Traceability-based reading technique. The result showed that software inspectors with different undergraduate degrees, undergraduate object oriented courses taken, and object-oriented work experience do not perform defect detection significantly differently.

1 INTRODUCTION

A majority of software inspection research has been on the technical view of software inspection. Software reading technique for defect detection was one of the main issues (Cataldo, 2009). These software reading techniques can be used on software artifact including software requirement, software design, and code.

Among the proposed reading techniques for object-oriented design documents, Ad hoc and Checklist-Based Reading (CBR) have been adopted in industrial practices, while the others, known as Scenario-based reading techniques including Defect-based reading (Andre B.de Carvalho, 2010), Perspective-Based Reading (PBR) (Chang Tiantian, 2009), Usage-based reading (Abel GarciaNajera, 2010), and Traceability-based reading (Chun Shan, 2014), have been validated empirically and experimentally.

A number of empirical researches comparing scenario-based reading techniques to Checklist-Based Reading (CBR) techniques or Ad hoc technique provided promising results (Elaine J, 2010). However, the results are not completely consistent. Most scenario-based reading techniques were claimed to provide more effectiveness of defect detection than Checklist-Based Reading (CBR) technique.

As defect detection in early stage in software development lifecycle is less cost to repair in later stages (Joel Lehman, 2011), software reading techniques can be applied to object oriented design and requirement for defect detection. Most of scenario-based reading techniques for

object-oriented design require some training on how to use the reading for the inspectors. Some experiments include a whole day subject training in the experimental protocol.

Object-oriented design specifications usually contain UML diagrams and descriptions, as well as descriptive requirement. Thus, object oriented software inspectors should understand the UML diagrams well enough. Traceability-based reading technique (Marco D'Ambros, 2012) allows inspectors to check whether the design corresponds to the requirements using vertical and horizontal reading. This reading technique provides software inspectors with steps on where and how to find defects in object oriented design documents.

Most students with software-related undergraduate degrees have taken at least one object-oriented course during their undergraduate study. These students should be ready to become software inspectors after graduation. With the software reading containing clear instructions, there should be a small amount of training on defect detection. With a readable reading technique such as Traceability-based reading technique, prior training and experience may not be necessary for a software inspector in defect detection on object-oriented design documents.

2 TRACEABILITY-BASED READING TECHNIQUE

This technique was specifically developed for object-oriented design document inspection (Nathalia Cristina Torres Mariani, 2014).

The main purpose of this technique is to ensure consistency and completeness of the work product. This reading technique covers horizontal readings, for checking the different design artifacts against each other, and vertical readings, checking whether the design corresponds to the requirements.

There are seven readings in a set of Traceability-based reading technique. Each reading asked inspectors to read specific design documents together, the steps to read them, and where and how to find defects in each step. The seven readings are:

Horizontal Review

- Class Diagrams with respect to Class Descriptions
- Class Diagrams with respect to State Machine Diagrams
- Interaction (Sequence) Diagrams with respect to State Diagrams
- Interaction (Sequence) Diagrams with respect to Class

Vertical Review

- Class Descriptions with respect to Requirements Description
- Interaction (Sequence) Diagrams with respect to Use cases
- State Diagrams with respect to Requirements Description and Use cases Diagrams

It was claimed that Traceability-based reading technique can help user without experience in software inspection through defect detection process in software reading.

3 EXPERIMENTAL PLANNING

3.1 *Goals of the study*

This study aims at discovering how inspectors of the different object oriented background perform defect detection using Traceability-based reading technique, namely OORT (Object-Oriented Reading Technique) (Chang Tiantian, 2009). The software inspector's object oriented backgrounds include undergraduate degree earned, undergraduate object-oriented courses taken, undergraduate degree performance, and object-oriented work experience.

3.2 *Subjects*

The subjects were 29 undergraduate students in Information Technology in Business Program at Huaiyin Institute of Technology. The students holding any Bachelor degree with at least 12 credits in information technology or statistics or mathematics were qualified for admission to this program. The students in this program have common interest in pursuing a career in Information technology with different background in their undergraduate degrees ranging from social science to sciences.

An object-oriented analysis and design course is one of the core courses in the first year in the program. Prior to this course, students either have some undergraduate courses in object-oriented programming, object-oriented analysis and design, or software engineering, or none. Some students might have been working prior to admission to the program or just graduated.

3.3 *Object-oriented design documents*

The design document in this study was a car rental application consisting of requirement specification, use case diagram and description, class diagram and class description, sequence diagram, and state diagram modified from the prior study (Nagalakshmi, 2012). There were 23 seed defects put in the design document.

As defect detection in software designs had identified a defect classification scheme for object-oriented design in 5 classifications (Song Qinbao, 2011).

Omission: One or more design diagrams that should contain some concepts from the general requirements or from the requirement document do not contain a representation for that concept.

Incorrect Fact: A design diagram contains a misrepresentation of a concept described in the general requirements or requirement document.

Inconsistency: A representation of a concept in one design diagram disagrees with a representation of the same concept in either the same or another design diagram.

Ambiguity: A representation of a concept in the design is unclear, and could cause a user of the document (developer, low-level designer, etc.) to misinterpret or misunderstand the meaning of the concept.

Extraneous Information: The design includes information that, while perhaps true, does not apply to this domain and should not be included in the design.

To determine appropriate seed defects in the design document for this study, twelve object oriented student group project design documents were inspected to analyze the highest frequently defect subtype occurrence in each defect classification. Incorrect fact and extraneous information were the defect classification found least from the twelve object oriented student group project design documents inspected. A variety of seed defects were placed into the design document in

Table 1. Number of seed defects in design document.

Defect classification	Number of seed defect
Omission	5
Incorrect Fact	1
Inconsistency	5
Ambiguity	5
Extraneous Information	1
Total	17

this study according to the subtype of each defect classification.

However, there were defects that can be detected due to the main defect. For example, when a class was missing from a class diagram, attributes, methods, and associations of the missing class were also missing from the design and thus detected as Omission and Inconsistency type of defect classification, respectively. In this study, a missing class defect leads to additional 4 Omission defects and 2 Inconsistency defects.

3.4 Instructions to defect detection

This study adopted the OORT (Object-Oriented Reading Technique) (Chang Tiantian, 2009) that was translated into Chinese language in prior study (Nathalia Cristina Torres Mariani, 2014). A comparison on the effectiveness of the subjects using a full set of readings and subject using only four readings of the reading technique, Sequence Diagrams with respect to class, Sequence Diagrams with respect to State Diagrams, class descriptions with respect to requirements description, and Sequence Diagrams with respect to use cases was conduct to offer a way to reduce inspection time (Joel Lehman, 2011). It was found that the number of defect detected using the full set of readings and the reduced set were not significantly different. It was also found that the efficiency rate per defect detected of subjects using the reduced set was higher (Song Qinbao, 2011). Therefore, the reduced set of the readings was used in this study.

The experiment was conducted at the beginning of first object-oriented course at the graduate level. At the beginning of the experiment, subjects were asked to answer demographic data questionnaire, such as their undergraduate degree title, the number of undergraduate object oriented courses taken, and years of work experience after graduation.

Each subject was given four sets of documents to work on: a set of design document, a set of reading instruction, a worksheet to record the defects detected, and a description of UML notation. The subjects were given to complete the task in their own time.

3.5 Hypothesis

This study was intended to discover the effect of the software inspector profiles on their performance in defect detection in object oriented design documents. The software inspector profile included undergraduate degree, undergraduate object oriented courses taken, undergraduate performance, and object oriented working experience.

The software inspector's performance in defect detection in object oriented design documents included the number of defect detected in object oriented design documents using Traceability-based reading technique.

Hypothesis 1: The software inspectors with software-related undergraduate degree perform differently in defect detection to the software inspectors with other undergraduate degrees.

Hypothesis 2: The software inspectors taken at least one object-oriented undergraduate course perform differently in defect detection to the software inspectors with no object-oriented undergraduate course.

Hypothesis 3: The software inspectors with higher undergraduate Grade Point Average (GPA) perform differently in defect detection to the software inspectors with lower undergraduate grades.

Hypothesis 4: The software inspectors with object-oriented work experience can detect different number of defects to the software inspectors with no object-oriented work experience.

3.6 Procedure

The subjects were given a questionnaire asking for their profiles on work experience, undergraduate degree, and courses taken. These questionnaires were filled out before any of the inspection tasks began. The subjects were given no time limitation to inspect the design documents. Once they detected a defect, they filled out the defect detection report indicating where (which design document) that defects were detected and description of the defects. The task were given out the subjects in the first class of object oriented analysis and design course offering in the first semester of the program. In this first class, only object-oriented concept was given to the subjects.

4 EXPERIMENTAL RESULT

The summary report of defect detection behavior from this study was shown in Table 2.

When looking at the number of defect detected by individual subjects by their degree earned (software-related and non software-related), the result

Table 2. Defect detection behavior of subjects in the study.

Defect classification	No of defects NOT detected by any subjects	Average percentage of subjects that were able to detect defects
1. Omission	2 out of 9	27.72
2. Incorrect Fact	0 out of 1	6.90
3. Inconsistency	2 out of 7	40.89
4. Ambiguity	0 out of 5	24.83
5. Extraneous Information	1 out of 1	0.00

Table 3. T-test analysis comparing the number of defect detected by the subjects undergraduate degree.

	N	Mean	Std. Dev	Sig (2-tail)
Software-related under-graduate degree holders	20	7.7	2.96	0.08
Non software-related under-graduate degree holders	9	5.44	3.36	

Table 4. The T-test analysis on the number of defect detected by the subjects with and without undergraduate object-oriented course taken.

	N	Mean	Std. Dev	Sig (2-tail)
At least 1 undergraduate OO course taken	18	7.61	3.35	.194
No undergraduate OO course taken	11	6.00	2.83	

Table 5. The t-test analysis on the number of defect delectated by the subjects with different undergraduate degree performance.

	N	Mean	Std. Dev	Sig (2-tail)
Undergraduate GPA higher than or equal to 3.0	10	8.7	2.91	0.036*
Undergraduate GPA under 3.0	19	6.11	3.05	

Table 6. T-test analysis on the number of defect detected by the subjects with and without undergraduate object-oriented experience.

	N	Mean	Std. Dev	Sig (2-tail)
Undergraduate GPA higher than or equal to 3.0	11	8.27	2.83	0.150
Undergraduate GPA under 3.0	18	6.59	2.93	

is shown in Table 3. It can be concluded that the undergraduate degree does not differ the subject performances in the number of defect detected in object-oriented design document inspection.

The number of defect detected for each individual subject was counted regardless of defect classification. The average number of defect detected by 29 subjects was 7.0 with the highest number of defect detected of 14 defects and the lowest number of defect detected of one defect.

The result in Table 3 also indicated that the defect detection performance of the subjects with software-related undergraduate degree might be different from the performance of the subjects with other undergraduate degree at the level of significance of 90%. Therefore, looking into whether the subjects had taken any object-oriented undergraduate course might be a better analysis to prove the hypothesis 1.

When comparing the number of defect detected by individual of the subject's taking undergraduate object-oriented courses (such as object-oriented programming, object-oriented analysis and design, object-oriented technology), it can be seen in Table 4 that there is no significant difference in defect detection performance between subjects with at least one undergraduate object-oriented courses to subject with no undergraduate object-oriented course. Further analysis confirmed that no matter how many undergraduate object-oriented courses the subjects took does not differ in the subjects' performance in defect detection.

The subjects participating in this study had an average undergraduate grade point average of 2.85. When comparing the number of defect detected by individual of the subject's undergraduate degree performance using grade point average, it can be seen in Table 5 that subjects with higher undergraduate degree performance (Grade Point Average or GPA of higher than or equal to 3.00) can detect defects in design document at a higher rate.

The subjects participating in this study had at least 3 years of experience with an average of 5.2 years. When counting the number of defect detected by subjects with the difference work experience (with object–oriented work experience and with no object-oriented work experience), it can be seen that the object-oriented work experience does not differ the subjects' performance in defect detection at the significant level of 95% as seen in Table 6.

Although only undergraduate GPA significantly indicates the difference in defect detection performance, work experience, undergraduate object oriented courses taken and software-related degree encourage higher average number of defect detected. It is yet to be further analyzed that how much of the object oriented work experience, undergraduate object oriented courses taken significantly improve defect detection performance.

5 CONCLUSION

It can be concluded that the software inspectors with software-related undergraduate degrees, undergraduate object oriented courses taken, and object-oriented work experience do not perform defect detection significantly differently. However, the software inspectors with higher undergraduate performance perform significantly better in defect detection on object oriented design documents.

Traceability-based reading technique can be used for the newly recruited software inspectors with or without prior object-oriented profiles during training. However, the average defect detection performance of these subjects in this study was not high, especially on extraneous information and incorrect fact classification of defects.

REFERENCES

Abel Garcia Najera, John A. Bullinaria. An improved multi-objective evolutionary algorithm for the vehicle routing problem with time windows. Computers and Operations Research. 2010 (1):96–105.

Andre B. de Carvalho, Aurora Pozo, Silvia Regina Vergilio. A symbolic fault-prediction model based on multiobjective particle swarm optimization. The Journal of Systems & Software. 2010 (5):167–178.

Cataldo, Marcelo, Mockus, Audris, Roberts, Jeffrey A, Herbsleb, James D. Software Dependencies, Work Dependencies, and Their Impact on Failures. IEEE Transactions on Software Engineering. 2009 (6):211–220.

Chang Tiantian, Liu Hongwei & Zhou Shuisheng Dept. of Applied Mathematics, Xidian Univ., Xi'an 710071, P.R. China. Large scale classification with local diversity AdaBoost SVM algorithm. Journal of Systems Engineering and Electronics. 2009(06):89–98.

Chun Shan, Long Huang, Xiao Lin Zhao. Software Structural Stability Evaluation Method Based on Motifs. Advanced Materials Research. 2014 (989): 330–335.

Elaine J. Weyuker, Thomas J. Ostrand, Robert M. Bell. Comparing the effectiveness of several modeling methods for fault prediction. Empirical Software Engineering. 2010 (3):104–112.

Joel Lehman, Kenneth O. Stanley. Abandoning Objectives: Evolution Through the Search for Novelty Alone. Evolutionary Computation. 2011 (2):246–257.

Marco D'Ambros, Michele Lanza, Romain Robbes. Evaluating defect prediction approaches: a benchmark and an extensive comparison. Empirical Software Engineering. 2012 (4):183–192.

Nathalia Cristina Torres Mariani, Rosangela Camara da Costa, Kassio Michell Gomes de Lima, Viviani Nardini, Luís Carlos Cunha Júnior, Gustavo Henrique de Almeida Teixeira. Predicting soluble solid content in intact jaboticaba Myrciaria jaboticaba (Vell.) O. Berg fruit using near-infrared spectroscopy and chemometrics. Food Chemistry. 2014.

Nagalakshmi, S., N. Kamaraj. Comparison of computational intelligence algorithms for loadability enhancement of restructured power system with FACTS devices. Swarm and Evolutionary Computation. 2012.

Song Qinbao, Jia Zihan, Shi Liu Jin. A General Software Defect-Proneness Prediction Framework. IEEE Transactions on Software Engineering. 2011 (3):105–116.

Civil, Architecture and Environmental Engineering – Kao & Sung (Eds)
© *2017 Taylor & Francis Group, ISBN 978-1-138-02985-9*

Webpage automatic summary extraction based on term frequency

Yangxin Yu & Liuyang Wang
Faculty of Computer and Software Engineering, Huaiyin Institute of Technology, Huai'an, China

ABSTRACT: Document automatic summary is an important research in the field of natural language understanding. Search engines are used nowadays by all Internet users. Keyword selection is a fast-growing industry in which different tools are used by companies to suggest their webpage's keywords. The research goal of the web automatic summarization technology is to solve this problem directly to provide a concise and comprehensive information page content summary to the user, in order to improve the efficiency of user access to information. It is important to understand how different search engines would choose a webpage in search results based on a user's query. The paper's aim is to propose a method that suggests the keywords of a webpage based on frequent terms. The method used in this paper is the term frequency for defining frequent terms. An experiment is executed to validate the method results, and the result of the new method is compared with the Google AdWord tool. The accuracy of the proposed method is 82.4%, which is considered to be a promising result. The experimental results show that the judgment and readability of web page content in this system were superior to the general web page design of automatic summarization.

1 INTRODUCTION

With the rapid development of the Internet, a variety of data on the Web has increased dramatically. The network has become a potential source of data warehouse and knowledge. Keyword selection is used in many applications as an example of preprocessing, text classification, web mining, semantic web, and sponsored search. Semantically, keyword extension would allow improving the quality of selected keywords.

The technology will automatically document content in a simple way, when the information retrieval technology is developed to a certain degree of natural extension. Online adverting depends mainly on text adverting since advertisers should aim to increase the volume of bid phrases and also chose the most relevant phrases for their products, otherwise online users would not click on the ad that transfers them to the purchase page of the product or the service being promoted.

The proposed approach is divided into four steps. First, loading and parsing the webpage into tokenizes (small pieces). Second, removing the stop words from the list of terms extracted from the parser, and then the third step is stemming the terms to return to the word's stem. Finally, the list of terms and their frequency is extracted from the proposed system.

The rest of the paper is organized as follows. First, the author describes the problem. Then, previous work related to keyword selection is listed.

After that, the proposed method is presented; the next part will describe the system snapshots. Finally, the proposed method's performance is evaluated on the available dataset and results discussed. The paper conclusion includes benefits of the proposed approach and directions for future work.

2 RELATED WORK

2.1 Keyword suggestion concept

The process of keyword suggestion is so important for different fields, e.g., semantic web, data mining, natural language process, and advertising. Adverting companies use different tools that suggest the keyword. This step is very important since the most appropriate keywords could maximize their profit using the click-through rates (A. Joshi, 2006).

Keyword suggestion can be classified into proximity search, query log mining, and meta-tag spidering. Proximity search methods extract terms from the search engine's result pages that already exist in the search term.

The second type is the query log mining method. This method suggests past queries containing the search term. The Google AdWords tool and the Yahoo Search Marketing tool are known examples of this method (Dongqing Zhu, 2010).

The last method which is the meta-tag spidering method uses the engine with the seed and extracts meta-tags from the best suggestion ranking.

This method is considered to be of a lower quality than other methods (G. Chen, 2008). Using the semantic meaning of different words and extending keywords by semantically related phrases could be used by the advertiser's website.

2.2 Keyword suggestion applications

Advertising companies use research studies to reach more profitable levels. Research studies investigate the suggestion keyword tools. Semantically related phrases can be extracted from any webpage and used to define the most important keyword to represent this webpage. It is very important to extract the keyword list that is less common but also most representative of a webpage. This can be achieved by defining the most similar words semantically by using the statistical occurrence methods.

The structure of the website can be improved by rearranging links between pages (H. Zahera, 2010). The text content can also be improved by identifying the most relevant keywords. These relevant keywords are extracted using Term Frequency-Inverse Document Frequency (TF-IDF).

The list of keywords that represent a specific webpage/document could be extracted not only by statistical methods but also using the conceptual similarity knowledge (Dongqing Zhu, 2010). In (J. Alpa, 2010) a novel approach proposed, called TermNet, the semantic relationships can be designed as a directed graph for defining the relevant terms extracted from a webpage.

Search engine results depend on a query or keyword; sometimes users do not use the right or specific keyword to reach the search goal (Jaime Arguello, 2011). Query recommendation systems are used to help the user reach his goal by suggesting some keywords that were used previously by the user or others (user log). Another way is to use the help of an information source or a thesaurus (J. Velasquez, 2004).

One of the well-known previous researchers who worked in online advertising proposed a method for phrase extraction for advertising purposes using HTML tags, TF-IDF, and query logs (L. Mostafa, 2009). Open Information Extraction (OIE) focuses on domain independent and scalable extraction of terms without requiring human input. S. Ravi, 2010 proposes a model of OIE over search query logs where clusters generated from the query logs can be very effective in web search.

2.3 Keyword suggestion tools

For the purpose of online advertising, different keyword suggestion tools also known as Bidterm suggestion tools (V. Abhishek, 2007) are used (Yuelin Li, 2010). Online advertisers bid on keywords through auctions. The winner of the auction can put his ad and links on the search result page of the search company when querying the Bidterm (L. Mostafa, 2009). However, these tools are commercial tools and sometimes their result is not related to the required target (Yuelin Li, 2010).

Also, keyword selection tools are used by different search engines to provide relevant search results as a response to a user query.

The Google AdWords tool can be used for keyword selection purposes (A. Joshi, 2006). However, the drawback of this tool is the proximity based searches in which the keyword extracted must be listed as a search term.

Some systems solve the problem by adopting a strategy similar to proximity search, but using the result pages to build a bag of word vectors for the seed, and then suggest keywords that have a similar vector (H. Zahera, 2010).

3 PROPOSED APPROACH

This study aims at discovering how inspectors of the different object-oriented background perform defect detection using traceability-based reading technique namely Object-Oriented Reading Technique (OORT) (L. Li, 2008). The software inspector's object-oriented backgrounds include undergraduate degree earned, undergraduate object-oriented courses taken, undergraduate degree performance, and object-oriented work experience.

The steps of the proposed method pass through three main modules, which are loader, parser, stopword remover, and stemmer, as shown in Figure 1. The input of the proposed approach is the webpage and the output is the list of words and their frequencies. The following figure shows the proposed approach framework.

3.1 Proposed approach phases

The parser module is the first step in which a parser is a program that breaks large units of data

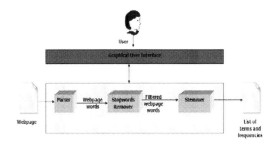

Figure 1. Proposed approach—three phases which include a parser, stopword remover, and stemmer.

into smaller pieces called tokenzs. After the webpage is fed to the system, the parser divides it into tokenzs. The output of the parser phase is the list of words in the webpage. The second phase is the stopword remover.

Stopwords are common words that carry less important meaning than the keywords. Usually, search engineers remove stopwords from a keyword phrase to return the most relevant result. Examples of stopwords are: the, an, a, are.

Different stopword removers can be used as PorterStemAnalyzer (L. Mostafa, 2009). The proposed approach uses Lucene English Stopwords removal (Lucene, 2011a) as it includes most of the existing stopwords list.

After removing the stopwords from the list of keywords, the relevant words can be discovered. The stemmer can then start working on these words.

Stemming is the process of reducing inflected or sometimes derived words to their stem, base or root form, e.g., education to educate. The process of stemming is important for search engines, query expansion or indexing, and natural language processing problems.

Different stemming algorithms can be used such as Porter stemming (2011b), Lovins stemming algorithm (2011c) and Krovetz stemming algorithm (2011d). One of the most used stemmers is the Snowball Stemmer (2011e). This stemmer is used in the proposed framework.

The output of the stemmer phase is terms with their frequencies. It is important to count the frequency of the terms so that terms with the highest frequency are considered as being one of the keywords of the webpage.

3.2 Approach tools

For preprocessing and extraction of keywords, an application was developed using the Java language environment that includes Java Integrated Development Environment (IDE), which is NetBeans and the Java virtual machine.

The stemming algorithm is a process for removing the commoner morphological and inflexional endings from words in English. Its main use is as part of a term normalization process that is usually done when setting up information retrieval systems. The stemmer used is Snowball Stemmer. The stopwords remover used is Lucene English stopwords removal (Lucien, 2011a).

4 PROPOSED APPROACH SYSTEM

The proposed framework is implemented using Java, a well-known object-oriented language.

Different modules were added, e.g., Snowball Stemmer. This section will describe the created system snapshots. The following figures will show the system's sequence.

The first step is feeding the webpage URL to the system, as described in the following Figure 2.

The second step as explained in the proposed approach section is parsing, stopword removal, and stemming. This step is shown in Figure 3.

Figure 2. The system screen in which the user can insert the webpage URL.

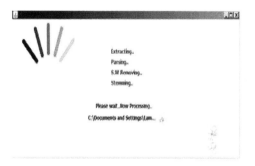

Figure 3. The loading screen that represents steps of Parsing, Removal, and Stemming.

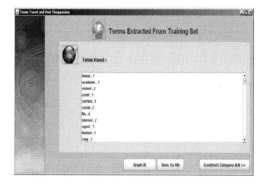

Figure 4. The last step of the proposed system that shows the output represented in a list of terms and their frequencies.

1295

Figure 4 shows the last step which contains the list of terms and their frequencies.

5 EVALUATION AND TEST RESULTS

5.1 Experiment design

The goal of the experiment is to validate the result of the proposed approach by measuring the F-measure and accuracy level. The proposed method compares its keywords suggestion to Google AdWords keywords. The method suggests three keywords that can be used for webpage search query.

The experiment design was mainly divided into two parts. The first part examines F-measure (precision and recall); the second defines the accuracy of the proposed approach.

The F-measure is a combination of precision (the percentage of positive predictions that are correct) and the recall (the percentage of positively labeled instances that were predicted as positive).

To calculate the F-measure, it is required to compute the precision and recall values. The following equations (1) are used for this purpose.

$$recall = \frac{number\ of\ relevant\ items\ retrieved}{number\ of\ relevant\ items\ in\ collection} \quad (1)$$

$$precision = \frac{number\ of\ relevant\ items\ retrieved}{total\ number\ of\ items\ retrieved} \quad (2)$$

$$F_1(r,p) = \frac{2rp}{r+p} \quad (3)$$

To accomplish the experiment design, a dataset for webpages should be used. The DMOZ2 dataset (Open Directory) is used for this purpose. DMOZ or The Open Directory Project is the largest, most comprehensive human-edited directory of the Web. It is constructed and maintained by a global community of volunteer editors. The Open Directory was founded in the spirit of the Open Source movement and is the only major directory that is 100% free. The dataset used consists of 50 webpages of shopping domain extracted from the DMOZ dataset.

5.2 Experimental results

For the purpose of defining the proposed approach, validation F-measure is used. Based on the 50 web pages in the dataset, the value of F-measure is 0.7 compared to the Google AdWords keywords as shown in Table 1.

Table 1. F-measure and accuracy values.

Precision	0.5467
Recall	0.8913
F-Measure	0.6777
Accuracy	82.467

The experiment compares each keyword suggested by the proposed approach with the Google AdWord tool. Each correct suggestion will affect the value of precision and recall. After calculating the F-measure, the accuracy level is calculated for the proposed approach validation.

6 CONCLUSION

The paper provides a framework for webpage keyword suggestion; the framework is ready for adding other modules in future work since it is implemented using one of the object-oriented languages, which is Java. The input of the approach is the webpage and the output is a list of keywords suggested.

The process of suggestion in the proposed model depends on the combination of the parser, stopword removal, and stemmer. After finishing this preprocessing step, the words and their frequency are calculated. The experiment depends on the comparison between each suggested word by the proposed approach and the Google AdWord tool. The accuracy of the proposed approach is measured along with the precision and recall values.

REFERENCES

Abhishek, V., K. Hosanagar. Keyword generation for search engine advertising using semantic similarity between terms. In ICEC '07, 2007:89–94.

Alpa, J., M. Pennacchiotti. Open Information Extraction from Web Search Query Logs. Technical Report YL, 2010:203–215.

Chen, G., B. Choi. Web page genre classification. Proceedings of the 2008 ACM symposium on Applied computing. 2008:203–215.

Dongqing Zhu, Ben Carterette. An Analysis of Assessor Behavior in Crowd sourced Preference Judgments. Proceedings of the SIGIR 2010 Workshop on Crowd sourcing for Search Evaluation (CSE 2010). 2010:68–76.

Http://lucene.apache.org/java/2_3_2/api/org/apache/lucene/analysis/standard/StandardAnalyzer.html. Retrieved:Mar.8, 2011a.

Http://snowball.tartarus.org.Retrieved: Mar. 8, 2011e.

Http://sourceforge.net/projects/stemmers.Retrieved: Mar. 8, 2011c.

Http://www.comp.lancs.ac.uk/computing/research/stemming/general/krovetz.htm.Retrieved: Mar. 8, 2011d.

Http://www.ils.unc.edu/~keyeg/java/porter/index.html. Retrieved:Mar. 8, 2011b.

Jaime Arguello, Fernando Diaz, Jamie Callan. Learning to aggregate vertical results into web search results. Proceedings of the 20th ACM international conference on Information and knowledge management. 2011:153–162.

Joshi, A., R. Motwani. Keyword generation for search engine advertising. Proceedings of Sixth IEEE-ICDM, 2006:123–129.

Li, L., S. Otsuka, M. Kitsuregawa. Query Recommendation Using Large-Scale Web Access Logs and Web Page Archive, 2008:134–141.

Mostafa, L., M. Farouk, M. Fakhry. An Automated Approach for Webpage Classification. ICCTA09 Proceedings of 19th International conference on computer theory and applications, 2009:89–99.

Ravi, S., A. Broder, E. Gabrilovich. Automatic generation of bid phrases for online advertising. In WSDM'10: Proceedings of the 3rd ACM International Conference on Web Search and Data Mining, 2010:341–350.

Velasquez, J., S. Ros, H. Yasuda. Identifying keywords to improve a web site text content. 6th International Conference on Information Integration and Web-based Applications & Services, 2004: 39–48.

Yuelin Li, Nicholas J. Belkin. An exploration of the relationships between work task and interactive information search behavior. J. Am. Soc. Inf. Sci. 2010 (9):105–126.

Zahera, H., G. Hady, W. Abdhed. Query Recommendation for Improving Search Engine Results. Proceedings of the World Congress on Engineering and Computer Science 2010 (WCECS 2010), 2010:243–246.

Civil, Architecture and Environmental Engineering – Kao & Sung (Eds)
© 2017 Taylor & Francis Group, ISBN 978-1-138-02985-9

Research and optimization of the layout of F company truck interior workshop based on GA

Wei Jiang, Yanhua Ma & Yuanchao Pan
Institute of Mechanical Science and Engineering, Jilin University, Changchun, China

ABSTRACT: Nowadays, the competition in the manufacturing industry is more and more fierce and a good workshop layout will have a significant impact on enterprise production in all aspects, especially in reducing the logistics intensity and logistics cost, and improving the production efficiency. Based on this background, this paper aimed at studying the improvement of the truck interior workshop layout in F company. During the study, a series of scientific research methods were adopted like F-D, SLP, GA, and AHP. When compared to the initial layout, the results obtained from the optimal layout indicate that the new layout scheme will bring about improvement in efficiency and reduction of resource waste. The process and results of this study have an important practical value for similar research objects.

1 SITUATION ANALYSIS OF F COMPANY TRUCK INTERIOR WORKSHOP

The main content of F company truck interior workshop is conducted truck cab assembly. The interior workshop is mainly composed of A, B, two parallel assembly lines and assembly tasks. The dimensions of the shop interior are 320 m × 50 m, the assembly line length is 236 m and width 4 m. The workshop uses towline car to be assembled and has the characteristics of flexibility. Now, the existing plant layout work units are divided into a total of 16 operating units. The truck interior workshop units are divided and the drawn interior shop layout diagram is as shown in Figure 1.

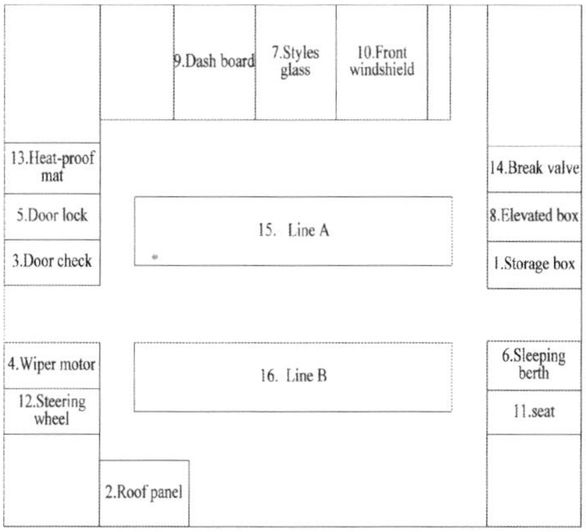

The truck interior workshop with each work unit size and truck interior part list requirement is presented in Table 1. By the analysis of F Company workshop existing truck interior planar layout, we found out the following problems:

1. The amount of logistics and distance between operating units 1 and the assembly line was relatively large.
2. The amount of logistics produced in the assembly line between operating unit 7 and operating unit 13 was not satisfactory, but far away.

Table 1. Truck interior workshop with each work unit size and truck interior part list requirement.

No.	Name of work unit	Size (m)	Area (m²)	Number/vehicle
1	Glove box	20 × 4	80	1
2	Top cover	10 × 8	80	1
3	Door stopper	10 × 4	40	1
4	Wiper motor	10 × 4	40	1
5	Door lock	10 × 4	40	2
6	Sleeper	20 × 4	80	2
7	Door glass	10 × 3	30	2
8	Elevated box	20 × 4	80	2
9	Dashboard	10 × 9	90	2
10	Front windshield	10 × 9	90	1
11	Seats	20 × 4	80	2
12	Steering wheel	10 × 4	40	1
13	Insulation pads	10 × 4	40	1
14	Brake valve	20 × 4	80	1
15	A line	236 × 4	944	
16	B line	236 × 4	944	

Figure 1. Truck interior existing plant layout.

2 RESEARCH EXISTING LAYOUT USING LAYOUT DESIGN

2.1 *The interior of the truck assembly plant in the existing layout analysis using the SLP method*

Since A, B, two assembly lines are arranged in two parallels, so the A and B line intensity of superposition analysis of logistics were collected and the logistic capacity is as shown in Table 2.

From the operating units integrated correlation diagram, a truck interior workshop unit position correlation diagram is obtained, as shown in Figure 2.

The combined operating units location map and associated operating units in a conventional layout in the actual area can be arranged in a number of reasonable layout programs. Several programs can be selected to evaluate the optimal solution, but the process of obtaining the candidate program approximately has great subjectivity and limitations and cannot guarantee the optimal layout of the global scope of the program, so this will be the only program as part of the initial population of the genetic algorithm. Based on this conclusion, this paper searched globally to get a real sense of the optimal solution.

Table 2. From the total amount of logistics to the table.

Level B \ Level C	B1	B2	B3	B4	Level C Ordered by the total weight
	0.1082	0.2532	0.0655	0.5731	
C1	0.0608	0.0695	0.0974	0.1111	0.0942
C2	0.3531	0.3484	0.5695	0.4444	0.4184
C3	0.5861	0.5821	0.3331	0.4444	0.4873

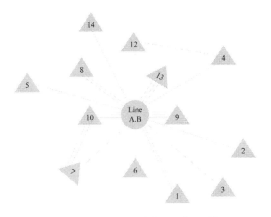

Figure 2. Truck interior workshop unit position correlation diagram.

2.2 *The main steps of the truck interior workshop layout optimization procedure using genetic algorithms*

1. The establishment of the fitness function
2. Design function constraints
3. The initial population design
4. Determining the coding method
5. Genetic operator design
6. Algorithm termination condition

2.3 *The use of MATLAB to achieve genetic algorithm*

By the MATLAB operation, this paper got two layout schemes in addition to the original layout, a total of three layout schemes. Next, it selected the best solution by conducting the comparative evaluation, where the original layout was named layout scheme 1, two programs by genetic algorithm are referred to as layout scheme 2 and scheme 3. Scheme 2 is as shown in Figure 3 and scheme 3 is as shown in Figure 4, with the original layout plan shown in Figure 5.

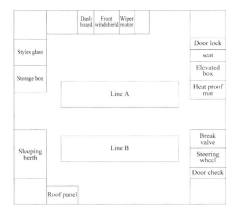

Figure 3. Layout scheme 2.

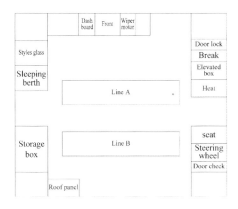

Figure 4. Layout scheme 3.

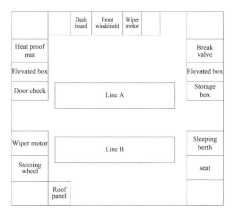

Figure 5. Layout scheme 1.

3 EVALUATION AND SELECTION OF THE BEST TRUCK INTERIOR WORKSHOP PLANAR LAYOUT PLAN

3.1 Evaluation and selection of the best truck interior workshop planar layout plan using the analytic hierarchy process

The paper selected the expert's questionnaires to build the judgment matrix and standards related target of judgment matrix as follows:

$$A = \begin{bmatrix} & B_1 & B_2 & B_3 & B_4 \\ & 1 & 1/3 & 2 & 1/5 \\ & 3 & 1 & 4 & 1/3 \\ & 1/2 & 1/4 & 1 & 1/7 \\ & 5 & 3 & 7 & 1 \end{bmatrix} \quad (1)$$

AHP total sort results shown in Table 3.

Calculated overall consistency ratio CR = 0.0117 <0.1 satisfies consistency. Total weight of the program to sort the results by the calculated layer can be seen. Total weight of the program C3> Total weight of the program C2> Total weight of the program C1.

It can be concluded that F company truck interior workshop layout scheme 3 is the optimal solution, but also an ideal layout is scheme 2. The two optimization programs are more excellent than the initial layout.

3.2 Comparative analysis results

By evaluating the safety factors, flexible layout, overall environment, the efficiency of logistics, and other aspects, the selected option 3 is the optimal layout, and integrated logistics cost from the following aspects before and after comparison with

Table 3. Total level sorting results.

Level B Level C	B1 0.1082	B2 0.2532	B3 0.0655	B4 0.5731	Level C Ordered by the total weight
C1	0.0608	0.0695	0.0974	0.1111	0.0942
C2	0.3531	0.3484	0.5695	0.4444	0.4184
C3	0.5861	0.5821	0.3331	0.4444	0.4873

more intuitive results were demonstrated before and after optimization.

First, this paper calculated the cost of the initial layout of logistics.

Then it used the following formula to solve the collected amount of logistics and logistics costs:

$$Z = c \sum_{i=1}^{n} \sum_{j=1}^{n} (f_{ij} \times d_{ij}) \quad (2)$$

where c is the unit cost of transport from the logistics, taking c = 1.

We solved for the initial layout logistics costs $Z1 = 3.3 \times 10^7$. Based on genetic algorithms, using MATLAB, we can get the optimal layout of the logistics cost $Z3 = 1.5 \times 10^7$. Optimal layout reduces logistics costs by 1.8×10^7 than the initial layout, a reduction of 55% of logistics cost.

So we can see that the selected planar layout scheme 3 through genetic algorithms and assessment method compared with the initial planar layout 1, greatly reduced the cost of logistics.

4 CONCLUSION

Firstly, this paper started collecting related data of the present layout. Accordingly, it adopted the F-D analytical method to get data to find out the deficiency of the present layout. Secondly, aiming to optimize the layout, SLP and genetic algorithm method were selected to get two approximate optimal solutions. In the end, the evaluating method AHP was used to choose the most excellent layout. As a result, the evaluation method got a consistent result. Thus the layout chosen is the best for F trucks interior workshop.

REFERENCES

Buffa E C. & Armour G C. 1964. Vollman T E. Allocating Facilities with CRAFT[J]. HarvardBus, Rev, 42:136–158.
Chang-Lin Yang, Shan-Ping Chuang. & Tsung-Shing Hsu. 2011. A genetic algorithm for dynamic facility

planning in jobshop manufacturing[J]. Advanced Manufacturing Technology, 52:303–309.

Ershi Qi. 2001. Logistics engineering [M]. Beijing: Chinese science and technology publishing house.

Lee. R.C. & J.M. Moore. 1967. CORELAP—Computerized Relationship Layout Planning[J]. Industrial Engineering, 18(3):195–200.

Madhusudanan Pillai V., Irappa Basappa Hunagund. & Krishna K. 2011. Design of robust layout for Dynamic Plant Layout Problems [J]. Computers & Industrial Engineering, 61:813–823.

Seomon Takakuwa. & Hiroki Takizawa. 2000. Simulation and Analysis of Non-automated Distribution Warehouse[A]. J.A. Joines. & R.R. Barton. & K. Kang. & P.A. Fishwick ed. Preceeding of the 2000 Winter Simulation[C]. New Jersey: Institute of Electrical and Electronics, 1177–1184.

Sunderesh S. & Heragu. 2006. Facilities design[M]. BWS iUniverse.

Civil, Architecture and Environmental Engineering – Kao & Sung (Eds)
© 2017 Taylor & Francis Group, ISBN 978-1-138-02985-9

The study and implementation of VoIP voice terminal system based on android platform

Gongjian Zhou
Xiamen University Tan Kah Kee College, Zhangzhou, Fujian, China

ABSTRACT: VoIP is a system which can realize real-time voice communication in IP network. Based on the research of VoIP related technology and protocol and Android platform, this paper designs and realizes the Android terminal VoIP voice communication system based on SIP protocol by applying Android NDK development framework and Java language integrated with SIPDroid. The test results show that the system is with multiple functions and in good performance, which can meet the practical application and marketing demand.

1 INTRODUCTION

In recent years, with the continuous development of Internet technology, IP-based real-time voice trans-mission technology has gradually become mature and VoIP (Voice Over Internet Protocol) telephone has been more and more widely used (Runsheng & Xiaorui 2004). The expansion of mobile communication services and the popularity of smart phones provide a strong market thrust for the application and promotion of soft-switching system which belongs to VoIP and achieves voice communication function through software. Given the fact that the Android system plays an absolute leading role in the mobile terminal market, this article uses Android NDK (Native Development Kit) integrated with SIPDroid (SIP open source protocol stack) as development environment after an in-depth study of the VoIP-related technology and Android platform, designs and realizes the VoIP communication system based on SIP protocol of Android terminal. The main achievements are as follows:

1. The paper analyses and studies VoIP technology, SIP protocol and Android technology, focusing on Android NDK development framework and SIPDoird architecture. We comprehensively understand and master the technological theories used by the system, which lays a solid theoretical foundation for the design and implementation of the system.
2. The paper uses MVC design pattern and Android NDK hierarchical structure, and divides the terminal system into four sub-modules, namely user interface, SIP message processing, voice processing, real-time transmission. Besides, the paper also uses Java language to develop and implement the four sub-modules.
3. The paper packages the developed programs to form apk, then downloads it in the cellphones. The paper also provides the corresponding experimental environment to test its functions, and uses Wireshark to analyze the package of voice packets and verify the completeness of the system's functions. The final results are correct, which achieves the design expectations.

2 SYSTEM-RELATED TECHNOLOGY AND PROTOCOL RESEARCH

2.1 *VoIP technology*

VOIP is a set of instant messaging technologies which transmit voice over IP network. It is not a specific agreement or standard, but a complex set of technologies (Runsheng & Xiaorui 2004). Its working principle can be showed in Figure 1.

The voice communication services can be realized as follows: convert voice signals to digital signals, compress coding and package the signals in groups, then turn the analog signals into IP telegrams with Internet as transmitting medium

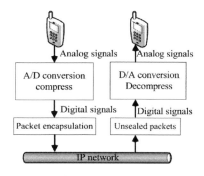

Figure 1. VoIP data processing flow chart.

which can be transmitted and exchanged on the Internet. Finally, turn IP telegrams back to analog voice signals through a series of reverse operations (Rosenbery 2002, Jiantao & Kaiyan 2007).

The key technologies related with VoIP include: signaling technology, coding technology, real-time transmission technology, Quality of Service (QoS) guarantee technology, network transmission technology, etc (Duanfeng & Xinhui 2005). After comprehensively considering various conditions, the system adopts the SIP (Session Initiation Protocol) as the signaling control protocol, Speex coding algorithm as the voice coding technology, RTP (Real time Transport Protocol) as the voice real-time transmission; RSVP (Resource Reservation Protocol) and RTCP (Real Time Transport Control Protocol) as the quality of service QoS guarantee. In terms of network transport layer protocol, the system selects UDP, supplemented by gateway interconnect, routing, network management and security certification and billing, mute detection, echo cancellation and other related network communication technologies.

2.2 SIP protocol

The SIP protocol is a text-based, application-layer signaling control protocol proposed by the Internet Engineering Task Force (IETF) in 1999. It is mainly used to create, modify and release one or more participants of the session. These sessions are a collection of signaling data and media data. Participants can communicate via multicast, unicast or even multicast (Rosenbery 2002, Jiantao & Kaiyan 2007).

2.2.1 Network entities of SIP

The SIP protocol adopts the Client/Server (C/S) model to complete the call establishment through the request and response between the client and the server. A typical SIP network usually contains two kinds of network entities (User Agent) and Network Server (Network Server). The network framework is shown in Figure 2.

User Agent (UA) is divided into UAC (User Agent Client) and UAS (User Agent Server). UAC's main task is to initiate SIP call request and UAS's main task is to respond to the call request. In many cases, the user agent undertakes the dual role of the user agent client and the user agent server.

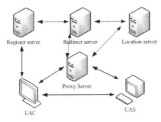

Figure 2. Network framework of SIP system.

The network server is composed of a proxy server, a register server, a redirect server, and other servers. Proxy server is mainly responsible for message routing. Registration server is mainly responsible for processing user registration information. Redirect server is mainly responsible for positioning. They can be distributed in different physical entities, and can also coexist in a device. According to actual needs, we will integrate the three servers into one through softwares.

2.2.2 SIP message mechanism

The messages in SIP are coded in text mode. They are composed of the start line, several message headers, blank lines and message bodies, and are divided into two types, namely request message and response message (Rosenbery 2002).

Generic-message = start-line
                    *message-header
                    CRLF
                    [message-body]

1. Start-Line gives the SIP version, the calling operation method, the current address of the invited user, the response type, and so on. The start line in the request message is the Request-Line, and the start line in the response message is the Status-Line.
2. Message header (Message-Header) is divided into four categories: general head, request head, response head and entity head. The symbol "*" indicates that the field can have more than one message head (Rosenbery 2002).
3. A blank line (CRLF) indicates the end of a message header field.
4. Message-Body generally uses the Session Description Protocol (SDP). It mainly includes the session description, and the cause and progress indication text may also be included in the response message.

The request message is used to activate SIP messages sent to the server, including INVITE, ACK, OPTIONS, BYE, CANCEL, and REGISTER, depending on particular operations.

The response message is used to respond to a request message indicating the success or failure status of the call. It contains a 3-bit integer status code to indicate the response of the called party to the request. The first bit indicates the response type, the last two bits indicate the specific response in the class. According to the different value, it is divided into the following types, 1xx, 2xx, 3xx, 4xx, 5xx, 6xx, and so on.

2.3 Android technology

Android is an open source operating system based on the Linux platform developed by Google and the Open Handset Alliance, and it is primarily used in mobile devices. Its essence is to add a Java virtual machine Dalvik in the standard Linux system, and build a JAVA application framework in Dalvik virtual machine. All applications are built on the JAVA application framework (Batyuk et al. 2009). Android system uses a layered architecture. From the top to the bottom, it is divided into four layers, namely, Linux kernel, system libraries and Android runtime, application program layer and application framework layer, which can be shown in Figure 3.

1. Linux kernel: Android core system services rely on Linux2.6 kernel, such as memory management, process management, network protocol stack and driver model. The Linux kernel also serves as an abstraction layer for the hardware and software stacks.
2. System library and Android runtime: The system library includes nine subsystems, namely, layer management, media library, SQLite, OpenGL, FreeType, WebKit, SGL, SSL and Libc. Android runtime includes the core library and Dalvik virtual machine, and the core library is compatible with most of the Java language functions. Dalvik virtual machine is a register-based java virtual machine, and its main functions include the life cycle management, stack management, thread management, security and exception management and garbage collection and other important functions (Batyuk et al. 2009).
3. Application framework layer: application framework layer includes the following ten parts, telephone manager, resource manager, location manager, notification manager, and other six parts. It is the basis for Android application development. In most cases, developers are dealing with the layer. On the Android platform, the developer has full access to the APIs used by the core application, and any application can publish its own functional modules, or use other functional modules that have been published by other applications to perform component replacement.
4. Application layer: This layer provides some core application packages, such as e-mail, SMS,

calendar, maps, browser and contact management. At the same time, developers can use the Java language to design and compile their own applications.

3 THE OVERALL DESIGN OF THE SYSTEM

3.1 *System layer module design*

According to the functional requirements of the system, the system uses a hierarchical modular design ideas, as shown in Figure 4.

1. Access layer: It mainly uses the existing PSTN, 3G, 4G and IP networks to achieve system access services, and analyzes the SIP message and SDP message transmitted from the Internet through the SIP signaling protocol stack.
2. Ability layer: It mainly provides the function module to deal with SIP signaling and RTP voice, which is the core part of the system. The SIP soft switch platform achieves the User Agent (UA). When it calls, it is regarded as UAC. When it is called, it is regarded as UAS. We also set other modules, such as coding, recording, RTP media, conference management, IVR, data engine and so on. Above those modules, we set a service engine in order to dock with business layer.
3. Business layer: It is a system function provided for end-users. In addition to achieving the basic point-to-point voice calls, it can also achieve switch board IVR voice processes, conference calls, voice mail and other value-added functions.
4. Display layer: It is a carrier to use system provided for end-users. At the mobile end, specialized APP is used and on the PC end SIP soft terminal softwares are used (eg X-Lite, Eyebeam, etc.).

3.2 *The protocol stack structure of the system*

The system is based on IP network voice communication and the protocol stack structure can be shown in Figure 5.

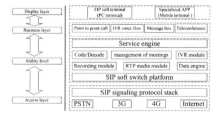

Figure 4. System hierarchical module diagram.

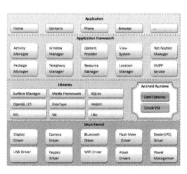

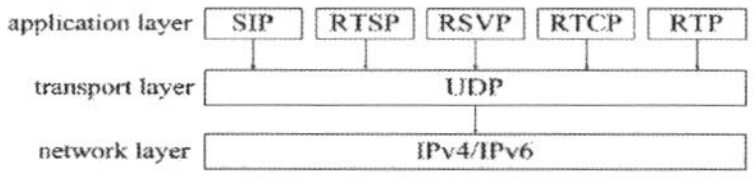

Figure 3. Android system architecture diagram.

Figure 5. System protocol stack structure.

At the application layer, SIP is used for voice signaling control, and RTSP is used to control "one-to-many" multimedia data flow. RSVP and RTCP are used to guarantee higher quality of service QoS. The former specifies the resource protection policy of the IP network and the latter is used to detect and potentially resolve transmission problems, thus monitoring the session quality and detecting the network. RTP is used to complete the end-to-end voice data real-time transmission services.

At the transport layer, we use the UDP protocol. On the one hand because the UDP does not require three handshake. The establishment process of SIP signaling has completed the functions which is equivalent to TCP three-way handshake. On the other hand, UDP can meet the real-time requirements of RTP while satisfying the requirement of SIP commonality (Stefano 2007).

3.3 Function modules' design of the system

Through the deep research of VoIP technology, SIP protocol and Android technology, we use NDK (Native Development Kit) framework which integrates open source SIPDroid under the Android platform as the development environment. The hierarchical functions planned by the system corresponds to the hierarchical modules in the NDK framework, which can be shown in Figure 6.

1. User Interface (UI) module: This module corresponds to the Java Application Layer of the NDK framework. It mainly uses the Java language to design and write the user interface, which is the part of the whole software system and user interaction. It implements the concrete SIP function by calling the interface of JNI layer (Java Native Interface Layer): for example, the initiation, establishment and termination of the session. Besides, it also includes many other subordinate function interfaces such as address book display, call parameter settings, dial-up and real-time calls.
2. SIP message processing module: This module corresponds to the NDK framework of the JNI layer and the native code layer (Native Code Layer), which is the core of the entire system. Android kernel itself does not provide SIP protocol stack, so in this module we use open-source SIP protocol stack SIPDroid SIP to process messages. Its source code has included specific methods to process the SIP messages, so we only need to design the appropriate JNI Interface function and directly use it in the UI module[3].
3. Voice processing module: This module corresponds to the NDK framework of the native code layer. The design of the layer are generally written in C language, which is mainly used to complete a series of functions such as the collection of voice data, compression, encoding and playback.
4. Real-time transmission module: This module corresponds to the NDK framework of the JNI layer and the native code layer. It is mainly used to transmit various messages of SIP, and RTP encapsulation and packing of voice data after voice processing, decapsulation of received voice data and so on (Batyuk et al. 2009). In SIPDroid framework, we design and implement the RTP transmission source code function, and use JNI to encapsulate and package calls.

4 THE SPECIFIC DESIGN AND IMPLEMENTATION OF THE SYSTEM

4.1 The design and implementation of User interface (UI interface)

UI interface is not only the interface of human-computer interaction, but also affects the user's overall experience of the application. In the framework of Android UI design, this system designs five UI interfaces, such as user login interface, parameter setting interface, dial interface, address book interface and real-time communication interface. The design mode of MVC (Model-View-Controller) can be shown in Figure 7.

The entire UI is built by View and View Group. Each visual interface uses Activity to achieve this component. Activity also carries a variety of different control elements. The use of XML file layout mechanism helps to define the user interface structurally and to ensure that the user interface is separated from the code definition. After completing the UI layout, the Android framework uses the event listener interface to listen to user interaction.

4.2 The design and implementation of SIP message processing module

SIP message processing module is the core module of the system, which can be realized by using open

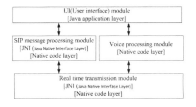

Figure 6.   Terminal system function module.

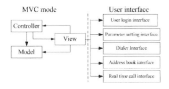

Figure 7.   UI interface design under MVC mode.

source VoIP protocol stack SIPDroid on Android platform. SIPDroid completely encapsulates the SIP message model, implements the RFC3261 protocol content, and employs high cohesive, low coupling software design architecture, which is easy for developers to customize their own new functions (Singhand & Schulzrinne 2006). The framework of SIPDroid protocol stack is shown in Figure 8.

1. UI layer: It is mainly responsible for the interface display and it provides users with the system interaction interface.
2. Core Engine Layers: This layer starts various services through SIPDroid Engine, initializes the software parameters, provides customized UI interfaces, and calls down the encapsulated SIP message processing interface, which is the core processing layer of the system. This layer contains several important files and classes: User Profile (User Profile) which is used to maintain the system configuration information and global variables; User Agent (user agent) which is used to handle a variety of user events and complete the SIP protocol stack call and session; Register Agent which is used to handle the registration event of the SIP message to complete the registration process of the client to the SIP server.
3. It mainly realizes the content of RTP and RTCP protocol, using UDP protocol to complete the data transmission and control. In addition, the layer also provides JSTN NAT penetration technology based on STUN protocol.

4.3 The design of voice processing module

We consider the design of voice processing module mainly from three aspects: the acquisition and playback of voice streams, compression and encoding and packaging and transmission the main process can be shown in Figure 9.

We use Audio development framework provided by Android media library to acquire and play voice

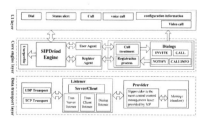

Figure 8. SIPDroid program framework diagram.

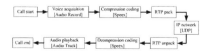

Figure 9. Voice processing module flow chart.

signals. Audio Record provided by Audio framework is used to record the voice and Audio Track is used to play the voice[4]. We also employ Speex Algorithm to compress and encode voice signals. RTP and UDP protocols are used to pack and transmit voice signals.

4.4 The design and implementation of real-time transmission module

The data transmitted in VoIP system can be broadly divided into two types: SIP signaling data and session audio data. Based on the architecture of SIPDroid, we design the above two kinds of data streams into two different transmission channels, which can be shown in Figure 10.

1. SIP signaling data transmission: The operations of the user at the UI layer operation (such as dial-up, answer, etc.) will be broadcast to the SIPDroid core engine. According to the types of operations, SIPDroid core engine will deliver them to the User Agent or Register Agent and they will produce different requests. SIPDroid Provider will analyze those requests and form corresponding SIP packages. The packages will be delivered to Udp Transport. After being handled, the packages will be delivered to Udp Provider. We can call Udp Socket and bind it with the target address and finally transmit the date through Datagram Socket of Java. When receiving SIP packets from the network, the data transfer order is reversed.
2. Session audio data transmission: the terminal compresses and codes the sampled and then delivers them to the SIPDroid core engine. Meanwhile, the backstage system will start two thread classes. One is used to send RTP packet virtual class RtpStreamSender, and the other is used to receive RTP packets of virtual Class RtpStreamReceiver. These two classes are Java thread classes, both of which have achieved Runnable interface and are in the audio session after the establishment of continuous operation. RtpStreamSender data will be delivered to the RtpSocket. Then RtpSocket will package the data packet into Rtp

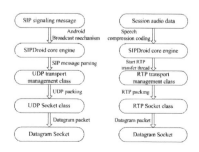

Figure 10. Two different real-time transmission data channels.

package, read the target IP address and the agreement of a good RTP port number, and bind those information with SIPDroid Socket. SIPDroid Socket can transmit the data by calling Java Datagram Socket. When receiving Rtp packets from the network, the data transfer order is reversed.

4.5 *Database design*

From the business logic point of view, the system selects open source, and operating system-independent SQLite as a database background, in which multiple tables are set up, including two most important tables, namely user information and contacts tables. The User Information Table includes SIP account, log-in password, permission level and other fields. The Contacts Table includes the contact name, SIP account number, E-mail address, and other fields.

5 OPERATION AND TESTING OF THE TERMINAL SYSTEM

After the development of all functional modules, we use ADT in Eclipse to package the program to generate the appropriate apk file and install it to the Android phone.

In the test, we choose the client application software developed in this paper as a caller, running on the HTC G23. Free VoIP soft phone X-Lite is chosen as the called party, running in the pc-side windows environment. The server uses the SIP server built by Asterisk and integrates the functions of the registration server and the proxy server. Wireshark uses the open-source network packet analysis software Wireshark to capture and analyze network packets during a call, which can be shown in Figure 11.

After comparing the data of Wireshark capturing packet with standard SIP call flow (see Figure 12), it is confirmed that the system is in good working condition and the basic function of voice terminal is complete. The main function modules can be used normally and the expected results are achieved.

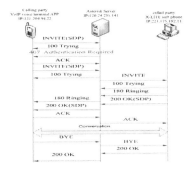

Figure 11.   VoIP basic call process.

Figure 12.   Wireshark capture graph.

6 SUMMARY

In this paper, we use open source development platform and tools to design and implement the VoIP voice terminal system based on the Android platform. Modularization method is also applied in this paper, so the system has the advantages of low development costs, complete voice communication functions, being convenient, practical, reliable, safe and good. The system can also be extended into the network interactive multimedia application with texts and video conferences to meet different application needs of the market, which has a good value for market application and promotion.

ACKNOWLEGEMENT

Fund Project: Application and Development of Mobile Voice Communication System of Smart Phones Based on SIP Protocol (Zhangzhou Natural Science Foundation of Science and Technology Project: ZZ2014J38).

REFERENCES

Batyuk L, Schmidt A D & Schmidt H G. (2009). Developing and benchmarking native linux applications on android. J. Mobile Wireless Middleware, Operating Systems, and Applications. 07, 381–392.

Duanfeng Si & Xinhui Han. (2005). The core technology and research progress of SIP standard. J. Journal of Software. 02, 239–249.

Faccin Stefano. (2007). IP multimedia services: Analysis of mobile IP and SIP interactions in 3G network. J. IEEE Communications Magazine, v42(1). 06, 85–87.

Jiantao Zhao & Kaiyan Wu. (2007). Design and Implementation of Soft Phone Based on SIP Protocol. J. Journal of North China Electric Power University. 01, 120–122.

Rosenbery. J (2002). SIP: Session Initiation Protocol. RFC 3261.

Runsheng Wang & Xiaorui Hu. (2004). Voip development in China. IEEE Computer. 37(9), 4–5.

Singhand K. & H. Schulzrinne. (2006). Interworking between SIP/SDP and H.323. J. Internet Engineering. 06, 22–28.

Civil, Architecture and Environmental Engineering – Kao & Sung (Eds)
© 2017 Taylor & Francis Group, ISBN 978-1-138-02985-9

Research on a malicious code behavior acquisition method based on the Bochs virtual machine

H.Y. Liu & Y.J. Cui
Department of Information Engineering, Academy of Armored Forces Engineering, Beijing, China

ABSTRACT: Bochs is an open source IA-32 (x86) emulator written in C++ that simulates the entire PC platform, including CPU, I/O devices, memory, and BIOS. This paper presents a method to acquire the behavior of a malicious code based on the Bochs virtual machine. It intercepts the instruction stream and data stream information conditionally when a malicious code is running in Bochs by redesigning the Bochs system, and then it records and parses the intercepted information. It also gets the system call information by linear address analysis so as to provide the executed system calls of a malicious code for the following behavior analysis. Experiments show that this method can effectively acquire the behavior characteristics of malicious codes.

1 INTRODUCTION

With the development of the Internet, the network has brought great convenience to people. However, the security problem ensues and has become more and more serious. Malicious code is one of the most widely used network attack methods. At present, there are millions of different types of malicious codes that have been discovered. Since network security has been upgraded to be an important factor of national security, which is related to a country's information infrastructure, the speeding up of detection technology of malicious code has become far more significant to social and national security.

Malicious code detection methods are generally divided into two types, i.e., static detection, and dynamic detection.

1. Static detection. The static detection is based on the signature of malicious code previously found, which searches the signature in the target file, the IP package, the mail attachment, etc. This method does not actually execute the code but analyzes malicious code by disassembling, decompiling, analyses file structure and other methods to get the required information. Due to various advanced self-protection technologies used in a malicious code, such as shell processing and obfuscation technology, it becomes more and more difficult for static detection method to discover malicious codes.
2. Dynamic detection. In this kind of method, the detection system discovers a malicious code when it is running by analyzing its behavior characteristics.

Behavior refers to the operation of the operating system such as process, memory, files, registry, and API functions. The basic idea behind malicious code detection is based on the behavioral analysis in that there is a clear distinction between the behavior that a malicious code exhibits when it is running on the system and the behavior that a normal program behaves when it runs so that it can be determined whether the code is malicious. Since there is no need for a predefined signature, malicious code detection based on behavioral analysis technology can discover unknown malicious code, which overcomes some of the limitations of static detection. With the high detection rate of this method, it has gradually become the focus of research in the field of malicious code detection. As such the research of behavior acquisition method has gradually become more and more important.

In this paper, the Bochs virtual machine is used in the acquisition of the behavior of a code, which provides the basis for modeling the behavior of a malicious code.

2 ANALYSIS OF THE PROGRAM BEHAVIOR ACQUISITION METHOD

At present, the main method to monitor the behavior of a running malicious code can be divided into three types, i.e., the environmental comparison method, debug method, and system call monitoring method.

Environmental comparison method, which is also called the black box observation method, acquires the behaviors of the malicious code by comparing different system views before and after the malicious code execution. However, this method cannot find specific reasons for the change of system views and cannot reflect the whole dynamic process of execution, which may result in the loss of some malicious code behavior.

The debug method executes the code step-by-step and tracks system changes by setting breakpoints through the debugger. This method can monitor code execution step-by-step and also execute a program fragment at a time. Although this method can fully monitor the execution process of the code, it is useless for those malicious codes using some anti-debugging techniques.

System call monitoring method intercepts the system calls during the execution of the code. The method can acquire the operation of code execution in real time, which can be used for the analysis of behavior characteristics as well as the detection of malicious code. To avoid being infected by the malicious code, the sandbox technology is generally used to execute the malicious code samples, and the virtual machine is a typical kind of sandbox.

The virtual machine simulates a physical host by the software. A virtual environment can protect the real host from being infected by malicious codes. The actual execution of the malicious code in the virtual machine will not cause infection and attack to the host. The virtual machine has the following advantages when used in the analysis of a malicious code. First, multiple operating systems can coexist on a single physical machine and are isolated from each other. Second, a virtual machine can provide an instruction set architecture which is different from the host computer. In other words, the operating system of the client can be different from its host. Third, virtual machines are usually highly available to maintain and provides some disaster recovery capabilities.

In this paper, the Bochs virtual machine is used to intercept the system call information when a malicious code is running.

3 BOCHS VIRTUAL MACHINE

At present, virtual machines can be divided into three categories according to their degree of virtualization, i.e., fully virtualization, semi virtualization, and simulator.

1. Fully virtualization. The most typical product of full virtualization is VMware. VMware fully virtualizes the underlying hardware by using a common method of PC hardware virtualization, while the guest OS of the virtual machine often has lower performance than the original host.
2. Semi virtualization. Semi virtualization makes some modifications to the kernel of the operating system so that some of the original need to directly implement the CPU protection tasks can be directly implemented in the modified

kernel which improves the efficiency of the virtual machine.
3. Simulator. The simulator reads the CPU instruction of the program, explains the instruction, and simulates the execution. It also simulates the state of various registers and state machines of the hardware.

Bochs is essentially a simulator by reading the CPU instructions and simulating the execution to achieve the virtualization effect of various hosts. It can effectively record and intercept the instruction stream information when a malicious code is running. Since any action that occurs while a program is running is implemented by a sequence of instructions, the instruction stream information can not only reflect the path of the static binary code execution but also reflect the various behaviors of the binary code.

Bochs is a GPL-licensed INTEL IA-32 virtual machine which provides the simulation of the entire computer hardware platform. It is a full system simulator including functional simulation of one or more processors and commonly used external devices. It fully explains and executes the operation of virtual machine instructions and memory access. Bochs provides a complete simulation of CPU behavior, including extraction, decoding, execution, and other CPU internal behaviors. Bochs can simulate a variety of operating systems, including Linux, DOS, Windows 95/98/NT/2000/XP, etc. Bochs project is an open source project, which is developed mostly by C++ programming language. Since its CPU simulation module is relatively independent and well designed, it is suitable for users to further design based on its source codes. This article will realize a virtual machine which can conditionally intercept the instruction stream by revising the source code of Bochs.

4 BEHAVIOR ACQUISITION AND ANALYSIS BASED ON BOCHS

4.1 Interception of instruction stream and data stream

4.1.1 Redesign of Bochs

Bochs program is mainly composed of a CPU module, memory module, interrupt module, and the peripheral module. The CPU module is the core of Bochs, which is responsible for simulation execution of instructions. After Bochs starts, it initializes the hardware modules and then enters the CPU module. The CPU module consists of a loop composed of interrupt processing, fetching instructions, decoding instructions, and executing instructions. The basic execution process of CPU is realized by the cpu_loop() function.

Based on the original code, we add the instruction stream intercepting module. The architecture of Bochs after adding the runtime interception module is shown in Figure 1.

It is no doubt that the efficiency of the virtual machine will be greatly reduced if each instruction is intercepted. So instructions can be intercepted conditionally. These interception conditions can be time, instruction linear address, instruction type, or operand. Conditional interception module can reduce the impact on the efficiency of the virtual machine brought by interception. This article sets conditions to be the linear address range of instructions, thus it only intercepts instructions in a specific memory address space. The linear address of the instruction is very important for instruction analysis. Taking the system call as an example, the import address table

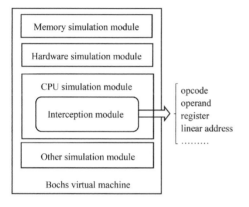

Figure 1. Bochs architecture with intercepting module.

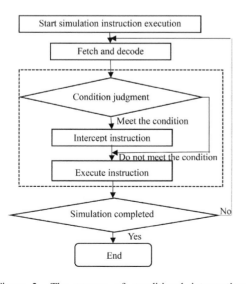

Figure 2. The process of conditional interception module.

of the system calls is known when the system is running, so which system call is now being executed can be known by checking the linear address in the import address table. The process of conditional interception module is shown in Figure 2.

4.2 Analysis of system call based on linear address

In Windows systems, the behavior of a malicious code is different from that of a normal program which can be described as a system call sequence. No matter how the malicious code evolves, the realization of its function must rely on system calls to be completed. So it can determine whether the code is malicious by combining the system call sequence with their parameters. Therefore, to get the executed system call sequence is the basis for identifying a malicious code.

To get the instruction stream and data stream in some readable form for further analysis, the intercepted instruction which is in binary format must be resolved to string format. The instruction stream after the analysis is shown in Table 1.

Table 1. The instruction stream after the analysis.

Linear address	Instruction	Register and memory value
[0 × 00542225]	MOV ECX, EDX	ECX[005421C1] EDX[00006BA4]
[0 × 00542225]	INC ESI	ESI[00527324]
[0 × 00542225]	SUB EAX, EBX	EAX[A16B9784] EBX[C60DFB26]
......	......	......
[0 × 00542225]	PUSH ECX	ECX[005421C1]
[0 × 00542225]	CALL 004CD4 AB	MW[0000000000: 0023FF2C] [005421C1]

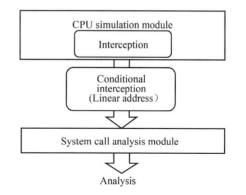

Figure 3. The overall structure of the system after redesigning.

The principle of system call resolving based on the linear address is as follows:

It determines the instruction function belongs to which system call according to the linear address of the instruction. For a particular operating system, the import linear address of a particular system call is usually invariant. Therefore, the analysis of a system call module can judge whether the function belongs to a system call according to the operating system information and the import linear address of the code block. For example, in the Windows PE 1.5 operating system, the import linear address of system call Kernel32.GetProcAddress is $0 \times 7C80AC28$. If an import linear address of the code block is $0 \times 7C80AC28$, the code block is determined as the system call Kernel32.GetProc Address. The analysis of specific system call is conducive to rapidly analyze the behavior characteristics of the function. The overall structure of the system after adding the interception function as well as the system call analysis function based on the linear address is shown in Figure 3.

5 EXPERIMENT AND RESULTS

In order to test whether the Bochs virtual machine after our redesigning can acquire the system call of a running code, we do the following experiment.

1. To install Windows XP operating system in a redesigned Bochs virtual machine. Bochs starts is shown in Figure 4.
2. To make the malicious code sample Win32. Kryptik.NX run in Windows XP.
3. To intercept and analyze the system calls of a sample. The analytical results acquired by the system are shown in Table 2.

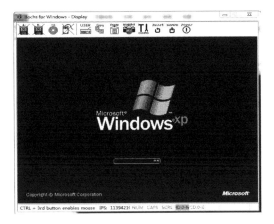

Figure 4. Bochs starts.

Table 2. The analytical results acquired by the system.

Linear address	System call
$0 \times 7C80B529$	Kernel32.GetmoduleHandleA
$0 \times 7C80AC28$	Kernel32.GetProcAddress
$0 \times 7C859B5C$	Kernel32.OutputDebugStringA
$0 \times 7C80AC28$	Kernel32.GetProcAddress
$0 \times 7C80AC28$	Kernel32.GetProcAddress
$0 \times 7C8017DD$	Kernel32.LoadLibraryA
$0 \times 7C809A81$	Kernel32.VirtualSlloc
$0 \times 7C859B5C$	Kernel32.OutputDebugStringA
$0 \times 7C80AC28$	Kernel32.GetProcAddress
$0 \times 7C80AC28$	Kernel32.GetProcAddress
$0 \times 7C859B5C$	Kernel32.OutputDebugStringA
$0 \times 7C859B5C$	Kernel32.CallWindowProcA

6 CONCLUSION

Bochs virtual machine can be used to execute malicious code in research to protect a system from being infected or attacked. Since it is an open source project, it can also be used to acquire runtime instruction information by redesigning the program, which is the basis for behavior analysis of malicious code. In this paper, an instruction interception module is added to the CPU simulation module to acquire the instruction executed when a malicious code is running. The work of this paper provides a basis for modeling the behavior of a malicious code.

REFERENCES

Chen, P. 2009. Research and implementation of obtaining malicious code behavior. Journal of Computer Application 29: 76–79.
Deng, C, G. 2011. A program understanding approach for stripped binary code. Computer Application. 28(10): 2608–2612.
Li, H. 2011. Technique of detecting malicious executable via behavioral and binary signatures. Application Research of Computers 28(3): 1127–1129.
Li, J, R. 2008. Differential analysis on dynamic binary and its application in malicious code analysis. Application Research of Computers 29(2): 654–660.
Li, Z, L. 2014. Hierarchical analysis method of malicious behavior based on API association. Computer Engineering and Design 35(11): 3730–3735.
Sun, M. 2012. A program understanding approach for stripped binary code. Computer Application. 28(10): 2608–2612.
Yu, Q. 2006. Understanding the source code of Bochs. Nanjing.
Zhao, X, J. 2015. Intrusion Detection Based on Malicious Code Behavior Analysis. Computer Simulation 32(4): 277–280.
Zhu, L, J. 2012. An Identification Method on Unknown malicious Code Based on Dynamic Behaviour. Journal of Shenyang University of Chemical Technology. 26(1): 77–80.

Civil, Architecture and Environmental Engineering – Kao & Sung (Eds)
© 2017 Taylor & Francis Group, ISBN 978-1-138-02985-9

Research on the network security equipment allocation problem based on Monte Carlo tree search and machine learning

Cheng Chen, Hengjun Wang & Nianwen Si
Zhengzhou Institute of Information Science and Technology, Zhengzhou, China

Yingying Liu
Henan University of Animal Husbandry and Economy, Zhengzhou, China

ABSTRACT: In this paper, research on the allocation of network security equipment is carried out thoroughly. We model the network security equipment with structured language. Based on this, in order to solve the equipment allocation problem, we combine Monte Carlo tree search with machine learning algorithm to design specific solutions for the problem. Furthermore, in the process of the model of this problem, we also implement the representation for specific knowledge, establishing the state space for the problem and designing for knowledge solving algorithm. The results indicate that the proposed scheme significantly improves the speed and accuracy of equipment allocation.

1 INTRODUCTION

In today's world, the Internet revolution is changing with each passing day, which has a profound impact on the development of international politics, economy, culture, society, military, and other fields. With the continuous development of network expansion, the problem of network security is also increasing. The current network security protection system has established a relatively mature and complete model and designed the scheme of network security solution for diversification under various conditions. The security equipment performance has become more and more perfect. However, the problem of how to allocate the network system security equipment and the corresponding mature solution is still very limited. At present, most of the allocations for communication network security equipment still rely on manual methods. How to improve the accuracy and timeliness of the manual allocation is still a complex problem which needs to be solved.

The traditional manual allocation has many defects in accuracy and timeliness, and these two points are also the key problems for the allocation of the security equipment problem. If the machine learning method is introduced to this problem, the system can realize the automatic allocation through self-learning and this will improve the safety and accuracy of the network from the urgent task to have a more positive significance. In this paper, we propose an effective scheme which combines the Monte Carlo tree search with reinforcement learning to solve the allocation of the network security problem.

2 MONTE CARLO TREE SEARCH FOR EQUIPMENT ALLOCATION

2.1 Preparing the new file with the correct template

Using search to solve the problem is to find a solution in a possible space. This space is the object of the search and each of the elements in the space represents a possible state. The state is a mathematical description of the progress at a certain time when the problem is solving. The path of description must be to keep the relevant information as far as possible to ensure the accuracy of the state representation and all the states of the data structure is called the state space. The search process is based on these states and the corresponding search algorithm aims to solve the problem. So the object of the search is to find the state in the formatted state space. In order to solve the problem by using the search algorithm, the model must be established. The state and the state space must be determined.

In the form of the structure, the state which is a set of ordered minimal variables $(q_0, q_1, q_2, \cdots, q_n)$ can be introduced to describe the difference between some things under different conditions. The form is defined as follows:

$$Q = \{q_0, q_1, \cdots, q_n\} \qquad (1)$$

Each of these elements is called a state variable. A specific state can be obtained by determining a set of values of the component.

The state space is an ordered set of all possible states $Q_1, Q_2, \cdots, Q_n$ as well as their relations. The form is defined as follows:

$$S = \{Q_1, Q_2, \cdots, Q_n\} \qquad (2)$$

Each of these elements is represented as a state of the problem. State space is in the form of a figure. The nodes correspond to the graph state and the edges between nodes correspond to the feasibility of state transition. The weight of the edge corresponds to the cost of transfer. The solution of the problem may be a state in the figure, or a path from the start state to the same states, or the cost of the target.

In problem solving using the search algorithm, we need to abstract various aspects of the problem for the computer, thus it can make way for it to be understood and also saved. The actual problem is ensured with the state of one correspondence, which is a model of the process for building. So this search problem can be established as follows.

Based on the known network **A**, the configuration of network security equipment, for example, required to configure each network node with the existing equipment, so as to ensure the dissemination of information in the safe range. If all the networks containing **N** nodes needs to configure the **M** warehouse equipment, these equipment are available for distribution and allocation, and these things can be set as follows:

P: The node to be allocated. The number of nodes is **N**. It can be expressed as a matrix $P = \{P_1, P_2, \ldots, P_n\}$, each element represents a network node.

E: The type of equipment available for allocation. The number of types of equipment is **M**. It can be expressed as a matrix $E = \{E_1, E_2, \ldots, E_m\}$. Each element represents a kind of type.

Number(Ex): The number of the class **X** equipment.

(S, Ex): In the state **S**, the next node to select the equipment **Ex**.

P(S, Ex): The probability choice function. The possibility of being selected in the **S** state of the equipment **Ex**.

T: The maximum simulation duration.

2.2 *Target of problem search*

According to the description of the problem, the solving process can be simplified in a **M*N** matrix (the number of equipment type is **M** and the number of the node is **N**). The problem is to find out the best path from the first node to the last node as shown in Figure 1.

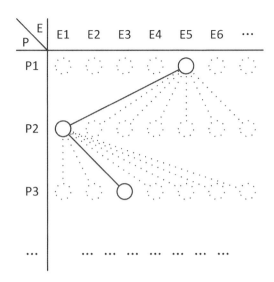

Figure 1.   The process of solving.

Table 1.   Frame representation.

\<Device name>	
\<Attribute 1>	\<Attribute value>\<Range>
\<Attribute 2>	\<Attribute value>\<Range>
......	......
\<Attribute n>	\<Attribute value>\<Range>

First, the simulation method used in the scheme is the Monte-Carlo tree search so that the machine can be the reinforcement learning and get the choice probability **P(S, E)** of the corresponding equipment in the node. Then through continuously repeated simulation, the equipment selection probability **P(S, E)** of each state can be more precisely found. In the probability selection, the function **P(S, E)** can be obtained after optimization which begins from an initial state. According to the **P(S, E)**, each state will choose a large number to compare with and reach the final state of **Sn**. Therefore, the key to the problem is to find the ideal choice probability **P(S, E)** value (the more numbers in the simulation experiments used, the greater the **T** value will be, and more accurate the results will be).

2.3 *Formal description of equipment and network node*

There are many kinds of network security devices and nodes, which can be used to frame representation

Table 2. Knowledge representation for device based on framework.

Device attribute	Representative	Range of values
Tag	Name and type	Text
Type	Type of security device	
Layer	Which layer is located on the network	"datelink", "network", "transport", "session", "presentation", "application"
Port	The type of port with device	"RJ45", "fiber", "RJ11", "AUI", "BNC"
Work_ with	what types of equipment can be interoperable with	Enumeration of devices that can communicate
Sum	Total number of devices	

Table 3. Knowledge representation for node based on framework.

Node attribute	Representative	Range of values
Number	Node ID	
Layer	Which layer is located on the network	"datalink", "network", "transport", "session", "presentation", "application"
Port	The type of port	"RJ45", "fiber", "RJ11", "AUI", "BNC"
Connect_with	Which nodes need to communicate with	Enumeration of nodes that must communicate with

to describe the characteristics of equipment and nodes. The first is to extract the feature and then summarize it as a number of attributes and determine the attribute name and value range of each attribute. The representation is as follows:

To build the framework which is required to extract the properties of the device, the extracted attributes should meet the following criteria: 1. Strong expression ability, which means it is able to express the required knowledge; 2. Easy to understand, which means it is easy to build data structure; 3. Easy to solve and reasoning, which means the knowledge of the symbolic structure and reasoning mechanism can support the knowledge base on the advanced search. This could quickly introduce the conclusion from the existing knowledge.

From the perspective of solving the target, the extraction of device properties can refer to the following features: the name and type of equipment, the work environment, the environmental adaptability, the technology performance, the adapter group, the quantity, and so on. Through the comparison of the above characteristics, the network security equipment can be extracted from the following aspects:

Similarly, the node can be extracted from the following aspects:

3 OPTIMIZED SEARCH ALGORITHM

The Monte Carlo algorithm is selected for the equipment allocation, which is ready to be configured by directly abandoning the low coincidence rate of equipment library selection and giving up the calculation of the exhaustive method which consumes a large amount of computing resources and specifying the main direction of calculation so that it has a high rate of choice for a more accurate calculation and analysis, especially in the selection strategy which aims to add more equipment-related professional knowledge. This makes the Monte Carlo search based equipment configuration algorithm to be a higher level of selection compared with the random search.

Since the Monte Carlo search is based on a large amount of reinforcement learning, the learning process is the focus of algorithm design and the variables are set as follows:

C_{ij}: The matching of the node and the equipment. It can be expressed by matrix $CAP=(C_{ij})m*n$, where C indicates that the i node is matched with the j equipment. If the equipment can be operated at that node, it is 1, else it is 0.

$V(S, E_x)$: The average accumulation rate. It represents the number of times $E_x(x \in m)$ has been selected and winning in the previous simulation of the state S.

$X(S, Ex)$: The prior knowledge function. In the current state S, according to configuration rules between equipment or between the node and equipment, the probability of the next step is to choose the equipment $E_x(x \in m)$.

$u(S, E_x)$: Knowledge selection probability function. The choice function is determined by known knowledge. It represents the probability of selecting $E_x(x \in m)$ in the state S.

$N(S, E_x)$: Recording the number of all the winning simulations in the state S to select $E_x(x \in m)$

$L(S, E_x, t)$: Recording that the (S, E_x) is to be accessed in the time of t simulation. It can be accessed to be 1, otherwise 0.

$R_t(S, E_x)$: Reward function. Recording the outcome of the game in the t simulation of the selected action (S, E_x). If it wins it is 1, otherwise 0.

The process of machine learning is to obtain a return function $R_t(S, E_x)$, which is combined with *a priori* knowledge. Constantly, it is also revised to the probability function $P(S, E_x)$. Finally, the optimal state transition path is found by $P(S, E_x)$.

In the beginning of the simulation process, due to the lack of experimental data, *a priori* knowledge is needed to remove some of the unreasonable combinations of the matching relationship between the equipment and the node (which can avoid such errors between the network layer equipment configuration and the physical layer). This not only reduces the state space but also improves the search speed. We can make the reasonable configuration that was earlier revealed, even if the data is not enough. Let $X(S, E_x)$ denote that the system is in the state S according to the *a priori* knowledge, then it computes the probability of the next step to select E.

$$X(S_i, E_x) = \frac{C_{i,E_x}}{\sum_{j=1}^{m} C_{ij}} \tag{3}$$

With the increasing number of simulations, the experimental data is gradually increasing and the reduction of the *a priori* knowledge is required to reduce the proportion of the *a priori* knowledge in the selection function. This will reduce the useless search and improve the convergence of the function, thus the choice of probability and *a priori* knowledge is proportional to the analog frequency which is inversely proportional. Let $u(S, E_x)$ denote that after the N round of simulations, the probability of the system is in the state S in the first step to choose E_x.

$$u(S, E) = \mu \frac{X(S, E)}{1 + N(S, E)} \tag{4}$$

At the end of each simulation process, the system will return a reward function based on the results of the simulation configuration. If such a configuration network is in accordance successfully, it returns back to **1**, otherwise **0**.

$$R_t(S, E_X) \begin{cases} 1 \\ 0 \end{cases} \tag{5}$$

Through constant repetition of the experiment, the reward function for 1 of the configuration is a steady stream of feedback. These "excellent" configurations were collected and used to determine the configuration of the next step with *a priori* knowledge. Let $N(S, E)$ denote the previous time step **t-1** in the simulation process in the state S

case, the number of times we will select the equipment **E** and the final equipment successfully.

$$N(S, E) = \sum_{i=1}^{t-1} L(S, E, i) R_i(S, E) \tag{6}$$

The use of reinforcement learning is to get feedback results as well as avoid information which is too much to be submerged into inspired information. Let $V(S, E_x)$ denote the probability of selecting equipment E_x under the state S, it is conducted according to the feedback of machine learning.

$$V_t(S, E) = \frac{1}{1 + N_t(S, E)} \sum_{i=1}^{t-1} L(S, E, i) R_i(S, E) \tag{7}$$

At the beginning of the simulation, the influence of *a priori* knowledge is relatively large and it can improve the convergence speed. With increasing number of simulations, the influence of the *a priori* knowledge is gradually reduced and influence of the feedback result of machine learning is increased so that the offset decreasing will make higher winning percentage allocation which is gradually displayed. Both of them are combined to get the probability choice function.

$$P(S, E) = (1 - \lambda) V(S, E) + \lambda u(S, E) \tag{8}$$

The methodology of Monte Carlo tree search with reinforcement learning can be described as follows:

Table 4. The steps of search.

Step 1	Set the number of simulations as **T**, the initial value is **1**
Step 2	Starting from the node P_1, through the probability choice function **P(S, Ex)** optimal configuration
Step 3	If P_1 is not the end node, then create the next node P_2, and select one of the configurations **Ex** according to **P(S, Ex)**
Step 4	Continuous simulation until the completion of all the node configurations, if the results of the configuration is in line with the requirements of the topic selected by the action of the reward function **Rt(S, Ex)** for 1, else 0
Step 5	According to the results of the last simulation update to the probability choice function **P(S, Ex)**, observe the simulation times **T** whether it reaches the upper limit, if not **T++** and return to **Step 1**, else the simulation is terminated

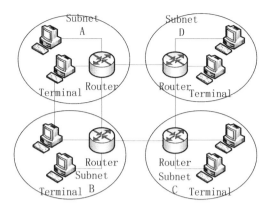

Figure 2. Network topology.

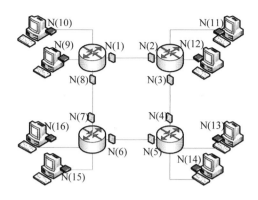

Figure 3. Schematic diagram of equipment distribution.

Table 5. Part of the results.

	N(1)	N(2)	N(3)	N(4)	N(5)	N(6)	N(7)	N(8)	N(9)	N(10)	N(11)	N(12)	N(13)	N(14)	N(15)	N(16)
1	EL1	EL1	EL1	EL1	EL1	EL1	EL2	EL2	EW1	EW1	EW3	EW3	EW2	EW2	EW3	EW3
2	EL1	EL1	EL1	EL1	EL1	EL2	EL1	EL2	EW2	EW2	EW3	EW3	EW1	EW1	EW3	EW3
3	EL1	EL1	EL1	EL1	EL2	EL1	EL1	EL2	EW1	EW2	EW3	EW3	EW2	EW1	EW3	EW3
4	EL1	EL1	EL1	EL2	EL1	EL1	EL1	EL2	EW2	EW1	EW3	EW3	EW2	EW1	EW3	EW3
5	EL1	EL1	EL2	EL1	EL1	EL1	EL1	EL2	EW1	EW2	EW3	EW3	EW1	EW2	EW3	EW3
6	EL1	EL2	EL1	EL1	EL1	EL1	EL1	EL2	EW2	EW1	EW3	EW3	EW1	EW2	EW3	EW3
7	EL2	EL1	EL1	EL1	EL1	EL1	EL1	EL2	EW3	EW3	EW1	EW1	EW3	EW3	EW2	EW2
8	EL1	EL1	EL1	EL1	EL1	EL2	EL2	EL1	EW3	EW3	EW1	EW2	EW3	EW3	EW1	EW2
9	EL1	EL1	EL1	EL1	EL2	EL1	EL2	EL1	EW3	EW3	EW1	EW2	EW3	EW3	EW2	EW1

...

4   EXPERIMENTAL RESULTS

The network topology of information system is shown in Figure 2. The network is made up of four subnetworks, each of which has two terminals. Each subnet was formatted as the backbone network.

Configuration target: The whole network is divided into two secure subnets: the implementation of the subnet **A** and subnet **C**, which is the terminal security communication. Among the subnet **B** and **D** and the subnet **A** and **C**, the subnet **B** and **D** is the terminal which is not connected.

Available security equipment: The security equipment **EL1** and **EL2** are in data link layer, each of which can communicate with others. There are **6** sets of **EL1** and **2** sets of **EL2**. Network layer security equipment is **EW1, EW2,** and **EW3**. The equipment **EW1** and **EW2** can communicate with each other, but the equipment **EW3** cannot communicate with others. There are **2** sets of **EW1, 2** sets of **EW2,** and **6** sets of **EW3**.

The name of the network node is shown in Figure 3.

Through the algorithm presented in the previous paper, it is concluded that there are 336 feasible solutions, that is, there are 336 states in the solution space and some results are shown in Table 1.

5   CONCLUSION

In this paper, the Monte Carlo search and reinforcement learning is applied to the selection of equipment configuration, which can give full play to intelligence, positive feedback, and cooperative characteristics of the algorithm. The experimental results show that the algorithm has better evolution ability and can obtain optimal allocation. In further research, the combination of two algorithms will be used to carry out the dynamic configuration of the equipment. In this way, we will make full use of the advantages of Monte Carlo search to find the solution in a wide range and this kind of algorithm will be more difficult to be strapped in the local optimum.

REFERENCES

Baird, L. (1995). *Residual Algorithms: Reinforcement Learning with Function Approximation*. Machine Learning Proceedings, 30–37.

Bellemare, M., Veness, J., & Bowling, M. (2013). *Bayesian Learning of Recursively Factored Environments*. International Conference on Machine Learning (pp.2248–2256).

Le, Q. V., Ngiam, J., Coates, A., Lahiri, A., Prochnow, B., & Ng, A. Y. (2011). *On Optimization Methods for Deep Learning*. International Conference on Machine Learning, ICML 2011, Bellevue, Washington, USA, June 28 - July (Vol.7, pp.67–105).

Metropolis, N., & Ulam, S. (1949). *The Monte Carlo Method*. Journal of the American Statistical Association, 44(247), 115–129.

Mnih, V., Kavukcuoglu, K., Silver, D., Graves, A., Antonoglou, I., & Wierstra, D., et al. (2013). *Playing Atari with Deep Reinforcement Learning*. Computer Science.

Nair, V., & Hinton, G. E. (2015). *Rectified Linear Units Improve Restricted Boltzmann Machines*. Proc Icml, 807–814.

Silver, D., Huang, A., Maddison, C. J., Guez, A., Sifre, L., & Van, d. D. G., et al. (2016). *Mastering the Game of Go with Geep Neural Networks and Tree Search*. Nature, 529(7587), 484–489.

Tianyilin. (2008). *The Research on Building Enterprise Knowledge Management Performance Evaluating Indicator System*. Modelling, Simulation and Optimization, International Workshop on (pp.25–29). IEEE.

Civil, Architecture and Environmental Engineering – Kao & Sung (Eds)
© 2017 Taylor & Francis Group, ISBN 978-1-138-02985-9

Robust automatic speech recognition based on neural network in reverberant environments

L. Bai, H.L. Li & Y.Y. He
National Computer Network Emergency Response Technical Team/Coordination Center of China, Beijing, China

ABSTRACT: The reverberant environment is still a big challenge to speech recognition. This paper presents a method of reverberant Automatic Speech Recognition (ASR) using front-end based methods and enhanced Voice Activity Detection (VAD). A 2-channel dereverberation method is adopted to achieve robust dereverberation under different reverberant conditions. Also a 2-channel spectral enhancement method is used where the gain of each frequency bin is controlled by acoustic scene, which is detected based on the analysis of full-band coherent property. We also use Deep Neural Network (DNN) as a feature extractor, and a DNN based VAD is also used to improve the ASR performance. The DNN based front-end allows a very flexible integration of meta-information. Bottle neck features are extracted in place of MFCC features used in the HMM-GMM system. Finally, we evaluate our methods on the data provided by REVERB challenge. On simulated data, the performance yields more than 33% relative reduction in Word Error Rate (WER).

1 INTRODUCTIONS

Improving the ASR performance in reverberant speech has been an important research topic for a long time. Lots of researches have proved that audio processing is helpful in improving the quality of the reverberant speech. Among the front-end signal processing technologies, three categories of dereverberation methods are generally applied: 1) beamforming using microphone arrays, 2) spectral enhancement, 3) blind system identification and inversion (Naylor & Gaubitch 2010). Spectral enhancement based dereverberation shows superiority due to its robustness in both reverberant and noisy environment (Yoshioka, Nakatani & Miyoshi 2009). Fractional time delay alignment filter is applied to the reverberant signal, and the acoustic scene is classified by analyzing the coherent component. Based on the acoustic scene, an appropriate spectral enhancing scheme is selected to eliminate the interference as much as possible while keeping the speech distortion always in a low level.

Recently, acoustic model based on the Deep Neural Networks (DNNs) has gained popularity with the consistent improvement in recognition performance over earlier Neural Network based front-ends, e.g. (Grézl *et al.* 2007). DNNs are either deployed as the front-end for standard Hidden Markov Model based on Gaussian Mixture Models (HMM-GMMs), or in a hybrid form to directly estimate state level posteriors. As noted in several publications (Larochelle *et al.* 2009, Seide *et al.* 2011, Liu *et al.* 2014), DNNs show general WER improvements on the order of 10–30% relative across a variety of small and large vocabulary tasks when compared with HMM-GMMs built on classic features (e.g. MFCC, PLP). Neural network based features have long been used successfully in speech recognition (Stolcke *et al.* 2005, Grezl *et al.* 2009, Hain *et al.* 2012). While the early research did not involve deep layers (Larochelle *et al.* 2009), the path towards deep learning was laid in the stacking of bottleneck networks (Grezl *et al.* 2009). A DNN is a conventional Multi-Layer Perceptron (MLP) with many internal or hidden layers. The BottleNeck (BN) features extracted from the internal layer with a relatively small amount of neurons have been shown to effectively improve the performance of ASR systems. It is possibly due to the limited number of units which creates a constriction in the network and further forces the information pertinent to classification into a low dimensional representation (Seide *et al.* 2011, Tüske *et al.* 2013). In many ASR systems, the neural network based features performs better than the cepstrum or spectrum based features (e.g. MFCC, PLP).

According to the REVERB challenge, the reverberant data is simulated or recorded in various rooms with different distances between source and microphones (Paul & Baker 1992, Robinson *et al.* 1995, Lincoln *et al.* 2005), and three kinds of utterance are provided: 1-channel, 2-channel and 8-channel. We choose the 2-channel dereverberation method for both Speech Enhancement (SE) task and ASR task. The work in this paper aims to extend and enhance the preliminary work in (Wang *et al.* 2015) with advanced DNN based feature extraction. We follow the standard strategies where the DNN includes one input layer, three hidden layers, another bottleneck layer and one output layer. The BN features are extracted from the bottleneck layer of DNN trained to predict the context-dependent clustered triphone states.

The rest of this paper is organized as follows. Section 2 introduces the algorithms of 2-channel dereverberation, VAD and DNN based feature extractor. Section 3 presents the experiment results. Finally, Section 4 concludes the paper and discusses the future work.

2 RECOGNITION OF REVERB SPEECH

2.1 *System overview*

Figure 1 shows the overall system proposed for REVERB challenge. It contains two basic modules, front-end module and back-end evaluation module. In the front-end module, different from our previous system, we adopt a DNN based feature extractor. For consistent comparison, we use the same recognizer provided by REVERB challenge organizer.

2.2 *Dereverberation and Noise Reduction*

Spectral enhancement methods show their priority on the robustness in the condition of both noise and reverberation. In this section, an acoustic scene aware dereverberation method is utilized for the purpose of achieving environmental adaptation (Wang *et al.* 2015). Two sensors are utilized since it is the basic topology of all microphone arrays and can be generalized to any other microphone arrays conveniently.

2.2.1 *Signal model*

Let $s(n)$ represent the target clean signal, and $x_m(n)$ is the time-aligned signal at sensor $m(m=1,2)$ $n_m(n)$ .is the noise of environment. $h_m(t)$ can be seen as the time-delayed Room Impulse Response (RIR) which is the convolution of real RIR from target signal to sensor $m(m=1,2)$ and alignment filter where the alignment filter is designed to steer the

direction of target speech signal. Then the observed signal each channel can be expressed as follows.

$$x_m(n) = h_m(t) * s(n) + n_m(n) \qquad (1)$$

Applying STFT to the 16 kHz time-aligned signal, we have sinal expression at lth frame and kth frequency bin in time-frequency domain.

$$x_m(l,k) = H_m(l,k)S(l,k) + N_m(l,k) \qquad (2)$$

2.2.2 *Spectral enhancement*

Spectral Enhancement method has a generalized form. The estimate of the amplitude spectrum of the target signal can be expressed as follows.

$$|\hat{S}(l,k)| = G(l,k)|\hat{X}(l,k)| \qquad (3)$$

$G(l,k)$ is the gain estimated on each frequency bin and $|\hat{X}(l,k)|$ is the amplitude spectrum of signal to be enhanced. Before overlap-and-add scheme, regardless of leakage between frequency bins caused by STFT, both $G(l,k)$ and $|\hat{X}(l,k)|$ should be chosen cautiously versus distortion to achieve robustness.

2.2.3 *Environmental adaptation based processing*

Reverberation, especially late reverberation, shows isotropic property as well as the environmental diffuse noise, while the direct sound shows strong coherent property (Kuttruff 2009). There are two main acoustic scenes of the room. One is the reflection condition which can be interpreted by reverberation time (T60) and the other one is the speaker-mic distance. By estimating the proportion of coherent component, the effects of the two are synthesized. All the diffuse part can be seen as noise to be filtered. We follow the Coherent-to-Diffuse energy Ratio (CDR) estimation in (Jeub *et al.* 2011), which is expressed as follows.

$$\varepsilon(e^{j\Omega}) = \frac{|\operatorname{sinc}(\Omega f_s d_{mic}/c)|^2 - |\Gamma_{x_1 x_2}(e^{j\Omega})|^2}{|\Gamma_{x_1 x_2}(e^{j\Omega})|^2 - 1} \qquad (4)$$

$\varepsilon(e^{j\Omega})$ is CDR estimator of each frequency bin and $\Gamma_{x_1 x_2}$ is the expression of complex coherence function (Mccowan & Bourlard 2003). And a Wiener filter can be formed based on the estimation of CDR which can filter the non-coherent part. According to (Jeub *et al.* 2011), it's more accurate when CDR is relatively large. Furthermore, We use global CDR, denoted by $\hat{\varepsilon}$, as a full-band acoustic scene aware controller indicating the level affected by reverberation and diffuse noise which can be interpreted as a direct sound activity detector to achieve environmental adaptation.

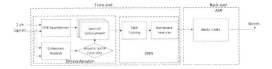

Figure 1.   System structure.

Up to now, lots of 1-channel and 2-channel dereverberation methods have proved the efficiency under the framework of spectral enhancement and achieved the robustness to noise compared with the methods using inverse filtering. Fixed beamforming, such as Delay-and-Sum Beamformer (DSB), helps to suppress the reverberation based on priori DOA information though its suppression ability is limited. Late reverberation suppressing method using generalized statistic model of reverberation (Habets *et al.* 2009) shows outstanding performance especially when reverberation is strong. Based on the controller mentioned above, we compare $\hat{\varepsilon}$ with three constants $\sigma_1, \sigma_2, \sigma_3$, (from large to small, chosen 15 dB, 10 dB, 5 dB).

Spectral enhancement strategy suggested above separates the acoustic scene into four cases. In the first case, the acoustic scene is ideal so that the speech signal recorded is very close to clean speech. In the second and third cases, a moderate trade-off between dereverberation and signal distortion is achieved. In the last case, more reverberation reduction means better performance when reverberation is strong enough.

To avoid music noise, both time recursive and adjacent frequency gain smoothing are conducted. The recovered signal is obtained by inverse STFT and overlap-and-add scheme. The phase of recovered speech signal equals the noisy phase of beamformer output. All the processing is with windows of 512 points and step-size of 256 points which means the result of a DFT with length 512 (32 ms) at a shift of 16 ms.

2.3 *DNN based VAD*

Voice Activity Detection (VAD) is important in speech processing. In the applications, the systems usually need to separate speech/non-speech parts, so that only the speech part can be dealt with. DNN, which proves its efficiency in speech recognition, has been widely used in recent years. We use simulated data to train a DNN based VAD. The input layer is the features of speech/non-speech parts, and the output layer is the posterior of speech/non-speech parts. We use Error Back Propagation (EBP) algorithm to train the model.

2.4 *DNN based features*

Figure 2 shows the architecture. It is similar to those in (Grézl *et al.* 2007, Grezl & Fousek 2008, Veselý *et al.* 2011). All DNNs are trained feedforward with the TNET[1] on GPUs. In a default

---

[1]http://speech.fit.vutbr.cz/software/neural-network-trainer-tnet

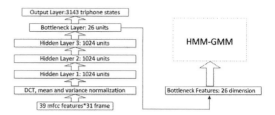

Figure 2. DNN module structure.

TNET setup, 31 adjacent frame MFCC features are decorrelated and compressed with DCT into a dimension of $624(31 \times 39 \rightarrow 16 \times 39)$. Global mean and variance normalization are performed on each dimension before feeding as the DNN input. The 6 layered DNN structure is shown in Figure 2. On average 10% data in training set is reserved for cross validation in DNN training. The training stops automatically when the improvement of frame-based target classification accuracy on cross validation set drops below 0.1%.

The bottleneck layer is placed just before the output layer, as in our initial experiments this topology gives the best performance. We set to 26 dimensions according to our experiments. DNNs are trained on classification targets of 3143 triphone states. In the bottleneck layer, linear BN features are extracted before the sigmoid activation.

3 EXPERIMENT

3.1 *Data and system set*

In the REVERB challenge, both simulated and real recorded data are provided. The simulated data (SimData) is convolved by clean utterance from WSJCAM0 corpus (Robinson *et al.*, 1995) with the recorded RIR in different rooms. The reverberation time of the rooms are 250 ms, 500 ms and 700 ms respectively. Recorded background noise is added to the reverberant data at a fixed Signal-to-Noise Ratio (SNR) of 20dB. The real recorded data (RealData), utterances from the MC-WSJ-AV corpus (Lincoln *et al.* 2005), consists of utterances recorded in a noisy and reverberant room with reverberation time of 700 ms. Both SimData and RealData include two types of distances between the speaker and microphone array (near = 50 cm and far = 200 cm). The test data are from the SimData and RealData databases under the following important assumptions. First, there is no drastic change in RIR within an utterance. Second, relative speaker-microphone position changes from utterance to utterance, which means the Direction of Arrival (DOA) of the target speech signal is uncertain, and this is essential to our dereverberation method. The recording 8-channel circular array has diameter of 20 cm and the 2-channel microphone distance, denoted by d_{mic}, which can be calculated.

Under the framework of HTK based recognizer, we retrain the acoustic model of "multi-condition" using HMMs structure and DNN respectively. The proper starting point is that the artificially distorted training signals are mismatch with the enhanced ones. Then the five possible cases are:

Clean+noEnh: "clean-condition" HMMs without dereverberation;

Multi+noEnh: "multi-condition" HMMs without dereverberation;

Clean+Enh: "clean-condition" HMMs with dereverberation;

Multi+Enh: "multi-condition" HMMs with dereverberation;

DNN+Multi+Enh: "multi-condition" HMMs with features from DNN.

3.2 *Baseline experiments*

For the ASR task, WER of test data is reported in Section 3.1. Baseline models of ASR task as well as retrained model are provided in this section. The "clean-condition" baseline system uses 39D Mel-Frequency Cepstral Coefficients (MFCCs) including Delta and Delta-Delta coefficients as features. As to acoustic models, it employs tied-state HMMs with 10 Gaussian components per state trained according to the maximum-likelihood criterion (Kinoshita *et al.*, 2013). All the training data for "clean-condition" HMMs is from WSJ-CAM0 corpus (Robinson *et al.*, 1995). Further, the model is retrained using the features of artificially distorted 7861 utterances to form the "multi-condition" HMMs. The utterances are in mixture with 24 kinds of RIRs and 6 kinds of noises. We test the enhanced signals on the two baseline systems both using and not using the unsupervised CMLLR model adaptation. We also retrain the

DNN based VAD to separate speech/non-speech parts.

Table 1 shows the WER results on SimData and RealData datasets respectively. It includes the ASR results of "clean-condition" model, "multi-condition" model. WER of near, far data and their average are reported separately. As we can see, "multi-condition" model performs better than "clean-condition" model. It achieves decrease on WER from 51.68% to 29.51% on average of SimData and from 88.53% to 56.94% on average of RealData. Consistent improvement across all recording conditions is achieved by using CMLLR which results in WER 25.25% on SimData and 48.85% on RealData.

3.3 *2-channel dereverberation*

Performance of dereverberation is examined using both "clean-condition" and "multi-condition" acoustic models. According to Table 2, recognizing the test set with "clean-condition" model without CMLLR adaptation, the dereverberation method achieves decrease on WER from 51.68% to 37.06% on average of SimData and from 88.53% to 72.62% on average of RealData. Consistent improvement across all recording conditions is achieved by using CMLLR which results in WER 27.93% on SimData and 62.12% on RealData. However, recognizing with "multi-condition" model without CMLLR adaptation, the performance (SimData: 34.22%, RealData: 59.72%) is worse than the "multi-condition" baseline (SimData: 29.51%, RealData: 56.94%) because the recognizing enhanced data is mismatched with the reverberant data used to train the "multi-condition" acoustic model, though using CMLLR gives a little improvement.

3.4 *Bottleneck features from DNN*

Table 3 shows the results for BN features based decoding. All the DNNs are trained with the

Table 1. Word error rate of baseline.

Test data		Word error rate(%)									
		SimData							RealData		
		Room 1		Room 2		Room 3			Room 1		
		Near	Far	Near	Far	Near	Far	Ave.	Near	Far	Ave.
Clean+noEnh	NOCMLLR	18.06	25.38	42.98	82.20	53.54	88.04	51.68	89.72	87.34	88.53
	CMLLR	14.81	18.86	24.63	64.58	33.77	78.42	39.16	82.31	80.76	81.53
Multi+noEnh	NOCMLLR	20.60	21.15	23.70	38.72	28.08	44.86	29.51	58.45	55.44	56.94
	CMLLR	16.23	18.71	20.50	32.47	24.76	38.88	25.25	50.14	47.57	48.85

Table 2. Word error rate of 2-channel dereverberation.

Test data		Word error rate(%)									
		SimData							RealData		
		Room 1		Room 2		Room 3			Room 1		
		Near	Far	Near	Far	Near	Far	Ave.	Near	Far	Ave.
Clean+noEnh	NOCMLLR	17.43	25.25	27.85	49.48	36.51	65.94	37.06	73.91	71.34	72.62
	CMLLR	14.47	19.47	21.19	34.86	27.16	50.50	27.93	62.66	61.58	62.12
Multi+noEnh	NOCMLLR	23.64	36.46	27.72	37.69	34.00	45.85	34.22	59.95	59.49	59.72
	CMLLR	16.93	20.04	19.91	26.84	23.95	34.33	23.66	44.87	45.81	45.34

Table 3. Word error rate of bottleneck features from DNN.

Test data		Word error rate(%)									
		SimData							RealData		
		Room 1		Room 2		Room 3			Room 1		
		Near	Far	Near	Far	Near	Far	Ave.	Near	Far	Ave.
DNN+Multi+Enh	NOCMLLR	17.67	23.36	23.56	29.39	30.35	39.36	27.28	59.60	58.71	59.16
	CMLLR	14.72	19.23	19.95	25.07	25.42	34.02	23.07	54.52	54.62	54.57

triphone state targets force-aligned using training set. The decoding is performed on the test set. We achieve different performance without CMLLR and with CMLLR. On "Multi+Enh" case, we get 33.43% relative decrease on SimData using BN features without CMLLR. But we only get 2.49% relative decrease with CMLLR. That is because we only use DNN as a front-end. During decoding process, we use the HMM-GMM structure of baseline system. CMLLR may not be so work for the DNN features. Another reason lies the feature dimensions, we use 39-dimension MFCC features for our baseline system, but only 26-dimension BN features adopted in this paper.

4 CONCLUSION

We have presented out a dereverberation approach to the REVERB challenge based on spectral enhancement. An acoustic scene aware technique is proposed to make dereverberation robust to different conditions. For ASR task, when it is combined with back-end ASR with matched training, it produces a significant decrease on WER.

The DNN based front-end was tested in the context of reverberant speech recognition. Experiments showed that on SimData, BN features gave an average 33.43% relative WER reduction over using MFCC features without CMLLR. Retrain-

ing did not give as much gain as we got on MFCC features. We also need to do further experiments on RealData about the performance difference between MFCC features and BN features.

REFERENCES

Grezl, F. & Fousek, P. 2008. Optimizing bottle-neck features for LVCSR. In: *2008 IEEE International Conference on Acoustics, Speech and Signal Processing, Las Vegas*, pp. 4729–4732.

Grezl, F., Karafiat, M. & Burget, L. 2009. Investigation into bottle-neck features for meeting speech recognition. In: *10th Annual Conference of the International Speech Communication Association, Brighton*, pp. 2947–2950.

Grézl, F., Karafiát, M., Kontár, S. & Cernocky, J. 2007. Probabilistic and bottle-neck features for LVCSR of meetings. In: *2007 IEEE International Conference on Acoustics, Speech and Signal Processing, Honolulu*, pp. IV757–760.

Habets, E.A.P., Gannot, S. & Cohen, I. 2009. Late reverberant spectral variance estimation based on a statistical model. *IEEE Signal Processing Letters. 16*, 770–773.

Hain, T., Burget, L., Dines, J., Garner, P.N., Grézl, F., El Hannani, A., Huijbregts, M., Karafiat, M., Lincoln, M. & Wan, V. 2012. Transcribing meetings with the amida systems. *IEEE Transactions on Audio, Speech, and Language Processing. 20*, 486–498.

Jeub, M., Nelke, C., Beaugeant, C. & Vary, P. 2011. Blind estimation of the coherent-to-diffuse energy ratio from noisy speech signals. In: *Signal Processing Conference, 2011 19th European, Barcelona*, pp. 1347–1351.

Kinoshita, K., Delcroix, M., Yoshioka, T., Nakatani, T., Sehr, A., Kellermann, W. & Maas, R. 2013. The reverb challenge: a common evaluation framework for dereverberation and recognition of reverberant speech. In: *2013 IEEE Workshop on Applications of Signal Processing to Audio and Acoustics, New Paltz*, pp. 1–4.

Kuttruff, H. 2009. *Room acoustics*. Boca Raton: CRC Press.

Larochelle, H., Bengio, Y., Louradour, J. & Lamblin, P. 2009. Exploring strategies for training deep neural networks. *Journal of Machine Learning Research. 10*, 1–40.

Lincoln, M., Mccowan, I., Vepa, J. & Maganti, H.K. 2005. The multi-channel wall street journal audio visual corpus (mc-wsj-av): specification and initial experiments. In: *2005 IEEE Workshop on Automatic Speech Recognition and Understanding, San Juan*, pp. 357–362.

Liu, Y.L., Zhang, P.Y. & Hain, T. 2014. Using neural network front-ends on far field multiple microphones based speech recognition. In: *2014 IEEE International Conference on Acoustics, Speech and Signal Processing, Brisbane*, pp. 5542–5546.

Mccowan, I.A. & Bourlard, H. 2003. Microphone array post-filter based on noise field coherence. *IEEE Transactions on Speech and Audio Processing. 11*, 709–716.

Naylor, P.A. & Gaubitch, N.D. 2010. *Speech dereverberation*. Springer Science & Business Media.

Paul, D.B. & Baker, J.M. 1992. The design for the wall street journal-based csr corpus. In: *Proceedings of the workshop on Speech and Natural Language, Morristown*, pp. 357–362.

Robinson, T., Fransen, J., Pye, D., Foote, J. & Renals, S. 1995. Wsjcamo: a british English speech corpus for large vocabulary continuous speech recognition. In: *1995 IEEE International Conference on Acoustics, Speech, and Signal Processing, Detroit*, pp. 81–84.

Seide, F., Li, G. & Yu, D. 2011. Conversational speech transcription using context-dependent deep neural networks. In: *12th Annual Conference of the International Speech Communication Association, Florence*, pp. 437–440.

Stolcke, A., Anguera, X., Boakye, K., çetin, Ö., Grézl, F., Janin, A., Mandal, A., Peskin, B., Wooters, C. & Zheng, J. 2005. Further progress in meeting recognition: the ICSI-SRI spring 2005 speech-to-text evaluation system. In: *International Workshop on Machine Learning for Multimodal Interaction, Edinburgh*, pp. 463–475. Heidelberg: Springer.

Tüske, Z., Schlüter, R. & Ney, H. 2013. Deep hierarchical bottleneck MRASTA features for LVCSR. In: *2013 IEEE International Conference on Acoustics, Speech and Signal Processing, Vancouver*, pp. 6970–6974.

Veselý, K., Karafiát, M. & Grézl, F. 2011. Convolutive bottleneck network features for LVCSR. In: *2011 IEEE Workshop on Automatic Speech Recognition and Understanding, Waikoloa*, pp. 42–47.

Wang, X.F., Guo, Y.M., Ge, F.P., Wu, C., Fu, Q. & Yan, Y.H. 2015. Speech-picking for speech systems with auditory attention ability. *Scientia Sinica Informationis. 45*, 1310.

Yoshioka, T., Nakatani, T. & Miyoshi, M. 2009. Integrated speech enhancement method using noise suppression and dereverberation. *IEEE Transactions on Audio, Speech, and Language Processing. 17*, 231–246.

Civil, Architecture and Environmental Engineering – Kao & Sung (Eds)
© 2017 Taylor & Francis Group, ISBN 978-1-138-02985-9

Application of association rule mining in electricity commerce

Han Yang
Department of Science and Engineering, Communication University of China, Beijing, China

Yecheng Pu
School of Business Administration, Dongbei University of Finance and Economics, Dalian, China

ABSTRACT: This paper proposed a new view of the application of data mining analysis in electricity commerce, and examined how to use k-means cluster and association rule mining as an entity to accurately explore the association rules between varieties of indexes, with whatever quantitative data or qualitative data. The online shops' indexes like product ratings, percentages of delivery speed, ranges among the whole industry, and regional distribution were first classified and clustered, and then applied or mined by Apriori algorithm, a method of association rule mining. The results with a comprehensive analysis will be helpful not only in instructing the merchants who run an online shop, but also in offering valuable information to the investigators related to electricity commerce.

1 MODELS AND BASIC ALGORITHMS

1.1 Cluster analysis

A cluster is dividing the dataset into multiple classes or clusters. The data points set are as similar as possible in the same cluster and as far as possible in a different cluster. Cluster analysis, as an important part of statistics and the related research, has a long history and is an important part of data mining and pattern recognition.

The cluster is applied in business most typically for it can help market analysts find different kinds of customers from the database and depict characteristics of different customer base by buying patterns. There are some other applications of the cluster including speech recognition and character recognition in pattern recognition, the data compression in image segmentation, image processing and information retrieval, genetic classification of biology, etc. The commonly used algorithms like K-means algorithm, DBSCAN algorithm, CURE algorithm, and COBWEB algorithm are applied for discriminative situations and conditions, especially on data. Due to a large amount of data in this paper, the accuracy of the cluster is not high, so K-means cluster was chosen here.

1.2 K-means cluster

The basic idea of K-means algorithm is to classify the objects nearest to them by clustering the k points in space. The iteration method is used to update the value of each cluster center until the best clustering result is obtained.

Assuming that the sample is divided into c categories, the algorithm is described as follows:

1. The initial center of c classes is chosen appropriately.
2. In the "k" th iteration for any sample find its distance, which is usually the Markov Distance to the c centers and classify the sample into the class where the center with the shortest distance to the sample locates.
3. The center value of the class is updated by the rest samples' average and the like.
4. For all c cluster centers if the value is not changing, then the iteration ends. Otherwise, the iteration is continued.

1.3 Association rule analysis

Association rule analysis was invented initially for the retail industry, shopping basket, in particular, to find the relationship between the merchandise of consumers and discover the association rules. This kind of analysis benefits the retail industry to appoint a marketing strategy. As a result of the relevance analysis having a wide range of applicability, many researchers have done in-depth research on association analysis including optimizing the algorithm and improving the efficiency of the algorithm. The improved algorithm for association rules has been proposed ever since.

Association analysis aims to find association rules between data. For example, the customers in a supermarket who purchase A often buy B at the same time. That is $A => B$, which is treated as an association rule. In 1993, *Agrawal et al.* firstly proposed mining association rules among items in the customer transaction database. The formal description is as follows.

Let I be a set of m different data items where the elements are called terms, the collection of items is called an item set, an item set containing k items is called a k-item set. Given a transaction D, that is the transaction database where each transaction T is a subset of data items, that is, $T \subseteq I$, there is a unique identifier TID. The transaction T includes the set X if and only if $X \subseteq T$. Then the association rule is an implication of "X = > Y". As for $X \subseteq I, Y \subseteq I, X \cap Y = \emptyset$, the records that satisfy the condition in X must also satisfy the Y as per association rule X = > Y in the transaction database, with support S and confidence C.

The transaction dataset D has a support degree S, that is, D, at least S% of the transaction contains $(X \cup Y)$, described as support $(X => Y) = P(X \cup Y)$. It appears that support (X = > Y) equals that of the number of simultaneous purchases of goods X and Y plus the total number of transactions at the same time in transaction dataset D with confidence C, that is, the items included in the transaction D contains at least C% of Y at the same time, which is described as confidence (X = > Y) = P (Y|X). It is like when someone purchases the goods X and the possibilities for purchasing goods Y, and confidence (X = > Y) equals the number of transactions in which the goods X and Y are purchased at the same time, plus the number of transactions in which the goods X is purchased.

Association rule mining is to mine the transaction database on the basis of the given "minimum support" and "minimum confidence" in order to find the strongest association rules which satisfy minimum support and minimum confidence. The mining of association rules includes one-dimensional Boolean association rule analysis, multi-level association rule mining, and constraint-based association rule mining. The typical algorithm of one-dimensional Boolean association rule analysis is Apriori algorithm. The core idea of this algorithm is to find the frequent item set firstly and the association rules can be generated by satisfying the minimum support and minimum confidence condition. The generation of frequent item sets is an iterative process. In the process of finding k frequent item sets every time a large item set is selected from a candidate set, the database will be scanned once, which greatly reduces the efficiency of the algorithm. If the database is a large database, each candidate set is very large and the deficiency of the algorithm is more obvious. Thus, how to improve the validity of Apriori algorithm is a key problem. There are a lot of revised versions of Apriori algorithms designed mainly to improve the efficiency of the original algorithm. The improved methods include compressing transaction sampling and the like to minimize the number of scans and number of scanning transactions.

1.4 Apriori algorithm

The property of Apriori (Combined by "a" and "priori") algorithm is called the priori principle that if an item set is frequent, then all of its non-empty subsets must also be frequent. Conversely, if an item set is infrequent, then all its supersets must be infrequent. If A does not satisfy the minimum support threshold s, then A is infrequent. If P (A) adds b to A, then the support of the resulting item set {A, b} will be less than A's support, {A, b} is not frequent.

The principle of the connection step: Suppose the element in the items and the items in the item set are sorted by the dictionary, if the first k-2 elements of I1 and I2 in the k-1 frequent item set are the same, then I1 and I2 can be connected to generate the new item set Ck.

Pruning principle: Ck is a superset of Ik, and scans the count of each item in the Ck in the total database to calculate the support. If it is greater than the support for the threshold, then it is the superset. If not, then it is discarded. If the number of item sets generated is too excessive to compute then we can use the Apriori property in this step: If the subset k-1 of the k item set is not included in the frequent k-1 item set, the candidate will not be frequent and will be deleted directly.

Algorithm steps:

Input: transaction database D, minimum support threshold.

Run the steps as:

1. Scan the database and generate C1 in a set of items.
2. Count prune C1 item set and then get frequent I1 item set.
3. According to the frequent I1 item set, the candidate item set C2 is generated.
4. Scan the database and discard the undercounted part from C2 to generate I2.
5. Use the Apriori property to generate and simultaneously prune (cut off the infrequent) getting C3.
6. Scan the database and discard the undercounted portion of C3 to generate I3; (n) Output: the final frequent item set according to the threshold.

2 EMPIRICAL ANALYSIS

2.1 About the data

The dataset includes 491 Taobao (a Chinese electronic commerce company like Amazon) electricity business shop information, mainly in the mobile phone industry. Shop information or indexes include shop name, the industry, the company, the punishment ratings, product rating, bad ratings, delivery speed scores, praises, dispute rates, credit rankings, and so on. The data coverage time is from June 18, 2008, to September 26, 2014. A part of the source data is in Table 1.

2.2 Data preprocessing

First of all, we removed some of the indexes that are not so significant such as shop creation time, store-owned platform, store name, etc., and then pre-processed the data. The data preprocessing includes judgment of the missing value and process of discretizing the various numerical indexes. For the shop data with a huge number of missing values, we discarded it directly. For shop data with a small number of missing values, we treated the missing value as 0 or the global average. The continuous data discretization is also divided into two categories:

1. For more evenly distributed data, we selected the quartile for the boundaries and the data divided into high, fair, acceptable, and low as four levels.
2. For data with different distributions and different levels, we used K-means clustering method to select the appropriate cluster number by several attempts and divide them into suitable n clusters (n is mainly 3, 4, 5). As shown in the following Table 2, it is part of the data preprocessing results.

2.3 Empirical analysis and visualization

There were 46,653 effective rules generated after initial mining. Some representative association rules are shown in Table 3. For the 46,653 valid rules, they revealed some negative correlation characteristics between the distribution of support and confidence shown in Figure 1.

Table 1. A small part of the source data.

create_date	dsr_zl_ grade_ percent_ 5 stars	seller_title	shop_id
2014/5/20	0.9474	GZ Telecom flagship store	110999741
2011/10/17	0.9382	HLJ Telecom flagship store	69265553
2011/11/14	0.931	SX Telecom flagship store	69643988
2013/9/10	0.8944	EP Official flagship store	106319694

Table 2. A small part of the preprocessed data.

dispute_rate	refund_rate	seller_dsr_fw	avgRefund
mild	easy	fair	fair
Fair	hard	high	fast
mild	hard	high	fast
mild	easy	fair	fast
mild	hard	fair	fast
mild	very hard	fair	fast
mild	very hard	high	fair

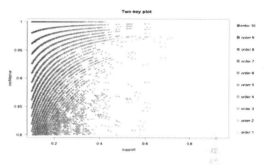

Figure 1. Distribution of initial mining support and confidence.

From Table 2 we can see some strong association rules:

1. For merchants with a low or even general delivery speed rating and low description rating, their shipments are ranked lower in the same industry. What's more, most of this part of the merchants are from the Taobao platform (type = b). These problems even appear in the sellers with few industry disputes.
2. For some of the sellers with a relatively low score in the index "conformity in products and description" and a less than general score in delivery rating in the same industry ranked lower or even the general ranking of the seller, they often have problems like low delivery speed scores and a part of the merchants with a low success rate of refund.

Then, 46,653 rules were selected for support and confidence, and 152 association rules with the highest degree of support and confidence were selected. As can be seen, the strong association rules of this 152 gather into a number of shops with lower quality scores, slower shipments, and lower rankings in the industry, as shown in Figure 2.

As can be seen from Figure 2, the shop with a relatively strong correlation is the industry-level ranking. One with the poor speed of delivery in the industry ranks and poor product ratings in the industry ranks often comes with the high industry

Table 3. A representative part of the association rule analysis results.

Lhs		rhs	support	confidence	lift
{dsr_fw_com=lower}	=>	{seller_dsr_fh_com=lower}	0.60	1.00	1.36
{seller_dsr_fw=fair}	=>	{seller_dsr_fh_com=lower}	0.54	1.00	1.36
{seller_dsr_fw=fair}	=>	{dsr_fw_com=lower}	0.54	0.99	1.58
{seller_dsr_fw=fair, dsr_fw_com=lower}	=>	{seller_dsr_fh_com=lower}	0.54	1.00	1.35
{seller_dsr_fh_com=lower, seller_dsr_fw=fair}	=>	{dsr_fw_com=lower}	0.54	0.99	1.58
{seller_dsr_fw=fair, dsr_zl_com=lower}	=>	{seller_dsr_fh_com=lower}	0.54	1.00	1.35
{seller_dsr_fw=fair, dsr_zl_com=lower}	=>	{dsr_fw_com=lower}	0.53	0.99	1.58
{seller_dsr_fw=fair, dsr_zl_com=lower, {dsr_fw_com=lower}	=>	{seller_dsr_fh_com=lower}	0.53	1.00	1.35
{seller_dsr_fh_com=lower, seller_dsr_fw=fair, dsr_zl_com=lower}	=>	{dsr_fw_com=lower}	0.53	0.99	1.58
{seller_user_type=b, seller_dsr_fw=fair}	=>	{seller_dsr_fh_com=lower}	0.53	1.00	1.35
{dsr_fw_com=lower}	=>	{seller_dsr_fh_com=lower}	0.43	1.00	1.35
{seller_dsr_zl_grade=bad}	=>	{seller_dsr_fh_com=lower}	0.42	1.00	1.35
{seller_dsr_zl_grade=bad, dsr_zl_com=lower}	=>	{seller_dsr_fh_com=lower}	0.42	1.00	1.36
{seller_dsr_zl_grade=bad, seller_dsr_fh_com=lower}	=>	{dsr_zl_com=lower}	0.42	1.00	1.38
{dsr_zl_com=lower}	=>	{dsr_fw_com=lower}	0.52	0.99	1.58

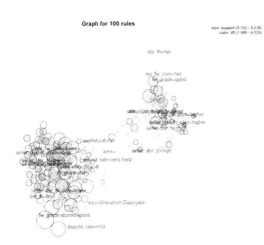

Figure 2. The rule distribution after screening.

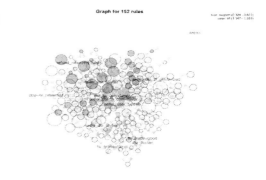

Figure 3. Regular distribution after secondary screening.

dispute rate and a slow refund rate. At the same time, we can also know, for sellers with "general" speed of delivery, the refund rate is often slow. Sometimes there are low scores towards their goods.

For those association rules of stores with good or at least general performance in product ratings, delivery speed, and industry ranking, by reducing the support we can obtain a certain degree of association with little weaker association rules and some more information about this type of store as shown in Figure 3.

What can be found from Figure 3 is that a number of sellers with fast delivery and delivery of

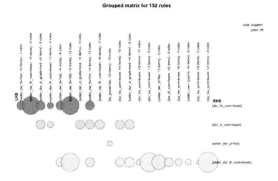

Figure 4. Strong association rules.

high ratings often get a high score in the "product and description consistent". At the same time, for some merchants with low rates of industry disputes, there is often a "shipping issue is controversial" issue occurred. Such merchants also make up a part of merchants located in Shenzhen, Guangdong Province, which can seem as a characteristic of the region. While in Shenzhen, Guangdong Province, the merchants also have a low delivery score, lower shipments in the industry rankings, which is just a conformity with the former conclusion.

3 CONCLUSIONS

3.1 *Text and indenting*

With the previous analysis, we conclude that:

1. Based on the characteristics of the source data, for most of the linear data, there is often an uneven distribution of the situation, especially because of the prevalent "Brush" phenomenon (merchants paid for increasing illusory high ratings and obliterating relatively lower ratings) in the industry, leading data distribution to be like the one with heavy head but light feet. For such data, if one wants to explore the correlation strength of indexes, the traditional contingency analysis or regression methods may cause something prone to the characteristics of the poor degree of fit, error, and poor predictive. The "cluster-association rule analysis" method used in this paper can enlarge the difference between data and make the distribution of processed data more even. The result is convincing and the error is controllable.
2. The relatively stronger association rules exist in the merchants or shops whose overall strength is considerably poor (shown in Figure 4). Thus, the problems of these merchants are much exhaustive like low product scores, slow delivery speed, low dispute rate, behindhand industry rankings, and poor refund performance, the competitiveness of which is the overall backwardness of the merchants. If they want to improve the business situation, they need to start with the whole system from product to shipment to solve all kinds of problems after the sale.
3. For those merchants with a high index such as those with high product ratings or fast delivery or good refunds, the association rules for each of these metrics are weaker indicating that for merchants of the industry when an index with outstanding performance occurs, the performance of other indexes are often not satisfactory. For these shops, the improvement needed is to avoid weaknesses and at the same time, without ignoring the "bucket effect" of the existence. Their inherent impressive performance need to be maintained. Relatively poor areas need to be further strengthened in order to search for diversified development.
4. As for the association rules for regional distribution, the most prominent one is the seller in Shenzhen, Guangdong, whose overall market performance is not high. It may be related to the development of the region itself, and merchants and governments in the region need to reflect on this.

As for the mining of association rules in this paper, the following problems exist:

1. Although there are 46,653 effective association rules in this paper, most of them have little-referenced value. For example, "some of the merchants with low score of delivery often come with low-ranking shipments in the industry", this kind of association rules that can consider as "nonsense with obvious truth" often takes a large part, and a great part of them are in the strong association rules, which makes the selection of the result very difficult.
2. The Apriori algorithm used in this paper is the most basic one. It has some shortcomings such as "needs to scan the database for many times which may waste too much time unnecessarily", "generates a large number of candidate sets" and "too much computation". There is a certain amount of space for further optimization.

REFERENCES

Tan P N, Steinbach M, Kumar V. Introduction to Data Mining: Global Edition, 2/E [M]. Pearson Schweiz Ag.

Wang A P, Wang Z F, Tao S G, et al. Common Algorithms of Association Rules Mining in Data Mining [J]. Computer Technology & Development, 2010(4): 105–108.

Wu X, Kumar V, Ross Quinlan J, et al. Top 10 Algorithms In Data Mining [J]. Knowledge and Information Systems, 2008, 14(1): 1–37.

Zhang B, Wu Z. Investigating on China's stock market volatility and macroeconomic Indexes [J]. Journal Of WUT (Information & Management Engineering), 2014(6): 843–847.

Zhao Y. R and Data Mining: Examples and Case Studies [M]. 2012.

Civil, Architecture and Environmental Engineering – Kao & Sung (Eds)
© 2017 Taylor & Francis Group, ISBN 978-1-138-02985-9

Research on a non-guided link discovery technology to obtain an early warning of financial risk

Chunyan Xue
Tan Kah Kee College, Xiamen University, Fujian Sheng, China

ABSTRACT: Early warning of financial risk is an important research direction in financial data mining. Financial risk warning is difficult because financial data are various, complex, and dynamic. Using data mining technology can discover the hidden abnormal transaction information from massive financial data, and then the technology can monitor and deal with it a timely manner. Early warning of financial risk can reduce the operational risk of financial institutions effectively. This paper introduces a method of discovering suspicious transaction information at risk with early warning. The method is based on the non-instructional link discovery technology in data mining. It can realize the hidden information found in massive data. The method first constructs the financial transaction network. Second, it finds the target node and path of the transaction, and then filters them. Then it calculates the transaction frequency of the selected path. Finally, it finds the outlier based on the distance-based outlier detection algorithm. These outliers are the suspicious transaction information. By processing these information, financial institutions can avoid financial risks.

1 INTRODUCTION

One of the main economic functions of the financial sector is to reduce and control the risks to investors and financiers in the process of social financing at an appropriate level. Early warning of financial risk refers to the supervision of the financial operation process and the warning on test results.

The system analyzes customer data including customer financial data, the credit contract information, the capital account transaction, the external judicial information, the financial information, and so on. Through the analysis, the system can detect suspicious transaction information and predict which customers in the future may be transformed into a bad customer.

Discovery of the suspicious transaction information is the most important part of the whole risk early warning system because bad customers' information is often more "alternative" than the normal customers. So it can infer the bad financial customers by discovering these "alternative" transaction information.

There are three main methods of detecting suspicious transaction information in the risk early warning system:

1. It uses manual calculation and manual monitoring to find the suspicious transaction information;
2. According to expert experience to solidify the knowledge into the model and then use the model to find the suspicious transaction information;
3. Using machine learning techniques to identify suspicious transaction information intelligently.

However, these three methods have their own shortcomings.

This paper presents a method for discovering suspicious transaction information in the financial risk early warning system. The idea is to discover non-coherent link discovery technology in data mining. The purpose of this method is to discover hidden information of abnormal data from massive data.

2 THE COMPOSITION AND STRUCTURE OF THE SYSTEM

The system has six modules:

1. Establishment of the transaction network module: the transaction data in the database are recorded separately in order to reflect the relationship between various transaction nodes and records. The transaction data in this module include account, transaction amount, an organization of account, transaction time, trading location, trading network model, and the link between them. The module will use this data graph theory to build a trading network map.
2. Discovery of the target node and transaction path module: This module is based on the transaction network module. The source node is the

given suspicious account. Then the module will find all the target nodes which have a direct relationship or indirect relationship with the source node and the transaction path between them;

3. Filter the target node and transaction path module: This module will provide five types of rules to filter the mass target node prior;
4. Extraction of the transaction path type module: The module's main function is to extract the path type from the filtered transaction path. It prepares for the frequency calculation module;
5. Frequency calculation modules: the module's main purpose is to quantify effect value between the goal node and each type of path. The value is the "frequency".
6. Discovery of suspicious transaction information module: Calculate the frequency between all target nodes and the transaction path and then construct all the frequencies as eigenvectors. The outlier detection algorithm is then used based on the distance to find outliers. The outliers are the suspicious transaction information that the users are interested in.

3 THE APPLICATION OF NON-GUIDED LINK DISCOVERY TECHNOLOGY IN THE PROCESS OF GAINING EARLY WARNING OF FINANCIAL RISK

3.1 *Non-guidance link discovery technology*

Non-guidance link discovery method is the expansion of link analysis method in data mining. The method builds some disconnected objects into a network. It finds useful patterns and trends in the network. The whole process is divided into the following three stages: to set up a model stage, abnormal trading path discovery, and abnormal trading node discovery. The key to the study is to determine the object type in the transaction model and the relationship between them.

3.2 *Financial risk early-warning process*

3.2.1 *Setting up the network module*
The graph G (V, E) represents the trading network. V is the nodes set. There are five type of nodes in the nodes set, including accounts, transaction amount, accounts organization, trading time, and trading locations. E is the edges set. The edges are the relationship among these six types of nodes shown in Figure 1.

Node type in the trading network:

- Pi account.
- Di transaction time.
- Ti virtual node represents one transaction.
- Si transaction amount.

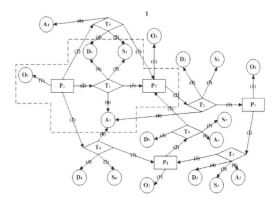

Figure 1.   Trading network diagram.

- Oi account organization.
- Ai transaction locations.

3.2.2 *Target node and transaction path discovery module*
The main function of this module is to select the account from the bad customer account database as the source node in the transaction network diagram and then use the directed graph traversal algorithm to discover the target nodes connected to the source node and all the transaction paths between them.

For example, account P1 is the source node, the target node which connected with P1 and all transactions path are shown in Table 1.

3.2.3 *Target node and transaction path filter module*
Since the number of object nodes involved in the transaction network graph is huge, when the source node is given, the amount of target node and transaction path data obtained from the above module will be huge too. It will affect the computational efficiency of subsequent modules. According to business domain knowledge and expert experience, some information out of range can be filtered in advance, so this module defines five filtering rules to filter the target nodes:

- The amount filter rules M: When the amount node M_i ($0 \leq i \leq n$) is the destination node, it will reserve the nodes whose amount is greater than or equal to M, and the others will be filtered.
- Time filtering rules (T_m, T_n): When time node T_i ($0 \leq i \leq n$) is the target node, it will reserve the nodes in the period in (T_m, T_n).
- Place filter rules {A_m}: When the location node A_i ($0 \leq i \leq n$) is the destination node, the nodes that are in the set{A_m} will be filtered.
- Filter rules to account organization {O_m}: When the account organization node O_i ($0 \leq i \leq n$) is

the destination node, the nodes who are in the set{O_m}will be filtered.

- Account filtering rules {P_m}: When the account node P_i ($0 \leq i \leq n$) is the destination node, the nodes who are in the set{P_m} will be filtered.

3.2.4 *Trading path type extraction module*

The main function of the module is to extract the transaction path according to the types. Since every node in the transaction model has a type, the transaction path composed of the nodes will also form different transaction path types. All paths in Table 1 can be merged into four types, as given in Table 2.

3.2.5 *Frequency calculation module*

Frequency is actually a relative measurement. It is established on the basis of different path types. It indicates that for the exchange path type tp, if S is the source node, count the probability of T which is the destination node". Frequency formula:

$$Frequency = N1/N2 \qquad (1)$$

N1 represents the total number of transaction paths between the given source node S and the target node T, which satisfies the path type tp;

Table 1. The path table as the source node is P1.

Source node-> Target node	Serial number	Trading path
P1→P2	Path 1	P1->T1->P2
	Path 2	P1->T5->P2
	Path 3	P1->T4->P4->T5->P2
	Path 4	P1->T1->P2->T2->P3-> T3->P4->T5->P2
P1→P3	Path 6	P1->T1->P2->T2->P3
P1→P4	Path 7	P1->T4->P4
	Path 8	P1->T1->P2->T2->P3->T3->P4

N2 represents the total number of transaction paths between the source node S and any destination nodes, which all match the path type tp.

According to the above description of the significance of frequency N1/N2, we can get a conclusion: when a source node terminates at a target node with a high probability, it means there are frequent trading path types between the source nodes and target nodes. So we think there is a close relationship between the source node and target node.

According to Figure 1, Table 1, and Table 2, N1, N2, and frequency values are calculated and given in Table 3.

3.2.6 *Obtaining suspicious trading information*

Quantification of the target nodes is the inputting condition. Outlier set {N} are obtained by using the outlier mining algorithm based on distance. The final result is suspicious transaction information that the users are interested in.

Outlier mining algorithm based on distance: If the data set {D} object has at least pct parts that the distance to the objects O is bigger than

Table 3. Frequency value table.

Node	Path type	N1	N2	Frequency
P1→P2	Type 1	2	3	2/3
	Type 2	1	2	1/2
	Type 3	0	1	0
	Type 4	1	1	1
P1→P3	Type 1	0	3	0
	Type 2	1	2	1/2
	Type 3	0	1	0
	Type 4	0	1	0
P1→P2	Type 1	1	3	1/3
	Type 2	0	2	0
	Type 3	1	1	1
	Type 4	0	1	0

Table 2. Path type whose source node is P1.

Path	Corresponding transaction path type	Merge paths
Path 1	Account->Virtual node->Account	1. Account->Virtual node->Account number
Path 2	Account->Virtual node->Account	2. Account->Virtual node->Account->Virtual node->Account number
Path 3	Account->Virtual node->Account-> Virtual node->Account	3. Account->Virtual node->Account->Virtual node->Account->Virtual node->Account number
Path 4	Account->Virtual node->Account-> Virtual node->Account->Virtual node->Virtual node->Account	4. Account->Virtual node->Accounts-> Virtual node->Account->Virtual node-> Account->Virtual node->Account number
Path 6	Account->Virtual node->Account-> Virtual node->Account;	
Path 7	Account->Virtual node->Account	
Path 8	Account->Virtual node->Account-> Virtual node->Account->Virtual node->Account	

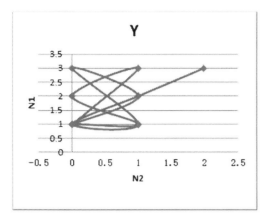

Figure 2.   Outlier point based on distance.

the dmin. So object O has outlier points based on parameters pct and dmin. In Figure 2, point A is the relative outlier to the other points. The data in Figure 2 is based on data in Table 3.

4   CONCLUSION

Non-guiding link discovery technology is efficient, time-saving, and labor-saving. The application of this technology can handle large amounts of data and improves the processing efficiency of risk information. The method can ensure finding suspicious transaction information results in a timely manner. The method reduces business manual calculation and time of manual monitoring. The technology is not based on expert experience entirely and ensures objectivity. It can find the unknown mode of suspicious transaction information.

REFERENCES

Aggarwal C, Yu P: Outlier Detection for High Dimensional Data [A]. ACM SIGM, 2001.

Jiawei Han, Michel Kamber, Jian Pei: Data Mining Concepts and Techniques Third Edition, Machinery Industry Press, 2012.

Liang Xun: Data Mining Algorithms and Application [M], Xiamen University Pressing, 2011.

Liu Fang, Fu Feng: Computer Engineering and Science, 2007, 29 (6): 77–80.

Ramaswamy S, Rastogi R: Efficient Algorithms for Mining Outliers from Large Date set. ACM SIGM, 2003.

Civil, Architecture and Environmental Engineering – Kao & Sung (Eds)
© 2017 Taylor & Francis Group, ISBN 978-1-138-02985-9

Effect of different operating media on the PVT system

Wei Pang, Yongzhe Zhang & Hui Yan
The College of Materials Science and Engineering, Beijing University of Technology, Beijing, China

Yu Liu
Chengdu Green Energy and Green Manufacturing Technology R&D Centre, Chengdu, China

Yanan Cui
National Research Center of Testing Techniques for Building Materials, National Center of Quality Supervision and Test for Building Materials, Beijing, China

ABSTRACT: The concept of nanofluid was forward and the innovative PhotoVoltaic-Thermal (PVT) systems design using nanofluid as the heat transfer medium was carried out. The overall efficiencies achieved were over 5% higher than the conventional media, water and air, due to the higher heat capacity of nanofluid. The phase change material was used to store latent heat, which was different from sensible heat. The thermal efficiency of the PVT system was further improved. This article gives a review of the efficiencies of the PVT systems with different media, including air, water, nanofluids, phase change materials, and other hybrid media. The performance of the PVT system was evaluated.

1 INTRODUCTION

Renewable energy is widely advocated by many countries in order to deal with the energy crisis. Solar energy is one of the most important resources because it is abundant, clean, non-toxic, etc. Photovoltaic modules (PV) are composed of series-parallel multiples of solar cells, which can convert solar energy directly into electricity and storage, but the major concern for developers and users is the temperature effect, that is, the output decline of PV with their temperature increasing during operation. Skoplaki et al. (Skoplaki E, 2009) had reviewed a large amount of literature and summarized the relationship value of ~0.5% between temperature and electrical output of silicon PV cells. Silicon PV panels are the protagonists in practical applications. Therefore, cooling the PV is necessary. The passive cooling method is commonly used through the setting fluid channel on the back of the PV to take away the heat and reduce its temperature. A hybrid PhotoVoltaic/Thermal (PVT) integration system (Chow T. T, 2010) arises at the historic moment, which can not only improve the efficiency of power generation but also make full use of heat. Because of the different heating methods, the PVT systems can be a flat plate or concentrated. They are also classified according to the type of working fluid used for PV cooling, including air (Tonui J. K, 2007), water (He W, 2006 & Tyagi V. V, 2012), nanofluid

(Chandrasekar M, 2010), phase change material (Tian Y, 2013), and so on. This paper focuses on the performances of the flat-plate PVT systems with different media and evaluates the advantages and disadvantages of them.

2 PERFORMANCE ANALYSIS OF A FLAT-PLATE PVT SYSTEM

The general structure of a flat-plate PVT system is shown in Figure 1. It mainly contains six parts: front cover, encapsulated PV, absorber plate, flow channel, thermal insulation, and aluminum alloy frame. In the flat-plate PVT system, the core components are PV and thermal collector. The silicon PV panels are commonly selected due to their low

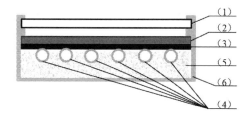

Figure 1. Cross-sections of a general PVT collector: (1) front cover; (2) encapsulated PV; (3) absorber plate; (4) flow channel; (5) thermal insulation; (6) aluminum alloy frame.

cost. Most of the thermal collectors are made up of copper and aluminum for their high thermal conductivity. The collector can work under different working media with different fluid flow in the channel.

2.1 Air

The design of the air-type flat-plate PVT system can be simplified in its structure, thereby reducing the cost of PV cooling with forced or natural flow as shown in Figure 2. Hegazy et al, (2000) compared four types of flat-plate PVT air collectors: air flowing over the absorber (Model I), air flowing under the absorber (Model II), air flowing on both sides of the absorber in a single pass (Model III), and air flowing on both sides of the absorber in a double-pass fashion (Model IV). It was found that the overall performance was dependent on the pass type of the model, the flow rate of the fluid, and the flow channel ratio under similar operational conditions. He also found the optimum ratio value is 2.5×10^{-3}, which effectively maximized the thermoelectric gains for such types of PVT air collectors. (Chandrasekar et al, 2015) verified experimentally that the cooling performance of flat PV modules could be enhanced by heat spreaders in conjunction with cotton wick structures. The electrical power of the PV module was increased by 14% with the developed air cooling methods. (Agrawal et al, 2013) comparatively analyzed different types of PVT air collectors. The overall annual thermal energy gain of hybrid PVT air collector is below 30% and nearly 60% without and with glass cover, respectively. (Raman et al, 2008) had not only evaluated annual thermal and exergy efficiency of a hybrid photovoltaic thermal air collector for different Indian climatic conditions but also analyzed the life cycle in terms of cost/kWh. It is observed that the exergy efficiency is 40% to 45% and the energy payback time and life cycle period were about 2 years and above 50 years, respectively.

2.2 Water

Due to the higher heat capacity of water, the water-type flat-plate PVT systems (Aste, N. del P. C, 2014), are used because they achieve higher overall efficiencies than air systems. (Zondag et al, 2003) also classified the water flat-plate PVT system into four types: sheet and tube, channel, free-flow,

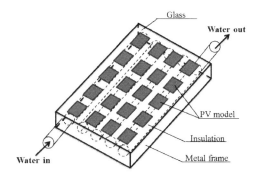

Figure 3.   Cross sections of a water PVT collector.

and two-absorber types. The channel-below-transparent-PV design and the PV-on-sheet-and-tube design were considered to be good alternatives because of their nearly annual efficiency, but the channel-below-transparent-PV design showed superior performance when compared with all types of collectors. (Saitoh et al, 2003) studied experimentally the performance of a single-glazed sheet-and-tube PVT collector under constant coolant supply temperature and flow rate and found that conversion efficiency of PV ranged from 10% to 13%, and thermal efficiencies were from 40% to 50%, respectively. (Shyam et al, 2016) evaluated the overall thermal energy gain and exergy gain performance of series connected PVT water collector. The pay-back time of overall thermal energy and exergy basis was found to be 1.50 and 14.19 years, which is shorter than the air PVT system.

2.3 Nanofluids

Considering the low thermal conductivity of air and water, (Choi, 1995) thermal conductivity was enhanced by adding nanoparticles in liquids called nanofluid. It is defined as a mixture of normal fluid with a very small amount of solid metallic or metallic oxide nanoparticles or nanotubes. As shown in Figure 4, a layer of nanofluid was inserted into the basic structure of the conventional PVT system. (Hussein et al, 2016) had given a comprehensive overview of the recent 10~15 years' advances related to the application of nanotechnology in different types of solar collectors. The thermal energy efficiency and overall energy efficiency were both enhanced about 5%~7%. (Otanicar et al, 2009) compared environmental and economic impacts of using nanofluids to enhance solar collector efficiency as compared to conventional solar collectors. The payback period of the nanofluid collector is slightly longer than the conventional collector. (Kasaeian et al, 2015) reviewed the applications of nanofluids in PVT systems. The

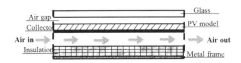

Figure 2.   Cross-sections of an air PVT collector.

advantages and disadvantages of nanofluids were summarized. Nanofluids had unique mechanical, optical, electrical, magnetic, and thermal properties, but increasing costs, heavy toxicity, instability, and environmental damage.

2.4 Phase change materials

Thermal energy storage is practiced in two ways: sensible heat and latent heat. The sensible heat storage is the quantity of heat of material changes caused by temperature gradient and latent heat storage is the quantity of heat of material changes caused by phase change. Phase Change Materials (PCM) (Tiwari G. N, 1998) have considerably higher thermal energy storage densities. The structure of a PVT system is shown in Figure 5. The layer of phase change material had settled between the back plate of the thermal collector and insulation layer. Koca et al. (Tiwari G. N, 2008) analyzed the energy and exergy of a solar collector with phase change material. It was observed that the average system energy and exergy efficiencies are 45% and 2.2% in October, respectively. (Shukla

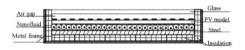

Figure 4.   Cross sections of a nanofluid PVT collector.

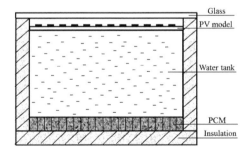

Figure 5.   Cross sections of the PVT system with phase change material.

et al, 2009) summarized the investigation and analysis of thermal energy storage incorporating with and without phase change material for use in solar water collectors. It was demonstrated that a large surface area for heat transfer could help enhance the thermal performance of solar water collectors by a phase change material. (Ma Tao et al, 2015) studied that electrical efficiency and thermal efficiency were both improved by using the phase change material. However, the PVT system with phase change material research is still in the laboratory. It will face attendant challenges for practical applications in the future.

2.5 The others

Tripanagnostopoulos introduced three alternative modes of PVT/bi-fluid collectors with improvement in thermal and electrical output features. The results showed that the integrative design is able to cut down payback time from 25 years to 15 years using a cheap diffuse reflector with lower concentration. Assoa et al. developed a PVT/bi-fluid collector that were able to reach thermal efficiencies approximately 80% for the specific collector length and mass flow rate. Jarimi et al. verified experimentally the model of a bi-fluid PVT solar collector. The primary energy saving efficiency of the collector was around 60% with air and water at the optimum flow rate. However, the design of this hybrid fluid PVT system is innovative but not practical owing to its high manufacturing difficulty.

3   RESULTS AND DISCUSSION

The performance comparisons of different PVT systems with different media were shown in Table 1, including cost, construction, toxicity, corrosivity, heat loss, operational temperature, and the environment. The nanofluids, PCMs, and other hybrid PVT systems were both of high cost and structurally complex, although they could work under high temperature for a long while or reduce the heat loss. In addition, the nanofluids PVT systems were not environment-friendly. Therefore, the air and water PVT systems would become the main force

Table 1.   Performance comparisons of different PVT systems with different media.

Medium	Cost	Construction	Toxicity	Corrosivity	Heat loss	Operational temperature	Environment
Air	Low	Simple	No	No	Yes	Low	Friend
Water	Low	Simple	No	No	Yes	Low	Friend
Nanofluids	High	Complex	Yes	Yes	No	High	Bad
PCMs	High	Complex	No	Portion	No	High	Friend
Others	High	Complex	Yes	Portion	No	High	Portion

of integrated photovoltaic buildings due to their low costs, simple structures, easy installation, and convenient use. In view of the thermal efficiency, the water PVT system would be highly promoted. The hot water would also be produced for residents and convenient for nomadic living.

4 CONCLUSION

The above review shows that great efforts have been demonstrated by several types of research. This paper presents a comprehensive analysis of the research carried out on flat-plate PVT systems operating with different media. The overall thermal energy gain and exergy gain performance had been evaluated and a comparative analysis of the cost input and payback time had been done on different flat-plate PVT systems. This review will help in designing a suitable PVT system for energy storage and building-integrated photovoltaics.

ACKNOWLEDGMENT

This project was supported by the National Science Foundation of China (NSFC No. 61574009) and the 14th Graduate Students of Science and Technology Fund of Beijing University of Technology (No. ykj-2015-11787).

REFERENCES

Agrawal, S. & Tiwari, G.N. 2013. Overall energy, exergy and carbon credit analysis by different type of hybrid photovoltaic thermal air collectors. *Energy conversion and Management* 65: 628–636.

Aste, N. del P.C. & Leonforte, F. 2014. Water flat plate PV-Thermal collectors: a review. *Solar Energy* 102: 98–115.

Chandrasekar, M. & Senthilkumar, T. 2015. Experimental demonstration of enhanced solar energy utilization in flat PV (PhotoVoltaic) modules cooled by heat spreaders in conjunction with cotton wick structures. *Energy* 90: 1401–1410.

Chandrasekar, M., Suresh, S., & Bose, A.C. 2010. Experimental studies on heat transfer and friction factor characteristics of Al$_2$O$_3$/water nanofluid in a circular pipe under laminar flow with wire coil inserts. *Experimental Thermal and Fluid Science*, 34(2): 122–130.

Chol, S.U.S. (1995). Enhancing thermal conductivity of fluids with nanoparticles. *ASME-Publications-Fed* 231: 99–106.

Chow, T.T. 2010. A review on photovoltaic/thermal hybrid solar technology. *Applied energy* 87(2): 365–379.

He, W., Chow, T.T., Ji, J., Lu, J.P., Pei, G., & Chan, L.S. 2006. Hybrid photovoltaic and thermal solar-collector designed for natural circulation of water. *Applied energy* 83(3): 199–210.

Hegazy, A.A. 2000. Comparative study of the performances of four photovoltaic/thermal solar air collectors. *Energy Convers Manage* 41: 861–881.

Hussein, A.K. 2016. Applications of nanotechnology to improve the performance of solar collectors–Recent advances and overview. *Renewable and Sustainable Energy Reviews* 62: 767–792.

Kasaeian, A., Eshghi, A.T., & Sameti, M. 2015. A review on the applications of nanofluids in solar energy systems. *Renewable and Sustainable Energy Reviews* 43: 584–598.

Koca, A., Oztop, H.F., Koyun, T., & Varol, Y. 2008. Energy and exergy analysis of a latent heat storage system with phase change material for a solar collector. *Renewable Energy* 33(4): 567–574.

Ma, T., Yang, H., Zhang, Y., Lu, L., & Wang, X. 2015. Using phase change materials in photovoltaic systems for thermal regulation and electrical efficiency improvement: A review and outlook. *Renewable and Sustainable Energy Reviews* 43: 1273–1284.

Otanicar, T.P. & Golden, J.S. 2009. Comparative environmental and economic analysis of conventional and nanofluid solar hot water technologies. *Environmental science & technology* 43(15): 6082–6087.

Raman, V. & Tiwari, G.N. 2008. Life cycle cost analysis of HPVT air collector under different Indian climatic conditions. *Energy Policy* 36(2): 603–611.

Saitoh, H., Hamada, Y., Kubota, H., Nakamura, M., Ochifuji, K., Yokoyama, S., & Nagano, K. 2003. Field experiments and analyses on a hybrid solar collector. *Applied Thermal Engineering* 23(16): 2089–2105.

Shukla, A., Buddhi, D., & Sawhney, R.L. 2009. Solar water heaters with phase change material thermal energy storage medium: a review. *Renewable and Sustainable Energy Reviews* 13(8): 2119–2125.

Shyam, Tiwari, G.N., Fischer, O., Mishra, R.K., & Al-Helal, I.M. 2016. Performance evaluation of N-PhotoVoltaic Thermal (PVT) water collectors partially covered by photovoltaic module connected in series: An experimental study. *Solar Energy* 134: 302–313.

Skoplaki, E. & Palyvos, J.A. 2009. On the temperature dependence of photovoltaic module electrical performance: A review of efficiency/power correlations. *Solar energy* 83(5): 614–624.

Tian, Y. & Zhao, C.Y. 2013. A review of solar collectors and thermal energy storage in solar thermal applications. *Applied Energy* 104: 538–553.

Tiwari, G.N., Fischer, O., Mishra, R.K., & Al-Helal, I.M. (2016). Performance evaluation of N-photovoltaic thermal (PVT) water collectors partially covered by photovoltaic module connected in series: An experimental study. *Solar Energy*, 134, 302–313.

Tiwari, G.N., Rai, S.N., Ram, S., & Singh, M. 1988. Performance prediction of PCCM collection-cum-storage water heater: quasi-steady state solution. *Energy conversion and management* 28(3), 219–223.

Tonui, J.K. & Tripanagnostopoulos, Y. 2007. Improved PV/T solar collectors with heat extraction by forced or natural air circulation. *Renewable energy* 32(4): 623–637.

Tyagi, V.V., Kaushik, S.C., & Tyagi, S.K. 2012. Advancement in solar PhotoVoltaic/Thermal (PV/T) hybrid collector technology. *Renewable and Sustainable Energy Reviews* 16(3): 1383–1398.

Zondag, H.A. De Vries, D.W. Van Helden, W.G.J. Van Zolingen, R.J.C. & Van Steenhoven, A.A. 2003. The yield of different combined PV-thermal collector designs. *Solar energy* 74(3): 253–269.

Civil, Architecture and Environmental Engineering – Kao & Sung (Eds)
© 2017 Taylor & Francis Group, ISBN 978-1-138-02985-9

A practical method for improving the coherence degree between SAR images

Wen Yu, Zhanqiang Chang, Xiaomeng Liu, Wei Wang & Jie Zhu
College of Resource, Environment and Tourism, Capital Normal University, Beijing, China
Key Laboratory of 3D Information Acquisition of Education Ministry of China, Beijing, China

ABSTRACT: The interferometric coherence is a key indicator of the quality of phase values in Interferometric Synthetic Aperture Radar (InSAR) processing. Accordingly, improving the coherence degree between SAR images will contribute to ensuring unwrapping smoothly and generating reliable intereferograms. In this paper, we proposed a practical registration method for improving the coherence degree between SAR images on the basis of systematically analyzing various decorrelation sources. A large number of experimental results indicate that the proposed method can obtain better quality interferograms.

1   INTRODUCTION

In the field of Microwave Remote Sensing, Interferometric Synthetic Aperture Radar (InSAR) is an earth observation technique and one of the more popular research tools. It has been widely applied in the field of topographic mapping, geodynamics, glacier excursion, forest investigation, and oceanic surveys. The interferometric coherence and the changes of coherence degree have also been used in several applications such as InSAR processing, forestry stock volume estimation, and so on. In the InSAR processing, the interferometric coherence is a very important indicator of the quality of the phase values in the interferograms. Generally, high-quality interferograms from which we extract reliable deformation information in the study area is based on fine interferometric coherence degree. The relevant research achievements have shown that it will only be possible for unwrapping the interferometric phase in conventional SAR interferometry if the ensemble average of the coherence between SAR images is equal to or greater than 0.3 or 0.4 (Rott H, 2006; Chang Z, 2007). Whereas, it is rather difficult to improve the coherence degree between SAR images in the InSAR application. Accordingly, we will try to find a feasible method to improve the coherence degree between SAR images.

2   ANALYSIS OF VARIOUS DECORRELATION SOURCES

The total decorrelation formula can be expressed as follows (Zebker H. A. & Villasenor J., 1992; Wang, C. Zhang, H., Liu Z., 2002)

$$\rho = \rho_{thermal}\rho_{spatial}\rho_{DC}\rho_{coreg}\rho_{temporal} \qquad (1)$$

where $\rho_{thermal}$ is the radar thermal noise decorrelation; $\rho_{spatial}$ denotes the decorrelation induced by spatial perpendicular baseline; ρ_{DC} is the decorrelation due to the difference of Doppler centroids between SAR images; ρ_{coreg} represents the decorrelation induced by co-registration error; and $\rho_{temporal}$ represents the decorrelation caused by physical changes on the ground surface over the time period between acquisitions.

Following this, we will analyze each influencing element respectively.

The correlation $\rho_{thermal}$ can be calculated according to the following formula (Zebker H. A. & Villasenor J., 1992):

$$\rho_{thermal} = \frac{1}{1 + SNR^{-1}} \qquad (2)$$

where SNR is the ratio of signal to noise of the SAR system.

The spatial decorrelation is induced by the perpendicular baseline $B_\perp$, Zebker and Villasenor gave the following formula (Zebker H. A. & Villasenor J., 1992)

$$\rho_{spatial} = 1 - \frac{2\,|B_\perp|\cos^2\theta}{r\lambda} \qquad (3)$$

The decorrelation induced by the frequency difference of Doppler centroid between SAR images can be calculated by the following formula (Wang, C. Zhang, H., Liu Z., 2002):

$$\rho_{DC} = 1 - \frac{\Delta f_{DC}}{B_A}|\Delta f_{DC} \le B_A) \qquad (4)$$

where B_A is the bandwidth of the signal; Δf_{DC} is the frequency difference of the Doppler centroids between SAR images.

The temporal decorrelation is very complicated and induced by many factors involving the motion of scatters, the changes of soil humidity and temperature, the vegetation growth, and so on. Given the probability distribution of the random scatters that obey Gaussian, it can be expressed (Zebker H. A. & Villasenor J., 1992) as

$$\rho_{temporal} = \exp\{-\frac{1}{2}(\frac{4\pi}{\lambda})^2(\sigma_y^2 \sin^2\theta + \sigma_z^2 \cos^2\theta)\} \quad (5)$$

We note that this formula only takes the motion of scatters into consideration.

The decorrelation induced by co-registration error in the range or azimuth direction is calculated using the following formulae (Just, D., Bamler, R., 1994)

$$\rho_{coreg,r} = \sin c(\mu_r) = \frac{\sin(\pi\mu_r)}{\pi\mu_r} \quad (6)$$

$$\rho_{coreg,a} = \sin c(\mu_a) = \frac{\sin(\pi\mu_a)}{\pi\mu_a} \quad (7)$$

where μ_r and μ_a, respectively, denote the co-registration errors in the range and azimuth directions.

Analyzing the various decorrelation sources above, we get to know that it is very tough to control most decorrelation sources in InSAR processing since they are previously determined or fixed by SAR sensor performances, satellite orbit, or the properties of ground scatters. However, the only decorrelation which can be handled is the decorrelation induced by co-registration errors. We discovered that the search window size can influence the co-registration accuracy to some extent.

According to equations (6) and (7), the decorrelation due to co-registration error can be calculated by

$$\rho_{coregistration} = \rho_{coreg,r} \cdot \rho_{coreg,a} \quad (8)$$

where $\rho_{coreg,r}$ represents the co-registration errors in range, $\rho_{coreg,a}$ represents the co-registration errors in azimuth.

In the past years, there has been some relevant research about the search window size. Ming Sheng Liao proposed a Multi-Stage Matching Method combined with least square matching, which has solved the issues of efficiency, precision, and reliability well. In his paper, he did some experiments about the least square matching by changing different search window sizes ($N = 17 \sim N = 71$).

Jingfa Zhang also did relevant work and said that 64×64 and 128×128 were the best-suited search window sizes in registration processing. In this paper, we improve the research results by adjusting the search window size based on summing up the results of previous research. As a result, we find a better search window size after doing a lot of experiments.

3 EXPERIMENTAL PROCEDURE

Based on the above theoretical analysis, we selected more than 10 ASAR images on Track 218 for experiment analysis, which are outlined in Table 1.

The registration was processed by using Gamma software, which was developed by Gamma Company. In practice, the experiments were carried out according to the following steps:

First, the ten images were clipped into the same sizes of 3200×14842; Second, we picked out the 20 pairs of images by baseline matching of SARscape. Third, the different fit standard deviations in range and azimuth were obtained by adjusting the search window sizes in the registration module of Gamma. Finally, the suitable search window size was selected according to the fit standard deviations that came from the co-registration errors and improved coherent images were obtained.

4 EXPERIMENTAL RESULTS

Figure 1 and Figure 2 illustrate respectively the fit standard deviations in range and azimuth directions for different search window sizes when generating the 20 pairs of interferograms. Generally speaking, with the increase in search window sizes, the fit standard deviations in range and azimuth directions will be getting smaller.

Table 1. The experience dataset and corresponding parameters.

Mission/mode	Orbit	Date	Track	Frame
Envisat ASAR/IM-2	28328	20070801	218	2799
Envisat ASAR/IM-2	28829	20070905	218	2799
Envisat ASAR/IM-2	29330	20071010	218	2799
Envisat ASAR/IM-2	29831	20071114	218	2799
Envisat ASAR/IM-2	30332	20071219	218	2799
Envisat ASAR/IM-2	31334	20080227	218	2799
Envisat ASAR/IM-2	31835	20080402	218	2799
Envisat ASAR/IM-2	32336	20080507	218	2799
Envisat ASAR/IM-2	32837	20080611	218	2799
Envisat ASAR/IM-2	33338	20080716	218	2799

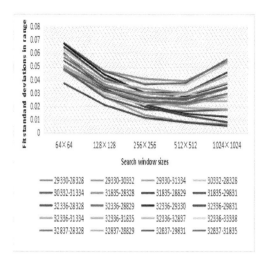

Figure 1. Comparison between fit standard deviations in range for different search windows sizes when generating interferograms.

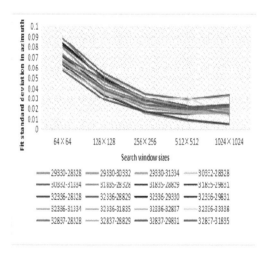

Figure 2. Comparison between fit standard deviations in azimuth for different search windows sizes when generating interferograms.

However, the 1024 × 1024 window size generates different results, which shows the instability of change. So the 1024 × 1024 is not the suitable search window size. Figure 3 shows that the real consumptive time has gotten longer with the increase of the search window. As a matter of fact, considering the fit standard deviations and the real consumptive time, we adopted 256 × 256 as the suitable search window size for co-registration and then implemented SAR interferometry to obtain

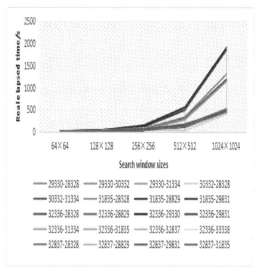

Figure 3. The real lapsed time for different search windows sizes about the 20 pairs of images.

the reliable deformation field in the study area. The search window 512 × 512 can be a candidate choice if we need higher co-registration precision.

5 DISCUSSION AND CONCLUSION

By statistically analyzing the experimental results, we found that reasonably increasing the search window size can significantly reduce the co-registration error, which is a feasible method to improve the coherence degree between SAR images.

As a result, higher quality interferograms can be generated, which set the stage for obtaining consequently high-quality interferometric products, such as Digital Elevation Model data, ground surface deformation information, and canopy height, etc.

More concretely, the experimental results indicate that the search window 256 × 256 is the optimal size for co-registration in view of both the fit standard deviations and the real consuming time.

Apart from that, the search window 512 × 512 can be a candidate if we need higher co-registration precision. Furthermore, the fit standard deviations in range and azimuth directions should be a comprehensive consideration due to their unique patterns.

In the future, we will be focusing on the mechanism of this method.

ACKNOWLEDGMENTS

This work was supported by the National Natural Science Foundation of China (Grant No. 41671417).

REFERENCES

Chang, Z., Zhang, J., Gong, H. et al. (2007). 'Maximal effective baseline' for conventional SAR interferometry. International Journal of Remote Sensing, 28(24), pp. 5603–5615.

Chen, E.X., Li, Z.Y. and Chen, X.J. 2000. Generating DEM from SAR interferometry and elevation error correction. High Technique Communication 10, pp. 57–63.

Just, D. and Bamler, R. 1994. Phase statistics of interferograms with applications to synthetic aperture radar. Applied Optics 33 (20), pp. 4361–4368.

Rott, H. and Nagler T. 2006. The contribution of radar interferometry to the assessment of landslide hazards. Advances in Space Research 37, pp. 710–719.

Stramondo, S., Moro, M., Tolomei, C., Cinti, F.R. and Doumaz F., 2005. InSAR surface displacement field and fault modelling for the 2003 Bam earthquake (southeastern Iran). Journal of Geodynamics 40, pp. 347–353.

Wang, C., Zhang, H. and Liu Z. 2002. Spaceborn Synthetic Apture Radar Interferometry (Beijing: Science Press), pp. 48–49.

Zebker, H.A. and Villasenor, J. 1992. Decorrelation in interferometric echoes. IEEE Transactions on Geoscience and Remote Sensing 30, pp. 950–959.

Civil, Architecture and Environmental Engineering – Kao & Sung (Eds)
© 2017 Taylor & Francis Group, ISBN 978-1-138-02985-9

Smart management and control of household appliances

Qiuyan Lu, Shizhe Chen & Tianran Li
School of Electrical and Automation Engineering, Nanjing Normal University, Jiangsu, China

ABSTRACT: In recent years, many traditional household appliances, like coal, oil, gas, and steam equipment, are gradually replaced by the electric power appliances. Subjective intention and objective requirement of residential users to participate in the demand side response are more and more strong. Through the simulation and analysis of the residential user's electricity behavior, the study is based on the principle of residential demand response of smart home appliances. In this study, the Smart Household Management (SHM) control strategy is proposed. The strategy provides feasible solutions for the residential customers to participate in the demand response plan, which in order to improve household electric economy, energy efficiency, and promote stability of distribution network, and to achieve mutual benefits for both supply and demand sides.

1 INTRODUCTION

At present, there are many researches on the load of residential housing. In the literature (Tsui, 2012; Du, 2011; Chen, 2012; Yi, 2013), the intelligent control strategy for the load of multiple resident houses is proposed, such as the peak filling valley of the same distribution network feeder, renewable energy involved in home user scheduling and so on. The control scheme is based on the actual situation of regional load. But the characteristics of the electric load of the residents in our country are obviously different, that is, there are obvious differences in the choice of equipment, use of time and frequency. For the reasonable participation in demand response of residential users, based on two-way friendship for electricity, the promotion of the experience of the intelligent residential power use, this paper proposes an intelligent residential load control strategy Smart Household Management (SHM).

Implementation of intelligent residential demand response needs the support of communication technology and control technology. The hardware foundation of SHM is the Home Area Network (HAN). HAN system realizes the wireless interconnection of the residential building internal power equipment and the direct control of the load and the intelligent socket. At the same time, the system can also be used to realize the information interconnection between the system, the community and the distribution network, so as to provide the electricity price and load information for the users to participate in the DR.

2 SHM LOAD INTEGRATED MODEL

2.1 SHM temperature model

1. Air conditioning load model
The control model of air conditioning and refrigeration operation state is as follows:

$$S_{C,t} = \begin{cases} 1 & T_{C,t} \geq T_{C,h} \\ 0 & T_{C,t} \leq T_{C,h} - \Delta T_C \\ S_{C,t-1} & T_{C,h} - \Delta T_C < T_{C,t} < T_{C,h} \end{cases} \tag{1}$$

$$I_C = \begin{cases} \dfrac{T_{C,h} - T_{C,t} - \Delta T_C}{\Delta T_C} & S_{C,t} = 1 \\ \dfrac{T_{C,h} - T_{C,t}}{\Delta T_C} & S_{C,t} = 0 \end{cases} \tag{2}$$

In the formula, $S_{C,t}$ represents the running state of the air conditioner (1 run, 0 stop, the same below); $T_{C,t}$ represents the internal temperature of the house; $S_{C,h}$ represents the highest temperature of the summer residence set; ΔT_C represents the range of residential temperature changes in summer; I_C is introduced as a residential temperature comfort criterion.

2. Heating load model
Floor heating operation state control model are as follows:

$$S_{F,t} = \begin{cases} 1 & T_{F,t} \geq T_{F,l} \\ 0 & T_{F,t} \leq T_{F,l} + \Delta T_F \\ S_{F,t-1} & T_{F,l} < T_{F,t} < T_{F,l} + \Delta T_F \end{cases} \tag{3}$$

$$I_F = \begin{cases} \dfrac{T_{F,t} - T_{F,l} - \Delta T_F}{\Delta T_F} & S_{F,t} = 1 \\ \dfrac{T_{F,t} - T_{F,l}}{\Delta T_F} & S_{F,t} = 0 \end{cases} \quad (4)$$

In the formula, $S_{F,t}$ represents the running state of the heating; $T_{F,t}$ represents the internal temperature of the house; $T_{F,l}$ represents the lowest temperature of the winter residence set; ΔT_F represents the range of residential temperature changes in winter; I_F is introduced as a residential temperature comfort criterion.

2.2 SHM uninterruptible load model

The running state control model of the uninterruptible load is as follows:

$$S_L = S_{PC} = S_{TV} = S_R = \begin{cases} 1 & I = -\infty \\ 0 & I = \infty \end{cases} \quad (5)$$

$$I_L = I_{PC} = I_{TV} = I_R = \begin{cases} \infty & R = 0 \\ -\infty & R = 1 \end{cases} \quad (6)$$

In the formula, S_L, S_{PC}, S_{TV}, S_R represents the running state of lamps, desktop computers, TV, refrigerator respectively. The formula introduces random parameter R (1 use, 0 nonuse, the same below). I_L, I_{PC}, I_{TV}, I_R represents the comfort criterion of lamps, desktop computers, TV, refrigerator respectively. The higher the value is, the higher the priority of the corresponding equipment will be.

2.3 SHM transferable load model

The control model of the load running state of the washing machine is as follows:

$$S_W = \begin{cases} 1 & I_W = -\infty \\ 0 & (T-1)_{W.s} + \Delta T_d > T_W > T_{W.s} + \Delta T_S \end{cases} \quad (7)$$

$$I_W = \begin{cases} \infty & R_W = 0 \\ -\infty & R_W = 1 \\ -(I_F + I_C) R_W & \neq 0,1 \end{cases} \quad (8)$$

In the formula, S_W indicates the running state of the washing machine; R_W is the introduction of random variables, which means that users need to use the device immediately for special reasons; T_W is the current time of washing machines; $T_{W.s}$ indicates the latest start running time of washing machine; ΔT_s indicates the single washing time; $(T-1)_{W.s}$ means the start running time of washing machine last time; ΔT_d indicates the operating interval frequency of washing machine in plan; I_W is priority criterion.

3 SHM LOAD CONTROL STRATEGY

Take the control of four types of load as an example, and the corresponding four order control functions are as follows:

$$N = I \cdot N_B + N_A = \begin{bmatrix} I_1 & I_2 & I_3 & I_4 \end{bmatrix}$$
$$\times \begin{bmatrix} N_{B1} & 0 & 0 & 0 \\ 0 & N_{B2} & 0 & 0 \\ 0 & 0 & N_{B3} & 0 \\ 0 & 0 & 0 & N_{B4} \end{bmatrix} \quad (9)$$
$$+ \begin{bmatrix} N_A & N_{A2} & N_{A3} & N_{A4} \end{bmatrix}$$

$$N_{Bi}(t) = \sum (m_i S_t, n_i P_t) \quad (10)$$

In the formula, N represents the priority control function; I represents device priority criteria for control; N_B is the basic control function; N_A is a modified control function based on price response or incentive response; S_t, P_t indicate the running state, running power of the equipment based on time; m_i, n_i indicate the model parameters based on different residents of residential users.

4 SIMULATION

Equipment detail and load type which participate in the simulation of demand response are as shown in Table 1.

4.1 Simulation 1

Simulation demand response data sampling period is set to 10 min. Residential load simulation global online load and all equipment priority changes are as shown in Figure 1.

Residential temperature, EV simulation results are shown in Figure 2.

Table 1. Load participating in simulation.

Device	Power/kW	Load type
Air conditioner	4.5	temperature control
Radiant floor heating	4.0	temperature control
Refrigerator	0.1	uninterruptible
Computer	0.5	uninterruptible
TV	0.3	uninterruptible
Lamps	0.2	uninterruptible
Washing machine	0.8	transferable
Electric vehicle	5.0	EV

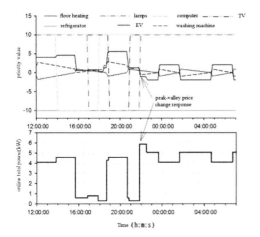

Figure 1. Simulation results of total power and priority.

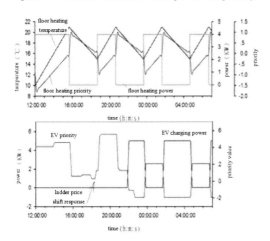

Figure 2. Floor heating and EV simulation results.

The results in Table 2 show that in the current power system price environment, the demand response of SHM control system has been able to realize the economic dispatch of the residential user load.

4.2 Simulation 2

Price incentive factors use Real-Time Pricing (RTP) system. The change degree of real time electricity price is drawn referring to the RTP curves in literature (Sheng, 2014; Zhang, 2004), as shown in Figure 3.

In this simulation, we increase the simulation sampling period to 3 mins. Residential load simulation of the global online load and all the device priority changes are as shown in Figure 4.

Residential temperature, EV control status are as shown in Figure 5.

Table 2. Comparison of the cost of electricity.

Without SHM	40.96 yuan
SHM control	30.98 yuan

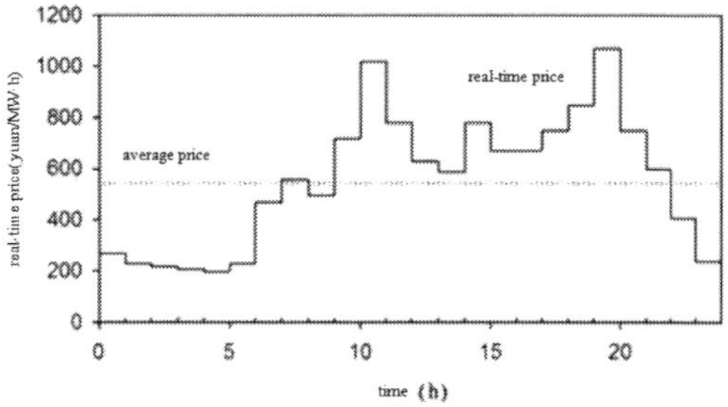

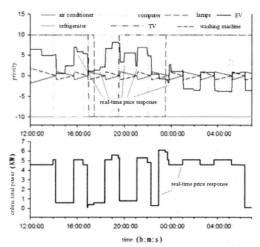

Figure 3. Real Time electricity Price (RTP) curve.

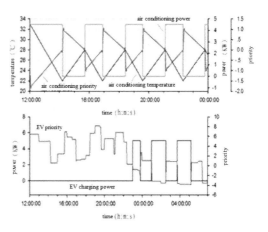

Figure 4. Simulation results of total power and priority.

Figure 5. Simulation results of temperature control equipment (air conditioning).

1345

Table 3. Comparison of the cost of electricity.

Without SHM	37.59 yuan
SHM control	32.73 yuan

The results in Table 3 show that the demand response level of SHM control system can be realized in the long-term electricity market RTP price environment, which can also realize cutting peak and filling valley and economic dispatch to the residential load.

5 CONCLUSION

This paper constructs a residential load control model based on demand side response, optimizes electrical characteristics of residential users and makes residents use electricity economically. The simulation results show that SHM economic dispatch strategy of cutting peak filling valley has been realized. Adding a new load EV significantly increases the residential housing demand response scheduling regulation capacity. And in the current and long-term power market environment, SHM control strategy is reliable and compatible. SHM control strategy on the one hand improves the use efficiency of electric energy of consumers, reduces the cost of electricity and improves the comfort of the use of electricity; on the other hand, it optimizes resident load, transfers the load smoothly, smooths the load curve and promotes the stability of power supply and distribution network.

REFERENCES

Chen Zhi, Wu Lei, Fu Yong. 2012. Real-Time Price-Based demand response management for residential appliances via stochastic optimization and robust optimization. *IEEE Transactions on Smart Grid* 3(4): 1822–1831.

Du Pengwei, Lu Ning. 2011. Appliance commitment for household load scheduling. *IEEE Transactions on Smart Grid* 2(2): 411–419.

Sheng Wanxin, Zhang Bo, Di Hongyu, et al. 2014. Application of an immune optimization algorithm based on dynamic antibody memory library in automatic demand response. *Chinese Journal of Electrical Engineering* (25): 4199–4206.

Tsui KM, Chan Sc. 2012. Demand response optimization for smart home scheduling under Real-Time pricing. *IEEE Transactions on Smart Grid* 3(4): 1812–1821.

Yi Peizhong, Dong Xihua, Iwayemi A, et al. 2013. Real-Time opportunistic scheduling for residential demand response. IEEE Transactions on Smart Grid 4(1): 227–234.

Zhang Xian, Wang Xifan, Wang Jianxue, et al. 2004. Application of block trading in power bilateral contract market. *Automation of electric power system* 28(11): 13–16.

Civil, Architecture and Environmental Engineering – Kao & Sung (Eds)
© *2017 Taylor & Francis Group, ISBN 978-1-138-02985-9*

Research and application of online monitoring and positioning systems for partial discharge of high-voltage cables

Ting Li
Jilin Institute of Electrical Engineering, Jilin Changchun, China

Xilin Zhang
State Grid Jilin Electric Power Company Limited & Changchun Power Supply Company, Jilin Changchun, China

Zhenhao Wang
Northeast Dianli University, Jilin Jilin, China

Pengyu Zhang
Jilin Institute of Electrical Engineering, Jilin Changchun, China

Yan Zhang
State Grid Jilin Electric Power Company Limited, Jilin Changchun, China

ABSTRACT: This paper discusses the configuration and main functions of "Online Monitoring and Positioning Systems for Partial Discharge of High-Voltage Cables (OMPSPDHVC)", the design of high-frequency current sensors, the schematic design of collector hardware, software design and detection test. Field operation proves that the method disclosed herein is correct and effective.

1 INTRODUCTION

The author studied the signal characteristics and online monitoring of Partial Discharge (PD) of XLPE cables, developed an OMPSPDHVC, designed a wideband current sensor that has a wider working frequency band and better amplitude-frequency characteristics, determined the optimal parameters through analytic comparison, and verified the performance of the wideband current sensor designed.

An OMPSPDHVC based on the two-end detection technology was developed, whose hardware equipment includes high-frequency current sensors, PD signal collectors and PD signal monitoring servers. To realize real-time monitoring and remote diagnosis of PD signal of XLPE cables, the software system designed herein is divided into a front-end control system and a remote diagnosis system. The paper mainly introduces the realization of collection control, data processing and reliability related with the front-end control system, as well as the realization of remote data transmission, database and diagnostic analysis mainly related with the remote diagnosis system. The correctness and practicability of the OMPSPDHVC has been

verified through XLPE cable PD signal detection tests.

The OMPSPDHVC based on the two-end detection technology can monitor, at any time, the safety status of field equipment and the insulation condition of live HV cables to avoid accidents or further deterioration of insulation. Therefore, promptly detecting and positioning PD signal of XLPE cables are an important means to timely find hidden troubles and ensure safe and reliable operation of power cables. By minimizing periodic artificial inspection, labor cost, workload and outage time and avoiding safety risks caused by repair operations or old working procedures, the OMPSPDHVC significantly improves work efficiency, power safety and economic benefit.

2 SYSTEM CONFIGURATION (Wang Zhenhao, 2015 & Pang Dan, 2014)

The OMPSPDHVC is divided into three tiers: sensor HFCT; PD signal collection terminal; PD signal monitoring concentrator. The system architecture is shown in Figure 1.

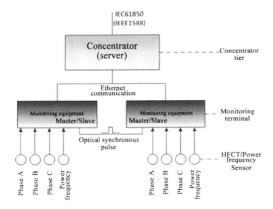

Figure 1. System architecture.

3 MAIN FUNCTIONS (Wang Zhenhao, 2015 & Pang Dan, 2014)

3.1. The synchronous multi-channel PD signal detection technology can realize synchronous detection of multiple phases of the same circuit and collect sufficient correlation information so as to increase the accuracy of detection.

3.2. By reconstructing the basic information of PD signal, it is possible to evaluate the amount of discharge and locate the source of discharge. By means of fuzzy classification algorithm and BP neural network algorithm, it is possible to identify the type of PD signal.

4 DESIGN OF HIGH-FREQUENCY CURRENT SENSORS

Soft magnetic materials are roughly divided into two types: metallic soft magnetic materials and ferrite soft magnetic materials. PD signals are micro current signals which cannot be detected by ordinary Rogowski coils. The purpose of using magnetic materials is to increase the coil's self-inductance coefficient "L", reduce the coil's size, number of turns, and stray capacitance "C", and thus improves its performance. The saturated magnetic induction strength, Curie temperature, basic magnetic permeability, and high-frequency eddy current loss of metallic soft magnetic materials are relatively high but their coercive force and bulk resistivity are relatively low, so such materials are suitable for low-frequency applications in most cases. Although the magnetic performance of ferrite soft magnetic materials is generally lower than that of metallic soft magnetic materials, their bulk resistivity is several orders of magnitude higher than that of the latter, and their high-frequency loss is small, so such materials are more suitable

for high-frequency applications. Furthermore, since ferrite soft magnetic materials are inexpensive, with high magnetic permeability, low loss, high saturated magnetic induction strength, high cutoff frequency, and high stability, they are widely used in the production of high-frequency inductor coil core so as to improve the coil's self-induction and reduce its volume.

Through comparative experiments and based on the existing conditions, the sensor parameters finalized herein are as follows: the magnetic core material is nickel zinc; the number of turns of the coil is about 15; the integral resistance is 100–150Ω.

5 HARDWARE SCHEMATIC DESIGN OF COLLECTOR (Wang Zhenhao, 2015; Pang Dan, 2014 & Wang Zhenhao, Pang, 2014)

Functionally, the PD signal collector can be divided into four parts: power management module, control processing module, data collection module, and signal conditioning module, as shown in Figure 2.

6 SOFTWARE DESIGN (Wang Zhenhao, 2015; Pang Dan, 2014; Wang Zhenhao Pang, 2014 & Zhang XiliN, 2014)

The upper computer software is installed in the tablet PC provided and communication with the collector is realized via the WiFi of the tablet PC. Test results are given after the wave forms captured by the collector are filtered and analyzed.

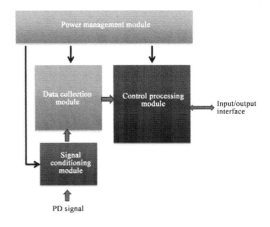

Figure 2. Hardware principle of collector.

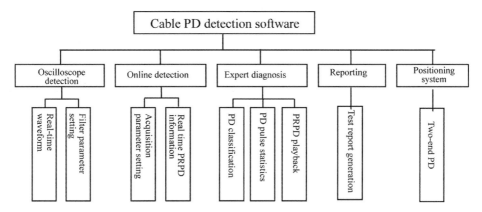

Figure 3. Functional block diagram of software system.

The software system is developed using QT development tools and adopts a modular design, the functional block diagram of which is shown in Figure 3.

7 DETECTION TEST

7.1 *In-situ investigation of cable tunnels and determination of electric power supply solutions*

The purpose is to understand the actual distribution of cable tunnels, determine the test plan and procedures, and solve the problem of power supply. Since the tunnel is not very long, power supply is led to the middle of the tunnel at the air vent. Where PD signal collectors work in the stand-alone operation mode, they are powered by batteries so as to facilitate on-site inspection.

7.2 *Propagation relationship test of PD signal in three-phase crossover interconnected cables*

Simulated PD signals are injected from different parts of the insulation joint, and an HFCT is respectively installed on the coaxial cable and the inner and outer shielding lines of the insulated cable connector to observe, via an oscilloscope, the propagation relationship of PD signal in three-phase crossover interconnected grounding wire. It is proved that installing HFCT on the cross connection box's grounding wire is feasible.

7.3 *In-situ PD signal detection test*

First, PD signal collectors are set in the stand-alone mode to perform signal sampling for all the 220kV cable insulation joints, with 50 sets of data collected for each joint. The upper computer software analyzes the collected signals and the background noise (intensity and type of interference) of the site to find out the cable joints where suspected PD signals exist. Then, according to the result of stand-alone inspection, "hand in hand" network detection is performed for the cables on both sides of the cable joint where there are obvious "PD signals". In-situ network is built, power supply system & communication optical fiber is laid, and PD signal collectors are set in the network operation mode. The next step is to further determine the location the PD source and the amount of PD.

7.4 *Analysis of the collected data and conclusion of the detection test*

Conclusion of the detection test is given based on the results of the above test and analysis so as to find out hidden troubles in time and provide a basis for ensuring the safe and reliable operation of power cables.

8 SUMMARY

Since put into use, the OMPSPDHVC has produced satisfactory results and proved the correctness and effectiveness of the said system constitution, main functions, high-frequency current sensor design, collector hardware schematic design, software design, and test method.

REFERENCES

Pang Dan, Zhang Xilin, Wang Zhenhao, etc.: Functions and theoretical basis of Partical Discharge On-line Mornitering System for high voltage cables [M]. *Applied Mechanics and Materials,* Vols. 672–674, 2014: 854–857.

Wang Zhenhao, Pang Dan, Zhang Xilin, etc.: Desing of High Voltage Power Cable Real-time Mornitoring System Function [M]. IIICEC 2015. 99–102.

Wang Zhenhao, Pang Dan, Zhang Xilin, etc.: The Hardware design of Partical Discharge On-line Mornitering System in High voltage Cable [M]. *Applied Mechanics and Materials,* Vols. 672–674, 2014: 822–826.

Zhang Xilin, Wang Zhenhao, Huang Zhongying: The Hardware design of Partial Discharge On-line Monitoring System in High voltage Cable High voltage cables partical discharge on-line mornitering system software design [M]. *Applied Mechanics and Materials,* Vols. 672–674, 2014: 810–813.

Civil, Architecture and Environmental Engineering – Kao & Sung (Eds)
© 2017 Taylor & Francis Group, ISBN 978-1-138-02985-9

The numerical analysis of rainfall erosivity irregularity in Zhejiang province

Fangchun Lu, Gang Li & Jinjuan Zhang
Zhejiang Institute of Hydraulics and Estuary, Hangzhou, Zhejiang, China

ABSTRACT: In this paper, new indexes are used to evaluate the temporal and spatial variation irregularity of rainfall erosivity, including the cumulative relative deviation rate (C_r), long-term variation rate (L_r), general deviation index (I_d) and special deviation index (D_i). The results showed that, since 1980s, the change process of rainfall erosivity in Zhejiang province was decrease-increase-decrease. But the change of rainfall erosivity was more complex in the early 21th Century. The Id and Di obviously showed the typical low value areas and high-value areas. In the spatial difference, the Id and Di of the eastern coastal area of Zhejiang were significantly larger than those of other regions ($p < 0.01$).

1 INTRODUCTION

Rainfall erosivity is the potential ability of soil erosion caused by rainfall, and it is also the most important external driving force. The rainfall erosivity R factor is an index to evaluate this kind of ability (Zhang et al. 2003). At present, the global temperature showed a significant increase. And this change has been profoundly effecting the global water distribution pattern by accelerating the water cycle process (Feng et al. 2006). The fluctuation of rainfall irregularity has already produced a significant impact on rivers, wetlands and natural vegetation (García-Barrón et al. 2011). And these have led to the change of the spatial distribution pattern of soil erosion, which is worth people's attention. The fourth assessment report of the UN Intergovernmental Panel on Climate Change (IPCC) noted that with the increase of heavy rainfall events frequency, if not in time to take measures to deal with global climate change, soil erosion in most areas will be further aggravated (IPCC 2007). Under the background of global climate change, the change of rainfall erosivity, caused by rainfall change (such as rainfall intensity, rainfall frequency and seasonal rainfall pattern changes), have a profound effect on soil erosion process (Zhang et al. 2005; D'Asaro et al. 2007). Therefore, the study on the spatial and temporal distribution of regional rainfall erosivity can accurately reveal the impact of climate change on regional soil erosion (Diodato and Bellocchi 2009).

2 DATA AND METHODS

According to the geomorphological features of Zhejiang province, the province was divided into five regions. These were the north of Zhejiang (plain area), northwest of Zhejiang (mountainous area), the central region of Zhejiang (hilly area), the southwest of Zhejiang (mountainous area) and eastern coast of Zhejiang (coastal area). The daily rainfall data of 84 meteorological stations in Zhejiang province during 1980–2009 were collected, and the power function model was used to calculate the rainfall erosivity (Zhang et al. 2002). And the long-term evolution characteristics of annual rainfall erosivity were analyzed by using the cumulative relative deviation rate (C_r), long-term variation rate (L_r), the general deviation index (I_d) and the special deviation index (D_i).

The calculation formula of the cumulative relative deviation rate is as follows (Qing et al. 2010):

$$C_{ri} = \Sigma^i_{j=1} (P_j/\mu - 1)$$

Where P_j is the rainfall erosivity of jth year. The μ is the average rainfall erosivity.

Long-term variation rate is defined as follows:

$$L_{ri} = \sigma_{(i,i-4)}/\mu_{(i,i-4)}$$

Where the $\sigma_{(i,i-4)}$ is the standard deviation of five consecutive year rainfall erosivity. The $\mu_{(i,i-4)}$ is the average of five consecutive year rainfall erosivity.

The general deviation index formula (Lana et al. 2000) is

$$I_d = \Sigma((P_{i+1} - P_i)^2/(n-1))^{0.5}/\mu$$

Where P_i is the rainfall erosivity of ith year. The μ is the average rainfall erosivity.

The specific deviation index proposed by García-Barrón is

$$D_i = ((P_i - P_{i-1})^2/2 + (P_{i+1}-P_i)^2/2)^{0.5}/\mu_{(i-1, i, i+1)}$$

Where P_i is the rainfall erosivity of ith year. And the $\mu_{(i-1, i, i+1)}$ is average rainfall erosivity of three consecutive years.

3 RESULTS AND ANALYSIS

3.1 The cumulative relative deviation rate

Figure 1 showed the variation of the C_r in different geomorphologic divisions. On the whole, the C_r reflected the processes of 4 significant change stages of rainfall erosivity, which was decrease-increase with fluctuations-sharply decrease-slow increase. The C_r was negative downward trend from 1980 to 1986 with the annual average decrease 0.12. However, from 1987 to 2002, the C_r was upward trend, the variation of 15 years increased by 0.94. In 2003 and 2004, the C_r showed a sharply decline, as well as in 2005 it began to rise year by year. From the analysis of the overall changes in different geomorphologic divisions.

In 1980s, the C_r in the northwest, the central region, the southwest and the eastern coast, showed decreased trends, as well as a fluctuant increased trend in the north. However, in 1990s, the C_r in the north, the northwest, the central region, and the southwest, showed increased trends, as well as a slightly fluctuant decreased trend in the eastern coast. To the beginning of 21th Century. The C_r in the north, the southwest and the eastern coast was decrease in the first five year and increase in the last five year. And it showed complete decline trends in the northwest and the central region. Thus, in early 21th Century, the Province affected by the global climate change and other factors, the change of rainfall erosivity was more complex.

3.2 The long-term variation rate

There were differences in the L_r of rainfall erosivity in different regions of Zhejiang province in

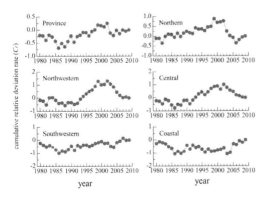

Figure 1.   The cumulative relative deviation rate.

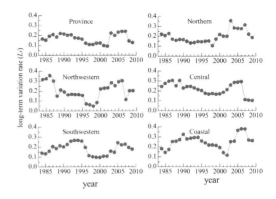

Figure 2.   The long-term variation rate.

recent 30 years (Figure 2). For the whole provincial annual average rainfall erosivity, in 1980s and 1990s, the L_r was shown to slight decreased trend. In early 21th Century, the L_r increased rapidly within a few years, followed by five of the balance period, but from 2007, the L_r again decreased significantly. The change process of the L_r in the north and the central region was similar with that of the province. By comparatively, the L_r in the northwest was decrease more obviously in 1980s and 1990s, nevertheless obvious increased in the first five years of 21th Century. But the change of the L_r in the southwest and the eastern coast were on the contrary.

3.3 The general deviation index

The statistical analysis indicated that the change range of the I_d was from 0.157 to 0.441, average 0.254 and variation coefficient 0.209. In the spatial pattern, the I_d had obvious regional characteristics (Figure 3). The I_d presented two typical low value regions and three typical high value regions. The two typical low value regions were located in the north and the south of Zhejiang, respectively. In which the I_d was less than 0.20. The northern low value region included Xiaoshan district (in Hangzhou city), the northwest of Keqiao district and the north of Zhuji County (in Shaoxing city). The three high value regions were located in the most north, the west and the eastern coast in which the Id could reach more than 0.3. The northern high value region included the north of Changxing County (in Huzhou city) and the northeast of Jiaxing city. The western high value region involved the middle parts of Qujiang district and Kaihua County (in Quzhou city). The eastern coastal high value region related to three regional centers, which were Fenghua-Xiangshan center (in Ningbo city), Sanmen-Linhai center (in Taizhou city) and Leqing-Yuhuan center (in Wenzhou city).

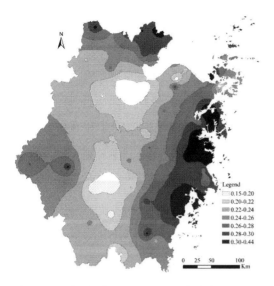

Figure 3. The distribution of general deviation index.

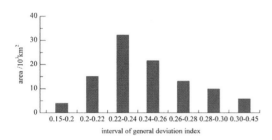

Figure 4. The area composition in general deviation index intervals.

Figure 4 indicated the area distribution of the I_d was positive skewness. In the range of 0.22–0.24, the largest area proportion was 31.70%. Secondly, the area proportion was 21.25% in the range of 0.24–0.26. Thus, I_d in 0.22–0.26 range, the area proportion accounted for 52.95%. The variance analysis of the I_d in different geomorphologic regions indicated that the I_d in the eastern coast (0.305) was significantly higher than in other regions ($p < 0.01$). The I_d in the northwest (0.234), the central region (0.237) and the southwest (0.228), was significantly lower than that in province (0.254). The I_d was no significant difference between in the north (0.252) and in the province.

3.4 The special deviation index

The min-value of the special deviation index (D_i) was 0.227 and max-value 0.537, average 0.372 and variation coefficient 0.18. The spatial distribution characteristic of the D_i was similar to that of the

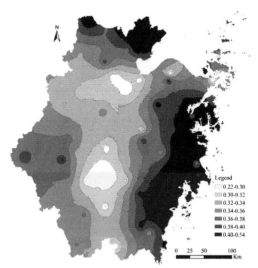

Figure 5. The distribution of special deviation index.

I_d (Figure 5). The D_i also presented two low value regions and three high value regions in the spatial distribution. The two low value regions were located in the north and the south. The northern low value region touched upon Xiaoshan district (in Hangzhou city), the west of Keqiao district and the north of Zhuji County (in Shaoxing city). The southern low value region covered the most region of Wuyi County, northeast of Suichang County (in Jinhua city), the northeast of Liandu district and the north of Songyang County (in Lishui city). Where the D_i was in the range of 0.22–0.30. The three high value regions were situated in the most north, the west and the eastern coast. The northern high value region included the north of Changxing County (in Huzhou city) and the whole northeast of Jiaxing city. The western high value region contained the central region of the kaihua County and Qujiang district (in Quzhou city). The eastern coastal high value region was a wide and Contiguous range.

It contained Ningbo city (Ninghai County, Xiangshan County), Taizhou city (municipal districts, Sanmen County, linhai County and Wenling County), and Wenzhou city (municipal districts, Yueqing County, Yongjia County, Ouhai District, Ruian County). The D_i in these regions was above 0.4.

According to the statistics (Figure 6), the D_i was mainly distributed in the 0.32–0.38 range. And the area proportion Accounted for 58.69%. The area proportion in the range of 0.32–0.34 and 0.34–0.36 was close to 20%. In addition, the area proportion of D_i greater than 0.40 accounted for 18.21%. through analysis of variance found that the D_i in

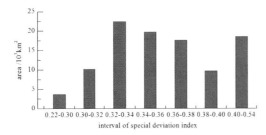

Figure 6. The area composition in special deviation index intervals.

the eastern coast (0.436) was significantly higher than in other regions ($p < 0.01$), and the D_i in the southwest (0.335) was significantly less than that in the province (0.372), but there was no significant difference between the north (0.378), the northwest (0.344), the central region (0.349) and the province.

4 CONCLUSIONS

In the study of the rainfall erosivity variation, most scholars used the standard statistics, such as anomaly value, standard deviation and variation coefficient to measure the variation of rainfall erosivity. In the study of the rainfall erosivity variation, most scholars used the standard statistics, such as anomaly value, standard deviation and variation coefficient, to measure the variation characteristic of rainfall erosivity. However, these standard statistics cannot describe the irregularity of rainfall erosivity. In this paper, the new indexes were used to analyze the irregularity of rainfall erosivity in Zhejiang province. The results of the study were as follows:

The cumulative relative deviation rate (C_r) of rainfall erosivity in the whole province showed a significant time variation feature. Since 1980s, the rainfall erosivity had experienced the process of decrease-increase-decrease. By the beginning of 21th Century, due to the impact of global climate change and other factors, the change of rainfall erosivity was more complicated. The change of long-term variation rate (L_r) had regional characteristics. The L_r, in the north, the northwest and the central region of Zhejiang, had experienced a slow decrease process in 1980s and 1990s. Sharply increase occurred during the 2000–2005. In recent years, there was a significant decline.

The general deviation index (I_d) and the special deviation index (D_i) have obvious spatial distribution characteristics, which were two low value regions and three high value regions. In the 52.95% region of the province, the Id of rainfall erosivity

was in the 0.22–0.26. And In the 58.69% region, the D_i of rainfall erosivity was in the 0.32–0.28. In the regional differences, the I_d and D_i in the eastern coast of Zhejiang were significantly higher than that in other regions ($p < 0.01$). The I_d, in the north, in the center region and in the southwest, was significantly smaller than that in the province. The D_i in the southwest of Zhejiang was significantly less than that of the province.

ACKNOWLEDGEMENT

The research work was supported by the Science and Technology Planning Project of Zhejiang Province, China under Grant No. 2015F50060.

REFERENCES

D'Asaro, F., D'Agostino, L. & Bagarello V. 2007. Assessing changes in rainfall erosivity in Sicily during the twentieth century. Hydrological Processes, 21(21):2862–2871.

Diodato, N. & Bellocchi G. 2009. Assessing and modelling changes in rainfall erosivity at different climate scales. Earth Surface Processes and Landforms, 34: 969–980.

Feng, S., Huang, Y. & Xu, Y.P. 2006. Impact of global warming on the water cycle in Xinjiang region. Journal of glaciology and geocryology, 28(4):500–505. (in Chinese).

García-Barrón, L., Aguilar, M. & Sousa, A. 2011. Evolution of annual rainfall irregularity in the southwest of the Iberian Peninsula. Theoretical and Applied Climatology, 103(1):13–26.

IPCC. 2007. Summary for policy makers of climate change 2007: The physical science basis contribution of working group I to the fourth assessment report of the intergovernmental panel on climate change. Cambridge: Cambridge University Press.

Lana, X. & Burgueno, A. 2000. Some statistical characteristics of monthly and annual pluviometric irregularity for the Spanish Mediterranean Coast. Theoretical and Applied Climatology, 65:79–97.

Qing, J.F., Wei, Z.Y., Ji, H.R., et al. 2010. Spatial pattern and evolution of annual precipitation irregularity in Xinjiang. Arid land geography. 33(6):853–860. (in Chinese).

Zhang, G.H., Nearing, M.A. & Liu, B.Y. 2005. Potential effects of climate change on rainfall erosivity in the Yellow River Basin of China. Transactions of the American Society of Agricultural Engineers, 48:511–517.

Zhang, W.B., Xie, Y. & Liu, B.Y. 2002, Rainfall erosivity estimation using daily rainfall amounts [J]. Scientia Gegraphica Sinica, 22(6):705–711. (in Chinese).

Zhang, W.B., Xie, Y. & Liu, Y.B. 2003. Spatial distribution of rainfall erosivity in China. Journal of Mountain Science, 21(1):33–40. (in Chinese).

Civil, Architecture and Environmental Engineering – Kao & Sung (Eds)
© *2017 Taylor & Francis Group, ISBN 978-1-138-02985-9*

Module design of container architecture

Ruitong Zhu

College of Civil Engineering, Tongji University, Shanghai, China

ABSTRACT: Container architecture, derived from freight containers, has an increasingly important position in the field of temporary buildings. With its modular and detachable structure, this newly emerging type of architecture shows a great development potential. However, container architecture has been confined to the absence of corresponding codes and technical modifications. In this article, a hockey rink in Huijia High School was taken as an example to illustrate a modification technology of container architecture, in which standard containers were displaced by redesigned modular units. Simplified models for engineering application were built through 3D3S in order to analyze the bearing capacity under vertical loads.

1 INTRODUCTION

1.1 *Origin of container architecture*

Container architecture is combined by modular units with full architectural functions. These units would be prefabricated in factories and then assembled at the final location.

Container architecture originated from a large number of empty containers piled up in ports around the world. The rapid development of the logistics industry has brought about a huge demand for containers and it is estimated that about 300 million of them are circulating around the world every year (Gong & Zhang 2010). Due to the lopsided pattern of international trade, a great number of containers are transported from the eastern to western countries loaded with raw materials. Yet despite the good condition of these containers, most of them would end up rusting at destination ports with no prospect of being transported back to Asia, which is much more expensive than purchasing new containers and therefore unacceptable to merchants. As a result, about 1 million Twenty-foot Equivalent Unit (TEU) were decommissioned worldwide (Xu 2008).

Container architecture came into being under such circumstances serving as temporary installations in ports initially. With the green concept rising in the mainstream of architecture, container architecture has received a lot of attention. For decades, the number of containers used in a building has been increasing, as well as the diversity of the combination.

1.2 *Modular tendency*

The domain of container architecture is also expanding. The basic unit of buildings, which used to be a standard container has turned into specially designed modular blocks. This new type of modular frames shares similar superiorities with the original container architecture. With all the modular blocks prefabricated in factories, little work needs to be done on site, which improves the construction efficiency a lot. Also, those highly industrialized blocks make it possible to retain the movability and provisionality of the architecture.

Barring the similarities, this new type of modular architecture presents with many improvements. The stress pattern of the container, of which the corrugated plates with higher stiffness bear much of the load (Lu 2014), is quite different from that of the framework. Therefore, orthodox container architecture has to keep those corrugated plates with restricted room in it. Yet after modification, the basic unit of the modular architecture is more of a frame. It means that the loads are to be carried by the columns at the corners leaving enough space to meet the requirements of a large bay. Due to the similarity of the modular blocks and boxes, this type of architecture is also known as box building (Li & Zhang 1992).

On June 1st, 2013, China issued *CECS 334:2013 Technical specification for modular freight container building* (CSCS 2013), the first national standard for reformation of modular containers. While traditional container architecture has been applied all over the world, however, the research over box building is still fumbling. Thus, whether to choose modular container as units, or design the modular block afresh, is indeed a question worth considering, and circumstances alter these cases. When it comes to a certain project, checking calculation and modeling analysis are necessary.

2 PROJECT AND BACKGROUND

The auxiliary building of the hockey rink in Huijia High School is a three-story office building which is 49.32 meters in length, 13.2 meters in width, and a total building area of 1953square meters. The owner is strict with the construction time and makes a request for the movability of the buildings, expecting it to be reassembled easily somewhere else. Meanwhile, as an office building for conference and reception, it has to be equipped with a large bay.

As is shown in Figure 1, the large bay of the typical floor makes it impossible to maintain the corrugated plates of modular freight containers, which means that all loads are to be imposed on the skeleton of those containers. Generally speaking, such frames can offer little support. So in order to demonstrate the viewpoint, a calculating model for the solitary modular container was built.

3 MODEL

3.1 *Model of universal container*

Two types of modular freight containers are needed in order to correspond with the architectural design and those two types of containers, according to the drawings in the collection, share the same width and height, 12192 mm × 2438 mm × 2591 mm for IAA and 9125 mm × 2438 mm × 2591 mm for IBB.

Construction drawings in Appendix A of the code, *GB/T 1413-2008 Series 1 freight containers-Classification, dimensions and ratings* (2008), provide a scientific basis for structural simplification, as is shown in Figure 2.

The model was built in 3D3S, a software for steel structure modeling. The aim of the modeling

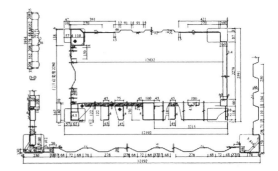

Figure 2. Structure diagram of universal containers IAA.

is to disprove the reliability of the frame for freight containers, so container IBB was chosen as the model. With a shorter span in length, the container IBB ought to be more capable than container IAA. This means if IBB was unable to satisfy the requirements of bearing capacity, then IAA would be automatically proven unsafe.

For the convenience of modeling, the distance between the secondary beams at the bottom of the container was unified to be 300 mm, and the minimum cross section of these beams was adopted to be on the safe side. Additionally, all sections of profiled steel sheets were approximated as fundamental or thin-walled sections to fit the function of the software. The detailed approximation is listed in Table 1, and their corresponding positions are marked in Figure 3.

A bracing system would be set up to enhance the horizontal strength and stiffness of the structure resisting lateral wind and seismic loads. Therefore, only vertical loads need to be considered in this model. As an office building, the dead load of its bottom was evaluated as 0.5 kN/m², and the live load of 2.0 kN/m². The load transferring pattern at the bottom resembles that of a one-way slab, which means being equally divided onto the secondary beams on both sides. The transferring pattern at the top of the frame, on the other hand, was like a two-way slab, considering the great distance it spans. Suspended ceilings were estimated to apply 0.5 kN/m² to the top as dead load.

Loads of the upper layers were passed through the columns at the four corners equally because of the monolithic bottom board. To simulate the limit state of the frame, the model was based on the ground floor, bearing the loads passed from the two floors above. Detailed information of loads was elaborated in Table 2. With the consideration of load combination, all the dead loads would be multiplied by 1.2 and live loads by 1.4 before calculating. Force transmission between layers was completed through corner fittings where there

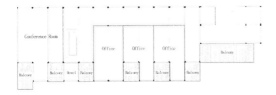

Figure 1. Typical floor plan.

Table 1. Component cross-references of the universal container.

Mark	Type of section	Size of section
GL1	Cold formed square	60 × 3
GL2	Cold formed rectangle	138 × 110 × 4
GL3	Cold formed square	60 × 4
GL4	Cold formed rectangle	140 × 50 × 4.5
GL5	Cold formed rectangle	152 × 60 × 4
GL6	Cold formed C steel	150 × 60 × 6
CL1	Cold formed C steel	122 × 45 × 4
GZ1	Cold formed C steel	230 × 40 × 6
GZ2	Cold formed rectangle	174 × 159 × 6

* All in mm.

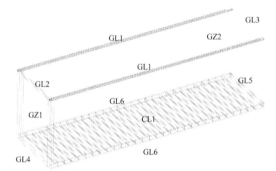

Figure 3. Approximated frame of a universal container.

Table 2. Load distribution of a universal container.

Position	Force transferring mode	Dead load	Live load
Bottom	One-way slab	0.5 kN/m²	2.0 kN/m²
Top	Two-way slab	0.5 kN/m²	/
Columns	Joint load	11.12 kN	22.25 kN

were hinge supports. Apart from the support, all the other joints were fixed.

The check of the model was based on *GB 50017-2003 Code for Design of Steel Structures*. Both strength and stability did not pass the examination and its strength-stress ratio was distinguished with different colors as shown in Figure 4.

Results show that over 5 of the components buckled and the maximum stress ratio of strength is up to 2.660. Moreover, slenderness ratio of some rods has gone beyond the limits. So far, it has been proven impracticable to take freight containers without corrugated wallboards as architectural units. In other words, the modular unit requires redesigning.

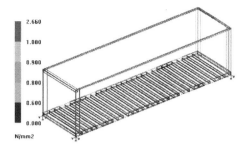

Figure 4. Distribution of stress ratio in the model of a universal container.

3.2 Model of a redesigned unit

The main principles of the design are to keep the feature of easy manufacturing and strengthen the frame of the module.

Still, two different types of modular units with identical width and height are necessary in order to match the architectural design. The difference lies in the phase of the conceptual design. The unit with greater length would be divided into the main body, which is exactly the same as the shorter one and the cantilevered part. For the sake of modular coordination, the units were designed to be 9020 mm × 3000 mm × 3100 mm and 12000 mm × 3000 mm × 3100 mm. The main body of the longer unit was 9.02 meters long while the rest of the part overhangs. Thus all the columns were in the same plane as is shown in Figure 5.

The longer unit would be taken as an example to illustrate how the frame was strengthened. In contrast with standard freight containers, universal beams and cold formed rectangles were chosen to form the frame, the cross sections of which are much larger. Also, secondary beams were added to the top, enhancing the stability as well as installing the ceiling.

The cantilevered part of the longer unit plays a negative role when calculating the structure, compared with the shorter unit. Therefore modeling longer unit is enough to ensure the safety of the units.

The second model is as shown in Figure 6, and detailed information is listed in Table 3.

On top and bottom of the unit, loading condition of this model in 3D3S was the same with the previous one, but the patterns of force transferring were both in the way of the two-way slab. Hinge supports were on the four corners of the main body. Based on what is mentioned above, the software can return the reaction force of the supports, which would be doubled and then added to the columns as the loads from the upper layers. Similarly, considering load combination, all the dead loads would time 1.2 and live loads would time 1.4. Detailed information was presented in Table 4.

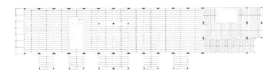

Figure 5.   Structural floor plan.

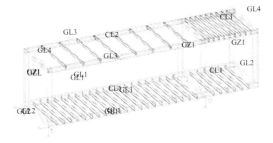

Figure 6.   Frame of a redesigned unit.

Table 3.   Components cross-references of the redesigned unit.

Mark	Type of section	Size of section
GL1	Cold formed rectangle	$250 \times 150 \times 8$
GL2	Universal beam	$250 \times 150 \times 6 \times 8$
GL3	Cold formed rectangle	$200 \times 150 \times 5$
GL4	Universal beam	$250 \times 150 \times 4.5 \times 6$
CL1	Cold formed rectangle	$250 \times 150 \times 6$
CL2	Cold formed rectangle	$150 \times 50 \times 3$
GZ1	Cold formed square	89×3

* All in mm.

Table 4.   Load distribution of the redesigned unit.

Position	Force transferring mode	Dead load	Live load
Bottom	One-way slab	0.5 kN/m²	2.0 kN/m²
Top	Two-way slab	0.5 kN/m²	/
Newels	Joint load	23.63 kN	24.13 kN
Front posts	Joint load	52.69 kN	47.87 kN

Judging from the outcome, the structure gets a sufficient safety margin with the maximum stress ratio of strength being 0.638 and stress ratio of stability being 0.608. As a consequence, the design of the modular unit is secure and feasible.

4   CONSTRUCTION MEASURES

Apart from the module design of the major structure, there are also some construction measures aimed at enhancing structural capacity.

Straight braces were arranged longitudinally along the center line at the top of the unit with 2 mm-thick steel plates as roof and side beams on top were welded together, on which a coat of sealant was applied.

All the ceiling joists should not be put on the top plates. Instead, these joists should be welded to the secondary beams or angle steel attached to them.

The bottom was covered with 28 mm thick wooden floor specifically for containers with some space reserved for the installation of doors and windows.

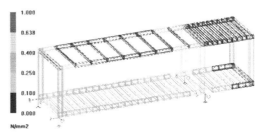

Figure 7.   Distribution of stress ratio in the model of the redesigned unit.

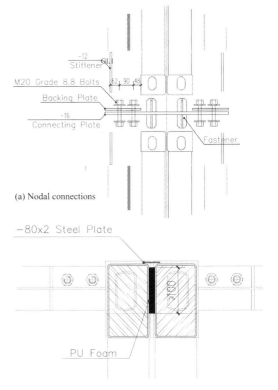

(a) Nodal connections

(b)   The connection between beams.

Figure 8.   The detailed structure of the junction.

There was a fitting on each corner of the unit through which the units were linked together. As is shown in Figure 8, the beams were aligned with corner fittings by backing plates. Two M20 grade 8.8 bolts were installed around each fitting along with the stiffener to avoid buckling.

Beams were connected through a steel lathing, welded with two contiguous beams. The inter-spaces were filled with Polyurethane Foam sealing agent (PU Foam) thicker than 100 mm.

Fittings on the top had a bolt hole on each of them for hoisting on the site, which would be filled with bolts upon completion.

The assembly had at most four members at the position of one beam or one column increasing the cross section of the beam or column tremendously. In this way, the frame would be strengthened and deflection decreased, influencing the usage and safety of the structure positively.

5 CONCLUSIONS

Modular buildings represented by the design in the article not only inherit the advantages like movability and prefabrication from the orthodox container architecture but also remove all the restraints for space. The diversity of modular units makes the style of the building more flexible. Such modular buildings are bound to set off a revolution in the field of temporary buildings.

REFERENCES

China Steel Construction Society (2013). Technical specification for modular freight container building. *CECS* 334: 2013.

Gong X.L. & Zhang Y.K. (2010). Transformation from container to building: an attempt of the sustainable building development. *J. World Architecture.* 10:124–127.

Li N.C. & Zhang X.Z. (1992). Design and construction of box buildings. *J. Architectural Journal.* 12:22–27, 1992.

Lu Y. (2014). Structural design technology of container buildings. *J. Industrial Constructions.* 02:130–136+97.

Series 1 freight containers-Classification. *GB/T* 1413–2008.

Xu Z.H. (2008). Final destination of secondhand container. *J. Containerization.* 19(5):34–35.

Civil, Architecture and Environmental Engineering – Kao & Sung (Eds)
© 2017 Taylor & Francis Group, ISBN 978-1-138-02985-9

Design of a reflective pulse oximetry sensing system based on the semi-active RFID tag

Liying Chen, Yang Wang & Hongwei Liu

School of Electronic and Information Engineering, Tianjin Polytechnic University, Tianjin, China

ABSTRACT: This paper describes the design of a reflective pulse oximetry sensing system based on the semi-active RFID tag. It can monitor human physiological signals continuously through the wireless sensor network. The system is composed of four modules: reflective pulse oximetry detection module, data processing module, RFID transceiver module, and PC interface. These modules can achieve the function of monitoring non-invasive, real-time, and long-distance wireless pulse oximetry. The reflectance probe is designed to collaborate with LEDs with wavelengths of 660 nm and 940 nm. The data processing module is designed to manage almost all of the internal processing. The RFID transceiver module is based on the EPC C1 GEN2/ISO 18000-6c protocol for communication. The RFID reader communicates with the RFID tag through the electromagnetic wave. The pulse and blood oxygen information measured by the sensor will be displayed in the upper computer interface. The experimental results show that the system has high accuracy and low power consumption, and will have a good application prospect in the direction of wearable medical wisdom research.

1 INTRODUCTION

The traditional health care philosophy is based on professional medical diagnosis and treatment institutions using professional diagnostic equipment to achieve unscheduled health care (Occhiuzzi, 2010; Peris-Lopez, 2011). This approach not only takes up excessive medical resources but also cannot grasp the body's health in real-time, even miss the best time to treat the disease, resulting in irreparable damage (Redondi, 2013). With the constant improvement of science and technology, networking technology, optical sensing technology, radio frequency identification technology, and smart micro-processing technology there has been rapid development. The portable intelligent medical monitoring devices begin to enter people's vision and towards the direction of the development of wireless communication and monolithic integrated chip (Catarinucci, 2013; Toh, 2009).

The real-time monitoring of vital signs of patients is a very important part in the new wisdom of the medical system (Ley, 2013). The medical point of human vital signs, such as oxygen saturation and pulse rate, are important indicators characterizing the degree of human health. So monitoring pulse oximetry safely and effectively plays an active role in the diagnosis of disease and health care aspects (Mendelson, 1988). As a kind of medical instrument, which can continuously and non-invasively collect the data of blood oxygen saturation, the pulse blood oxygen measuring instrument can be divided into two types: transmission type and reflection type. Because of the limitation of the transmission type instrument to the human body collection position, the reflection type sensor has been paid more and more attention by people. At present, there are few reports about the reflectance pulse oximeter clinical monitoring data (Kirk, 2003). For the reflection type sensors, the design of hardware, software, and measurement methods are worthy of further exploration.

In this paper, the measurement of pulse oximetry employs a reflective optical measurement method according to the principle of infrared spectrum measurement. The pulse and blood oxygen saturation detection system based on DCM03 reflective sensor and a microprocessor with RFID technology is introduced in detail. The system uses advanced digital signal processing technology to effectively restrain the influence of human movement and weak perfusion measurement and provides a fast and accurate method for measuring blood oxygen saturation and pulse rate.

2 RELATED WORKS

Pulse is the number of times the heart beats in a minute. Blood oxygen saturation is the percentage of total content of hemoglobin (HbO2) in the blood, which can be combined with oxygen. Blood oxygen saturation is defined by the formula:

$$SPO_2 = \frac{HbO_2}{HbO_2 + Hb} \times 100\%$$ (1)

SpO_2 represents oxygen saturation, HbO_2 and Hb represent the content of hemoglobin and oxygenated hemoglobin in the blood, respectively. Measurement of oxygen saturation is based on the Lambert-beer law as the theoretical basis, using the principle of the spectrum effect combined with digital signal processing algorithms to achieve it. Blood in the veins of humans will be regarded as a liquid medium in the analysis process due to the absorption rate of different wavelengths of light for different components in the blood. So we can calculate the content of different components in blood by measuring the attenuation degree of light through blood.

Measurement of pulse oximetry is based on the principle that light absorption of ripple component in arterial blood will change with the contraction or expansion of the artery, while the light absorption of other organizations is almost constant. For the arterial blood vessel pulse, the light absorption of the arterial blood will change, which is known as the AC component. The light absorption of skin, muscle, bone, blood and non-pulsatile blood will be constant, known as the straight flow. Cardiac contraction or expansion will affect the volume of the artery. Change in the optical path will affect the absorbance according to the Lambert-Beer law and the ripple component of the artery is known as photoplethysmography (PPG). Based on the photon diffusion equation and time-resolved spectroscopy technique, reflective light intensity can be obtained by the formula shown below.

$$R(\rho,t) = (4\pi Dc)^{-3/2} z_0 t^{-5/2}$$
$$\exp(-\mu_a ct) \exp\left(-\frac{\rho^2 + z_0^2}{4Dct}\right)$$ (2)

The relationship between the variation of the reflected light intensity and the absorption coefficient is derived as given below.

$$W = -\mu_a c$$ (3)

Connecting the absorption coefficient and absorption material concentration c, so that we can obtain the measurement method of dual wavelength blood oxygen saturation.

The following formula is obtained by the double beam method.

$$\frac{W_{\lambda_1}}{W_{\lambda_2}} = \frac{\varepsilon_{Hb}{}^{\lambda_1} C_{Hb} + \varepsilon_{Hbo_2}{}^{\lambda_2} C_{Hbo_2}}{\varepsilon_{Hb}{}^{\lambda_2} C_{Hb} + \varepsilon_{Hbo_2}{}^{\lambda_1} C_{Hbo_2}}$$ (4)

The formula of saturation of blood oxygen is obtained by selecting the wavelength of λ_2 as an equal absorption point.

$$SpO_2 = \frac{\varepsilon_{Hb}{}^{\lambda_1}}{\varepsilon_{Hb}{}^{\lambda_1} - \varepsilon_{Hbo_2}{}^{\lambda_1}} - \frac{\varepsilon_{Hb}{}^{\lambda_2}}{\varepsilon_{Hb}{}^{\lambda_1} - \varepsilon_{Hbo_2}{}^{\lambda_1}} \cdot \frac{W_{\lambda_1}}{W_{\lambda_2}}$$ (5)

The absorption of two kinds of hemoglobin in the red spectral region varies dramatically, but the difference of absorption in the near-infrared region of the spectrum is subtle, so the light absorption degree of different oxygen saturation depends on the content of two kinds of hemoglobin in the blood. According to the Lambert-Beer law, the analysis of light detector to the red and infrared volume pulse wave, hemoglobin concentration, and oxygen saturation can be calculated by analyzing the PPG of the red and infrared light.

3 SYSTEM DESIGN

The reflective pulse oximetry sensing system based on semi-active RFID tags is composed of four modules: the reflective pulse oximetry detection module, data processing module, RFID transceiver module, and the PC interface.

3.1 Pulse oximetry sensing system

The pulse oximetry detection module, data processing module, and RFID tag are respectively responsible for the collection, processing, and transmission of pulse oximetry information. The RFID reader is connected to the host computer through the USB interface. They are jointly responsible for coordinating the management of the pulse oximetry sensing system, obtaining real-time sensing information of all nodes, planning and coordinating the behavior of each module. The RFID reader communicates with the RFID tag through the electromagnetic wave. On the basis of power and precision, making the normal work of each module and establishing a stable communication are key to the design of the system. The overall design of the reflective pulse oximetry sensing system is shown in Figure 1.

In order to verify the feasibility of the system, we firstly select the appropriate hardware to build the system. Among them, the pulse oximeter sensor uses the APM's latest evaluation kit for the reflective SDPPG sensor, including DCM03

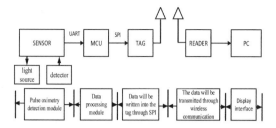

Figure 1. Structural block diagram of the reflective pulse oximetry sensing system.

Figure 2. The platform of the pulse oximetry sensing system.

reflection-type probe and a built-in 32-bit Cortex-M3 processor. The DCM03 reflective sensor used in this design is composed of a red light emitting diode, infrared light emitting diode, and a high-performance photodiode. It uses 660 nm red light and 940 nm near-infrared light as the incident light source. The light source and detector are located on the same side of the body tissue. A part of light emitted from the source is absorbed through human tissue, another part of the reflected light is absorbed by the detector. It can not only measure the PPG signal but also measure the SDPPG signal. Data processing module uses the STC-12C5A60S2 of STC MCU Limited. The reader adopts the RLM300 UHF RFID introduced by RAY-LINKS. RLM300 works within the range of 840~960 MHz based on the EPC C1 GEN2/ISO 18000-6C communication protocol. The design is based on advanced RFID analog circuit combined with the digital signal processing technology. The tag selects the IDS-SL900A used in the same EPC UHF G2 tag based on 18000-6C. The platform of the pulse oximetry sensing system is built as shown in Figure 2.

The four modules of the system are:

1. The pulse oximetry detection processing module comprising a reflection type sensor and a microprocessor. The light source device emits red light and infrared light of two different wavelengths under the control of a microprocessor. The light is received by an optical detector after reflection from human tissues, and carry the oxygen pulse information in the optical signal. The light signal is converted into an electrical signal by a light sensitive element and fed into a microprocessor. The electrical signal will be processed with amplification, filtering, and A/D conversion, and then separated into DC component and AC signal.

2. Data processing module includes high-speed A/D conversion module, six digital tube, central processing unit, and data memory EEPROM. Its input and output support UART and SPI, two kinds of interface mode. The pulse oximetry detection processing module connects data processing module through the UART serial port, so pulse and blood oxygen information can be displayed on the digital tube. Communication between data processing module and the tag through the SPI protocol achieves the configuration of running parameters of the tag so that the analog signal will be written into the tag's EEPROM in the form of a digital signal after processing.

3. The RFID transceiver module includes two parts: RFID tag and RFID reader. RFID tags are generally composed of a tag chip and RF antenna. Each tag chip has a different identity code called EPC code. It is stored in a specific storage area of the tag memory, and can also operate and identify multiple tags through the EPC code at the same time. The RFID reader communicates with RFID tags through the transmission and reception of radio frequency signals. The microprocessor in the reader controls the work of the whole reader. It is responsible for decoding and encoding the received signal and communication with the host computer.

4. The PC interface includes a host computer and the interface written by VC++ MFC programming. The reader connects directly to the host computer through the USB interface and data obtained through wireless communication will be displayed on the interface. They can obtain real-time physiological information such as pulse and blood oxygen saturation. The pulse and blood oxygen saturation monitoring system are connected as shown in Figure 3.

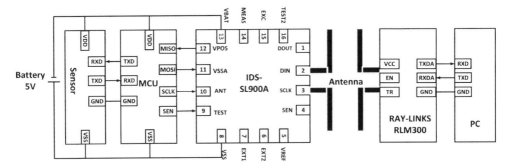

Figure 3. Diagram of the reflective pulse oximetry sensing system.

3.2 Data processing module

After running the entire system program, the hardware and software environment will firstly be initialized and then the host computer sends the sensor information to the reader. The reader will wait for query instructions of the sensor information sent by the host computer. If the reader does not receive any query from the host computer, the system remains in the initial state. Once the query instruction is received from the host computer, the reader will send selection, inventory, and access instructions to the tag through the non-modulated RF carrier, so that it will establish a connection with the label and complete the configuration of the parameters of the tag. When the tag receives the reader inquiry instruction, the data processing module transmits the enabling signal to the blood oxygen pulse sensor. The sensor begins to work after receiving the signal. Information will be sent to the data processing module after completion of the corresponding data acquisition. The data processing module communicates with the tag through the SPI interface, the received data will be written into the internal register of the tag after being extracted and analyzed. After the reader identifies the tag, the data will be transmitted to the internal register of the reader and the tag through wireless communication. Finally, the real-time data received by the reader will be displayed in the PC interface. The system flow diagram is shown in Figure 4.

3.3 Transceiver communications

Communications between the interrogator and tag are based on EPC C1 GEN2/ISO 18000-6c protocol. The parameter of the interface signals for RF communications, such as prescribed frequency and modulation, data coding, RF envelope, and data rate will be set up under the protocol. Tags shall receive power from and communicate with interrogators within the frequency range from 860 MHz to 960 MHz.

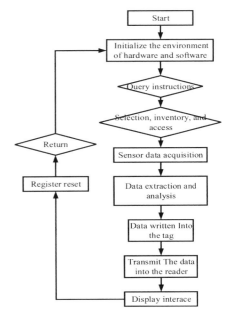

Figure 4. System flow diagram.

An interrogator manages tag populations using three basic operations: Select, Inventory, and Access. Tags shall implement a reply state. Upon entering a reply, a tag shall backscatter an RN16. If the tag receives a valid acknowledgment (ACK), it shall transition to the acknowledged state, backscattering the reply of its PC, EPC, and CRC-16. If the tag fails to receive an ACK or receives an invalid ACK or an ACK with an erroneous RN16, it shall return to arbitrate.

3.4 PC display interface

This design involves two kinds of PC display interface. One kind of interface can display the waveform of the measured data, which is obtained from the host computer connected directly with the pulse

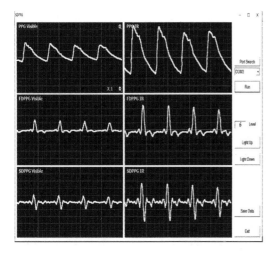

Figure 5. Information obtained directly from the pulse blood oxygen sensor.

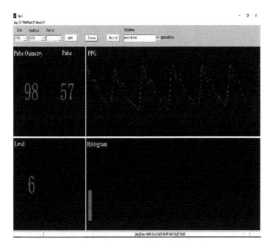

Figure 6. Data collected through the sensing system.

blood oxygen sensor module through the USB interface. The other can display data measured by the pulse blood oxygen sensor after processing by the data processing module and RFID transceiver module. The two pulse blood oxygen display interfaces are as follows:

1. The display interface of the pulse blood oxygen module can not only display the red and infrared PPG signals directly but can also display the red and infrared SDPPG signals in the interface as shown in Figure 5.

2. The real-time pulse blood oxygen information collected through the sensing system can be displayed by the waveform diagram, histogram,

and numerical method with the interface as shown in Figure 6.

4 EXPERIMENTAL VALIDATION

After debugging the software and hardware of the system, the clip type for adult probe BSJ09001C is selected from Shanghai Berry for contrast. Data measured by the oximeter will be compared with the data measured by the sensing system in order to verify further feasibility and accuracy of the system. Twenty individuals of the same age are selected for testing. Each person must have no strenuous exercise before the test and all subjects must have rest for at least 5 minutes before the signal was extracted. During the measurement, the room should be quiet and the result will be recorded when the measured data is stable.

The data measured by the pulse oximetry sensing system was analyzed and compared to the data obtained from the pulse oximeter. The results of blood oxygen saturation and pulse rate are shown in Figure 7 (a), and 7 (b). Series 1 represents the data measured by the blood oxygen sensor, and series 2 represents the data measured by the system.

The results of error analysis of blood oxygen saturation and pulse measurement are shown in Figure 8. The series 1 and 2 respectively represent the percentage error of the pulse rate and oxygen

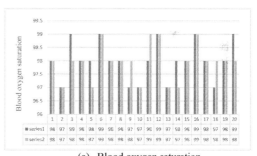

(a) Blood oxygen saturation

(b) Pulse rate

Figure 7. Results of blood oxygen saturation and pulse rate.

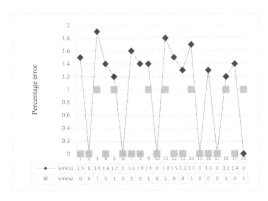

Figure 8. The results of error analysis of the system.

saturation. After calculation, the average error of blood oxygen saturation is 0.35% and the maximum error is 1%. The average error of pulse rate is 1% and the maximum error is less than 2%. Overall, the system has good stability. It shows that this method can be applied to real-time monitoring of pulse and blood oxygen saturation basically to meet the design requirements.

5 CONCLUSION

In this paper, a reflective pulse oximetry sensing system is proposed based on active RFID tags. A sensor combined with a microprocessor and wireless communication technology can realize the function of real-time monitoring of pulse blood oxygen information in the middle and long distance. The experimental results prove the feasibility of the research project and the system has the characteristics of a stable operation, high accuracy, and low power consumption, and so on. Along with the development of science and technology, the improvement of medical treatment level, the characteristics of the miniaturization, and wireless real-time monitoring have good application prospects in the direction of the wearable medical wisdom.

ACKNOWLEDGMENTS

This research was funded by the Tianjin Research Program of Application Foundation and Advanced Technology (No. 15 JCYBJC16300), and the Tianjin Science and Technology Commissioner Project (No. 16 JCTPJC45500).

REFERENCES

Catarinucci, L., S. Guglielmi, V. Mainetti, An energy-efficient MAC scheduler based on a switched-beam antenna for wireless sensor networks, Journal of Communication Software and Systems, 9(2), 117–127 (2013).

Kirk, V., S. Bohn, W. Flemons, and J. Rammers, Comparison of home oximetry monitoring with laboratory polysomnography in children, Amer. College Chest Phys, 124, 1702–1708(2003).

Ley, S., D. Laqua, P. Husar, Simulation of Photon Propagation in Multi-layered Tissue for Non-invasive Fetal Pulse Oximetry, The 15th International Conference on Biomedical Engineering, 43, 356–359(2013).

Mendelson, Y., Noninvasive pulse oximetry utilizing skin reflectance photo plethysmography, IEEE Transactions on Biomedical Engineering, 35, (10), 798–805(1988).

Occhiuzzi, C., G. Marrocco, The RFID Technology for Neurosciences: Feasibility of Limbs' Monitoring in Sleep Diseases, IEEE Transactions on Information Technology in Biomedicine, 14(1), 37–43(2010).

Peris-Lopez, P., A. Orfila, A. Mitrokotsa, and J.C.A. van der Lubbe, A comprehensive RFID solution to enhance inpatient medication safety, International Journal of Medical Informatics, 80(1), 13–24(2011).

Redondi, A., M. Chirico, L. Borsani, An integrated system based on wireless sensor networks for patient monitoring, localization, and tracking, Ad Hoc Networks, 11(2), 39–53(2013).

Toh, S., K. Do, Health decision support for biomedical signals monitoring system over a WSN, in Proc. 2nd Int. Symp. Electron. Commerce Security, 605–608(2009).

Research of the long-range greenhouse environment monitoring system based on the cloud technology

Hao-wei Zhang
School of Electronics and Information Engineering, Tianjin Polytechnic University, Tianjin, China
Engineering Research Center of High Power Solid State Lighting Application System, Tianjin, China

Ping-juan Niu
School of Electrical Engineering and Automation, Tianjin Key Laborary of Advanced Electrical Engineering
and Energy Technology, Tianjin Polytechnic University, Tianjin, China
Engineering Research Center of High Power Solid State Lighting Application System, Tianjin, China

Hai-tao Tian
School of Electrical Engineering and Automation, Tianjin Key Laborary of Advanced Electrical Engineering
and Energy Technology, Tianjin Polytechnic University, Tianjin, China

Wei-fang Xue
School of Electronics and Information Engineering, Tianjin Polytechnic University, Tianjin, China

ABSTRACT: The purpose of the greenhouse environment monitoring is to provide a good growing environment for plants, increase crop production, cut down the growth cycle and reduce the unconscious influence on the greenhouse. Combine the smart city, plant factories, large data, other hot evaluation, and design a system of long-range greenhouse environment monitoring based on the cloud technology combine the smart city. The system is mainly made up of cloud technology platform, STM32 microprocessor, sensor module, wireless module and monitoring terminal (mobile phones and personal computers), which realizes the long-range wireless environment monitoring and controlling. The System of STM32 microprocessor connects the sensor module and wireless module to collect the data of greenhouse environment (such as temperature, humidity, light, etc.), and transmits them to the cloud platform. Monitoring terminal visits the cloud platform to get and display the greenhouse environment data through TCP/IP. This system adopts the connection of cloud platform technology and Wi-Fi technology, which is more convenient than the traditional network and Bluetooth technology. The results show that the data collected by the system is stable, more real-time and has lower packet loss rate.

1 INTRODUCTION

The greenhouse environment monitoring and control obtains the optimal growth conditions for crops through changing the greenhouse environment factors (such as temperature, humidity, light intensity, etc.) (Zhang, 2013; Guo, 2010; Chen, 2011; Mittal, 2012), so as to prevent crop disease, improve crop quality and increase economic benefits. Our country is a large agricultural country with vast area, and every year serious losses are often caused by natural disasters to crops (Robert, 2008; Zhang, 2013; Zhao, 2012). Deploying the sensor nodes within the farm, access to climate conditions in real time and make corresponding measures are helpful for improving the yield of plants. Many scholars have conducted the research

in this field: the greenhouse environment monitoring system based on GSM message designed by Zhao Liyan et al. (Zhao, 2009) realized the real-time monitoring of environmental temperature, humidity and light intensity. The greenhouse environment monitoring system based on Bluetooth technology designed by Li Li et al. (Li, 2006) realized the wireless greenhouse environmental information collection through the Bluetooth. The greenhouse environmental data and control system based on ARM designed by Miao Fengyun, et al. (Miao, 2015) realized the greenhouse environment control in view of the damage of dew condensation on crop surface.

Greenhouse is a kind of advanced facility agriculture (Li, 2005), and the farmers can control environmental conditions to increase crop yield and avoid the influence of climatic conditions on crops (Liu, 2015; Li, 2009). Greenhouse also plays an important role in anti-season vegetables and flowers. Traditional greenhouse monitoring system is mostly implemented by the wired multipoint system with complex wiring, difficult installation and poor anti-interference ability. The wireless transmission can solve the shortage of the traditional greenhouse monitoring, but Bluetooth and Wi-Fi can only achieve the short transmission control. GSM and GPRS can realize the wireless transmission control, but the SMS cost and traffic cost are high.

The research aims to design and implement a kind of remote wireless real-time greenhouse environment monitoring system based on Wi-Fi technology and cloud platform technology, so as to solve the above-mentioned problems existing in the greenhouse monitoring. STM32 is the main control chip, combined with the sensor module, realizes the collection of greenhouse environment, and uses ESP8266 Wi-Fi module to implement the wireless transmission of environment factor data; the cloud platform realizes the data storage, monitoring terminal realizes the data display and control, as well as the mass data processing in greenhouse, and reduces the influence of staff on the greenhouse environment. Also, the management personnel can timely understand the change of the greenhouse environment, control the greenhouse environment and increase the crop yield and quality.

2 SYSTEM HARDWARE PLATFORM DESIGN

This system mainly consists of the power module, sensor module, wireless module, mobile phone APP and cloud platform, etc. The development board and LED of power module provides the voltage for operation; sensor module is used for real-time acquisition of greenhouse environment parameters; wireless module is used for the acquisition of terminal access networks and wireless data transmission to the cloud platform; mobile phone APP is used for the remote access to environmental parameters and issuing control instructions; cloud platform is mainly used for data processing and storage. The schematic diagram is as shown in Figure 1.

The system main control chip uses the Corter-M3 microprocessor chip of ST Company with the working frequency up to 72 MHZ, cache of 512 bytes, SRAM of 64 K bytes and rich I/O interface, and the interrupt processing can satisfy the requirement of high real-time performance. The chip has rich resources and perfect software

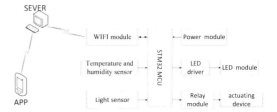

Figure 1.   General schematic diagram of the system.

library, can simplify the system hardware, greatly shorten the development cycle and reduce development cost; therefore, it is widely used in various kinds of electronic products, such as the electronic measuring instrument, environmental monitoring, hand held measuring tools, etc.

2.1 Wireless module design

The system uses the wireless access networks and adopts the wireless module highly integrated by ESP8266 chip with the built-in TCP/IP protocol, 32-bit microprocessor and a set of complete and self-contained Wi-Fi solution, which can carry the software application, and uninstall all network functions through another processor. The module has powerful processing and storage capabilities through GPIO port integrated sensor and other application specific equipment, so in the process of development, it occupies less system resources. Due to the low power consumption of system, it is widely used in mobile devices and wearable products. The module in this system is used to connect cloud server platform, realize the environment factor data transmission of collection terminal, and the connection with the MCU is as shown in Figure 2.

2.2 Sensor module design

The system environment temperature and humidity is collected using DHT11 sensor containing a resistor type moisture component, a NTC thermometer component and high-performance single chip microcomputer for processing data with the standard calibration function. The standard calibration coefficient is stored in the OTP memory, therefore, the sensor has high reliability and stability. It has widely measuring range, high precision, so it can satisfy the measurement task of this system. Its internal structure is as shown in Figure 3.

Light intensity of the system is collected using the TSL2560 chip sensor module of TAOS Company, and the module can convert light intensity analog signal into digital signal, which is a kind of high conversion rate, low power consumption,

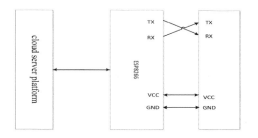

Figure 2.   Wi-Fi module connection schematic diagram.

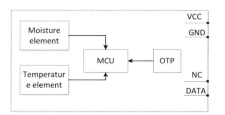

Figure 3.   Internal structure diagram of DHT11.

wide range digital conversion chip. There are two channels inside the module, containing two photosensitive diodes, and the channel 0 is sensitive to visible light and infrared ray, and only channel 1 is sensitive to infrared light. The current through the photosensitive diode accumulates via A/D converter, and is translated into digital quantity stored in the chip. The chip has the programmable light intensity threshold, when the actual light exceeds the threshold, the interrupt signal will be sent, advantageous for the constant illumination function and satisfy the light acquisition task. Its internal structure is as shown in Figure 4.

2.3   Relay control circuit

The system environment control module has the relay module and single chip. The single chip receives the control commands from the upper machine, and executes the command to control the relay, so as to realize the control of heating system, ventilation system and the environment inside the greenhouse. Relay control circuit diagram is as shown in Figure 5 below.

2.4   PWM dimming modules

According to the special voltage and current characteristics of LED, it adopts the constant current drive mode, and the LED driver uses the PT4115 chip design. The chip is a step-down constant current source of continuous inductor current conduction mode, used to drive the one or more series

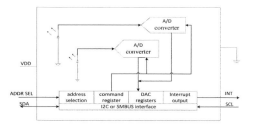

Figure 4.   Internal structure diagram of TSL2560.

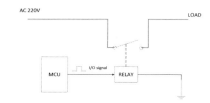

Figure 5.   Relay connection schematic diagram.

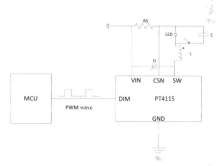

Figure 6.   The connection schematic diagram of PWM dimming module and MCU.

LED. It has independent PWM control interface, and the connection with MCU is as shown in Figure 6.

PWM wave produced by STM32 is used to adjust the light intensity, the GPIO interface outputting the PWM wave is connected to the LED driver module, the output of LED driver is connected to the LED module, and MCU controls the brightness of LED lights by controlling the output of PWM wave.

3   SYSTEM SOFTWARE DESIGN

System software design includes the design of collection software, cloud platform building and monitoring terminal software design. The software and corresponding hardware platform work together, so as to realize the system function.

3.1 Collection software design

STM32 is the main control chip of collection end as the core of intelligent environmental factors acquisition and control. On the one hand, controlling the ESP8266 module access network realizes the connection with the cloud platform; on the other hand, the sensor module realizes the real-time acquisition and processing of greenhouse environmental parameters; it should complete the data transmission and execute commands from the PC, and start the actuators.

After the system is powered on, initialize the serial port and Wi-Fi module, guarantee that the equipment is connected to the network; after joining the network, the equipment will automatically send the IP address to cloud platform. Then initialize the sensor module to collect the environment temperature and humidity, light intensity, environmental parameters, and send to cloud storage platform, and then wait for the outside wireless packet. When receiving the wireless packet, detect the data packet, determine whether it is the device's IP address; if that, parse the packet; if not, discard the packet. When the accepted packet is the control command (such as the control of pump and humidifier, etc.), start the executing agency, execute the corresponding action, and complete the corresponding function.

3.2 Cloud platform building

Cloud platform technology is a new kind of data processing technology with a large volume of data, various data types, low value density, high commercial value and processing speed (Lu, 2016; Liu, 2016). Greenhouse environmental information collection and control contains huge amount of information. The cloud platform technology can help realize the greenhouse environment information acquisition and processing in different regions, providing accurate and timely data for the crop growth research.

The cloud platform of this system has perception layer, transport layer, cloud service layer and access layer. Perception layer is at the front end of the cloud platform, responsible for remote acquisition and control of greenhouse environment parameters; the transport layer is responsible for the transmission of greenhouse environmental data, user preferences and related commands. The cloud services layer is the core of the greenhouse environment monitoring system, used to provide the cloud infrastructure; the cloud application is based on the cloud platform infrastructure, the core of the cloud layer, which allows the user to view environmental parameters and control it anytime and anywhere, share logs, experience farming

fun; the access layer is used for the mobile terminal of platform, and provide the APP way into the cloud through the cloud computing center.

3.3 Mobile terminal software design

Mobile terminal system is the equipment of obtaining the environmental information and sending control command with the mobile phone as the mobile terminal equipment, making it convenience for users to access to information. Android is the open source mobile operating system developed by Google, including the interface development, application development, etc. This system uses the cloud platform as the intermediate connecting device, and the client development uses the Java language of Android, combined with scoket for network communication and data interaction, and produces the APK installation file. It can be installed directly on the Android mobile phone; compared to the traditional wireless environment monitoring system, this system is not subject to the region, time, distance and other environmental factors. The software design of mobile terminal is as shown in Figure 7.

After the system client software starts, first of all, initialize the interface, enter the interface button and click the QR code scanning to scan the acquisition terminal equipment, the equipment system will automatically generate identifiable ID (only IP and port number), and

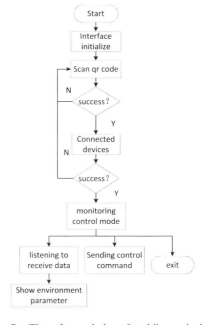

Figure 7. The software design of mobile terminal.

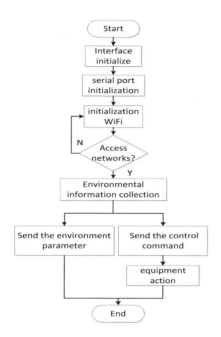

Figure 8. The software design of acquisition side.

the device ID is stored in the cloud platform; connect the devices into environmental monitoring control mode, send packets to the cloud platform and start listening mode, receive the environmental parameters from the cloud platform, and send the corresponding control command to the device through the control button. The software design of acquisition side is as shown in Figure 8.

4 SYSTEM FUNCTIONAL TEST AND ANALYSIS

In order to verify the feasibility of the designed monitoring system application and other performances, the whole system should be operated and tested.

4.1 System functional test

The test environment of this system is the smart plant growth chamber produced by Haiyu Company of Tianjin Polytechnic University; the monitoring system is installed in intelligent plant growth box with serial number in each growth chamber. Monitoring terminal uses the GALAXY Grand 7106 mobile phone produced by Samsung; inside the mobile phone is the designed APP; open the APP and scan QR code on the growth chamber, and make do a specific identification for each growth chamber for convenient control and management of growth chamber.

Supply the power to the system, the system accesses to the network and begins to collect environment parameter, open the phone App, enter the environment parameters monitoring page, and you can see the current plant growth parameters, such as temperature, humidity and light intensity. The pages are as shown in Figure 9 above. Enter the settings page, set time, plant growth and other information as shown in the Figure below. Different buttons can be used to realize the control of collection relay, and the control of environmental parameters. The interface is as shown in Figure 10 below.

4.2 Analysis of system performance

To test and verify the packet loss rate of system, send the temperature parameters request command to different growth chambers on the PC

Figure 9. Environmental monitoring page.

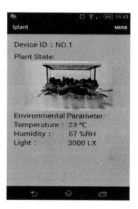

Figure 10. Environmental control page.

Table 1. Unicast request packet loss rate.

Number	Number of sending data	Back packets	Packet loss rate (%)
1	2000	1995	0.25
2	2000	1996	0.20
3	2000	1996	0.20
4	2000	1993	0.35
5	2000	1995	0.25
6	2000	1994	0.30
7	2000	1991	0.45
8	2000	1992	0.40

side in the form of unicast, so as to test the system packet loss rate. The unicast packet loss rate is as shown in Table 1.

The system adopts cloud platform technology and Wi-Fi technology with no distance limit. No matter when and where it is, there is network in the phone, it can realize the real-time monitoring of greenhouse environment. In the case of 2G network, it can also refresh the page quickly.

5 CONCLUSION

This paper designs and implements a kind of remote wireless greenhouse monitoring system combined with cloud platform technology and Wi-Fi technology, so it effectively overcomes the traditional problems of greenhouse environment monitoring and management. It combines the cloud platform and Internet technology that the users are allowed to obtain and control the greenhouse environment parameters through App, so it effectively avoids the crop losses due to the improper environment parameters in greenhouse. Also, it realizes the unified management of the greenhouse environment monitoring, gets rid of the traditional individual greenhouse environment monitoring system, effectively realizes the multiple remote monitoring and management of greenhouse. It eliminates the defects of traditional greenhouse management, thus exploring the development of greenhouse environment monitoring and providing strong theoretical basis.

REFERENCES

Ai, Q., Chen, C., "Green House Environment Monitor Technology Implementation Based on Android Mobile Platform", IEEE Conferece Publications. Pages (s):5584–5587, 2011.

Guo Wenchuan, Cheng Hanjie, Li Ruiming, et al. Greenhouse environmental information monitor system based on wireless sensor network [J]. Journal of agricultural machinery, **41**, 181–185 (2010).

Li Haisheng. Remote monitoring system based on GSM short message [D]. Yanshan University, (2005).

Li Li, Liu Gang, Design of greenhouse environment monitoring system based on Bluetooth technology[J]. Journal of agricultural machinery, **39**, 97–100 (2006).

Li Yuan. Design and research of wireless environment monitoring system based on ZigBee [D]. Hunan University, (2009).

Liu Lufeng. Cloud platform construction research for university library discipline service under big data environment [J]. Library Journal, 03: 99–101+108 (2016).

Liu Tong, Li Yao, He Hongwei, Ma Jianshe. Closed LED plant factory monitoring and control system based on ZigBee[J]. Journal of Agricultural Mechanization, **37**, 75–81 (2015).

Lu Hui, Gao Hongbo, Zhang Fengman, et al. Job scheduling algorithm based on resource prediction in Hadoop cloud platform [J]. Computer application research, 08:1–6 (2016).

Miao Fengjuan, Tao Bairui, Liu Tongkai, et al. Design of greenhouse environment data acquisition and control system based on ARM [J]. Journal of Agricultural Mechanization Research, **37**, 138–141 (2015).

Mittal, M., Tripathi, G., "Green House Monitor and Control Using Wireless System Network", VSRD-IJEECE, Vol. 2 (6), 2012, 337–345, 2012.

Yunseop (James) Kim, Robert G. Evans, and William M. Iversen, "Remote Sensing and Control of an Irrigation System Using a Distributed Wireless Sensor Network," IEEE Transactions on Instrumentation and Messagement, vol. 57, No. 7, July 2008.

Zhang Meng, Fang Junlong, Han Yu. Design of remote monitoring and control system based on ZigBee and Internet for greenhouse group [J]. Journal of Agricultural Engineering, **29**, 171–176 (2013).

Zhang Xingwei. Research and design of greenhouse environment monitoring system based on WSN [D]. Zhengzhou University, (2013).

Zhao Chunjiang, Qu Lihua, Chen Ming, et al. Design of image sensor node for greenhouse environment monitoring based on ZigBee [J]. Journal of agricultural machinery, **43**, 192–196 (2012).

Zhao Liyan, Xu Liang. Greenhouse environment monitoring system based on GSM short message [J]. Electronic design engineering, **17**, 29–31 (2009).

Civil, Architecture and Environmental Engineering – Kao & Sung (Eds)
© 2017 Taylor & Francis Group, ISBN 978-1-138-02985-9

Negative emotion speech classification for a six-leg rescue robot

Xiaolei Han
College of Information and Computer Science, Shanghai Business School, Shanghai, China

ABSTRACT: This paper aims at providing an approach for a user-dependent emotion recognition system of affective speech based on multiple features using prosodic information and Higher Order Spectra (HOS) analyses. Prosodic features including speech rate, short-term energy, and pitch-related features are extracted to indicate the effect of linear aspects, and HOS analyses are utilized to indicate the impact of nonlinear aspects of affective speech signal. Some significant representatives are taken to form the reduced feature set and build the classifiers for emotion recognition. The proposed algorithm is implemented on our six-leg robot, which is designed for search and rescue tasks in real disaster sites using the predictive accuracy of dynamic time warping. The effectiveness of this algorithm in both emotional speech signal recognition and feature dimensionality reduction is illustrated with some experimental results. An important finding is that the proposed algorithm is shown to be effective for identifying negative-type emotion with excellent predictive capability.

1 INTRODUCTION

A huge earthquake with the magnitude 9.0 hit Fukushima, Japan on March 11th, 2011 and has raised the alarm of emergency measure problems both in information and robotics technology for the case of large-scale disasters. The advantage of utilizing rescue robots in such dangerous situations is that they are able to carry out high-risk tasks without exposing rescue crews to danger (Valero, 2010). This means that the rescuers' lives are less at risk and that survivors can be more easily and quickly located. If the robot cannot bring the survivors out on its own, it should be able to detect and transmit survivors' vital signs and emotional states to rescuers and doctors and keep the surviving life a friendly company while rescuers try to reach them.

Under such emergency conditions, the trapped survivors easily plunge into desperation and their audio signals normally have aspects of negative emotion (e.g., fear, sadness, panic, stress, hopelessness, and despair). With the ability of cognitive and affective reaction, once the rescue robot perceives a negative emotion through survivors' speech, it could try to ease and encourage them as rescuers or psychologists do. Recent developments in speech-based emotion recognition enable us to use them for this case (Koolagudi, 2012). In fact, negative emotion classification can not only be used to improve the naturalness and effectiveness of interaction between robots and humans, other applications can also benefit such as life-support systems,

in-car driver interfaces, call-center management, educational software, games, etc. (Koolagudi, 2012; Sravros, 2012; Ververidis, 2006; Chen, 2007).

Our motivation is to develop an emotional speech detection system for survivors and rescue robots in the context of disasters (earthquake, fire, building collapse, etc.). With emotional speech recognition ability, a teleoperated rescue robot could detect, analyze, and feed a survivor's vital signs and emotional states back to the control base outside the disaster area, and it could also be a friend to accompany trapped people while rescuers are trying to reach them. The main novelty is that we adopt Higher Order Spectra (HOS) to analyze nonlinear aspects of affective speech signals, and try to search for embedding dimension and associated features that best show the emotional class structure of our data.

2 PREPARATIONS

2.1 Six-leg rescue robot

The first version of our six-leg teleoperated rescue robot (Figure 1) is developed for executing rescue missions at disaster areas. With the help of rubber suction cups, it is capable of surmounting obstacles, operating valves, etc. in disaster areas. The sensory equipment includes auditory, tactile, ultrasonic, and vision sensors, which allow the rescue robot to behave autonomously and interact with rescuers. For protection consideration, all processing and control systems are located in the robot's body.

Figure 1. The six-leg rescue robot and a simulated scene of operating the valve.

After locating the survivors, the rescue robot can recreate and model its surroundings, and feed this information to rescuers waiting outside. Meanwhile, the rescue robot friend accompanies the survivors until rescuers arrive.

2.2 Difficulties of emotional speech recognition in disaster areas

The current ability of the six-leg rescue robot's emotional speech recognition is not satisfied due to noise from the disaster environment. Rescuers can communicate with survivors even under quite noisy conditions. However, such noise is critical for the rescue robot. Meanwhile, the lack of corpus and studies dealing with emotions in the disaster environment encouraged us to build the SJTU corpus, which is a multi-language emotional database including seven emotion states.

Meanwhile, survivors' emotional expressions occurring in the disaster area may have some particularities. Though much remains to be discovered, survivors' emotions seem to be more intense and biased toward certain types of emotion states, such as inner strain and destructiveness of sadness, and fear controlled by the intensity of roiling emotions that block recovery. Some of these particularities are beneficial for emotion recognition because the recognition model can be built to interact specifically with survivors.

2.3 SJTU emotional database

Adapting any general emotion recognition system for a particular individual requires speech samples and prior knowledge about their emotional content. However, in real disaster scenarios, annotated emotional speech date is unavailable to train or adapt the models. To address this problem, this paper introduces an ongoing recorded SJTU corpus which is a syllable-based multi-language

Table 1. Number of utterances available per emotion type.

	Ch_M		En_M		Ch_F		En_F	
Emotion	Tr	Te	Tr	Te	Tr	Te	Tr	Te
Happiness	90	60	191	130	297	100	180	120
Sadness	180	120	401	270	147	200	450	300
Fear	135	90	296	200	222	150	315	210
Neutral	45	30	87	60	72	50	90	60
Total	450	300	975	660	738	500	1035	690

emotional database (English, Chinese, German, and Italian) including seven emotions: neutral, surprise, sadness, disgust, fear, happiness, and anger. The corpus was recorded in a sound-proof room (lower than 35 dB) where it can block any outside noise. So far, a total of 10 college students (5 males 25.6 years ± 3.65 and 5 females 25.8 ± 2.49) without any history of voice disorders or larynx diseases participated in this research. Emotional stimuli used in this study are audio-visual film clips, which were examined for their appropriateness and effectiveness through a preliminary experiment.

Considering the factors such as lower computational power needed for speech processing, the 16-bit recordings were taken with a sampling frequency of 16 kHz. Emotional audio data are saved in. wav format. For our emotion detection system, we selected 5348 utterances through subjective evaluation tests by 20 listeners (students from Shanghai Jiao Tong University in China) including four emotions: happiness, sadness, fear, and neutral (Table 1). For this research, we assume sadness and fear as negative-type emotions. The meanings of the abbreviations in Table 1 and the rest of this paper are shown as follows:

- Ch_M, Ch_F: Chinese utterances by male speakers and female speakers.
- En_M, En_F: English utterances by male speakers and female speakers.
- Tr, Te: The number of utterances for training and testing.

3 ACOUSTIC FEATURES

It is well known that the performance of emotional speech recognition algorithms is greatly influenced by some fundamental issues in which parameters are selected as emotional features. Some researchers combined tens or even thousands of parameters to enhance performance in recognition of emotional states through speech. The most common features are prosodic features, qualitative features, vocal tract features, and features based

Table 2. List of extracted parameters.

Type	Abbr.	Description	No.
Prosodic parameters	Duration	Ratio of voicing against unvoiced segments	1
	Pause	Ratio of pauses against speech duration	1
	MeanF0	Mean pitch	1
	VarF0	Pitch variance	1
	StdF0	Standard deviation of pitch	1
	MedianF0	Median pitch	1
	MaxF0	99% value of pitch	1
	MeanEn	Mean of short time energy	1
	VarEn	Short time energy variance	1
	StdEn	Standard deviation of short time energy	1
	Median En	Median short time energy	1
	Max En	99% value of short time energy	1
	Min En	1% value of short time energy	1
Nonlinear Parameters	HOS	Proposed HOS parameters	128

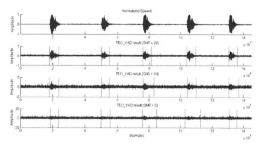

Figure 2. The TEO-based VAD result of the noisy speech signal with neutral emotion at SNR of 20, 10, and 5 dB.

to create noisy signals at specific SNR 20, 10, and 5 dB (Figure 2). Then, each incoming speech signal is divided into 30 msec. long with 10 msec. overlapping duration. The features of all frames in one emotion syllabic unit are clustered to form a feature vector. All features are then standardized with zero mean and standard deviation of one.

3.1 Prosodic features

As prosody is believed to be the primary indicator of a speaker's emotional state (Lee, 2005; Ingale, 2012), most of the works adopt prosodic features (Koolagudi, 2012). In this work, we extracted 14 prosodic parameters as listed in Table 2. Duration means the ratio of voicing against unvoiced segments. Pause here means the ratio of pauses against speech. Pitch (F0) contains information about emotion because it depends on the tension of the vocal folds and the subglottal air pressure, measured in Hz (Ververidis, 2006; Chen, 2007; Lee, 2005; Ingale, 2012). Short-term energy is based on the speech signal's amplitude, which is related to the arousal level of emotions (Ververidis, 2006).

For emotion classification, not all parameters contribute. Some of the parameters may be highly correlated with redundant information and hence may not be optimal. We use the absolute value of Pearson's correlation coefficient to calculate the linear correlation between different parameters, p_i and p_j:

$$\rho_{i,j} = \left| \frac{\text{cov}\left(p_i, p_j \right)}{\sigma_i \sigma_j} \right| \tag{1}$$

From Cauchy-Schwarz inequality, we know $\rho_{i,j}$ is less than or equal to 1. We find $\rho_{4,5}$ $\rho_{4,5}$, $\rho_{8,11}$, and $\rho_{9,10}$ presents significant linear correlation with minimum value varying from 0.91 to 0.96. It is crucial to discard irrelevant and redundant features for efficient computation and accurate

on the Teager energy operator (Sravros, 2012; Ververidis, 2006).

Teager suggested that the true source of sound production is actually the vortex-flow interactions, which are nonlinear. Therefore, nonlinear speech features are necessary to classify negative emotion speech from neutral. HOS offers a novel set of techniques, algorithms, and methodologies for the study and analysis of signals, especially it can detect and characterize nonlinearities in time series and can suppress additive Gaussian noise of unknown spectral characteristics. Emotional speech has significant nonlinearity, thereby making HOS a potential candidate in detection, parameter estimation, and classification problems. The objective of this study is to investigate whether or not bispectral analysis, a particular form of HOS, may be utilized for analyzing the emotion states through speech. As long as there is no common agreement on the best features to use, we chose to combine the most widely investigated prosodic parameters and the proposed HOS parameters as emotional feature sets which are listed in Table 2.

After pre-emphasis, we use the Teager Energy Operator (TEO)-based Voice Activity Detection (VAD) method to segment the utterances into syllabic units (Chen, 2007). To verify the performance of TEO based VAD method, we vary the testing condition. Noise is added to the clean speech corpus

classification. According to the average value of $\{\rho_4, \rho_5, \rho_8, \rho_9, \rho_{10}, \rho_{11}\}$ it is equal to $\{0.48, 0.46, 0.57, 0.45, 0.47, 0.52\}$. We eliminate the pitch variance, mean and standard deviation of short time energy (index 4, 8, and 10) to reduce the higher correlation.

3.2 HOS features

Let $X(n)$ represents a syllabic unit of digital discrete-time emotional speech. Assuming the emotion expression of $X(n)$ is stable within a word and its moment up to order k exists, the n-th moment function of $X(n)$ is defined by

$$
\begin{aligned}
&m_k\left(\tau_1, \tau_2, \cdots, \tau_{k-1}\right) \\
&= E\left\{x(n)x(n+\tau_1)\cdots x(n+\tau_{k-1})\right\}
\end{aligned} \tag{2}
$$

where $E\{\cdot\}$ stands for the expected-value operator. m_k only depends on the time differences $\tau_1, \tau_2, \cdots, \tau_{k-1}$. Bispectral analysis is related to the third moment (skewness) of $X(n)$. The relationship between the 3rd-order cumulant and the moment $m_3(\tau_1, \tau_2)$ is given by

$$
c_3\left(\tau_1, \tau_2\right) = m_3\left(\tau_1, \tau_2\right) \tag{3}
$$

Let $\tau_1 = \tau_2 = 0$ in equation(2) and (3). The skewness γ_3 is obtained by

$$
\gamma_3 = c_3\left(\tau_1, \tau_2\right) = E\left\{x^3(n)\right\} \tag{4}
$$

The bispectrum is the Fourier transform of the third order correlation of $X(n)$, which can be estimated by

$$
B(f_1, f_2) = E\left\{X(f_1)X(f_2)X^*(f_1 + f_2)\right\} \tag{5}
$$

where * denotes the complex conjugate. X(f) is the Fourier transform of $X(n)$. For more details of HOS please refer to (David, 1991; Mendel, 1991).

In order to observe the bispectrum of emotional speech, we do 128 samples of FFT for each syllabic unit. The contours plots of the magnitude of the indirectly estimated bispectrum for the "Sh-owl" (this order in Chinese will let the horse step back) by the same speaker in different emotions are shown in Figure 3. We notice that the bispectral analysis is capable of distinguishing different emotions. Each bispectrum is a 128×128 array, which is too large to be used in emotional speech recognition algorithm. To this end, the eigenvector and eigenvalue of the bispectrum are extracted. Through experiments, we found that each bispectrum matrix only has one eigenvalue. We only take the first column of the eigenvector,

Figure 3. The bispectra of neutral, surprise, sadness, disgust, fear, happiness, and anger on bifrequency plane for "Sh-owl".

V, into consideration, which is a 128×1 vector. V is further normalized as follows:

$$
\widehat{V} = |V| / \sqrt{\left(\Sigma V^2\right)} \tag{6}
$$

where $\widehat{V}$ is the emotional feature of HOS analysis.

Compared to the dimension of $\hat{V}$, the size of training data is small, which will direct influence of the accurate rate of recognition. Therefore, we introduce Nonlinear Iterative Partial Least Squares (NIPALS) algorithm to handle this problem. Moreover, from Pearson's correlation coefficient, we find the significant correlation period of $\hat{V}$ is 5, and values of $\{\rho_{i,i+5}\}$ ($i = 1, 2,\ldots, 123$) are higher than 0.9. This means $\hat{V}$ contains a high correlation. The Nonlinear Iterative Partial Least Squares (NIPALS) algorithm has been developed by H. Wold, first for PCA and later on for Partial Least Squares (PLS). It can handle data matrices with a high degree of collinearity and perform well when the size of training data is small. NIPALS has been successfully applied in speech and visual related works. Thus, in this work, we apply the PLS method which is based on the NIPALS algorithm to training HOS feature vectors. More details about NIPALS can be seen in (Bowden, 2010; Alin, 2009).

4 EMOTION CLASSIFICATION AND RESULTS AND DISCUSSION

As an emotion classifier, we utilize the NIPALS method for HOS parameters and use a Dynamic Time Warping (DTW) method for prosodic features, since DTW is well proven although it has some imperfections. The results from emotion detection are classified into three categories: positive-type emotion (denoted as Pt), Negative-type emotion (Nt), and Neutral emotion (Ne). If a happiness emotion is detected from the speech, the system classifies it as a positive-type emotion.

If we detect fear or sadness, the system classifies it as a negative-type emotion. Otherwise, the system assumes a neutral emotion. As a result, if a negative-type emotion is detected, the rescue robots should take appropriate countermeasures to console and encourage survivors.

4.1 Results of emotion recognition performance

We compare the emotion recognition output from the system with labeled emotions. The results of the comparison are shown in Table 3. The "Average" means the average accuracy rate for emotion states. As a result, the average accuracy of happiness, sadness, fear, and neutral emotion detection from the speech is 73.03%, 82.43%, 67.59%, and 87.58%, respectively.

4.2 Results of recognition performance by using different features

Furthermore, we observe the effects of emotion category recognition and compare the performance of proposed features with some of the most commonly investigated emotional features:

- F_1: HOS parameters.
- F_2: Set 1, Set 2, and Set 3.
- F_3: F_1 and F_2.

F_1 is the proposed HOS feature, which presents the nonlinear aspect of the emotional speech signal. F_3 is the proposed feature vector which can also be used in Table 3, and the final emotion category for F_3 is generated by the ballot. In order to prove the effectiveness of this new emotional feature, the most investigated features of F_2 is: 1) preprocessed by grouping Set 1, Set 2, and Set 3 together as one feature vector; 2) the final emotion category is generated by the ballot of sets of features with more than half of votes cast. Table 4 to Table 6 show the results of emotion category recognition (Pt: positive-type emotion, Nt: negative-type emotion, and Ne: neutral emotion). The "average" represents the rate that the system output correctly matched the labeled emotion states for all emotion states. As a result, the average accuracy by F_1, F_2, F_2-vote,

Table 3. Accuracy rate of emotion classification (%).

	Happiness	Sadness	Fear	Neutral
Ch_M	71.67	92.50	64.44	93.33
En_M	58.46	78.89	70.50	80.00
Ch_F	92.00	76.00	57.33	92.00
En_F	70.00	82.33	78.10	85.00
Average	73.03	82.43	67.59	87.58
	77.66			

Table 4. Accuracy rate of emotion category recognition by using F_1 (%).

	Pt	Nt		Ne
	Happiness	Sadness	Fear	Neutral
Ch_M	86.67	88.33	91.11	76.67
En_M	87.69	72.59	79.50	75.00
Ch_F	96.00	93.00	90.00	84.00
En_F	92.50	96.67	92.38	73.33
Average	90.72	87.65	88.25	77.25
	85.97			

Table 5. Average accuracy of emotion category recognition by using F_2 with votes cast (%).

	Pt	Nt		Ne
	Happiness	Sadness	Fear	Neutral
Ch_M	96.67	100.00	96.67	96.67
En_M	92.31	90.00	94.50	75.00
Ch_F	96.00	100.00	84.67	86.00
En_F	75.00	87.00	100.00	80.00
Average	90.00	94.25	93.96	84.42
	90.66			

Table 6. Average accuracy of emotion category recognition by using F_3 (%).

	Pt	Nt		Ne
	Happiness	Sadness	Fear	Neutral
Ch_M	96.67	100.00	96.67	100.00
En_M	90.00	90.00	94.50	85.00
Ch_F	100.00	100.00	86.00	94.00
En_F	79.17	87.00	100.00	85.00
Average	91.46	94.25	94.17	91.00
	92.72			

and F_3 is 85.97%, 79.57%, 90.66%, and 92.72%, respectively.

4.3 Discussion

The performance of proposed features with votes cast (Table 6) is better than the one of prosodic features with votes cast (Table 5), increasing from 0% to 6.58%. The results illustrate that combining the existing emotion features with the proposed HOS feature can enhance the performance to classify emotion states and the ability of emotion category detection.

Although the average accuracy of using the proposed emotion detection algorithm for happiness, sadness, fear, and neutral from speech is unsatisfied (77.66%), the one for emotion category recognition significantly increases the precision of the system (15.06%), that is to say the nonlinear aspect affects emotion expression and the HOS feature has a good ability to distinguish differences among positive-type emotion, negative-type emotion, and neutral emotion.

Since the prosodic features with votes cast perform better than the one without votes cast, increasing from 5.64% to 14.17%, the voting system is one of the important steps in emotion recognition. Meanwhile, some individual accuracy rates of Table 6 are lower than that of Table 5 and Table 3 because prosodic features, 3/4 of votes are the major factor that affects the final result of emotion category detection. On the other hand, the SJTU corpus was recorded by Chinese speakers, which will lower the naturalness of emotion expression in English. This may also be the reason why the accuracy rate of English is lower than Chinese.

The average accuracy of the negative-type emotion recognition is higher than the positive-type (2.75% in Table 6) and neutral emotion (3.21% in Table 6). The negative-type emotion recognition works well for this particular experiment because the proposed emotion recognition system is trained and evaluated under low noise conditions. The noise from a realistic disaster environment and the rescue robot itself will probably decrease the performance. Thus, for practical and effective use, we should improve the anti-noise ability of negative-type emotion detection performance in a real disaster situation. Our next step is to post-process the utterances by adding the noise of the rescue robot and disaster environment. The disaster environment noise can be acquired from the real disaster environments and disaster films, which are realistic and near to the situations defined by our application field.

5 CONCLUSION

This paper presented an approach to improve the interaction between rescue robots and survivors based on emotion recognition, particularly, negative-type emotion. The experiment results revealed that HOS parameters are essential for this approach. Experiments were carried out with the SJTU database. Finally, this method was implemented in our six-leg rescue robot. However, the present research is being done in laboratory circumstances and does not consider the influence by disaster environmental noise. In the next step, we would like to achieve higher recognition accuracy and enhance the capability of anti-noise to make rescue robots work in the real world.

ACKNOWLEDGMENTS

This research was partially supported by the Training Foundation for The Excellent Youth Teachers of Shanghai Education Committee (Grant No. sxy15002).

REFERENCES

Alin, A. *Statistical Papers*, **50**, pp. 711–720 (2009).

Bowden, J.C. *e-Science Workshops, Sixth IEEE International Conference*, pp. 25–30 (2010).

Chen, M.Y., H. Li. *Journal of Chongqing Institute of Technology (Natural Science Edition)*, **21**, pp. 112–114 (2007).

David, R.B. *Statistica Sinica*, **1**, pp. 465–476 (1991).

Ingale, A.B., D.S. Chaudhari. *International Journal of Soft Computing and Engineering*, **2**, pp. 2231–2307 (2012).

Koolagudi, S.G., K.S. Rao. *International Journal of Speech Technology*, **15**, 2, pp. 99–117 (2012).

Lee, C.M., S.S. Narayanan. *IEEE Transaction on Speech and Audio Processing*, **13**, pp. 293–303 (2005).

Mendel, J. *Proc. IEEE*, **79**, pp. 278–304 (1991).

Sravros, N., F. Nikos. *IEEE Transactions on affective computing*, **3**, pp. 116–125 (2012).

Valero, A., C. Saracini, P de la Puente, et al. *Proceedings Second International Symposium on New Frontiers in Human-Robot Interaction*, pp. 120–127 (2010).

Ververidis, D., C. Kotropoulos. *Speech Communication*, **48**, pp. 1162–1181 (2006).

Civil, Architecture and Environmental Engineering – Kao & Sung (Eds)
© 2017 Taylor & Francis Group, ISBN 978-1-138-02985-9

Nonlinear vibration of a commensurate fractional-order happiness system

Zhongjin Guo

College of Mechanical Engineering, Beijing University of Technology, Beijing, China
School of Mathematics and Statistics, Taishan University, Taian, P.R. China

Wei Zhang

College of Mechanical Engineering, Beijing University of Technology, Beijing, China
Beijing Key Laboratory on Nonlinear Vibrations and Strength of Mechanical Structures, Beijing, P.R. China

ABSTRACT: In this work, nonlinear vibration of a happiness model with commensurate fractional-order derivative is studied analytically and ascertained by numerical ones. The influences of the impact factor of memory on the curves of external frequency, excitation amplitude, and nonlinear damping coefficient versus maximal amplitude of one's response to external events are presented and compared by introducing the polynomial homotopy continuation technique. The result shows that when one's experiences have more and more influences on his/her present and future life, the vibration amplitude of one's response to external events becomes bigger and bigger under the same external events, and some strong nonlinear phenomena such as multiple solutions, bifurcation and peak amplitude with the increasing impact factor of memory are illustrated. In addition, we can find that all the stable solution branches obtained match well with the numerical ones.

1 INTRODUCTION

An interesting dynamic model of happiness composed of a 3-order differential equation has recently been reported by Sprott (Sprott, 2005). This model describes the time variation of happiness (or sadness) displayed by individuals under certain external circumstances, and the governing equation is given as

$$\frac{d^3R}{dt^3} + \alpha\frac{d^2R}{dt^2} + \beta(1-R^2)\frac{dR}{dt} + R = f(t) \qquad (1)$$

where α and β are the parameters of the model and β denotes the nonlinear damping coefficient. $f(t)$ is a time-dependent forcing function representing one's external circumstance. R represents that what others presume one's happiness to be based on one's circumstances.

Fractional-order derivative (Oldham, 1974), a generalization of the traditional integer order, has a history of over 300 years. It is an adequate tool for the study of the so-called "anomalous" social and physical behaviors, in reflecting the non-local, frequency and history-dependent properties of these phenomena (Monje, 2010). The integral-order derivative of a function is only related to its nearby points, while the fractional derivative has a relationship with all of the function history

information. As a result, a model described by fractional-order equations possesses memory. In fact, real world processes generally or most likely are fractional-order systems, namely dynamical systems governed by the fractional order derivative equations. As we all know, the emotion of happiness has a relationship with all of the past factors, that is, it is influenced by memory. Thus, the characteristics of fractional-order models and the practical emotion of happiness coincide with each other well. So it is more proper to describe the happiness model with fractional-order equations rather than with integer-order equations. For instance, Song et al. (Song, 2010) demonstrated and exhibited various behaviors of a dynamical model of happiness described through fractional-order differential equations via numerical simulations. Ahmad and Khazali (Ahmad, 2007) examined the fractional-order dynamical model of love and obtained strange chaotic attractors under different fractional orders by using numerical simulations. Sun et al. (Sun, 2012) used the predictor-corrector scheme to demonstrate the dynamical analysis of a driven nonlinear happiness model.

This study aims to introduce methods of harmonic balance (Leung, 2016; Guo, 2016; Ju, 2015) and polynomial homotopy continuation (Sommese, 2005) to study the periodic vibration and effect of Impact Factor of Memory (IFM) of an individual (fractional-order) on steady-state response of a happiness model with commensurate fractional-order derivative.

2 FRACTIONAL-ORDER MODEL OF HAPPINESS

The fractional-order derivative features the memory effect of past states. In equation (1), replacing the integer-order derivative with a fractional-order $0 < \lambda \le 1$ in the sense of Caputo definition, a commensurate fractional-order happiness system is given as below.

$$D_t^{3\lambda}R + \alpha D_t^{2\lambda}R + \beta(1 - R^2)D_t^{\lambda}R + R = f(t) \qquad (2)$$

In state space, equation (9) can be rewritten as

$$\begin{cases} D_t^{\lambda}R = E \\ D_t^{\lambda}E = H \\ D_t^{\lambda}H = -R - \beta(1 - R^2)E - \alpha H + f(t) \end{cases} \qquad (3)$$

where α and β are model parameters, $f(t)$ denotes a time-dependent forcing function and D_t^{λ} is the fractional derivative in the Caputo sense.

From Ref (Song, 2010), R represents one's responses to external events, namely, what others presume one's happiness depending on one's circumstance. E means one's true feelings. The parameter pair (α, β) is to categorize people with different personality, i.e., different (α, β) represents a different kind of individuals, and the order λ represents the impact factor of the memory of an individual, that is, as the value of λ increases from 0 to 1, the impact factor of the memory of individual increases correspondingly. It is noteworthy that the conception of the impact factor of memory is proposed here to denote a measurement of how the influence of an individual is by his/her past experiences. When one's IFM is low, his/her past experiences may have little influence on his/her present and future life; when the IFM is high, it might be difficult for him/her to escape from past experiences despite nightmares or sweet memories.

In fact, one's mood will change due to different external circumstances. We assume the external periodic event $f(t) = q \cos(\omega t)$ and then equation (3) becomes

$$\begin{cases} \dfrac{d^{\lambda}R}{dt^{\lambda}} = E \\ \dfrac{d^{\lambda}E}{dt^{\lambda}} = H \\ \dfrac{d^{\lambda}H}{dt^{\lambda}} = -R - \beta(1 - R^2)E - \alpha H + q\cos(\omega t) \end{cases} \qquad (4)$$

3 STEADY-STATE RESPONSE ANALYSIS

Since the periodic solutions are of interest, the steady-state response of equation (4) can be expanded in the Fourier series

$$R(t) = \sum_{k=0}^{n} a_k \cos(k\omega t) + \sum_{k=1}^{n} b_k \sin(k\omega t) \qquad (5)$$

where a_i, b_i are the unknown Fourier coefficients. n is an integer representing the maximum order which retained harmonics.

Substituting equation (5) into equation (4) and using the Galerkin procedure, it results in the following harmonic balance equations

$$\begin{aligned} R_k^c&(a_0, a_1, \cdots, a_n, b_1, \cdots, b_n) \\ &= \frac{2}{\pi}\int_0^{\pi} \Phi(D_{\tau}^{3\lambda}R, D_{\tau}^{2\lambda}R, D_{\tau}^{\lambda}R, R) \\ &\quad \times \cos(i\,\tau)\mathrm{d}\,\tau, i = 1, 2, \cdots, n \end{aligned} \qquad (6)$$

$$\begin{aligned} R_k^s&(a_0, a_1, \cdots, a_n, b_1, \cdots, b_n) \\ &= \frac{2}{\pi}\int_0^{\pi} \Phi(D_{\tau}^{3\lambda}R, D_{\tau}^{2\lambda}R, D_{\tau}^{\lambda}R, R) \\ &\quad \times \sin(i\,\tau)\mathrm{d}\,\tau, i = 1, \cdots, n \end{aligned} \qquad (7)$$

Where

$$\begin{aligned} \Phi(D_{\tau}^{3\lambda}R\omega^{3\lambda}, D_{\tau}^{2\lambda}R, D_{\tau}^{\lambda}R, R) &\overset{\Delta}{=} D_{\tau}^{3\lambda}R + \alpha\omega^{2\lambda}D_{\tau}^{2\lambda}R \\ &+ \beta\omega^{\lambda}(1 - R^2)D_{\tau}^{\lambda}R - q\cos(\tau). \end{aligned}$$

After carrying out the integration, it yields

$$[K(f, \mu)]\{f\} - \{F\} = \{R_1\} = 0 \qquad (8)$$

where $[K(f, \mu)]$ is the total stiffness matrix, $\{R_1\} = [R_0^c, R_1^c, \cdots, R_n^c, R_1^s, \cdots, R_n^s]^T$, $\{f\} = [a_0, a_1, \cdots, a_n, b_1, \cdots, b_n]^T$ Given the parameter values of μ, equation (8) is solved using the method of harmonic balance in combination with homotopy continuation algorithm. Further, all the steady-state analytical solutions are obtained.

4 RESONANCE RESPONSE AND DISCUSSION

Considering the third-order harmonic case, all the steady-state solutions are obtained by the method of residue harmonic balance in combination with polynomial homotopy continuation in this section. The influence of the impact factor of memory is examined by external frequency, excitation amplitude, and nonlinear damping coefficient versus maximal amplitude of one's response to external events. The dynamics and some simple nonlinear phenomena are shown. In the following figures, stable and unstable branches are indicated, respectively, by solid and broken lines. The circles correspond to the results of numerical simulation.

4.1 External frequency versus maximal amplitude

In this subsection, we focus on the effect of IFM on the relation of steady-state frequency versus maximal amplitude of one's response and true feeling. From Figure 1, we can find that (i) When the IFM $\lambda = 0.6$, the maximal amplitudes of response and true feeling slightly increase first and then decrease. The solutions are always stable, that is when one's past experiences have little influence on his/her present and future life, his/her response and true feelings to external events in a steady state only appear with some slight vibration. (ii) When one's experiences have more and more influence on his/her present and future life, the vibration amplitude

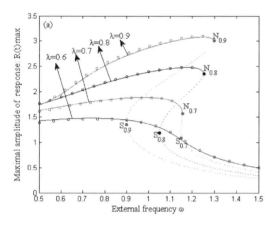

Figure 1(a).   Effect of IFM on external frequency versus maximal amplitude of steady response.

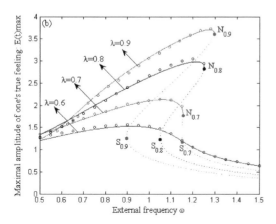

Figure 1(b).   Effect of IFM on external frequency versus maximal amplitude of one's true feeling.

of response to external events becomes bigger and bigger and the model system sequentially presents quasiperiodic motion with increasing external frequency. When IFM $\lambda = 0.7$, 0.8, 0.9 respectively, the change of stable maximal amplitudes of vibration motion to one's response and feelings becomes more and more evident as increases in external frequency to $N_{0.7}(\omega \approx 1.155)$, $N_{0.8}(\omega \approx 1.251345)$, and $N_{0.9}(\omega \approx 1.296356)$ sequentially. In addition, the model system presents strong nonlinear phenomena such as multiple solutions, quasiperiodic motion, etc. When the IFM $\lambda = 0.7$, one's response and feeling to external events present three coexisting periodic solutions in the region of $[S_{0.7}, N_{0.7}] \approx [1.147962, 1.155]$, and as further increases in external frequency from point $N_{0.7}$, the vibration presents in quasiperiodic motion. Similarly, some key points for IFM $\lambda = 0.8$ and 0.9 are $S_{0.8} = 1.05191$, $N_{0.8} = 1.251345$, $S_{0.9} = 0.896894$, $N_{0.9} = 1.296356$. As a result, it is evident that, with the increase in IFM, the region of existing periodic motions gradually become broad and frequency of quasiperiodic motion occurring sequentially becomes big. (iii) All the stable branch solutions are in excellent agreement with numerical simulation ones.

4.2 Excitation amplitude versus maximal amplitude

In this subsection, the relation of excitation amplitude versus maximal amplitude of one's response and the effect of IFM are illustrated and shown. From Figure 2, we can find that: (i) As the excitation amplitude increases, the stable vibration amplitude of one's response slightly increases but change is not evident. (ii) When one's experiences have more and more influence on his/her present and future life, the vibration amplitude becomes bigger and bigger under the same external events. (iii) When IFM $\lambda = 0.7$, the model system first presents

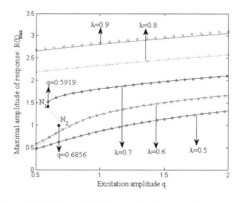

Figure 2.   Effect of IFM on excitation amplitude versus maximal amplitude of one's response.

quasiperiodic motion, multiple solutions, and then develops into a stable response branch with increasing excitation amplitude. That is, when one's past experiences have a moderate influence on his/her present and future life and the excitation amplitude is smaller, one's response to external events will present a swing. The corresponding saddle-node points are $N_1(q \approx 0.5919)$ and $N_2(0.6856)$. (iv) All the present analytically stable solutions match well with the numerical ones.

4.3 Nonlinear damping coefficient versus maximal amplitude

This subsection illustrates the effect of IFM on the relation between nonlinear damping coefficient and maximal amplitude of response in the steady state. From Figure 3, it is evident that (i) When one's past experiences have little influence on his/her present and future life such as IFM $\lambda = 0.6, 0.7$, the maximal amplitude of one's response to external events in the steady state keeps decreasing as the nonlinear damping coefficient from −2 to 0. (ii) As the IFM increases, for instance when $\lambda = 0.8$, the response to external events first increases and then decreases and presents a peak amplitude of nonlinear damping coefficient varying from −2 to 0. (iii) When one's experiences have much influence on his/her present and future life, we can see

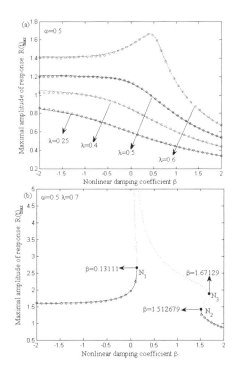

Figure 4. Nonlinear damping coefficient versus maximal amplitude curve and effect of IFM with $\omega = 0.5$.

that the maximal amplitude of one's response to external events abruptly become very big when the nonlinear damping coefficient tends to zero as in Figure 3(b). (iv) When one's experiences have more and more influence on his/her present and future life, the vibration amplitude becomes bigger and bigger under the same nonlinear damping coefficient. In addition, all the present analytical solutions match well with numerical ones. In Figure 4, we expand the nonlinear damping coefficient from −2 to 2 and take external frequency $\omega = 0.5$ and IFM $\lambda = 0.25, 0.4, 0.5, 0.6$, and 0.7, respectively. Similar responses are shown in Figure 4. However, from Figure 4, we can see that (i) when the IFM $\lambda = 0.6$, the peak amplitude of the curve of nonlinear damping coefficient versus maximal amplitude of response has emerged for nonlinear damping coefficient varying from −2 to 2. (ii) When IFM $\lambda = 0.7$, the system presents some strong nonlinear phenomena such as multiple solutions, saddle-node bifurcation.

5 CONCLUSION

A model of happiness with commensurate fractional-order derivative has been explored in detail by using the method of harmonic balance

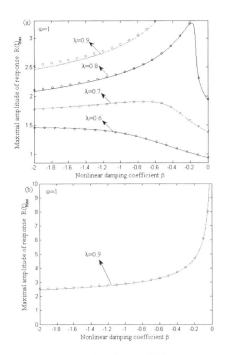

Figure 3. Nonlinear damping coefficient versus maximal amplitude curve and effect of IFM with $\omega = 1$.

in combination with homotopy continuation. The influence of the impact factor of memory on curves of external frequency, excitation amplitude, and nonlinear damping coefficient versus maximal amplitude of one's response to external events is examined. The results show that when one's experiences have more and more influence on his/her present and future life, the vibration amplitude of the response to external events becomes bigger and bigger under the same external events. The model system sequentially presents quasiperiodic motion with the increasing of external frequency and the peak amplitude of curve of nonlinear damping coefficient versus maximal amplitude of one's response to external events will emerge as the IFM increases for the same parameters. In addition, all the stable solution branches have been verified and are in agreement with numerical ones.

ACKNOWLEDGMENTS

The authors gratefully acknowledge the support of the National Natural Science Foundation of China through Grant Nos. 11290152, 11427801 and 11502160, the Natural Science Foundation of Shandong Province, China through Grant Nos. ZR2014 JL002 and ZR2014 AQ028, Beijing Postdoctoral Research Foundation through Grant No. 2015zz-18, the Higher Educational Science and Technology Program of Shandong Province, China through Grant No. J15 LI13 and the Funding Project for Academic Human Resources Development in Institutions of Higher Learning under the Jurisdiction of Beijing Municipality. The authors would also like to thank the anonymous referees for their helpful comments.

REFERENCES

Ahmad, W.M., R. EI-Khazali, Chaos, Solitons Fractals **33** (2007).

Guo, Z.J., W. Zhang, Appl. Math. Model. **40** (2016).

Ju, P.J., Appl. Math. Model. **39** (2015).

Leung, A.Y.T., Z.J. Guo, Commun Nonlinear Sci Numer Simulat **16** (2011).

Monje, C.A., Y. Chen, B.M. Vinagre, et al., *Fractional-order systems and controls: fundamentals and applications*. London: Springer-Verlag, (2010).

Oldham, K.B., J. Spanier, *The fractional calculus*, Academic Press, New York, (1974).

Sommese, A.J., C.W. Wampler, *Numerical solution of systems of polynomials arising in engineering and science*, World Scientific Press, Singapore, (2005).

Song, L., S.Y. Xu, J.Y. Yang, Commun Nonlinear Sci Numer Simulat **15** (2010).

Sprott, J.C., Psychology, and Life Sciences **9** (2005).

Sun, Y.X., X.H. Qiao, B.C. Bao, J. Circuits Syst. **17** (2012).

Civil, Architecture and Environmental Engineering – Kao & Sung (Eds)
© *2017 Taylor & Francis Group, ISBN 978-1-138-02985-9*

Simulation analysis of airdrop cargo extraction posture

Hong Wang, Xu-hui Chen, Ya-hong Zhou & Xiang Li
Air Force Airborne Academy, Guilin, China

ABSTRACT: Based on the ANSYS/LS-DYNA software platform and taking certain types of heavy equipment airdrop systems for example, the method of numerical analysis was used to study the microcosmic changing process of cargo extraction posture in different situations. The change in the overturn angle and speed as time changes was analyzed. The best binding cargo barycenter position was determined, which provided guidance to the design of the airdrop system and the actual airdrop operation.

1   INTRODUCTION

In the airdropping process of heavy equipment, the binding cargo barycenter position has a great influence on cargo extraction posture. At present, cargo binding and moorage solutions are basically relying on experiences. It not only wastes a lot of manpower and material resources but also may cause damages to cargo. Therefore, it is necessary to study this problem. Based on ANSYS/LS-DYNA numerical simulation and analysis system (Li, 2012; Qi, 2014), aiming at certain types of heavy equipment airdrop systems, simulation analysis is made on the changing process of cargo extraction posture in this paper. Taking the cargo overturning speed and angle as a reference standard (Li, 2013), the best binding cargo barycenter position is determined at an extraction speed of 3 m/s, which provides guidance to the design of airdrop system and actual binding and moorage operation in airdropping.

2   SETTING UP A SIMULATION MODEL

2.1   *Simulation object*

Aiming at certain types of heavy equipment airdrop systems in this paper, a simulation study is done on airdropping heavy equipment, the time when the cargo leaving the cabin and the parachute cord of the main chute are completely straightened, the cargo at an extraction speed of 3 m/s and the relationship between barycenter position and the biggest overturn angle and speed (Zhang, 2014; Cheng, 2013). Heavy equipment airdrop system is generally composed of a cargo platform and parachute. Cargo platform is the cargo carrier. Here we take cargo and cargo platform as a rectangle overall, where the length, width, and height are 3.65 m,

2.4 m, and 2 m; respectively, and the cabin floor is 4 m × 2.4 m × 0.2 m cuboid.

Considering the characteristics of the airdrop process and the main purpose of numerical analysis, in order to reduce simulation difficulties and calculating time, the airdrop simulations are simplified as follows:

1. Choose the cargo extraction time from when the cargo platform geometric center reaching the cabin edge to the parachute cord of the main chute are straightened. This period of time is less than 3 s, so the simulation time is identified as 3 s.
2. When airdropping, keep the aircraft in a level flight and the aircraft cabin at floor level.
3. Take the cargo and cargo platform as a whole, use cuboid to substitute the actual model. Do not consider the moorage problem.
4. Take the aircraft as motionless, the cargo platform and cargo at an extraction speed of 3 m/s relative to the aircraft.

Based on the assumptions mentioned above, the changing process of the cargo extraction posture is simulated to find out the relationship between the barycenter position and the overturn angle and speed. According to actual airdrop experiences, 3 kinds of barycenter positions are chosen, simulation calculations in 3 kinds of working conditions (see Table 1) are made to seek the best binding cargo barycenter position. Coordinates of barycenter y and z are the model geometric centers in the directions, and the positions of x-direction are shown in Table 1.

2.2   *Structure model*

According to the cabin structure, the composition of heavy equipment airdrop system, geometric size of each part, assumption of simulation conditions, and actual process of airdropping cargo extraction

Table 1. Simulation calculation of the working condition.

Working conditions	Extraction speed (m/s)	Barycenter x coordinates (m)
Working condition 1	3	2.1132
Working condition 2	3	2.0172
Working condition 3	3	1.9211

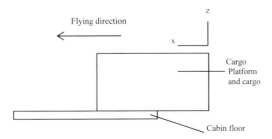

Figure 1. Structure model of simulation analysis of the changing process of airdropping cargo extraction posture (side elevation).

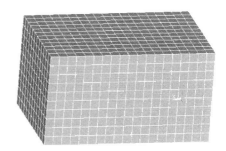

Figure 2. Finite element models of cargo platform and cargo.

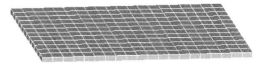

Figure 3. Finite element model of cabin floor.

(Chen, 2009), the structure model of simulation analysis of the changing process of airdropping cargo extraction posture is determined as shown in Figure 1.

Based on that, grid division on the finite element is done. Finite element models of cargo platform, cargo and cabin floor are set up as shown in Figure 2 and Figure 3.

2.3 *Definition of physical conditions*

In order to describe the whole extraction process of the cargo platform and cargo completely, physical conditions are defined as follows, such as material parameters, interface, boundary, and so on.

1. Definition of material parameters
Cabin floor, cargo platform, and cargo are all defined as linear elastic material. The elasticity modulus is 210 GPa, Poisson ratio is 0.3, the density of cabin floor is 7800 Kg/m³, density of cargo got from cargo quality and volume is 475 Kg/m³.

2. Definition of interface
The contact-impact problem is one of the most difficult nonlinear problems, because of which the response in the contact-impact problem is not smooth. When the impact occurs, the speed perpendicular to the interface is transient and discontinuous. As for Coulomb friction model, when there is a viscous glide action, tangential speed along the interface is also discontinuous. These characteristics of contact-impact problem bring obvious difficulties at the time integral of discrete

equations. Therefore, the selection of suitable methods and algorithm are vital to the success of numerical analysis. In this paper, there is a problem of contact pairs of cargo platform and cabin floor in the simulation analysis. According to the characteristics of the cargo platform gliding in the cabin, the contact is defined as a separable automatic surface contact model.

3. Definition of original and boundary condition
Load: The gravity to cargo platform and cargo is 81340N, which is added to the barycenter position.

Speed: The extraction speed of cargo platform and cargo is given as 3 m/s.
    Boundary condition: Displacement and turning degree of freedom on the cabin floor are all constrained in X, Y, and Z directions.

3 RESULTS AND ANALYSIS

By numerically calculating the simulation, motion pictures are used to visualize the extraction process of the cargo platform and cargo in different working conditions. Various parameters in the process are gained. By comparing the changing curve of displacement, a difference of nodes 807 and 808 in the Z direction in 3 s as time changes, we know the cargo overturning conditions. Since the distance of node 807 to 808 is 3.65 m, if the displacement difference of the two points reaches the peak value of 3.65 m, it shows that the cargo overturns to the vertical direction; if the peak value does not reach 3.65 m in 3 s, it shows that the cargo overturning angle in 3 s is less than 90 degrees. In 3 s, the less the peak value, the smaller the overturning angle is.

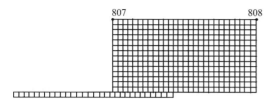

Figure 4. Diagram of nodes 807 and 808.

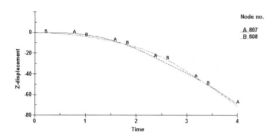

Figure 5. Displacement curve in the Z direction of nodes 807 and 808 in working condition 1.

3.1 Analysis of cargo posture changing process in working condition 1

Figure 5 shows the barycenter position x = 2.1132 m, that is the displacement curve in the Z direction of nodes 807 and 808 in working condition 1.

Figure 6 shows the barycenter position x = 2.1132 m, that is the displacement difference curve in the Z direction of nodes 807 and 808 in working condition 1.

Figure 5 and Figure 6 show that the cargo platform and cargo are mainly in a steady state before 0.2 s; it begins to overturn right after 0.2 s, the overturn angle is not big; it begins to overturn left after 1 s, it reaches the vertical direction at about 2.5 s and then overturns right again.

3.2 Analysis of cargo posture changing process in working condition 2

Figure 7 shows the barycenter position x = 2.0172 m, that is the displacement curve in the Z direction of nodes 807 and 808 in working condition 2.

Figure 8 shows the barycenter position x = 2.0172 m, that is the displacement difference curve in the Z direction of nodes 807 and 808 in working condition 2.

Figure 7 and Figure 8 show that the cargo platform and cargo are mainly in a steady state before 0.2 s and it begins to overturn right after 0.2 s, the overturn angle reaches the biggest at about 1.5 s, but does not reach the vertical direction and then it begins to overturn left until near about 2.5 s.

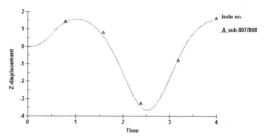

Figure 6. Displacement difference curve in the Z direction of nodes 807 and 808 in working condition 1.

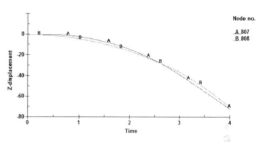

Figure 7. Displacement curve in the Z direction of nodes 807 and 808 in working condition 2.

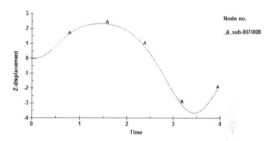

Figure 8. Displacement difference curve in the Z direction of nodes 807 and 808 in working condition 2.

3.3 Analysis of cargo posture changing process in working condition 3

Figure 9 shows the barycenter position x = 1.9211 m, that is the displacement curve in the Z direction of nodes 807 and 808 in working condition 2.

Figure 10 shows the barycenter position x = 1.9211 m, that is the displacement difference curve in the Z direction of nodes 807 and 808 in working condition 3.

Figure 9 and Figure 10 show that the cargo platform and cargo are mainly in a steady state before 0.2 s; it begins to overturn right after 0.2 s, then the overturn angle becomes bigger and bigger with the overturning left and right angles reaching the biggest and in the vertical direction at about 2.4 s, then it begins to overturn left till it is in a steady state.

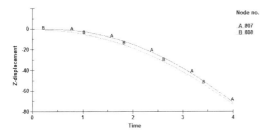

Figure 9. Displacement curve in the Z direction of nodes 807 and 808 in working condition 3.

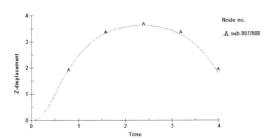

Figure 10. Displacement difference curve in the Z direction of nodes 807 and 808 in working condition 2.

Analysis results in 3 kinds of working conditions show that: at the beginning, cargoes in 3 kinds of working conditions are all overturning right, cargo overturns to vertical direction in working condition 1 at 2.5 s; the same as in working condition 3 at 2.4 s; only in working condition 2, it overturns to vertical direction at 3.5 s and the overall overturning speed is comparatively gentle and slow, and the overturning angle is not big. Therefore, all-in-all, at an extraction speed of 3 m/s, the choice of the barycenter position as in working condition 2 is better.

4 CONCLUSIONS

By numerical simulation analysis carried out in this paper, we obtained the changing curve of cargo at an extraction speed of 3 m/s in different barycenter positions with overturning conditions as time passes. The result indicated that the binding barycenter position of cargo had an obvious influence on the overturning speed and angle after the cargo extraction, which determined the steady state of cargo extraction. As for certain types of heavy equipment airdrop systems in the paper, the best cargo barycenter position was x = 2.0172 m. This was mainly coherent with the monitoring result of cargo overturning in actual airdropping. This leads to the identification of using numerical simulation calculation methods that can provide a certain theoretical guidance to the study of how to control the cargo barycenter when airdropping heavy equipment, and at the same time this verifies the application probability of the numerical simulation method in this field.

REFERENCES

Chen J, Shi Z.K. Aircraft Modeling and Simulation with Cargo Moving Inside [J]. Chinese Journal of Aeronautics, 2009, 22(2):191–197.

Li Da-dong, Sun Xiu-xia, Dong Wen-han, et al. Improved Heavy-weight Airdrop Model Considering Many Influence Factors [J]. Systems Engineering and Electronics, 2013, 35(2):447–451.

Li Liang-chun, Huang Gang, Li Wen-sheng, et al. Simulation Analysis of New Type Landing Cushion Airbag Based on ANSYS/LS-DYNA [J]. Packaging Engineering, 2012, 33(5):16–20.

Qi Ming-si, Liu Shou-jun, Zhao Qi, et al. Simulation Research of the Cushioning Airbag during the Landing Process Based on Ansys/LS-DYNA. Packaging Engineering, 2014, 35(11):13–17.

Zhang Heng-ming, Cheng De-feng, Hui Du-yi. Influences Factors of the Cargo Overturn during Airdrop. Journal of Sichuan Ordnance [J], 2014, 35(3):37–40.

Zhang Heng-ming, Cheng De-feng. Influences of the Barycenter Position on Pose Angle during Airdrop [J]. Aeronautical Science and Technology, 2013(5):27–29.

Civil, Architecture and Environmental Engineering – Kao & Sung (Eds)
© 2017 Taylor & Francis Group, ISBN 978-1-138-02985-9

Research on the comparison between the compressed UF-tree algorithm and the UF-Eclat algorithm based on uncertain data

Chaoquan Chen & Xiujuan Guo
College of Information Science and Engineering, Guilin University of Technology, Guilin, China

Jiahuan Huang
Jieshun Science and Technology Industry Company, Shenzhen, Guangdong, China

Wenqi Liao
College of Civil Engineering and Architecture, Guilin University of Technology, Guilin, China

ABSTRACT: In the existing mining algorithm based on uncertain data, we change the rules of building the UF-tree as follows: we merge the data item to the branch once the data item in transaction matched the tree node in a branch; otherwise, we create a new branch which is composed of the current item and subsequent items from the unmatched tree node. Then, the compressed UF-tree algorithm is proposed, while the UF-Eclat algorithm is proposed by transplanting the classic vertical mining algorithm-Eclat and applying it to uncertain data. It builds the probability vector of a single data item and calculates the degree of support for candidate data items to mine the frequent items. The results indicate that the compressed UF-tree algorithm is more efficient.

1 INTRODUCTION

With the continuous progress of technology, uncertain data research has gradually received wide attention, and mining frequent items has become one of the key research questions (Zhou, 2009). According to uncertain data mining of a frequent item, Chui et al. proposed U-Apriori algorithm (Chui, 2007) transplanted on Apriori. It got a large number of candidate items which led to large execution time and system consumption. Leung et al. proposed the UF-growth algorithm based on an FP-growth algorithm (Aggarwal, 2009), which used a tree structure to mine the frequent items. Chaoquan Chen et al. compressed the UF-tree and proposed the compressed UF-tree algorithm (Chen, 2014). To a certain extent, it could reduce the memory consumption and improve the feasibility and efficiency of the algorithm. In 2000, J. Zaki proposed a depth-first algorithm: Eclat (Liu, 2012), can efficiently mine frequent items in a vertical dataset. So the Eclat algorithm is transplanted and is used for uncertain data, then the UF-Eclat algorithm is proposed.

2 RELATED DESCRIPTION AND DEFINITION

In uncertain datasets, the data items of a transaction appear in a certain probability, then we use T_i to represent a transaction, x represents a data item, then $p(x \in T_i)$ indicates the existence probability of T_i in x, if it does not exist x in T_i, then $p(x \in T_i)$.

Definition 1. The degree of support for uncertain data items in the dataset is $\sup(x) = \sum_{i=1}^{k} p(x \in T_i)$, where k is the number of transactions in a dataset.

Definition 2. If the support for any data item x in the dataset satisfies the minimum support threshold value *min Support*, x is a frequent term, that is, $\sup(x) > min Support*k$.

Definition 3. We define the vector of $1 \times k$, the existence probability of a single data item x in transaction T_i is p, then $V_{(1,i)} = p$ in V_x, the vector V_x obtained by this rule is the probability vector of x.

Definition 4. We filter out the useless candidate items from the search tree of the item set, define a decimal B as the pruning rate, which is between 0 and 1, and clip the extension of the tree structure to restrain the meaningless growth of the tree structure.

Definition 5. Define a decimal A as a matching rate which is between 0 and 1 and used to outline the probability of the shared node or branch when a transaction is added to the tree structure.

According to the possibility of the world model, there are two possibilities for the data item x and transaction T_i: (1) when x exists in transaction T_i, we call it W1, $p(W_1) = p(x \in T_i)$ represents the possibility. (2) If x does not exist in transaction T_i, we call it W2, $p(W_2) = 1 - p(x^\in T_i)$. For a multiple data item xy, its probability in T_i can be represented as $p(x \in T_i)*p(y \in T_i)$ (HAN, 2007). According to the Definition 2, $\sup(xy) = \sum p(x \in T_i)*p(y \in T_i)$ is the degree of support for data item xy.

3 INTRODUCTION OF THE ALGORITHM

The UF-tree compression algorithm works in the form of a level dataset and its main research object is the transaction in the dataset, while the UF-Eclat algorithm works in the vertical form of datasets, the main research goal is the data item. The level of uncertain dataset is composed of TID: (Item: probability), as shown in Table 1.

We assume that the support threshold value is 0.08. According to the Definition 3, the minimum support value is $0.08*6 = 0.48$. Next, we count the degree of support for a single data item and sort the data items according to the size of the degree of support and update the dataset. 1. We sort each transaction of dataset according to the order of data items; 2. We delete the data item which does not meet the minimum support for the dataset (Chen, 2014). The dataset is shown in Table 2.

The form of vertical datasets is Item: (TID: probability). In order to objectively compare the two algorithms, the vertical dataset is produced by the level dataset, as shown in Table 3.

From Table 3 we can see that the form of vertical datasets is closer to the calculation process of the degree of support for a single data item.

Table 1. The uncertain dataset.

TID (transaction ID)	Item: probability
1	A:0.8 B:0.7 C:0.7 D:0.2
2	A:0.7 B:0.2 D:0.7
3	A:0.4 B:0.7 C:0.1 D:0.6
4	A:0.5
5	B:0.3 C:0.8 D:0.3
6	D:0.7 F:0.1 G:0.5

Table 2. The updated level dataset.

TID (transaction ID)	Item: probability
1	A:0.8 D:0.2 B:0.7 C:0.7
2	A:0.7 D:0.7 B:0.2
3	A:0.4 D:0.6 B:0.7 C:0.1
4	A:0.5
5	B:0.3 C:0.8
6	D:0.7 G:0.5

Table 3. The vertical dataset.

Item	TID: probability
A	1:0.8 2:0.7 3:0.4 4:0.5
B	1:0.7 2:0.2 3:0.7 5:0.3
C	1:0.7 3:0.1 5:0.8
D	1:0.2 2:0.7 3:0.6 6:0.7
G	6:0.5

3.1 The UF-Eclat algorithm

The UF-Eclat algorithm is proposed by combining the Apriori algorithm and Eclat algorithm. It uses the depth first search process and takes an effective pruning strategy. The UF-Eclat algorithm uses a structure called the search tree of the item set to generate a candidate item. We extend the search tree and use the probability vector dot product operation to obtain the degree of support for the extension node. As the node of the search tree does not contain any data and the degree of support is calculated by the probability vector dot product, so the node of the search tree is a frequent item. An intuitive description of the UF-Eclat algorithm of construct for the item set search tree is as follows: (1) It scans the uncertain vertical database Ud, calculates the degree of support for the data item and uses the minimum support threshold to filter out the non-frequent items and sorts the dataset according to the degree of support in descending order. (2) It creates a root node, the single frequent item nodes are like the child nodes of the root node. The first child node of the root node is used as the current node. (3) If the number of sibling nodes of the current node is 1, then the only sibling node is a suffix node, do(4). If the parent node of the current is the root, that is the end. Otherwise, the next sibling node of the parent node is the current node, do(3). If the sibling number is greater than 1, the followed sibling nodes are the suffix node one by one, do(4). If the child node number of the current node is greater than 1, the first child node of the current node is used as the current node, otherwise, the next sibling node of the current is used as the current node, do(3). (4) Finally, it reduces prefix connection to the data items of the parent node and suffix node and gets the candidate item and calculates its degree of support. We compare it with the minimum degree of support, if the former is bigger than the latter, the frequent item is added to the search tree of the item set as the child node of the parent node. Return (3).

We take the depth-first strategy to build the search tree of the item set and prune the data from Table 3 without taking the *a priori* knowledge of the Apriori algorithm. The size of the set search tree is enlarged and a lot of lower layer nodes are useless, as shown in Figure 1.

In Figure 1, four nodes of the A node is useless. The UF-Eclat algorithm builds a complete search tree of the item set for data from Table 3, as shown in Figure 2.

3.2 The compressed UF-tree algorithm

When dealing with a large database, the UF-growth algorithm will take up a lot of memory to build the UF-tree and spend a lot of time. Therefore, the compressed UF-tree algorithm improves the structure of the tree based on UF-growth algorithm:1. It will match the current data item in a transaction with data items of the corresponding nodes of all tree branches. If they are matched with each other, the data item is merged into the branch and it shares the branch node; if not, a new branch is opened from the parent node of the matching node, and the new branch consists of the matching data item and its post data item.2. It saves the transaction's TID for the nodes of the last data item in each transaction. In addition, the tree node does not save the probability of the corresponding data item.

The compressed UF-tree constructed by the data from Table 2 is shown in Figure 3.

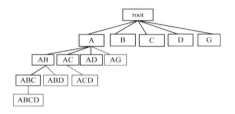

Figure 1. The subtree constructed by the depth first search from node A.

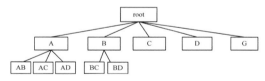

Figure 2. The search tree of the data item.

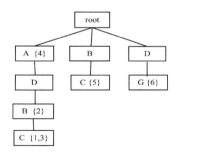

Figure 3. The compressed UF-tree.

The second step is the generation of candidate items and the process is as follows. (1) The leaf node is the start node to search tree nodes along the branch, the transaction number stored in the start node is transmitted to the search node. The candidate data items are composed of the initial node and the nodes from the start node to the node on the path of the search node. (2) The transaction number set saved by the initial node is passed to the generated candidate data items. It merges the transaction number sets for the same candidate data items. (3) The degree of support for candidate data items is calculated based on the probability vector of the single data item and the transaction number set of the candidate data.

4 EXPERIMENTAL ANALYSIS AND RESULTS

The experimental environment is Intel(R) 2.3 GHz CPU, 2 GB memory, 32 bit Windows 7 system PC, MyEclipse 8.5, Java language. The thirteen simulated datasets used in the experiment are generated by the modified IBM data generator. They meet the conditions: the name form of dataset is "sj+ transactions". In the transaction, the number of individual data items is randomly generated and controlled within 14, and the number of attributes is 14. The attribute name is simplified to the letter from a to n, and the probability value is controlled from 0 to 1. Algorithm 1 is the compressed UF-tree algorithm, algorithm 2 is the UF-Eclat algorithm.

The first experiment: the support threshold is fixed to 0.08, the dataset, in turn, takes sj10, sj30, sj50, sj80, sj100, sj300, sj500...... sj5000, sj10000. Run UF-Eclat algorithm and compression UF-tree algorithm, the number of frequent items of the two algorithms is the same as shown in Table 4.

Table 4. The number of frequent items (support threshold 0.08).

Dataset (transaction number)	Frequent item number
10	60
30	35
50	44
80	37
100	30
300	39
500	41
800	43
1000	45
3000	45
5000	45
8000	45
10000	45

The performance difference between the two algorithms is as shown in Figure 4.

The execution time of the algorithm increases with increase in the number of datasets, while the UF-Eclat algorithm increases slowly. The reason is that the compressed UF-tree algorithm scans the database two times while the other only scans one time; the compressed UF-tree algorithm generates the candidate item by traversing the tree structure. It screens the frequent items. While in the UF-Eclat algorithm, the node generated by expanding the search tree is a frequent item. The UF-Eclat algorithm is better for the mining of frequent items.

The second experiment: dataset is fixed to sj1000 and the support threshold value respectively is 0.01, 0.02, 0.03,..... 0.09. Run the two algorithms and the operation results are the same as we can see from Table 5. We compare the performance of the two algorithms, as shown in Figure 5.

Figure 5 shows the comparison of performance between the two algorithms in the same dataset. The gradually increasing support threshold has little effect on the execution time of the compressed UF-tree algorithm. The UF-Eclat algorithm shows a gradually decreasing trend. The support threshold affects the number of frequent items and then determines the execution time of the UF-Eclat

algorithm. The results of the above two experiments have verified that the mining results of the two algorithms are the same.

The third experiment: with the same experimental environment of the first experiment, the performance of the two algorithms is as shown in Figure 6.

In Figure 6 and Figure 7, the performance of the compressed UF-tree algorithm is better in most

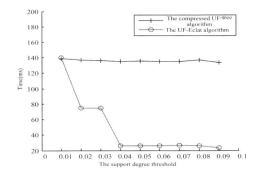

Figure 5. The performance difference between the two algorithms (dataset sj1000).

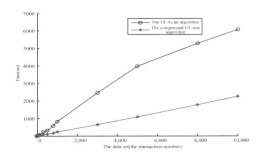

Figure 6. The performance of the two algorithms (support degree 0.08).

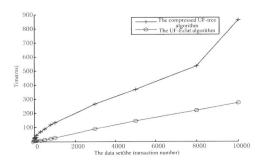

Figure 4. The performance difference between the two algorithms (support threshold 0.08).

Table 5. The number of frequent items (support threshold 0.08).

Support threshold value	Frequent item number
0.01	375
0.02	165
0.03	165
0.04	47
0.05	45
0.06	45
0.07	45
0.08	45
0.09	37

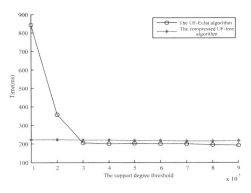

Figure 7. The performance of the two algorithms (dataset sj1000).

cases. The conclusion is just the opposite of the first two experiments, that is, the effect of different properties of different datasets on the performance of the two algorithms is different and the specific analysis is as follows:

The number of transactions affects the performance of the compressed UF-tree algorithm, while it has no effect on the UF-Eclat algorithm; when the number of attributes is large, the performance of the UF-Eclat algorithm is greatly affected. In addition, the pruning rate and matching rate, to some extent, affect the performance of the algorithm. The high matching rate can reduce the consuming time of the compressed UF-tree algorithm and the low pruning rate can increase the running time of the UF-Eclat algorithm.

5 CONCLUSION

This paper studied the comparison between the compressed UF-tree algorithm and the UF-Eclat algorithm for uncertain data. Through experiments, we know that when we deal with a large database, the performance of the UF-tree algorithm is better; but there are still some limitations, for example, in the UF-tree compression algorithm. It takes a long time to generate the candidate items by traversing the tree structure and the process is more complicated. While the transaction number of datasets or attribute numbers is too much, it is a huge challenge for the memory to build the UF-tree or the item set search tree. Therefore, how to balance between memory and performance is also worth studying in the future.

ACKNOWLEDGMENT

This work was supported by the Guangxi Universities key Laboratory Fund of Embedded Technology and the National Natural Science Foundation of China (Grant No. 61540054).

REFERENCES

Aggarwal C C. Li Y. Wang J Y. et al. Frequent pattern mining with uncertain data. Proc of the 15th ACM SIGKDD International Conference on Knowledge Discovery and Data Mining [C]. New York: ACM Press, (2009): 29–38.

Aoying Zhou. Cheqing Jin. Guoren Wang. A Survey on the Management of Uncertain Data. CJOC, **32**, 1(2009): 1–16.

Chaoquan Chen. Jiahuan Huang. Yunhui Jiang. Mining frequent items in uncertain dataset using compressed UF-tree. AROC, **31**, 3(2014): 716–719.

Chui C K. Kao B. Hung E. Mining frequent itemsets from uncertain data. Proc of the 11th Pacific-Asia Conference on Knowledge Discovery and Data Mining, (2007): 47–58.

Han J W. Kamber M. Data mining: Concepts and Technologies Second Edition (2007): 155–156.

Ian H. Witten. Eibe Frank. Data Mining: Practical Machine Learning Tools and Technologies Second Edition. (2005): 76–81.

Leung C K S. Carmichael C L. Hao B. Efficient mining of frequent patterns from uncertain data. Data Mining Workshops, (2007). ICDM Workshops 2007. Seventh IEEE International Conference on. IEEE, (2007): 489–494.

Lixin Liu. Xiaolin Zhang. Yimin Mao. Efficient mining probabilistic frequent itemset in uncertain database. AROC, **29**, 3(2012):841–843.

Civil, Architecture and Environmental Engineering – Kao & Sung (Eds)
© 2017 Taylor & Francis Group, ISBN 978-1-138-02985-9

Research on the sync technology of the ARINC659 bus redundancy fault-tolerant system

Yu Li & Xin Li
AVIC The First Aircraft Institute, Xi'an, China

Gang An
Xi'an Jiaotong University, Xi'an, China

ABSTRACT: The ARINC659 bus is suitable for high security and high reliability requirements of the integrated and modular aircraft airborne computer communication system design. This paper analyzes the structure of the ARINC659 bus, window activity and sync mode, introduces the system architecture, the bus bridge mode, and the working principle of four redundancy computers which is composed of the ARINC659 bus as the core. Sync is the basis and core function of redundant fault-tolerant computer system data transmission and redundancy management. This paper analyzes the cause for clock drift and asynchronous degree in the process of initialization sync and resynchronization, designs task sync algorithm of system application level, and message sync algorithm of bus communication level.

1 INTRODUCTION

Under the premise of the basic reliability index of components, in order to improve the reliability of the airborne system, the key components such as sensors, controllers, actuators, computers, and so on are usually designed with the technology of redundancy. Different units perform the same tasks in a system. The advantage of this design is that the fault unit can be effectively isolated and the task switched to the normal unit to continue execution. However, to ensure the reliability of the system, it has brought the management question of the redundant fault-tolerance system: the system will face how to ensure the coordination between various computing units.

Sync is the basis and key technology, which directly affects the function and operation of the system. It ensures that redundant modules keep step with work, an action to accomplish the same task at the same time. It can shield a few failures in accordance with the principle of the majority of the same to accurately switch to the normal module. The realization of the structure of the reconfiguration is such that the fault tolerant system can be sustained, correct and a reliable operation was done. In this way, the voting process of the fault tolerant system is significant.

ARINC659 bus from Honeywell's SAFEBus: In 1993, the SAFEBus was adopted by the Airlines Electronic Engineering Committee and became the standard and promulgated for the ARINC659

bus specification. The ARINC659 bus is a dual redundancy system based on time-triggered architecture, which supports robust time and space partitioning. It is a key technology of integrated and modular avionics system. The bus technology has been applied to the Airplane information management system of the Boeing's B777. The Chinese latest and most advanced, the largest military transport aircraft also used the bus, as the communication bus between high security and high-reliability redundancy fault-tolerant computer internal function modules of communication backplane bus and the computer.

2 REDUNDANT FAULT-TOLERANT COMPUTER SYSTEM

2.1 BUS structure

Each BIU (Bus Interface Unit) is connected to two buses. Each LRM (On Line Replaceable Module) has two bus interface units, BIUx and BIUy, transmission through the X and Y bus. Each bus is driven by an independent transceiver in the LRM to prevent a single failure from adversely affecting one or more buses.

Through cross validation between BIUx and BIUy and with the four buses cross-checked, LRM has a dual self-test capability. The use of serial transmission line reduces the hardware, simplifies the whole concurrency control, and improves the basic reliability of the bus. The cross-validation

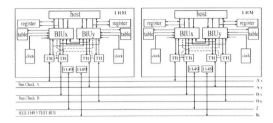

Figure 1.   ARINC659 bus architecture.

between the four buses improves the reliability and availability of data. Bus architecture, interface, and transmission line structure and connection relations are shown in Figure 1.

2.2 *Bus activity*

ARINC659 bus activity is composed of alternating message and gap, each window occupying the relevant LRM command table specified fixed time period.

The window can contain a data message, sync information, or free message. Each window can have a unique transmitter or a limited set of backup transmitters, which are programmed in a table memory. The source and destination addresses of each message are included in the table memory, not by bus. System designers can organize several types of intermodule message structures it supports: module to module (point to point) communication, a module to a set of modules (broadcast) communication. There are 2 types of messages: the basic message and the master/backup message. Basic messages for a single source and single purpose or a number of purposes. The master/backup message is used for a plurality of backup sources and for a single purpose or multiple purposes, and the system arbitration mechanism only allows one of the primary or backup sources to access the bus. A backup source is only the main and other priority than its high backup transmitter in a predetermined time period to remain silent on the bus to send.

2.3 *Computer system*

The greatest advantages of the ARINC659 bus are the time deterministic in the transmission process, dual self-test capability, fault detection, and isolation. So it is very suitable for the design of a redundant fault-tolerant computer system. In this paper, the computer system with the maximum redundancy level of aircraft is designed as the goal. This paper introduces the structure of a redundant computer system based on ARINC659 bus technology, as shown in Figure 2.

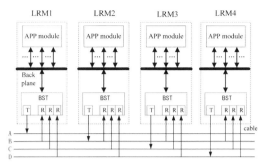

Figure 2.   Four redundancy computer system.

Each LRM is a separate control channel. The channel is composed of application module, ARINC659 backplane, and Bus Bridge Transfer (BST) module. ARINC659 backboard bus is the core of the channel. Each function module is an intelligent node, which contains a processor or microcontroller. The signal on the function board is regulated and monitored by the processor or the microcontroller. Each function module can carry out information exchange, resource sharing, and resource reconstruction. Finally, the system can achieve the resource dynamic reconfiguration and fault-tolerance of the four redundant system.

BST board exchanges and shares the ARINC659 bus data between the computers. BST converts the backplane mode to cable mode for forwarding to other LRM. BST is a continuation of the bus signal and also is the backplane signal at the computer level conversion and transmission, and can realize the redundant channels between CCDL (Cross Communication Data Link).

ARINC659 bus specifies the physical layer and data link layer, and the data communication sync of ARINC659 bus mainly works on the data link layer. After the computer power is switched on, the BST is responsible for the initialization of the CCDL sync to achieve the data sync between channels. Then, each BST initiates the sync request to the ARINC659 bus in the channel and realizes the data sync between the internal function modules. In the task period, the BST is responsible for maintaining the bus sync and channel sync.

3   RESEARCH ON THE SYNC METHOD

3.1 *Sync mode*

For the system c, if the ideal clock time is T, and the node's clock time set as $c_c(T)$, for the other two systems p, q, if satisfied $|c_p(T)-c_q(T)| \le \delta$max ($\delta$max which is the maximum degree of the asynchronous clock), the system p and q sync. According to different sync periods, sync technology can be divided into clock sync, loose sync, and task level sync.

Clock sync is a distributed system which uses the same clock source to drive the nodes. Each node is synchronized to the clock cycle, so each node in the system is running at the same time, as shown in Figure 3. Clock sync is mainly based on hardware, the advantages of sync are strict, no extra time overhead, but the requirements of the clock is very high, which often requires the use of a fault-tolerant clock.

Loose sync is a time sync method based on time slice, as shown in Figure 4. Every time slice is composed of a plurality of clock cycles. In each time slice, it allows each node in the system to have an asynchronous degree. As long as the asynchronous degree is in the allowable range, each node in the system synchronization will be resynchronized in the next time slice. In this way, it is usually implemented with a logic clock sync software. The main purpose of loose synchronization is to periodically align the clocks of each node. This can reduce or eliminate the asynchronous degree between nodes and reduce the error accumulation of the clock. Loose sync can establish the same conditions for each node to complete the same task so that the suppression of the fault caused by the asynchronous module spread can be restrained. There is a certain deviation in the sync of loose sync, which can eliminate the influence of some instantaneous faults, but the time cost is large.

Task sync, which is a kind of loose sync, is the basic unit of time for the task period, and each task period includes task execution time and task idle time. The resynchronization cycle is composed of multiple task cycles. Every node performs tasks repeatedly in each task cycle. One of the nodes of the clock or all the nodes of the clock is set to a reference clock. If there is a node of the clock in asynchronous degree greater than the maximum asynchronous, the standard clock corrects the standard for all nodes of the logical clock so that each node in the system returns to the status of sync, as shown in Figure 5. The clock of each node can be operated independently and the realization is simple and easy to implement.

3.2 Application level task sync design

Application level task sync refers to the task sync process which is based on the specific application requirements of the service system including initialization sync and resynchronization.

Figure 6 shows the initialization sync. LRM1, LRM2, LRM3, LRM4 began the duty cycle of the moment, respectively to t_a, t_b, t_c, t_d. The $t_a > t_b > t_c > t_d$, for several moments are not overlapping. In the first period, LRM1 did not receive the time information of the other three LRM, operating according to a predetermined time delay T_d and other LRM, although it can receive time information, but the time offset calculation is not yet complete. Therefore, it is according to the same predetermined delay time T_d of operation. In the second period, the time of LRM1 is taken as the baseline time of the whole network, and it is still in the time of the delay time T_d, and the other three LRM in accordance with the calculation of the time delay, respectively T_{db}, T_{dc}, T_{dd}.

If LRM1 and LRM2 in the M + 1 period achieved sync, then the formula (1) was established:

$$T_s + T_d + T_c + \mathbf{M}*\left(T_s + T_c + T_{da}\right) + \left(t_a - t_b\right) \\ = T_s + T_d + T_c + \mathbf{M}*\left(T_s + T_c + T_{db}\right) \tag{1}$$

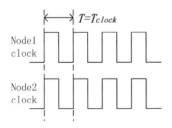

Figure 3.   Clock sync.

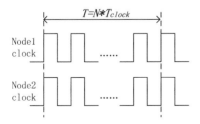

Figure 4.   Loose sync.

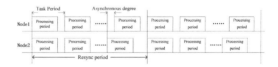

Figure 5.   Task sync.

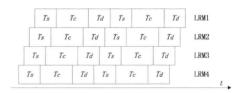

Figure 6.   Task period initialization.

If LRM1 and LRM3 in the N + 1 period achieved sync, then the formula (2) was established:

$$T_s + T_d + T_c + \mathbf{M}^*\left(T_s + T_c + T_{da}\right) + \left(t_a - t_c\right)$$
$$= T_s + T_d + T_c + \mathbf{M}^*\left(T_s + T_c + T_{db}\right) \quad (2)$$

If LRM1 and LRM4 in the P + 1 period achieved sync, then the formula (3) was established:

$$T_s + T_d + T_c + \mathbf{M}^*\left(T_s + T_c + T_{da}\right) + \left(t_a - t_d\right)$$
$$= T_s + T_d + T_c + \mathbf{M}^*\left(T_s + T_c + T_{dd}\right) \quad (3)$$

Therefore, LRM2, LRM3, and LRM4 were synchronized with LRM1 after three task periods. The delay times of the four LRM are $(t_a-t_b)/(T_{db}-T_d)$, $(t_a-t_c)/(T_{dc}-T_d)$, $(t_a-t_d)/(T_{dd}-T_d)$, and finally the initialization sync is realized.

Resynchronization: Initialize sync is complete because each LRM hardware clock source, clock drift problem, and periodic task also exist with subtle differences. With the passage of time, clock drift differences continue to accumulate and showed larger asynchronicity. If the asynchronous degree exceeds the maximum allowed, it is considered that the four LRM is not in the sync state, as shown in Figure 7. After a period of time, compared to the LRM2, LRM3, LRM4, and LRM1, respectively, appeared the δ_{ab}, δ_{ac}, δ_{ad} synchronous, and $|\delta_{ad}|$ being the greatest, with the passage of time, δ_{ad} must first achieve the allowed maximum asynchronous degree δ_{max}. At this point, you need to reduce central processing unit D delay, delay T_{dd} operation in accordance with the newly calculated, making A and D achieve resynchronization, B and C resynchronization process with the process being similar.

Figure 2 shows the four LRM in the redundant fault-tolerant computer system, which is expressed as W, X, Y, Z. In the system dynamic operation process, each one of the ID of LRM is W, and the rest of the ID is X, Y, Z. So the corresponding relationship between the four LRM and W, X, Y, Z is as follows:

LRM1: W − LRM1, X − LRM2, Y − LRM3, Z − LRM4

LRM2: W − LRM2, X − LRM4, Y − LRM3, Z − LRM1

LRM3: W − LRM3, X − LRM1, Y − LRM2, Z − LRM4

LRM4: W − LRM4, X − LRM1, Y − LRM2, Z − LRM3

The sync process can be designed as shown in Figure 8.

3.3 *Bus communication level sync design*

Before the execution of the frame period, the user can define the window on the bus. The frame is composed of the sync, the data and the free window of the cycle period, and the frame period is equal to the sum of the length of each window of the frame.

Table memory can store the frame command sequences with different lengths. Frame switching is implemented by the frame switching command. There are two different types of frames: the version and non-version. In a version of the frame, all BIU on the bus should have the same table version number. The frame switching mechanism of the version ensures that all the sync of the BIU verifies its own version number according to the received frame switch messages in the version of the data. If the version number of the table is not consistent, it will lose sync with the bus.

In a non-version frame, the version of the table is ignored. As long as a BIU in the LRM can be synchronized with the bus, it is able to participate in the non-version of the frame. In addition, all messages in the non-version frame should be divided by the maximum time gap (9-bit time). For master/slave messages, the maximum step time gap (10-bit time) must be used.

When BIU is in an asynchronous state, it will attempt to recover sync with the active BIU.

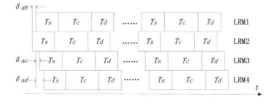

Figure 7. Task period resynchronization.

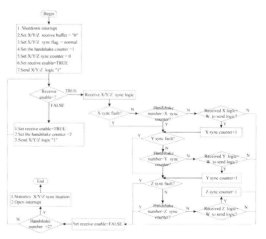

Figure 8. Sync process.

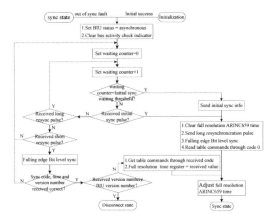

Figure 9. Flow chart of the frame level sync process.

Initialization sync limit wait time refers to the length of time that the BIU searches for sync messages (such as short message, long synchronization message, or initialization synchronization message) before the BIU decides to send the initialization sync pulse. If the BIU has not yet searched a sync pulse when the BIU exceeds the initialization sync limit wait time, the BIU will transmit an initialization sync pulse. If the BIU receives an initialization synchronization message (sent by another module or by itself), it will execute an inherent command immediately to receive a non-version into the sync message. Message synchronization code is "0" and the full resolution time value is "0".

4 CONCLUSION

The ARINC659 bus has the advantages of large data throughput, high coding efficiency, transmission time determination, strict fault isolation, fault tolerance and so on. Communication system design for aircraft airborne computer, controller and so on are suitable for high security, high reliability, and high comprehensive degree. It can be designed to become the backplane data bus in a computer, but also to meet the communication needs of multiple computers. The four redundancy fault-tolerant computer system based on

ARINC659 bus is the basis of data transmission and redundancy management of the computer internal function module. In this paper, the characteristics of the bus, the characteristics of four redundancy fault-tolerant computer architecture, sync transmission mode, clock drift, and asynchronous degree are analyzed. The application task and bus communication two-level sync algorithm are designed to ensure that the data from the application level to the data link level are transmitted synchronously.

REFERENCES

Airlines Electronic Engineering Committee, Arinc specification 659 Backplane data bus [C]. Dec. 27, 1993.

Feng Fulai. Comments on several data bus applications in avionics system. Measurement and control technology. 1999, 18(5):42–44.

Gao Jinyuan, Xia Jie. Principle of computer control. Beijing: Tsinghua University press, 2006.

Huo Man. Prospects for the development of integrated avionics technology. Avionics technology, 2003(3): 12–17.

Liu Lin, Guo Enyou. Aircraft control system sub system. Beijing: National Defence Industry Press. 2003.

Ren Suzhong, Feng Xiaofeng. Structural design of integrated modular avionics system for Boeing B777 aircraft. Electronic mechanical engineering. 1998(6).

Wang Yong, Yu Hongkun. Airborne computer system. Beijing: National Defense Industry Press. 2008.

Wang Yunming. Standards, processes, methods, and tools for embedded software development. Journal of computer technology and development. 2006.10.

Wen Chuanyuan. Modern flight control. Beijing: Beihang University press. 2004.

Wu Wenhai. Flight integrated control system. Beijing: Aviation Industry Press. 2007.

Xu Yanling, Zhang Xinguo, Wang Lin. Redundant digital fly by wire control system input information redundancy management strategy and its application. Measurement and control technology. 1000–8829 (2008) 11-0011-03.

Xu Zezhong, Lei Xun, Hu Rong. A study on the development of the new generation integrated avionics system. Avionics Technology. 2001, 32(4):11–18.

Yang Wei. Fault tolerant flight control system. Xi'an: Northwestern Polytechnical University press. 2007.

Zhi Chaoyou, Airborne data bus technology and its application. National defense industry press. Jan. 2009.

Civil, Architecture and Environmental Engineering – Kao & Sung (Eds)
© 2017 Taylor & Francis Group, ISBN 978-1-138-02985-9

Identity recognition approach based on the improved phase congruency of the gait energy image

Ling-yao Jia, Chao Liang & Dong-cheng Shi
School of Computer Science and Engineering, Changchun University of Technology, Changchun, China

ABSTRACT: The use of phase congruency for marking features has significant advantages over gradient-based methods. It is a dimensionless quantity that is invariant to changes in image brightness or contrast, hence it provides an absolute measure of the significance of feature points. In this paper, identity recognition based on improved Log-Gabor phase congruency is proposed. The improved phase congruency algorithm is based on the improved local energy calculation method, frequency spread, and noise compensation tactics. The feature of improved phase congruency is of good location and recognition, and then the LDA method is used to project features into a low-dimensional space. The nearest neighbor classifier based on Euclidean distance is tested in the CASIA gait database. The experimental results show that our approach outperforms other state of the art automatic algorithms in terms of recognition accuracy.

1 INTRODUCTION

Gait recognition (Chai, 2011; Nixon, 2004) is the process of identifying an individual by the way they walk. This is a less unobtrusive biometric, which offers the possibility to identify people at a distance. Moreover, gait recognition offers great potential for recognition of low-resolution videos where other biometric technologies may be invalid because of insufficient pixels to identify human subjects.

2 RELATED WORK

Approaches for human recognition by gait can be divided into two categories: model-based approaches and model-free approaches. The former's main focus is the human body and uses parameters of the model for recognition, which results in a complex model. The computational time, date storage, and cost are extremely high due to its complex searching and matching procedures. The latter generally differentiates the whole motion pattern of the human body by a concise representation such as a silhouette without considering the underlying structure. Due to that reason, a large number of model-free approaches are proposed in recent years.

CR Li et al. (Li, 2012) presents an effective gait recognition method using the magnitude and the phase of Quaternion Wavelet Transformation (QWT). I Rida et al. (Rida, 2014) describe a supervised feature selection method to select relevant features for human recognition, which can mitigate the impact of covariates (such as clothing, carrying conditions). HP Mohan Kumar et al. (Mohan, 2014) provides a new representation technique, which is able to capture variations in gait due to changes in cloth, carrying a bag, and different instances of normal walking conditions more effectively. YY Zhang et al. (Zhang, 2013) use three different local textural features of the Gait Energy Image (GEI) to represent the local space distribution of pixel brightness. In this paper, the identity recognition based on the improved Log-Gabor phase congruency of the GEI is proposed.

3 THE GAIT RECOGNITION SYSTEM BASED ON THE GEI

The structure of the gait recognition system based on GEI is shown in Figure 1. First, gait frames are extracted from gait videos and then the frames undergo silhouette pre-processing, morphological filtering, and binarization, which estimates the gait cycle and computes GEI.

Figure 1. The flow chart of gait recognition system based on the gait energy image.

Given a size-normalized and horizontal-aligned human walking binary silhouette sequence $B(x, y, t)$, the grey-level GEI $A(x, y)$ is defined (Han, 2006) as follows.

$$A(x,y) = \frac{1}{N}\sum_{i=1}^{N} B(x,y,t) \tag{1}$$

where N is the number of frames in complete cycles of the sequence, t is the frame number of the sequence, and x and y are values in the 2D image coordinate. The examples of normalized and aligned silhouette frames in walking sequences and GEI (the rightmost image) are shown in Figure 2.

3.1 Feature extraction of phase congruency

The process of feature extraction of improved phase congruency is as described previously (Kovesi, 1999). In this paper, the Log-Gabor is used for the calculation of phase congruency. On the linear frequency scale, the Log-Gabor function has a transfer function of the form

$$g(w) = e^{\frac{-(\log(w/w_0))^2}{2(\log(k/w_0))^2}} \tag{2}$$

where w_0 is the filter's center frequency. To obtain constant-shape ratio filters, the term k/w_0 must also be held constant for varying w_0. Let I denote the signal and M_n^e and M_n^o denote the even-symmetric and odd-symmetric wavelets at a scale n, and then the responses of each quadrature pair of filters as forming a response vector,

$$[e_n(x), o_n(x)] = [I(x) * M_n^e, I(x) * M_n^o] \tag{3}$$

The amplitude of the transformation at a given wavelet scale is given by

$$A_n(x) = \sqrt{e_n(x)^2 + o_n(x)^2} \tag{4}$$

and the phase is given by

$$\phi_n(x) = a\tan 2(e_n(x), o_n(x)) \tag{5}$$

Each point in the signal will have a series of response vectors, each of which has a response vector of the filter response. One of the difficulties in the extraction of phase congruency is the processing of the noise response of the filter. According

Figure 2. Gait sequences and gait energy image.

to related description, the improved expression for energy can be got by

$$E = \sqrt{(\sum_n e_n)^2 + (\sum_n o_n)^2} \tag{6}$$

where e_n and o_n are the outputs of even and odd symmetric filters on scale n.

According to relevant theory, it can be concluded that amplitude distribution of energy vector obeys the Rayleigh distribution of the form

$$R(x) = \frac{x}{\sigma_G^2} e^{\frac{-x^2}{2\sigma_G^2}} \tag{7}$$

The mean and variance of the Rayleigh distribution are as follows.

$$\mu_R = \sigma_G \sqrt{\frac{\pi}{2}} \tag{8}$$

$$\sigma_R^2 = \frac{4-\pi}{2}\sigma_G^2 \tag{9}$$

where σ_G^2 is the variance of the Gaussian distribution describing the position of the total energy vector.

The noise radius can be obtained by

$$T = \mu_R + k\sigma_R \quad k \in [2,3] \tag{10}$$

If the estimated noise effect is subtracted from the local energy before normalizing it by the sum of the wavelet response amplitudes, spurious responses to noise can be eliminated. So the expression for phase congruency can be got by

$$PC(x) = \frac{\lfloor E(x) - T \rfloor}{\sum_n A_n(x) + \varepsilon} \tag{11}$$

where $\lfloor \rfloor$ denotes that the enclosed quantity is equal to itself when its value is positive, and zero otherwise.

Phase congruency of feature points has important significance to identify only over a wide frequency range. Frequency expansion expression is as follows.

$$s(x) = \frac{1}{N}\left(\frac{\sum_n A_n(x)}{\varepsilon + A_{max}(x)}\right) \tag{12}$$

where N represents the total number of wavelet scales, $A_{max}(x)$ is the amplitude of the filter pair having a maximum response at x, ε is also in order to prevent the denominator from being zero. A phase congruency weighting function can then be constructed by applying a sigmoid function to the filter response spread value, namely

$$W(x) = \frac{1}{1 + e^{\gamma(c - s(x))}} \qquad (13)$$

where c is the cut-off value of filter response spread below which phase congruency values become penalized, and γ is a gain factor that controls the sharpness of the cutoff. Thus, the calculation of phase congruency can be modified as follows.

$$PC(x) = \frac{W(x)\lfloor E(x) - T \rfloor}{\sum_n A_n(x) + \varepsilon} \qquad (14)$$

For the fuzzy image, the positioning perform-ance of phase congruency is relatively weak. For this reason, the calculation for E(x) is modified by

$$\Delta\Phi(x) = \cos(\phi_n(x) - \bar{\phi}(x)) - |\sin(\phi_n(x) - \bar{\phi}(x))| \qquad (15)$$

$$E(x) = A_n(x)\Delta\Phi(x) \qquad (16)$$

Using this new measure of phase deviation, $\Delta\Phi(x)$, a new measure of phase congruency can be defined as

$$PC(x) = \frac{\sum_n W(x)\lfloor A_n(x)\Delta\Phi_n(x) - T \rfloor}{\sum_n A_n(x) + \varepsilon} \qquad (17)$$

In order to be able to make the direction and possible connection of wavelet filter features, which are of equal importance, the extraction of phase congruency of one image can be defined as

$$PC2(x) = \frac{\sum_o \sum_n W_o(x)\lfloor A_{no}(x)\Delta\Phi_{no}(x) - T_o \rfloor}{\sum_o \sum_n A_{no}(x) + \varepsilon} \qquad (18)$$

where o denotes the index over orientations. In order to achieve the effect of uniform coverage of the spectrum in all filtering orientations, the inter-val between the angle direction of the filter can be calculated by the following formula:

$$G(\theta) = e^{-\frac{(\theta - \theta_0)}{2\sigma_\theta^2}} \qquad (19)$$

where θ_0 represents the filter orientation and σ_0 is the standard deviation of Gauss distribution function.

3.2 *Individual recognition*

In this paper, the features extracted from the gait energy image are subjected to the problem of excessive dimensionality. To reduce their dimensionality, the multi-class linear discriminant analysis (Theodoridis, 2009) is used to project the features into a low-dimensional space. The GEIs used in this paper are normalized and aligned so that they can be clustered by the distance measure. The classic Euclidean distance and nearest neigh-bor classifier are adopted to discriminate different gait patterns.

4 EXPERIMENTAL RESULTS AND ANALYSIS

Our approach has been applied to the CASIA Dataset B database consisting of 124 different subjects with variations in walking status (normal, in a coat, or with a bag) at the lateral view. Each person has 10 GEIs with a resolution of 240 * 240 and they are divided into training and test set with the ratio 1:1. The related parameters are set as follows: number of scales, 5; number of orienta-tions 8; scaling factor between successive filters 2.1; ratio of the standard deviation of Gaussian 0.55; number of standard deviations of noise energy beyond the mean at which we set the noise thresh-old point 2.0; the fractional measure of frequency spread below which phase congruency values get penalized 0.5; and the sharpness control of transi-tion in sigmoid function used in weight phase con-gruency for frequency spread 10. One gait energy image and phase congruency using the jet color-map is shown in Figures 3 and 4, respectively, and the filter response of one scale in eight orientations is shown in Figure 5.

The cumulative matching characteristic and receiver operating characteristic are used to assess

Figure 3. Gait energy image.

Figure 4. The phase congruency.

Figure 5. Log-Gabor filter response.

Table 1. Recognition rate of CMC in rank 1.

Approach	QWT-Magnitude-Phase[3]	GEI-PCA-MDA[4]	GEI-RBL-LBP[5]	GEI-SDOLB (Zhang, 2013)	Proposed
0	68.33%	87.50%	81.25%	93.89%	99.58%
54	82.50%	73.13%	45.42%	95.56%	99.58%
90	89.17%	89.38%	90.42%	91.67%	97.08%
126	88.33%	67.50%	81.67%	62.22%	99.17%
180	80.00%	90.00%	90.83%	88.33%	100.0%
Mean	81.67%	81.50%	77.92%	86.33%	99.08%

Table 2. Verification rate of ROC in FAR = 1%.

Approach	QWT-Magnitude-Phase[3]	GEI-PCA-MDA[4]	GEI-RBL-LBP[5]	GEI-SDOLB[6]	Proposed
0	79.17%	91.25%	82.08%	83.33%	98.33%
54	90.00%	80.00%	40.42%	85.00%	99.17%
90	95.83%	95.00%	90.83%	89.44%	95.00%
126	90.83%	63.75%	71.25%	83.89%	98.33%
180	84.17%	93.75%	90.83%	85.56%	98.75%
Mean	88.00%	84.75%	75.08%	85.44%	97.92%

the performance of the proposed approach. The related experimental data is shown in Tables 1 and 2.

It can be seen from Tables 1 and 2 that the proposed approach, as well as comparison with other approaches to individual recognition by gait, has the highest recognition rate of CMC in Rank 1 and verification rate of ROC in FAR = 1% practically. It can be concluded that the proposed approach renders the best recognition effect of the five approaches. In order to assess the overall performance of the five approaches, the equal error rate is computed and is shown in Figure 6.

The smaller the equal error rate, the better the overall performance of one algorithm. It can be seen from Figure 6 that the equal error rate of the proposed approach is the least.

The CMC and ROC curves of CASIA gait database at 0, 54, and 90 angles are shown in Figures 7–12, respectively.

In the CMC and ROC curves, the best performance of one approach should be close to the upper left corner as far as possible. It can be seen from Figures 7–12 that our approach (GEI-PC) is closest to the upper left corner. Usually, the area under the CMC and ROC curves can reflect the feature identification performance and algorithm verification performance, respectively. It can be seen from Figures 7–12 that the area under the curve of the proposed approach is the largest. The experimental results on CASIA gait database exhibit the effectiveness of the proposed approach.

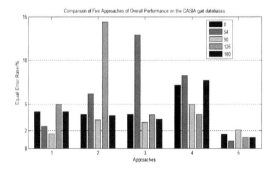

Figure 6. The equal error rate of five approaches.

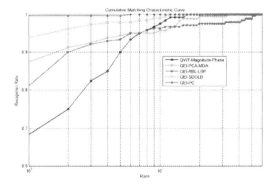

Figure 7. Recognition performance of CMC curves at 0 angle.

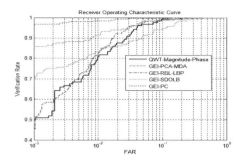

Figure 8.   Verification performance of ROC curves at 0 angle.

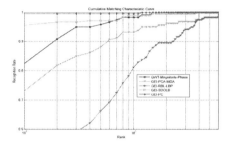

Figure 9.   Recognition performance of CMC curves at 54 angle.

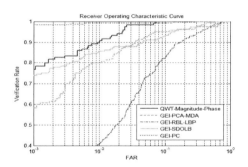

Figure 10.   Verification performance of ROC curves at 54 angle.

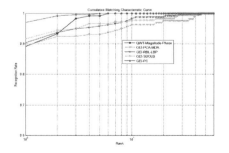

Figure 11.   Recognition performance of CMC curves at 90 angle.

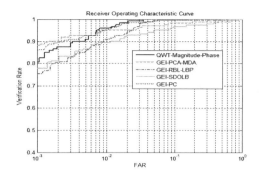

Figure 12.   Verification performance of ROC curves at 90 angle.

5   CONCLUSIONS AND FUTURE WORK

In this paper, a new way of calculating phase congruency through the use of wavelets is adapted to extract features of the gait energy image. It shows that the extracted gait features are good localization and recognition. Experimental results on CASIA gait database show that the proposed approach has the characteristic of a high correct recognition rate. Further research will focus on the following aspects: a) explore a more effective method to extract gait features; b) seek a more effective feature dimension reduction method; c) seek a more effective method of similarity measurement; d) seek a more effective classifier; e) build more large-scale, more status of walking gait database to test the performance of the algorithm. In addition, it is important to find strategies to overcome some gait status under the interference of unexpected factors such as stopover and occlusion in the real scene.

ACKNOWLEDGMENT

This work was in part supported by the Education Department of Jilin Province Science and Technology Research Project in the 13th Five-Year in 2016 under contract No. 349. The CASIA gait database used in this paper was from the Institute of Automation of the Chinese Academy of Sciences. The authors would like to express their gratitude for their support.

REFERENCES

Chai Y, Ren J, Han W, et al. Human gait recognition: Approaches, datasets and challenges [C]. International Conference on Imaging for Crime Detection and Prevention. IET, 2011:3-3.

1405

Chao-Rong Li, Jian-Ping Li, Xing-Chun Yang, et al. Gait recognition using the magnitude and phase of quaternion wavelet transform [C]. International Conference on Wavelet Active Media Technology and Information Processing. 2012:322–324.

Han J, Bhanu B. B. Individual recognition using gait energy image [J]. IEEE Transactions on Pattern Analysis & Machine Intelligence, 2006, 28(2):316–322.

Kovesi P. Image Features From Phase Congruency [J]. Journal of Computer Vision Research, 1999, 1(1): 115–116.

Mohan Kumar H P, Nagendraswamy H S. LBP for gait recognition: A symbolic approach based on GEI plus RBL of GEI [C]. International Conference on Electronics and Communication Systems. 2014:1–5.

Nixon M S, Carter J N. Advances in automatic gait recognition [C]. IEEE International Conference on Automatic Face & Gesture Recognition. 2004:11–16.

Rida I, Almaadeed S, Bouridane A. Improved gait recognition based on gait energy images [C]. ICM 2014. 2014:40–43.

Theodoridis S, Koutroumbas K. Pattern Recognition (Fourth Edition) [M]. 2009.

Zhang Y Y, Jing L I, Jiang S M, et al. Local features of gait energy image through the method of texture analysis [J]. Jilin Daxue Xuebao, 2013, 43:193–198.

Civil, Architecture and Environmental Engineering – Kao & Sung (Eds)
© *2017 Taylor & Francis Group, ISBN 978-1-138-02985-9*

Fractional-order generalized augmented Lü system and its application in image encryption

Hong-yan Jia & Yong-jun Wu

Department of Automation, Tianjin University of Science and Technology, Tianjin, China

ABSTRACT: Based on the fractional-order generalized augmented Lü system, a kind of double encryption algorithm method is designed to realize image encryption. The method mainly refers to the scramble of the pixel position and the transformation of the pixel value. The effectiveness of the double encryption method is verified by encrypting and decrypting the image. The experimental results indicate that this method not only has a better effect of image encryption and decryption, but also ensures the security of the encrypted image. The results from the histogram analysis and correlation analysis also verify the security and effectiveness of the proposed encryption method.

1 INTRODUCTION

Since Lorenz first proposed the famous Lorenz chaotic system in 1963, many scholars have put forward some new chaotic systems such as the Chen system (Lorenz, 1963), Qi system (Chen, 1999), Lü system (Hu, 2002). Subsequently, based on the Lü system, Qiao et al. proposed a new 3D autonomous chaotic system called the generalized augmented Lü system (Xiao, 2009), which is obtained by adding a linear state inverse controller in the third equation of the augmented Lü system. Although there are only three system parameters in the system, it has 5 balance points so that it can generate fan-shaped four-wing chaotic attractors without any need for external excitation.

Since the idea of chaotic encryption was firstly proposed by Matthews, the research of chaotic cryptography and chaotic secure communication has become a hot research topic (Zai, 2014; Li, 2016; Akram, 2016; Yi, 2016). At first, the chaotic encryption technology is mostly implemented by the low dimensional chaotic system, which has the advantages of a simple model and fast calculation. At the same time, the disadvantages that the key space is too small and the chaotic sequence is too simple are very distinct, which may lead to the low security of the encryption technology. Therefore, high dimensional chaotic systems, such as three-dimensional Lü system, some hyperchaotic systems are always used for image encryption because of the advantages of large enough key space of the encryption and high security of the encryption system.

Based on the fractional-order generalized augmented Lü system (Wei, 2014), this paper firstly analyzes the chaotic characteristics of the system and gives Lyapunov exponent spectrum and bifurcation diagram of the 2.7 dimensional generalized augmented Lü system. In the second section, a simple but secure and efficient image encryption algorithm is designed based on the fractional generalized augmented Lü system. The results from the numerical analysis, histogram analysis, and correlation analysis verify the security and effectiveness of the proposed encryption algorithm.

2 FRACTIONAL GENERALIZED AUGMENTED LÜ SYSTEM

The fractional-order generalized augmented Lü system is described as (Wei, 2014)

$$\begin{cases} \dfrac{d^q x}{dt^q} = -\left[ab/(a+b)\right]x - yz \\ \dfrac{d^q y}{dt^q} = ay + xz \\ \dfrac{d^q z}{dt^q} = bz + xy + cx \end{cases} \quad (1)$$

where $0 < q < 1$, a, b, and c are constants, $a < 0$ and $b < 0$. By utilizing the time-frequency domain method, we obtain the attractors of the system (1) when $q = 0.9$, $a = -10$, $b = -6$, $c = 0.5$. The phase portraits are depicted in Figure 1.

From the phase portraits shown in Figure 1, we can see that chaotic attractors exist in the system (1) with $q = 0.9$, $a = -10$, $b = -6$, and $c = 0.5$. In addition, for the 2.7-order generalized augmented Lü system, the bifurcation diagrams and Lyapunov

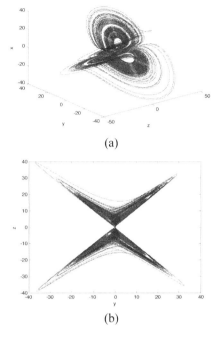

(a)

(b)

Figure 1. Phase portraits of the system (1) with q = 0.9, a = −10, b = −6, and c = 0.5: (a) x-y-z plane; (b) y-z plane.

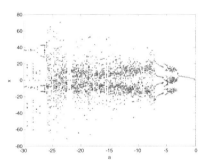

Figure 2. Bifurcation diagrams of system (1).

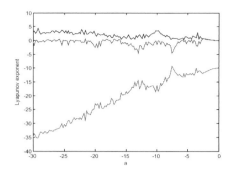

Figure 3. Lyapunov exponent diagrams of system (1).

divergent. The above numerical analysis results from bifurcation diagrams and the Lyapunov exponents further verify that chaotic attractors exist in the system (1).

3 FRACTIONAL GENERALIZED AUGMENTED LÜ SYSTEM

3.1 Image encryption algorithm

During the process of image encryption, the color image is firstly decomposed into three tricolor components (R, G, B) that are red, green, and blue monochrome images. The selected image is regarded as an m × n matrix and each dot of the matrix presents a pixel value of the image. Before encrypting the image, we firstly transform the m × n matrix into a one-dimensional vector with the length of m × n. The sequence x of chaotic sequences is sorted in ascending order and one can use it to scramble pixels of the image of R, G, and B components to obtain the scrambled image. Next, we implement the XOR operation on three components of the numerical matrices by using the 3D chaotic sequences respectively to complete the image encryption. Image decryption process is the inverse process of encryption. The three-color separations are firstly decomposed from the encrypted image, and the XOR operation is operated on three components of the numerical matrices by using chaotic sequences. Then the reduction of image pixel value scramble is realized by using the sequence x. Thus, decryption of the image is implemented finally.

3.2 Image encryption and decryption results

We select the image *Einstein. jpg* as the research object and its size is 512 × 512. The experiment was implemented in MATLAB 2010, with the initial values of state variables being 0.1, 0.2, and 0.3. In this manner, we obtained the encrypted images by

exponents are also shown in Figure 2 and Figure 3, respectively. In the range from −30 to 0 of parameter *a*, bifurcation diagrams in Figure 2 show that there exist abundant and complex dynamical characteristics in the 2.7-order system. Moreover, the Lyapunov exponents represent the characteristics of the system movement. Two of the Lyapunov exponents in Figure 3 are always negative indicating that the adjacent orbits in the attractor of the system are average convergent in direction. The biggest Lyapunov exponent in Figure 3 is positive when $a \in (-25, -8)$ explaining that the adjacent orbits in the attractor of the system are average

using the fractional-order generalized augmented Lü system shown in Figure 4. The decrypted images of *Einstein.jpg* are shown in Figure 5. A comparison of the encrypted image and decrypted image shows that the encrypted image is completely different from the original image, while the decrypted image and original image are identical. It illustrates that the image encryption algorithm is practical and effective and has good encryption effectiveness.

(a) Original image   (b) Sequence encrypted image

(c) Spatial encrypted image

Figure 4. Encrypted images: (a) original image; (b) sequence encrypted image; (c) spatial encrypted image.

(a) Encrypted image   (b) Spatial decrypted image

(c) Decrypted image

Figure 5. Decrypted images: (a) encrypted image; (b) spatial decrypted image; (c) decrypted image.

4 FRACTIONAL GENERALIZED AUGMENTED LÜ SYSTEM

4.1 *Initial value sensitivity analysis*

Selecting the system parameters as $q = 0.9$, $a = -10$, $b = -6$, $c = 0.5$, and the initial state value of the system $(x_0, y_0, z_0) = (0.1, 0.2, 0.3)$ as the encryption key, after the encryption, if you enter the correct key, you can get a successful decryption image as shown in Figure 5. When the initial value is increased by 0.000000001, we get the wrong decrypted image as shown in Figure 6. One can also change the parameters a, b, or c by a small numerical value, such as 0.000000001, obtaining the error decrypted image just as the image shown in Figure 6. It can be concluded that even a very small change in the key can also lead to the failure of decryption, which shows that the encryption system has high sensitivity to the encryption key.

4.2 *Histogram analysis*

Histogram of images can clearly reflect the distribution of characteristics of the images (Nooshin, 2012), so it is a practical and effective method for people to correctly evaluate the overall quality condition of the image. The histograms of *Einstein.jpg* are shown in Figure 7. We can find that most differences between histograms of the encrypted image and the original image are that the former is better than the latter in histogram distribution. That is to say, the encrypted image effectively masks the basic information of the original image, which indicates that it is able to resist the attack of statistical analysis very well.

4.3 *Correlation analysis*

The correlation coefficient among adjacent pixels of the encrypted image and original image are shown in Table 1. As can be seen from Table 1, the correlation coefficient of adjacent pixels in the original image is close to 1, which indicates

error decrypted image

Figure 6. Error decrypted image.

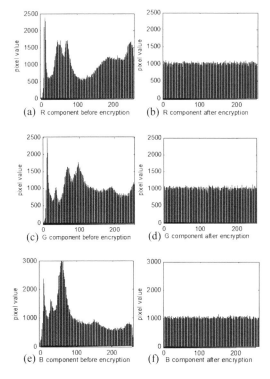

Figure 7.   Histograms of original image and encrypted image: (a) R component for original image; (b) R component for encrypted image; (c) G component for original image; (d) G component for encrypted image; (e) B component for original image; (f) B component for encrypted image.

Table 1.   Correlation coefficient.

Image	Correlation coefficient			
	Horizontal direction		Vertical direction	
R	0.99468	0.0020953	0.99547	0.0040664
G	0.99334	0.00094717	0.99433	0.0023295
B	0.99320	0.0023591	0.99419	0.0022439

that the adjacent pixels are highly correlated, while the correlation coefficient of adjacent pixels in the encrypted image is close to 0, which shows that the effect of the image encryption is better. The results verify that the encryption algorithm can well resist malicious attacks.

5   CONCLUSIONS

This paper adopts an effective and secure method to implement the image encryption operation. The method is based on the chaotic sequences generated by the fractional-order generalized augmented Lü system and combines the double encrypted technique including the pixel position scrambling of image and pixel value transform of the image. The experimental results and performance analysis show that this method not only has good encryption results, but also ensures high security of the encrypted image.

REFERENCES

Akram Belazi, Ahmed A. Abd El-Latif, Safya Belghith, J. Signal Process. **128** (2016) 155–170.
Chen G R, Ueta T, Yet another chaotic attractor, J. Int J Bifurcat Chaos. **9** (1999) 1465–1466.
Jinhu, L, C Guanrong, J. Int J Bifurcat Chaos. **12** (2002) 659–661.
Lorenz E N, J. J Atmos Sci. **20** (1963) 130–141.
Nooshin B, Youset F, Karim A, J. Comput Electr Eng. **38** (2012) 356–369.
Wei, X, X Jingkang, C Shijian, J Hongyan, J. Chin. Phys. B. **23** (2014) 060501-1-060501-8.
Xiaohua Q, Bocheng B, The 3D four wings generalized augmented Lu system. (2009)
Xiaowei Li, Chengqing Li, J. Signal Process. **125** (2016) 48–63.
Yi, Q, W Hongjuan, W Zhipeng, J. Opt Laser Eng. **84** (2016) 62–73.
Zaiping, P, W Chunhua, J. Acta Phys Sin-Ch Ed. **63** (2014) 240506-1-240506-10.

Civil, Architecture and Environmental Engineering – Kao & Sung (Eds)
© 2017 Taylor & Francis Group, ISBN 978-1-138-02985-9

Ground operation efficiency analysis of a large airport airfield based on computer simulation

Xiong Li
Planning and Design Institute, China Airport Construction Group Corporation of CAAC, Beijing, China

Xiaoqing Chen
Aviation Industry Development Research Center of China, Beijing, China

Dongxuan Wei
Highway School, Chang'an University, Xi'an, China

ABSTRACT: Airfield ground operation efficiency is an important basis to judge whether an airport's planning and design is reasonable. Using Simmod simulation software, taking the Beijing New Airport as an example, in this study, the airfield simulation models were established. The airport ground operation efficiency, arrival/departure flight delay time, and peak hour movements in southward and northward operation directions of the airport were explored in this study. The simulation results indicate that under the current construction scheme, the Beijing New Airport is able to meet the needs of the target year. In the main and secondary operation directions, the average ground delay time of departure flight is 2.73 minutes and 2.63 minutes, respectively, and that of arrival flight is less than 1 minute. The peak hour movements are about 90 flights.

1 INTRODUCTION

The continuous growth of social economy has driven China air transport on a fast-growing track, because of which China civil aviation has become the second largest air transport system in the world. Needless to say, the soaring aviation market leaves the airport systems with not many choices but to enlarge the airport system scale in order to tackle the booming demand of capacity. Under such a context, as an alternative to relieve the operation pressure of the Beijing Capital International Airport and to better the experience of air travelers in Beijing-Tianjin-Hebei region, construction of Beijing New Airport (hereinafter referred to as BNA) is imperative. With the current plan of four runways (three vertical and one horizontal), BNA is the first airport to have a non-parallel runway configuration in China. Given the complicated runway configuration and characteristic design of terminal building in BNA, and to better direct the planning, design, and operation of the new airport, it is necessary to conduct an in-depth exploration of ground operation efficiency of BNA's airfield using computer simulation.

In general, there are three approaches to evaluate airport capacity and operation efficiency: (1) data statistics, (2) mathematical theory, and (3) computer simulation. (1) Method of data statistics is used to estimate the airport capacity by drawing a capacity envelope graph based on the existing statistical information of the airport flow. As a frequently used approach for capacity analysis in early stage, it involves certain disadvantages; it only evaluates the established operation pattern of the existing runway system in busy airport, which lacks prospectiveness (Gilbo, 1993). (2) The approach of mathematical theory can be harnessed to calculate the runway capacity by proposing a proper assumption and establishing mathematical equations of airport ground operation parameters. It is mainly used for macroscopic capacity evaluation, which lacks a consideration of operation details and commonality (Neufville, 2002; Yu, 2002; Chen, 2005; Chen, 2007). (3) Approach of computer simulation is at present the most popular solution for evaluation of airport capacity and efficiency. By establishing the operating environment and control rule models using the simulation software, the airport's operation status can be reflected with reality, and can be reached to the airport capacity based on the analysis of the simulated operation data (Wang, 2008; Gao, 2010; Huang, 2010; Yu, 2009). Simmod and TAAM are the mainstream simulation modeling software for airport capacity evaluation (Huang, 2004).

Employing BNA as the research object and applying Simmod for computer simulation and modeling, in this paper, a comprehensive analysis has been performed on airport ground operation efficiency under different conditions, and the recommended schemes have been put forward. Conclusions in this paper can be used as a reference and instructions for planning, construction, and operation pattern selecting of multi-runway systems in China's large airports.

2 AIRFIELD LAYOUT AND OPERATION RULES OF BNA

In the current phase, the BNA construction project is to build four runways for a yearly passenger throughput of 72 million. The layout of its airfield is shown in Fig. 1. The planned terminal building, targeting a yearly passenger throughput of 45 million, applies a centralized combinational configuration, with the five concourses distributed in a "centrally radial" shape. Taking the main landing direction as an example, the runway operation pattern employed is West 1 Runway (W1) and North 1 Runway (N1) only used for take-off; West 2 Runway (W2) and East 1 Runway (E1) mainly used for landing; and E1 used for assisted take-off in the morning departure peak. The runways will operate independently.

Taxiing routes should be reasonably and appropriately setup, based on the airport operation, by the principle that main taxiways such as parallel taxiways, western and eastern contact taxiways, etc. should be operated in a monodirectional fashion. Taxiway system of airfield could be divided into four areas based on the average taxiing speed

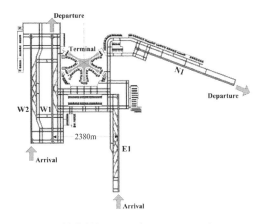

Figure 1. Airfield layout and runway operation pattern of BNA.

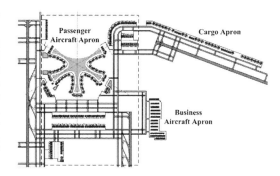

Figure 2. Apron distribution.

(from 5 km/h to 46 km/h): push-out taxiway, apron taxiway, main taxiway, and high-speed taxiway.

As shown in Fig. 2, there are 322 gates in airfield, among which 197 are for passenger plane, 40 for cargo aircraft, and 85 for corporate aircrafts. Based on the differences among plane types, the corresponding boarding and alighting times are set accordingly in the scope from 30 to 80 minutes.

3 AIRPORT SIMULATION MODELING

Employing the simulation software Simmod Plus 7.6, the ground operation efficiency of BNA was simulated and evaluated. Simmod, launched in 1978 by the U.S. Federal Aviation Administration (FAA), is a microscopic, dynamic, and comprehensive airport simulation software. After constant upgrading and perfecting, it is now one of the most popular airport and airspace simulation software [9,10]. Based on apron operation pattern (a), the simulation modeling mainly takes two operation directions (southward and northward) into consideration. Then, on that basis, and taking the southward operation as an example, further simulation and comparison on patterns (b) and (c) is conducted.

Landing interval is controlled mainly based on the wake vortex separation, as shown in Table 1. In addition, extra safe redundancy is taken into account. According to the differences between the aircrafts, departure separation of each runway is arranged at 80–140 seconds. Moreover, given the influence of the Beijing Capital International Airport on BNA in terminal airspace, departure interval of 3 minutes is considered for western runways in the morning time of 7:30–8:30, and a same departure interval is considered for eastern runways in the morning time of 8:30–9:30. When a single runway is in alternate operation, and the distance from the landing flight to the runway is less than 8 km, departure flights are forbidden to taxi out from the waiting point.

Based on the predicted flight volume of BNA and requirement of targeted yearly transportation volume of the terminal building, three typical daily flight schedules are planned in simulation, as shown in Table 2. The simulated operation demonstration is shown in Fig. 3.

Table 1. Wake vortex separation of landing flights.

Latter aircraft	Former aircraft		
	A380	Heavy	Medium
A380	8 km	6 km	6 km
Heavy	12 km	8 km	6 km
Medium	13 km	10 km	6 km

Table 2. Typical daily take-off and landing flights.

Yearly passenger throughput (person-time)	Typical daily aircraft movements	Aircraft movements in peak hour
30 million	808	80
45 million	1210	95
60 million	1600	115

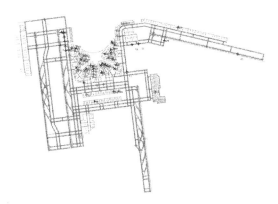

Figure 3. Simulated operation demonstration.

4 OPERATION RESULTS AND ANALYSIS

Table 3 gives out the average ground delay times under different daily flight volumes when the airport is running in northward and southward directions. In the calculation of the comprehensive delay time, the proportion of operation in northern and southern directions is temporarily considered as 9:1. When the passenger throughput is 45 million person-times, average ground delay time of departure and arrival flights is 2.72 minutes and 58 seconds, respectively, which stands for a high operation efficiency. When the passenger throughput reaches 60 million person-times, the average ground delay time of departure flights is 4.41 minutes, which is still a sound efficiency of ground operation.

For departure flights using boarding gates, the total delay time consists of three parts: (1) take-off waiting, including waiting for take-off at runway entrance and delay of push-out; (2) delay in angle area of main terminal buildings, including waiting in apron taxiway and waiting caused by interference due to push-out of other flights; and (3) taxiing delay occurring in other areas. Table 4 presents the details and proportion of each part. With an increase of flight volume, the take-off waiting and taxiing delay in other areas grows evidently, whereas delay in angle areas grows slowly, with a slight change of delay time. When passenger throughput exceeds 45 million person-times, proportion of delay time in angle area is relatively low, accounting for 1/5 or 1/6 of total delay time; this indicates that it is not the bottleneck of airport ground operation.

Fig. 4 gives the aircraft movement in daily time frames under a passenger throughput of 45 million person-times. The total movement is 1210 flights, with 90 flights in each peak hour.

5 CONCLUSIONS

In this paper, a systematic analysis on the ground operation efficiency of the airfield of BNA was conducted. Utilizing the simulation software Simmod, a simulation study on ground operation efficiency, delay time, and peak hour movements

Table 3. Average ground delay time.

Yearly passenger throughput (person-time)	Departure flight average ground delay		Arrival flight average ground delay	
	Northward	Southward	Northward	Southward
30 million	1.32'	1.33'	0.57'	0.49'
45 million	2.73'	2.63'	0.97'	0.83'
60 million	4.43'	4.21'	1.42'	1.15'

Table 4. Constitution of ground delay of departure flights using boarding gates.

Yearly passenger throughput (person-time)	Airport operation direction	Total ground delay of departure flight	Take-off waiting	Delay in angle area	Taxiing delay in other areas
30 million	Northward	1.61	0.83	0.68	0.1
	Southward	1.51	0.82	0.6	0.09
	Mean value	1.6	0.83	0.67	0.1
	Proportion		51.8%	42.0%	6.2%
45 million	Northward	3.41	1.83	0.75	0.83
	Southward	3.36	1.88	0.72	0.76
	Mean value	3.41	1.84	0.75	0.82
	Proportion		53.9%	21.9%	24.2%
60 million	Northward	5.47	3.05	0.85	1.57
	Southward	5.25	3.02	0.91	1.32
	Mean value	5.45	3.05	0.86	1.55
	Proportion		55.9%	15.7%	28.4%

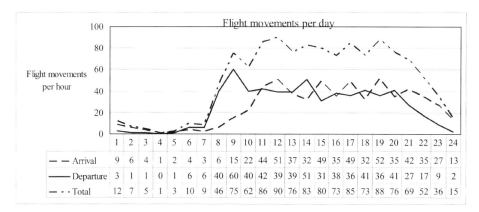

Figure 4. Distribution of aircraft movements in daily time frames.

was performed. Results of all the above-mentioned endeavors are as follows:

1. When the yearly passenger throughput is 45 million person-times, as for the construction scheme in current stage of BNA, the average ground delay time of departure flights is about 2.72 minutes, and that of arrival flights is about 58 seconds. This indicates that the airport is running in high efficiency, and that the target yearly transportation volume can be reached.
2. As for flights using boarding gates, over half of ground delay of departure flights is caused by take-off waiting and only 1/5 occurs in angle area between main buildings. Furthermore, this part accounts for a smaller proportion when the flight volume grows, proving that it is not the bottleneck of airport operation.
3. In BNA, the extreme value of take-off flight is in the morning. The peak hour movements are 90 flights, appearing at noon.

However, it should be noted that the simulation study in this paper only works for the capacity and delay time of the airport under normal operation. Furthermore, flight delay in simulation is normally less than that in actual cases, since influencing factors such as air traffic flow control, bad weather, airlines, military actions, and aviation events have been excluded in the study.

ACKNOWLEDGEMENTS

This work was supported by two research projects:

1. China civil aviation security ability building funds—Research on application of simulation technology in civil airport planning and design; and
2. National Key Technologies R&D Program—Research and demonstration on key technology of green airport construction (2014BAJ04B02).

REFERENCES

Chen, X. Research on capacity evaluation and optimization methods at airport airside, PhD thesis, Nanjing University of Aeronautics & Astronautics (2007).

Chen, Y., Cao, Y. and Zhou, Y. Estimation of Beijing international airport capacity, *Flight Dynamics*, 23(4), pp. 86–89, (2005).

Gao, W. and Huang, C.W. Research on capacity evaluation for Chongqing *TMA, Journal of Civil Aviation Univ. of China*, 28(5), pp. 1–4, (2010).

Gao, W. and Jiang, S.Y. Simulation study on closely spaced parallel runway analysis using SIMMOD plus, in *Proc. 2010 International Conference on Intelligent Computation Technology and Automation*, (Los Alamitos, CA, USA, 2010).

Gilbo, E.P. Airport capacity: representation, estimation, optimization, *IEEE Transactions on Control Systems Technology*, 1(3), pp. 144–154, (1993).

Huang, A.S., Schleicher, D. and Hunter, G. Future flight demand generation tool, in *Proc. AIAA 4th Aviation Technology, Integration and Operations (ATIO) Forum*, (Chicago, IL, USA, 2004).

Neufville, R.D. and Odoni, A.R. *Airport systems planning, designs, and management* (McGraw-Hill, New York, 2002).

Wang, C., Zhang, X.Y. and Xu, X.H. Simulation study on airfield system capacity analysis using SIMMOD, *in Proc. 2008 International Symposium on Computational Intelligence and Design*, (Wuhan, China, 2008).

Yu, H.Y., Hu, H.Q. and Yao, J.J. Computer emulation methods for aviation capacity evaluation of airports, *Air Traffic Management*, 2, pp. 21–25, (2009).

Yu, J. and Pu, Y. Runway capacity probability model and capacity curve, *Journal of Traffic and Transportation Engineering*, 2(4), pp. 99–102, (2002).

Civil, Architecture and Environmental Engineering – Kao & Sung (Eds)
© *2017 Taylor & Francis Group, ISBN 978-1-138-02985-9*

Matrix search algorithm of robust common quadratic Lyapunov function for uncertain switched systems

Xiaoyu Zhang, Ping Li & Bin Shen
School of Electronics and Information Engineering, North China Institute of Science and Technology, Beijing, China

ABSTRACT: Considering the stability problem of uncertain linear switched systems, this article puts forward a searching algorithm of Common Robust Quadratic Lyapunov Function (CRQLF). The sufficient conditions of CRQLF are given when the stable matrix set is not involuntary. The recurrence formula of CRQLF and the search algorithm in terms of Linear Matrix Inequality (LMI) are given. The simulation results verify the existence of the CRQLF for the stability of uncertain linear switched system under arbitrary switching signal.

1 INTRODUCTION

The stability of Switched Systems (SS) is a significant problem for practice. Many research works about it have been done since the early years. For example, the basic control analysis was carried out (Beldiman & Bushnell 1999). A Lie algebraic condition was proposed about the stability of SS (Liberzon & Morse 1999). The Lie-algebraic stability conditions for nonlinear switched systems subsequently were considered in (Margaliot & Liberzon 2006). Branicky showed solicitude for stability of SS in early years (Branicky 1994, Branicky 1997, Branicky 1998) and made excellent contribution to the field.

It is known that the subsystems stability does not guarantee the stability of the whole SS. An important stability analysis method is related to the existence of Common Lyapunov Functions (CLF). If there is a Lyapunov function whose time derivative is negative all the time along with every subsystem, the stability of the whole SS is reached. In (Narendra & Balakrishnan 1994), two subsystems who comprise SS are studied. It is well-known that the commutativity of the two system matrices is a sufficient condition for a Common Quadratic Lyapunov Function (CQLF). Therein an algorithm of NB Algorithm was born, that is for finding such a CQLF. Ooba and Funahashi present triangle conditions for the existence of a CQLF concerning a pair of linear systems (Ooba & Funahashi 1997). Furthermore, a sufficient condition for the global asymptotic stability of a switched nonlinear system is proposed by the analysis of the Lie bracket of any pair of each sub system. The pairwise

commutation of the vector fields is sufficient for the existence of a CLF (Mancilla-Aguilar 2000). Fierro et al. tried to find different way to validate a hybrid system that is composed of continuous subsystem, and certified that the method of CLF is conservative by contrast of cases (Fierro et al. 1996). How to find the CQLF? Cheng et al. expanded the NB-Algorithm and present a necessary and sufficient condition for the existence of a CQLF of a finite set of stable matrices. They also proved that a set of block upper triangular matrices share a CQLF, if each set of diagonal blocks share a CQLF (Cheng et al. 2003, Cheng et al. 2006). More particularly, the introduction about CLF refers to (Margaliot 2006, Liberzon 2003, Lin & Antsaklis 2009).

Recently, the CLF method continues to be concerned by many researchers. A switched linear copositive Lyapunov function method for switched positive systems is presented in (Liu 2009), and the existence problem of a Common Copositive Lyapunov Function (CCLF) for switched positive linear systems with stable and pairwise commutable subsystems is investigated in (Tong et al. 2013). The present results demonstrated that a CCLF can be constructed for the underlying system whenever its subsystems are continuous time, discrete-time or the mixed type. Further, For the finite-time stability problem of switched positive linear systems, Chen and Yang present a sufficient condition for finite-time stability using the CCLF and multiple copositive Lyapunov functions (Chen & Yang 2014). Meanwhile a computational method for vector functions used to construct the CLF is proposed. A Common Diagonal Lyapunov Function (CDLF) is constructed for a special linear SS (Bykkoroglu 2012). Contradistinctively, a method of computing a CQLF for a stable linear time-invariant system, whose system matrix is tridiagonal (Yu & Gao 2009). Many CLF research work is carried out for nonlinear switched systems. A CLF is based on generalized Kullback-Leibler divergence or Csiszar's divergence between the state and equilibrium is presented in (Abou Al-Ola et al. 2011). Aleksandrov addressed the stability analysis of a class of switched nonlinear systems with uncertain nonlinear functions constrained in a sector set, which are called admissible sector nonlinearities (Aleksandrov et al. 2011). The existence condition in terms of linear inequalities of a CLF and a constructive algorithm based on the modified Gaussian elimination procedure are given. And for a family of nonlinear mechanical systems with one degree of freedom, He proposed the existence condition of a CLF once again (Aleksandrov & Murzinov 2012). In (Wang et al. 2014), sufficient conditions for the existence of a CQLF are given for two classes of switched linear systems which possess negative row strictly diagonally dominant and diagonalizable stable state matrices, respectively. Some kinds of different switched systems are concerned about their CQLFs or stability problems via CQLF. For example, planar linear switched system (Ben Salah et al. 2015), descriptor system (Gu et al. 2009, Zhai 2009) and switched Persidskii-type system (Chen et al. 2011). The non-existence of a CQLF for SS is also a interesting problem. A sufficient condition for the non-existence of a CQLF for N systems is presented and the determining algorithm is designed based on particle swarm optimisation (Duarte-Mermoud et al. 2012). The research work about CQLF is also extended to discrete switched system (Sun et al. 2011, Liu & Zhao 2015).

Finding a CLF is still a challenging problem for SS study. Although many results of CLF have been proposed during several past decades, the conditions of its existence and the finding algorithm are difficult yet. Furthermore, it is closely related with the control analysis and design. Majority of results about finding CLF are always based on the assumption that every subsystem matrix A_i is stable, i.e. $\mathbb{A} = \{A_1, A_2,\ldots, A_N\}$ is a stable matrix set. When the matrices of the subsystems are not commutive, the NB Algorithm is not applicable. Under the situation Zhu et al. presented the method of constructing CQLF (Zhu et al. 2007). However, there is no specific applicable algorithm to find a CQLF in their method. And the conditions are difficult to be testified.

The purpose of the present paper is to provide a recurrence searching algorithm of a CQLF for uncertain switched system. It will discuss the CQLF problem, i.e. determination for uncertain

switched linear system stability. If considered individually the stability of linear switched systems with uncertainty, namely, the problem of robust stability and constructing a CQLF will be more difficult. In order to overcome this difficulty, this paper presents Robust Stable Matrix Set (RSMS) concept, and further releases the determinant and constructing theorems of a CQLF for a RSMS. The results make sense for robust control problem of uncertain switched linear systems under arbitrary switching rule.

2 PROBLEM FORMULATION

Consider the following uncertain switched system:

$$\dot{x}(t) = \left(A_\sigma + \Delta A_\sigma\right)x(t) \tag{1}$$

where $x(t) \in \mathbb{R}^n$ denotes the system state variable, ΔA_σ reflects the uncertainties of the parameters, $\sigma(t): \mathbb{R} \to \mathbb{N} \cong \{1,2,\cdots,N\}$ is a piecewise constant function of the time t, also referred to as the switching signal (rule). Then we define a switching sequence:

$$Q \triangleq x_0; (i_0,t_0), (i_1,t_1),\cdots,(i_k,t_k),\cdots,$$
$$\forall i_k \in \mathbb{N}, k \in \mathbb{Z}^+$$

i.e., the i_k-th subsystem operates when $t \in [t_k, t_{k-1}]$.

For the switching signal $\sigma(t) = i, i \in \mathbb{N}$, the i-th subsystem parameters can be denoted as: $A_\sigma \triangleq A_i, \Delta A_\sigma \triangleq \Delta A_i$. Accordingly, for the k-th switching, when $t_k \le t < t_{k+1}$. We can set $\sigma(t) = i$, i.e., $i_k = i \in \mathbb{N}$. According to (1), the system can be described as:

$$\dot{x}(t) = \left(A_i + \Delta A_i\right)x(t) \tag{2}$$

The switching system as described in (1) can satisfy the following assumptions.

Assumption 1 The uncertainty parameters, ΔA_i satisfy the following relationships:

$$\Delta A_i = H_{a,i}F_{a,i}(t)E_{a,i} \tag{3}$$

where $H_{a,i} \in \mathbb{R}^{n \times r_a}$ and $E_{a,i} \in \mathbb{R}^{r_a \times n}$ are all the known constant matrices, and the unknown time-varying matrices $F_{a,i}(t)$ satisfy: $F_{a,i}^T(t)F_{a,i}(t) \le I$.

Generally, the stability problem of (1) is significant for control. The problem of this paper is: if A_i, $i \in \mathbb{N}$ comprise a stable matrix set, i.e., $\mathbb{A} = \{A_1, A_2,\cdots,A_N\}$, then finding a CQLF guarantees the stability of uncertain switched system (1). Firstly two lemmas and a definition that are used in the present study will be introduced.

Lemma 1 (Schur Complement) Given a symmetric matrix $S = \begin{bmatrix} S_{11} & S_{12} \\ S_{21} & S_{22} \end{bmatrix}$, the following three conditions are equivalent.

i. $S < 0$;
ii. $S_{11} < 0, S_{22} - S_{21}S_{11}^{-1}S_{12} < 0$;
iii. $S_{22} < 0, S_{11} - S_{12}S_{22}^{-1}S_{21} < 0$.

Lemma 2 Assuming that H and E are the matrices consisting of real constants with approximate dimension, $F(t)$ satisfies $F^T(t)F(t) \le I$, the following relation is true for any constant $\varepsilon > 0$,

$$HF(t)E + E^T F^T(t)H^T \le \varepsilon^{-1}HH^T + \varepsilon E^T E. \quad (4)$$

Lemma 3 If $\forall \varepsilon_i > 0$, the Linear Matrix Inequality (LMI)

$$\begin{bmatrix} P_i A_i + A_i^T P + L_i & P_i H_{a,i} \\ * & -\varepsilon_i I \end{bmatrix} < 0, \forall i \in \mathbb{N} \quad (5)$$

has a positive definite symmetric solution matrix P_i, every subsystem (2) is quadratically robust stable.

Proof: Select a Lyapunov function of every subsystem (2)

$$V_i(x(t)) = x^T(t)P_i x(t).$$

And find the time derivative along (2),

$$\dot{V}i(x(t)) = x^T(t)[P_i A_i + A_i^T P_i + P_i \Delta A_i + \Delta A_i^T P_i]x(t)$$

According to Assumption 1 and Lemma 2, $\forall \varepsilon_i > 0$, the inequality

$$P_i \Delta A_i + \Delta A_i^T P_i \le \varepsilon_i^{-1} P_i G_i P_i + L_i$$

holds with

$$G_i = \varepsilon_i^{-1} H_{a,i} H_{a,i}^T, L_i = \varepsilon_i E_{a,i}^T E_{a,i} \quad (6)$$

are matrix items produced by the system uncertainty. Substitute it into $\dot{V}_i(x(t))$, we have

$$\dot{V}_i(x(t)) \le x^T(t)[P_i A_i + A_i^T P_i + P_i G_i P_i + L_i]x(t)$$

According to Lemma 1, if (5) is satisfied,

$$\dot{V}_i(x(t)) \le -x^T(t)Q_i x(t)$$

hold, where Q_i is some positive definite matrix. Then every subsystem (2) is quadratically robust stable.

Suppose that every subsystem matrix A_i satisfies the following Assumption.

Assumption 2 Every matrix A_i in stable matrix set $\mathbb{A} = \{A_1, A_2, \cdots, A_N\}$ is quadratically robust stable, i.e. satisfies Lemma 3.

To underlie the following analysis, define

$$P_{ij} = A_i^T P_j + P_j A_i, U_{i,j} = P_j G_i P_j + L_i, \forall i,j \in \mathbb{N}. \quad (7)$$

3 PRELIMINARIES OF CQLF

3.1 Known results

A great deal of research was conducted on the construction of a CQLF of a stable matrix set. The sufficient condition as well as important conditions for constructing a CQLF of a matrix set was elaborated in detail in Section 1. As described in (Zhu et al. 2007), the researchers probed into the methods for searching the CLF and concluded several conditions for the existences of CQLF. Based on the research results in (Zhu et al. 2007), when any two matrices in a stable matrix set $\mathbb{A}$ are not interchangeable but satisfy the involution conditions, the CQLF can be constructed according to the following lemmas.

Assuming a stable matrix set $\mathbb{A} = \{A_1, A_2, \cdots, A_N\}$ with their corresponding Lyapunov matrix P_i, and defining $[A_N, A_i] = \sum_{k=1}^{N} \gamma_i^k A_k, \forall i = \mathbb{N}$, in which $\gamma_i^k > 0$ is a scaling coefficient parameter. Given a randomly selected $P_{N-1} > 0$, we set

$$P_{N,N} = P_N A_N + A_N^T P_N = -P_{N-1}.$$

Lemma 4 Given a stable matrix set $\mathbb{A} = \{A_1, A_2, \cdots, A_N\}$, if $\forall i = 1, \cdots, N-1$ the following conditions can be satisfied:

$$\begin{cases} \max(\gamma_i^i) < 2\min|\text{Re}\,\lambda(A_N)|, \\ -P_{i,N-1} + \gamma_i^N P_{N-1} - \sum_{k=1,k\ne i}^{N-1} \gamma_i^k P_{k,N} > 0, \end{cases} \quad (8)$$

where $P_{i,j} = A_i^T P_j + P_j A_i, \forall i,j \in \mathbb{N} = \{1,2,\cdots,N\}$, P_N is a CQLF of $\mathbb{A}$.

Still, we should randomly select $P_{N-1} > 0$ and set $P_N A_N + A_N^T P_N = -P_{N-1}$, and moreover,

$$C_{N,j} = [A_N, A_j], \forall j \in \mathbb{N} \quad (9)$$

According to the results in [32], the following Lemma can be deduced.

Lemma 5 Given a stable matrix set $\mathbb{A}$, defining $C_{N,j}, \forall j \in \mathbb{N}$ as similar to (9), if the following inequality can be satisfied:

$$P_{i,N-1} + P_N C_{N,i} + C_{N,i}^T P_N < 0, i = 1, \cdots, N-1 \quad (10)$$

1419

P_N is a CQLF of $\mathbb{A}$.

3.2 Robust stable matrix set

Defining a robust stable matrix set

$$\mathbb{A} = \{A_1, A_2, \cdots, A_N\},$$

in which each stable matrix is of robust stability, each matrix can satisfy Lyapunov Equation:

$$P_i A_i + A_i^T P_i + P_i G_i P_i + L_i = -Q_i, \quad \forall i = 1, 2, \cdots, N \tag{11}$$

where G_i and L_i are the matrices of Lyapunov Equation induced by the system uncertainty; i.e. defined by (6).

If the matrix set $\mathbb{A}$ is involutory, we still assume that

$$[A_N, A_i] = \sum_{k=1}^{N} \gamma_i^k A_k, \forall i = \mathbb{N},$$

where $\gamma_i^k > 0$ is the scalar coefficient parameter. To random select $P_{N-1} > 0$ and set the fowling equation:

$$P_N A_N + A_N^T P_N + U_{N,N} = -P_{N-1}. \tag{12}$$

4 MAIN RESULTS

The following corollaries can be deduced based on Lemma 4 and Lemma 5.

4.1 Common robust quadratic Lyapunov function

Corollary 1 Given a robust stable matrix set $\mathbb{A} = \{A_1, A_2, \cdots, A_N\}$, which can satisfy (11), if $\forall i = 1, \cdots, N-1$ the following conditions can be satisfied:

$$\begin{cases} \max(\gamma_i^i) < 2\min|\mathrm{Re}\,\lambda(A_N)|, \\ -P_{i,N-1} + \gamma_i^N P_{N-1} - \displaystyle\sum_{k=1,k\neq i}^{N-1} \gamma_i^k P_{k,N} + \gamma_i^i U_{i,N} \\ + \gamma_i^N U_{N,N} + U_{i,N} A_N + A_N^T U_{i,N} - U_{N,N} A_i \\ \qquad - A_i^T U_{N,N} > 0, \end{cases} \tag{13}$$

P_N is a common robust quadratic Lyapunov function of $\mathbb{A}$, i.e, they can satisfy the Riccati Inequality:

$$P_N A_i + A_i^T P_N + P_N G_i P_N + L_i < 0, \forall i = 1, 2, \cdots, N. \tag{14}$$

Proof:

$$\left(P_{i,N} + U_{i,N}\right)\left(A_N + \frac{\gamma_i^i}{2}I\right) + \left(A_N + \frac{\gamma_i^i}{2}I\right)^T \left(P_{i,N} + U_{i,N}\right)$$
$$= \left(P_N A_i + A_i^T P_N + U_{i,N}\right) A_N + A_N^T \left(P_N A_i + A_i^T P_N + U_{i,N}\right)$$
$$\quad + \gamma_i^i \left(P_N A_i + A_i^T P_N + U_{i,N}\right)$$
$$= P_N \left(A_N A_i - \sum_{k=1}^{N} \gamma_i^k A_k\right) + A_i^T P_N A_N + A_N^T P_N A_i$$
$$\quad + \left(A_i^T A_N^T - \sum_{k=1}^{N} \gamma_i^k A_k^T\right) P_N + \gamma_i^i \left(P_N A_i + A_i^T P_N + U_{i,N}\right)$$
$$\quad + U_{i,N} A_N + A_N^T U_{i,N}$$
$$= \left(P_N A_N + A_N^T P_N\right) A_i + A_i^T \left(P_N A_N + A_N^T P_N\right)$$
$$\quad - \gamma_i^N \left(P_N A_N + A_N^T P_N\right) - \sum_{k=1,k\neq i}^{N-1} \gamma_i^k \left(P_N A_k + A_k^T P_N\right)$$
$$\quad + U_{i,N}\left(A_N + \gamma_i^i\right) + A_N^T U_{i,N}$$
$$= -P_{i,N-1} + \gamma_i^N P_{N-1} - \sum_{k=1,K\neq i}^{N-1} \gamma_i^k P_{k,N} + U_{i,N}\left(A_N + \gamma_i^i\right)$$
$$\quad + A_N^T U_{i,N} - U_{N,N}\left(A_i - \gamma_i^N\right) - A_i^T U_{N,N}.$$

According to the second condition in (13), if the inequality

$$-P_{i,N-1} + \gamma_i^N P_{N-1} - \sum_{k=1,k\neq i}^{N-1} \gamma_i^k P_{k,N} + U_{i,N}\left(A_N + \gamma_i^i\right)$$
$$+ A_N^T U_{i,N} - U_{N,N}\left(A_i - \gamma_i^N\right) - A_i^T U_{N,N} > 0,$$

Is valid $\forall i = 1, 2, \cdots, N-1$, the product of $P_{i,N} + U_{i,N}$ and the transposed matrix of $A_N + \gamma_i^i/2\,I$ is positive definite. According to the first inequality in (13), it can be inferred that $A_N + \gamma_i^i/2\,I$ is stable, and the following inequality is true:

$$P_N A_i + A_i^T P_N + P_N G_i P_N + L_i < 0, \forall i \in \mathbb{N}. \tag{15}$$

It means that P_N is a common quadratic Lyapunov function of $\mathbb{A}$; moreover, for each matrix A_i in $\mathbb{A}$. P_N satisfies the Riccati Inequality that contains the uncertain item $U_{i,N}$ (as described in (15)). Therefore, P_N is a CRQLF of $\mathbb{A}$.

Assuming a robust stable matrix set $\mathbb{A} = \{A_1, A_2, \cdots, A_N\}$, in which each stable matrix A_i is of robust stability and thus satisfies Lyapunov equation as described in (11). We still define (6) and (12), but the matrix $\mathbb{A}$ is not involutory. Based on the Lemma 5 on the search of the CQLF, the following Corollary 2 can be deduced.

Corollary 2 Given a robust stable matrix set $\mathbb{A} = \{A_1, A_2, \cdots, A_N\}$, which satisfies (11), we define $C_{N,j}$ as described in (9). If $\forall i = 1, 2, \cdots, N-1$, P_N satisfies

$$P_{i,N-1} + P_N C_{N,i} + C_{N,i}^T P_N + U_{N,N} A_i$$
$$+ A_i^T U_{N,N} - U_{i,N} A_N - A_N^T U_{i,N} < 0, \tag{16}$$

P_N is a CRQLF of $\mathbb{A}$, i.e. they can satisfy the Riccati inequality:

$$P_N A_i + A_i^T P_N + P_N G_i P_N + L_i < 0, \forall i = 1, 2, \cdots, N \tag{17}$$

Proof:

$$\left(P_{i,N}+U_{i,N}\right)A_N + A_N^T\left(P_{i,N}+U_{i,N}\right)$$
$$=\left(P_N A_i + A_i^T P_N + U_{i,n}\right)A_N + A_N^T\left(P_N A_i + A_i^T P_N + U_{i,N}\right)$$
$$= P_N\left(A_N A_i - C_{N,i}\right) + A_i^T P_N A_N + A_N^T P_N A_i$$
$$\quad +\left(A_i^T A_N^T - C_{N,i}^T\right)P_N + U_{i,N}A_N + A_N^T U_{i,N}$$
$$=\left(P_N A_N + A_N^T P_N\right)A_i + A_i^T\left(P_N A_N + A_N^T P_N\right)$$
$$\quad - P_N C_{N,i} - C_{N,i}^T P_N + U_{i,N}A_N + A_N^T U_{i,N}$$
$$= -\left(P_{i,N-1} + P_N C_{N,i} + C_{N,i}^T P_N + U_{N,N}A_i \right.$$
$$\quad \left. + A_i^T U_{N,N} - U_{i,N}A_N - A_N^T U_{i,N}\right).$$

If the inequality as described in (16) is satisfied, $\left(P_{i,N}+U_{i,N}\right)A_N + A_N^T\left(P_{i,N}+U_{i,N}\right)>0$ is true for $\forall i =1,2,\cdots,N-1$, suggesting that the product of $P_{i,N}+U_{i,N}$ and the transposed matrix of A_N is positive definite. As stated above, A_N is stable, and therefore, $P_{i,N}+U_{i,N}<0$, $i=1,\ldots,$ $N-1$ is true, *i.e.* the following inequality is true:

$$P_N A_i + A_i^T P_N + P_N G_i P_N + L_i < 0, \forall i \in \mathbb{N}. \quad (18)$$

Accordingly, P_N is a CQLF of $\mathbb{A}$, and moreover, for each matrix A_i in $\mathbb{A}$, P_N satisfies the Riccati Inequality that contains the uncertain item $U_{i,N}$ (as described in (18)). Therefore, P_N is a *CRQLF* of $\mathbb{A}$.

4.2 *Recursive algorithm of CRQLF*

Based on Lemma 5, the recursive algorithm for searching a CQLF of the stable matrix set can be constructed.

Theorem 1 Given a stable matrix set $\mathbb{A}=\left\{A_1,A_2,\cdots,A_N\right\}$, if $\mathbb{A}$ has a CLF, the matrix $P_k\left(\forall k\in\mathbb{N}\right)$ that can satisfy the following conditions:

$$\begin{cases} P_{k,k} < 0, \\ A_i^T P_{k,k} + P_{k,k}A_i - P_k C_{k,i} - C_{k,i}^T P_k > 0 \end{cases} \quad (19)$$

can be regarded as the CQLF matrix of the matrix set $A_k = \{A_1, A_2, \ldots A_k\}$, in which

$$C_{i,j} = A_i A_j - A_j A_i.$$

Proof:

$$P_{i,k}A_k + A_k^T P_{i,k}$$
$$=\left(P_k A_i + A_i^T P_k\right)A_k + A_k^T\left(P_k A_i + A_i^T P_k\right)$$
$$= P_k\left(A_k A_i - C_{k,i}\right) + A_i^T P_k A_k + A_k^T P_k A_i$$
$$\quad +\left(A_k A_i - C_{k,i}\right)^T P_k$$
$$=\left(P_k A_k + A_k^T P_k\right)A_i + A_i^T\left(P_k A_k + A_k^T P_k\right)$$
$$\quad - P_k C_{k,i} - C_{k,i}^T P_k$$
$$= A_i^T P_{k,k} + P_{k,k}A_i - P_k C_{k,i} - C_{k,i}^T P_k.$$

If the inequality in (19) can be satisfied, obviously, $P_{i,k}A_k + A_k^T P_{i,k} > 0$. Since A_k is a stable matrix,

$$P_{i,k} = A_i^T P_k + P_k A_i < 0, \forall i = 1,\cdots,k$$

is true. Therefore, P_k is a CQLF of

$$\mathbb{A}_k = \left\{A_1, A_2, \cdots, A_k\right\}.$$

We assume a robust stable matrix set $\mathbb{A}$ in which each stable matrix is of robust stability and can satisfy Lyapunov equation as described in (11). Still, we define (7) but don't define (12), and the matrix set $\mathbb{A}$ is not involutory. Based to the above-described Theorem 1 on the common quadratic Lyapunov function, the following algorithm for recursively constructing the CLF can be deduced.

Theorem 2 There is a CLF in robust stable matrix set $\mathbb{A} = \left\{A_1, A_2, \cdots A_N\right\}$. For arbitrary positive integer $\varepsilon_1, \varepsilon_2, \cdots, \varepsilon_k$ and $\forall i = 1, \cdots, k-1, k \in \mathbb{N}$ the positive definite symmetric matrix $P_k\left(\forall k \in \mathbb{N}\right)$ that satisfies the following conditions:

$$\begin{cases} P_{k,k} + U_{k,k} < 0, \\ A_i^T P_{k,k} + P_{k,k}A_i - P_k C_{k,i} - C_{k,i}^T P_k \\ \quad + U_{i,k}A_k + A_k^T U_{i,k} > 0 \end{cases} \quad (20)$$

can be regarded as the CRQLF matrix of the matrix set $A_k = \{A_1, A_2, \ldots, A_k\}$ And P_k satisfy the Riccati inequality:

$$P_k A_i + A_i^T P_k + U_{i,k} < 0, \forall i = 1,2,\cdots,k \quad (21)$$

where $C_{i,j} = A_i A_j - A_j A_i$.

Proof:

$$\left(P_{i,k}+U_{i,k}\right)A_k + A_k^T\left(P_{i,k}+U_{i,k}\right)$$
$$=\left(P_k A_i + A_i^T P_k + U_{i,k}\right)A_k$$
$$\quad + A_k^T\left(P_k A_i + A_i^T P_k + U_{i,k}\right)$$
$$= P_k\left(A_k A_i - C_{k,i}\right) + A_i^T P_k A_k + A_k^T P_k A_i$$
$$\quad +\left(A_k A_i - C_{k,i}\right)^T P_k + U_{i,k}A_k + A_k^T U_{i,k}$$
$$=\left(P_k A_k + A_k^T P_k\right)A_i + A_i^T\left(P_k A_k + A_k^T P_k\right)$$
$$\quad - P_k C_{k,i} - C_{k,i}^T P_k + U_{i,k}A_k + A_k^T U_{i,k}$$
$$= A_i^T P_{k,k} + P_{k,k}A_i - P_k C_{k,i} - C_{k,i}^T P_k$$
$$\quad + U_{i,k}A_k + A_k^T U_{i,k}.$$

If the inequality in (20) can be satisfied, obviously, $\left(P_{i,k}+U_{i,k}\right)A_k + A_k^T\left(P_{i,k}+U_{i,k}\right)>0$. Since A_k is a stable matrix,

$$P_{i,k} + U_{i,k} = A_i^T P_k + P_k A_i + U_{i,k} < 0, \forall i = 1,\cdots,k-1$$

is true. Therefore, P_k is a CRQLF of
$A_k = \{A_1, A_2,\ldots, A_k\}$.

5 NUMERICAL SIMULATION

A switched system (1) consisting of three subsystems $\left(N \triangleq \{1,2,3\}\right)$ were simulated when the system state is measurable. The parameters of each subsystem are listed below.

Subsystem 1:

$$A_1 = \begin{bmatrix} -1 & 1 \\ 2.25 & -2.75 \end{bmatrix}, \Delta A_1 = \begin{bmatrix} 0.38\sin 4\pi t & 0 \\ 0 & 1.1e^{-t} \end{bmatrix},$$

Subsystem 2:

$$A_2 = \begin{bmatrix} -5.6 & 0.4 \\ 3.4 & -2.6 \end{bmatrix}, \Delta A_2 = \begin{bmatrix} 0.65\cos 7\pi t & 0.2\cos 7\pi t \\ 0 & 0.56e^{-2.2t} \end{bmatrix},$$

Subsystem 3:

$$A_3 = \begin{bmatrix} -0.8 & -7.6 \\ 2.2 & -0.6 \end{bmatrix}, \Delta A_3 = \begin{bmatrix} 0 & 0 \\ 1.7\sin 2\pi t & 0.76e^{-t^2} \end{bmatrix}.$$

According to Assumption 1, the uncertainty of every subsystem can be decomposed as the following.

$$H_{a,1} = \begin{bmatrix} 1 & 0 \\ 0 & 1 \end{bmatrix}, F_{a,1} = \begin{bmatrix} \sin 4\pi t & 0 \\ 0 & e^{-t} \end{bmatrix}, E_{a,1} = \begin{bmatrix} 0.38 & 0 \\ 0 & 1.1 \end{bmatrix},$$

$$H_{a,2} = \begin{bmatrix} 1 & 0 \\ 0 & 1 \end{bmatrix}, F_{a,2} = \begin{bmatrix} \cos 7\pi t & 0 \\ 0 & e^{-2.2t} \end{bmatrix},$$

$$E_{a,2} = \begin{bmatrix} 0.65 & 0.2 \\ 0 & 0.56 \end{bmatrix},$$

$$H_{a,3} = \begin{bmatrix} 0 & 0 \\ 1 & 1 \end{bmatrix}, F_{a,3} = \begin{bmatrix} \sin 2\pi t & 0 \\ 0 & e^{-t^2} \end{bmatrix},$$

$$E_{a,3} = \begin{bmatrix} 1.7 & 0 \\ 0 & 0.76 \end{bmatrix}.$$

We can testify that A_i, $i \in \mathbb{N}$ comprise a RSMS using Lemma 3. Namely Assumption 2 is satisfied. LMIs (21) was solved by Theorem 2 recursively. Let $k = 2$, 3 and the resolving results are following respectively.

A CRQLF of subsystem 1, 2 is

$$P_2 = \begin{bmatrix} 744.1 & 713.5 \\ 713.5 & 1162.7 \end{bmatrix}.$$

A CRQLF of subsystem 1, 2, 3 is

$$P_3 = \begin{bmatrix} 146220 & 21680 \\ 21680 & 474580 \end{bmatrix}.$$

Thus, we find a CRQLF of the system (1). It is showed that the system (1) is robust stable under arbitrary switching rule. The initial state value is

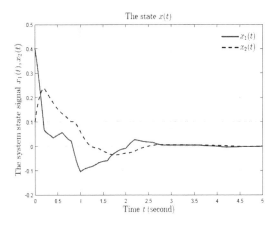

Figure 1. The system state signal $x(t)$.

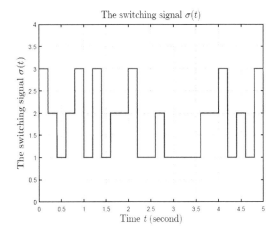

Figure 2. The system switching signal $\sigma(t)$.

selected as $x(t_0) = [0.4 \, 0.1]^T$. The system state running curves are simulated under arbitrary rule, which is in Fig. 1. The switching signal is in Fig. 2. They are showed that the switched system with uncertainty is robust stable under arbitrary switching rule.

6 CONCLUSIONS

The CQLF finding problem for uncertain switched linear systems is discussed. If the subsystem matrices constitute robust stability matrix set, the CQLF is called Common Robust Quadratic Lyapunov Function (CRQLF). A CRQLF determination method and its recursive constructing algorithm are proposed. Construction algorithm is in terms of recursive Linear Matrix Inequality (LMI) form, which brings convenience to facilitate the solution and strong practicability. The numerical example and simulation verify the feasibility of the presented results.

REFERENCES

Abou Al-Ola, O.M., Fujimoto, K., Yoshinaga, T., 2011. Common Lyapunov function based on kullbackcleibler divergence for a switched nonlinear system. *Mathematical Problems in Engineering* 2011(3): 1–12.

Aleksandrov, A.Y., Chen, Y., Platonov, A.V., Zhang, L., 2011. Stability analysis for a class of switched nonlinear systems. *Automatica* 47(10): 2286–2291.

Aleksandrov, A.Y., Murzinov, I.E., 2012. On the existence of a common Lyapunov function for a family of nonlinear mechanical systems with one degree of freedom. *Nonlinear Dynamics & Systems Theory* 2(2): 137–144.

Beldiman, O., Bushnell, L., 1999. Stability, linearization and control of switched systems, in: *Proc. American Control Conference*, pp. 2950–2955.

Ben Salah, J., Valentin, C., Jerbi, H., 2015. Quadratic common Lyapunov function design for planar linear switched system stabilization, in: *Control Conference (ECC), 2009* European.

Branicky, M., 1994. Stability of switched and hybrid systems, in: *Proc.33rd IEEE Conf. Decision and Control*, pp. 3498–3503.

Branicky, M., 1997. Stability of hybrid systems: state of the art, in: *Proceedings of the 36th Conference of Decision and Control*, pp. 120–125.

Branicky, M., 1998. Mutiple Lyapunov functions and other analysis tools for switched and hybrid systems. *IEEE Trans. Automatic Control* 43(4): 475–482.

Bykkoroglu, T., 2012. Common diagonal Lyapunov function for third order linear switched system. *Journal of Computational & Applied Mathematics* 236(15): 3647–3653.

Chen, G., Yang, Y., 2014. Finite-time stability of switched positive linear systems. *International Journal of Robust & Nonlinear Control* 24(1): 179–190.

Chen, Y., He, Z., Aleksandrov, A.Y., 2011. Existence of common Lyapunov function based on KKM lemma for switched Persidskii-type systems, in: *Electric Information and Control Engineering (ICEICE), 2011 International Conference on*, pp. 371–374.

Cheng, D., Guo, L., Huan, J., 2003. On quadratic Lyapunov functions. *IEEE Trans. Automatic Control* 48(5): 885–890.

Cheng, D., Zhu, Y., Hu, Q., 2006. Stabilization of switched systems via common Lyapunov function, in: *Proceedings of the World Congress on Intelligent Control and Automation (WCICA)*, pp. 183–187.

Duarte-Mermoud, M.A., Ordonez-Hurtado, R.H., Zagalak, P., 2012. A method for determining the non-existence of a common quadratic Lyapunov function for switched linear systems based on particle swarm optimisation. *International Journal of Systems Science* 43(11): 1–19.

Fierro, R., Lewis, F., Abdallah, C., 1996. Common, multiple, and parametric Lyapunov functions for a class of hybrid dynamical systems, in: *Proc. the 4th IEEE Mediterranean Symp. New Directions in Contr. and Automat.*, pp. 77–82.

Gu, Z., Liu, H., Liao, F., 2009. A common Lyapunov function for a class of switched descriptor systems, in: *Informatics in Control, Automation and Robotics, 2009. CAR '09. International Asia Conference on*, pp. 29–31.

Liberzon, D., 2003. Switching in Systems and Control. Birkhauser Boston. Liberzon, D., Morse, J., 1999. Stability of switched systems: a lie algebraic condition. *Systems & Control Letters* 37(3): 117–122.

Liberzon, D., 2003. *Switching in Systems and Control.* Boston: Birkhauser.

Lin, H., Antsaklis, P.J., 2009. Stability and stabilizability of switched linear systems: A survey of recent results. *IEEE Transactions on Automatic Control* 54(2): 308–322.

Liu, X., 2009. Stability analysis of switched positive systems: A switched linear copositive Lyapunov function method. *IEEE Transactions on Circuits & Systems II Express Briefs* 56(5): 414–418.

Liu, X., Zhao, X., 2016. Stability analysis of discrete-time switched systems: a switched homogeneous Lyapunov function method. *International Journal of Control* 89(2): 297–305.

Mancilla-Aguilar, J., 2000. A condition for the stability of switched nonlinear systems. *IEEE Trans. Automatic Control* 45(11): 2077–2079.

Margaliot, M., 2006. Stability analysis of switched systems using variational principles: an introduction. *Automatica* 42(12): 2059–2077.

Margaliot, M., Liberzon, D., 2006. Liealgebraic stability conditions for nonlinear switched systems and differential inclusions. *Systems & Control Letters* 55(1): 8–16.

Narendra, K., Balakrishnan, J.A., 1994. A common Lyapunov function for stable lti systems with commuting a-matrices. *IEEE Trans. Automatic Control* 39(12): 2469–2471.

Ooba, T., Funahashi, Y., 1997. Two conditions concerning common quadratic Lyapunov functions for linear systems. *IEEE Trans. Automatic Control* 42(5): 719–721.

Sun, C., Fang, B., Huang, W., 2011. Existence of a common quadratic Lyapunov function for discrete switched linear systems with m stable subsystems. *IET Control Theory & Applications* 5(3): 535–537.

Tong, Y., Zhang, L., Shi, P., Wang, C., 2013. A common linear copositive Lyapunov function for switched positive linear systems with commutable subsystems. *International Journal of Systems Science* 44(11): 1994–2003.

Wang, C., Shen, T., Ji, H., 2014. Common quadratic Lyapunov function for two classes of special switched linear systems. *IEICE Transactions on Information & Systems* E97.D(2): 175–183.

Yu, M., Gao, Y., 2009. The computation of a common standard quadratic Lyapunov function for tridiagonal system. *Adv. Differ. Equ. control Process* 3(1): 19–31.

Zhai, G., 2009. A common Lyapunov function approach to analysis and design of switched linear descriptor systems, in: *Proceedings of the Japan Joint Automatic Control Conference*, pp. 155–155.

Zhu, Y., Cheng, D., Qin, H., 2007. Constructing conmmon Lyapunov functions for a class of stable matrices. *Acta Automatica Sinica* 33(2): 202–204.

Civil, Architecture and Environmental Engineering – Kao & Sung (Eds)
© 2017 Taylor & Francis Group, ISBN 978-1-138-02985-9

HFSS simulation of RCS based on MIMO system

Yangyang Han & Lu Sun
School of Electromechanical Engineering, Xidian University, Xi'an, Shaanxi, P.R. China

Cong Hu
School of Electronic Engineering and Automation, Guilin University of Electronic Technology, Guilin, GuangXi, P.R. China

ABSTRACT: RCS (Radar Cross Section) of an object is an important parameter in military action. The accuracy and stability of typical single radar can not be applied to many huge targets. With the development of MIMO (Multiple-Input and Multiple-Output) technology, a new way of RCS simulation based on MIMO system is formulated in this paper. It introduces the simulation theory of RCS which can be simulated by HFSS. After synthesizing the multiple radar effect, it's verified that MIMO system has a better property comparing with single radar.

1 INTRODUCTION

Nowadays, it's expected to alleviate the impact for targets RCS with the variation of the observation angle (Mark, 2014). Comparing with the traditional radar, MIMO radar has the advantage of parallel multi-channel access to information, so it has broad application prospect. MIMO radar uses the transmitted and received signals simultaneously. As a result, many signals can be separated among the time domain, spatial domain and polarization domain. It has the advantages of higher processing dimension, making full use of scattered and received apertures, and higher angular resolution. MIMO radar uses irrelevance of echo signal raised by spatial diversity of target scattering, keeps average energy received of echo wave approximate to constant and the air target RCS smoothly, improves the target RCS fluctuation, and increases the detection performance and spatial resolution.

Nowadays, it becomes convenient for us to estimate the RCS by means of electromagnetic simulator. With the development of high speed calculation methods, other types of software such as HFSS employing FEM algorithm have been used on electromagnetic simulation (Yoshihide, 2014).

2 FUNDAMENTAL THEORY

Based on array antenna structure, MIMO radar with M emitters and N receivers can radiate orthogonal signals at the same time, these waveform signals can be scattered by the target and received by N array (Ghotbi, 2014). The model of statistical MIMO radar is shown in figure 1.

RCS is a key subject that reflects the observability of a target (Levent, 2013). It can be defined and calculated from the target-scattered field caused by an incident plane wave hitting on the object:

$$\sigma = \lim_{R \to \infty} 4\pi R^2 \frac{|E_s|^2}{|E_i|^2} \tag{1}$$

where R is the distance between the radar transmitter and the target, and **Es** and **Ei** are the scattered and incident electric fields. The unit of RCS is square meter (m2) or dBsm (Doren, 2008).

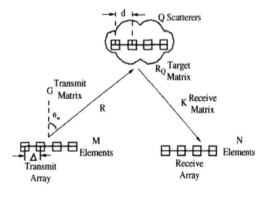

Figure 1. Statistical MIMO radar target model.

HFSS is a high-performance full-wave electromagnetic (EM) field simulator for arbitrary 3-dimension passive device modeling which takes advantage of the familiar Microsoft Windows graphical user interface. Ansoft HFSS employs the Finite Element Method (FEM), adaptive meshing, and brilliant graphics to solve Maxwell's equations.

HFSS-IE uses MOM (method of moments) solution type to solve large, open, radiating or scattering analyses. It is a new Integral Equation solver technology in the HFSS desktop. So it's used to simulate the RCS of dihedral angle structure below.

In this paper, the length and width of dihedral angle (perfect electrical conduct) are all 5.6088λ, the rotated angle θ is 90°. The frequency of incident wave is 9.4 GHz with vertical polarization. In HFSS operation, the excitation is set to be 1V/m and the boundary is set to be perfect electricity boundary (Sheng, 2013). Fig. 3 shows the model of dihedral angle.

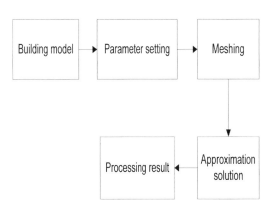

Figure 2. HFSS simulation process.

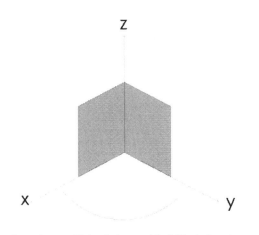

Figure 3. HFSS simulation model of dihedral angle.

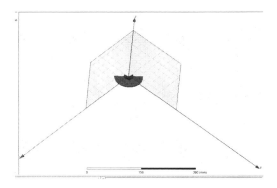

Figure 4. Mono-static simulation model with incident wave phi (–45°~135°).

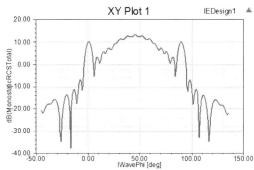

Figure 5. Mono-static RCS simulation result.

3 MONO-STATIC SIMULATION

In the simulation of mono-static situation, the direction of incident wave is the same as that radiation wave. The rotated angle θ is 90° and azimuth angle φ sweeps from –45° to 135°, the scan interval is 1°. Fig. 4 shows the simulation model. The simulation result of RCS is revealed in Fig. 5 in which primary sweep is incident wave Phi and vertical axis unit is dB (RCS Total). For comparison, we also use the theoretical calculation values from reference article (Timothy, 1987).

Comparing the simulation result with correlative literature datum, it's obvious that simulation and calculation results of mono-static RCS match well with the change of azimuth angle.

4 MIMO BI-STATIC SIMULATION

In the MIMO system, four special angles are used to simulate RCS. The simulation model of bi-static is as same as mono-static. The bi-static RCS simulations of a target are obtained from

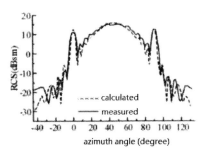

Figure 6. Calculated result compared with measured values of reference.

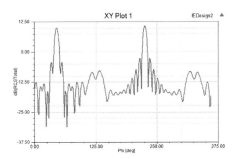

Figure 7. Simulation results of total RCS.

the amplitudes of received signal strength for each of the transmitting—receiving paths (Ebrahimi, 2007). Through four angles' data processing, MIMO radar's result can be revealed clearly.

4.1 Case 1

The electric field direction is set to be $\mathbf{E} = (0,0,1)$ and electromagnetic vector is $\mathbf{k} = (-1,-1,0)$. The incident angle is $\varphi = 45°$ $\theta = 90°$ while XOY plane as radiating direction. There is no change in the size of model. Fig. 7 reveals the simulation results and primary sweep is φ.

4.2 Case 2

The electric field direction is set to be $\mathbf{E} = (0,0,1)$ and electromagnetic vector is $\mathbf{k} = (1,-1,0)$. The incident angle is $\varphi = 135°$ $\theta = 90°$ while XOY plane as radiating direction. Fig. 8 reveals the bi-static simulation results.

4.3 Case 3

When the electric field direction is $\mathbf{E} = (0,0,1)$ and electromagnetic vector is set to be $\mathbf{k} = (1,1,0)$. The incident angle is $\varphi = 225°$ $\theta = 90°$ and XOY plane as radiating direction. Through the simulation and Fig. 9 reveals the results.

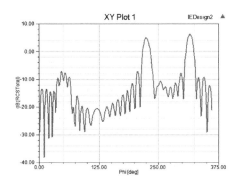

Figure 8. Bi-static RCS simulation results with $\varphi = 135°$, $\theta = 90°$.

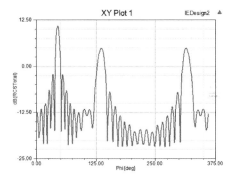

Figure 9. Total RCS simulation results with $\varphi = 225°$, $\theta = 90°$.

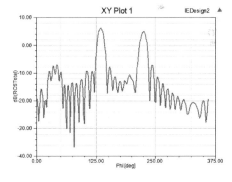

Figure 10. RCS simulation results with $\varphi = 135°$, $\theta = 90°$.

4.4 Case 4

The electric field direction is set to be $\mathbf{E} = (0,0,1)$ and electromagnetic vector is $\mathbf{k} = (-1,1,0)$. The incident angle is set to be $\varphi = 315°$ and $\theta = 90°$ while XOY plane as radiating direction. Fig. 10 reveals the simulation result.

Table 1. RCS values of four cases.

	dB(RCSTotal) [] - Freq='9.4GHz' Theta='90deg'			
2	-18.78023444	-20.57046238	-12.84608912	-19.50732391
3	-15.84578653	-28.36036746	-12.41774309	-19.36478537
4	-13.98312929	-25.17221935	-12.60227861	-20.04073729
5	-13.03298791	-18.9827261	-13.48318029	-21.70878804
6	-12.89435109	-15.76007549	-15.17521463	-24.60534561
7	-13.58012011	-14.05034369	-17.6267675	-27.43782516
8	-15.24297693	-13.38447543	-19.34289118	-24.96788831
9	-18.28540226	-13.6449251	-17.65694232	-20.9904872
10	-23.49605221	-14.93439317	-14.82810168	-18.23221189
11	-26.00670251	-17.6930394	-12.76545403	-16.56923598
12	-20.51384068	-23.56211475	-11.66480103	-15.80448392
13	-17.02705894	-41.50268394	-11.50463977	-15.87080635
14	-15.2530221	-21.97653451	-12.3616674	-16.81825705
15	-14.6099679	-16.84283207	-14.50335837	-18.83179434
16	-14.76647793	-14.37368644	-18.36465009	-22.23143984
17	-15.32841829	-13.3538688	-20.95049847	-26.21139363

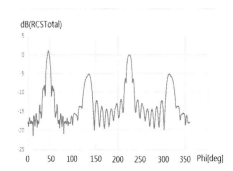

Figure 11. Average RCS results of last four cases.

5 RESULT ANALYSIS

Through extracting the data of four kinds of cases above, Some data listed in the following table 1. After average processing, the total result shown in Fig. 11.

From the above result, it can be found that the direction of the incident wave RCS is bigger than that of the others. Comparing MIMO radar with typical method, the fluctuation of RCS was improved efficiently by MIMO radar. This method can be used to forecast electromagnetic scattering characteristics of large complex targets with higher frequency (like airplane and ship) faster and more accurate.

6 CONCLUSION

RCS is one of the important features of radar target signal. Although the electromagnetic theory has its integrity which can analyze the scattering mechanism of some objects, it's difficult for complex targets to calculate the RCS value using existed methods. Now RCS can be simulated by the electromagnetic simulation software. Traditional single radar can not be applied in some complex objects, so we will apply the MIMO radar to RCS simulation.

From dihedral angle RCS simulation in HFSS, it's verified that the result of HFSS simulation agrees well with the theoretical result. By RCS numerical treatment of four kinds of angles, we obtain the RCS curve under the MIMO system. It can be concluded that the value of the target RCS is related to the angle of incident wave. This study has successfully demonstrated the application of MIMO theory in radar signal receiving and target detection, especially laid the foundation of MIMO radar in the electromagnetic stealth technology of military equipment.

ACKNOWLEDGEMENT

The authors would like to thank the editors and the reviewers for their valuable and insightful comments. This research was supported by the following three funds: Common Technology Funds under Grant No. 90406150008, National Natural Science Foundation of China under Grant No. 61501344 and Automatic Detection Technology and Instrument in Guangxi Key Laboratory Open Fund under Grant No. YQ15205.

REFERENCES

Doren H., W, LAPC, 37(2008).
Ebrahimi-Tofighi, N., M. ArdebiliPour, M. Shahabadi, S. Rajabi, ICACT, 1748(2007).
Ghotbi, S., M. Ahmadi, MA. Sebt, SPC, 1701(2014).
Levent, S., R. Zubair, M. Irfan, APM, **55**, 278(2013).
Mark F., T, S. Kyle B, M. Ninoslav, J. Joel T, ITEAS, **50**, 1595(2014).
Sheng Chih, C., L. Kun-Chou, AAWPL, 12, 937(2013).
Timothy, G., B. Constantine A, TOAAP, **AP-35**, 1137(1987).
Yoshihide, Y., N. Michishita, D. Nguyen Quoc, ATC, 69(2014).

Civil, Architecture and Environmental Engineering – Kao & Sung (Eds)
© 2017 Taylor & Francis Group, ISBN 978-1-138-02985-9

Object rough localization of tongue images via clustering and gray projection

Weixia Liu
Straits Institute, Minjiang University, Fuzhou, China

Zuchang Zhang, Yeyu Lin, Danying Shenand & Zuoyong Li
Fujian Provincial Key Laboratory of Information Processing and Intelligent Control (Minjiang University), Fuzhou, China

ABSTRACT: Tongue diagnosis is one of the most important diagnosis methods in Traditional Chinese Medicine (TCM). Tongue image segmentation is a crucial step in developing automatic tongue diagnosis system. Rough localization of tongue body is a useful preprocessing step in tongue image segmentation, which can eliminate the adverse effect of strong edges from neighboring tissues such as face and lip when extracting tongue body contour. After exploring the existing tongue rough localization method based on gray projection, we propose a modified method combining clustering with gray projecting in this paper. The proposed method first conducts clustering on image pixels' hue components in HSI (Hue, Saturation and Intensity) space to seek multiple thresholds. Then, image thresholding and morphological operations are conducted to generate a binary image, and the largest object region is taken as the initial localization result. Finally, the localization result is refined by gray projection on pixels' red components. Experiment results on a variety of tongue images show that the proposed method significantly improves the accuracy of tongue rough localization in comparison with the existing gray projection method.

1 INTRODUCTION

Tongue diagnosis (Kirschbaum, 2000) is one of widely used diagnostic methods in Traditional Chinese Medicine (TCM) due to its virtues such as effectiveness, painlessness, simplicity and immediacy. Tongue diagnosis has a history of at least 3000 years, and its practitioners have accumulated very rich clinical experiences on drawing physiological and pathological information according to the features of tongue body, such as color, texture, shape and coating. Eight principles of tongue diagnosis reveal that different sub-regions of tongue body can reflect health statuses of different human organs such as heart, lung, spleen and stomach. For example, tongue's appearance is the most useful gauge for monitoring the improvement or deterioration of a patient's health status. The color and texture features of tongue coating, which are called TCM syndromes, often reflect many diseases and human health conditions such as inflammation and infection (Hsu, 2003).

In modern Western Medicine, tongue has increasingly been taken as an extension of the upper gastrointestinal tract that can provide important clues to human health status. Some researchers in Western Medicine have also taken tongue diagnosis as a helpful method for clinical decision making (Anastasi, 2009). For example, tongue coating is a risk indicator for aspiration pneumonia in edentate patients, as it is associated with a number of viable salivary bacteria (Abe, 2008). It has also been reported that the amyloidosis of tongue may be a diagnostic manifestation of plasmacytoma (Hoefert, 1999). These researches exhibit the potential for inferring systemic disorders from the tongue in clinical diagnosis.

The conventional tongue diagnosis lacks quantitative and robust diagnosis accuracy due to its high dependence on practitioners' experience. Now, powerful computer processors have made it possible to develop automatic computer-aided tongue diagnosis system (Pang, 2004) via image processing and pattern recognition techniques. This automatic tongue diagnosis system usually first extracts tongue body in a tongue image by image segmentation technique, then calculates the features of the tongue body by feature extraction technique, and finally uses a classifier to obtain the final diagnosis result. Hence, tongue image segmentation (i.e., extracting tongue body) is a crucial step. Some researchers have presented some algorithms (Ning, 2012; Shi, 2013) to resolve this problem. However, to date, tongue image segmentation is still a challenging task due to large personal variation of tongue body on shape, color, coating, texture, and weak edges caused by similar color between tongue body and its neighboring tissues.

To improve accuracy of tongue image segmentation, Zhang and Qin (Zhang, 2010) proposed an image preprocessing method for tongue image segmentation, which uses gray projection technique to achieve rough localization of the object (i.e., tongue body) in a tongue image. The goal of tongue rough localization is to find the rectangle region located by the tongue body. Tongue rough localization is helpful for subsequent tongue segmentation since it can avoid strong image edges out of tongue body to be erroneously regarded as a part of tongue contour. After exploring the limitation of the existing tongue rough localization method based on gray projection (Zhang, 2010), we propose a modified rough localization method combining clustering and the gray projection. Experimental results on a series of tongue images with large variation on tongue size, shape, color, coating and texture demonstrate the effectiveness of the proposed method.

2 TONGUE ROUGH LOCALIZATION BASED ON GRAY PROJECTION

The tongue rough localization method based on gray projection is presented according to two prior knowledge. The first knowledge is that tongue root region is usually darker than other regions on human face when generating tongue images. The second knowledge is that there may exist dark regions near tongue contour due to the blocked light by stretched tongue body when generating tongue images. These dark regions cause those involved image rows and image columns have lower average gray values than other rows and columns, respectively.

For a given tongue image I with the size of $M \times N$, let $I(i, j)$ denote the gray level of pixel $p_{i,j}$ at the i-th row and j-th column. The detailed process of the gray projection based tongue rough localization is as follows:

1. Calculate average gray value of each image row, where the average gray value of i-th row is defined as

$$R(i) = \sum_{j=1}^{N} I(i, j). \qquad (1)$$

Similarly, the average gray value of j-th column is defined as

$$C(j) = \sum_{i=1}^{M} I(i, j). \qquad (2)$$

2. Find the row with the lowest R value among the front half of image rows, which can be formulated as

$$r_{up} = \min_{1 \le i \le M/2} R(i). \qquad (3)$$

Similarly, the row with the lowest R value among the latter half of image rows can be determined as

$$r_{down} = \min_{M/2 < i \le M} R(i). \qquad (4)$$

3. Find the column with the lowest C value among the front half of image columns, which can be formulated as

$$c_{left} = \min_{1 \le j \le N/2} C(j). \qquad (5)$$

Similarly, the column with the lowest C value among the latter half of image columns can be determined as

$$c_{right} = \min_{N/2 < j \le N} C(j). \qquad (6)$$

4. Use r_{up}, r_{down}, c_{left} and c_{right} to confine a rectangle region as the rough localization result of tongue body.

To intuitively show tongue rough localization results, we exhibit them as blue rectangles in original tongue images as shown in Figures 1 (b) and (d). After exploring the gray projection method, we found that it usually fails to obtain satisfactory result on those acquired original tongue images, but can obtain satisfactory result on those clipped tongue images. Figure 1 gives an example to explain the above conclusion, where Figures 1 (b) and (d) exhibit the rough localization results obtained by applying the gray projection method to an acquired original tongue image in Figure 1 (a) and its clipped version in Figure 1 (c), respectively. From Figure 1, one can observe that the gray projection method obtains satisfactory rough localization result on the clipped tongue image, but obtains unsatisfactory result on the original tongue

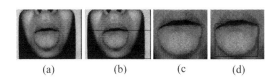

| | | | |
| (a) | (b) | (c | (d) |

Figure 1. Rough localization results obatined by the gray projection method (Zhang, 2010): (a) original tongue image, (b) result of (a), (c) the clipped image of (a), (d) result of (c).

image. This demonstrates the limitation of the gray projection method on tongue rough localization.

3 THE PROPOSED METHOD

After exploring the gray projection method (Zhang, 2010), we found that it is invalid for acquired original tongue image as illustrated in Figure 1. To obtain satisfactory rough localization result, the method need clip original tongue image, which is time-consuming and inconvenient. To resolve this issue, we propose a modified rough localization method based on clustering and gray projection. The detailed rough localization process is as follows:

1. Color space transformation: this step maps an image in RGB color space into HSI color space, and the mapping is formulated as

$$H = \begin{cases} \theta, & G \geq B \\ \theta - 2\pi, & G < B \end{cases} \quad (7)$$

$$S = 1 - \frac{3}{R+G+B} \min\{R,G,B\}, \quad (8)$$

$$I = \frac{1}{3}(R+G+B), \quad (9)$$

where

$$\theta = \arccos\left\{ \frac{[(R-G)+(R-B)]/2}{[(R-G)^2+(R-B)(G-B)]^{1/2}} \right\}. \quad (10)$$

In Eqs. (7)–(10), R, G and B indicate the red, green and blue components of an image, respectively.

2. Thresholds determination based on clustering: this step uses a clustering scheme to seek three thresholds for subsequent image thresholding. Here, the H component of an image in HSI color space is used for clustering. Figure 2 (b) exhibits the H component of the used example in Figure 1 (a). In addition, the class number is empirically set to 3, and the three initial class centers are set to the minimum value, the median value and the maximum value of all image pixels' H values. The centers of three classes are updated iteratively until they are unchanged. At each iteration, we first classify each pixel's H value into one class according to the minimum absolute difference of H values, and then update each class center with the average H value of the class. After the clustering process on pixels' H values, we gather three final class centers into a set called T_{set} with ascending order.

3. Image thresholding and morphological operations: this step first uses the elements in T_{set} to conduct image thresholding, where the image thresholding result B_{img} is defined as

$$B_{img}(i,j)$$
$$= \begin{cases} 1, & if\ Tset(1) < I(i,j)\ or\ I(i,j) > Tset(3) \\ 0, & otherwise \end{cases} \quad (11)$$

where $T_{set}(i)$ denotes the i-th element in T_{set}. To further refine the thresholding result B_{img} as RB_{img}, we sequentially apply morphological operations (i.e., open operation and image filling operation) to B_{img}, where the structure element of the open operation is shown in Figure 2. Take the original tongue image used in Figure 1 (a) as an example, Figures 3 (c) and (d) exhibit initial image thresholding result B_{img} and the refined image thresholding result RB_{img} after the morphological operations, respectively.

0	0	1	1	1	1	1	0	0
0	1	1	1	1	1	1	1	0
1	1	1	1	1	1	1	1	1
1	1	1	1	1	1	1	1	1
1	1	1	1	1	1	1	1	1
1	1	1	1	1	1	1	1	1
1	1	1	1	1	1	1	1	1
0	1	1	1	1	1	1	1	0
0	0	1	1	1	1	1	0	0

Figure 2. Structure element of morphological open operation.

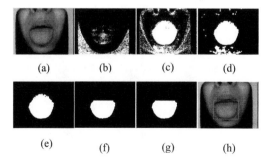

(a)　　(b)　　(c)　　(d)

(e)　　(f)　　(g)　　(h)

Figure 3. Tongue rough localization based on the proposed method: (a) original tongue image, (b) the H component in HSI color space, (c) image thresholding result of (b), (d) the refined image thresholding result after morphological operations, (e) the largest object region, (f) the remained object region after clipping operation according to row localization result, (g) the largest object region of (f), (h) final rough localization result.

1431

4. Selection of object region: this step selects the largest object region (white region) in the refined image thresholding result RB_{img} as possible tongue body region, and records the largest object region in a binary image called O_{img}, where 1 and 0 denote object pixel and background pixel, respectively. Figure 2 (e) shows the image O_{img}, which is generated from Figure 2 (d).

5. Gray projection based row localization: this step first conducts row clipping by gray projection on those rows containing object (white) pixels in the binary image O_{img}, where we take the red component of an image in RGB color space for gray projection. In detail, we first calculate the average gray value of object pixels on those rows containing object pixels, and then find the row with the lowest average gray value among those image rows containing object pixels as the upper bound (r_{up}) of the rectangle located by the tongue body. In addition, we take the maximum row number of the object region in O_{img} as the down bound (r_{down}) of the rectangle located by the tongue body. Because we think that the tongue body should be located between the r_{up}-th row and the r_{down}-th row, we remove the object pixels on those rows out of the row range [r_{up} r_{down}], i.e., setting corresponding object pixels' values in O_{img} as 0 s. Figure 3 (f) exhibits the remained object region after removing the object pixels on those rows out of [r_{up} r_{down}].

6. Column localization based on the largest object region: this step first finds the largest object region in Fig. 3 (f), and then the left and the right bounds of the object region are determined as the left bound (c_{left}) and the right bound (c_{right}) of the rectangle located by tongue body.

7. Finish the tongue rough localization by using the row bounds (i.e., r_{up} and r_{down}) and the column bounds (i.e., c_{left} and c_{right}). To intuitively exhibit the rough localization result obtained by the proposed method, we use green rectangle to represent it in an original tongue image as shown in Figure 3 (h).

4 EXPERIMENTAL RESULTS

To evaluate the performance of the proposed method on tongue rough localization, we have applied it to a variety of tongue images with sizes of 640×480. The rough localization results yielded by the proposed method were compared with those results obtained by the gray projection method (Zhang, 2010). To quantitatively measure the accuracy of tongue rough localization, we take the rectangle region obtained by each method as the object region in a binary segmentation result. In

this case, accuracy of rough localization can be quantitatively evaluated via Misclassification Error (ME) (Yasnoff, 1977) of the corresponding binary segmentation result. ME regards image segmentation as a pixel classification process. It reflects the percentage of background pixels incorrectly classified into foreground, and conversely, foreground pixels erroneously assigned to background. For a two-class segmentation, ME can be simply formulated as

$$ME = 1 - \frac{|B_O \cap B_T| + |F_O \cap F_T|}{|B_O| + |F_O|}, \quad (12)$$

where B_O and F_O are the background and foreground of the manual ideal segmentation result, B_T and F_T the background and foreground of the binary rough localization result, and $|\cdot|$ cardinality of a set. The value of ME varies between 0 for a perfectly classified image to 1 for a totally erroneously classified one. A lower value of ME means better accuracy of rough localization. All experiments were performed on a notebook PC with 1.7G Intel Core i5–3317U CPU and 4G RAM.

Our experiments are divided into four groups. Testing samples in the first group of experiments contain eight tongue images with large difference of tongue body size. The eight testing samples are shown in the first row of Figure 4, where the front four images have small tongue bodies, meanwhile the latter four images have large tongue bodies. The second row in Figure 4 exhibits manual segmentation results of tongue images, where tongue body contours are indicated by cyan curves. The third row in Figure 4 shows the rough localization results obtained by the gray projection method (Zhang, 2010), where the localization results are indicated by blue rectangles. The last row in Figure 4 exhibits the rough localization results obtained by the proposed method, where the localization results are indicated by green rectangles. From Figure 4, one can observe that the gray projection method can effectively find locations of tongue roots, but cannot correctly find locations of tongue tips and tongue bodies' left and right boundaries. Adversely, the proposed method successfully finds four boundaries of tongue bodies and obtains satisfactory rough localization results.

To further explore the performance of the proposed method on object rough localization of tongue images with large difference on tongue body's shape, color, coating and texture. Figure 5 exhibits visual rough localization results on eight tongue images with large tongue shape difference, where the front three images have square

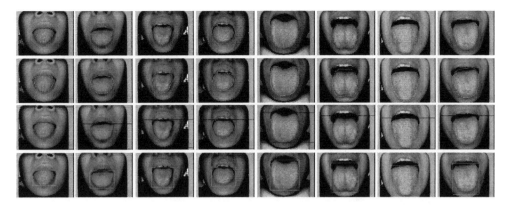

Figure 4. Visual rough localization results of tongue images with large difference of tongue size from up to down: original images, manual segmentation results, localization results of the gray projection method (Zhang, 2010), localization results of the proposed method.

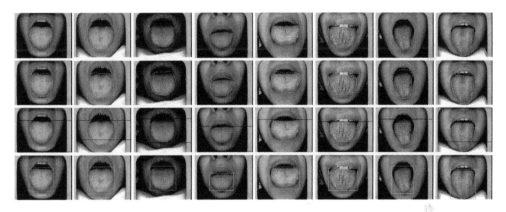

Figure 5. Visual rough localization results of tongue images with large difference of tongue shape from up to down: original images, manual segmentation results, localization results of the gray projection method (Zhang, 2010), localization results of the proposed method.

tongue bodies, the middle three images have flat tongue bodies, and the last two images have vertical tongue bodies. Figure 6 exhibits visual rough localization results on eight tongue images with large tongue color difference, where the front four images have tongue bodies with slight white color, and the latter four images have tongue bodies with deep red color. Figure 7 exhibits visual rough localization results on eight tongue images with strong tongue coating and texture, where the front four images have strong tongue coating, and the latter four images have strong texture. From Figures 5–7, one can observe that the proposed method successfully achieves object rough localization on the above tongue images with large difference on tongue shape, color, coating and texture of tongue body. However, the gray projection method usually only finds the locations of tongue roots, and fails to successfully achieve tongue rough localization.

After performing the above qualitative comparison on tongue rough localization results, we further conduct quantitative comparison on four groups of tongue images. We used Misclassification Error (ME) to quantitatively evaluate those rough localization results. The average ME values corresponding to four groups of tongue images are listed in Table 1. From the table, one can conclude that the average ME value of the proposed method on each group is obviously lower than that of the gray projection method. This further demonstrates the advantage of the proposed method over the gray projection method.

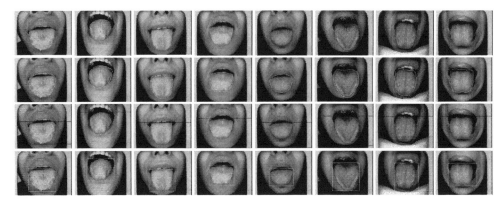

Figure 6.   Visual rough localization results of tongue images with large difference of tongue color from up to down: original images, manual segmentation results, localization results of the gray projection method (Zhang, 2010), localization results of the proposed method.

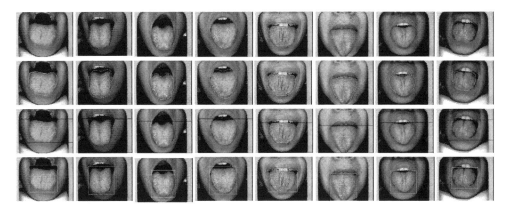

Figure 7.   Visual rough localization results of tongue images with strong coating and texture from up to down: original images, manual segmentation results, localization results of the gray projection method (Zhang, 2010), localization results of the proposed method.

Table 1.   Average ME values obtained by different methods on four groups of experiments.

Group number	Gray projection (Zhang, 2010)	Proposed method
1	0.482	0.067
2	0.442	0.063
3	0.419	0.067
4	0.414	0.070

5   CONCLUSIONS

Tongue image segmentation is a crucial step in developing automatic tongue diagnosis system. Rough localization of tongue body is a useful preprocessing step in tongue image segmentation. The existing rough localization method based on gray projection fails to obtain satisfactory results on original tongue images. To resolve this issue, we presented a new method based on clustering in HSI color space and gray projection. Experimental results on a variety of tongue images with large difference on tongue size, shape, color, coating and texture shows the superiority of the proposed method over the existing gray projection method.

ACKNOWLEDGMENT

This work is partially supported by National Natural Science Foundation of China (61202318), Program for New Century Excellent Talents in Fujian Province University (NCETFJ), Technology Project of Provincial University of Fujian Province (JK2014040), Fuzhou Science and

Technology Project (2015-PT-91, 2016-S-116, and 2015-G-60), Technology Project of Education Department of Fujian Province (JA15424, JA15425 and JAT160391).

REFERENCES

Anastasi, J.K., L.M. Currie, G.H. Kim, Understanding diagnostic reasoning in TCM practice: tongue diagnosis, Altern. Ther. Health M., **15(3)**, 18–28 (2009).

Abe, S., K. Ishihara, M. Adachi, K. Okuda, Tongue-coating as risk indicator for aspiration pneumonia in edentate elderly, Arch. Gerontol. Geriatr, **47(2)**, 267–275 (2008).

Hoefert, S., E. Schilling, S. Philippou, H. Eufinger, Amyloidosis of the tongue as a possible diagnostic manifestation of plasmacytoma, Mund Kiefer Gesichtschir, **3(1)**, 46–49 (1999).

Hsu, C.H., M.C. Yu, C.H. Lee, T.C. Lee, S.Y. Yang, High eosinophil cationic protein level in asthmatic patients with Heat Zheng, AM. J. Chin. Med., **31(2)**, 277–283 (2003).

Kirschbaum, B. *Altas of Chinese tongue diagnosis.* Seattle, WA: Eastland (2000).

Ning, J., D. Zhang, C. Wu, F. Yue, Automatic tongue image segmentation based on gradient vector flow and region merging, Neural Comput. Appl., **21(8)**, 1819–1826 (2012).

Pang, B., D. Zhang, N. Li, K. Wang, Computerized tongue diagnosis based on Bayesian networks, IEEE Trans. Biomed. Eng., **51(10)**, 1803–1810 (2004).

Shi, M., G. Li, F. Li, C2G2FSnake: automatic tongue image segmentation utilizing prior knowledge, Sci. China Inf. Sci., **56(9)**, 1–14 (2013).

Yasnoff, W. A., J.K. Mui, J.W. Bacus, Error measures for scene segmentation, Pattern Recognit., **9(4)**, 217–231 (1977).

Zhang, L, Qin J. Tongue-image segmentation based on gray projection and threshold-adaptive method, Journal of Clinical Rehabilitative Tissue Engineering Research, **14(9)**, 1638–1641 (2010).

Multi-sensor attitude algorithm design for a low-cost strap-down system based on the direction cosine matrix

Jin Du
North University of China Science and Technology on Electronic Test and Measurement Laboratory, Taiyuan, China

Jie Li
North University of China Science and Technology on Electronic Test and Measurement Laboratory, Taiyuan, China
Key Laboratory of Instrumentation Science and Dynamic Measurement (North University of China) Ministry of Education, Taiyuan, China

Chenjun Hu
Suzhou Fashion Nano-Technology Co. Ltd., China

Kaiqiang Feng
North University of China Science and Technology on Electronic Test and Measurement Laboratory, Taiyuan, China
Key Laboratory of Instrumentation Science and Dynamic Measurement (North University of China) Ministry of Education, Taiyuan, China

ABSTRACT: In order to satisfy the urgent needs of low-cost, high-performance measurement system for the attitude angle of the carrier, this paper presents a multi-sensor data fusion algorithm based on Direction Cosine Matrix (DCM). Aiming at the problem of inaccurate measurement of the attitude incurred by low-cost MEMS gyroscope drifting, the study constructs a multi-axis complementary filter, proposes reference vectors to detect gyroscope bias error in each axis and uses PI controller as feedback to compensate the drift error of gyros. Static experiment on a three-axis turntable shows that the static accuracy of the attitude angle is better than 0.2°, and dynamic vehicle experiment shows that the dynamic accuracy is better than 4°, meeting the application requirements of low-cost attitude determination system.

1 INTRODUCTION

As the first step to realize attitude control of the carrier, attitude measurement results play a crucial role in many navigation and control systems. The current trend towards miniaturization and real-time control has been fuelling research in the field of attitude measurement (Dong, 2016). The development of microelectronics gave birth to MEMS sensors, which possess the advantages of small size, low power consumption, and low cost (Robert, 2013). This made the realization of strap-down attitude angle determination system based on MEMS sensors feasible.

A variety of methods are adopted to get the three-dimensional attitude information for the strap-down attitude angle determination system. Some scholars use GPS carrier phase signals to determine the body's attitude (Park, 2004). This kind of method does not accumulate errors over time; however, the installation of the antenna increases the complexity of the system (Gaygysyz, 2010). Some scholars have used images from the airborne camera to identify the horizon or sea-sky line. The roll angle and pitch angle of the aircraft can be measured quickly according to the obtained linear information (Mosavi, 2013; Hu, 2016; Ji, 2016). However, this method is limited in the complex terrain.

The mainstream method is an INS-aided system, and the key to this method lies in how to carry out data fusion. Typical data fusion algorithms can be roughly divided into two categories. One is from the frequency domain that distinguishes and eliminates noise, such as complementary filter (Vaibhav, 2014). The other is to use the state space method to design the filter in the time domain, such as Kalman Filter (KF), Extended Kalman

Filter (EKF), Unscented Kalman Filter (UKF), and so on (Stefano, 2010; Chingiz, 2016). As we know, Kalman filter is the optimal filter to estimate a linear system. However, many real-world systems are nonlinear in nature. EKF is developed to help account for these nonlinearities. The method is convenient to implement, and it can also meet the requirements of the application of nonlinear system in many cases. Using UKF, estimated accuracy and rate of convergence are apparently enhanced in comparison to EKF, but the problem of computational burden cannot be solved effectively with this method.

Research is currently being carried out in many laboratories for simple and effective filtering techniques that result in both linear and non-linear systems. A new method of multi-axis complementary filter based on Direction Cosine Matrix (DCM) is proposed in this paper. Experimental results show that the static accuracy of the attitude angle is better than 0.2°, and that the dynamic accuracy is better than 4°.

2   COORDINATE SYSTEM AND DCM-BASED ATTITUDE ALGORITHM

2.1   *Definition of coordinate system*

Geographical coordinate system is fixed to the inertial coordinate system and is linked to the Earth referential. The orthogonal axis of geographical coordinate frame is called North, East, Down (NED) frame, and the axes of x_n, y_n and z_n represent the axes of north, east and down, respectively, which satisfy the right-hand rule.

Body coordinate system is fixed to the airplane and represents a coordinate system that is attached to the carrier. The origin is the gravity centre of carrier; the x_b-axis points in the forward direction, and it is aligned with the roll axis; the z_b-axis (yaw) points to the bottom of the MIMU; and the y_b-axis (pitch) represents the direction of aircraft starboard. And rounds up the right-handed orthogonal coordinate system.

Figure 1 shows the relationship between the geographical coordinate system and the body coordinate system.

2.2   *The direction cosine matrix*

There are many ways to describe the orientation relation of the moving coordinate system with respect to the reference coordinate system, such as by using the Euler angle, the quaternion approaches and the DCM. Using the DCM to solve the attitude angle is more intuitive than the quaternion approaches, which provides the ability to work in the entire range of attitude angles (Chiemela, 2014).

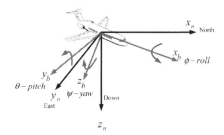

Figure 1.   Definition of a coordinate system.

The transformation from the body coordinate system to the geographical coordinate system can be realized by three times rotation. The relationship between the navigation coordinate system and the body coordinate system can be expressed by a matrix, that is, the DCM. As a result of the multiplication of three rotation matrices, each column of the DCM represents a coordinate system in another coordinate system. Therefore, DCM has two characteristics: orthogonal and unit.

2.3   *Updating of DCM and extraction of attitude angle*

Neglecting the influence of the earth self-rotation rate, the change of heading, pitch and roll angle tend to be zero when the time interval t tends to be infinity. The expression of DCM updating with time is as follows:

$$C_b^n(t+dt) = C_b^n(t)\begin{pmatrix} 1 & -d\psi & d\theta \\ d\psi & 1 & -d\phi \\ -d\theta & d\phi & 1 \end{pmatrix} \tag{1}$$

$$= C_b^n(t)\begin{pmatrix} 1 & -\omega_z dt & \omega_y dt \\ \omega_z dt & 1 & -\omega_x dt \\ -\omega_y dt & \omega_x dt & 1 \end{pmatrix} \tag{2}$$

The derivative of DCM is obtained as follows:

$$\dot{C}_b^n = C_b^n \Omega_b^b \tag{3}$$

$$\Omega_b^b = \begin{pmatrix} 0 & -\omega_z & \omega_y \\ \omega_z & 0 & -\omega_x \\ -\omega_y & \omega_x & 0 \end{pmatrix} \tag{4}$$

Ω_b^b represents the projection of the angular velocity of the body coordinate system with respect to the inertial coordinate system. ω_x, ω_y and ω_z represent the angular velocity of X, Y and Z axis in the body coordinate system, respectively.

ω_x, ω_y, and ω_z are measured by three gyro mounted on the carrier. The structured flowchart of DCM updating is shown in Figure 2.

By using the measured data of the gyroscope, the system realizes the real-time updating of DCM and also the real-time updating of attitude angle calculation. After getting the DCM, it is easy to deduce the expression of pitch angle, heading angle and roll angle:

$$\theta = \arcsin(-C^n_{b31})$$
$$\psi = \arctan(\frac{C^n_{b21}}{C^n_{b11}}) \quad (5)$$
$$\phi = \arctan(\frac{C^n_{b32}}{C^n_{b33}})$$

Yaw and roll angle have multiple values, and we need to judge the quadrant after working out the results. Due to space limitations, we have not provided a detailed description here.

3 COMPLEMENTARY FILTERING AND MULTI-SENSOR DATA FUSION

The attitude angles are obtained by integrating the angular rate of the gyro's output, which is accurate and reliable in short time. Due to the drift error of the MEMS gyroscope itself and the influence of various external noises, gyros and other sensors are often used in combination. The method of data fusion enables the system to provide accurate attitude information for a long time (Liu, 2014; Hu, 2016).

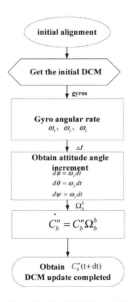

Figure 2. The flowchart of DCM updating.

Due to orthogonality and unit constraints, any pairs of rows or columns in the matrix are orthogonal to each other. However, owing to the accumulation of numerical rounding error, gyro drift and other reasons, the lengths of the matrix are not equal to each other in the process of constant updating of the DCM; therefore, the original orthogonal matrix is tilted, as shown in Figure 3.

Non-orthogonal change of the DCM is bound to affect the calculation accuracy of the attitude angle. The modification of the DCM is an urgent problem that needs to be solved in the process of obtaining the accurate attitude angle. It is known to us all that for the same set of vectors, with different coordinate systems, the mode and direction of the vectors must be the same. As mentioned earlier, there is a bias in the DCM. In this paper, we have modified the rotation matrix by calculating the deviation, and have further achieved the purpose of correcting the attitude angle.

In this study, we have selected two direction reference vectors. One is the vector of gravity field, measured by the accelerometer, and the other is the course vector, which is measured by the electronic magnetic compass. The gravity field vector is used to detect the deviation of pitch angle and roll angle, and the direction vector is used to detect the deviation of the heading angle. The direction reference vectors and the vectors obtained by the DCM are vectors of the same direction in theory. Their outer product is defined as the error vector we are seeking. The model value of the external product is directly proportional to the angle of the two vectors.

3.1 Calculation of pitch angle and roll angle error vector based on gravitational field

Set $\left[v_x, v_y, v_z\right]^T$ as the projection of gravity field $[0,0,1]^T$ under the body coordinate system. The following formula can be derived from the DCM:

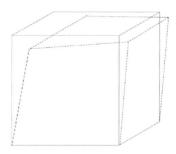

Figure 3. Schematic diagram of DCM non-orthogonal change.

$$
\begin{bmatrix} v_x \\ v_y \\ v_z \end{bmatrix} = \begin{bmatrix} \cos\theta\cos\psi & \cos\theta\sin\psi & -\sin\theta \\ \sin\phi\sin\theta\cos\psi - \cos\phi\sin\psi & \sin\phi\sin\theta\sin\psi + \cos\phi\cos\psi & \sin\phi\cos\theta \\ \cos\phi\sin\theta\cos\psi + \sin\phi\sin\psi & \cos\phi\sin\theta\sin\psi - \sin\phi\sin\psi & \cos\phi\cos\theta \end{bmatrix} \begin{bmatrix} 0 \\ 0 \\ 1 \end{bmatrix} \tag{6}
$$

$$
\begin{bmatrix} v_x \\ v_y \\ v_z \end{bmatrix} = \begin{bmatrix} -\sin\theta \\ \sin\phi\cos\theta \\ \cos\phi\cos\theta \end{bmatrix} \tag{7}
$$

$[v_x, v_y, v_z]^T$ obtained here is the projection of the gravity field in the body coordinate system, which are the direction reference vectors that we mentioned before. $[a_x^b, a_y^b, a_z^b]^T$ is the three axes force information measured by accelerometers under the body coordinate system. The main component is the projection of gravity in the three axes, which we set as $[g_x^b, g_y^b, g_z^b]^T$. However, due to the motion of the vehicle, the three axes force information also includes the motion acceleration and the centrifugal acceleration of the vehicle, which we set as $[\dot{v}_x^b, \dot{v}_y^b, \dot{v}_z^b]^T$ and $a_{centrifugal}$, respectively.

Therefore, the projection of gravity in the three axes can be described by the following formula:

$$
\begin{bmatrix} g_x^b \\ g_y^b \\ g_z^b \end{bmatrix} = \begin{bmatrix} a_x^b \\ a_y^b \\ a_z^b \end{bmatrix} - \begin{bmatrix} \dot{v}_x^b \\ \dot{v}_y^b \\ \dot{v}_z^b \end{bmatrix} - a_{centrifugal} \tag{8}
$$

Among them, the calculation formula of the centrifugal acceleration is as follows:

$$
a_{centrifugal} = \omega_{gyro} \times v = \begin{pmatrix} \omega_y v_z - \omega_z v_y \\ \omega_z v_x - \omega_x v_z \\ \omega_x v_y - \omega_y v_x \end{pmatrix} \tag{9}
$$

$V = [v_x, v_y, v_z]^T$ is the velocity of the vehicle in the body coordinate system, which can be measured by the GPS receiver. V_x represents the forward direction speed of the carrier, which is much faster than the other two axes. Therefore, we think V_y, V_z approach zero. The expression of the centrifugal acceleration is as follows:

$$
a_{centrifugal} = \begin{pmatrix} 0 \\ \omega_z v_x \\ -\omega_y v_x \end{pmatrix} \tag{10}
$$

$[g_x^b, g_y^b, g_z^b]^T$ can be simplified as

$$
\begin{bmatrix} g_x^b \\ g_y^b \\ g_z^b \end{bmatrix} = \begin{bmatrix} a_x^b \\ a_y^b \\ a_z^b \end{bmatrix} - \begin{bmatrix} \dot{v}_x^b \\ \omega_z v_x \\ -\omega_y v_x \end{bmatrix} \tag{11}
$$

Therefore, the error vector used to modify the pitch angle and the roll angle can be calculated by the outer product of $[v_x, v_y, v_z]^T$ and $[g_x^b, g_y^b, g_z^b]^T$. We set this error vector as $error_{pitch,roll}$. The expression of $error_{pitch,roll}$ is as follows:

$$
error_{pitch, roll} = \begin{bmatrix} v_x \\ v_y \\ v_z \end{bmatrix} \times \begin{bmatrix} g_x^b \\ g_y^b \\ g_z^b \end{bmatrix} \tag{12}
$$

3.2 Calculation of heading angle error vector based on geomagnetic field

Magnetoresistive sensor is used for heading angle measurement. By measuring the horizontal component of the geomagnetic field, we can calculate the angle between the carrier and the magnetic north pole. When the magnetic resistance sensor is installed, the three sensitive axes of the magnetoresistive sensor become coincident with the body coordinate system.

The magnetic field of the x axis is b_x, and the magnetic field of the z axis is b_z. Similar to the method of using the gravity field as the reference vector to compensate the accelerometer, we select the theoretical value of the magnetoresistive sensor as the reference vector, which can be described as $[b_x\ 0\ b_z]^T$. Now, we calculate the error vector to correct the heading angle.

Through the DCM, the output of the magnetoresistive sensor in the body coordinate system can be converted to the horizontal coordinate system. The conversion process is as follows:

$$
\begin{bmatrix} h_x \\ h_y \\ h_z \end{bmatrix} = \begin{bmatrix} \cos\theta\cos\psi & \sin\phi\sin\theta\cos\psi - \cos\phi\sin\psi & \cos\phi\sin\theta\cos\psi + \sin\phi\sin\psi \\ \cos\theta\sin\psi & \sin\phi\sin\theta\sin\psi + \cos\phi\cos\psi & \cos\phi\sin\theta\sin\psi - \sin\phi\sin\psi \\ -\sin\theta & \sin\phi\cos\theta & \cos\phi\cos\theta \end{bmatrix} \begin{bmatrix} m_x \\ m_y \\ m_z \end{bmatrix} \tag{13}
$$

In the formula, $[m_x, m_y, m_z]^T$ represents the output of the magnetoresistive sensor in the body coordinate system. The value of the vector modulus measured by the magnetoresistive sensor must be the same in the XOZ plane in the body coordinate system. Therefore, the following formula can be deduced:

$$b_x = \sqrt{(h_x^2 + h_y^2)} \qquad (14)$$

The two methods get the same size of the vector in the Z-axis, so we get an equation: bz = hz. Now, we obtain the theoretical value of the magnetoresistive sensor in the navigation coordinate system, that is $[b_x \quad 0 \quad b_z]^T$. We use this vector to deduce backward to obtain the output of the magnetoresistive sensor in the navigation coordinate system, and we get the following vector $[m_{bx}, m_{by}, m_{bz}]^T$. The deduction process is as follows:

$$\begin{bmatrix} m_{bx} \\ m_{by} \\ m_{bz} \end{bmatrix} = C_n^b \begin{bmatrix} b_x \\ 0 \\ b_z \end{bmatrix} \qquad (15)$$

At this point, the error vector used to correct heading angle can be obtained from the outer product of $[m_x, m_y, m_z]^T$ and $[m_{bx}, m_{by}, m_{bz}]^T$. We name this error vector as $error_{yaw}$, and its expression is as follows:

$$error_{yaw} = \begin{bmatrix} m_x \\ m_y \\ m_z \end{bmatrix} \times \begin{bmatrix} m_{bx} \\ m_{by} \\ m_{bz} \end{bmatrix} \qquad (16)$$

In summary, we set the total error vector as $e = [e_x, e_y, e_z]^T$, which can be obtained by the following formula:

$$e = error_{pitch,roll} + error_{yaw} \qquad (17)$$

3.3 Fusion strategy based on explicit complementary filter

Gyroscopes have an outstanding performance in dynamic response characteristics, but accumulate errors with time. Accelerometer and magnetoresistive sensors have no accumulated errors, but have disadvantages of poor dynamic response (Sousa, 2009). Therefore, they complement each other in the frequency domain. Due to the slow drift of the gyro, the noise has mainly low frequency, and the accelerometer noise mainly exists in the high-frequency band. The complementary filter can effectively filter out the interference of low-frequency and high-

frequency noise, and get close to the true value of the reconstructed signal (Roberto, 2016).

We constructed a multi-axis complementary filter, proposed a reference vector to detect gyroscope bias error in each axis and used PI controller as feedback to compensate the drift error. Our test system uses four kinds of sensors, including gyroscope, accelerometer, magnetic sensor and GPS receiver. As the accelerometer cannot interpret the difference between motion acceleration and acceleration of gravity, we had to use GPS to get the speed and then the acceleration of motion. The acceleration of motion is subtracted from the acceleration measured by the accelerometer, and the induction of the gravity field is obtained. The error vector is used as the input of the filter, and the corrected value of the gyroscope is the output of the filter. The structure of the multi-axis complementary filter fusion algorithm is shown in Figure 4.

The input error vector is $e = [e_x, e_y, e_z]^T$. The output of the system after PI adjustment can be expressed as follows:

$$u(k) = K_p e(k) + K_i \sum_{j=0}^{k} e(j) \qquad (18)$$

The expression of the gyro angular rate ω after the correction is

$$\omega = \omega_g + Ku(k) \qquad (19)$$

where ω_g is the angular rate measured by the gyroscope; K_p is the proportional coefficient; K_i is the integral coefficient; K is the new coefficient of the system after data fusion. In $K = 2\pi f t_s$, f is the cut-off frequency of the system, and t_s is the sampling period. The appropriate choice of filter parameters is equivalent to determining the cut-off frequency of the filter. Low cut-off frequency setting leads to relying on the gyro slightly more. High cut-off frequency setting leads to depending on the accelerometer and the magnetoresistive

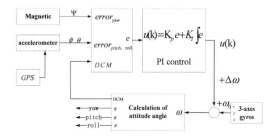

Figure 4. The structure of the multi-axis complementary filter fusion algorithm.

1441

sensor slightly more. Due to the influence of vibration and other factors, the high frequency noise interference is more in the practical application. Complementary filter cut-off frequency should select a smaller value in a reasonable range. Specific parameters need to be determined according to the level of noise in the actual application.

4   EXPERIMENTS AND RESULTS

In order to verify the effectiveness of the proposed method, this paper set up static and dynamic experiments.

4.1   *Static experiment on a three-axis turntable*

In order to obtain the attitude angle calculation accuracy of the attitude determine system at different positions, we set up the static experiment and compared the results between ECF algorithm and turntable feedback data.

The static experiment on a three-axis turntable was to verify the attitude angle calculation accuracy under the static condition. In order to increase the number of experiment samples and make the experiment more convincing, we used the high precision three-axis turntable as a static experiment platform. By setting the program, we controlled the rotation of the turntable and made it stable at different locations. The experimental system collected the data from the sensor at different angles under the static condition. The real-time feedback angle of three-axis turntable was used as the reference value. The accuracy and reliability of the algorithm were verified by comparing the attitude angle obtained by the proposed algorithm with the reference value.

The hardware system used in the experiment is an invention of our research group. Each axis is equipped with a gyroscope, an accelerometer and a geomagnetic sensor, and all the three are MEMS devices. A single antenna GPS receiver module is also integrated in the system. In order to facilitate the outdoor experiment record data, the system integrates the 8 GB chip data storage. The physical diagram of the static experiment is shown below.

The experimental steps of the static experiment are as follows:

1. Turn on the turntable, power on the experiment system and make its pitch angle at the following position:
   $0°$, $30°$, $45°$, $60°$, $75°$, $0°$, $-30°$, $-60°$, $0°$
2. Make the three-axis of the turntable return to zero position, subsequently;
3. Make its roll angle at the following position:
   $0°$, $30°$, $45°$, $60°$, $75°$, $0°$, $-30°$, $-60°$, $0°$

Figure 5.   Static experiment on a three-axis turntable.

4. Get the feedback data of turntable and calculation results of the experiment system.

The magnetoresistive sensor could be largely affected owing to the iron material of the turntable; therefore, no heading angle test was carried out. Comparing the feedback angle of the turntable with the result from the attitude algorithm based on the ECF algorithm, we could obtain the following experiment result.

According to the military standard GJB 729–1989, attitude angle accuracy assessment standards of the static experiment are as follows:

$$RMS = \sqrt{\frac{1}{n}\sum_{i=1}^{n}\frac{1}{m_i}\sum_{j=1}^{m_i}\Delta E_{i,j}^2} \qquad (20)$$

$\Delta E_{i,j}$ stands for the attitude angle error of number j sampling time during number i experiment. m_i stands for data sampling points of the number i experiment. n is the number of the effective experiments. From the above formula, attitude angle calculation accuracy under the static condition can be calculated such that RMS_{pitch} is $0.1648°$ and RMS_{roll} is $0.1059°$. From the experimental results, it can be seen that when the carrier is placed in a stable position, the attitude angle calculation accuracy is high, and the attitude angle error is less than $0.2°$.

4.2   *Dynamic vehicle experiment*

In order to verify the calculation accuracy of attitude angle under dynamic conditions and the heading angle accuracy, which is not verified in the static experiment, we proposed a dynamic vehicle experiment. Our test system and the high-precision inertial navigation system (considered as reference) were installed side by side on the test

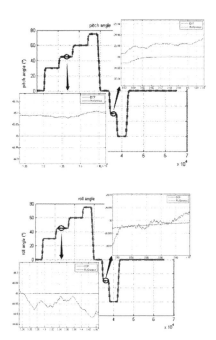

Figure 6.   Experimental result of the static experiment.

Figure 7.   Test platform of the dynamic vehicle experiment.

platform, which was connected with the car body. The motion information of the vehicle body was collected by the two systems at the same time, and the test time was about 180 seconds. The test platform of the dynamic vehicle experiment is shown in Figure 7.

The output of attitude angle from our test system was compared with that from the high-precision inertial navigation system, and the results are shown in Figure 8.

It can be seen from Figure 8 that the attitude angle deviation due to the gyro drift is evidently reduced. Under the dynamic environment of vehicle experiment, the attitude angle obtained by the proposed ECF algorithm is highly consist-

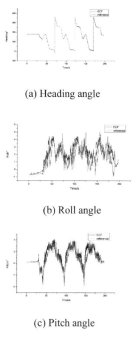

(a) Heading angle

(b) Roll angle

(c) Pitch angle

Figure 8.   Attitude angle of the dynamic vehicle experiment.

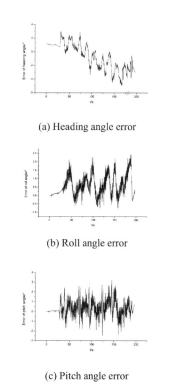

(a) Heading angle error

(b) Roll angle error

(c) Pitch angle error

Figure 9.   Attitude angle error curve.

ent with the attitude information provided by the high-precision inertial navigation system. The new algorithm has the performance of real-time tracking of dynamic changes.

From Figure 9, it can be seen that the heading angle error is the largest in the three-axis attitude angle error, reaching 4°. This is because the determination of heading angle requires geomagnetic sensor, and the sensor itself is vulnerable to interference. Fortunately, after using the proposed ECF algorithm, the heading angle error divergence is suppressed, and the error value becomes stable.

Roll angle error and pitch angle error are relatively small. Roll angle error is less than 2.3°, and the pitch angle error is less than 3.1°. The error curve has no tendency to diffuse. From the above experimental data, the RMSE of heading angle, roll angle and pitch angle can be calculated, and the values are 3.2682°, 1.3257°, and 1.5954°, respectively.

5 CONCLUSION

In this paper, a new complementary filter algorithm for the attitude angle based on DCM is proposed. In this method, we constructed a multi-axis complementary filter, proposed reference vectors to detect gyroscope bias error in each axis and used PI controller as feedback to compensate the drift error of gyros. The proposed algorithm was tested through three-axis turntable and actual vehicle experiment. The experiment result showed that the static accuracy of the attitude angle is better than 0.2°, and the dynamic accuracy is better than 4°, satisfying the application requirements of low-cost attitude determination system.

ACKNOWLEDGMENTS

The authors wish to thank Jie Li for helpful discussions. The authors would like to appreciate the assistance of all co-workers in the Science and Technology on Electronic Test & Measurement Laboratory for their contribution to experiments. This work was supported by the National Natural Science Foundation of China (No. 51575500, No. 50905169).

REFERENCES

Anonymous. Multi MEMS sensor development board [J]. Electronics Weekly (2013).

Chiemela Onunka, Glen Bright, Riaan Stopforth. USV attitude estimation: an approach using quaternion in direction cosine matrix [J]. Robotica (2014).

Chingiz Hajiyev, Demet Cilden, Yevgeny Somov. Gyro-free attitude and rate estimation for a small satellite using SVD and EKF [J]. Aerospace Science and Technology (2016).

Development of sewer pipe measurement system by vehicle equipped with low-priced MEMS sensor [J]. Mechanical Engineering Journal (2016).

Francesco Cappello, Roberto Sabatini, Subramanian Ramasamy. A low-cost and high performance navigation system for small RPAS applications [J]. Aerospace Science and Technology (2016).

Gaoge Hu, Shesheng Gao, Yongmin Zhong, Bingbing Gao, Aleksandar Subic. Matrix weighted multisensor data fusion for INS/GNSS/CNS integration [J]. Proceedings of the Institution of Mechanical Engineers (2016).

Gaygysyz Jorayev, Karol Wehr, Alfonso Benito-Calvo, Jackson Njau, Ignacio de la Torre.

Ji Hyoung Ryu, Ganduulga Gankhuyag, Kil To Chong, Hana Vaisocherova. Navigation System Heading and Position Accuracy Improvement through GPS and INS Data Fusion [J]. Journal of Sensors, 2016, 2016.

Joon Goo Park, Jang Gyu Lee, Chan Gook Park. SDINS/GPS in-flight alignment using GPS carrier phase rate [J]. GPS Solutions (2004).

Liu, Yu, Xiang, Gaolin, Cao, Yang, Wang, Ruijie, Gong, Dawei. Research on the Temperature Compensation Algorithm of Zero Drift in MEMS Gyroscope Based on Wavelet Transform and Improved Grey Theory [J]. Sensors & amp; Transducers (2014).

Mosavi, M. R., M. Soltani Azad, I. EmamGholipour. Position Estimation in Single-Frequency GPS Receivers Using Kalman Filter with Pseudo-Range and Carrier Phase Measurements [J]. Wireless Personal Communications (2013).

Robert Bogue. Recent developments in MEMS sensors: a review of applications, markets and technologies [J]. Sensor Review (2013).

Sousa, R., P. Oliveira. Joint Positioning and Navigation Aiding Systems for Multiple Underwater Robots [J]. IFAC Proceedings Volumes (2009).

Stefano Corbetta, Ivo Boniolo, Sergio M. Savaresi. Attitude estimation of a motorcycle via Unscented Kalman Filter [J]. IFAC Proceedings Volumes (2010).

Vaibhav G. Awale, Hari B. Hablani. Fusion of Navigation Solutions from Different Navigation Systems for an Autonomous Underwater Vehicle [J]. IFAC Proceedings Volumes (2014).

Yiqun Dong, Youmin Zhang, Jianliang Ai. Full-altitude attitude angles envelope and model predictive control-based attitude angles protection for civil aircraft [J]. Aerospace Science and Technology (2016).

Zhongxu Hu, Barry Gallacher. Extended Kalman filtering based parameter estimation and drift compensation for a MEMS rate integrating gyroscope [J]. Sensors & amp; Actuators: A. Physical (2016).

A study on personalized information push service based on context awareness and spatio-temporal correlation

Xingchao Wang
School of Information Science and Engineering, Yunnan University, Kunming, China

Lingyun Yuan
School of Computer Science and Technology, Yunnan Normal University, Kunming, China

Guili Ge
Library, Yunnan University of Traditional Chinese Medicine, Kunming, China

ABSTRACT: Personalized information service has become an inevitable trend with the development of the current information technology. Aiming at the defects of today's personalized information service as well as the growing requirement of personalized information, a personalized information push service model based on context awareness has been put forward in this paper. In this model, Internet of things and context awareness have been used to build a context information acquisition environment and to gather the context information of users. Furthermore, a user-personalized interest model has been constructed and combined with the spatio-temporal correlation to catch the relationship among the user groups, so that an information push mechanism based on spatio-temporal correlation can be achieved. Our method provides a more efficient and precise information push service.

1 INTRODUCTION

With the user at the core, the personalized information service is devoted to push more effective and pointed personalized information service to the user by studying some context information such as the user's behavior, environment, and emotions. This makes it possible to obtain the user's demand information automatically and decrease the user's search decision. Personalized information service needs to provide for real time and individual demands of the users and fulfill their personalized needs, which means that the service should possess the characteristic of self-adaptability. With the conventional information push technology, the user demands are maintained stably, but lack individuation, real time, and renewability; therefore, the traditional methods of catching user message are no longer in use. With the boom of the Internet Of Things (IOT), any information that is linked to the user's conduct, target, or the surrounding environment is all considered as an item that has some interactive relation with the user and the pervasive computing environment; thus, combining the context data with the user's behavior would be a great way to figure out the user's interest preference. Using the IOT technology to gain relative context information and considering a user's behavior on

the Internet, we can learn about the interests of the user more comprehensively, and then be able to offer personalized information service to the user independently.

The context awareness technology is derived from the context calculation; therefore, the "context" also inherits the advantage of "who, when, where, what" of pervasive computing. Recently, context information has been deemed as the foremost part of the Recommendation and push System (RS) (Feng, 2014). The traditional information push mined user information through the Internet and pushed the information after completing its analysis and processing. With the dramatic development of the IOT, letting the context information join the conventional two-dimensional push system "user-item" and building a three-dimensional push system based on the form of "user-item-context" has already become a mainstream trend of information push service. JinHai Feng proposed using indoor positioning technology to track user activity in the market, in line with the historical information that is recorded, which contains information about where the user came from and what commodities the user browsed to estimate the user's hobbies and favorites, and then accurately recommending goods to the user that they might be interested in (Wang, 2012). LiCai Wang has been trying to judge whether the requirements of mobile users would suffer from the context by computing the volatility of mobile users' behavior that is under constraint of the context, and to make sure of the extent of the influence by making use of the volatility (Gong, 2011). Combining mobile Internet and QRCode, XinWen Gong designed a mobile learning platform based on context awareness, in which the known context information of the users can be used as an important reference to provide the corresponding learning content (Li, 2012). Based on context awareness technology, XiaoYan Li established three application scenarios to provide shopping reminders for users automatically by combining user opinion with the context to mine users' shopping decisions and underlying demands and accordingly sending shopping reminders to the users automatically (Adomavicius, 2005). Champiri and his coworkers hold the view that there are three conditions of context information: user context, file context, and environment background. They also emphasized combining the concept of context with user opinion to make academic recommendations in the academic field; therefore, a digital library recommendation system based on context awareness has been designed and implemented effectively (Baltrunas, 2011). Gavala and Kenteris used context information such as the current location, time, and condition to extend collaborative filtering technology and provide corresponding mobile travel guidance according to the user's mobile device (Gavalas, 2011).

Based on the influence and function of context awareness in the process of personalized information push, with the IOT as a basic technology to collect and handle the context information, a personalized information push service model based on the context awareness has been proposed in this paper. In addition, the spatio-temporal correlation semantics technology has been added to the procedure of information push in order to push more delicate information. The rest of the paper is composed of the following: the personalized information push model framework has been constructed in section 2; the user personalized interest has been described and presented in section 3; in section 4, the information push mechanism based on spatio-temporal correlation has been proposed; and finally, the conclusions have been drawn.

2 THE CONSTRUCTION OF PERSONALIZED INFORMATION PUSH MODEL

Personalized information push uses context awareness as the main way to capture user behavior. According to the user interest and preference, a user model can be constructed. By matching the context information and user interest, the information that users want can be pushed to the users, by which we can provide adaptive adjustment according to user feedback, thus achieving a personalized information push. Hence, a personalized information push system mainly includes the capturing and handling of context information, the user interest modeling and information push. The framework of the personalized information push model based on context awareness is as shown in Figure 1.

The sensing layer of the whole framework of information push mainly captures information, including explicit and implicit information. The data layer organizes and manages the captured data information and builds user interest model with the data information stored orderly. The transport layer conducts information match of the newly captured information and user model. Lastly, the application layer combines time and space relation with similar user interest to push correlated information to the users. The work process of information push system is as following: firstly, get context information (environment information, user message) through the wireless sensor; use context-aware computing and mining technology to catch user interest and similar users; build database to store user information and data information (real-time information and history information); then build the context model to mine different users or the preference attributes of different conditions for the same users, by building the match rules, proposing the favor of users, extracting the information that matched with the users, and sending it to them; and finally, get user feedback across the context and make some adjustments and optimizations.

3 THE DESCRIPTION AND REPRESENTATION OF PERSONALIZED USER INTEREST

3.1 *User interest modeling based on context awareness*

With the foundation of research groups' prior results, a user interest model based on context awareness has been constructed in the paper, as shown in Figure 2. The construction of this model mainly includes four parts: information retrieval, information processing, model building, and model matching. Data flow is captured by context awareness and mining of network data. Here, the information is divided into explicit information and implicit information. By adaptive adjustment of the weight of the two kinds of information, interest tags that match with the information through

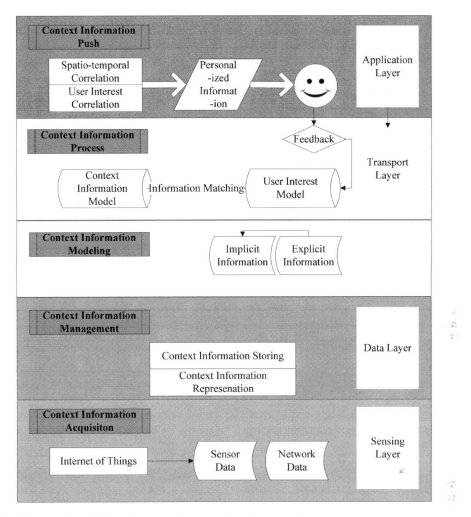

Figure 1.   The personalized information push framework based on context awareness.

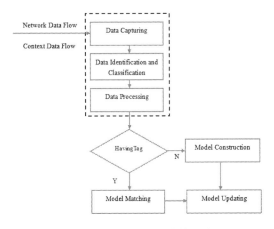

Figure 2.   M-C-W user interest model based on context.

similar algorithms are found, and the users are judged with or without the model according to the user tags. If the model fails in case of interest information while finding out the matched interest tags, it will build the user interest model directly; otherwise, it will update the original user model when successful. The process of constructing user context interest model is shown in Figure 2 (Ge, 2016).

3.2   *The representation of the user interest model*

The context information captured by the sensor is modeled in an angle of "user-item-context", which contains the user's context, behavior, and item, and is expressed in the form of recorded data. The expression form can be summarized as: *context* (*user, condition, location*). In this formula, context means the perceptive context information in the system; user means the user object; condition means the current situation of the user; and location means the current location of the user.

The user interest can be expressed as follows (1):

$$M_i = f(U_i, C_i, W_i) \qquad (1)$$

where M_i is the user interest model; U_i is the object of the user; C_i is the context information of the user; and W_i is the degree of user interest.

4 THE INFORMATION PUSH MECHANISM BASED ON SPATIO-TEMPORAL CORRELATION

The IOT environment comprises certain kind of temporal and spatial correlation among things and humans. As time goes by, the number and the relation of things changes as well. For instance, effectiveness of space is a necessary characteristic of IOT data, and it is also a valuable property when considering the relation among people; people and things; and things. In order to understand the user interest better, we can carry out comprehensive calculation in the user model about the information of adjacent users and similar items; collect the users that have the same behavior; put them in one set based on the temporal and spatial similarity; and then push the spatio-temporal correlation information.

4.1 *User context information acquisition based on spatio-temporal correlation*

The context properties are divided into three classes: time, location, and relation. The acquisition of mobile user information is mainly done via tracking and learning about the user preference information and conducting the correlation calculation of user information by considering different properties. In terms of the angle of "user-item-context", it needs to get the context of users, task, and environment and combine them to get the information push, as shown in Figure 3.

For example, a smart library and museum has a vital significance for acquiring the relation among the users according to the same location and reading time. Time and place is the temporal and spatial definition about the affairs the user's joined or implemented item; therefore, we mainly use the spatio-temporal correlation of context environment to catch user environment context and to combine, link, and match the user context, task, and environment, in order to obtain the similar

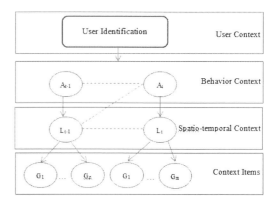

Figure 3. User context information acquisition based on spatio-temporal correlation.

user groups by the combination of time and space situations and to push the collaborative information at the end. The collaborative push based on spatio-temporal correlation mainly fulfills the similarity between time and space and promotes the adaptive matching of situation between the needs of users and the push services of system.

4.2 *User interest similarity criterion based on spatio-temporal correlation*

4.2.1 *Spatio-temporal similarity calculation*
User interest change should be described from the perspective of change of space, time, and property, in which the change of state, space, and interest property of users within different periods are included. In the semantic net, spatio-temporal semantics is deemed as the common repression of time semantics, space semantics, and property semantics. We use a relationship like "user-item-context" to display the interest of spatio-temporal correlation. Among the user objects that have similar interests exists certain special space and time links. In the link of spatio-temporal correlation, users have a special space and time correlation emphasizing the relation of time at which the affairs occurred; the qualitative relation is shown in Figure 4.

It is assumed that the data range of temporal and spatial similarity is between *0* and *1*, where *0* means totally unrelated in time-space, and *1* means related perfectly. Therefore, the spatio-temporal similarity criterion under the influence of the space and time is equal. Except the time relation, the spatio-temporal correlation also emphasizes space relation and spatial timeliness. For example, when the space similarity is in the state of *0*, and even the time similarity is *1*, the state of space similarity would also be *0*, which means that they are still dissimilar in terms of time-space character.

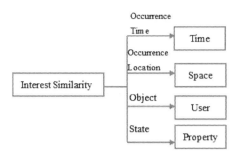

Figure 4. The space and time relationship of user interest.

Some corresponding relations exist among time, place, and property. The occurrence of a behavior property must occur in a special period of time and space. In spatio-temporal correlation, user A's state of interest in time T and space P would be similar to user B's interest at a similar time T or a similar space location P.

Let's suppose that M is a collection of property, where $M = \{m_1, m_2, ..., m_n\}$; T is a collection of time, where $T = \{t_1, t_2, ..., t_n\}$. t is regarded as a point in time or a period of time; m_n and t_n can't be empty, and t is the only number, which means that the occurrence of property must be in a mere period of time, but a point in time or a period of time has many properties. The relation between property and time, and property and place is both $n:1$, and $1:1$ is the relation between time and space. Therefore, we can make sure of one user's property with time and place.

The similarity measure of time-space series should contain spatial similarity measurement and time series similarity measurement. The object distance of spatial point can be measured in many ways. In this paper, we use the Euclidean Distance of two-dimensional space computing, for two points $P_i\ (X_{i1}, X_{i2},...,X_{in})$ and $P_j\ (X_{j1}, X_{j2},...,X_{jn})$ in R^n. The general form of generalized distance is as follows (2):

$$d_n\left(p_i, p_j\right) = \left[\sum_{k=1}^{n}\left(x_{ik} - x_{jk}\right)^n\right]^{1/n} \tag{2}$$

The space object is usually expressed in the two-dimensional or three-dimensional geographic spaces; therefore, n is always $1, 2, 3$. We mainly think about the two-dimensional space, using Euclidean Distance to carry out the spatial similarity calculation.

When $n = 2$, it is shown as follows (3):

$$d_{ij} = \left[\sum_{k=1}^{n}\left(x_{ik} - x_{jk}\right)^2\right]^{1/2} \tag{3}$$

To express the close relationship between the distances more directly, we use similarity factor to measure the level of similarity of variables, to get the degree of correlation between the spaces, in which the distance variable needs to be transformed as shown below:

$$sim\left(p_i, p_j\right) = \frac{1}{1 + d\left(p_i, p_j\right)} \tag{4}$$

The time correlation measurement is based on the linear correlation coefficient of time series; it is carried out by calculating within the same period of time and conducting similarity computation of two equal length time series. Assuming that u and v are two time series of the same length; the correlation coefficient between u and v can be described as follows (5):

$$\rho(u,v) = \frac{\sum_{i=1}^{n}\left(u_i - \bar{u}\right)\left(v_i - \bar{v}\right)}{\sqrt{\sum_{i=1}^{n}\left(u_i - \bar{u}\right)^2}\sqrt{\sum_{i=1}^{n}\left(v_i - \bar{v}\right)^2}} \tag{5}$$

In this formula, n represents the length of time-space series. The values of $\rho(u,v)$ range from $[-1,1]$, in which $\rho(u,v) > 0$ means positive correlation, while $\rho(u,v) < 0$ means a negative correlation, and the absolute value of $\rho(u,v)$ refers to the level of the degree of correlation. When $\rho(u,v) = 1$, it means a completely positive correlation, which shows that there is an evident relation between the two series, whereas $\rho(u,v) = 0$ means a totally negative correlation, which means that there is no relation between the two series.

To measure the degree of temporal and spatial similarity, we need to calculate time similarity and space similarity separately. If the time-space possesses similarity, then we judge the relation between users through temporal and spatial similarity, and then judge the interest relation between users by capturing the users' time-space interest through the user group.

4.2.2 Spatio-temporal correlation calculation

Due to the advantage of probabilistic method in memory requirements that can judge the interest relation between users, we used the Bayesian method for the calculation. On getting the conclusions, Bayesian computation should consider not only the current observed sample information, but also the past experience and knowledge reasoned. Provided that every temporal and spatial characteristic property is conditionally independent, the deduction based on Bayesian computation is as follows (6):

$$P\left(y_i \mid x\right) = \frac{P(x \mid y_i)P(y_i)}{P(x)} \tag{6}$$

The denominator is constant for all categories; therefore, it is only necessary to maximize the molecular. Meanwhile, the attributes are independent of the conditions, and thus the computing is according to the following formula (7):

$$P(y_i \mid x) = \frac{P(y_i)\prod_{j=1}^{m} P(x \mid y_i)}{P(x)} \quad (7)$$

In formula (7), $P(y_i \mid x)/p(x)$ is called "probability function". As an adjustment factor, it makes the estimated probability closer to the real probability. First, it estimates a prior probability, and then it combines the experimental results to observe if the experiment enhances the prior probability, so that we can get the posterior probability that is close to reality. Here, when $P(y_i \mid x)/p(x) > 1$, it means that the prior probability is enhanced, the possibility of occurrence of event y becomes bigger, and user has more interest in y. When the probability function is 1, the occurrence of affair x does not judge the possibility of A. If the probability function is less than 1, it means that the prior probability is weakened, and the probability of event A becomes small (Shang, 2012).

To judge the spatio-temporal correlation, we need to judge the influence relation between the time property and space property. Here, we set event properties to be $Ei(l_i, t_i)$. It can be expressed by spatio-temporal correlation as follows (8):

$$P(t_i \mid l) = \frac{P(l)\prod_{i=1}^{m} P(l \mid t_i)}{P(t_i)} \quad (8)$$

In the model of personal interest, there is an important relationship among the interest, behavior, and score of users. In the environment of context, user interest always has vital relation with time and space, because the time-space relation between the user and himself is always $1{:}1$, whereas the relation between time and space is $n{:}m$; therefore, when judging the user interest in a certain time and space, we need to judge the time-space relation of ourselves. If the relationship between time and space is related to a certain time and space, we need to judge the possibility of $P(t_i \mid l)$, if the time or time period that occurs when an event occurs in the L space does exist.

4.3 The information push mechanism based on spatio-temporal correlation

Starting with the angle of "user-item-context", we can conduct information push by combining the background of time-space, the users, and

the project. The information push mechanism is shown in Figure 5. Known from the relation of time-space, if user A and B have a similar relation in time and space, it can be inferred that user A and user B have similar spatio-temporal relationships; therefore, they might have similar interests, and then we can consider pushing the interest of user A to user B.

When judging the spatio-temporal correlation of user groups, if similarity relation in time and space exists, then it is thought that this user has a similar relation with others, and a similar interest push can be made, combined with the spatio-temporal relationship in order to determine the user's personal interest model. If the level of spatio-temporal correlation is high, we can get user interest tags from the user interest model to push the information. Its approximate process is shown in Figure 6.

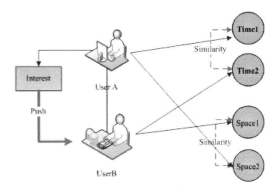

Figure 5. User interest criterion based on spatio-temporal correlation.

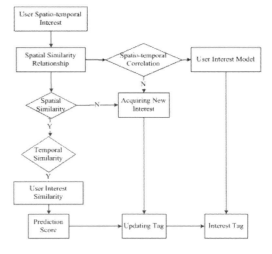

Figure 6. User information push mechanism based on spatio-temporal correlation.

5 CONCLUSIONS

The context awareness technology has the ability to percept user needs in an intelligent and invisible manner. It can reduce the interaction and participation of users and capture their personalized interest smartly. Traditional information push can no longer satisfy user commands consistently, in real-time, and with personalization. Combining the technology of context awareness and information push, a personalized information push model based on context awareness and spatio-temporal correlation was proposed in this paper, in which spatio-temporal correlation was considered to build a new information push method, in order to realize the combination of intelligent input and output, and then to provide a true meaning of personalized information push service. In the future work, we will apply the model to a digital ethnography museum to validate the efficiency further.

REFERENCES

Adomavicius, G., R. Sankaranarayana, S. Sen. Incorporating contextual information in recommender systems using a multidimensional approach. ACM Transactions on Information Systems, 23(1): 104–145 (2005).

Baltrunas, L. et al. Context-aware places of interest recommendations for mobile users. A Springer-Verlag Berlin Heidelberg (2011).

Feng, J.H. Personalized WeChat recommendation system based on indoor WLAN localization. Computer Engineering and Science, 36(10):1925–1931 (2014).

Gavalas, D., M. Kenteris. A web-based pervasive recommendation system for mobile tourist guides. Personal Ubiquitous Computing, 15(7):759–770 (2011).

Ge, G.L., L.Y. Yuan, X.C. Wang. Personalized user interest modeling based on context aware. Application Research of Computer, 33: 1–6 (2016).

Gong, X.W. The design and implementation of mobile learning platform based on context awareness. Tianjin: TianJin University, 38–45 (2011).

Li, X.Y. The research and design of mobile shopping assistant based on context awareness. Dalian: Dalian Maritime University (2012).

Shang, C.L. The study on time series feature based on bayesian spam filtering. Guangzhou: South China University of Technology (2012).

Wang, L.C., X.W. Meng, Y.J. Zhang. Context-aware recommender systems. Journal of Software, 23(1): 1–20 (2012).

Design of a new in-flight entertainment terminal

Hairong Xu, Hong Zhou & Hui Yang
College of Air Transportation, Shanghai University of Engineering Science, Shanghai, China

ABSTRACT: A new design of In-Flight Entertainment terminal is presented in this paper. Qualcomm's quad-core processor APQ8016 is used as an application processor to process multimedia, games, and applications. This design integrates advanced OLED display technology and intelligent detection technology. The new terminal will provide passengers with more intelligent and personalized service through the Android operating system and big data analysis technology. It will be more intelligent and greatly reduce power consumption.

1 INTRODUCTION

In-Flight Entertainment (IFE) system is an important component of civil aircraft cabin systems and an important way to improve the quality of service. It connects to cabin communication system through wired or wireless connection. It can achieve cabin announcements and entertainment services. The basic idea behind the IFE system is to provide passengers with comfort during long flights; therefore, these services were initially based on delivering food and beverages to passengers. As passengers' demand for services increased, accompanied with an increase in airlines competition and technology advancement, more services were introduced, and modern electronic devices began playing a remarkable role in the same. This caused a change in the basic concept behind the IFE system. It became more than just delivering physical comfort and providing food. It has been extended to provide interactive services that allow passengers to participate as a part of the entertainment process as well as providing business-oriented services through connectivity tools. Moreover, it can provide means of health monitoring and physiological comfort.

Flight entertainment began before the First World War by the Graf Zeppelin. Starting from 1960, IFE system began to attract attention; it was basically a pre-selected audio track that could be accompanied with a film projector. The IFE system was gradually developed into a powerful system. Cabin telephones allowed passengers to make phone calls during the flight. The system became interactive and allowed passengers to select their own services, while in the past, passengers had to follow fixed services. Web-based internet services allow passengers to use some services such as

emails and SMS messaging. IFE's data transmission system has been gradually upgraded from the analogue mode to the digital mode (Kang, 2012). The IFE system consists of a server, the passenger terminal, and the network. Passenger terminal includes a multimedia processing circuit board, display panel, and passenger control unit. It is connected to the server by wired or wireless connection, and the server provides audios, videos, games, and advertisements to the passengers.

A new design of passenger terminal system is presented in this paper. In this design, powerful hardware will meet all kinds of requirements of the passengers. The software system will provide various services and applications to facilitate passengers' travel experience through advanced operating system. With big data analytics, airline companies can analyze a mix of complex datasets from various sources in order to gain important insights into customer behavior and use such feedback to provide better customer service. This design can push personalized media, games, and advertisements to the terminal from an on-board server according to various passengers' demands.

In the new design, passenger terminal system will integrate cutting edge technology of electronics, communications, computers, and big data analysis. These advanced technologies will provide more personalized service and increase the economic benefits of the airline companies. Passenger terminal system will place more emphasis on the comfort of the seats because the passengers spend most of their travel time in their seats. Audios and videos will be saved in local storage instead of the server, which will reduce the dependence on the server and the network (Akl, 2011).

2 HARDWARE DESIGN OF THE SYSTEM

Terminal system of the IFE system has a monitor, a circuit board, buttons, a touch panel, and accessories such as a headset. A typical terminal uses LCD or CRT as the display device, which has a big size and high power consumption. The central processor in the traditional terminal is not powerful enough to run certain games. Buttons and touch panel are used as the input, which are installed behind the backrest or installed on the roof of the cabin.

In the new design, each terminal is an independent information carrier and multimedia processing unit. Large-capacity data storage in the terminal can store various games, movies, and applications. The central processor can handle various applications and games. The terminal can retrieve content from the server and update passenger's data to the server through wired or wireless connection (Alamdari, 1999).

The new design uses a powerful multimedia processing chip APQ8016 (Qualcomm's processor) to handle audios, videos, and games. High-resolution Organic Light-Emitting Diode (OLED) display panel is used as the display device, and a capacitive touch screen is used as an input device. Passengers can play games with joystick connected to the USB host connector. The terminal has data acquisition interface to connect to the seat belt sensor system and passengers' health sensor system (Z, 2016). It can automatically acquire real-time data and send it to the server. This kind of data can also be saved in local massive storage. The new terminal is more intelligent and powerful compared to the traditional device. The hardware design block diagram is shown in Figure 1.

2.1 CPU subsystem design

The CPU subsystem is the core of the entire hardware system. It is made up of application processor, power management chip, and memory chip. Qualcomm Snapdragon 410 processor APQ8016

is used as the multimedia processor. APQ8016 has quad 64-bit ARM Cortex-A53 MPcore Harvard Superscalar core, LP-DDR2 / LP-DDR3 SDRAM interface, Hexagon QDSP6, 13.5 MP camera support, 400 MHz Adreno 306 GPU, 1080p video encode/decode, gpsOneGen 8 A with GLONASS, Bluetooth 4.0, OpenGL ES 3.0, DirectX, OpenCL, Renderscript Compute, and FlexRender support. APQ8016 can support a variety of multimedia applications and game applications.

The PM8916 mixed-signal HV-CMOS device is used as a power management chip. It has 20 low dropout linear regulators, which can power external devices. System power on and off can be controlled by APQ8016 through SPMI interface. There is an audio codec inside APQ8016, and one stereo head phone driver is supported.

The memory chip (MCP) includes 1GB LPDDR3 and 8GB solid state storage, which is used for the operating system and different applications. It is connected to the main processor through high-speed data bus. Extended SD card can be used to save audios, videos, and games. The CPU subsystem design diagram is shown in Figure 2.

2.2 Man–machine interaction subsystem

Man–machine interaction subsystem is the main part of the design. This subsystem has an input section and an output section. The input section will receive the passenger's command, and the output section will display the passenger's favorite content on the screen.

Traditional display device uses Liquid Crystal Display (LCD) panel as the output. A backlight system is used for display module, as LCD does not produce light by itself. LCD needs illumination (ambient light or a special light source) to produce a visible image.

In this design, 720p OLED display panel is used. OLED is a kind of new display technology with better power efficiency and thickness, and it has

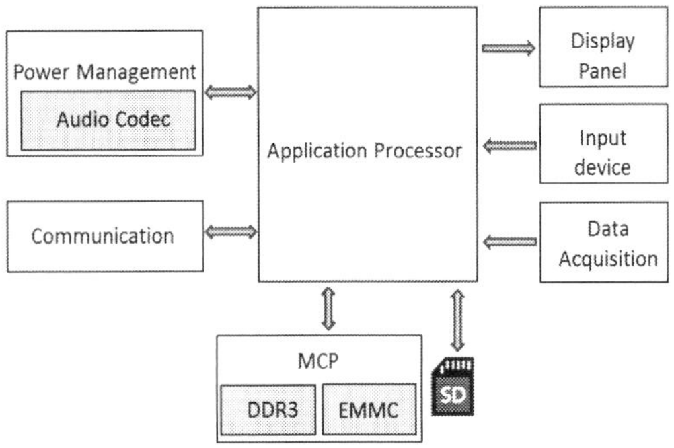

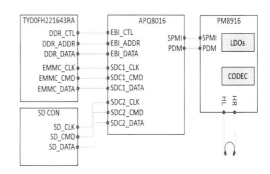

Figure 1. Hardware design.

Figure 2. CPU subsystem design.

much faster response time than the LCD. OLED display panel is connected to the core processor via MIPI interface. Since OLED display works without a backlight system, power consumption is very low, and heating phenomenon does not exist in the display module; thus, it can improve the economic efficiency and improve the airline safety.

Normal design of the terminal uses resistive touch screen as the input that does not support multi-touch with fingers. Capacitive touch screen is used in this design, which can recognize two or more points of contact on the surface concurrently. It is extremely good for object manipulations—touch, drag, and zoom in and out, which greatly facilitate the interaction between the passengers and the device; it is also good for certain games.

Flight conditions may cause the cabin environment to be tough, especially for people who can face ill conditions. Flight duration, dehydration, pressure, engine noise, and other factors can be the reasons of physical and/or psychological problems. New intelligent seat with the seat belt monitoring system, health monitoring system, and other sensors can help get passengers' status during their flights (Z, 2016). All these monitoring systems in the seat can be connected to device via I2C interface. Passenger's status data can be saved in the terminal and be sent to on-board server. Cabin crew can monitor the status of all passengers on the server side; thus, it will reduce the workload of the cabin service and greatly improve the efficiency of passengers' services. When the aircraft would experience some turbulence, cabin crew won't need to walk through the cabin and do a seatbelt check one by one for the passengers. A sensory system integrated in the IFE system can provide a way to sense bad health conditions for passengers with health problems and can either inform the crew members or perform an action to reduce the effect. The server will save all the data that is acquired from the passengers, and this data will be analyzed by big data processing technology on the local server after landing.

Capacitive touch screen and the heartbeat, seat belt monitoring system can be connected to the device via I2C interface.

The terminal works as a media center and game station. The joystick must be supported, as it is a principal control device for games; it can be connected to the device via a USB host connector.

Man–machine interaction subsystem design diagram is shown in Figure 3.

2.3 *Communication system design*

Nowadays, wired networks are the principal technology of implementing IFE systems. Ethernet is currently the standard for wired communication in different fields. Compared with wired connection,

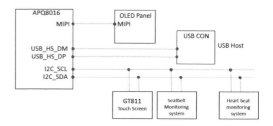

Figure 3.   Man–machine interaction subsystem design.

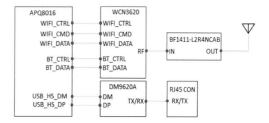

Figure 4.   Communication subsystem design.

wireless connection can reduce the costs of network inside the cabin. Wired and wireless connections interface are both implemented in the device. Peer-to-peer network can be set up to distribute video content over the network instead of traditional client-server architecture. The peer-to-peer approach will allow IFE units to monitor, store, and serve media content with each other.

In this new design, wired and wireless connections are supported. The device uses RF transceiver WCN3620 to support WIFI and Bluetooth. WCN3620 integrates two different radio technologies: WLAN, compatible with IEEE 802.11b/g/n specification; and Bluetooth, compatible with BT version 4.1 specification. DM9620 A is USB to 10/100Mbps fast Ethernet controller that is used to implement wired connection between the on-board server and the terminal.

The terminal can communicate with the on-board server via wireless or wired connection. The server can push flight information, shopping information, and news to the passengers according to their specific requirements and preferences. A Bluetooth headset can be connected to the device to listen to music. Gaming systems can be networked to allow interactive playing by multiple passengers and enable high-quality gaming in an aircraft cabin environment. The communication system design is shown in Figure 4.

3   SOFTWARE DESIGN OF THE SYSTEM

In order to support various applications and games on the terminal, Android is used as the operating system. It is a mobile operating system developed by Google, based on the Linux kernel, designed primarily for touchscreen device such as smartphones and tablets. Android's source code is released by Google under open source licenses.

Applications that extend the functionality of devices are written using the Android Software Development Kit (SDK). One customized Android system is required because there is a big difference between tablets and terminals. Applications used in the terminals are not same as the applications used in the mobile devices and tablets. These applications must be based on cabin services and entertainment.

The software system consists of login applications, cabin information applications, flight information applications, online shopping applications, settings, and various games. It also has certain background services that get data from the seat and communicate with the on-board server. Basic diagram of the whole software architecture is shown in Figure 5.

There are four main software layers in the whole system as shown in the architecture diagram.

1. Linux kernel

At the bottom of the layers is Linux kernel. It provides a level of abstraction between the device hardware and contains all the essential hardware drivers such as OLED display, capacitive touch panel, I2C, Joystick, WIFI, Bluetooth, and Ethernet. The kernel works as the abstract layer between software and hardware. It hides all the detailed information of the hardware and provides a uniform interface for upper layers.

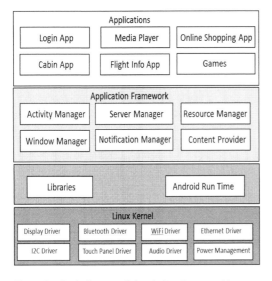

Figure 5.   Basic diagram of the whole software architecture.

2. Libraries and Android runtime

It encompasses those libraries that are specific to audio and video playback, game graphics, drawing, and database access.

Android runtime provides a key component called Dalvik Virtual Machine (VM), which is a kind of Java Virtual Machine specially designed and optimized for Android. The Dalvik VM enables every Android application to run in its own process, with its own instance of the Dalvik VM.

3. Application Framework

The application framework layer provides many higher level services to applications in the form of Java classes.

Server manager is the interface for the data transmission between terminal and server. Applications and services use this manager to get data from the server and push data to the server.

Notification manager is the interface that is used to send the cabin information message to the terminal.

4. Applications

You will find all the Android applications at the top layer. Examples of such applications are media player, cabin information, flight information, online shopping, settings, and games. Media and commodities may be different according to different passengers, since they may have various differing requirements. All the data pushed to the passengers is highly customized according to the analysis result of big data processing.

4   BIG DATA PROCESSING

There are two ways to collect data from passengers: one way is from the internet, and another way is from the terminal. Passengers' hobbies will be gathered if they use the internet. Passenger's heart beat data, seat belt status data, and tracking data in each application will be sent to on-board server and then to the local server after landing. All the related data can be analyzed by big data processing technology to create a profile for each passenger. This profile is a snapshot of the passenger. The content of the profile includes the personal information of the passengers, their favorite food and beverages, and so on. This profile is used to configure applications and games in the terminal and provide customized content to the passengers. Each passenger will be able to access different applications, videos, and game; therefore, the passenger would not need to waste time in selecting items during the flight, and the profile of the passenger would be used for future travels. For health services, automatic pop-up reminders can be used to stop passengers from being stuck to the entertainment content.

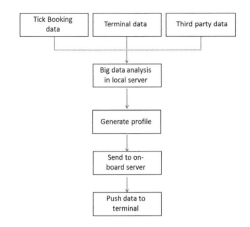

```
┌──────────────┐  ┌──────────────┐  ┌──────────────┐
│ Tick Booking │  │ Terminal data│  │Third party   │
│    data      │  │              │  │    data      │
└──────────────┘  └──────────────┘  └──────────────┘
                         ↓
               ┌──────────────────┐
               │ Big data analysis│
               │  in local server │
               └──────────────────┘
                         ↓
               ┌──────────────────┐
               │ Generate profile │
               └──────────────────┘
                         ↓
               ┌──────────────────┐
               │  Send to on-     │
               │  board server    │
               └──────────────────┘
                         ↓
               ┌──────────────────┐
               │  Push data to    │
               │    terminal      │
               └──────────────────┘
```

Figure 6. Working flow of big data processing.

Profile data and media data will be updated and pushed to the passenger terminal before the plane takes off.

The working flow is shown in Figure 6.

5 CONCLUSION

At the beginning, the IFE system was targeting just the comfort of the passengers. With time, the system began to reveal another dimension of services to support crew members and airline companies in order to facilitate crew tasks and increase airline revenue. Compared with the traditional terminal design, the new design can greatly reduce the weight of the terminal system and improve the economic efficiency of airline companies. This design makes the passenger the prime focus, and as a digital intelligent system, provides personalized services to end users. The new design integrates big data analysis technology, latest display technology, and electronic technology. Multi-touch panel and joystick are supported, which greatly improve the interaction and operability of the terminal.

IFE system is still in the development phase and under research. Although the development of the IFE system made a great leap in the past years, there are still various issues that need further research. These developments range from enhancing current systems to adding new components and services. As the technology improves, more advanced devices can be used to enhance current components such as increasing network bandwidth, using more accurate contactless sensors, wireless devices, and lighter components. There is no limit for new services that can be added to IFE systems. Using 3D displaying devices can introduce a new sensation to IFE entertainment. Furthermore, hologram images can be used to present safety instructions instead of crew members doing the same.

REFERENCES

Akl, A., T. Gayraud, P. Berthou, *International Journal on Advances in Networks and Services*, **4**, 159 (2011).

Alamdari, F. *Journal of Air Transport Management*, **5**, 203 (1999).

Jiang-Lan, Z., B. Ying-ying, L. Sen, Process Automation Instrumentation, **36**, 6 (2016).

Kang, W. *Communications Technology*, **45**, 7 (2012).

Civil, Architecture and Environmental Engineering – Kao & Sung (Eds)
© 2017 Taylor & Francis Group, ISBN 978-1-138-02985-9

The countdown traffic light error detection method based on union color spaces and fuzzy PCA

Shaoqing Mo, Rui Zhang & Yuexiang Wen
School of Automotion and Transportation, Tianjin University of Technology and Education, Tianjin, China

ABSTRACT: Traffic lights are very important for traffic management and control. If there are errors in traffic lights, managing the traffic becomes difficult. Therefore, it is highly essential to recognize the state of traffic lights on line. This paper proposes a detection method of errors in countdown traffic lights based on color union spaces and fuzzy Principal Component Analysis (PCA). In this method, first, pixels are classified into color pixels and achromatic pixels in RGB color space. Then, the principal color of the color pixels is extracted by histograms in HVS space, and the color feature is recognized based on the H component of the principal color. The color countdown characters are converted into gray according to the principal color, and then the training samples of the countdown characters are divided into subsets in terms of fuzzy degree; then, their corresponding PCA subspaces are constructed. Lastly, the character is recognized in the specific subspaces, which are chosen by the character's fuzzy degree. The experiment results show that our method has a more favorable comprehensive performance than other algorithms, and that it can better meet the demands of precision and real-time processing simultaneously.

1 INTRODUCTION

Traffic lights are highly significant for traffic control and management, and they are indispensable for city roads, which are increasingly becoming congested. Traffic lights not only help vehicles and people pass through intersections from different directions in an orderly fashion, but also ensure higher traffic flow and traffic safety (Jensen, 2016). The countdown on traffic lights displays the residual time of the current phase. Drivers can make decision in advance according to the residual time, which can improve the traffic capacity of a given road and ensure greater traffic safety. Therefore, the countdown traffic lights are becoming increasingly popular in city traffic systems. As traffic lights operate in the exterior environment, they break down easily, and their state is difficult to maintain regularly manually, as they are scattered across the city. When the traffic lights are faulty, the traffic management department is unable to channel the traffic flow timely due to less information, which causes traffic jams and even causes traffic accidents. Therefore, it is necessary to recognize state of the traffic lights automatically.

The main errors in the traffic lights are the following: a) inaccurate color of traffic lights; b) phase color conflict; c) countdown character errors such as incomplete characters display or messy code; d) countdown time conflict. Based on the above-mentioned errors, the core issues of error detection in countdown traffic lights are the color features and countdown character recognition.

The existing traffic lights error detection methods can be classified into two classes: methods based on hardware detection (Kong, 2015; Zou, 2012), and methods based on video recognition (Diaz, 2015; Omachi, 2009; Gomez, 2014; Jie, 2015). The former is based on the measurement of internal current and voltage of traffic lights. Due to difference in internal structures of traffic lights, curves and amplitudes of current and voltage are not similar; therefore, methods based on hardware are not universal. Moreover, the countdown characters cannot be recognized by the amplitudes of the current and voltage. The methods based on video recognition always use the color, shape, and texture. However, these features are difficult to extract due to the change of lighting condition.

This paper proposes a new countdown traffic light error detection method based on union color space and fuzzy Principal Component Analysis (PCA). First, we classify pixels into color pixels and achromatic pixels according to the maximum difference of R, G, and B components in RGB color space. Then, the principal color of color pixels is extracted by histograms in HVS space, and the color feature of traffic light is recognized based on the H component of the principal color. The color countdown characters are converted to gray images according to the principal color, and the fuzzy degree of the gray image is computed by using triangular norm and non-fuzziness cardinality. Then, training samples of the count-down characters are divided into subsets in terms of fuzzy degree, and their corresponding PCA subspaces are constructed. Lastly, the character is recognized in the specific subspaces, which are chosen by the character's fuzzy degree. The experiment's results show that our method has a more favorable comprehensive performance than other algorithms; it can better meet the demands of precision and real-time processing simultaneously.

2 COLOR FEATURE RECOGNITION

The color of traffic lights has a large dynamic range due to reflection and backlight, as shown in Fig. 1. In Fig. 1, a) shows two images for traffic lights on black; b) shows two images for traffic lights on red; c) shows two images for traffic lights on green; and d) shows two images for traffic lights on yellow. The color of left image in each group is distorted heavily and has a halo, and the color of the right image is also different than the standard color. The color feature of red is always represented by 2R – G – B, the green is represented by 2G – R – B, and the yellow is represented by R + G – 2B, in RGB color space. In HSV color space, the color feature is always represented by the H component. Fig. 2 shows histograms of images in Fig. 1; a) shows histograms of 2R – G – B, rep-

resenting the red component; b) shows histograms of 2G – R – B, representing the green component; c) shows histograms of R – G – 2B, representing the yellow component; and d) shows histograms of the H component. In each figure, black, red, green, and yellow curves depict histograms for a), b), c), and d) in Fig. 1 respectively; solid lines are for the left images and the dotted lines are for the right images of a) to d). As shown in Fig. 2, black, red, green, and yellow for traffic lights can't be discriminated in RGB space and HVS space, which results in difficulty of color feature recognition.

By statistical analysis, the reason for no obvious division of colors is interference of achromatic pixels in HVS color space. Therefore, we classify pixels into color pixels and achromatic pixels according to the maximum difference of R, G, and B components in RGB color space, and then recognize the color feature in HVS space for color pixels.

For image *I* in RGB color space, the maximum difference of R, G, and B, denoted by $md(i, j)$ is computed by equation (1):

$$md(i, j) = max(|R - B|, |R - G|, |G - B|) \quad (1)$$

where R, G, and B are the red, green, and blue values of pixel *(i, j)*, respectively.

According to $md(i, j)$, pixels are classified into color and achromatic, as is described in equation (2):

$$C(i, j) = \begin{cases} 1, md(i, j) > T_1 \\ 0, md(i, j) \le T_1 \end{cases} \quad (2)$$

$C(i, j) = 1$ denotes that pixel (i, j) is color; otherwise, pixel *(i, j)* is achromatic, and T_1 is a given threshold.

Then, image *I* is transformed from RGB space to HVS space. The *H* component histogram of color pixels is defined as equation (3):

$$hist(x_i) = \frac{S(x_i)}{\sum C(i, j)}, s.t. H(i.j) = x_i \& C(i, j) = 1 \quad (3)$$

We denote the peak value of histogram as *MH*. The principal color, (Pcolor), is defined as the weighted mean of *H* histograms, whose deviation from *MH* is less than the given threshold T_3:

$$PColor = \sum_{d=-T_3}^{T_3} (MH + d) * hist(MH + d) / \sum_{d=-T_3}^{T_3} hist(MH + d) \quad (4)$$

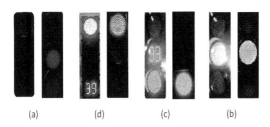

(a)   (d)   (c)   (b)

Figure 1.   Traffic lights under different conditions.

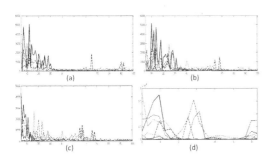

(a)   (b)

(c)   (d)

Figure 2.   Histograms for images of Fig. 1 respectively.

Then, the distances between the principal color and the standard red ($H = 0$), standard yellow ($H = 1$), and standard green ($H = 2.1$) respectively, are computed by equations (5).

$$distR = Pcolor$$
$$distY = |Pcolor - 1| \qquad (5)$$
$$distG = |Pcolor - 2.1|$$

If $minDist = \min\{distR, distY, distG\} < T_4$, then the color of traffic light is the color whose distance is minimum. If $minDist \geq T_4$, then the traffic light is not red, green, and yellow, which is in error.

Fig. 3 shows histograms of color pixels for images in Fig. 1, respectively. When the traffic light is black, there are no color pixels, as shown in Figure 3(a). H component of red centers around 0 or 2π; that of green centers 2.8; and that of yellow is about 1, as shown in Figure 3(b), (c), and (d), respectively. The principal color of same color traffic lights is nearer, and the distance of principal colors for red, green, and yellow is longer. That is to say, the distance of same colors reduces; otherwise, the distance of different colors gets enhanced, which is good for the color feature recognition of traffic lights.

3 COUNTDOWN CHARACTER RECOGNITION

3.1 Fuzzy degree

In order to highlight the characters and reduce the background and noise, the character images are translated into gray image according to equation (6):

$$GI(i,j)$$
$$= \begin{cases} 255 - \dfrac{255*|H(i,j) - Pcolor|}{\max\{(|H(i,j) - Pcolor|)'\}} & C(i,j) = 1 \\ 0 & C(i,j) = 0 \end{cases} \qquad (6)$$

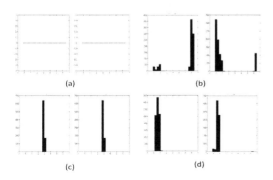

Figure 3. The H histogram of color pixels.

To the gray image GI with K pixels, V_{ij} denotes the $k*k$ neighborhood of pixel (i, j), and its gray feature is defined by equation (7).

$$\mu_G(i,j) = \frac{\left[0V(M(i,j) + m(i,j) - 2GI(i,j))\right]^2}{\left[M(i,j) - m(i,j)\right]^2} \qquad (7)$$

where $M(i, j)$ and $m(i, j)$ are the maximum and minimum gray values, respectively. If $M(i, j) = m(i, j)$, $\mu_G(i, j) = 0$.

The variance feature of V_{ij} is described as

$$\mu_D(i,j) = 1 \wedge \left[01(D(i,j) - D_1)\right] / (D_2 - D_1) \qquad (8)$$

where D_2 and D_1 are the given thresholds, and $D_2 > D_1 > 0$; $D(i, j)$ is gray variance, such as

$$D(i,j) = \sum_{(s,t) \in V^{ij}} \left(GI(s,t) - \overline{GI(i,j)}\right)^2 \qquad (9)$$

where $\overline{GI(i,j)} = \sum_{(s,t) \in V^{ij}} GI(s,t) / k^2$

The local contrast feature $\mu_L(i,j)$ is acquired based on the triangular module operation of $\mu_G(i,j)$ and $\mu_D(i,j)$.

$$\mu_L(i,j) = T(u_G(i,j), \mu_D(i,j)) \qquad (10)$$

where T is triangular module.

Then, the fuzzy degree of image I is

$$FuzzyD = 1 \wedge [lgK - lg \sum Count(L)) / F_0 \qquad (11)$$

where $\sum Count(L) = \sum_{i,j} \mu_L(i,j)$, and F_0 is fuzzy threshold, ranging from 4 to 8.

The fuzzy degree, $FuzzyD$, ranges from 0 to 1; the bigger the value, the more blurred the image.

3.2 PCA classification method

PCA is a statistical method that reduces multiple indicators into a few composite indicators (Abdelmalek, 2007; Oja, 1983; Purnima, 2012). PCA method simplifies data by projecting high dimensional data into low dimensional space with less information loss. According to PCA method, PCA subspace classifier extracts salient features of each class and represents it by a template set. The core of PCA subspace classifier is that each class has its own feature set, and the feature set changes according to the class.

Suppose there is a sample space set containing K subclasses, the number of samples for each subclass is N_i, and the test sample is x. The PCA classification method can be described as follows.

1. Arranging original images into feature vector by rows or columns, and denoting the jth sample vector of the ith class as $\boldsymbol{x}_j^i$; then computing the covariance matrix of subclass.

$$R_i = \frac{1}{N_i} \sum_{j=1}^{N_i} (\boldsymbol{x}_j^i - \boldsymbol{M}_i)(\boldsymbol{x}_j^i - \boldsymbol{M}_i)^T \qquad (12)$$

where $\boldsymbol{M}_i$ is the mean value of the ith class samples. N_i is the number of samples.

2. Computing eigenvectors $\boldsymbol{\mu}_j^{(i)}$ $(j = 1,2,...,p)$ and eigenvalue $\lambda_j^{(i)}$ according to SVD;
3. Arranging eigenvalues in descending order $(\lambda_1^{(i)} \geq \lambda_2^{(i)} \geq ... \geq \lambda_p^{(i)})$, and determining dimension m of subspaces.

$$\sum_{j=1}^{m} \lambda_j^{(i)} / \sum_{j=1}^{p} \lambda_j^{(i)} \geq r \qquad (13)$$

where $0 < r < 1$, and m is the smallest integer of equation (13).

The characteristic matrix $U^{(i)}$ is

$$U^{(i)} = \left(u_1^{(i)}, u_2^{(i)}, ..., u_m^{(i)} \right) \qquad (14)$$

and the subspace L^i is

$$L^i = \left(u_1^{(i)}, u_2^{(i)}, ..., u_m^{(i)} \right) \qquad (15)$$

4. Constructing the project matrix for each subspace;
5. Computing the project residual error distance:

$$g_s(\boldsymbol{x}, L^i) = \|\boldsymbol{x} - \boldsymbol{M}_i\|^2 - \|P_i(\boldsymbol{x} - \boldsymbol{M}_i)\|^2 \qquad (16)$$

The minimum residual error distance $\min D = \min(g_s(\boldsymbol{x}, L^i))$ the class which has $minD$ is $i = \min\{g_s(\boldsymbol{x}, L^i)\}$. If $minD < T_s$, then the character is the ith class; otherwise, the character is error.

3.3 PCA based on fuzzy degree

PCA classifier demands samples of each subclass to have similar features. If samples of each subclass differ from each other, in other words, if the subclass has bigger intra-subclass distances, the feature of the subclass is no longer salient. Then the character matrix derived from PCA method can't describe the subclass sufficiently, which will result in missorting. Due to the change of light and vibration, countdown characters of traffic lights are blurred to some extent, which augments the intra-subclass distance and reduces the inter-subclass distance; as a result, there is high error rate for using PCA classifier directly. In order to get higher recognition rate, the intra-subclass distance must be reduced and the inter-subclass distance must be enhanced. Therefore, we group samples into subsets according to their fuzzy degree. Samples of each subset are more similar to each other, and the intra-subset distance becomes smaller. Then, we build the subspace of each subset and recognize characters in the corresponding subspace according to the character's fuzzy degree. The complete procedure can be stated as follows:

a. Classifying total training samples into three subsets S_1, S_2, and S_3. Ensuring that the testing sample locates to the appropriate subset, the three subsets overlap each other: the fuzzy degree of S_1 is [0, 0.35], that of S_2 is [0.3, 0.65], and that of S_3 is [0.6, 1];
b. Constructing projection matrix $\{P_i\}^s$ for every character of each subset, where $i = 1,2,...,K$, denoting the indices of characters, $s = 1,2,3$, denoting the indices of subsets;
c. Selecting appropriate $\{P_i\}^s$ for recognition according to the $FuzzyD$ of the testing sample. The selection rule is

If $FuzzyD < 0.325$, select $\{P_i\}^1$;
If $0.325 \leq FuzzyD < 0.625$, select $\{P_i\}^2$
If $0.625 \leq FuzzyD$, select $\{P_i\}^3$

4 EXPERIMENTS

To evaluate the performance of our method, we tested on six video sequences acquired under difference conditions. Video 1 was acquired in the morning with strong reflection; video 2 was acquired at noon with normal light; video 3 was acquired at afternoon with backlight; video 4 was acquired at night; video 5 was acquired in rain; and video 6 was acquired in fog day. Total 54000 countdown characters ranged from 99 to 0, half of the characters were training samples, and the remaining were testing samples. The color recognition result is shown in Table 1, and the countdown character recognition result is shown in Table 2.

Table 1. Experiment results of color recognition.

Video	Recognition rate		
	Our method	HVS method	RGB method
Video 1	0.92	0.85	0.83
Video 2	0.94	0.93	0.92
Video 3	0.93	0.87	0.87
Video 4	0.93	0.89	0.88
Video 5	0.92	0.90	0.89
Video 6	0.93	0.89	0.90

Table 2. Experiment results of countdown character recognition.

Method	Recognition rate	Recognition time (ms)
Our method	0.94	4.2
Traditional PCA method	0.87	3.8

According to Table 1, our method outperforms the HVS method and RGB method in general. Under normal lighting condition, our method has recognition rate similar to the HVS method and the RGB method; however, when the illumination condition is not ideal, such as reflection, backlight, rainy, and foggy, and so on, the color of traffic lights distorts to some degree, and our method has higher recognition rate than the other methods.

According to Table 2, our method outperforms the traditional PCA method as well. The reason is that the intra-subclass distance decreases and the inter-subclass distance gets enhanced after classifying the samples into subsets according to fuzzy degree.

5 RESULTS

This paper presented a robust detection method for countdown traffic light errors based on union color spaces and fuzzy PCA classification. Based on the union color spaces of RGB and HSV, our method eliminates the effect of color tone shifting and halo disturbances and exhibits excellent performance in color recognition. Classifying countdown characters into different subsets according to the image quality enhances the intra-similarity and increases the inter-difference, and the character recognition rate is improved. The experiment results of six video sequences with different time and conditions show that our method has a favorable comprehensive performance, which can better meet the demands of precision and real-time processing simultaneously.

ACKNOWLEDGEMENTS

This work was supported by TUTE (XJKC031310) and NSFC (61503284).

REFERENCES

Abdelmalek Zidouri. PCA-based Arabic Character feature extraction, *in Proc. 9th ISSPA*, 1–4 (2007).

Diaz, M., P. Cerri, G. Pirlo, M. Ferrer, and D. Impedovo, A survey on traffic light detection," in *Proc. New Trends ICIAP Workshops*. 201–208 (2015).

Gomez, A.E. F.A.R. Alencar; P.V. Prado. Traffic lights detection and state estimation using Hidden Markov Models. *in Proc. IEEE Intell. Veh. Sym.*, 750–755 (2014).

Jensen, M., M. P. philipsen, A. Mogelmose, et al. vision for looking at traffic lights: issues, survey, and perspectives. IEEE Trans. Intell. Transp. Syst. **17(7)**: 1800–1815 (2016).

Jie, Y., T. Jin, L. Lei. Traffic light detection based on similar shape searching for visually impaired person. *in Proc. 6th int. conf. intell. con. inf.*, 376–380 (2015).

Kong, L. X. Xu, Z. Shi. Design and implementation of traffic lights fault-detection system. measurement & control technology, **4**, 51–54 (2015).

Oja E. Subspace methods of pattern recognition. England. *Research Studies Press* (1983).

Omachi, M., S. Omachi. Traffic light detection with color and edge information. *in Proc. IEEE 2nd Int. Conf. Comp. Sci. Inf. Tech.*, 284–287 (2009).

Purnima Kumari Sharma; Mondira Deori; Balbindar Kaur; Chandralekha Dey; Karen Das Radon Transform and PCA based feature extraction to design an Assamese Character Recognition system, *in Proc. 3rd NCETACS*, 46–51 (2012).

Zou, X., J. Bao, Q.Hu, S. Sun. An fault-monitoring design for system of traffic signal lights. Computer measurement & control. 8, 2024–2027 (2012).

Civil, Architecture and Environmental Engineering – Kao & Sung (Eds)
© 2017 Taylor & Francis Group, ISBN 978-1-138-02985-9

Study of the optimization evaluation algorithm of improved genetic-BP neural network in the monitoring and diagnosis of shortwave transmitting system

Yong Luo & Jun Gao
Naval University of Engineering, Wuhan, China

ABSTRACT: In the process of monitoring the running state of shortwave transmitting system, it is necessary to realize the intelligent discrimination and analysis of the monitoring data and enable users to obtain the real-time health status of the system quickly and accurately. Considering the large amount of information in monitoring data and the highly difficult analysis, the initial structure of BP neural network is optimized with genetic algorithm based on the calculation evaluation with traditional BP neural network algorithm, and then the optimal initial weight value is obtained by the optimizations such as improvement of chromosome coding, selection of fitness function and design of related operators. Based on the simulation training of a large number of monitoring data, the improvement effect of the algorithm is experimentally verified and analyzed. After this improvement, the algorithm optimization evaluation in the monitoring and diagnosis of shortwave transmitting system gets strong convergence ability and high diagnostic accuracy, which greatly increases the use efficiency of the monitoring system and lays a good practical foundation.

1 INTRODUCTION

Today, with the deepening of artificial intelligence research, various intelligent algorithms have been widely used in various fields, leading to a series of revolutionary technological and scientific research innovations. It has no exception in the intelligent monitoring research of shortwave transmitter. Compared with the traditional manual monitoring disposal, the automatic monitoring control and intelligent state discrimination and disposal also need to be introduced. Artificial intelligence is a very broad field, involving a lot of technical concepts and research directions. One important aspect is the artificial neural network. Artificial neural network can well solve the problem of pattern recognition which is difficult to be distinguished by traditional means because of its own features such as parallelism, self-learning, self-organization and associative memory. Among the monitoring technologies based on neural network, BP neural network is a widely used model. In the intelligent monitoring of the state of shortwave transmitter, due to many types of equipment state, unapparent information difference and so on, relying solely on BP algorithm may cause problems, such as slow convergence rate of discrimination model, easy to fall into local extremum. Therefore, it is necessary to optimize and adjust BP neural network to a certain degree. In this paper,

the genetic algorithm with excellent global search ability is used to optimize and improve the BP neural network model, and it can realize the quick discrimination and accurate diagnosis of the transmitter state better.

2 TRADITIONAL BP CALCULATION EVALUATION METHOD

As shown in Figure 1, the classic BP neural network consists of three parts, namely input layer, hidden layer and output layer. The number of hidden layers is adjusted according to the need of

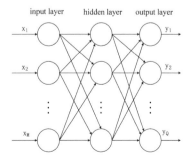

Figure 1. Typical structure diagram of BP neural network.

actual calculation. The number of nodes in each layer of the network structure is set according to the types of input and output data as well as related requirements of network construction, and the number of nodes on the network is the number of neurons. Data enters into the network in the input layer and moves forward along with the arrow direction, then moves forward after calculated at each node, and finally the result is output on the output layer.

In general, BP neural network on the third layer is common. Assuming the number of neurons on each layer is respectively M, N and Q, $x_j (1 \leq j \leq M)$ is the input of the j-th node on the input layer, $o_k (1 \leq k \leq Q)$ is the output of the k-th node on the output layer, and y_k is the desired output of the k-th node on the output layer. The threshold value of the i-th node on the hidden layer is $\theta_i (1 \leq i \leq N)$, the threshold value of the k-th node on the output layer is a_k; w_{ij} is the weight from the j-th node on the input layer to i-th node on the hidden layer; v_{ki} is the weight from i-th node on the hidden layer to the k-th node on the output layer; g(x) and f(x) are respectively the transfer functions of the hidden layer and output layer.

The input and output of the i-th node on the hidden layer are:

$$s_i = \sum_{j=1}^{M} w_{ij} x_j + \theta_i \qquad (1)$$

$$b_i = g(s_i) = g\left(\sum_{j=1}^{M} w_{ij} x_j + \theta_i \right) \qquad (2)$$

Similarly, the input and output of the k-th node on the output layer are:

$$t_k = \sum_{i=1}^{N} v_{ki} b_i + a_k$$
$$= \sum_{i=1}^{N} v_{ki} g\left(\sum_{j=1}^{M} w_{ij} x_j + \theta_i \right) + a_k \qquad (3)$$

$$o_k = f(t_k) = f\left[\sum_{i=1}^{N} v_{ki} g\left(\sum_{j=1}^{M} w_{ij} x_j + \theta_i \right) + a_k \right] \qquad (4)$$

Then for each sample p, the error function between the actual output value o_k and the ideal expected value y_k of the neural network is:

$$E_p = \frac{1}{2} \sum_{k=1}^{Q} (y_k - o_k)^2 \qquad (5)$$

The total error function for P training samples is:

$$E = \frac{1}{2} \sum_{p=1}^{P} \sum_{k=1}^{Q} (y_k^p - o_k^p)^2 \qquad (6)$$

The essence of the BP neural network algorithm is taking the total error function as the objective function. By adjusting the weights and thresholds of each layer with the gradient descent method, the final modified network output can approach the expected value, so as to find the minimum error. In the initial stage of training neural network, the actual output value is largely deviated from the expected value, and the relative value of the error function is large, with a larger space to adjust down. Thus, a substantial adjustment of the error function will increase the convergence rate of the network; as the number of training samples and the number of training times increase gradually, the output value gradually approaches the expected value, the relative value of the error function continues to decrease and the decline space will continue to decrease. At this time, the microadjustment of the error function will slow down the convergence rate of the network. Therefore, in the practical application, the convergence rate of traditional BP algorithm in the later period of training will be slow. As the optimization is gradually carried out in the process of gradual decline of the error function, when the surface space of the error function is multi-dimensional, it is possible to make the minimum error found in training a local minimum rather than the global optimum.

3 FIGURES AND TABLES OPTIMIZATION AND IMPROVEMENT OF BP NEURAL NETWORK BASED ON GENETIC ALGORITHM

As the genetic algorithm is based on group optimization, the global research from multiple samples can effectively avoid the search being trapped into the local optimal solution, thus better finding the global minimum error or the good suboptimal solution. Therefore, in order to improve the accuracy of BP network algorithm and the accuracy of state recognition and to overcome the drawback of easily falling into local minimum, the genetic algorithm is used to optimize the structure parameters of neural network. For the BP neural network, the essence of the algorithm is an optimization problem of the complex function. Its weight distribution and size contain the network's all ideas. Therefore, the genetic algorithm can be used to optimize the weight of neural network, so as to better improve the overall performance of the system. Its basic structure thought is to divide the optimization of neural network structure into two steps:

1. Using genetic algorithm to achieve fast global optimization, namely building the structure parameters of the neural network as the chromosome of genetic algorithm, using genetic and mutation recombination for global optimization of weights and thresholds of the network, and quickly converging the network structure to the global optimal solution or its vicinity through fitness screening;
2. Taking the optimal network structure found by the genetic algorithm as the starting point of BP algorithm studying, using BP algorithm for further local search, and again optimizing the network weights and thresholds. In this way, the network structure can be quickly adjusted to the optimal solution or satisfactory solution.

Here we mainly deal with the following three problems: chromosome coding improvement, fitness function selection and related operator design. The genetic algorithm is used to improve the structure of BP neural network and improve the convergence effect of the algorithm.

3.1 Chromosome coding design

The key step of genetic algorithm optimization is to reconstruct the chromosomes in the group by genetic variation, and gradually search the individuals with good fitness in the group, so as to achieve the goal of finding the optimal solution by gradually increasing the number through mutation. In order to achieve reasonable genetic variation, it is necessary to encode the parameters in the problem space and convert the solutions in the space that can be processed by the genetic algorithm for optimization. At present, the typical encoding method is binary coding. However, in dealing with the problem of large data volume, binary coding will increase the length of the chromosome and increase the search space, resulting in lower search efficiency. As the overall performance of BP network is closely related to the weights and thresholds in network structure, the method of floating-point coding is used in this paper. BP neural network adopts three-layer structure. The parameters to be optimized are all weights and thresholds, and the number is $(L+1)M+(M+1)N$. All the parameters constitute a chromosome, and the sorting order of genes in the chromosome is from input layer to hidden layer weight, hidden layer to output layer weight, hidden layer threshold, output layer threshold. The mapping relationship of code is:

$$X = \{w_{11}, w_{12}, \ldots, w_{NM}, v_{11}, v_{12}, \ldots, v_{QN}, \quad (7)$$
$$\theta_1, \theta_2, \ldots, \theta_N, a_1, a_2, \ldots, a_Q\}$$

Thus, each individual constituted corresponds to the parameter on the network structure. It can be considered that the characteristics of the individual represent the structural characteristics of the neural network.

3.2 Construction of fitness function

The fitness function is also called as the evaluation function, which is often used in the genetic algorithm to measure the degree of excellence that the individual can reach or approach the optimal solution in the optimization calculation. The value generated by the fitness function can also be considered as the "scoring" or "judgment" of the individual and it is a criterion to distinguish between good and bad individuals in a group. The convergence rate of genetic algorithm and the choice of optimal solution depend largely on the selection of fitness function. The design principles are non-negativity, continuity, uniqueness, rationality and simplicity. The key is rationality and simplicity, which requires that the fitness function should really reflect the advantages and disadvantages of the corresponding solution, and the calculation process should be as simple as possible to reduce the time and complexity of calculation as well as the cost of calculation. It is often difficult to reach these conditions. In this paper, the fitness function is chosen as the reciprocal of the mean square error function. When the individual error is larger, the corresponding fitness is smaller. When the error is smaller, the corresponding fitness is larger.

$$f(x_p) = \frac{1}{1+E_p} = \frac{1}{1+\frac{1}{2}\sum_{k=1}^{Q}(y_k - o_k)^2} \quad (8)$$

3.3 Selection Algorithm

According to the basic genetic concept, it is necessary to select the superior and eliminate the inferior individuals. This process is achieved by selection algorithm in calculation. The fitness of the individual is taken as a genetic standard. If the fitness is high, the probability of being inherited to the next generation of group is large. In this way, the individuals with higher fitness will be retained as far as possible, so that the fitness of individuals in the group constantly approaches the optimal solution. A typical selection algorithm is the roulette selection algorithm. In order to better avoid the loss of useful genetic information, improve the global convergence and computational efficiency, the best reservation selection algorithm is used as a supplement of the typical roulette selection algorithm.

1467

In the typical roulette selection algorithm, the probability p_c of the individual being passed on to the next generation can be calculated by the Equation (9):

$$p_c = f\left(x_p\right) / \sum_{p=1}^{P} f\left(x_p\right) \qquad (9)$$

All the individuals in the group are placed on a disc, and the circular area is divided into the corresponding fan-shaped areas according to the number of individuals. Each area represents an individual. The size of the area is proportional to the individual fitness. When the disk is rotated at random, it is clear that the larger the fan-shaped area is, the larger the probability of the pointer stopping in this area is, namely, the larger the probability of individual selected corresponding to the fan-shaped area is. However, due to limited size of the group and the randomness of operation, there may be some discrepancies between the actual number of individuals selected and the expectation. Therefore, the selection error of this original roulette algorithm is relatively large, or even some optimal solutions may be lost. In order to avoid this omission, the best reservation method is adopted, that is, executing the selection operation of genetic algorithm with the roulette selection method, and then retaining the individuals with the highest fitness in the current group to the next generation, thus ensuring that the iterative termination results are the individuals with the highest fitness in the past generations.

3.4 Cross Algorithm

The inheritance and evolution of natural organisms are mainly achieved through mating and reconstructing. The genetic algorithm uses cross for imitation. Two individuals are selected from the group based on a certain probability, and new genetic individuals are generated by exchanging part of genes. This exchange method is called as cross algorithm. In general, there is single point, multiple points, order, cycle cross and so on. When encoding chromosomes, the sort order of genes is from input layer to hidden layer weight, from hidden layer to output layer weight, hidden layer threshold, output layer threshold. Different kinds of neural network structure parameters play different roles in calculation. Therefore, when considering crossing genes, it is necessary to distinguish them. In this paper, for parameters of neural network, the multi-point cross method is used, as shown in Figure 2:

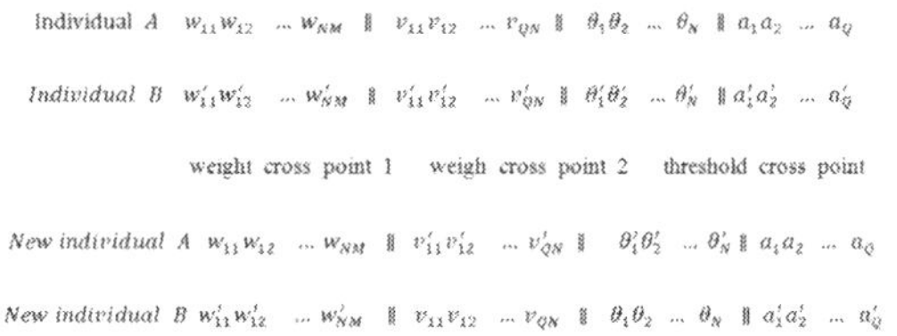

Figure 2. Structure diagram of multi-point cross method.

3.5 Mutation Algorithm

Mutation is also a way to generate new individuals. In genetic algorithm, the method of replacing some gene values in the individual chromosome coding by other values is used to achieve mutation simulation. In the process of inheritance, the mutation is usually randomly generated. Compared with selection and crossover operation, the individual's local mutation, as the supplement of the new individual method, can avoid the loss of some information and ensure the effectiveness of the genetic process. Genetic algorithm achieves a global search by the crossover operation and improves local random search ability by mutation operation. In this paper, the method of uniform mutation is used, namely setting the gene locus in the chromosome coding as the mutation point, and randomly selecting a numerical value to replace the original gene value in the value range of the gene corresponding to the mutation probability p_m, thus generating new individuals at a lower probability.

4 MONITORING DATA SIMULATION AND COMPARISON

For shortwave transmitter, the power status data is a very important technical indicator, directly affecting the distance and reliability of communication. At present, the data that can be directly monitored and analyzed in the working process of transmitter are the transmitter's input power, output RF power, antenna transmission forward power and reverse power. In the working process of the transmitter, the transmitter mainly experiences four states, namely waiting for transmission, transmission beginning, transmission and transmission ending. Transmission beginning and transmission ending are two dynamic processes, and the change of monitored power data is big. The role of the neural network is to classify and identify the monitoring data, and to correctly identify the specific working status of the transmitter.

Table 1. Typical input and output samples.

| Number | Sample output after normalization processing | | | | | | | | | | Sample output |
	T1	T2	T3	T4	T5	T6	T7	T8	T9	T10	
1	0			0.158		0.667		0.932		0.988	1
	0.935			0.961		0.935		0.961		0.935	
2	0.854			0.854		0.879		0.854		0.888	2
	0.854			0.879		0.859		0.879		0.854	
3	0.967			0.967		0.948		0.967		0.963	3
	0.965			0.969		0.666		0.145		0	
4	0			0.146		0.677		0.879		0.683	7
	0.686			0.705		0.686		0.705		0.683	
5	0.447			0.591		0.879		0.889		0.695	8
	0.641			0.553		0.472		0.563		0.528	
6	0.699			0.688		0.686		0.688		0.670	9
	0.591			0.594		0.578		0.145		0	

The main faults of transmitter include abnormal input power, abnormal out RF power, and abnormal antenna transmission forward power and reverse power. To verify the recognition performance and convergence of BP neural network improved by genetic algorithm, two models of conventional BP network and improved BP network are established. For the different states of the transmitter in the working process, 5000sets of monitoring data are sampled to constitute the training sample. The typical training sample is shown as follows:

The correspondence between the sample output of the neural network and the fault state of the transmitter is shown in the following table.

The convergence curves of two BP algorithms are shown in Figures 3, 4.

10 sets of test data are randomly selected as the test samples, as shown in Table 3. The results obtained from the simulation of two kinds of neural are shown in Table 4.

From the simulation results it can be seen that:

1. The convergence rates of traditional BP network and genetic BP network are relatively fast in the early stage of training and learning, with the same effects. However, the difference is obvious in later period. The traditional BP network converges very slowly in the later period of training, difficult to meet the requirements after long time of training; while the genetic BP network converges still relatively fast in the later period of training, able to meet the training requirements faster.

2. After early optimization of improved BP neural network by genetic algorithm, the initial network structure finds the optimal solution. After later training, the recognition and analysis of

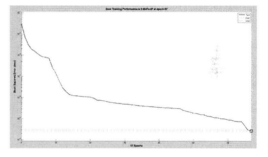

Figure 3. Convergence process curve of traditional BP network.

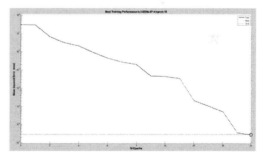

Figure 4. Convergence process curve of improved BP network.

monitoring data can be accurately and comprehensively realized. Therefore, the agreement of recognition and analysis of test samples by genetic BP network with the real result is obviously higher than by traditional BP network, and the recognition accuracy is greatly improved.

1469

Table 2. Correspondence between the sample output and the fault state of the transmitter.

Sample output	State description	Fault analysis
1, 4	When transmission starts, the RF power output is normal, the power input is normal, and the transmission system is working properly	Null
2, 5	During the transmission, the RF power output is normal, the power input is normal, and the transmission system is working properly	Null
3, 6	When transmission ends, the RF power output is normal, the power input is normal, and the transmission system is working properly	Null
7, 10	When transmission starts, the RF power output is abnormal, the power input is abnormal, and the transmission system is working improperly	Transmitter fault
7, 4	When transmission starts, the RF power output is abnormal, the power input is normal, and the transmission system is working improperly	Antenna fault
8, 11	When transmission is being, the RF power output is abnormal, the power input is abnormal, and the transmission system is working improperly	Transmitter fault
8, 5	During the transmission, the RF power output is abnormal, the power input is normal, and the transmission system is working improperly	Antenna fault
9, 12	When transmission ends, the RF power output is abnormal, the power input is abnormal, and the transmission system is working improperly	Transmitter fault
9, 6	When transmission ends, the RF power output is abnormal, the power input is normal, and the transmission system is working improperly	Transmitter fault

Table 3. Randomly selected test samples.

Number	\multicolumn{10}{Sample output after normalization processing}										Sample output
	T1	T2	T3	T4	T5	T6	T7	T8	T9	T10	
1	0	0.913	0.238	0.915	0.780	0.913	0.965	0.962	0.913	0.913	1
2	0.922	0.915	0.922	0.922	0.937	0.922	0.922	0.905	0.922	0.922	2
3	0.967	0.969	0.967	0.666	0.948	0.145	0.967	0.963	0	0.965	3
4	0	0.705	0.146	0.686	0.677	0.705	0.879	0.683	0.683	0.686	7
5	0.447	0.553	0.591	0.472	0.879	0.563	0.889	0.695	0.528	0.641	8
6	0.699	0.594	0.688	0.578	0.686	0.145	0.688	0.670	0	0.591	9

5 CONCLUSION

For the difficulties such as large amount and difficult analysis of monitoring data of shortwave transmitting system, the initial structure of the BP neural network is optimized with genetic algorithm based on the calculation evaluation with traditional BP neural network algorithm, and the simulation experiment is applied to verify the results. By genetic algorithm, we can realize the global optimization in the solution space, quickly find a more optimized network structure, quickly converge to the vicinity of the global optimal solution, and then we use BP algorithm for local calculation,

Table 4. Summary of simulation results.

Number	Output of traditional BP network	Output of genetic BP network	Actual state	State description
1	1.0210	1.0107	1	It works normally when transmission starts
2	2.0045	2.0042	2	It works normally during the transmission
3	3.0401	2.9989	3	It works normally when transmission ends
4	6.9875	7.0035	7	It works abnormally when transmission starts
5	7.9986	8.0010	8	It works abnormally during the transmission
6	8.9797	8.9968	9	It works abnormally when transmission ends

effectively avoiding traditional BP network from being easily trapped in the problem of local optimum, and effectively improving the training effect and accuracy of neural network. After the BP network model improved by genetic algorithm is effectively used in the monitoring of shortwave transmitter, the accuracy of optimized evaluation algorithm becomes higher and the effect of intelligent evaluation and judgment becomes more ideal.

REFERENCES

Cabanas, M.F.; Pedrayes, F.; Melero, M.G.; Rojas, C.H.; Orcajo, G.A.; Cano, J.M.; Norniella, J.G.: Insulation fault diagnosis inhigh voltage power transformers by means of leakage flux analysis. Prog. Electromagn. Res. 114, 211–234 (2011).

Chaturvedi; D.K.; Kumar, R.; Mohan, M.; Kalra, P.K.: Artificial Neural Network learning using improved Genetic algorithm. J. IE (I), EL 82 (2001).

Gill, J.; Singh, B.; Singh, S. (2010) Training back propagation neural networks with genetic algorithm for weather forecasting. Proceeding of 8th IEEE International Symposium on Intelligent Systems and Informatics. 465–469, Ludhiana, India, September 10–11.

Ning Lu; Jianzhong Zhou;, Particle Swarm Optimization-Based RBF Neural Network Load Forecasting Model, Power and Energy Engineering Conference, APPEEC 2009, 27–31 March 2009. Asia-Pacific, pp. 1–4.

Sudhansu Kumar Mishra; Ganapati Panda and SukadevMeher; RitanjaliMajhi, "Comparative Performance Study of Multiobjective Algorithms for Financial Portfolio Design" International Journal of Computational Vision and Robotics, Inderscience publisher. Vol. 1, No. 2, pp. 236–247, 2010.

Yin F.; Mao H.J.; Hua L.: A hybrid of back propagation neural network and genetic algorithm for optimization of injection molding process parameters. Mater. Des. 32(6), 3457–3464 (2011).

Zhang, Y.; Ding, X.; Liu, Y.; Griffin, P.J.: An artificial neural networkapproach to transformer fault diagnosis. IEEE Trans. Power Deliv. 11(4), 183–184 (1996).

Civil, Architecture and Environmental Engineering – Kao & Sung (Eds)
© 2017 Taylor & Francis Group, ISBN 978-1-138-02985-9

Loyalty model in the networked environment based on catastrophe

Qiaoge Liu
Agricultural Bank of China, Beijing, P.R. China

Chenguang Yang
Chinese Electronic Equipment System Corporation Institute, Beijing, P.R. China

ABSTRACT: Loyalty is an important index to evaluate a customer. Understanding the dynamic process of loyalty and its impact factors is very useful for service system. On the basis of the catastrophe theory, a loyalty model influenced by satisfaction and switching cost is studied. In particular, we analyzed the loyalty model with word-of-mouth.

1 INTRODUCTION

A service system is very complex as it includes many different roles, relations, and interactions. Customers and service providers are two vital parts in a service system. A service provider always wants to hold a large amount of loyal customers and then bring the maximal profit to them. In this paper, we will focus on modeling the relationship of customers and service providers from the viewpoint of customer loyalty.

Customer loyalty and his repeated purchase intention are closely related (Wang, 2005). A loyal customer does not likely change his choice because of once dissatisfaction with the service. A satisfied customer does not mean he will choose the service again. Therefore, loyalty is regarded as an important goal in the service system, and it has attracted numerous studies on the principle of loyalty (Yang, 2004). Although the phenomenon of positive influence of service satisfaction to loyalty is well known and many conceptual models have been proposed, how satisfaction affects loyalty is not clear, especially in the modern society with popular networks.

Oliva (1992) used the catastrophe model to describe the loyalty model. This model showed how satisfaction degree and switching cost could influence loyalty.

On the basis of the basic model in Oliva (1992), we will combine the analysis of economics, psychology, sociology, catastrophe, complex network, and control theory to evaluate loyalty. By modeling the satisfaction degree, we will show how satisfaction and thus loyalty is affected in the networked environment. Because of the interaction between customers, loyalty will not change in a large range. Judging a person's loyalty is not limited in his own behavior, but largely affected by his surroundings.

This paper is organized as follows. Loyalty model based on the catastrophe theory is presented in Section 2. Analysis of loyalty in the networked environment by modeling satisfaction is presented in Section 3. Section 4 presents some simulations results showing the influence of network on loyalty. Conclusion is given in Section 5.

2 LOYALTY MODEL BASED ON THE CATASTROPHE THEORY

2.1 Overview of the catastrophe theory

Issues surrounding nonlinearities, discontinuities, and multivaluedness have presented difficult problems for researchers. The catastrophe theory is such a technique for estimating a class of nonlinear dynamic systems.

The catastrophe theory was proposed by Thom 4 in 1975 and developed by Zeeman 5. It describes how small and continuous changes in independent variables can have a sudden, discontinuous effect on a dependent variable. The elementary catastrophe theory focuses on seven models and the cusp catastrophe model is widely used. The cusp catastrophe model is (1) given by:

$$z^3 + yz - x = 0 \qquad (1)$$

where z represents the dependent variable, and the independent variables x and y are control variables. The surface defined by (1) is shown in Figure 1.

Variable x exhibits a right and left movement, y shows a back and front movement, and z exhibits a vertical movement. Changing the control variables x and y can affect the dynamic of z. If y is large, smooth change of z occurs and is proportional to

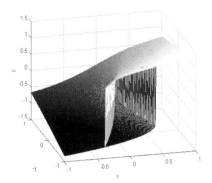

Figure 1.   Cusp catastrophe model.

that of x. If y is small, change of x will produce only a relatively small change of z until x reached a threshold value. Then, z will have a sudden discontinuous shift. When x is increasing and z is at the bottom of the surface, z will not increase quickly and will keep moving in the bottom but not in the upper part until x is large enough. When x is decreasing and z is at the top, z will stay at the top with only a slight decrease. Until x drops low enough, z will fall suddenly to the bottom and reduce as x does. Thus, the dynamics of z is controlled by x and y. z cannot change fluently, and there are some values that it cannot approach when y is large.

The catastrophe theory has been applied in many studies, such as physics, biology, and finance, social sciences, work, and organization behavior. For example, the cusp catastrophe model has been applied to explain not only job turnover of employees but also loyalty in services [3].

2.2   Loyalty model of catastrophe

The loyalty model based on the cusp catastrophe model was defined as (1). z, x, and $-y$ represent customer loyalty, customer satisfaction degree, and switching cost, respectively.

Customer satisfaction has always been treated as a determinant component in many customer retention programs. However, recent studies have found that although satisfaction is a necessary component of loyal or secure customers, the mere quality of being satisfied does not necessarily make customers loyal [1]. Sudden and discontinuous shifts are very familiar phenomena in the reality. For example, a customer's emotional response may change from one side to the opposite side without any intermediate process. When a sudden shift occurs, it is difficult to return to the former side. The relationship between customer satisfaction and customer loyalty is stronger for customers with high switching costs than for those with low switching cost.

3   LOYALTY IN THE NETWORKED ENVIRONMENT

There are always many customers focusing on or buying one service from one service provider. The customers will communicate information with each other by many ways, such as face to face, website, or news. Especially with the popularization of the Internet, it is much easier for a customer to collect information widely. Any evaluation about one service or some service provider is easily searched or spread. Thus, opinion of a customer is not only that of him, but also influenced by others. Then, to manage the satisfaction and loyalty of customers is difficult than before.

This section will model satisfaction degree and analyze how the interaction influences a customer's loyalty.

3.1   Word-of-mouth

The interactions between customers in a networked environment have been studied a lot from different aspects. [67]

Word-of-Mouth (WOM) is a pervasive and intriguing phenomenon. Both satisfied and dissatisfied consumers tend to spread positive and negative WOM, respectively, regarding the services they use. WOM spreads through the customer network. Nowadays, massive quantities of data on large social networks are available from blogs, social networking sites, newsgroups, chat rooms, and so on. Therefore, the size of the customer network is enlarged dramatically. Most of the customers can easily know others' opinion.

WOM has been studied using complex theory widely [8]. In this paper, we choose a simple decision function for each node in the customer network:

$$y_i(k+1) = w_i \cdot \left(\sum_{j=1}^{N_i} y_j(k) \right) \Big/ N_i + (1 - w_i) \cdot y_i(k) \quad (2)$$

where $y_i(k)$ represents the satisfaction degree of i th node at time k, N_i is the number of nodes connecting with node i, $\left(\sum_{j=1}^{N_i} y_j(k) \right) \Big/ N_i$ denotes the influence of conjoint nodes, and w_i shows the characteristic of each node. Some customers may only rely on others' opinions ($w_i = 1$), some customers may not be influenced by others ($w_i = 0$), some customer may consider others' opinions ($0 < w_i < 1$), and so on.

1474

3.2 Satisfaction degree

As defined in 9, customer satisfaction degree can be explained as the difference between the level of service performance expected and perceived by the customer. The perceived service performance is positive relative to the service quality provided by the service provider. The expected service is influenced by word-of-mouth, personal needs, and past experience, as proposed by 10 and popularly recognized by most of the researchers. Therefore, on the basis of the definition in 9, its quantitative description is given by:

$$Sd = S_P - S_E \qquad (3)$$

where S_E is the expected service, S_P is the perceived service, and Sd is the satisfaction degree.

Expected Service (S_E) is influenced by past experience (P_E), word-of-mouth (W_{OM}), and personal need (P_N). Suppose the function is (4):

$$S_E = f_{es}\left(P_E, P_N, W_{OM}\right) \qquad (4)$$

Both expected service and perceived service are the psychological results of the customer. Considering that service degree is the difference between them, we propose a quantitative method to calculate satisfaction degree, on the basis of the thought of control system framework. That is, the satisfaction degree can be thought as the degree of personal need being satisfied, which is influenced by past experience and word-of-mouth. Satisfaction degree is an accumulation variable 9, so the influence of past experience can be viewed as an integral part. The integration result is averaged and influenced by WOM.

According to the satisfaction degree, the service provider will make improvement and provide service to satisfy the customer's need. Then, a service cycle is presented in such control framework.

The control framework of service providing is shown in Figure 2. The service provider makes improvement according to the satisfaction degree and provides service to satisfy the customer's need.

3.3 Switching cost

Switching cost is the cost involved in changing from one service to another. It is a key factor in service systems, which always plays an important role when customers want to change from one service provider to another. As a lock-in method, switching cost can help prevent customers from switching. Therefore, it is very important for service providers to manage switching cost, especially in the competitive environment.

How to evaluate switching cost is difficult which has been shown in many articles. First, switching cost is the collective result of economic, psychological, and physical analyses. Second, different types of service have different features. We have done some research on modeling switching cost mathematically in other papers. Here, for simplicity, switching cost is supposed not to change a lot.

4 SIMULATION RESULTS

We choose a random network as the customer network and three types of nodes mentioned above. The data are randomly generated and scaled to zero mean-centered value. Figures 3–10 respectively show the dynamic influenced from WOM or not.

Figures 3–6 show curves that are not influenced from WOM.

Figure 3 shows the dynamic curves of average satisfaction degree and average loyalty. Figure 4 is the 3D dot plot of the catastrophe model and the relationship curves of loyalty, satisfaction degree, and switching cost. Because of limited line or dot types, we use only one type of lines or dots to represent the traces of all the customers.

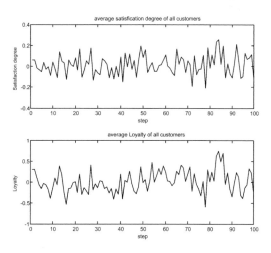

Figure 3. Average of loyalty and satisfaction degree without influence from WOM.

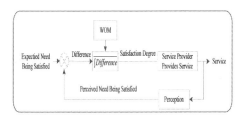

Figure 2. Control framework of service providing.

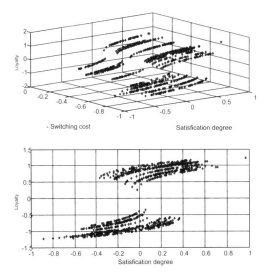

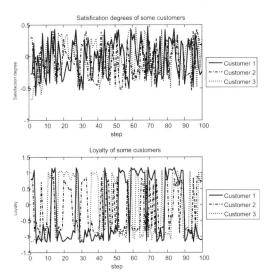

Figure 4. Relation of loyalty, satisfaction degree, and switching cost without influence from WOM.

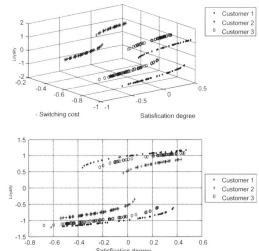

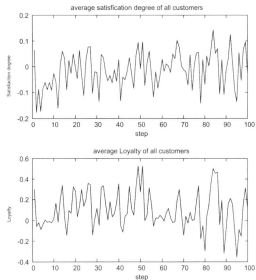

Figure 6. Relation of loyalty, satisfaction degree, and switching cost without influence from WOM—three types of customers.

Figure 5. Dynamic of loyalty and satisfaction degree without influence from WOM—three types of customers.

Figure 7. Average of loyalty and satisfaction degree with influence from WOM.

Figure 5 shows the loyalty and satisfaction degree of three customers. Each of them represent a different type of node, and are marked with three types of lines. The solid line represents "Customer 1" who will consider others' opinions ($0 < w_i < 1$). The dash–dot line denotes "Customer 2" who relies only on others' opinions ($w_i = 1$). The dotted line denotes "Customer 3" who is not influenced by others ($w_i = 0$).

Figure 6 shows the relationship curves of loyalty, satisfaction degree, and switching cost of these three types of customers. Different dots

represent a different type of customer. The dots marked with "." are "Customer 1" who are not influenced by others. The dots marked with "+" are "Customer 2" who will consider others' opinions. The dots marked with "o" are "Customer 3" who relies only on others' opinions.

Figures 7–10 are curves influenced from WOM.
Figure 7 is the dynamic curves of average satisfaction degree and average loyalty. Figure 8 is the

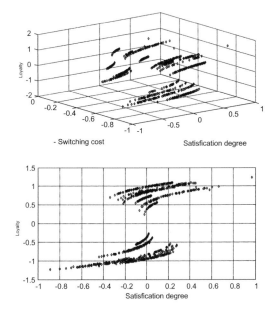

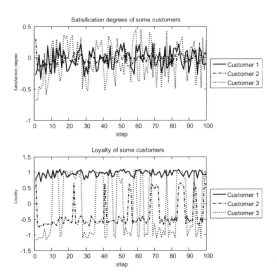

Figure 8. Relationship between loyalty, satisfaction degree, and switching cost with influence from WOM.

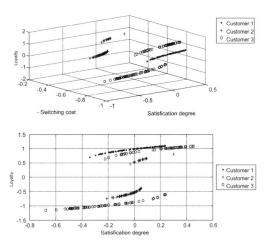

Figure 10. Relationship between loyalty, satisfaction degree, and switching cost with influence from WOM—three types of customers.

of lines. The solid line represents "Customer 1" who will consider others' opinions ($0 < w_i < 1$). The dash–dot line denotes "Customer 2" who only relies on others' opinions ($w_i = 1$). The dotted line denotes "Customer 3" who is not influenced by others ($w_i = 0$).

Figure 10 is the relationship curves of loyalty, satisfaction degree, and switching cost of these three types of customers. Different dots represent a different type of customer. The dots marked with "." are "Customer 1" who are not influenced by others. The dots marked with "+" are "Customer 2" who will consider others' opinions. The dots marked with "o" are "Customer 3" who only relies on others' opinions.

Simulation results show that the interaction between customers has a great effect on satisfaction degree and loyalty. It makes the satisfaction degree and loyalty around the middle status. Compared with no interaction, satisfaction degree and loyalty will float smaller and without much diversity. Average curves in Figure 7 change with a smaller range than those in Figure 3. Customers satisfaction and loyalty shown in Figure 5 seem to change more steadily with the effect of the customer network than in Figure 9.

For different types of customers, those who are not influenced by other's opinions are more smooth than others. Therefore, these types of customers can easily become loyal with several satisfied service. However, once a customer becomes disloyal, service providers must pay more attention to improve his/her satisfaction. Because employing word-of-mouth effect cannot make sense, for the customers who can listen to others' opinions, utilizing customer interaction is a good method

Figure 9. Dynamic of loyalty and satisfaction degree with influence from WOM—three types of customers.

3D dot plot of catastrophe model and the relationship curves of loyalty, satisfaction degree, and switching cost. Because of limited line or dot types, we use only one type of lines or dots to represent the traces of all the customers.

Figure 9 shows loyalty and satisfaction degree of three customers. Each of them represent a type of node, respectively, and are marked with three types

to increase his/her satisfaction and loyalty. While for those who are prone to make decision relying on others, introducing a group of some loyal customers to him is an easy way.

Therefore, service providers may consider pushing the process to increase the network effect. Especially, when the customers have a low level of loyalty, try to increase the word-of-mouth to find loyal customers easily. Also clearly grouping the customers types can help service providers make detailed plans aiming at different customers.

5 CONCLUSION

Loyalty is so important for service providers that it will influence the retention of customers to repurchase. In this paper, the loyalty model was studied. By modeling satisfaction and analyzing word-of-mouth and switching cost, we showed how the loyalty is influenced in the networked environment. Besides improving the service quality to attract more customers, well utilizing the network interaction is very useful to gain loyalty. Understanding such complex model of customers will help service providers to predict the response of the delivered service among customers.

REFERENCES

Daniel Birke, G.M. Peter Swann. Network Effects, network structure and consumer interaction in mobile telecommunications. ITS: 16th European Regional Conference Porto, September 2005.

Hahm, J., W. Chu, and J.W. Yoon. "A Strategic Approach to Customer Satisfaction in the Telecommunication Service Market", Elsevier Science, 1997, 33, pp. 825–828.

Jacob Goldenberg, Barak Libai, Eitan Muller, "Talk of the Network-A Complex Systems Look at the Underlying Process of Word-of-Mouth", Marketing Letters, 12(3), 2001, pp. 211–223.

Oliva TA, Oliver RL, MacMillan IC. "A catastrophe model for developing service satisfaction strategies". Journal of Marketing, 1992, 56, pp. 83–95.

Parasuraman, A., Valarie A. Zeithaml, Leonard L. Berry. "A Conceptual Model of Service Quality and Its Implications for Future Research", Journal of Marketing, 1985. 49(4), pp. 41–50.

Thom, R. "Structural Stability and Morphogenesis", Benjamin-Addison Wesley, New York, 1975.

Wang, L.H., M.F. Chen, "Using switching barriers as moderator to explore the relationship between customer satisfaction and customer loyalty", Master Thesis, Tatung University, 2005.

William A. Brock. Discrete Choice with Social Interactions. Review of Economic Studies, 2001. 68: 235–260.

Yang, Z.L., R.T. Peterson, "Customer perceived value, satisfaction, and loyalty: the role of switching cost", Psychology & Marketing, 2004, 21(10), pp. 799–822.

Zeeman, E.C. "Catastrophe Theory: Selected Papers (1972–1977)", Addison-Wesley, New York, 1977.

Civil, Architecture and Environmental Engineering – Kao & Sung (Eds)
© 2017 Taylor & Francis Group, ISBN 978-1-138-02985-9

Research on obstacle avoidance of a mobile robot based on visual information

Jin Jiang

Department of Information Engineering, Jiangsu Polytechnic College of Agriculture and Forestry, Jurong, Jiangsu, China

ABSTRACT: Active obstacle detection and avoidance behavior control has been the key research topic in the field of robot control. In this paper, we propose a visual information acquisition scheme on the basis of visual information and then analyze the visual information into path information source. According to the established route, when encountering obstacles, an algorithm is used to adjust the moving track in real time to meet the goal of intelligent control of mobile robots. The simulation results show that the integration of the visual sensing information can not only obtain the obstacle information, but also guarantee the real-time accuracy of the robot movement control.

1 INTRODUCTION

In recent years, with the rapid development of science and technology, robots have been widely used in various fields. At the same time, the user requires higher degree of robot intelligence. How to select and adjust the real-time route as well as how to take the initiative to avoid obstacles in an unknown environment for the process of robot movement will be undoubtedly one of the key issues to be solved (Khansari, 2012).

At present, the use of ultrasonic, infrared, stereo vision, and other sensor technologies is an important means to detect obstacles. Avoiding strategy after an obstacle is found as one of the key issues of the current robotic exploration. A large number of studies have been carried out in the academic circles. For example, Kim (2015) adopted a neural network-based method to find the optimal path, Huang (2015) adopted a method based on the comprehensive analysis of the grid method means to find the shortest movement trajectory, and Montiel (2015) adopted the artificial field potential method to dynamically adjust the movement trajectory of the robot. Therefore, in this paper, we choose the mode to detect obstacles based on visual information and focus on the strategy and path adaptive control of obstacle avoidance in the process of robot moving.

2 VISUAL INFORMATION CONVERSION

In order to avoid obstacles, the robot needs to follow the collected visual information to carry out the following process. To effectively reduce the volume and weight of the robot, this paper uses a pinhole camera to collect the information of the surrounding environment.

As shown in Figure 3, according to the imaging principle of the pinhole camera, the pixel coordinates in the image plane can be collected by the camera, and the 3D coordinates of the robot coordinate system can be transformed into the coordinate system (Cherubini, 2014). Assuming that the camera is fixed on the robot's height H, the origin of parameters O_L for the robot coordinate system is (X_L, Y_L, Z_L), the lens optical center O_c in the robot coordinate system parameter is (O_{cx}, O_{cy}, O_{cz}), the origin of parameters of image center O_l in camera coordinate system is on the Z_c axis, (I_x, I_y) is one of the feature points in the image and (x_r, y_r) is the point coordinate on the ground in the robot coordinate system.

According to the principle of imaging and the internal and external parameters of the camera, the coordinates of the points on the ground can be determined by using the coordinates of points extracted from the image (X_L, Y_L). The space position relation between the object and the object in the robot coordinate system is obtained.

The camera captures image information, and an ideal image should be a digital image without no distortion, small noise, and clear image. The general optical imaging system is far away from the optical axis, the image will have a larger distortion, including barrel distortion and the distortion of the shape of the pillow.

Considering the radial and tangential distortion in the imaging optical system to model distortion

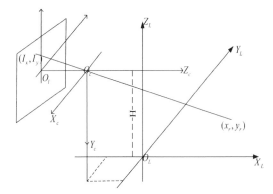

Figure 1.   Camera imaging model.

and distortion parameters through the appropriate existing system, it can be better to correct the distortion caused by the error.

Image distortion model coefficients are usually defined as the normalized image coordinates of the ideal linear pinhole model. No image distortion is normalized to (x_n, y_n), $r = \sqrt{x_n^2 + y_n^2}$ denotes an image point from the optical axis of the camera, and the amount of distortion is represented by δ. The tangential distortion is generally expressed as:

$$\begin{bmatrix} \delta x^{(t)} \\ \delta y^{(t)} \end{bmatrix} = \begin{bmatrix} 2p_1 x_n y_n + p_2(r^2 + 2x_n^2) \\ p_1(r^2 + 2y_n^2 + 2p_2 x_n y_n) \end{bmatrix} \quad (1)$$

The radial distortion generally uses the following formula:

$$\begin{bmatrix} \delta x^{(t)} \\ \delta y^{(t)} \end{bmatrix} = \begin{bmatrix} x_n(k_1 r^2 + k_2 r^4 + k_3 r^6 + \ldots) \\ y_n(k_1 r^2 + k_2 r^4 + k_3 r^6 + \ldots) \end{bmatrix} \quad (2)$$

where k_i represents the order coefficient. Because of the different calculation model, the calculated mean number of parameters and parameter values are very different, and different calculation models may lead to different errors.

3   FRAME OF ROBOT OBSTACLE AVOIDANCE BASED ON VISION

Vision-based robot obstacle avoidance framework involves two kinds of systems: visual information collection and mobile control. The design principles and ideas of the framework can be described as follows.

First, by installing the camera on the robot to detect environmental information, the information is transmitted to the robot behavior control system

in order to carry out the analysis of environmental information. Meanwhile, according to the camera used to capture the image data, the operation range is determined.

Then, the behavior control system to receive the video data, after the image processing, traffic information will be collected, obstacle information and target position data are used to construct the corresponding two-dimensional map. Then, using the path planning module, we generate a path from the initial point to the target point and then the obstacle avoidance algorithm is used to control the behavior of obstacles.

In the process of moving the control system, robot will be used to obtain the coordinate values, speed, steering angle, and other parameters, in order to guide the movement of the robot. The robot obstacle avoidance framework is shown in Figure 2.

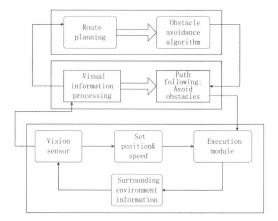

Figure 2.   Frame of robot obstacle avoidance based on vision information.

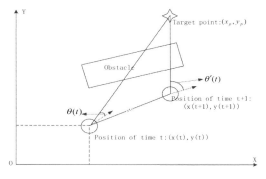

Figure 3.   Representation of parameters in a coordinate system.

4 DESIGN OF ROBOT OBSTACLE AVOIDANCE ALGORITHM

Figure 3 shows the establishment of a two-dimensional reference coordinate system for the controlled object and robot movement. For the convenience of calculation, it is assumed that there is no sliding between the wheel and the ground (Yang, 2016). The initial coordinate system is *XOY*; the target point coordinates are (x_p, y_p); at any moment t, the robot's line speed is $v(t)$, and the robot's current position state is $(x(t), y(t), v(t))$.

Assuming that current heading angle of the robot and the connection point between the centroid target is $\theta(t)$ and the steering angle is $\theta'(t)$, the following relationship exists between the various parameters:

$$v(t) = \arctan \frac{|\Delta y|}{|\Delta x|} = \frac{|y(t+1) - y(t)|}{|x(t+1) - x(t)|} \quad (3)$$

$$D = \sqrt{(x_p - x(t))^2 + (y_p - y(t))^2} \quad (4)$$

$$\theta(t) = \arctan \frac{|y_p|}{|x_p|} - v(t) \quad (5)$$

When the robot departs from the starting point, it will move to the target point in accordance with established routes. In the process of moving, if the camera sensor does not detect the obstacle information or detects obstacle distance, according to the detected robot vision-based orientation sensor and the target heading to the robot connection to the steering by angle $\theta(t)$, it moves to the target point with the speed of $v(t)$. In the course of motion, if the vision sensor detects the obstacles, which have a relatively close distance, the steering control is carried out according to the direction of the obstacle relative to the robot and the direction of the target, by avoiding obstacles at the speed of $v(t)$. The lower the distance of the obstacle, the greater the value of $\theta'(t)$, so as to avoid collision. After the obstacle is avoided, the robot will move to the target point based on the angle $\theta(t+1)$ measured by the azimuth detector.

The steps involved in moving the robot according to the obstacle avoidance algorithm are as follows:

Step 1: Determine whether the current position is the target point; if it is stopped, enter the next step.

Step 2: To determine whether the current collision is possible, if any, then go to the next step, or jump to step 4.

Step 3: According to the algorithm to control the robot dynamic obstacle avoidance, avoid obstacles.

Step 4: Toward the target point movement and after a decision cycle, jump to step 1.

5 SIMULATION OF OBSTACLE AVOIDANCE MODEL

In order to facilitate the simulation, four types of parameters need to be set. The first is to set the starting position of the robot. The second is to set the target position of the robot. Once again, set up multiple obstacles. Finally, set up obstructions distance (when the distance between the obstacle and the car is greater than the distance, it is not affected by the obstacles). In order to facilitate the representation, assuming that the starting position and the ending position are parallel to the coordinate axes, the upper and the lower edges of the two connecting lines are respectively provided with two obstacles, and the middle of the connecting line is provided with an obstacle.

According to the above design ideas and algorithms, this experiment uses MATLAB to simulate. There from, the robot obstacle avoidance trajectory model can be simulated as shown in Figure 4.

Similarly, in accordance with the above design ideas, set four types of parameter information.

Different from the first simulation model, the starting position and the target position are set on the two axes of the segmentation line, and the two coordinate axes are distributed with 45°. Meanwhile, the mobile robot designed in this paper adopts two wheel drives, which are divided into linear motion and circular motion. If the line speed on both sides of the wheel is consistent, the robot moves along a straight line. If the line speed of the two wheels is not consistent, the robot will make a certain radius of circular motion. According to the previous algorithm, the robot obstacle avoidance trajectory model is simulated as shown in Figure 5.

Using the simulation model of the robot by controlling the speed and direction of movement of the line angle, we can achieve the purpose of controlling a mobile robot. Because different parameter settings will produce different results, we cannot achieve the desired goal, which may be related to the obstacles in the distance setting, and the initial coefficient is not suitable for the relevant.

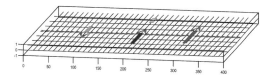

Figure 4. Trajectory model of robot obstacle avoidance.

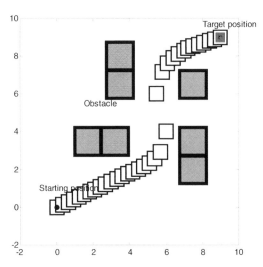

Figure 5. Trajectory model of robot obstacle avoidance.

As can be seen from the above model, the robot uses the information collected by the vision sensor to be sent to the control center for processing and meets the requirements of real-time obstacle avoidance. Simulation results prove that the algorithm is feasible and obstacle avoidance control process has certain stability.

6 CONCLUSIONS

The dynamic environment information sensed by the mobile robot is the key to avoid obstacles. Meanwhile, the vision sensor not only needs to be large enough to detect the scope, but also needs a higher acquisition real-time performance. Therefore, it is an urgent problem to be solved in the process of robot control to avoid obstacles in the dynamic environment with static and moving objects.

In this paper, we discussed the algorithm of obstacle avoidance of a mobile robot in the pro-cess of acquiring, using visual environment obstacle information, then according to the direction of movement and obstacle area dynamic obstacle avoidance processing, effectively avoiding obstacles and moving to goal. However, there are still problems such as real-time system problems, diversification of the target characteristics, and uncertainty of the road surface in the process of robot control, which need to be studied in the future.

ACKNOWLEDGMENT

This work was financially supported by the Science and technology project of Jiangsu Polytechnic College of Agriculture and Forestry (2016kj015).

REFERENCES

Cherubini A, Spindler F, Chaumette F. Autonomous visual navigation and laser-based moving obstacle avoidance [J]. IEEE Transactions on Intelligent Transportation Systems, 2014, 15(5): 2101–2110.

Huang R, Liang H, Chen J, et al. An intent inference based dynamic obstacle avoidance method for intelligent vehicle in structured environment [C] //2015 IEEE International Conference on Robotics and Biomimetics (ROBIO). IEEE, 2015: 1465–1470.

Khansari-Zadeh S M, Billard A. A dynamical system approach to realtime obstacle avoidance [J]. Autonomous Robots, 2012, 32(4): 433–454.

Kim C J, Chwa D. Obstacle avoidance method for wheeled mobile robots using interval type-2 fuzzy neural network [J]. IEEE Transactions on Fuzzy Systems, 2015, 23(3): 677–687.

Montiel O, Orozco-Rosas U, Sepúlveda R. Path planning for mobile robots using Bacterial Potential Field for avoiding static and dynamic obstacles [J]. Expert Systems with Applications, 2015, 42(12): 5177–5191.

Yang H, Fan X, Shi P, et al. Nonlinear Control for Tracking and Obstacle Avoidance of a Wheeled Mobile Robot With Nonholonomic Constraint [J]. IEEE Transactions on Control Systems Technology, 2016, 24(2): 741–746.

Civil, Architecture and Environmental Engineering – Kao & Sung (Eds)
© *2017 Taylor & Francis Group, ISBN 978-1-138-02985-9*

Multifocus image fusion based on defocus map estimation and NSCT

Ruixia Shi, Fanjie Meng, Dalong Shan, Pingping Zeng & Yanlong Wang
Institute of Intelligent Control and Image Engineering, Xdian University, Xi'an, China

ABSTRACT: A novel method based on the defocus map estimation and the Non-Subsampled Contourlet Transform (NSCT) is proposed in this paper. First, the degree of defocus of source images is estimated. Second, source images are divided into the left focused area, right focused area, and uncertain area, according to the defocus map estimation of source images. At the same time, high-frequency coefficients and low-frequency coefficients of source images are obtained after the NSCT is employed for source images. For uncertain area, Sum of Modified Laplace (SML) and Spatial Frequency (SF) are employed to fuse low-frequency coefficients and high-frequency coefficients, respectively. For the left focused area and the right focused area, the corresponding focused areas of source image are selected directly. Finally, through the combination of the fused uncertain area, the left focused area and the right focused area of the fused image are achieved. Compared with the traditional methods, experiment results show that the fusion effect is improved obviously in vision and parameters.

1 INTRODUCTION

Because of the limitation of focused range of optical lens, getting an image in which all targets are clear in one scene is difficult. Usually there are parts of focused and defocused targets in an image. Therefore, in order to achieve an image that all targets are clear, multifocus images must be fused.

Usually, multifocus image fusion methods are based on the pixel-level fusion and region-level fusion. Traditional pixel-level fusion (Li, 2004) is mainly based on multiscale transform, which includes pyramid transform (Liu, 2001), wavelet transform (Amolins, 2010), contourlet transform (Nencini, 2007), and NSCT transform (Cunha, 2006). These methods only concerned multiscale images, but not analyzed region characteristics of the source images. Because of the limitation of rules of pixel-level fusion, it cannot select all coefficients from the focused area, which will decline the quality of the fused image. Region-level fusion methods are mainly based on region segmentation and regional block segmentation.

The traditional segmentation methods are K-means clustering segmentation (Pena, 1999), entropy rate superpixel segmentation (Li, 2013), 3D doctor (Bindu, 2014), and regional block segmentation. However, the traditional segmentation method makes some blocks containing both focused and unfocused regions, thereby declining the quality of the fused image. However, region block fusion is very likely to cause the blocking artifacts in the fused image.

In order to solve the above problems, this paper proposes a fusion algorithm on the basis of region-level fusion and pixel-level fusion. The focused region is fused based on region level, and the uncertain region is fused based on pixel-level. This method can make focused area more accurate and the utilization of useful information of the source image maximum.

2 REGIONAL DIVISION BASED ON DEFOCUS MAP ESTIMATION

In recent years, the focused region is detected by calculating the Root Mean Square Error (RMSE) (Yang, 2015) between initial fused image and source images, or Symantec Security Information Manager (SSIM) between initial fused images and source images (Ning, 2015). Then, a mathematical morphological filter is employed to filter the initial region detection map. Finally, the final regional detection map is achieved. However, the methods of region detection based on RMSE and SSIM rely on initial fusion rules, different initial fused images can be obtained with different fusion rules, and different region detection maps are obtained. Therefore, these algorithms have higher requirement for the quality of the initial fused image; otherwise, we cannot obtain the accurate region detection map, leading to choosing the wrong region of the source image as the region of fused image.

Currently, the detection of fuzzy region has made some achievements. Zhuo (2011) used a

Gaussian kernel to re-blur input source image and calculate the ratio of the gradients of input to re-blurred images. By propagating the blur amount at edge location to the entire image, a full defocus map can be obtained. The robustness of the method is higher and not susceptible to noise. The higher the luminance value of the region in defocus depth estimation map, the more blur the corresponding region is. The lower the luminance value of the region in defocus depth estimation map, the clearer the corresponding region is, as shown in Fig. 1.

For multifocus images, by estimating the degree of defocusing of source images, the defocus estimated map of source images is obtained, as show in Figure 2(c) and (d). By comparing the defocus estimated map of image A and image B, a binary map C is achieved:

$$C(i,j) = \begin{cases} 1 & A1(i,j) - B1(i,j) \geq T \\ 0 & A1(i,j) - B1(i,j) < T \end{cases} \quad (1)$$

where A1 represents the defocus estimated map of the right focused image, B1 represents the defocus estimated map of the left focused image, C represents the initial region detection map, and T is a threshold; generally T is 0 or (–0.5 to 0.5).

Then, a mathematical morphological filter is employed to the initial detection map to remove isolated dots and small-area regions. However, the dividing line between the focused region and defocused region is not very accurate, if the fuse

(a)   (b)

Figure 1.   (a) Source image and (b) defocus map of (a).

(a)   (b)   (c)   (d)

(e)   (f)   (g)   (h)

Figure 2.   (a) Left focus image, (b) right focus image, (c) defocus map of (a), (d) the defocus map of (b), (e) the consequence of Ref. (Wang, 2012), (f) the consequence of ref. (Sun, 2013), and (g) and (h) consequences of our method.

source image is obtained directly from the initial detection map. It will result in a wrong choice of local border region, which reduces the quality of the fused image. Therefore, volatile operation is employed for white areas and black areas of C, and then the gray region, namely the uncertain region, is achieved, as shown in Figure 2(g). The region detection map, which is divided into left focused area, right focused region, and uncertain region, is shown in Fig. 2(h).

3   THE NSCT AND FUSION ALGORITHM

3.1   Introduction of the NSCT

The NSCT is based on Non-Downsampling Pyramid (NSP) and no-subsampled directional filter banks to implement multiscale and multidirectional decomposition, respectively. It is proposed based on improved contour transform. The structure of NSCT is similar to that of contour transform, but comparable to that of contour transform. It not only has the advantages of retaining multiresolution, localization, and multidirection, but also has the advantage of translation invariance, which can eliminate the Gibbs phenomenon. It overcomes the limitation of direction in traditional wavelet transform and variation of translation in contour transform, and can better express the detailed features of the image.

3.2   Fusion algorithm based on the NSCT

1. Estimate the degree of defocusing of the source image A and B, and then defocus the estimated map of A and B to achieve A1 and B1.
2. By comparing A1 and B1, a binary map C is achieved. Then, the morphological filter is employed to filter map C; finally, we can obtain the zoning map C1, which is divided into left focused area, right focused area, and uncertain area.
3. The NSCT is employed for source images, selecting fusion rules of the uncertain area.
4. According to the zoning map C1, select the corresponding area of left focus image A as the left focused area of fused image and select the corresponding area of right focus image B as the right focused area of fused image.
5. Combining the fused uncertain area, the left focused area, and the right focused area, the final fused image F is achieved.

The fusion flowchart is as follows:

The coefficient in the low-frequency sub-band represents the approximate information of source images and contains the highest energy of the source images. However, traditional average fusion

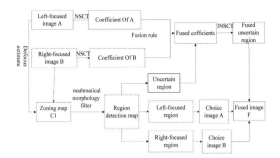

Figure 3. Schematic diagram of the proposed image fusion algorithm.

methods not only decrease the overall brightness of image, but also make the fused image fuzzy.

Therefore, considering that the visual characteristics of the human eye and SML can evaluate degree of focus better, the fusion rule of low-frequency coefficients presented in this paper employs the principle of SML on the basis of local visual contrast. Local features of an image are not determined by a single pixel, but are characterized by a plurality of pixels in a common region. Thus, we calculate the region's LV of each low-frequency coefficient in equation (2). The size of window is $n \times n$, where $n = 3$.

It is defined as follows:

$$LV^x(i,j) = \begin{cases} \dfrac{SML^x(i,j)}{\overline{L^x(i,j)}^{1+a}} & if\ \overline{L^x}(i,j) \neq 0 \\ SML^A(i,j) & otherwise \end{cases} \quad (2)$$

$$LV1^x(x,y) = \sum_{p=-(n-1)/2}^{p=-(n+1)/2} \sum_{q=-(n-1)/2}^{q=-(n+1)/2} LV^x(i+p,j+q) \quad (3)$$

where a = 0.6, $\overline{L^x}(i,j)$ is the mean value region of the pixel (i, j), and x is the source image A or B. The final fused rule is defined as follows:

$$L^F(i,j) = \begin{cases} L^A(i,j) & LV1^A(i,j) \geq LV1^B(i,j) \\ L^B(i,j) & LV1^A(i,j) < LV1^B(i,j) \end{cases} \quad (4)$$

where $L^A(i,j)$ and $L^B(i,j)$ represent the low-frequency coefficients of image A and B, respectively, and $L^F(i,j)$ is the fused low-frequency coefficient.

After the NSCT transformation of source image, the high-frequency coefficient can fully reflect the rich details information of source image, such as edge profile. Spatial Frequency (SF) reflects fundamental features used by the human visual system, so for image fusion, salient information and more detailed information can be extracted from source images effectively via SF. Our fusion method is defined as follows:

The Row Frequency (RF) and the Column Frequency (CF) of an image are expressed as:

$$CF^2 = \sum_{i=1}^{m} \sum_{j=0}^{n} (f(i,j) - f(i-1,j))^2$$
$$RF^2 = \sum_{i=0}^{m} \sum_{j=1}^{n} (f(i,j) - f(i,j-1))^2 \quad (5)$$

where $f(i,j)$ represents the images of size $M \times N$. The total spatial frequency is calculated by Equation (6):

$$SF = \sqrt{\frac{1}{M \times N}(RF^2 + CF^2)} \quad (6)$$

Therefore, regional spatial high frequency is computed by equation (7), with the size of window n = 3:

$$S^x(i,j) = \sum_{p=-(n-1)/2}^{p=-(n+1)/2} \sum_{q=-(n-1)/2}^{q=-(n+1)/2} SF_i(i+p,j+q) \quad (7)$$

$S^x(i,j)$ is regional spatial frequency of pixel (i, j) and x represents the left focused image A or the right focused image B. The fused rule is defined as follows:

$$H^F(i,j) = \begin{cases} H^A(i,j) & S^A(i,j) \geq S^B(i,j) \\ H^B(i,j) & S^A(i,j) < S^B(i,j) \end{cases} \quad (8)$$

where $H^A(i,j), H^B(i,j)$, and $H^F(i,j)$ represent the high-frequency coefficients of images A, B, and fused image F, respectively.

Finally, according to image C1, the corresponding area of left focus image A is selected as the left focused area of the fused image, and the corresponding area of right focus image B is selected as the right focused area of fused image. Combining the fused uncertain area, the left focused area, and the right focused area, the final fused image F is achieved.

4 EXPERIMENTS

Because there are no reference images, the noise or false information may be introduced in the process of fusion, and the parameters of entropy, spatial resolution, and gradient evaluation cannot evaluate the quality of fused image correctly. Therefore, the Mutual Information (MI) and the edge-dependent (Q_{abf}) are used to measure how much information of the original images the fused image F contains.

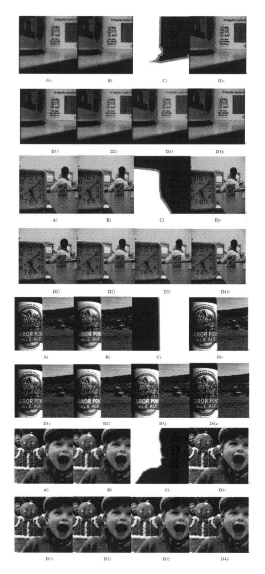

Figure 4. Fusion results of images. A) Left focus image, B) right focus image, C) region detection map, and D) fusion results of our method: D1)–D4) fusion results of Wang (2012), Sun (2013), Maruturi (2014), and Liu (2015), respectively.

The larger the value of MI and Q_{abf}, the better the fusion result is. Compared with the fusion methods proposed recently (Maruturi, 2014; Liu, 2015; Luigi, 2009; Lillo, 2008), the superiority of the method proposed in this paper is verified.

Figure 4 shows that the DCT algorithm has the block effect and cannot describe the information correctly. Pixel-level fusion algorithms based on NSCT cannot obtain focus area accurately. The proposed method can not only extract the focus

Table 1. Objective evaluation of different fusion methods for different multifocus images.

| Image | Evaluation index | Multifocus image fusion methods | | | | |
		Ref (Maruturi, 2014)	Ref (Liu, 2015)	Ref (Luigi, 2009)	Ref (Lillo, 2008)	Ours
Images 1	MI	6.9547	7.0390	7.1023	8.2346	**8.8162**
	Q_{abf}	0.7517	0.7183	0.5854	0.7755	**0.7827**
Images 2	MI	7.2076	7.0077	7.4177	8.7226	**8.8343**
	Q_{abf}	0.6984	0.6428	0.5154	0.7430	**0.7457**
Images 3	MI	6.2639	5.1960	5.6477	8.2204	**8.3232**
	Q_{abf}	0.7806	0.6355	0.3941	0.8018	**0.8023**
Images 4	MI	6.3877	6.0883	6.5551	8.1998	**8.2313**
	Q_{abf}	0.7134	0.6209	0.4952	0.7432	**0.7447**

area better, but also fully utilize the useful information of source image.

Table 1 shows the objective evaluation index of four group images. It can be seen from the table that the indexes of the method proposed in this paper are higher than those of other methods; the fused image in this paper can not only extract useful information fully from the source image but also make the spectrum information of fused image loss minimum.

ACKNOWLEDGMENTS

This work was supported by the National Natural Science Foundation of China under grant 61305040 and the Fundamental Research Funds for the Central Universities under Grant JB161305.

REFERENCES

Aili Wang, Changyan Qi, et. al. Multifocus Image Fusion Based on Nonsubsample Contourlet Transform [C]. Strategic Technology (IFOST), 7th, 1–4, 2012.

Amolins, K., Y. Zhang, P. Dare. Multifocus image fusion based on redundant wavelet transform [J]. Image Processing, 2010, 4(4): 283–293.

da Cunha, A. L., J. Zhou, and M. N. Do. The nonsubsampled contourlet transform: Theory, design, and applications [J]. IEEE Trans. Image Process., 2006, 15(10): 3089–3101.

De Lillo, F., F. Cecconi, G. Lacorata, A. Vulpiani, EPL, 84 (2008).

Hima Bindu, C., K. Veera Swamy. Medical Image Fusion using Content Based Automatic Segmentation [C]. Recent Advance & Innovations in Engineering, pp. 1–5. 2014.

Li, Q.P., J.P. Do, C. Wang. et.al. Region-based Multifocus Image Fusion Using the Local Spatial Frequency [C]. Control and Decision Conference, 25th, 3792–3796, 2013.

Li, S., J.T. Kwok, I.W. Tsang, Y. Wang. Fusing images with different focuses using support vector machines [J]. Neural Networks, 2004, **15**(6): 2004.

Liu, Z., K. Tsukada, K. Hanasaki, Y.K. Ho, and Y.P. Dai. Image fusion by using steerable pyramid [J]. Pattern Recognition Letters, 2001, **22**(2): 929–939.

Liu Cao, Longxu Jin, et.al. Multi-Focus Image Fusion Based on Spatial Frequency in Discrete Cosine Transform Domain [J]. IEEE Signal Processing Letters, 2015, **22**(2): 220–224.

Luigi, T. De Luca, Propulsion physics (EDP Sciences, Les Ulis, 2009).

Maruturi Haribabu, C.H. Hima Bindu Dr. K. Satya Prasad. Image Fusion with Biorthogonal Wavelet Transform Based On Maximum Selection and Region Energy [C]. International Conference on Computer Communication and Informatics, 03–05, 2014.

Nencini, F., A. Garzelli, S. Baronti, and L. Alparone. Remote sensing image fusion using the curvelet transform [J]. Information Fusion, 2007, **8**(2): 143–156.

Pena, J.M., J.A. Lozano, and P. Larranaga. An empirical comparison of four initialization methods for the K-means algorithm [J]. Parrern Recognition Letters, 1999, **20**: 1027–1040.

Shaojie Zhuo Terence Sim. Defocus Map Estimation from a Single Image [J]. Preprint submitted to Pattern Recognition, 2011, **44**(9): 1852–1858.

Xiangda Sun 1, Junping Du 1, Qingping Li, 1. et al. Improved Energy Contrast Image Fusion based on Nonsubsampled Contourlet Transform [C]. Industrial Electronics and Applications (ICIEA), **8**th, 2013.

Yang Ning Ou, Ning Zou, Tong Zhang. et.al. Multi-focus Image Fusion Based on NSST and Focused Area Detection, [J]. Computer Application, 2015, **35**(2): 490–494.

Yong Yang, Song Tong, Shuying Huang. et.al. Multifocus Image Fsuion Based on NSCT and Focused Area Detection [J]. IEEE Sensors Fournal, 2015, **15**(5): 2824–2838.

Civil, Architecture and Environmental Engineering – Kao & Sung (Eds)
© 2017 Taylor & Francis Group, ISBN 978-1-138-02985-9

A new and effective image retrieval method based on representative regions

Dalong Shan, Fanjie Meng, Ruixia Shi, Yanlong Wang & Pingping Zeng
Institute of Intelligent Control and Image Engineering, Xdian University, Xi'an, China

ABSTRACT: In this paper, we propose a new image retrieval system based on representative regions. First, image is segmented to several categories using K-means and Affinity Propagation (AP) clustering methods, and the largest region of every category is found as the representative region. Then, color and texture features of the representative regions are obtained through the HSV color histogram and Local Binary Pattern (LBP). Representative Region Matching (RRM) algorithm is used for calculating the distance between the query image and database images combining color, texture features with location, and area weight. Experimental results show that the proposed method is more prominent than retrieval using some of the existing methods.

1 INTRODUCTION

With the development of computer technology and network, information disseminating through digital images is becoming more and more general. To find the target images from the big date, the Content-Based Image Retrieval (CBIR) (Flickner, 1995; Stricker, 1995) technology attracts extensive attention. However, it is also faced with some problems, especially "the semantic gap" (Chen, 2005), which refers to the gap between the low-level visual feature and human semantic interpretation of an image. Region-Based Image Retrieval (RBIR) technology is the most common method, which can solve this problem by segmenting image into regions and choosing one or multiple regions to participate in the matching; the way of matching mainly includes individual region matching and integrate region matching. Li et al. (2001) from Stanford University proposed the SIMPLIcity system, which extracts color, texture, shape, and location characteristics of all regions in the image. The Integrated Region Matching (IRM) algorithm is used to allow one region matching with many regions belonging to another image, so the influence on the retrieving result due to image segmentation error is reduced; in other words, the robustness of the system is improved. Yuber et al. (Li, 2001) proposed an RBIR system, in which a binary coding method is used to accelerate the running of system by coding regions of images in the database, and to reduce the influence of irrelevant regions, the user can select regions from the query image that represent their interest. In the integrate region matching system such as SIMPLIcity, regions are assigned different weights. However,

the similarity of regions is ignored and the similar regions are matched repeatedly, which result in the interference to the weights of other regions. In individual region matching system, there are mainly two methods to obtain regions: artificial selection and salience method. The former causes the neglect of background and low-level salient regions in the image, whereas the latter can bring huge workload for the user.

To solve these problems, this paper proposes an image retrieval method based on representative regions. Image is segmented into independent regions by using K-means and Affinity Propagation (AP) (Frey, 2007) clustering segmentation methods. The largest region of every object is recorded as the representative region participating

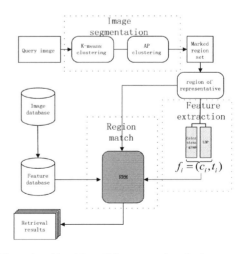

Figure 1. Algorithm of the proposed method.

in matching to ensure the integrity and avoid repetition of matching. Then, the color and texture features of the representative regions are obtained through the HSV color histogram and Local Binary Pattern (LBP), respectively. Representative Region Matching (RRM) algorithm is used for calculating the image distance between query image and database images combined with color, texture features, location, and area weight. Experimental results show that the proposed method provides better retrieving result than retrieval using some of the existing methods. The outline of the proposed framework is illustrated in Fig. 1.

2 PROPOSED METHOD

A formal description of the proposed retrieval framework based on representative regions is given in this section.

2.1 Extraction of representative regions

In order to find blocks representing different substance in an image precisely, K-means and AP clustering segmentation methods are used. In K-means algorithm, n objects are divided into K clusters and classified depending on the similarity of data. The number of clusters (K) need to be set in advance in K-means; however, the number of objects in different image are not equal, so setting a confirmed value K is inaccurate. K-means is combined with AP algorithm in our method to identify the number of clusters automatically. The strategy we adopt is over clustering and clustering again; categories of objects are usually less than 10 in an image, and accordingly, images are operated by the K-means clustering, where K is set to 10, and then, the results are operated by AP clustering in order to merge the similarity regions to the same category. The reason for not adapting AP along is that the algorithm is complex and time-consuming facing a big database.

The largest region of every category is found as the representative region used for matching with assigned different weights. Our method can describe image feature more accurately than the methods focusing on salient region detection and those based on fixed block.

2.2 Feature extraction

We calculate the color features represented regions using HSV histogram statistics, and H, S, and V color channels are divided into 18, 3, and 3 sections, respectively. The LBP method is used to calculate the texture features of regions and then distribute the LBP results into 24 sections and normalized:

$$c(h,s,v) = \frac{N(h,s,v)}{N_{total}} \quad (1)$$

$$t(lbp) = \frac{N(lbp)}{N_{total}} \quad (2)$$

$$lbp = \sum_{r=0}^{7} l(g_r - g_c)2^r \quad (3)$$

$$l(x) = \begin{cases} 1, x \geq 0 \\ 0, x < 0 \end{cases} \quad (4)$$

where c, t are color and textual features, respectively; $N(h, s, v)$ represents the number of pixels whose value is h, s, v in H, S, V channel, respectively; N_{total} means the number of pixels in the area; $N(lbp)$ indicates the pixels number when the LBP value is lbp in regions; g_c is the average pixel value of 3×3 block, and g_r is the value of pixel r.

2.3 Representative region matching

IRM algorithm is used to allow one region matching with many regions of another image, so the influence to the retrieving results due to image segmentation error is reduced. However, the similarities of regions are ignored, and similar regions were matched repeatedly, which result in the interference of weight

Figure 2. (a) Original image, (b1–b3) regions participated in matching in Krishnamurthy (2015), (c1–c5) regions participated in matching in Rashno (2015), (d1–d3) regions participated in matching in the proposed method.

Figure 3. Five images and their segmentation results of K-means combined with AP clustering method.

1490

of other regions; At the same time, regions weight is assigned just by its area in IRM algorithm, which cannot represent the interest of regions precisely. Therefore, RRM algorithm with combined area and location weight is proposed to calculate the distance between images precisely.

For computing the color and texture features and the distance between representative regions in I_1 and I_2, we choose the region b_j in I_2 with the minimum distance between a_i in I_1 as the matched region of a_i. The distance between a_i and b_j is defined as:

$$D_{ij} = d(c_i, c_j) + d(t_i, t_j) \qquad (5)$$

where i and j are the ordinals of representative regions in I_1 and I_2; c_i, c_j, t_i, and t_j are, respectively, the color and texture feature vectors of region a_i and b_j. The distance between I_1 and I_2 is defined as:

$$D = \sum_{i=1}^{m} \min_{j=1}^{n}(D_{ij}) * W(i) \qquad (6)$$

where W is the weight vector of representative regions in query image:

$$W(i) = 1 / (\sum_{i \in r} \| f(n) - (0.5, 0.5) \| / N_i) + sqrt(N_i / N_{total}) \qquad (7)$$

$$f(n) = [n_x / row, n_y / col] \qquad (8)$$

where m and n are the numbers of representative regions in images I_1 and I_2; N_i, N_{total} respectively, are the number of pixels in representative region i and the total image; n_x, n_y represent the coordinates of pixel n in i; row and col denote the length and breadth of image. The weight of representative regions vary directly as the area of regions and inversely as the average distance between pixels in regions and the center of the image, that is, the region closer to the center of image and with a larger area will have a high weight.

3 EXPERIMENTAL RESULTS

The algorithm has been implemented using MATLAB-2014a in Windows 7 and run on CPU 3.20GHz PC. The database used in our experiments is Wang database, which has 1000 images and categorized into 10 categories (African, beach, building, buses, dinosaurs, elephants, roses, horses, mountains, and food). In order to indicate the validity of our method, precision rate is used to evaluate the result, and is defined as:

$$P = n_k / K \qquad (9)$$

where K is the number of retrieved images and n_k is the number of relevant images in the retrieved

images. The average precision of the images belonging to the category q is given by:

$$\bar{P} = \sum_{q=1}^{10} P_q / 10 \qquad (10)$$

In the proposed algorithm, multiple regions with different weights participating in matching not only ensure the integrity of image features, but also avoid repeated matching. As shown in Fig. 4, no matter images with complex scenes like beach and building which do not have an obvious salient region or the images have clear objects, such as flowers and horses, both show excellent retrieval performance. Table 1 shows a comparison of our method and others, from which we can find that our method is better with six types of images: bus, dinosaur, elephant, flower horse, and food. However, our method does not have prominent performance when dealing with images that have complex background

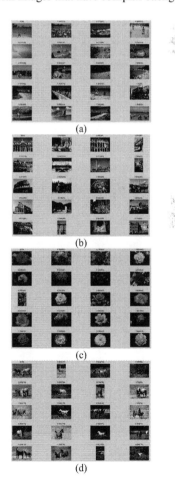

(a)

(b)

(c)

(d)

Figure 4.   Some image retrieving results.

1491

Table 1. Precision of the retrieval by different methods when K = 20.

Category	CTD CIRS (2011)	Vimina (2013)	Rashno (2015)	Lin (2009)	Proposed method
African	56.20	**71.52**	44.80	68.30	62.00
Beach	53.60	43.60	47.20	**54.00**	52.50
Building	**61.00**	53.55	53.40	56.20	60.00
Buses	89.30	85.30	73.40	88.80	**90.50**
Dinosaurs	98.40	99.55	**99.80**	99.30	**99.80**
Elephants	57.80	59.10	56.80	65.80	**75.00**
Roses	89.90	90.95	87.50	89.10	**94.50**
Horses	78.00	92.40	70.70	80.30	**94.50**
Mountains	51.20	38.35	39.30	**52.20**	47.50
Food	69.40	72.40	61.00	73.30	**75.00**
Average precision	53.24	70.67	63.39	72.70	**75.13**

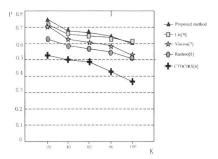

Figure 5. Average precision of different methods.

such as Africa, beach, and mountain; this indicates that our method is suitable for the images that have obvious salient regions. Generally speaking, the average retrieval precision of 75.13% is more prominent than the other method mentioned.

The comparison of average precision obtained by the proposed system with other retrieval systems when K is 20, 40, 60, 80, and 100 is shown in Fig. 5. We can observe that the average precision of our method is not highest when K = 100. These results clearly show that the performance of the proposed method is better than that of the other methods.

4 CONCLUSIONS

We propose a new image retrieval system based on representative regions. First, the image is segmented into several categories using K-means and AP clustering methods. The largest region of every category is found as the representative region. RRM algorithm is used for calculating the distance between query and database images combining color, texture features with location, and area weight. Experimental results show that the proposed method is more prominent than the existing methods. The aim of our next work is to further improve the accuracy of segmentation and make the method to have better applicability in the images with complex background.

ACKNOWLEDGMENT

This work was supported by the National Natural Science Foundation of China under Grant 61305040 and the Fundamental Research Funds for the Central Universities under Grant JB161305.

REFERENCES

Chen, Y., J.Z. Wang, and R. Krovetz: CLUE: Cluster-based retrieval of images by unsupervised learning. IEEE Transactions on Image Processing, 14(8): pp. 1187–1201 (2005).

Flickner, M., H. Sawhney, W. Niblack, J. Ashley, Q. Huang, B. Dom et al.: Query by Image and Video Content: The QBIC System, IEEE Computer, vol. 28, pp. 23–32 (1995).

Frey, Brendan J. and Delbert Dueck: Clustering by passing messages between data points. Science 315.5 814: pp. 972–976 (2007).

http://wang.ist.psu.edu/docs/related/

Krishnamurthy, Lakshmi, and Jayanthi Kesavan: An improved Content Based Image Retrieval using three region color and straight line signatures of the image. Communications and Signal Processing (ICCSP), International Conference on. IEEE (2015).

Li, J., J.Z. Wang, and G. Wiederhold: Simplicity: Semantics-sensitive integrated matching for picture libraries. IEEE Transactions on Pattern Analysis and Machine Intelligence, vol. 23, no. 9, pp. 947–963 (2001).

Lin, C.-H., R.-T. Chen, and Y.-K. Chan: A smart content-based image retrieval system based on color and texture feature. Image Vis. Compute. vol. 27, no. 6, pp. 658–665 (2009).

Rao, M.B., B.P. Rao, and A. Govardhan: CTDCIRS: Content based Image Retrieval System based on Dominant Color and Texture Features. International Journal of Computer Applications, Vol. 18–No.6, pp. 0975–8887 (2011).

Rashno Abdolreza, Sadri Saeed, and Sadeghian Nejad Hossein: An efficient content-based image retrieval with ant colony optimization feature selection schema based on wavelet and color features. In: 2015 International symposium on artificial intelligence and signal processing (AISP). IEEE. pp. 59–64 (2015).

Stricker, M. and M. Orengo: Similarity of Color Images. SPIE Storage and Retrieval for Image and Video Databases, pp. 381–392 (1995).

Vimina, E.R. and K.P. Jacob: A sub-block based image retrieval using modified integrated region matching. in IJCSI International Journal of Computer Science Issues, vol. 10, no. 2, pp. 686–692 (2013).

Wang, J.Z., J. Li, and G. Wiederhold, SIMPLIcity: Semantics-sensitive Integrated Matching for Picture LIbraries, IEEE Transactions on Pattern Analysis and Machine Intelligence, vol. 23, no. 9, pp. 947–963. (2001).

Civil, Architecture and Environmental Engineering – Kao & Sung (Eds)
© *2017 Taylor & Francis Group, ISBN 978-1-138-02985-9*

Semantic similarity research on case retrieval based on ontology

Yiling Liu, Hong Duan & Lei Luo
College of Information System and Management, National University of Defence Technology, Changsha, Hunan, China

ABSTRACT: Case retrieval is an important link in Case-Based Reasoning (CBR). In this paper, we choose ontology as a tool to describe cases and construct case library. On this basis, we analyze the features and retrieval requirements to choose appropriate similarity measurement and design the case retrieval algorithm. We proposed a new similarity measurement called Matched Genealogy Measure (MGM), considering not only elements separately but also cases as a whole. Then, we took combat simulation as an example to illustrate the ontology-based algorithm and designed experiments to evaluate it by comparing with some existing measurements. Furthermore, we performed user studies to obtain the ideal retrieval results and compared the retrieval results of different algorithms using Window Distance with the ideal results. The results suggest that MGM could perform better in matching human intuition and obtain reasonable retrieval results. Finally, we adjusted the two coefficients in MGM and demonstrated that MGM could meet different preferences of users.

1 INTRODUCTION

Case-Based Reasoning (CBR) is one of the three reasoning methods, together with Rule-Based Reasoning and Model-Based Reasoning. The main idea of CBR is to take advantage of historical experiences to solve new problems. Thus, we could know the importance of case retrieval by determining which cases are similar to the new one and could help to solve the problem. Most of the cases are recorded in natural text, and the traditional retrieval algorithms would match them by keywords. However, the keyword-based matching could result in more outlying information. Thus, more and more studies focus on knowledge expressing and semantic retrieval.

Ontology (Gruber, 1993; Asuncion, 1999; Xie, 2009) is a specification of a conceptualization, and it is very important for knowledge expressing and organizing, which facilitates essential semantic knowledge describing and understanding in computers. The recently studied ontology modeling makes use of these features to organize knowledge and takes some effects; for example, Gruninger (Gruninger, 1995) put forward a method to design and describe ontology and constructed a frame to evaluating the completeness of ontology.

In this paper, we proposed a new similarity measurement called Matched Genealogy Measure (MGM), which is based on the chosen Genealogy Measure and affected by matching ratio. The GM method described semantic similarity between concepts, but it is short of presenting cases or

collections as a whole. In fact, when we consider two similar cases, they are apparently not only similar in concepts, but also their construct or scale will have something in common. The MGM will present these features and try to perform better on meeting human intuition.

Taking combat simulation as an example to illustrate the ontology-based algorithm, we describe some combat simulation cases based on ontology and design experiments on the case library. We conducted user studies to get an ideal list of cases ordered by their similarity compared with a given case. Different algorithms will get different orders of these cases and compare these lists to know how well these algorithms meet human intuition.

The remainder of this paper is organized as follows: Section 2 introduces some related work on similarity measurements. Section 3 describes the case retrieval algorithm based on ontology. Sections 4 and 5 present the experiments on combat simulation cases and results. Section 6 provides some conclusion and proposes some further research issues.

2 RELATED WORK

In particular, case retrieval refers to calculating similarity between cases on which to base the choice of some similar cases to the current one from library, according to a certain similarity measurement. There are several studies on concept or case similarity.

Bernstein A proposed "ontology distance" (Bernstein, 2005) to describe the shortest path with a common ancestor node of two objects or the shortest common path linking two objects with a common descendant node. It can be predicted that the similarity will be higher at a smaller distance. The ontology distance considered both the distance between two concepts in the tree and the information described by the depth of concepts in the tree. It fits more to measure semantic similarity of concepts in the same ontology, but is difficult to measure similarity between objects or collections.

Al-Mubaid and Nguyen (Al-Mubaid, 2006) combined path length and common trait to get concept similarity, where path length is the length of the shortest path between two concepts, and common trait depends on the depth of the Least Common Subsumer of two concepts.

Traditionally, objects are considered as collections of concepts when calculating similarity, for example, Dice's similarity formula (Ganesan, 2003):

$$sim_{Dice}(X,Y) = \frac{2 \times |X \cap Y|}{|X| + |Y|}$$

Some other methods also abstract concepts or elements into dimension in vector space, while objects as vectors, which is the most widely used model for information retrieval (Salton, 1983). An apparent benefit of this model is that the weight of different elements could be adjusted. In vector space model, similarity between two objects is normally measured by the cosine value of the two corresponding vectors. Another method measures the similarity by Pearson's correlation coefficient, which is commonly used in consensus filter, such as GroupLens' research (Resnick, 1994).

Ganesan P (Xie, 2009) improved the vector space model by redefining dot product of leaves with the depth of the Least Common Ancestor, calling it as Generalized Co-sine-Similarity Measure (GCSM). However, considering that objects could be quite different from each other, the vectors could be so sparse that the data redundancy is serious. Another algorithm proposed by Ganesan P is Genealogy Measure, which introduced hierarchical domain structure and measured similarity with the depth of the Least Common Ancestor. The authors also put forward some improvements for multiple occurrences. The four algorithms in Ganesan's thesis fit different situations: GCSM does well in long query, especially when containing some similar key words; OGM focuses more on similarity of objects while ignore differences; choice between BGM and RGM depends on whether the user focuses on coverage or distribution of elements. Besides, RGM could balance the coverage and distribution of elements by adjusting parameters.

3 CASE RETRIEVAL ALGORITHM BASED ON ONTOLOGY

3.1 Genealogy measure for similarity

Ontology is actually a specification of conceptualization, which could be shown by many methods. Assume that ontology O could be depicted by a concept tree U, which has a prior weighting function W for its leaves. Then, the cases could be expressed as a collection of leaves of the concept tree. For given leaves l_1 and l_2, we can define the Lowest Common Ancestor (LCA) as the deepest ancestor node of l_1 and l_2, notated as $LCA(l_1, l_2)$. In general, the Genealogy Measure is based on the depth of LCA.

Consider two objects C_1 and C_2, defining introduced trees as T_1 and T_2. Define $LCA_{T_1,T_2}(l_1)$ for every leaf l_1 in T_1, expressing the deepest ancestor of l_1 appeared in T_2. Then, we determine the similarity of the two objects following Balanced Genealogy Measure (BGM):

1. Find a matching in T_2 for every leaf l_{1i} in T_1, denoted as l_{2i}; enlarge the *match_count* value for l_{2i}, which records how many times l_{2i} matched leaves in T_1.
2. Define:

$$optleafsim_{T_1,T_2}(l_{1i}) = \frac{depth(LCA_{T_1,T_2}(l_{1i}))}{depth(l_{1i})}$$

then we have:

$$leafsim_{T_1,T_2}(l_{1i}) = optleafsim_{T_1,T_2}(l_{1i}) \times \beta^{match\_count(l_{2i})-1}$$

3. Similarity between C_1 and C_2 is defined as the weighted average for *leafsim* values of all the leaves in T_1:

$$sim(C_1,C_2) = \frac{\sum_{l_{1i} \in C_1} leafsim_{T_1,T_2}(l_{1i}) \times W(l_{1i})}{\sum_{l_{1i} \in C_1} \times W(l_{1i})}$$

3.3 Matched genealogy measure

Intuitively, two main factors affect similarity among objects:

1. How many elements could be matched in another object, called overlap ratio;
2. To what extent are the elements matched to each other, called element similarity.

BGM focused more on element similarity, but ignored the overlap ratio. Define a match collection:

$$match_{T_1,T_2}(l_1) = \{l_2 \in C_2 \mid LCA(l_1,l_2) = LCA_{T_1,T_2}(l_1)\}$$

which includes all the elements in C_2 that could be matched with the elements in C_1 following the rule $LCA(l_1, l_2) = LCA_{T_1, T_2}(l_1)$. To understand intuitively, $match_{T_1, T_2}(l_1)$ is a collection of all the similar elements of l_1 in C_2. By denoting $|match_{T_1, T_2}(l_1)|$ as the number of elements in $match_{T_1, T_2}(l_1)$, we define matching ratio for l_1 as:

$$\xi_1 = \frac{|match_{T_1, T_2}(l_1)|}{|C_2|}$$

Then, ξ_1 is the element's percentage in C_2 that could match l_1. As for a whole introduced tree, we define collection $match_{T_1}(T_2)$ similarly to include all the elements in C_2 that could match any elements in C_1, and the matching ratio for T_1 and T_2 is given by:

$$\xi_{T_1, T_2} = \frac{|match_{T_1}(T_2)|}{|C_2|}$$

In a weighted model, we have:

$$\xi_{A,B} = \frac{\sum_{l_i \in match_{T_1}(T_2)} W_{l_i}}{\sum_{l_j \in C_2} W_{l_j}}$$

where W_{l_i} is the weight value of the corresponding leaf.

Apparently, matching ratio is a real number less than 1, which presents overlap ratio of two collections or cases.

The matching ratio and similarity calculated by BGM present the two aspects of similarity between cases. Thus, we define a new similarity measurement called Matched Genealogy Measure (MGM), which integrates both the factors:

$$sim_{MGM}(A, B) = a \times sim_{BGM}(A, B) + b \times \xi_{A,B}$$

where a and b present the importance of BGM similarity and the matching ratio when considering the similarity between two cases. In other words, the coefficients describe which of the two factors do the case library or user emphasize, element similarity or overlap ratio. They are prior knowledge of the retrieved library or field, and could be assigned by experts or users.

4 EXPERIMENTS

4.1 Describe combat simulation cases

A combat simulation case mainly contains three parts: combat entities, combat scenarios, and simulation resources. Thus, the ontology is also

constituted by three parts: entity ontology, mission ontology, and simulation ontology. The logistic structure of combat simulation case ontology is shown in Figure 1.

These concepts include more specific aspects, and are finally developed into a concept tree of ontology. Case retrieval based mainly on scenario tables (Kolonder, 1993), which describe entities, missions, and other elements related to the tasks in a combat simulation, regulated mainly by entity ontology and mission ontology. Figure 2 gives part of scenario table, including some specific attributes of a combat simulation case, hinting the relationships between the attributes and the three ontologies. Contents regulated by simulation ontology, however, are usually stored in resource tables as solutions and results of cases, such as data tables of simulation experiments and experiment scheme tables. For example, we have a case described as (A_obj, battle, F117A,...), where all the elements are actually leaves of the concept tree, referring to a simulation of a battle aimed at A_obj with F117A being the main platform.

4.2 Evaluation algorithm

The most direct way to evaluate a retrieval algorithm is testing to what extent the algorithm meets user' requirements. It is not reasonable to expect users to come up with absolute similarity scores between cases. Instead, we conducted user studies, asking them to rank cases according to their similarity to a given case.

Figure 1. Logistic structure of case ontology.

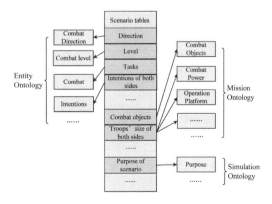

Figure 2. Part of scenario tables.

1495

We described 16 cases based on some typical battles or combat simulations in history as the case library. We asked some "users" to sort the cases by similarity with a given case (case X), and get an "ideal order" (L) of the 16 cases on the basis of statistic results.

Assume that we can rank a list of all cases in the designed library with a given retrieval algorithm M, denoting the rank as $L(M)$. If $rank_M(A)$ is the ranking value of case A in rank $L(M)$, then we express the Window Distance (Ganesan, 2003) between $rank_M(A)$ and the ideal list (L) as:

$$WindowDist_{L,M,I,K}(X) = \frac{\sum_{i=I}^{I+K-1} |rank_M(L[i]) - i|}{K}$$

where L is the ideal list of the given cases, I is the start position, and K is the size of the window.

Intuitively, the Window Distance gives a quantitative description of differences between the list to be evaluated and the ideal list. For example, case A is listed at the third position in the testing list while at the first position in the ideal one; then, it will contribute 2 to the Window Distance of the testing list.

5 RESULTS

5.1 Results of user study

We choose 100 people to participate in our user study and get "ideal list" from the statistic results. For example, the statistic results of the first place in the list are given in Table 1; we regard Case 2 as the most similar one to X.

5.2 Evaluation results

We compared our modified algorithm (MGM) with the original algorithms: OGM, BGM, and RGM. Considering the volume of library, we set the size of window as $K = 5$

As we claimed before, the two coefficients of MGM could be adjusted to meet different preferences of users. First, we set $a = b = 0.5$. Figure 3 shows the Window Distance of different algorithms as a function of window position. We can conclude that the Window Distance of MGM is generally smaller than others', which suggests that it matched human intuition better, especially in the first few ones. In other words, the retrieval

Table 1. Example of user studies results.

Case number	1	2	5	7	8	10
Proportion (%)	12	76	3	6	2	1

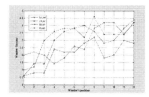

Figure 3. Window distance of different algorithms.

Table 2. First seven cases of result list.

Ideal list	OGM list	BGM list	RGM list	MGM list
2	2	1	2	2
1	1	2	1	1
12	16	16	12	12
15	15	15	16	16
9	12	12	15	15
16	13	13	3	9
13	5	6	13	13

results of MGM match users' intuition better, especially for the most similar ones.

It also shows that the first three or five distances of all the algorithms are acceptable but will get larger as the position moves on. That means, these measurements could describe similarity well when the cases actually share something in common under the given conditions, but the original algorithms have no idea to differ the "not-that-similar" aspects. Thus, the performance gets worse as the window position moves to the end. Table 4 shows that cases in the first seven collections are substantially same in the list of MGM, but the GMs have some disagreements about case 13, which also proved that results of MGM match human tuition better.

5.3 Adjusting for different users' preference

When $a = b = 0.5$ in MGM, we get a relatively good result, as we showed before. It is important to note that a certain group of coefficients could not fit all case libraries or users. The users could also adjust the coefficients to meet different requirements. For instance, some users might focus only on the first few cases of the list, while others might concern on the whole list.

If the users focus more on the first few cases of the list, we expect that the first three window distances could be as smaller as possible. We find that when $a = 0.61$ and $b = 0.39$, we could meet the expectation, and the results are shown in Figure 4 and Table 3.

If users want the algorithm to exhibit a better performance over the whole library, which means that we expect the average distance of the whole

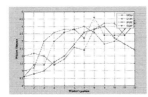

Figure 4.　Window distances when $a = 0.61$, $b = 0.39$.

Table 3.　Average distance of the first three windows.

	OGM	BGM	RGM	MGM
Average distance	1.7333	2.0667	1.1333	0.8

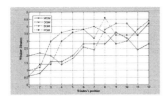

Figure 5.　Window distances when $a = 0.61$, $b = 0.39$.

Table 4.　Average distance when $a = 0.61$, $b = 0.39$.

	OGM	BGM	RGM	MGM
Average distance	3.1667	2.8500	3	2.2500

list, we have $a = 0.65$ and $b = 0.35$ to meet the requirement, and the results are given in Figure 5 and Table 4.

Briefly, different users could determine different values and get different results. However, we suggest $a = b = 0.5$ as a balance if users have no idea about them.

6　CONCLUSIONS AND FUTURE WORKS

Case retrieval is an important link for Case-Based Reasoning, and semantic retrieval is popular in recent researches for case retrieval.

The main contribution of this paper is proposal of a new semantic similarity measurement between cases based on ontology, called the Matched Genealogy Measurement (MGM), which is proved to perform well in matching human intuition. We took combat simulation as an example to illustrate ontology-based algorithm and described some typical combat simulation cases with the method as case library. The conclusion is supported by a user study and analysis of Window Distance. MGM could also meet different users' requirements by adjusting coefficients. It is actually a general measurement

of semantic similarity, and could be used widely in information retrieval and recommendation systems.

Further research will focus on the semantic information presented by relations of different level of the ontology. A possible direction may be constructing a concept tree with weighted edges. The depth of LCA could develop to the distance from the root.

We also need to pay some attention to the influence of the volume of cases. Considering that the Genealogy Measures are essentially average value of leaves' similarity, the similar or different features might be submerged by a large number of elements. It may be solved by assigning significant elements. However, it is difficult to say which elements are more important than others, and it might change with different user of application.

REFERENCES

Al-Mubaid H, Nguyen H A. A cluster-based approach for semantic similarity in the biomedical domain [C]. Engineering in Medicine and Biology Society, 2006. EMBS'06. 28th Annual International Conference of the IEEE. IEEE, 2006: 2713–2717.

Asuncion Gomez Perez, V. Richard Benjamins. Overview of Knowledge Sharing and Reuse Components: Ontologies and Problem-Solving Methods, Proceedings of the IJCAI-99 workshop on Ontologies and Problem-Solving Methods (KRR5) Stockholm, Sweden, August 2, 1999.

Bernstein A, Kaufmann E, Bürki C, et al. How similar is it? towards personalized similarity measures in ontologies [M]. Wirtschaftsinformatik 2005. Physica-Verlag HD, 2005: 1347–1366.

Ganesan P, Garcia-Molina H, Widom J. Exploiting hierarchical domain structure to compute similarity [J]. ACM Transactions on Information Systems (TOIS), 2003, 21(1): 64–93.

Gruber T R. A translation approach to portable ontology specifications [J]. Knowledge acquisition, 1993, 5(2): 199–220.

Gruninger, M. and Fox, M.S. Methodology for the design and evaluation of ontologies [C], Workshop on Basic Ontological Issues in Knowledge Sharing, International Joint Conference on Artificial Intelligence 1995, Montreal.

Kang Xiao-yu1, 2, Deng Gui-shi1. Overview of Military Scenario Research in Warfare Simulation [J]. Journal of System Simulation, 2009 (10): 2797–2800.

Kolonder J. Case-Based Reasoning [M]. San Mateo, California: Morrgan Kaufmann, 1993.

Resnick P, Iacovou N, Suchak M, et al. GroupLens: an open architecture for collaborative filtering of netnews [C]. Proceedings of the 1994 ACM conference on Computer supported cooperative work. ACM, 1994: 175–186.

Salton G, McGill M J. Introduction to modern information retrieval [J]. 1983.

Xie Hong-wei, Li Jian-wei. Research of case-based reasoning model based on ontology [J]. Application Research of Computers, 2009, 26(4): 1422–1424.

Civil, Architecture and Environmental Engineering – Kao & Sung (Eds)
© 2017 Taylor & Francis Group, ISBN 978-1-138-02985-9

Vibration energy acquisition and storage management system based on MSMA

Qingxin Zhang, Kai Lin & Jikun Yang
Automation Institute, Shenyang Aerospace University, China

ABSTRACT: Load identification is a major concept in the field of smart homes and smart grids. Nonintrusive Load Monitoring (NILM) method is applied to solve this problem, which is performed by analyzing the total current and voltage signal of the main distribution board to estimate the energy consumption of individual appliance and turning on/off or other operation. In this paper, we used the theory of NILM to identify household electric load. By analyzing the total current signal, extracting related features, and using Genetic Algorithm (GA) and Support Vector Machine (SVM), we identify different electric loads. We also use the BLUED data set (Anderson et al. 2012) as the experimental data set. Finally, rationality and effectiveness of the proposed method was verified by MATLAB simulation.

1 INTRODUCTION

MSMA is a new kind of intelligent material. Studies have shown that the Martensitic transformation appears when exogenous magnetic field and stress was applied to the MSMA. In theory, it can produce 10% linear deformation rate and 18% bent deformation rate (Aljanaideh, 2013). Magnetic control shape memory characteristics can be reused. On the one hand, under the action of external magnetic field deformation, it can produce the output force. On the other hand, its vibration under the excitation of the magnetic properties changes and then produces larger electromagnetic signal changes. Studies have also shown that the excellent performance of MSMA is particularly applicable to the collection of micro energy. Thus, MASA has a broad prospect for development (Zhang, 2013; Zhang, 2015).

The author analyzes the MSMA principle of vibration energy harvesting and establishes the equivalent circuit model of vibration energy harvesting of MSMA. In order to collect and store the weak AC voltage, the author designed the MSMA vibration energy acquisition power management circuit.

2 MSMA ENERGY HARVESTING DEVICE

2.1 Principle of MSMA energy harvesting device

MSMA is a new kind of intelligent material, which is used in the field of vibration energy acquisition. In this paper, we use the MSMA material of frame size $5 \times 5 \times 20$ mm purchased from the AdaptaMat company of Finland (Zhang, 2015; John, 2000). The principle of vibration energy acquisition in a device is shown in Figure 1.

When MSMA is subjected to external stress, its magnetization will change such that the magnetic flux of the induction coil changes. According to Faraday's law of electromagnetic induction, the magnetic induction intensity will change too. The experiment indicates that the instantaneous value of induction voltage can reach 46 V (Wen, 2012; Zhang, 2016).

As shown in Figure 1, when a mechanical vibration force is applied to the stretching rod, the material would be deformed by the force. According to the Villari effect (when MSMA is affected by an external force, such as vibration), the coefficient of magnetic conductivity will be proportional to the deformation of material. Under a bias magnetic field, magnetization will also be changed with the stretch and compression of the MSMA material, and the maximum change range could be up to 0.7T. If the magnetic flux through the internal material

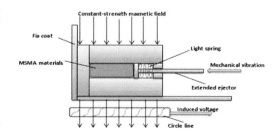

Figure 1. MSMA energy harvesting device working principle diagram.

permeability changes, the constant magnetic field forces around the MSMA will change. Meanwhile, the magnetic flux through the coil will change too. The induction voltages could be obtained as follows:

$$V = -\frac{\partial B}{\partial t} NS \tag{1}$$

where V is the induction voltage in the coil (V); B denotes the magnetic induction intensity caused by the change of the magnetic intensity in the material (T); $B = \mu_0(H + M)$, where H is the imposed magnetic field (A/m); M is the magnetization of the material (A/m); μ_0 is the permeability of vacuum (H/m); N is coil turns (turns); and S is the cross-sectional area of the coil (m^2).

2.2 Circuit model of MSMA vibration energy collector equivalent

In this section, we establish an equivalent model to calculate the corresponding induced voltage. The equivalent circuit and magnetic circuit of the vibration energy acquisition system are shown in Figure 2 while ignoring the eddy effects, saturation effect, and hysteresis effect.

$$\Phi_c = \frac{R_{mPM}\Phi_{PM}}{R_{mPM} + R_{mc} + \frac{(R_{mG} + R_{MSMA})R_{mS}}{R_{mG} + R_{MSMA} + R_{mS}}} \tag{2}$$

where Φ_{PM} is produced by a constant permanent magnet flux; R_{mpM} represents bias air gap reluctance; R_{mc} is the core reluctance; R_{ms} is the magnetic reluctance through the material leakage; R_{mG} is the air gap reluctance between the MSMA material and iron core; and R_{MSMA} is the reluctance of MSMA material. The shape of the MSMA material changes with the air gap and material width. Therefore, R_{mG} and R_{MSMA} depend on the deformation state of the MSMA material. In addition, R_{MSMA} also depends on the internal magnetic field (H_{MSMA}, B_{MSMA}) of the MSMA material. By calculating following equality, we can get the magnetic flux of the coil:

When the external vibration causes the deformation of the MSMA, the magnetic flux of the coil is changed, and the corresponding induction voltage can be obtained as:

$$u_e = N\frac{d\Phi_c}{dt} = \frac{N}{l_{MSMA}}\frac{d\Phi_c}{d\varepsilon}V \tag{3}$$

where l_{MSMA} is the length of the material and V is the speed of change of the shape of MSMA materials. From (3), we know that the induced voltage is influenced by the geometry of drives and material parameters as well as the speed of MSMA material.

3 POWER MANAGEMENT SYSTEM BASED ON MSMA

The author proposes a case of MSMA vibration energy-based power management system to collect and store the MSMA vibration energy output voltage AC. It is composed of six parts: AC/DC rectifier circuit, the voltage-control circuit, super-capacitor circuits, DC/DC converters based on LTC3526L boost regulator circuit, a temporary storage circuit, and lithium-ion battery charging circuit based on MAX1811. The schematic is shown in Figure 3.

When the force disappears, the charge generated by the MSMA vibration energy collector disappears immediately. Therefore, we need an external capacitor to storage the charges. As shown in Figure 3, the collective electric from MSMA vibration energy collection device is stored in the super-capacitor after passing the rectifier circuit. When the control module to monitor the voltage of the supercapacitor meets the set value, turn on the DC/DC chip LTC boost voltage regulator circuit. Then, the stable supplying DC voltage will be stored in a supercapacitor. When the voltage of the capacitor reaches the requirement of the input voltage of the

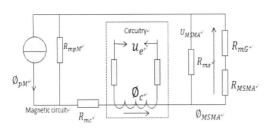

Figure 2. MSMA collector–magnetic equivalent circuit diagram.

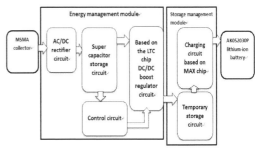

Figure 3. MSMA vibration energy harvesting power management.

MAX1811 charging control terminal, the charging circuit begins to work for the lithium battery, and the power management of the energy is realized.

3.1 Supercapacitor storage

In micro-vibration energy harvesting systems, supercapacitor is regarded as energy storage components. The output of the system would be instantaneous pulse DC. The supercapacitor will discharge rapidly when the capacitor voltage reaches the discharge level; however, the discharge process is difficult to control. Therefore, it is difficult to adopt supercapacitor direct power to supply the electrical equipment directly. In this paper, the supercapacitor is used as a temporary energy storage element. 0.01f supercapacitor is generally used as a temporary storage element, and the circuit is shown in Figure 4.

3.2 LTC3256L chip introduction

LTC3526L is a kind of DC/DC converter with synchronous output disconnection function and fixed frequency. The chip uses a startup voltage of 0.68V, and once it starts, it will operate at voltages as low as 0.5V. In the intermittent operation mode, LTC3526L implements burst mode operation under light load conditions so as to exhibit high efficiency under a wide load range. Antiringing control circuit can reduce EMI (Electro Magnetic Interference) by damping the sensor of the intermittent mode of operation. Figure 5 is the peripheral circuit.

3.3 Simulation analysis based on MSMA energy management circuit

The author used LTspice IV software for simulation analysis to verify the stability of the circuit module design. The circuit schematic diagram is shown in Figure 6.

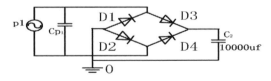

Figure 4. Supercapacitor storage emulation.

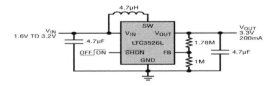

Figure 5. LTC3526L peripheral circuit diagram.

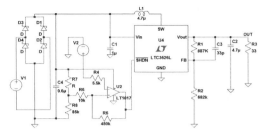

Figure 6. MSMA energy management circuit schematic diagram.

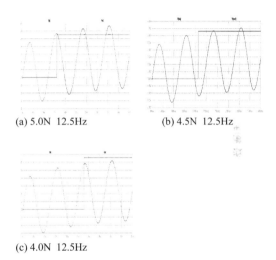

(a) 5.0N 12.5Hz   (b) 4.5N 12.5Hz

(c) 4.0N 12.5Hz

Figure 7. Circuit output simulation diagram under different input stress amplitudes.

We could obtain different input voltages in different vibration conditions. In order to analyze the stability of the simulating circuit, we observe the effects of different amplitude and frequency on the output response. In this paper, we adopt the method of inputting analog value, observing the output state of the simulating circuit by inputting different sine waves. First, keeping frequency of the stress constant f = 12.5 Hz and changing the size of the amplitude, the results change with output-induced voltage in Figure 7. The greater the external force, the shorter time required for the output voltage to reach a stable state. Simulation results show that the circuit can output stable output DC voltage in different stress amplitudes.

Second, we tried to control the amplitude of stress always being 4.5N. By changing the frequency of stress, we observed the effects of different frequency to output response. With the increase of the frequency of the external force, the output voltage gradually reaches steady state at shorter times. Simulation results show that the circuit can output stable output DC voltage in different stress frequencies.

Different vibration conditions produce different voltages. Regardless of the amplitude or frequency of change, the results show that the design management circuit can output a stable 3.3V DC voltage.

3.4 *Energy storage management module*

MAX1811 is a high-performance charging management chip produced in the United States. The input voltage is 4.35–6.5V; the maximum charging voltage is 4.1 or 4.2V; the maximum voltage of error range is 0.5%; and the charging current can be controlled with 100 or 500 mA. It has a built-in temperature control circuit and a battery voltage detector. There is no necessity to be controlled by a microprocessor with low power consumption. MAX1811 pin diagram is shown in Figure 9.

The operating voltage of the designed management charging chip is 4.35–6.5V. However, when the 3.3V circuit acquisition and management is lower than 4.35V, the chip cannot work. In order to save energy and improve the charging efficiency, the acquisition power is stored in supercapacitors C1 temporarily. The power management chip cannot work until the input voltage reaches more than

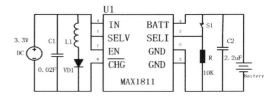

Figure 10. MSMA energy storage management module.

4.35V. Meanwhile, the temporary electric energy storage in supercapacitors can also supply energy to MAX1811. The light-emitted diode VD1 is lit. Once the voltage on the super capacitor is less than the threshold value of the chip, in order to reduce the dissipated energy, MAX1811 will automatically turn off. The output voltage of MAX1811 will be kept on a constant value as long as the voltage exceeds 4.35V, regardless of the fluctuation of the voltage of the supercapacitor.

In the test, the storage circuit input voltage is a stable DC voltage of 3.3V, and the SELV pin of the MAX1811 and SELI pin are set high and low, respectively, to make the lithium-ion charging current 100 mA and the final charge voltage 4.2V. In the charging experiment, a capacity of 250 mAh lithium battery-AK052030P is selected. Results show that the AK052030P lithium battery is fully charged in 2.8 h at a charging current and voltage are 100 mA and 4.2V, respectively.

4 CONCLUSION

In this paper, we designed a power management system based on MSMA. By utilizing the MSMA power management system, the mechanical energy in the environment can be transferred into electrical energy. The feasibility and superiority of the system are proved by experiment results. The advantages of the proposed method are small, green, and pollution-free. With the improvement of material properties and the progress of microelectronics and MEMS techniques, there is every reason to believe that integrated energy storage device will appear in the future.

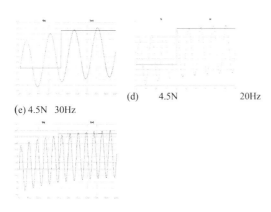

(d)   4.5N   20Hz

(e) 4.5N   30Hz

(f) 4.5N   40Hz

Figure 8.   Circuit output simulation diagram under different input stress frequencies.

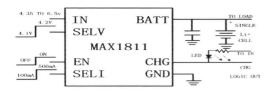

Figure 9.   MAX1811 pin map.

ACKNOWLEDGMENTS

This work was supported by the National Nature Science Foundation under Grant 51277126, Natural Science Foundation projects of Liaoning Province 2014024014, Excellent Talent Supporting Project of Higher Education Institution in Liaoning Province LR2013007, and Aviation Science Foundation under Grant 2014ZB54010.

REFERENCES

Aljanaideh O, AI janaideh M, Rakheja S, et al, Compensation of rate-dependent hysteresis nonlinearities in a magnetostrictive actuator using an inverse Prandtl-Ishlinskii model [J]. Smart Materials and Structures, Vol. 22, No. 2, 27–36, 2013.

John Kymissis, Clyde Kendall, Joseph Paradiso, Neil Gershenfeld. Parasitic Power Harvesting in Shoes [C]. the Second IEEE Internetional Conference on Wearable Computing, IEEE Computer Societu Press, 2000, 132–139.

Qingxin Zhang, Jian Li, Li Yu, Luping Wang. Thermo Magneto Mechanical Model and Simulation of Vibration Generator of Magnetically Controlled Shape Memory Alloy. The 27th Chinese Control and Decision Conference (2015 CCDC), May 23–25, 2015, Qingdao, China. pp. 4979–4983.

Wen Yumei, Wu Hanzhong, Li ping, Yin Wenjian, A self-powered power supply management circuit using frequency conversion network [J]. Acta Electronica Sinica, Vol. 40, No. 11, 2324–2329, 2012.

Zhang Q, Li J, Yu L, et al, Thermo magneto mechanical model and simulation of vibration generator of magnetically controlled shape memory alloy [C]. Control and Decision Conference (CCDC), 2015 27th Chinese. IEEE, 2015.

Zhang Qingxin, Yu Li, Gao Yunhong. "Vibration energy gathering power management system based on DC/DC boost converter circuit", 2016 Chinese Control and Decision Conference (CCDC), 2016.

Zhang, Qing Xin, Lu Ping Wang, Yu Huan Xie, and Zhan Bo Cui. "Vibration Energy Harvesting System Based on Magnetically Controlled Shape Memory Alloy", Advanced Materials Research, 2013.

Civil, Architecture and Environmental Engineering – Kao & Sung (Eds)
© 2017 Taylor & Francis Group, ISBN 978-1-138-02985-9

Architecture of the on-chip debug module for a multiprocessor system

Kexin Zhang & Jian Yu
Changzhou College of Information Technology, Changzhou, Jiangsu Province, China

ABSTRACT: This paper proposes an architecture description of the on-chip debug module aiming at a multiprocessor system. Some special function registers were used for setting the debugging mode and debugging flow control. Through the internal priority register setting, different processors can be set with or without different priory level, by which the different debugging structure can be realized. Internal arbitration mechanism handles the competition between different processors and decides the access order for different processors. Finally, all these techniques are integrated to make the on-chip debugging operation more efficient and flexible.

1 INTRODUCTION

With the development of integration technology, more and more electrical systems consist more than one highly integrated processor (Benini, 2005; Tan, 2012; S, 2014). With such a development, the complexity of the whole system and the increasing levels of integration result in novel challenges in the development of debugging methods.

For system with a single processor, some dedicated circuits would be used to gain access to some of the processor internal state, and the information will be communicated via serial interface such as JTAG (Maier, 2003; Portelagarcía, 2011); then, the on-chip debug function is realized.

Currently, on-chip debug method is mainly applied for single-processor systems, which cannot respond to the development of multiprocessor system. This paper proposes a high-performance on-chip debug architecture, which not only includes on-chip debug functions (such as internal or external breakpoint and single-stepping) for single processor, but also presents some specific features for multiprocessor systems. Such architecture has been developed to support on-chip debug for multiprocessor system, and the essential advantage is based on such architecture. The multiprocessor debug system with a high performance and flexibility would be built and used easily.

2 ON-CHIP DEBUG SUPPORT REALIZATION

2.1 Structure of on-chip debug module for single-processor system

The typical debugging device uses some dedicated facilities to gain access to some of the processor's internal state, and these devices generally use a serial debug interface (like JTAG) to communicate between the processor and the host computer (Zhang, 2013; Wang, 2011).

By contrast, as Figure 1 shows, this paper first presents the basic structure of on-chip debug module for single-processor systems, which is composed of three function units. First, the submodule of communication is used as an intermediate for data transmission between host PC and on-chip debug module. Second, the module configuration register submodule controls the operation modes of the entire on-chip debug module. Third, the control unit is responsible for controlling the entire debugging flow. The last unit of data register is used for temporary data storage.

Through this method, only some SFR (Special Function Register) and some control logic realize the main debugging functions. The definition of such SFRs determines the debugging mode, which decides under which conditions the debugging function will be triggered.

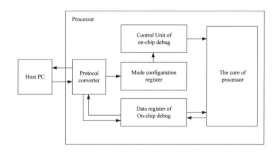

Figure 1. On-chip debug module for single-processor system.

2.2 Structure of on-chip debug module for multiprocessor system

Moving from single-processor to multiprocessor system, one critical problem needed to be solved is how to handle the competition of more than one processors sending debugging request to the same target processor. In this paper, the master-slave structure was adopted from the ARM protocol (Seys, 2009)

The on-chip arbitration is designed to be used as a central multiplexer. On the basis of the internal arbitration mechanism, all the master processors (e.g., microprocessors) drive out the debugging, requesting singles, and debugging commands, indicating that the debugging transfer they wish to perform and the on-chip arbiter of slave processor determine which request is granted. Figure 2 illustrates the structure required to implement the on-chip debugging aiming at multiprocessor system with three master processors and one slave processor.

Once the presetting conditions are satisfied, the operation of processor is paused and the system is moved into debugging mode, and the control of the system will be transferred to the on-chip debug module. During the debugging mode, the information required would be recorded and sent out. After the debugging work finished, some external stimulus is sent to microprocessors to make it work continually from the paused point. The additional pins would remain hidden otherwise and be controlled by the software developer.

2.3 Definition of debugging special function register

The on-chip debug system constructs the integrated entity that comprises hardware and software

together. The on-chip debug capability provides functions as breakpoint, single stepping, and external breaking. Either internal or external breaking is treated as a special form of interruption.

When implemented, the on-chip debug logic is part of the actual microprocessor silicon, the on-chip debug system provides the means to set the internal or external breakpoints, checking the internal states of the processors, and single step through coding the special function registers.

The definition of special function register for on-chip debug includes the following three main parts:

- Enabling/disabling on-chip debug system;
- On-chip debug mode setting;
- Debugging priority setting.

The format of the command, including all the above information that is sent from master processors, is depicted in Figure 3.

The debugging mode decides which debugging function is selected among the internal breakpoint, external breakpoint, single stepping, and other debugging functions. If internal breakpoint mode is selected, the source between special SFR or particular address comparing with target value set in breakpoint register can be further selected. Finally yet importantly, flag of priority level gives a clear sign of priority level, and the principle is the priority level decreases as the number increases, which means all zero get the highest priority.

2.4 Archiving of on-chip debug

When coming into debugging mode, the processor compares the corresponding signals with certain conditions, which were set beforehand. If the value matches one of the targeted values, the processor will come into breakpoint debugging mode. This is called the function of breakpoint.

The advantage is the breakpoint conditions could be set either external or internal. External breakpoint is the PC breakpoint, namely it comes into debugging mode when PC index reaches the given value. In the same way, on-chip debug module can also choose internal SFR comparing with target value.

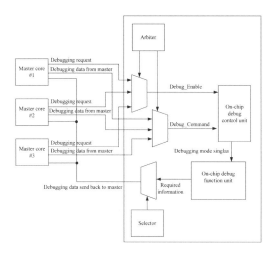

Figure 2. Structure of on-chip debug for multiprocessor.

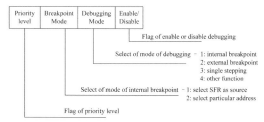

Figure 3. Format of debugging command.

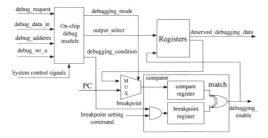

Figure 4. Hardware description of on-chip debug module.

As shown in Figure 4, as the debugging condition is satisfied, the matched signal will be sent out as debug-enabled signals, which indicates the clock module to pause the running of the clock, and then the entire processor will consequently stay in the pause state.

3 PRIORITY SETTING AND ARBITRATION

3.1 *Arbitration mechanism*

The arbitration mechanism is critical for multiprocessor system because the processors may run independently. There are inevitably some cases that more than one processor send debugging request to the same target microprocessor.

In our design, the internal on-chip arbitration mechanism is proposed to ensure that only one master has access to the target processor at any one time.

The main principles of the judgment are as follows:

- The internal on-chip arbiter performs the function of observing the debugging request from different master microprocessors and deciding which one is the highest priority master.
- Requests with different priorities: the request with a higher priority wins the control of the target microprocessor, the request with a lower priority should wait until the higher priority finished.
- On the contrary, if the request with the same priority, that is, internal Round-Robin mechanism, is applied, then the request would get the equal opportunity to get control of the target microprocessor. If one request wins, the others should wait until the debugging is finished.

The Round-Robin switch selects the request from different microprocessors having equal priority one by one. After the judgment, if a higher priority exists, the other debugging request should

be forced to keep in the waiting state; if any request having higher priority exists, the cycle repeats.

3.2 *Priority setting*

Select the system topology as the first step in designing a multiprocessor system because the performance of multiprocessor system depends heavily on the strategy of selecting the structure of the multiprocessor system.

Different applications based on the multiprocessor system leads to different structure requirements; thus, the primary target of our design provides a flexible on-chip debug module, which would be adjusted according to the multiprocessor system architecture. In our design, one agile method—different priority setting—was used.

The overview of processing of different priority settings is as follows:

- First, all the processors are regarded as equal before the priority setting; therefore, from the perspective of debugging, now the whole system is flat, no master processors or slave processors;
- Then, through the internal priority register setting, the processors under debugging can be divided into different groups with different debugging priorities. Apparently, for processors with lower priority, the higher one acts as master, which control the debugging processing;
- After priority setting, from the perspective of debugging, the whole system becomes a hierarchy now. One issue arises when more than one processors with the same priority sending debugging requests to the same slave processor. In such cases, some arbitration mechanisms are needed, as described in the "arbitration" section;
- The number of bits predefined in debugging special function register decides how many possible combinations on priority levels we have in our multiprocessor system debugging process. For instance, if three bits were used, then we would have eight possible priority levels, that is, from 3'b000 to 3'b111.
- With debugging priorities setting, different processors can get different or the same priority level; therefore, the debugging structure can be freely adjusted according to the practical applications.

4 DEBUGGING OPERATION

4.1 *Overview of the debugging operation process*

From the perspective of debugging, microprocessors in a multiprocessor system can be categorized into two types—debugging processor or target processor. The debugging processor act as master and the target processor acts as slave in the system. The protocol is designed to be used with a central multiplexor interconnection scheme. Using this scheme, all the debugging processors (masters) drive out the control information, indicating the debugging work they wish to perform to the central arbiter; then, through the internal mechanism, the arbiter determines which master gets the control of the debugging of the slaves.

As Figure 5 shows the detailed operation process of debugging, the complete debugging process has five steps:

- Step1: If the required debugging conditions are triggered, the normal operation of the debugging processor (as master in the whole debugging process) is paused, and the system gets into debugging mode;
- Step 2: Through the setting of debugging mode, the debugging control information becomes available, which includes the address of target processors (slave), and is drove out to the arbiter;
- Step 3: Before the debugging work commence, the debugging processor must be granted. This process is carried out by the debugging processor by asserting a debugging request to the central arbiter, whose role is to control which master accesses to the target processor (slave). For a detailed description of arbitration scheme, the reader is referred to section "Arbitration mechanism design".
- Step 4: A granted debugging processor starts the debugging work by driving the debugging control information to the target processor. The control information leads the target processor into debugging mode such as single-stepping, breakpoint of SFR, and breakpoint of address;

- Step 5: The significant advantage of on-chip debug is that it provides an efficient way to know the real-time state of processor under debugging. After the target processor is forced into debugging mode, the operation of processor is paused, and the specific area, such as some SFRs and memory, is recorded. These information are sent back to the debugging processor, which decides the next step.

5 SIMULATION RESULT

The experimental multiprocessor system platform was implemented containing four processors.

The simulation result in Fig. 6 shows that if debugging request is sent with a different priority level to the same slave, then the microprocessor with a higher priority level will win the competition; and if priority levels are the same, then the debugging request followed the round-robin order one by one.

We can conclude from Fig. 7 that when the predefined condition is satisfied (in this case, it means when the value of ACC matches the value of the breakpoint), the on-chip debug function is triggered, and the debugging process begins.

6 CONCLUSIONS

Debugging is a critical issue in the multiprocessor system, requiring a high–quality, efficient way to investigate the internal running state of the system. Current on-chip debug module is focused on the single-processor system. How to upgrade

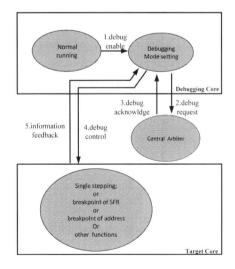

Figure 5.   Debugging operation process.

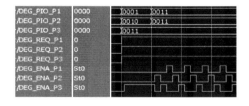

Figure 6.   Arbitration test.

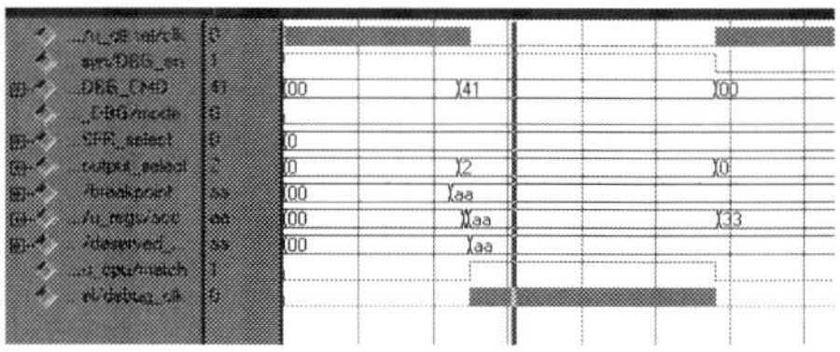

Figure 7.   On-chip debug function test.

such module to adjusting the ever increasing multiprocessor system and overcome the limitation of the debugging management and communication between multiprocessor systems is the main problem that needs to be addressed. From the available results, it is possible to conclude that the proposed on-chip debug module is an efficient and flexible solution for verifying and validating a multiprocessor system.

REFERENCES

ARM, AHB-Lite Protocol Specification, http://infocenter.arm.com

Benini, L., D. Bertozzi, A. Bogliolo, F. Menichelli, M. Olivieri, MPARM: Exploring the Multi-Processor SoC Design Space with SystemC, 《Journal of Vlsi Signal Processing》, pp. 169–182 (2005).

Maier, K.D. On-chip debug support for embedded Systems-on-Chip, International Symposium on Circuits & Systems, V-565–V-568 vol.5 (2003).

Portelagarcía, M., C. Lópezongil, M.G. Valderas, L. Entrena, Fault Injection in Modern Microprocessors Using On-Chip Debugging Infrastructures, IEEE Transactions on Dependable & Secure Computing, pp. 308–314 (2011).

Rusu S., Muljono H., Ayers D., et al. Ivytown: A 22 nm 15-core enterprise Xeon processor family, IEEE International Solid-State Circuits Conference Digest of Technical Papers, pp. 102–103, (2014).

Seys, S., B. Preneel. ARM: anonymous routing protocol for mobile ad hoc networks, International Journal of Wireless & Mobile Computing pp. 145–155 (2009).

Tan Hai, Zhou Xinqin, Tan Chengzhu, A new 3D network-on-chip for many-core system: China, 201210077519 [P]. (2012).

Wang Gang, Zhang Shengbing. On-chip debug architecture for MCU-DSP Core based system-on-chip, Computer Science and Automation Engineering (CSAE), IEEE International Conference on 2011, pp. 605–608 (2011).

Zhang Peng, Fan Xiaoya, Huang Xiaoping. An on-chip debugging method based on bus access, Signal Processing, Communication and Computing (ICSPCC), IEEE International Conference on 2013, pp. 1–5. (2013).

Civil, Architecture and Environmental Engineering – Kao & Sung (Eds)
© 2017 Taylor & Francis Group, ISBN 978-1-138-02985-9

Load balancing algorithm for computing cluster using improved cultural particle swarm optimization

Weihua Huang, Zhong Ma, Xinfa Dai, Yi Gao & Mingdi Xu
Wuhan Digital Engineering Institute, Wuhan, China

ABSTRACT: Aiming at defect of premature convergence in particle swarm optimization algorithm, an improved cultural particle swarm optimization algorithm is put forward for load balancing of computing cluster. Firstly, main population space of particle swarm optimization algorithm and knowledge space of cultural algorithm are combined to form the mechanism of "double evolution and double promotion", which can improve global searching capacity of algorithm and operational efficiency. Then evolution mechanism of genetic algorithm is adopted to enhance evolution operation of knowledge space. Finally, the algorithm is applied to load balancing problem of computing cluster to find solution. Experimental results show that the proposed algorithm increases resource utilization rate of computing cluster and makes load more balanced, proving to be an effective and reliable load balancing algorithm.

1 INTRODUCTION

Load balancing for computing cluster is to achieve equitable distribution of loads under certain constraint of real computation environment. Its purpose is to minimize response time of tasks for application programs, and it is proved to be a typical NP-hard problem (Z, 2016). Scholars have done extensive research in this field. Load balancing algorithm is mainly divided into static and dynamic way (Li, 2015). As static load balancing algorithm cannot correctly reflect current load on any node, its application is restricted (Jiang, 2016). Dynamic load balancing algorithm, which takes into consideration the current state of node load, becomes the main balancing algorithm employed to handle load balancing issues for computing cluster. Existing balancing algorithms optimize connection numbers and response time. Though they have good performances for small cluster system, they cause low efficiency and load imbalance for large cluster system (Wang, 2008; Sun, 2011).

Aiming at NP-hard feature of load balancing, some scholars put forward many inspirational algorithms, such as genetic algorithm (Wang, 2016), simulated annealing algorithm (Su, 2016), ant colony algorithm (Ghumman, 2015) and particle swarm optimization algorithm (Zhao, 2015). Due to the advantages of strong swarm intelligence and search capability, those algorithms can effectively optimize load balancing and improve resource utilization rate of CPU (Zhao, 2016; Jain, 2016; Kanimozhi, 2015). Among these algorithms, particle swarm optimization algorithm is a

swarm intelligence algorithm successfully applied to load balancing problem for computing cluster (Wu, 2010). However, particle swarm optimization algorithm can easily fall into local optimization and premature, causing certain inconsistency between optimization results and ideal results (Huang, 2012).

Cultural Particle Swarm Optimization (CPSO) (Deng, 2016; Wu, 2010) algorithm is an intelligent algorithm that integrates cultural algorithm into particle swarm optimization algorithm. During the evolution of CPSO, the particles can be updated by tracking two goals, namely global extreme value and individual extreme value (Yan, 2012). The continuous iteration of present global optimal solution up to now form a trajectory, which will be stored and considered as global knowledge space for global iterative search. Thus, acquired knowledge is transmitted to the next generation, guiding the individuals towards perceived global optimal solution and providing a systematic method for self-evolution. At the same time, greater global searching capacity of algorithm is achieved through double evolution and mutual effect of PSO space and knowledge space (Qin, 2016). However, due to the local optimum deficiency of knowledge space in cultural algorithm, knowledge space cluster cannot achieve desirable effect during self-evolution, reducing influence on lower-layer main cluster space.

To avoid above phenomenon, an **I**mproved **C**ultural **P**article **S**warm **O**ptimization algorithm (ICPSO) is put forward in this paper. Experimental results demonstrate that the proposed algorithm is an effective optimization algorithm for balancing computing cluster.

The rest of paper is organized as follows: Section II describes the load balancing problem. In Section III, the improved algorithm is introduced. Section IV explains in detail the working process of the algorithm. Section V gives the experimental settings and the result discussions. Finally, conclusions and further research directions are given in Section VI.

2 LOAD BALANCING PROBLEM

Load balancing for computing cluster performs reasonable load distribution so as to improve resource utilization rate of cluster system and accelerate response time of user requests.

Suppose cluster system consists of n computing nodes, $\{N_1, N_2, \cdots, N_n\}$, and within a period of time, there are m concurrent requests, then load index definition of the i-th node is as follows:

$$Load(N_i) = \omega_1 * Lcpu(N_i) + \omega_2 * Lmemory(N_i) + \omega_3 * Lio(N_i) + \omega_4 * Lqtime(N_i) \quad (1)$$

where weighted value is $\sum \omega_i = 1$.

Suppose that T_i is the time needed when computing node N_i processes user requests, then the optimum balancing scheme for computing cluster is to find out the minimum value of $\sum T_i$, which can be denoted by:

$$\min \sum \tilde{T}_i \quad (2)$$

3 IMPROVED CULTURAL PARTICLE SWARM OPTIMIZATION ALGORITHM

Cultural algorithm obtains useful knowledge and information through the evolution space of micro-level, main population space, and reserves it in the evolution space of macro-level, belief space. By incorporating population evolution mechanism of genetic algorithm into self-evolution process of knowledge space, the proposed algorithm can evolve and update knowledge space cluster through selection, crossover and mutation, so as to improve global searching capacity and operational efficiency of its evolutionary operations. The framework of ICPSO algorithm is as shown in Fig. 1.

Based on population evolution mechanism of genetic algorithm, the evolution and update of knowledge space is improved to evolve through three operations: selection, crossover and mutation.

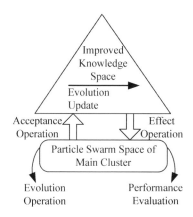

Figure 1. Framework of ICPSO algorithm.

3.1 Selection operation

Step 1. Calculate fitness value of each individual: $f(i)$, $i = 1, 2, \cdots, n$.

Step 2. Confirm individual selection probability $P(C_i)$. Individual sorted from small to large according to adaptability is marked as: $\{C_1, C_2, \cdots, C_n\}$, and then individual selection probability is:

$$P(C_i) = \frac{1}{n}\left\{\sigma^g - \frac{\sigma^a - \sigma^b}{n-1}(i-1)\right\} \quad (3)$$

In the above equation, i is ordinal number of individual; σ^g is expected value of optimal individual C_1 after selection operation and $\sigma^g = n \times P(C_1)$; σ^b is expected value of the worst individual C_n after selection operation and $\sigma^b = n \times P(C_n)$. Generally, we require $1 \leq \sigma^g \leq 2$ and $\sigma^b = 2 - \sigma^g$. When $\sigma_g = 2$ and $\sigma^b = 0$, the expected number for the worst individual surviving in next generation is 0 and selection probability of optimal individual is obviously greater than that of other individuals, which will lead the algorithm to converge too early. When $\sigma^g = \sigma^b = 1$, selection pattern (Liu, 2016) becomes distributed random selection. In real practice, $\sigma^g = 1.1$.

Then individual probability is calculated according to equation (3), namely, recording relatively superior individual in accordance with rules of roulette wheel selection method.

3.2 Crossover operation

This paper adopts single-point crossover method to perform operation. A single crossover point is randomly selected for both father generations. Then all priority value beyond that point is

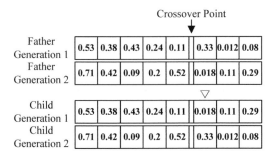

	Crossover Point ↓							
Father Generation 1	0.53	0.38	0.43	0.24	0.11	0.33	0.012	0.08
Father Generation 2	0.71	0.42	0.09	0.2	0.52	0.018	0.11	0.29
					▽			
Child Generation 1	0.53	0.38	0.43	0.24	0.11	0.018	0.11	0.29
Child Generation 2	0.71	0.42	0.09	0.2	0.52	0.33	0.012	0.08

Figure 2. Crossover operation.

swapped between two parent organisms, as shown in Fig. 2, to form two new child generations.

3.3 Mutation operation

Mutation operator performs operations based on mutation probability P_m. The priority value (Bonyadi, 2016) of some individual is selected according to P_m, and then their original values are replaced by values generated randomly between interval (0,1) according to uniform distribution.

4 PROCEDURE OF LOAD BALANCING BASED ON ICPSO

The load balancing algorithm for computing cluster using improved cultural particle swarm optimization works as follows:

Step 1. Initialize parameters, such as the maximum iterations of main population space and knowledge space respectively, G_1 and G_2, inertia weight ω, population size N and so on;

Step 2. Calculate particle fitness value of each cluster, and record the present optimal position and present optimal fitness value;

Step 3. Calculate new speed and position for particles in knowledge space according to equation (4) and (5), and meanwhile, limit the range for new position and speed of each particle respectively,

$$v_{id}^{t+1} = \omega v_{id}^t + c_1 r_1 (p_{id}^t - x_{id}^t) + c_2 r_2 (p_{gd}^t - x_{id}^t) \quad (4)$$

$$x_{id}^{t+1} = x_{id}^t + v_{id}^{t+1} \quad (5)$$

In the above equations, ω represents inertia weight; c_1 and c_2 are learning factors; r_1 and r_2 are random numbers distributed uniformly within range of (0,1); vector x_i^t and v_i^t respectively represent the position coordinate and velocity of the i-th particle in t-th generation; p_i^t represents

present optimal location of individual; p_g^t represents global optimal location of population;

Step 4. Perform the evolution and update of knowledge space using improved cultural particle swarm optimization. Replace the present optimal location of the particle with new fitness value if fitness value is greater than the optimal location of the particle; Update present global optimal location to the new fitness value if fitness value is greater than the global optimal location. Thus, the evaluation index of the premature convergence for the particle swarm is computed;

Step 5. Compare fitness value of each particle with extreme value of cluster, and if it is more superior to extreme value of cluster, then replace extreme value of cluster with fitness value of the particle, and replace optimal location of cluster with location of the particle;

Step 6. Judge whether acceptance function is true. If it is true, it means the output value of function is larger than theoretical sum. This output value describes the adaptive degree of global optimal particle for both population space and knowledge space. Then replace the particle with the worst fitness in knowledge space with particle that has the best fitness value in the population space;

Step 7. When maximum allowable number of iteration is exceeded or the searched optimal location satisfies the minimum threshold value, the search should stop and optimal location and optimal fitness value will be output. Otherwise, return to step 3 to continue the search;

Step 8. Decode optimal location node for task allocation and obtain optimal scheme of load balancing for computing cluster.

5 EXPERIMENTS

5.1 Experimental environment

In order to verify the performance of the proposed algorithm, experiments are carried out. The framework of computing cluster is as shown in Fig. 3.

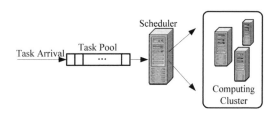

Figure 3. Framework of computing cluster.

The overall cluster environment is configured with CentOS 6.6 operating system, one test terminal, one management node and 30 computing nodes.

The performance of the algorithm will be evaluated by the following three indexes: average utilization rate of CPU, average response time to user requests, and average throughput capacity of cluster. Meanwhile, Generic Algorithm (GA) and standard Cultural Particle Swarm Optimization algorithm (CPSO) will also be tested for contrast analysis. Parameters for GA are specified as: population size is 20; crossover rate is 0.8; mutation rate is 0.02. Parameters of ICPSO is specified as: population size is 50; inertia factor ω is 1.2; c_1 and c_2 are 2; all algorithms are iterated 200 times.

5.2 Results and analysis

Regarding experiments on GA, CPSO and ICPSO, CPU utilization rate is sampled every minute. CPU utilization rate of five groups of experiments is shown in Fig. 4; average response time for user requests is shown in Fig. 5; throughput capacity of computing cluster is shown in Fig. 6.

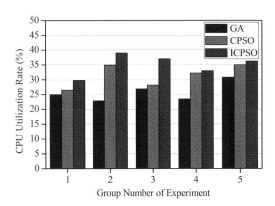

Figure 4.   Comparison of CPU utilization rate.

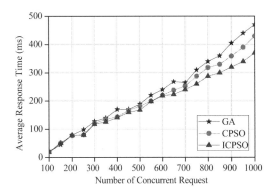

Figure 5.   Comparison of average response time.

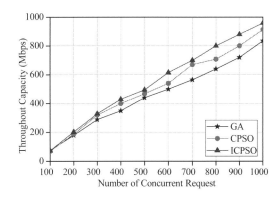

Figure 6.   Comparison of throughput capacity of computing cluster.

From Fig. 4, CPU utilization rate of GA algorithm is the lowest, and it is mainly because of local optimum and premature defect in GA algorithm. CPU utilization rate of CPSO algorithm is higher than that of GA algorithm, and it is mainly because cultural algorithm and particle swarm optimization algorithm complement each other to form a more superior computing load dispatch scheme. CPU utilization rate of ICPSO is the highest. Comparison results show that the "double evolution and double promotion" mechanism of the main population space and knowledge space can improve global searching capacity and operational efficiency. The evolution mechanism of GA is adopted to avoid the problem that knowledge space of cultural algorithm is likely to be trapped into local optimization. Thus, global optimization solution can be obtained and overall utilization rate of computing cluster is improved.

Fig. 5 shows that as the number of user requests increases, average response time of GA, CPSO and ICPSO increases correspondingly. The time increasing degree for GA is the greatest, followed by CPSO and ICPSO has the smallest increasing range. Compared with GA and CPSO algorithm, ICPSO algorithm accelerates average response for user requests. Computing cluster using ICPSO has good performance of load balancing so as to take respond to user request more quickly. This feature is favorable especially for large scale computing cluster with massive user requests.

It can be observed from Fig. 6 that throughput capacity of ICPSO algorithm is obviously greater than that of GA and CPSO algorithm because of the capability of global optimization. This means load scheduling for computing cluster through ICPSO algorithm is more applicable and effective to large scale computing cluster.

From the above analysis, the conclusion can be reached that ICPSO outperforms GA and CPSO.

6 CONCLUSION

Aiming at deficiency existing in standard particle swarm optimization algorithm, an improved cultural particle swarm optimization algorithm for load balancing is put forward. Experimental results show that the proposed algorithm is an effective and reliable load balancing algorithm that can improve resource utilization rate and make load more balanced for computing cluster. However, further work should be dedicated to establishing the theory model of parameter setting and figuring out how it influences the final solution.

ACKNOWLEDGMENT

This work is sponsored by National Science Foundation of China grant No. 61502438.

REFERENCES

Bonyadi, M., Z. Michalewicz, Stability Analysis of the Particle Swarm Optimization Without Stagnation Assumption, *IEEE Transactions on Evolutionary Computation*, 814–819 (2016).

Deng, C., Y. Liu, L. Xu, J. Yang, J. Liu, et al. A MapReduce-based Parallel K-means Clustering for Large-scale CIM Data Verification, Concurrency and Computation Practice and Experience, 28, 11, 3096–3114 (2016).

Ghumman, N., R. Kaur, Dynamic Combination of Improved Max-min and Ant Colony Algorithm for Load Bbalancing in Cloud System, 2015 6th International Conference on Computing, Communication and Networking Technologies (ICCCNT), 1–5 (2015).

Huang, H., J. Guo, B. Wang, An Improved KNN Algorithm Based on Adaptive Cluster Distance Bounding for High Dimensional Indexing, 2012 Third Global Congress on Intelligent Systems (GCIS), 213–217 (2012).

Jain, A., R. Kumar, A Multi Stage Load Balancing Technique for Cloud Environment, 2016 International Conference on Information Communication and Embedded Systems, 1–7 (2016).

Jiang, D., Z. Xu, Z. Lv, A Multicast Delivery Approach with Minimum Energy Consumption for Wireless Multi-hop Networks, Telecommunication Systems, 62, 4, 771–782 (2016).

Kanimozhi, T., K. Latha, An Integrated Approach to Region Based Image Retrieval Using Firefly Algorithm and Support Vector Machine, Neurocomputing, 151, 7, 1099–1111 (2015).

Li, X., Z. Lv, J. Xu, B. Zhang, LY. Shi, et al. Xearth: A 3d Gis Platform for Managing Massive City Information, 2015 IEEE International Conference on CIVEMSA, 1–6 (2015).

Liu, J., X. Li, An Analysis of the Inertia Weight Parameter for Binary Particle Swarm Optimization, *IEEE Transactions on Evolutionary Computation*, 666–681 (2016).

Lv, Z., T. Yin, H. Song, and G. Chen, Virtual Reality Smart City Based on WebVRGIS, IEEE Internet of Things Journal, 6, 99, 1–1 (2016).

Qin, Q., S. Cheng, Q. Zhang, L. Li, Y. Shi, Particle Swarm Optimization With Interswarm Interactive Learning Strategy, *IEEE Transactions on Cybernetics*, 2238–2251 (2016).

Su, N., A. Shi, C. Chen, E. Chen, Y. Wang, Research on Virtual Machine Placement in the Cloud based on Improved Simulated Annealing Algorithm, *2016 World Automation Congress (WAC)*, 1–7 (2016).

Sun, L., S. Yoshida, Y. Liang, A Support Vector and K-Means Based Hybrid Intelligent Data Clustering Algorithm, Ieice Transactions on Information and Systems, 94, 11, 2234–2243 (2011).

Wang, B., J. Li, Load balancing task scheduling based on Multi-Population Genetic Algorithm in cloud computing, *2016 35th Chinese Control Conference (CCC)*, 5261–5266 (2016).

Wang, Q., H. Chen, Y. Shen, Decision Tree Support Vector Machine based on Genetic Algorithm for Fault Diagnosis, IEEE International Conference on Automation and Logistics, 2668–2672 (2008).

Wu, S., G. Wei, High dimensional data Clustering Algorithm Based on Sparse Feature Vector for Categorical Attributes, 2010 International Conference on Logistics Systems and Intelligent Management, 973–976 (2010).

Wu, Y., X. Gao, X. Huang, K. Zenger, A Cultural Particle Swarm Optimization Algorithm, *2010 Sixth International Conference on Natural Computation* (ICNC), 2505–2509 (2010).

Yan, J., W. Li, W. Chen, W. Luo, C. Zhang, et al. Cultural Algorithm for Engineering Design Problems, IJCSI International Journal of Computer Science Issues, 9, 6, 53–61 (2012).

Zhao, J., Z. Ma, Virtual Network Mapping Algorithm based on Load Balancing Multi-objective Particle Swarm Optimization, 11th International Conference on Wireless Communications, Networking and Mobile Computing (WiCOM 2015), 1–5 (2015).

Zhao, J., K. Yang, XH Wei, Y. Ding, L. Hu, et al. A Heuristic Clustering-based Task Deployment Approach for Load Balancing Using Bayes Theorem in Cloud Environment, IEEE Transactions on Parallel and Distributed Systems, 27, 2, 305–316 (2016).

Civil, Architecture and Environmental Engineering – Kao & Sung (Eds)
© 2017 Taylor & Francis Group, ISBN 978-1-138-02985-9

An analysis of open source operating systems based on complex networks theory

Denghui Zhang
School of Mechanical Engineering, Shandong University, Jinan, China

Zhengxu Zhao
School of Information Science and Technology, Shijiazhuang Tiedao University, Shijiazhuang, China

Yiqi Zhou
School of Mechanical Engineering, Shandong University, Jinan, China

Yang Guo
School of Information Science and Technology, Shijiazhuang Tiedao University, Shijiazhuang, China

ABSTRACT: Open Source Software (OSS) is the software which grants access to its source code. A better understanding of how OSS functions may help developers take a more effective means for the system development. Studying software systems by the complex network theory can potentially provide useful insights into the diversity and success of OSS. However, research efforts on OSS concentrates on relationships among software systems. Little attention has been paid to operating systems themselves. In this paper, a data collection framework is proposed, and the distribution network of Linux operating systems is constructed. The network models distributions as nodes and dependencies among them as edges according to data crawled from DistroWatch. It is found that the distribution network is a scale-free and small world network similar to those identified in other fields. The inconsistency in share of distributions and targeted users is revealed and discussed. It is expected the constructed model can be a guide for distribution development in the future.

1 INTRODUCTION

In recent years, the dramatic growth to identify and classify network has been witnessed in a wide variety of fields. Rather different from those found in regular networks or simple random networks, it has been discovered that the underlying structures of these networks—including scientific collaboration networks (Wang, 2013), movie actor collaboration in sociology, and Power Grids (Pagani, 2013) and engineering informatics (Zhao, 2008) in engineering—share many scale-free and small-world qualities. The small-world characteristic of complex networks reflects "six degrees of separation" phenomenon in a real social network, while its scale-free feature bears the Matthew effect in the field of economy.

Software systems have become the core of the information-based world and take an important part in modern society. The rapid development in the Open Source Software (OSS) domain gives researchers opportunity to access kinds of software systems and collect data easily. Studying open source software systems from a perspective of complex networks theory contributes to manage its functional complexity and high evolvability (Zheng, 2008). While research results of modeling software system as complex networks in turn conduce to a better understanding of other forms of complex networks.

Studies of software systems based on complex networks theory nowadays are mainly focusing on dependency of software modules and packages. Christopher Zachor (2013) examined the structure, function, and evolvability of software collaboration networks. All of them reveal scale-free and/or heavy-tailed degree distributions, which implies software systems represent another important field which complex networks theory can contribute to the quantifiable measures. One disadvantage of Christopher's work is that only a few hundred software packages are analyzed. Xiaolong Zheng et al.(2008) modeled the whole package network of a Linux distribution Gentoo. They developed two new network growth models which take into consideration aging effect of old nodes to better explain empirical results.

Social network theory provides another powerful tool to model individuals as nodes and relationships among them as edges. Orcun Temizkan et al. (2015) modeled open source projects from SourceForge as self-organizing and collaborate social networks. They argued open source movement is preferentially connected networks. Through the social networks analysis of virtual communities for OSS projects, S.L. Toral et al. (2010) discern the major contributors in OSS projects. The results suggested a small brokers network plays an important role in projects. All of works is to analyze relationships among packages and individuals. However, little of them have focused on the relationship among operating systems underneath OSS.

In this paper, the Linux distributions dependency is analyzed by modeling it as a complex network based on the data crawled from the DistroWatch website. It is hypothesized that Linux distributions display preferential attachment in its structure. The empirical analysis suggests this is the case. The rest of this paper is structured as follows. In Section 2, a framework of data collection is presented. The distribution network is constructed by modeling distributions as nodes and dependencies among them as edges. In Section 3, an empirical analysis of the Linux distribution network is taken on. Results show that the Linux distribution network complies with a small-world and scale-free network similar to those identified in other fields. The research provides another example of complex networks in the real world. At last, the conclusions are summarized and possible future research is discussed.

2 METHODOLOGY

The Linux kernel was first proposed by Linus Torvalds in 1991. Prior to that, Richard Stallman found the Free Software Foundation (FSF) and the GNU project to contribute to various GNU programs. After continuous outstanding developers joining the GNU project, they created the Linux, also known as GNU/Linux system. Different Linux distributions are used for different purposes ranging from embedded devices and personal computers to powerful supercomputers. The distributions come in all shapes and sizes. They can be divided into two categories, one is a commercial company maintenance release with Red Hat as the representative, and the other is entirely community-driven distributions. Debian is representative of the latter.

A network which has parts or all of the features of self-organization, self-similar, attractor, small-world and scale-free is called a complex network (Ren, 2016). The complex network theory aims to reveal the principles of network systems forming, and remaining robust and adaptable when evolving. The rise of the open source software movement gives researchers sufficient data and chances to apply complex networks theory to. For this study, data was gathered from the DistroWatch, a web-based Linux distributions popularity ranking project. It provides information for over 200 distributions. It is noted that not all distributions are listed at DistroWatch. However, given the popularity of the site, it is rational to suppose distributions at DistroWatch could be representative of Linux distributions.

The data collection architecture is shown in Figure 1. Scrapy, a python-based crawler module, drives the data flow. Scrapy can be adopted to extract information from a site like DistroWatch which does not provide an API or other programmable access mechanism. Spiders schedule the first URL to crawl based on CrawlerRules. Scheduler sorts URL requests into a queue, and then sends them to the Downloader. Once the webpage is downloaded completely, Downloader sends the response content to Spiders, and then Spiders transfers the response to a HTML Filter for further process. Spiders returns new requests to the Scheduler at the same time. HTML Filter is the key of the architecture. It extracts the distribution dependency from a webpage. Due to the asynchronous network, the dependency is first saved in an intermediate file for each distribution webpage. After all of dependencies are collected, they are converted into a GraphML file which models distributions as nodes and dependencies among them as edges for the next complex network analysis. The process repeats until there are no more requests from the Scheduler.

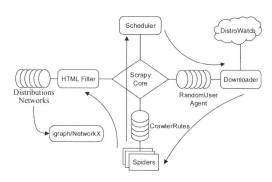

Figure 1.   Data collection architecture of DistroWatch.

It can be noted that there are two middlewares CrawlerRules and RandomUserAgent in the architecture. The RandomUserAgent is customized to prevent the crawler being identified or blocked by servers because of the generic user agent being frequently used. In order to improve the efficiency of the crawler, CrawlerRules defines a set of rules to only crawl pages containing distribution-specific information rather than all of them. Considering small size of the network, either igraph or NetworkX (Akhtar, 2014) module can be used to analyze characteristics of distribution networks.

3 RESULTS AND DISCUSSION

At last, after removing a self-loop edge, a complex network with 286 nodes and 318 edges was constructed as of the last 12 months. The average degree is about 1.12. The number of edges is 40755 in a complete graph with the same number of nodes, which is more than 128 times bigger. That is, the distribution network is sparse, which is similar to the Gentoo network.

In this section, properties of the distribution network are analyzed. The degree distribution and clustering coefficient are mainly focused.

Power-law Distribution: The real-world networks tend to show deviation from randomly constructed graphs with two non-trivial properties: power-law distribution and small-world effect. That is, the probability of a vertex having m edges decays with respect to the constant a. If the node degree of a network follows the power-law distribution, it can be said the network is a scale-free network, while random networks follow Poisson distribution basically.

Let $p(x)$ be the probability distribution function. If its histogram is a straight line on log-log, that is, $p(x) = -a \ln x + c$ where a and c are constants. With the maximum likelihood method, the exponent a of the degree distribution could be calculated as follows:

$$a = 1 + n \left[\sum_{i=1}^{n} In \frac{x_i}{x_{min}} \right] \quad (1)$$

where x_{min} is the minimum value above which power-law only follows at real-world networks. So the exponent of a network could be got from a set of n values x_i.

The degree distribution of the network can be seen in Figure 2. The plot shows that the degree of the distribution network follows power-law distribution. That is, the network is a scale-free network. It declines rapidly with the dashed line with

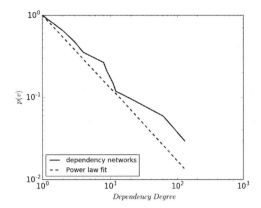

Figure 2. Probability density function of the distribution network.

a slope a, sharing same properties with other real networks.

Clustering Coefficient: Let $\lambda_G(V_i)$ be the number of all closed triplets (3 edges and 3 vertices) including v_i in graph G, Let $\tau_G(v)$ be the number of all open triplets (2 edges and 3 vertices) including v_i in G. The clustering coefficient $\bar{C}$ is defined below:

$$\bar{C} = \frac{1}{n} \sum_{i}^{n} C(i) = \frac{1}{n} \sum_{i=1}^{n} \frac{\lambda_G(v_i)}{\tau_G(v_i) + \lambda_G(v_i)} \quad (2)$$

In a random network, there is $\bar{C} \sim 1/n$. However, it has been shown that nodes have a strong tendency to form groups, which means real networks has a bigger clustering coefficient compared to random networks. Especially, a network with highly clustering coefficient like regular networks and small average path length like random network is called small world networks.

The clustering coefficient for the distribution network is about 0.34, while the clustering coefficient of a corresponding random graph is $\bar{C} = M / N(N-1) = 0.0039$. So the clustering coefficient of the distribution network is 87 times higher than the random network. For the corresponding Barabási–Albert model with same number of nodes and $m = 2$, a network with 569 edges is obtained, where m is the number of new edges in each step. The clustering coefficient of the network is 0.12, which is in the same order of magnitude as that of the distribution network. Considering its average path length is 2.36, it could be said that the distribution network is a small world network, which provides further evidence that complex networks is widespread in the real world.

Different with other social or biological networks like the actor-movie network, where over 90% of actors belong to a giant cluster, the largest cluster with Ubuntu and debian as its core comprises of 52% of distributions, which indicates that the distribution network is not as connected as other complex networks. One of the reason may be the diversity of distributions. There are 36 distributions which do not belong to any clusters. Two quintessential examples that should be cited are Smoothwall which is a family to Internet security products and SliTaz which is designed to run speedily on limited hardware. These distributions are not as universal as other desktop distributions, so they have to been developed independent.

The second largest cluster is a Red Hat focused network. The two clusters are representative of Linux distributions in desktop and server operating systems. As Diomidis etc. (2012) proposed that the function to fix or modify operating systems based on users' particular needs makes OSS a choice of competitive advantage for companies. Although there in only about 1.6% desktop computers using Linux, distributions oriented toward desktop use occupy the majority. It can be explained by the diverse needs of users for desktop environment. Anyone may create a distribution according to his flavor. While Linux is a leading operating system on servers, distributions oriented toward server use are relatively less. The reason may be that Linux servers have to run for years without failure. The downtime could have disastrous consequences for users. So only well-known server distributions survive.

4 CONCLUSION

With the rapid development of big data and cloud computing technology, open source operating systems are playing an increasingly important and irreplaceable role. In this paper, a distribution network with data crawled from DistroWatch was constructed. Different with previous random networks, the work has for the first time discovered that the distribution network is a scale-free and small world network. A giant cluster comprising of 52% distributions exists in the network. The results also reveal marked inconsistencies in share of distributions and targeted users. Desktop users who contribute most of the distributions have a priority need of customized systems, while server users focus on the stability of the system. This may help explain the success of OSS movement. That is, do one thing well and provide users with alternatives.

The study mainly concentrates to degree distribution and clustering coefficient of a complex network. Further studies could involve other features of a complex network, such as closeness centrality, betweenness centrality, degree growth rate, etc., which will provide useful insights into open source operating systems in the future.

ACKNOWLEDGMENTS

The authors would like to thank editors and reviewers for their detailed reviews and constructive comments. This study is supported by National Key Technology R&D Program of the Ministry of Science of China under Grants. 2015BAF07B04 and High-level Personnel Training Plan in Hebei Province of China under Grants. Z1100903.

REFERENCES

Akhtar, N., Social Network Analysis Tools, *Communication Systems and Network Technologies (CSNT), 2014 Fourth International Conference on*, IEEE, 388–392, (2014).

Pagani, G.A., and M. Aiello, The Power Grid as a complex network: A survey, Phys. Stat. Mech. Its Appl., **392**, 2688–2700, (2013).

Ren, J., H. Wu, R. Gao, G. Huang, and J. Dong, Identifying Important Nodes in Complex Software NETWORK Based on Ripple Effects, ICIC Express Lett. Part B Appl. Int. J. Res. Surv., **7**, 257–264, (2016).

Spinellis, D., and V. Giannikas, Organizational adoption of open source software, J. Syst. Softw., **85**, 666–682, (2012).

Temizkan, O., and R. L. Kumar, Exploitation and Exploration Networks in Open Source Software Development: An Artifact-Level Analysis, J. Manag. Inf. Syst., **32**, 116–150, (2015).

Toral, S.L., M.R. Martínez-Torres, and F. Barrero, Analysis of virtual communities supporting OSS projects using social network analysis, Inf. Softw. Technol., **52**, 296–303, (2010).

Wang, D., C. Song, and A. L. Barabási, Quantifying long-term scientific impact, Science, **342**, 127–132, (2013).

Zachor, C., and M. H. Gunes, Software Collaboration Networks, *Complex Networks*, Springer Berlin Heidelberg, 257–264, (2013).

Zhao, Z., and L. Z. Zhao, Small-world phenomenon: toward an analytical model for data exchange in Product Lifecycle Management, Int. J. Internet Manuf. Serv., **1**, 213–230, (2008).

Zheng, X., D. Zeng, H. Li, and F. Wang, Analyzing open-source software systems as complex networks, Phys. Stat. Mech. Its Appl., **387**, 6190–6200, (2008).

Civil, Architecture and Environmental Engineering – Kao & Sung (Eds)
© 2017 Taylor & Francis Group, ISBN 978-1-138-02985-9

Design of fitness information recording and network monitoring system based on BDS/GPS and Arduino Yun

Chao Jiang
School of Electronics and Information Engineering, Shanghai University of Electric Power, Shanghai, China
School of Communication and Information Engineering, Shanghai University, Shanghai, China

Jun Yu Wu & Yun Xuan Tu
School of Electronics and Information Engineering, Shanghai University of Electric Power, Shanghai, China

Yu Yun Hu
School of Mechatronic Engineering and automation, Shanghai University, Shanghai, China

ABSTRACT: In this paper, we propose a fitness information recoding and network monitoring system prototype based on BDS/GPS and Arduino Yun. UM220-III NL as a BDS/GPS of China independent intellectual property rights affords accurate coordinate position. Arduino Yun, integrating Atmega 32U4 and Linino AR9331, is used as the controller and Wi-Fi connecting to Yeelink, which offers powerful cloud service. The experimental results show that our proposed system prototype has not only realized fitness information recoding and network monitoring, but also provided accurate and reasonable description by utilizing its low-cost, convenient, and controllable characteristics.

1 INTRODUCTION

People vigorously carry out the national fitness program and promote the balanced development of recreational sports and competitive sports. The Rio Olympic Games showed a big grand feast to people all over the word. Athletes and fitness enthusiasts have been always deeply concerned about their fitness training (Noh, 2010). Besides professionals, people of different ages have got into fitness activities (Wen, 2013). Other than traditional styles, GPS, ECG, body area sensor network, walking route navigation system, and many other methods (Choi, 2013; Varatharajah, 2013) have been used to assist fitness activities, such as walking, hiking, and outdoor fitness equipment and medical monitoring (Pitman, 2012; Komninos, 2015). Even some merchants develop several fitness applications for smartphones (Buttussi, 2010; Altini, 2014). The construction of urban and rural public sports and exercise facilities are accelerated, and the national fitness program has taken root.

With the motivation of the above literature review, a prototype of fitness information recording and network monitoring system based on BDS/GPS and Arduino Yun is proposed in this paper. The main goal of this work is to design and implement a low-cost, commonly used, convenient, and controllable prototype.

In Section II, we review related works about BDS/GPS, Arduino Yun, and Yeelink. Section III presents our proposed system. Experimental results of the proposed system are shown in Section IV. Finally, conclusions are summarized in Section V.

2 RELATED WORKS

2.1 BDS/GPS

Global Navigation Satellite Systems (GNSS) provides significant benefit to improve satellite geometry, accuracy, integrity, continuity, and availability. There are four operational GNSS: Global Positioning System (GPS), GALILEO, GLONASS, and BeiDou System (BDS). The BDS navigation satellite system is an important part of GNSS, which is independently built by China. According to its overall planning schedule, the BD system is planned to be established completely and provide global service by 2020. Then, it will consist of five Geostationary Earth Orbit (GEO) satellites and 30 Non-Geostationary satellites (Ren, 2015). Moreover, the interoperability between BDS and other navigation satellite systems is expected to further enhance the accuracy of district positioning contribution based on BDS/GPS navigation satellite system (Basiri, 2014). Positioning technology for fitness purpose has been widely studied by many scholars (Peng, 2016).

Navigation is becoming more and more important from national strategy to civilian facilities. Smart wearable devices with BDS/GPS chip are emerging. At present, there are mainly four BDS/GPS companies in China: Icofchina in Hangzhou, Unicorecomm in Beijing, CEC Huada Electronic Desgin Co., Ltd. in Beijing, and MenXin Technology in Wuhan.

UM220-III NL of Unicorecomm is a BDS/GPS dual-system module designed for auto aftermarket, such as vehicle monitoring and navigation. It is also a good choice for handheld devices. Here, we choose UM220-III NL as our BDS/GPS navigation. As shown in Figure 1, UM220-III NL is the third generation of UM220 series module based on Unicore Low power GNSS SoC (HumbirdTM)1, and is by far the smallest domestic BDS/GPS module in the market. UM220-III NL is small and lightweight, requires ultralow power (120mW), less expensive, has high precision of 0.1m/s (RMS), data update rate of 1Hz, and independent intellectual-property rights.

Table 1 shows the RMC protocol description of NMEA message for UM220-III NL, the format of which is $--RMC, time, status, Lat, N, Lon, E, spd, cog, date, mv, mvE, mode*cs, such as $GPRMC, 123400.000,A,4002.217821, N, 11618.105743,E,0.026,181.631,180411„E,A*2C.

2.2 *Arduino*

In this paper, the Wi-Fi, which has built-in Arduino Yun Board, allows only the authorized user to interact with the Internet.

By using wireless communication, the data packets and signal are transferred in a dynamic environment between the user and system device. Arduino is an open-source simple tool that can sense, monitor, store, and control more applications than desktop computers.

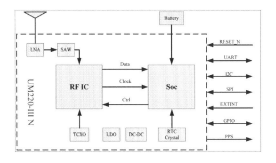

Figure 1.   Structure diagram of UM220-III NL.

As shown in Figures 2 and 3, Arduino Yun is a microcontroller board based on the ATmega32u4 and the Atheros AR9331, which supports a Linux distribution based on OpenWrt named OpenWrt-Yun. The board has built-in Ethernet and Wi-Fi support, a USB-A port, micro-SD card slot, 20 digital input/output pins, a 16 MHz crystal oscillator, a micro-USB connection, and an ICSP header.

Table 1.   RMC protocol description of NMEA message.

Parameter name	Type	Description
–	STR	GP(GPS); BD(BDS); GN(GPS and BDS)
time	STR	hhmmss.sss
status	STR	V: invalidity; A: validity
Lat	STR	ddmm.mmmmmm
N	STR	N: north latitude; S: south latitude
Lon	STR	ddmm.mmmmmm
E	STR	E: east longitude; W: west longitude
spd	Double	Land speed
cog	Double	Land direction (clockwise from north)
date	STR	Ddmmyy
mv	Double	magnetic declination (always vacant)
mvE	STR	magnetic declination direction (E)
mode	STR	Location: N(not); A(single); D(double)
cs	STR	Checksum (from '$' to '*')

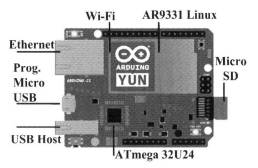

Figure 2.   Arduino Yun PCB board.

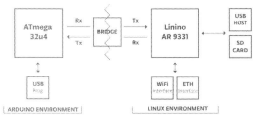

Figure 3.   Structure diagram of Arduino Yun.

1522

The Bridge library facilitates communication between the two processors, giving Arduino sketches the ability to run shell scripts, communicate with network interfaces, and receive information from the AR9331 processor. The USB host, network interfaces and SD card are not connected to the 32U4, but the AR9331 and the Bridge library also enable the Arduino to interface with those peripherals.

2.3 *Yeelink*

Yeelink is an open-source hardware and intelligent equipment service company, which provides the sensor cloud services. Through real-time data processing, Yeelink platform provides users with safe and reliable condition monitoring. The developer can use it by following: Step1, Registering users; Step2, Adding device; Step3, Adding sensors; Step4, Uploading data; Step5, Retrieving data, and then repeating Step4 and Step5. The Yeelink API key is coded by JSON (JavaScript Object Notation), of which ① POST(upload), ② PUT(edit), ③ GET, and ④ DELETE are the four formats.

①: curl --request POST --data-binary @datafile.txt
    --header "U-ApiKey:
    YOUR_API_KEY_HERE"
    http://api.yeelink.net/v1.0/device/12/sensor/3/
    datapoints
②: curl --request PUT --data-binary @datafile.txt
    --header "U-ApiKey:
    YOUR_API_KEY_HERE"
    http://api.yeelink.net/v1.0/device/12/sensor/3/
    datapoint/2016-10-15T17:53:16
③: curl --request GET --header "U-ApiKey:
    YOUR_API_KEY_HERE"
    http://api.yeelink.net/v1.0/device/12/sensor/3/
    datapoint/2016-10-15T17:53:16
④: curl --request DELETE --header "U-ApiKey:
    YOUR_API_KEY_HERE"
    --http://api.yeelink.net/v1.0/device/12/sensor/3/
    datapoint/2016-10-15T17:53:16

3 SYSTEM DESCRIPTION

3.1 *Hardware system*

As shown in Figures 4 and 5, the proposed prototype of fitness information recording and network monitoring system mainly consists of BDS/GPS, Arduino Yun, and ASR (Auto Speech Recognize) voice module. Fitness information including coordinate position, moving speed, and distance can be recorded and monitored by using smartphone Yeelink APP or Website in computer network (Hu, 2015; Yang, 2015).

ASR voice module, which can translate voice to data as well as translate information to speaking, is the interaction between person and system prototype. BDS/GPS gains coordinate position of itself and offers longitude, latitude, and speed by its self-calibrating algorithm. Data information from ASR and BDS/GPS input Arduino Yun by its SoftwareSerial function. The Bridge library of Arduino Yun facilitates communication between its own processors Atmega 32U4 and AR9331. By using Wi-Fi, the encoded fitness information data are transmitted to Yeelink, which could provide cloud service. One can analyze the real-time fitness information using a smartphone or PC as well as acquire them in SD card.

3.2 *Software description*

Figure 6 shows a brief flowchart for system software. Arduino Yun, including Atmega 32U4 and AR9331, ASR voice module, and BDS/GPS reset first. ASR interaction module translates voice speaking to data. BDS/GPS transmits coordinate position of itself, including longitude and latitude, and speed by its self-calibrating algorithm to Arduino Yun by system reset configuration. Here,

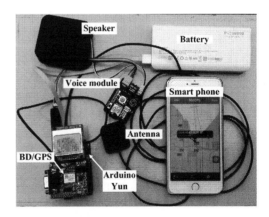

Figure 4. Fitness information recording and network monitoring system prototype.

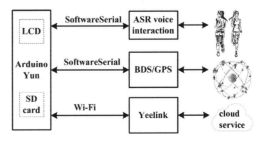

Figure 5. Block diagram of system prototype.

1523

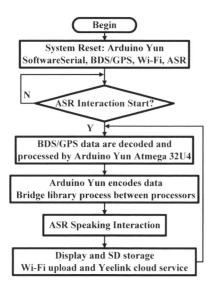

Figure 6. Brief flowchart of system software.

```
void Location_Post() {//Latitude_data, Latitude_data and Speed_data
  Process POST;
  POST.runShellCommand("curl --request POST -d '{"
    + String("\"value\":{\"lat\":")
    + Latitude_data
    + ",\"lng\":"
    + Longitude_data
    + ",\"speed\":"
    + Speed_data
    + ",\"offset\":\"yes\"}"
    + "}' --header \"U-ApiKey: fe8bace70b110a3c0a1233f0d130a3ea\"-
    + http://api.yeelink.net/v1.0/device/'
    + DEVICEID
    + "/sensor/"
    + BD
    + "/datapoints");
  Serial.flush();
}
```

Figure 7. Bridge process function based on API key.

we choose a frequency of 1Hz. Atmega 32U4 decodes the RMC protocol of NMEA message for UM220-III NL according to Table 1. The Bridge library of Arduino Yun transmits encoded data to AR9331 according to Yeelink API key, whose time interval should be greater than 10s. Figure 7 is the bridge process function based on Yeelink API key. The fitness information can be provided on Yeelink terminal, including smartphone APP and website online, and can also be gained in SD card offline.

4 EXPERIMENTAL RESULTS

UM220-III NL as a BDS/GPS of China independent intellectual property rights affords accurate coordinate position. Arduino Yun, integrating Atmega 32U4 and Linino AR9331, is used as the

controller and Wi-Fi connecting to Yeelink, which offers powerful cloud service.

Figure 8 shows smartphone APP interface and a three-circle normal walking playground test coordinate position. Figures 9 and 10 are the speed and distance curves for the test. Figure 11 shows the SD card recoding data description including coordinate position and speed.

The experimental results of our proposed system prototype show that the BDS/GPS coordinate position accuracy, at a reasonable speed of 4.7 km/h and the three-circle playground distance more or less based on BDS/GPS 1Hz data sample frequency and Yeelink 10 s time interval.

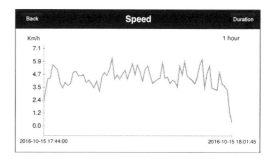

(a)　　　　(b)　　　　(c)

Figure 8. Smartphone APP interface and playground test.

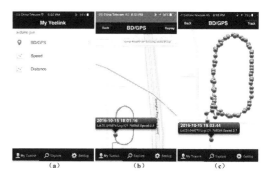

Figure 9. Speed curve of playground test.

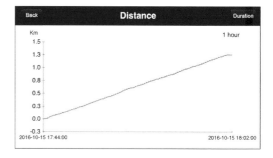

Figure 10. Distance curve of playground test.

```
192.168.1.104 - PuTTY                                    —  □  ×
root@Arduino:/mnt/sda1# cd 20161015
root@Arduino:/mnt/sda1/20161015# ls
17441510.txt   17481510.txt   17521510.txt   17561510.txt   18001510.txt
17451510.txt   17491510.txt   17531510.txt   17571510.txt   18011510.txt
17461510.txt   17501510.txt   17541510.txt   17581510.txt   18021510.txt
17471510.txt   17511510.txt   17551510.txt   17591510.txt   18031510.txt
root@Arduino:/mnt/sda1/20161015# cat 17451510.txt
{"value":{"lat":31.051483,"lng":121.785010,"speed":5.0,"offset":"yes"}}
{"value":{"lat":31.051418,"lng":121.785160,"speed":5.1,"offset":"yes"}}
{"value":{"lat":31.051371,"lng":121.785200,"speed":3.9,"offset":"yes"}}
{"value":{"lat":31.051327,"lng":121.785390,"speed":3.5,"offset":"yes"}}
root@Arduino:/mnt/sda1/20161015#
```

Figure 11.　SD card recoding data description.

5　CONCLUSION

Our proposed fitness information recoding and network monitoring system prototype based on BDS/GPS, Arduino Yun, and Yeelink has not only realized fitness information recoding and network monitoring, but also provided accurate and reasonable description, with a utilization of low-cost, convenient, and controllable character. Furthermore, an in-depth study should be conducted based on the proposed system prototype.

REFERENCES

Altini M, Vullers R, Van Hoof C, et al. Self-calibration of walking speed estimations using smartphone sensors[C]// Pervasive Computing and Communications Workshops (PERCOM Workshops) 2014 IEEE International Conference on. IEEE, 2014: 10–18.

Basiri A, Lohan E S, Silva P F E, et al. Overview of positioning technologies from fitness-to-purpose point of view[C]// International Conference on Localization and Gnss. 2014: 1–7.

Buttussi F, Chittaro L. Smarter phones for healthier lifestyles: An adaptive fitness game [J]. IEEE Pervasive Computing, 2010, 9(4): 51–57.

Choi K S, Yong S J, Kim S K. Automatic exercise counter for outdoor exercise equipment[C]// IEEE International Conference on Consumer Electronics. IEEE, 2013:436–437.

Hsueh-Wen C. Outdoor fitness equipment in parks: A qualitative study from older adults' perceptions [J]. Bmc Public Health, 2013, 13(1):1–9.

Hu Jiao, Sun Jian, Wang Xiaowei, Shen Shu, Zou Zhiqaing. The cloud water environment monitoring systems based on WSNS. Microcomputer & Its Application [J], 2015, 34(11):60–64.

Komninos A, Dunlop M D, Rowe D, et al. Using degraded music quality to encourage a health improving walking pace: BeatClearWalker[C]// Pervasive Computing Technologies for Healthcare (Pervasive Health), 2015 9th International Conference on. IEEE, 2015: 57–64.

Noh Y S, Han Y M, Yoon U J, et al. Development of sports health care system suitable to the fitness club environment[C]// Biomedical Engineering and Sciences (IECBES), 2010 IEEE EMBS Conference on. IEEE, 2010: 93–96.

Peng Zhenzhong, Li Qianxia, Xu Zhiqiu, Yang Bo, Xia Linyuan. Beidou and GPS Integrated High Precision Positioning and Monitoring Analysis of Bridges [J]. Tropical Geography, 2016, 36(4): 717–726.

Pitman A, Zanker M, Gamper J, et al. Individualized Hiking Time Estimation [J]. 2012: 101–105.

Ren Ye, Li Xiaohui, Wu Haitao, Xu Longxia. Effects on dual-system interoperability performance with system time offset[C]// 2015 IEEE 12th International Conference on Electronic Measurement & Instruments (ICEMI 2015), 674: 677.

Varatharajah Y, Karunathilaka N, Rismi M, et al. Body area sensor network for evaluating fitness exercise [C]// Wireless and Mobile NETWORKING Conference. IEEE, 2013: 1–8.

Yang Q, Zhou G, Qin W, et al. Air-kare: A Wi-Fi based, multi-sensor, real-time indoor air quality monitor [C]// Wireless Symposium. IEEE, 2015.

Civil, Architecture and Environmental Engineering – Kao & Sung (Eds)
© 2017 Taylor & Francis Group, ISBN 978-1-138-02985-9

A knowledge integration method for innovation teams based on social tagging in an open environment

Xinmiao Li & Xuefeng Zhang
School of Information Management and Engineering, Shanghai University of Finance and Economics, Shanghai, China

ABSTRACT: In today's global economic integration, the direction of global innovation is moving toward open innovation, and in the open innovation environment, cross-organizational, cross-regional open innovation teams have become the main carrier of knowledge innovation. In the open team knowledge innovation process, it is one of the most important factors in improving the knowledge innovation capability and innovation efficiency of the team that distributed knowledge is acquired accurately and integrated rapidly. In this paper, a knowledge integration method for innovation teams based on social tagging in an open environment was proposed. Further, a corresponding model describing this technique was constructed. This method is classified into tacit knowledge integration method and explicit knowledge integration method. Based on the knowledge integration method for innovation teams based on social tagging in an open environment, the high level of match between the knowledge and the knowledge requirement can be selected for the team member.

1 INTRODUCTION

In an era of knowledge-driven economy, innovation is regarded to have replaced efficiency and quality as the core competitiveness for companies (Schumaker, 2013). In the open innovation environment, cross-organizational, cross-regional open innovation teams have become the main carrier of knowledge innovation (Wei, 2014). Knowledge is one of the most important factors for innovation. In an open innovation environment, explicit knowledge and experts with tacit knowledge are distributed in different geographical locations and different organizations. In the open innovation environment, it is the key factor to acquire required knowledge quickly (including explicit knowledge and tacit knowledge possessed by experts) and integrate distinctive knowledge effectively in order to complete the innovation task collectively.

Since an innovation task is creative, the knowledge required by the task usually cannot be clearly addressed and described by some words. In addition, the knowledge required by the innovation team members is changing constantly as the task is progressing. Because knowledge requirements of innovative tasks have the characteristics of vagueness and dynamics, the existing knowledge search and knowledge organization methods can hardly support the acquirement and integration of knowledge in an open environment for the innovation team.

Knowledge search based on key words is the most common approach (Li, 2009). This approach

is both convenient and inexpensive. However, this approach does not build a close connection between key words. In addition, various people may have different cognitions on key words and a person may generate distinctive understandings at different time. Thus, the key word used by a person can't be necessarily recognized by the others. Therefore, the method of knowledge search based on key words is low accuracy and efficiency (Ning, 2008).

Some scholars have studied methods like knowledge maps (Hao, 2010). However, methods of knowledge map need the knowledge demander to classify and search knowledge based on the categories they are not familiar with (Lin, 2009). Some knowledge demanders have difficulty in recognizing classification and some even do not agree with such classification standards. In that case, innovation team members are not willing to use such methods for knowledge coordination and management. What's more, the standards of knowledge classification are so fixed that they are not suitable for the characteristics of dynamics and vagueness of innovative task's knowledge requirements. In that case, the method of knowledge classification could hardly support knowledge acquirement and integration for innovation team members in an open environment.

Social tagging is one of the most widely used method in Web 2.0 (Gabriel, 2014). This paper has come to a conclusion that social tagging has the following characteristics (Esteban-Gil, 2012): (1) social tagging allows knowledge users to label knowledge based on their own cognitions. Since social tagging is open for sharing, knowledge sharing and interaction in social tagging system is more convenient. (2) One piece of knowledge can be labelled with different tags, which reflect the knowledge's different characteristics. In that case, knowledge users can search knowledge according to their own knowledge requirements from different perspectives instead of based on defined and fixed classification and key words. Therefore, the method of social tagging enhances the efficiency of knowledge searching. (3) There is correlation between tags in social tagging system, which is beneficial for integrating collective wisdom and creating new knowledge. (4) Tags will change dynamically as knowledge and persons' cognitions change, which is suitable for the characteristics of dynamics of innovative task's knowledge requirements and knowledge integration. (5) Tags have the characteristics of vagueness in describing knowledge, which meets the demand of knowledge innovation.

Based on the above characteristics of social tagging, the method of social tagging is suitable for knowledge integration for open innovation teams. Even though social tagging has the above advantages, there is a lack of research on Tag's application in supporting open team knowledge

innovation. R. Arakji, R. Benbunan-Fich and M. Koufaris (2009) studied the motivation for users in using tags in social network. E. Tsui, W. M. Wang, C. F. Cheung and A. S. M. Lau (2010) established a hierarchical structure for a social tagging system. W-T. Hsieh, J. Stu, Y-L. Chen and S-C. T. Chou (2009) built a tag system of team coordination in order to manage group knowledge.

Based on our review, this paper proposed a knowledge integration method for innovation teams based on social tagging in an open environment and constructed the corresponding model. This model focuses on the recognition and integration of experts (tacit knowledge) and explicit knowledge in an open environment, which can support for open team knowledge innovation.

2 KNOWLEDGE INTEGRATION METHOD FOR INNOVATION TEAMS BASED ON SOCIAL TAGGING IN AN OPEN ENVIRONMENT

In this paper, we proposed a knowledge integration method for innovation teams based on social tagging in an open environment, as shown in Figure 1.

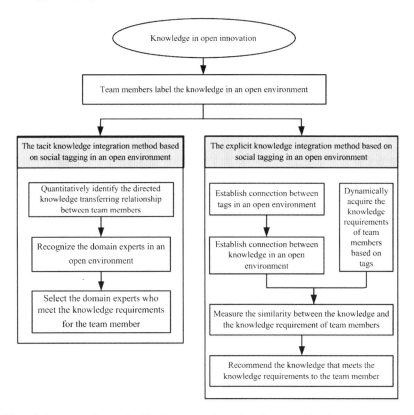

Figure 1.   A knowledge integration method for innovation teams based on social tagging in an open environment.

In Figure 1, first, team members label the knowledge in an open innovation environment. Second, the knowledge integration method for innovation teams based on social tagging in an open environment can be classified into two parts: tacit knowledge integration method and explicit knowledge integration method based on social tagging in an open innovation environment. In the tacit knowledge integration method, first, domain experts are identified by directed knowledge transfer relationship between team members. After the domain experts are recognized in an open environment, the domain experts who meet the knowledge requirements of the team member can be selected for the team member. In the explicit knowledge integration method, first, relationship between tags is established based on tags' co-occurrence. Second, the knowledge requirement of the team member is acquired dynamically based on tags labeled by him. Third, the similarity between the knowledge and the team member's knowledge requirement is measured. Finally, the knowledge that meets the knowledge requirements of the team member can be selected for the team member.

3 TACIT KNOWLEDGE INTEGRATION METHOD BASED ON SOCIAL TAGGING IN AN OPEN ENVIRONMENT

Based on team member's labeling behavior in an open environment, according to the direct knowledge transfer relationship between team members, the domain experts can be recognized and selected. The Tacit knowledge integration method based on social tagging in an open environment is based on the following assumption:

Assumption 1: In an open environment, a team member will label the knowledge if the member considers the knowledge is valuable to complete the innovation task. On the contrary, the member will not label the knowledge.

Based on the above assumption, we suppose that $TEAMMATE_i$ is the ith member, $KNOWLEDGE_k$ is the kth term of knowledge, and TAG_t is the tth tag. We suppose that the team member $TEAMMATE_j$ released the term of knowledge $KNOWLEDGE_{kj}$. If the team member $TEAMMATE_i$ read and labeled the term of knowledge $KNOWLEDGE_{kj}$, then $f_{TEAMMATE_i \rightarrow KNOWLEDGE_{kj}} = 1$, which represents the team member $TEAMMATE_i$ label the term of knowledge $KNOWLEDGE_{kj}$ once. On the contrary, $f_{TEAMMATE_i \rightarrow KNOWLEDGE_{kj}} = 0$.

The tacit knowledge integration method based on social tagging is based on the following rule of team members' knowledge labeling behavior to identify domain experts with tacit knowledge in an open environment.

Rule 1: In the process of open team knowledge innovation, if the member $TEAMMATE_i$ always reads and labels the knowledge of other people released, furthermore the other members never label the knowledge released by $TEAMMATE_i$, $TEAMMATE_i$ can be considered as a junior knowledge learners in the open team. With more and more people label the knowledge released by $TEAMMATE_i$, $TEAMMATE_i$ gradually becomes a knowledge transmitter from a knowledge acquirer, even a knowledge creator from a junior knowledge learner. When $TEAMMATE_i$ becomes a domain expert, he will seldom label the knowledge released by the other team members. In the meantime, more team members will label the knowledge released by him.

Based on Rule 1, the knowledge transfer relationship between $TEAMMATE_i$ and $TEAMMATE_j$ is established as follows:

$$R(TEAMMATE_i, TEAMMATE_j)$$
$$= \frac{\sum_{k=1}^{n} \frac{f_{TEAMMATE_i \rightarrow KNOWLEDGE_{kj}}}{\lg (Month(CurrentDate - LabeledDate_{kj}) + 1)}}{\sum_{p=1}^{m} \frac{f_{TEAMMATE_j \rightarrow KNOWLEDGE_{pi}}}{\lg (Month(CurrentDate - LabeledDate_{pi}) + 1)}} + b_1$$
(1)

Where $R(TEAMMATE_i, TEAMMATE_j)$ represents the knowledge transfer relationship between $TEAMMATE_i$ and $TEAMMATE_j$. $KNOWLEDGE_{kj}$ is the kth term of knowledge released by $TEAMMATE_j$. $TEAMMATE_j$ released n terms of knowledge. $KNOWLEDGE_{pi}$ is the pth term of knowledge released by $TEAMMATE_i$. $TEAMMATE_i$ released m terms of knowledge. $LabeledDate_{kj}$ is the labeled date of $KNOWLEDGE_{kj}$. $CurrentDate$ is the current date. $Month()$ is the number of months converted from the difference between the current date and the labeled date. b_1 is a constant used to adjust $R(TEAMMATE_i, TEAMMATE_j)$.

The labeled date of knowledge influences on current team knowledge innovation in an open environment. The influence rule is as follows:

Rule 2: The smaller the difference between $LabeledDate$ and $CurrentDate$ of the term of knowledge is, the greater influence the term of knowledge has on current team knowledge innovation. On the contrary, if $LabeledDate$ of the term of knowledge is far away from $CurrentDate$, the term of the knowledge has a minor influence on the current team knowledge innovation, which should be given a smaller weight to the term of knowledge.

Based on Rule 2, the function of $\frac{1}{\lg(x+1)}$ is a monotone decreasing function. $\frac{1}{\lg(x+1)} \geq 1$, the

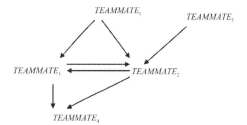

Figure 2. The knowledge transfer relationship between team members in an open innovation team.

first derivative $\left(\dfrac{1}{\lg(x+1)}\right)' < 0$ and the second derivative $\left(\dfrac{1}{\lg(x+1)}\right)'' > 0$, which show that $\dfrac{1}{\lg(x+1)}$ decreases with the independent variable x increases, and the decreasing rate gradually slows down. Therefore, the function of $\dfrac{1}{\lg(x+1)}$ can be used to fit the influence of labeled time of knowledge on current team knowledge innovation. In (1), we use $\dfrac{1}{\lg(Month(CurrentDate - LabeledDate)+1)}$ to represent the influence of labeled time of knowledge on current team knowledge innovation.

The knowledge transfer relationship between team members is shown in Figure 2. In Figure 2, the arrow points to those members who acquire knowledge. The more upside located a team member, the more senior he is in the domain. In Figure 2, $TEAMMATE_1$ and $TEAMMATE_5$ are both on the top. Moreover, there is no arrow between them. This means that $TEAMMATE_1$ and $TEAMMATE_5$ are likely to be in different knowledge domains. $TEAMMATE_2$ acquires knowledge from $TEAMMATE_1$ and $TEAMMATE_5$, which indicates $TEAMMATE_2$ is likely to create new knowledge by combining the knowledge in two domains. $TEAMMATE_4$ is a junior knowledge acquirer. Therefore, based on (1), we can recognize the domain experts according to the knowledge transfer relationships between team members in an open environment. After the domain experts are recognized in an open environment, the domain experts who meet the knowledge requirements of the team member can be selected for the team member.

4 EXPLICIT KNOWLEDGE INTEGRATION METHOD BASED ON SOCIAL TAGGING

4.1 Establishing the relationship between tags

Knowledge in an open innovation environment is distributed in different geographical locations and different organizations. It is difficult to quickly acquire the demanded knowledge and effectively integrate knowledge. In this paper, we integrate knowledge by establishing the relationship between tags.

We establish the relationship between tags through calculating the correlation degree based on co-occurrence between tags. The algorithms of the correlation degree between tags can be mainly divided into the symmetric algorithm and the asymmetric algorithm. The symmetric algorithm for the correlation degree between TAG_a and TAG_b is as (2):

$$R(TAG_a, TAG_b) = \frac{\sum f(TAG_a, TAG_b)}{\sum f(TAG_a) + \sum f(TAG_b)} \quad (2)$$

Where TAG_a and TAG_b are two different tags in an open innovation environment. $\sum f(TAG_a, TAG_b)$ is the co-occurrence frequency of TAG_a and TAG_b. $\sum f(TAG_a)$ is the occurrence frequency of TAG_a. $\sum f(TAG_b)$ is the occurrence frequency of TAG_b.

In (2), if one of the dividers is extremely large or small, a big deviation will be produced in the symmetric algorithm. Therefore, we use the asymmetric algorithm to calculate the correlation degree between TAG_a and TAG_b. The asymmetric algorithm for the correlation degree between TAG_a and TAG_b is as (3):

$$R(TAG_a, TAG_b) = \frac{\sum f(TAG_a, TAG_b)}{\sum f(TAG_a)} \quad (3)$$

In accordance to the correlation degree, the relationship between tags can be established. Based on the relationship between tags, the relationship between knowledge is established in an open innovation environment.

4.2 Measuring the similarity between the knowledge and the knowledge requirements

Team members' knowledge requirements are multidisciplinary and integrated. And team members' knowledge requirements are changing with the progress of the innovation task dynamically, which brings difficulties to the knowledge integration for an open innovation team. The explicit knowledge integration method based on social tagging in an open environment is based on the following assumption:

Assumption 2: In an open innovation environment, team members' labeling behavior for knowledge really reflects the process of members' acquiring, sharing and creating knowledge.

Based on Assumption 2, we use the tags labeled by $TEAMMATE_i$ to represent for the knowledge requirements of $TEAMMATE_i$. Therefore, the similarity between the tags labeled by $TEAMMATE_i$ and other knowledge' tags can be used to represent for the similarity between the knowledge requirements of members and other knowledge. Based on the similarity, the knowledge meets the requirements of the team member can be identified and selected for the team member.

We suppose $TEAMMATE_i$ is the target team member. All tags that $TEAMMATE_i$ labeled are extracted from an open environment. The repetitive tags are removed. $Tag_{TEAMMATE_i}$ is the set of tags labeled by $TEAMMATE_i$. $Tag_{TEAMMATE_i}$ is used to express the knowledge requirements of the target member $TEAMMATE_i$.

All tags of the kth term of knowledge $KNOWLEDGE_k$ are extracted. The repetitive tags are removed. $Tag_{KNOWLEDGE_k}$ is the tags set of $KNOWLEDGE_k$. $Tag_{KNOWLEDGE_k}$ is used to express the features set of $KNOWLEDGE_k$.

$Sim(Tag_{TEAMMATE_i}, Tag_{KNOWLEDGE_k})$ is the similarity between $Tag_{TEAMMATE_i}$ and $Tag_{KNOWLEDGE_k}$. The larger the similarity, the higher level of match between $KNOWLEDGE_k$ and the knowledge requirement of $TEAMMATE_i$.

$Tag_{TEAMMATE_i} = \{TAG_{i1}, TAG_{i2}, ..., TAG_{im}\}$ is the set of knowledge requirements of the team member $TEAMMATE_i$. $Tag_{KNOWLEDGE_k} = \{TAG_{k1}, TAG_{k2}, ..., TAG_{kn}\}$ is the features set of $KNOWLEDGE_k$. The similarity between $Tag_{TEAMMATE_i}$ and $Tag_{KNOWLEDGE_k}$ is as (4):

$$Sim(Tag_{TEAMMATE_i}, Tag_{KNOWLEDGE_k})$$
$$= \frac{1}{m \times n} \sum_{x=1}^{m} \sum_{y=1}^{n} Sim(TAG_{ix}, TAG_{ky}) \quad (4)$$

$$Sim(TAG_{ix}, TAG_{ky})$$
$$= \left(\frac{\sum f(TAG_{ix}, TAG_{ky})}{\sum f(TAG_{ix})} + \frac{\sum f(TAG_{ix}, TAG_{ky})}{\sum f(TAG_{ky})} \right) \div 2 \quad (5)$$

Where TAG_{ix} is the xth tag labeled by $TEAMMATE_i$. TAG_{ky} is the yth tag of $KNOWLEDGE_k$.

The created date of knowledge influences on current team knowledge innovation in an open environment. The influence rule is as follows:

Rule 3: The newer the knowledge is, the greater the knowledge influences on current team innovation. The influence of knowledge's created time shows that new knowledge is usually more valuable than old one. Therefore, new knowledge should be given greater weight. On the contrary, the knowledge created earlier has minor influence on team knowledge innovation and should be given smaller

weight. That is to say, the smaller the difference between $CreatedDate$ and $CurrentDate$ of the term of knowledge is, the greater influence the term of knowledge has on current team knowledge innovation. On the contrary, if $CreatedDate$ of the term of knowledge is far away from $CurrentDate$, the term of the knowledge has a minor influence on the current team knowledge innovation, which should be given a smaller weight to the term of knowledge.

Based on Rule 3, the function of $\frac{1}{\lg(x+1)}$ is a monotone decreasing function. $\frac{1}{\lg(x+1)} \geq 1$, the first derivative $\frac{1}{\lg(x+1)}' < 0$ and the second derivative $\frac{1}{\lg(x+1)}'' > 0$, which show that $\frac{1}{\lg(x+1)}$ decreases with the independent variable x increases, and the decreasing rate gradually slows down. Therefore, the function of $\frac{1}{\lg(x+1)}$ can be used to fit the influence of the created time of knowledge on current team knowledge innovation.

In this paper, we use

$$\frac{1}{\lg(Month(CurrentDate - CreatedDate) + 1)}$$ to

represent the influence of the created time of knowledge on current team knowledge innovation.

$$\frac{1}{\lg(Month(CurrentDate - CreatedDate) + 1)}$$ is used

as the time influence factor to modify the similarity between the tags labeled by $TEAMMATE_i$ and other knowledge' tags. Therefore, (4) is changed to (6) as follows:

$$score = \frac{Sim(Tag_{TEAMMATE_i}, Tag_{KNOWLEDGE_k}) * b_2}{\lg(Month(CurrentDate - CreatedDate) + 1)} \quad (6)$$

Where $CreatedDate$ is the date that the knowledge is created. b_2 is a constant used to adjust $score$.

Based on (6), the bigger $score$, the higher level of match between $KNOWLEDGE_k$ and the knowledge requirement of $TEAMMATE_i$. Therefore, the knowledge with the bigger $score$ should be recommended to $TEAMMATE_i$.

5 CONCLUSIONS

Social tagging is one of the most widely used concepts of the Web2.0 era. In this paper, a knowledge integration method for innovation teams based on social tagging in an open environment was proposed. This method is classified into tacit knowledge integration method and explicit knowledge integration method. Based on the tacit knowledge integration method for innovation teams based on social tagging in an open environment, the high level of match between the domain experts and the knowledge requirement can be selected for the team member. Based on the explicit knowledge integration method for innovation teams based on social tagging in an open environment, the high level of match between the knowledge and the knowledge requirement can be recommended to the team member.

ACKNOWLEDGMENT

This study was supported by Shanghai Municipal Natural Science Foundation of China (NO.14ZR1413400).

REFERENCES

Arakji, R., R. Benbunan-Fich, M. Koufaris, Decision Support Systems, **47**, 3 (2009).

Esteban-Gil, A., F. Garcia-Sanchez, R. Valencia-Garcia, J.T. Fernandez-Breis, Expert Systems with Applications, **39**, 9715–9722 (2012).

Gabriel, H-H., M. Spiliopoulou, A. Nanopoulos, Expert Systems with Applications, **41**, (2014).

Hao, J-X., R. C-W. Kwok, R. Y-K. Lau, A.Y. Yu, Decision Support Systems, **48**, 4 (2010).

Hsieh, W-T., J. Stu, Y-L. Chen, S-C.T. Chou, Expert Systems with Applications, **36**, 5 (2009).

Li, G., Chen Li, J. Feng, L. Zhou, Information Sciences, **179**, 21 (2009).

Lin, F-R., J-H. Yu, Decision Support Systems, **46**, 4 (2009).

Ning, X., H. Jin, H. Wu, Information Processing & Management, **44**, 2 (2008).

Schumaker, R.P., Decision Support Systems, **54**, 3 (2013).

Tsui, E., W.M. Wang, C.F. Cheung, A.S.M. Lau, Information Processing & Management, **46**, 1 (2010).

Wei, K., K. Crowston, N.L. Li, R. Heckman, I & M, **51**, 3 (2014).

Civil, Architecture and Environmental Engineering – Kao & Sung (Eds)
© 2017 Taylor & Francis Group, ISBN 978-1-138-02985-9

A study on the classification of Yunnan folk songs based on ensemble learning

Dong Wang
School of Art, Southwest Forestry University, Yunnan Province, China
Key Laboratory of Educational Informatization for Nationalities Ministry of Education, Yunnan Province, China

Yan Zhang & Dan-Ju Lv
School of Computer and Information, Southwest Forestry University, Yunnan Province, China

ABSTRACT: Nowadays, music classification has become an important application field for classification algorithms. There is only little research about folk songs classification. This paper used the original folk songs in Yunnan Province as research objects, extracting their CELP audio characters on a 11-dimensional scale, comparing and analyzing the classified experimental data from different proportions of marked training samples, which are processed by a single classifier and an ensemble classifier (such as Bagging, AdaBoost, and MCS). Among them, AdaBoost is the most efficient one with an accuracy rate of higher than 85%. The result showed that the ensemble classifiers could mean the effectiveness of the study methodology.

1 INTRODUCTION

As an important auxiliary measure to searching music information and audio processing, music classification plays a potential critical role in musical research theory and practicing. At present, music classification studies mainly focused on fields such as musical emotion, musical genres, instruments, and traditional dramas, whereas it is seldom applied in ethnical music style research (Sun, 2014).

Ethnic music has very high cultural and artistic value (Hong, 2014). The various folk music, which spread by oral teaching of different ethnic groups in Yunnan Province, show obviously different music styles in performance form, singing ways, melody, scale, mode, and rhythm, being originated from different geographical environment, ethnic origin, languages, religious faith, life style, and customs (Zhang, 2006). This paper classified and summarized the characters of different folk songs in Yunnan Province by modern digital audio processing techniques and built a database of those musical characters. This study provided a new way for digital processing of music, which could also be a reference and methodology for music producers.

2 ENSEMBLE LEARNING

In contrast to ordinary learning approaches, which try to construct on one type of learner from training data, ensemble methods try to construct a set of learners and combine them. Ensemble learning is also called committee-based learning, or learning multiple classifier systems (Zhou, 2012).To build an effective ensemble system, three strategies need to be considered, including data sampling or selection; training member classifiers; and combining classifiers. In ensemble learning, the key issue is how to design the base classifiers with stronger generalization and diversity (Zhang, 2012).The success of ensemble learning lies in achieving a good trade-off between the individual performance and diversity. Therefore, it is desired that the individual learners should be accurate and diverse. How to measuring diversity in classifier ensembles are diversity for building the ensemble are presented in detail in literature (Kuncheva, 2004).The Bagging and Boosting algorithms are popular ensemble strategies in the application.

2.1 Bagging

Bagging (Breiman, 1996) was proposed by Breiman. It uses bootstrap sampling to obtain the data subsets for training learners and adopts the most popular voting strategy for aggregating the outputs of each individual learner. Therefore, it should be used with unstable learners such as neural network and decision trees. That is, the more unstable, the larger the performance improvement. If the base classifiers are stable, the improved performance of the bagging is not obvious. The bagging algorithm (Zhang, 2006) is summarized as follows.

Input: Training data set L, learning method H, number of base classifiers N

Output: H_{final}
For i = 1 to N
$L_i = bootstrap\ sample\ from\ L;$
$H_i = H(L_i);$ //train a classifier H_i on L_i with H;
End For

$$H_{final} = \arg\max_{y \in Y} \sum_{i=1}^{N} I(Hi(x) - y)$$

2.2 Boosting

Boosting is a general method for improving the performance of a weak learner. It works by training a set of learners sequentially and combining them for prediction, where the later learners focus more on the mistakes of the earlier learners.

AdaBoost (Adaptive Boosting) is a popular ensemble algorithm that improves the simple boosting algorithm via an iterative process. The amount of focus is quantified by a weight that is assigned to every pattern in the training set. Initially, the same weight is assigned to all the patterns. In each iteration, the weights of all misclassified instances are increased, whereas the weights of correctly classified instances are decreased. As a consequence, the weak leaner is forced to focus on the difficult instances of the training set by performing additional iterations and creating more classifiers. Furthermore, a weight is assigned to every individual classifier. This weight measures the overall accuracy of the classifier and is a function of the total weight of the correctly classified patterns. Thus, weights are used for the classification of new patterns (L. Rokach, 2010).

2.3 Ensemble with different-type base classifiers

Different types of classifiers are trained by the same training data and combined with some strategies. The final classifier classifies new data and gives their label by predictions. The multi-classifier system is built by m classifiers with m learning algorithms, and the model is shown in Fig 1.

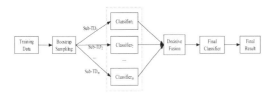

Figure 1. Model of the multi-classifier ensemble.

To measure ensemble diversity, a classical approach is to measure the pairwise dissimilarity between two learners and then average all the pairwise measurements for the overall diversity (Kuncheva, 2003). Given a data set D = {(x1, y1),...,(xm, ym)}, for two classifiers hi and hj, some characters are denoted:

$$a = \sum_{k=1}^{m} I(h_i(x_k) = y_k \cap h_j(x_k) = y_k) \tag{1}$$

$$b = \sum_{k=1}^{m} I(h_i(x_k) = y_k \cap h_j(x_k) \neq y_k) \tag{2}$$

$$c = \sum_{k=1}^{m} I(h_i(x_k) \neq y_k \cap h_j(x_k) = y_k) \tag{3}$$

$$d = \sum_{k=1}^{m} I(h_i(x_k) \neq y_k \cap h_j(x_k) \neq y_k) \tag{4}$$

Two approaches need to be adopted: correlation coefficient and kappa-statistic.

1. Correlation coefficient:

$$\rho_{ij} = \frac{ad - bc}{\sqrt{(a+b)(a+c)(c+d)(b+d)}} \tag{5}$$

This is a classic statistic for measuring the correlation between two binary vectors. The value indicates that the smaller the measurement, the larger the diversity.

2. Kappa-statistic
It is defined as:

$$Kp = \frac{\theta_1 - \theta_2}{1 - \theta_2} \tag{6}$$

where θ_1 and θ_2 are the probabilities that the two classifiers agree and agree by chance, respectively. The probabilities for hi and hj can be estimated on the data set D.

$$\theta_1 = \frac{a+d}{m} \tag{7}$$

$$\theta_2 = \frac{(a+b)(a+c)+(c+d)(b+d)}{m^2} \tag{8}$$

The value indicates that the smaller the measurement, the larger the diversity.

2.4 Random forest

Random Forest (RF) (Breiman, 2001) is an extension of bagging, which consists of a combination of tree classifiers, where each classifier is generated using random vector sampled independently from the original training examples. Given a training set size N with M attributes, the bagging method generates a number of new training sets each of size n (n < N) by randomly drawing samples with replacement from the original training set. For each node of the tree, m (m < M) attributes are randomly chosen to provide the base for calculating the best split at that node. Once the random forest is formed, each sample is classified to the class taking the most popular votes from all the tree predictors in the forest.

3 EXPERIMENTAL DATA AND METHOD

The experimental data are acquired from live recording in villages of Yunnan Minority, with 44.1 sampling rate, 16 bits, and stereo-track. The folk song data include five types of minority, such as Daizu, Hanizu, Nuzu, Wazu, and Yizu. The song's audio length amounts to almost 7 min. The denoising is executed in the preprocessing stage.

3.1 Feature extraction

The feature extraction is executed based on the bit-stream through the G.723.1 data encoding on the Matlab 7.1 platform. The receiving terminal unpacks the bit-stream as frame-steam. LPC features are extracted at each bit-frame 10 order coefficients from 0~23bit (LPC0~LPC2), which consists of 10 dimensions of LPC features (Zhang, 2014). The process of exacting feature is shown in Fig 2.

3.2 The method of experiment

The experiment is carried out on the platform of development Weka. At first, the feature data are converted to an ARFF format file. Then, the ARFF format file can be obtained, and the

module is classified in Java development based on the Weka. The experiments compared the performances of the single classifier and ensemble classifiers.

The single classifier includes traditional classifying methods such as decision tree J48 and neural network, while Bagging, AdaBoost, and random forest are involved in the ensemble strategies.

4 ANALYSIS OF THE EXPERIMENTAL RESULTS

In the experiment, the data are selected from the total according to a third of the total amount of data in each category. It is 10 times sampling randomly with 67% as the training examples and 33% as test samples. Five classes of music samples are shown in Table 1.

4.1 Single-classifier experiment

The single-classifier algorithm adopts the decision tree (J48), Neural Network (NN), Naïve Bayes (NB), Radial Basic Function (RBF), and Random Forest (RF). The results of classification are listed in Table 2. According to the table, the random forest outperforms others, whereas the Naïve Bayes and RBF show worse performance. BP is secondly to RF, but better than J48. Accuracy of each class and Overall Accuracy (OA) in three better single classifiers are shown in Table 3.

4.2 Ensemble with multiple classifiers

4.2.1 Select diverse base classifiers
There are five base classifiers above the experiment. In order to select the larger diversity classifiers involved in the ensemble, two methods are

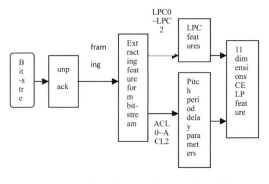

Figure 2. Extracting of CELP features from folk songs audio.

Table 1. Data set of training and test.

Class	Train data#	Test data#	Total
daizu	580	290	870
hani	600	300	900
nuzu	600	300	900
wazu	600	300	900
yizu	600	300	900
total	2980	1490	4470

Table 2. Classification accuracy of single classifiers.

Classifier	J48	BP	NB	RBF	RF
Accuracy	0.7040	0.7973	0.4718	0.5416	0.8208

Table 3. Classification accuracy of better single classifiers.

Classifier	daizu	hani	nuzu	wazu	yizu	OA
Decision tree	66.21%	75%	68%	68.67%	74%	70.40%
Neural network	83.45%	87.67%	76%	77.43%	74.33%	79.73%
Random forest	80%	89.33%	78.67%	77.67%	84.67%	82.08%

Table 4. Difference value of each classifier.

Classifier	Kp	Correlation coefficient
Decision tree	0.1232	0.1313
Neural network	0.1233	0.1385
Radial basic function	0.1766	0.1940
Random forest	0.1611	0.1846
Naïve Bayes	0.1680	0.1921

Table 5. Classification accuracy of ensemble methods.

Class	Bagging			AdaBoost			MCS J48 BP RF
	J48	BP	RF	J48	BP	RF	
daizu	74.14%	86.90%	83.45%	80%	86.55%	90%	83.79%
hani	86.67%	90.33%	89%	91%	92.67%	93%	92.33%
nuzu	78.33%	77.67%	83.33%	84.67%	84.33%	88%	81.33%
wazu	76.67%	84%	82.67%	82.67%	84.33%	83%	82%
yizu	76.67%	89.33%	91.67%	88.33%	88%	91.67%	86%
OA	80.60%	85.64%	86.04%	85.37%	87.18%	89.13%	85.10%

adopted to measure the diversity candidate classifiers. The process is as follows.

Step 1: To calculate the index of difference of each pairwise classifier;
Step 2: To construct the difference matrix;
Step 3: To calculate the value of difference of each classifier and sort the results in ascend order;
Step 4: To select the top ones for the ensemble classifiers.

In the experiment, the difference value of each classifier is calculated as illustrated in Table 4. According to the smaller difference value and larger diversity, three out of five classifiers are selected including, decision tree, neural network, and random forest (bold font in Table 4), which are involved in the ensemble multiple classification.

4.2.2 Ensemble classifiers

In the experiment of Bagging and AdaBoost, decision tree, neural network, and random forest are used as the base classifier, and the performances are summarized in Table 5.

It is obvious that the performances of Bagging, AdaBoost, and MCS are superior to the base classifier for the music classification in terms of overall

accuracy. Especially, that of AdaBoost is better than others. In addition, the classifiers have shown the different performance on specific classes, indicating that the classifier performing well for one class may be poor for other classes. Class "hani" and "yizu" obtained better performance with most classification methods, while classes "nuzu" and "wazu" had relatively low accuracy. In this case, the MCS is only better than Bagging with J48, but inferior to the other base classifiers in Bagging and AdaBoost. In the three base classifiers, RF has the highest accuracy, whereas BP has a lower accuracy and J48 has the lowest accuracy. In both Bagging and AdaBoost, the accuracy of the base classifiers ranked in the pattern as using the base classifiers. Furthermore, J48, BP, and RF showed better accuracies in AdaBoost than in Bagging.

5 CONCLUSIONS

Yunnan folk music is a treasure of ethnic culture and art and hence its classification is significantly valuable. Ensemble learning, constructing a set of classifiers, and then classifying new data via taking a vote of their predictions is a good way to achieve high accuracy in classifying. This research is about classifying five types of Yunnan folk music through extracting their main features. This research contains experiments of using single classifier, and ensemble methods including Bagging, AdaBoost, MCS, and random forest to analyze music data and classify them. From the results, better performance is found in ensemble methods. An effective way of classification can be provided by ensemble methods. Among these methods, the AdaBoost performs the best. More studies about combining ensemble learning with semi-supervised learning to build learning model with high accuracy are on underway.

ACKNOWLEDGMENTS

This study was supported by the National Nature Science Fund Project (61262071), National Science and Technology Support Program (2013BAJ07B02), and Key Project of Applied Basic Research Program of Yunnan Province (2016FA024).

REFERENCES

Breiman, L: Bagging predicitors. Machine Learning Vol. 24(2) (1996), p.123–140.

Breiman, L: Random forests. Machine Learning. Vol. 5–32 (2001), p. 45.

Cha Zhang, Yunqian Ma: Ensemble machine learning methods and application. Springer (2012).

Chao-Hui Hong: Music structure in Yunnan minority music aesthetic. Popular Literature (2014).

Ke Sun: The Study of Feature Extraction and Automatic Classification for Chinese Folk Music. Master Dissertation, Dong Hua University (2014).

Kuncheva L.I: Combining Pattern Classifications: Methods and Algorithms. John Wiely & Sons, Hoboken, NJ (2004), in press.

Kuncheva, L.I., C.J. Whitaker: Measures of diversity in classifier ensembles. Machine Learning Vol. 148–207 (2003), p. 51.

Rokach, L: Pattern Classification Using Ensemble Methods. World Scientific. Singapore (2010).

Xing-Rong Zhang: Original folk Music of Yunnan. CCOM Publisher (2006).

Yan Zhang: The Application of Ramdom Forest for Classifying Environmental Audio Data. Internation Conference on ComputerScience and Software Engineering. (2014) p. 190–196.

Zhi-Hua Zhou: Ensemble Methods Foundations and Algorithms. CRC Press (Taylor & Francis Group 2012).

Civil, Architecture and Environmental Engineering – Kao & Sung (Eds)
© *2017 Taylor & Francis Group, ISBN 978-1-138-02985-9*

RFID-based elevator positioning algorithm by the virtual reference label

Sheng Zhang & Xiu-Shan Zhang
Department of Computer Science, Naval University of Engineering, Wu Han, China

ABSTRACT: In this paper, a localization algorithm based on RFID is proposed to solve the problem of elevator positioning in the centralized monitoring system. The algorithm follows Euclidean distance by the received signal strength value (RSSI) between the reference tag and unknown node, adopts the dichotomy to rapidly approximate to determine the corresponding positioning interval, inserts the virtual label according to the required positioning precision dynamically, and estimates the unknown node positioning using the least-squares method. Performance analysis results indicate that the RFID-based system has an error of less than 0.1 m and improves the accuracy approximately 50% comparing with the common centroid algorithm. The results indicate that the algorithm can effectively locate the elevator and have a good positioning accuracy.

1 INTRODUCTION

The elevator safety problem is widely concerned nowadays. Therefore, it is necessary to monitor the elevator effectively to guarantee the safe operation of the elevator. The single-vendor solution cannot meet the actual demand because of the different hardware interface of different brand elevators (Jiang, 2015). To solve the unified supervision of different brand elevators, much research has been done by scholars from all fields, and several achievements have been made, such as network monitoring equipment, video overlay, and complete sets of elevator monitoring sensor system. In the above-mentioned solution, real-time acquisition of the elevator position information is a key problem. Therefore, it can be divided into two types according to the way of elevator position information acquisition. One is through the elevator own level sensor to achieve location information collection. The implementation method is mainly to install optical or magnetic induction sensor, which can provide a switching value when the elevator moves through a flat baffle. Then, we can know the position information about the elevator. The biggest shortcoming in this way is that the positioning accuracy is in floors, so that the system cannot detect the potential problems in the elevator motion process. The other is to collect the absolute position information of the elevator directly by laser, infrared, and other methods. It has higher positioning accuracy, which can be in the scale of centimeters. However, because of the mobility of car and the more dust and shelter in the environment of elevator shaft, the robustness

of the system is low and the implementation cost is rather high (Hui, 2013).

In order to solve the shortcomings of the above two schemes, this paper proposes an elevator positioning algorithm based on RFID technology, which combines the characteristics of elevator movement on the basis of RFID indoor positioning technology. It can realize low-cost positioning of the elevator and have a better positioning accuracy.

2 BACKGROUND KNOWLEDGE

In this section, we introduce the RFID-based location technology using RSSI and the weighted centroid algorithm, which are the foundations of our study.

2.1 *RFID-based location technology using RSSI*

RFID is a technology to achieve information in noncontact ways through wireless transmission of signals. Thus, according to the theory of wireless communication, the signal in the process of transmission will attenuate with the increase of distance. The path loss has the following empirical formula in indoor environment (De Angelis, 2010):

$$PL(d) = PL(d_0) + 20N \lg \frac{d}{d_0} + X_\sigma \qquad (1)$$

where $PL(d)$ represents the path loss of the unknown node, which is at a distance of d from the reader, $PL(d_0)$ is the path loss at reference distance

d_0, N is the pass loss factor of the links, which depends on the signal transmission environment, and X_σ is an environment variable, which is a normal distribution random variable with mean 0.

Assuming that the power of the signal after the antenna gain of the reader is P_0, the signal RSSI value can be expressed as:

$$RSSI(d) = P_0 - PL(d) = RSSI(d_0) - 20N \lg \frac{d}{d_0} - X_\sigma \tag{2}$$

Therefore, we can get the value of the path loss factor N by collecting a large number of sample points and estimate X_σ according to the Gaussian distribution method proposed by Zhi-feng Lin in (2011).

2.2 Weighted centroid algorithm

Considering the actual situation of the elevator movement, we set the vertical axis of the reference tag as the reference axis and the vertical axis of the elevator movement as the track axis. Under the ideal circumstances, we can determine the location of the current node A when distances r_1 and r_2 are the distances between the conference nodes and the unknown point A is given (Tao, 2014). However, in practice, the calculated distance between node A and the reference axis may be greater than r_0, as shown in Figure 1.

To solve the problem of calculation of point A in the real environment, the following processing is performed. First, the two-dimensional coordinate system is constructed, and the reference axis is the y-axis; therefore, the coordinates of the two reference nodes are $(0, h1)$ and $(0, h2)$. Then, set the two reference nodes for the center points and construct two circles R_1 and R_2 with radii of r_1 and r_2, respectively. A is the intersection point in the limited quadrant (here is the first quadrant). By substituting r_0 into the equation of R_1 and R_2, we can get the intersection part of the line $x = r_0$ and two circles. Let the points B and C be the intersection points given by the line $x = r_0$ and two circles which are the two closest pints to A. Now, A is on line BC. The traditional centroid algorithm takes the BC midpoint as the actual coordinate of point A. However, in the weighted centroid algorithm, the RSSI value is used to determine the size of the center of mass decision, that is, the larger the node RSSI value, the greater the impact on the position of the center of mass. Let $B(r_0, y_b)$ and $C(r_0, y_c)$, respectively, take the value of point A in the weighted centroid algorithm as:

$$\begin{cases} x = r_0 \\ y = \dfrac{\dfrac{y_c}{r_1} + \dfrac{y_b}{r_2}}{\dfrac{1}{r_1} + \dfrac{1}{r_2}} \end{cases} \tag{3}$$

3 ALGORITHM IMPROVEMENT

Because the RSSI value is greatly affected by the environment, and when the entity label measurement error is larger, the estimation result will have a bigger error, so we need to improve the algorithm.

3.1 Use dichotomy to estimate elevator node position

Throughout the elevator positioning algorithm, the reader may receive the reflected signals from more than two tags. According to formula (2), we always select two reference nodes that are the closest to the unknown nodes to reduce the influence of environmental factors. That is to say, the greater the density of the reference labels, the smaller the distance of the positioning interval. Therefore, the distances between the given nodes and unknown node are lower and the positioning accuracy is higher. However, in the actual deployment process, we found that the density of the reference label can not only increase the hardware cost of the system but more importantly lower the positioning accuracy because of the signal interferences between the tags while the density is too large. In this regard, we take the method of inserting virtual tags to improvement.

There is always a point D_{max} on the reference axis, which is r_0 away from the elevator in the case of the

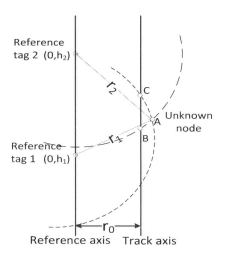

Figure 1. Principle of weighted centroid algorithm.

special trajectory of the elevator. The RSSI value of D_{max} should be maximum, and we think D_{max} is the projection of the elevator node on the reference axis whose y-coordinate is the same as the elevator node. If we ignore the influence of environmental factors, the RSSI values at two points are the same when the distances of two points to D_{max} on the reference axis are equal. On the basis of this feature, the dichotomy process of calculating the elevator position is as follows:

1. Assuming the reader selects the physical reference labels B and C as the two ends of the positioning interval, we read the RSSI values of two points.
2. We take the midpoint D of the line BC as the D_{max} and ΔRSSI as the difference between the RSSI values of the two points B and C. If ΔRSSI $\in$ $[-\delta, +\delta]$, we think that the signal strength values of B and C are equal. The midpoint of the line BC is the actual position of D_{max}, where δ is the error tolerance setting in consideration of the influence of environmental factors.
3. When ΔRSSI $\notin$ $[-\delta, +\delta]$, according to formula (2) we can select one with a larger RSSI value (we take B here). Then, make D replace C as new C.
4. Repeat Step 2, let d1 be the current distance of CD. Here, d the distance between C and the imaginary elevator node is $\sqrt{d_i^2 + r_0^2}$, and then through formula (2) to calculate the current C-point RSSI value.

In this way, we can acquire the estimating coordinates (r_0, y), where y is the height information of lift by a certain number of recursive processes.

3.2 *Estimate the elevator node using the least-squares method*

In Section 3.1, the idea of recursion of limit by dichotomy was used to estimate the position of elevator nodes. However, the uncertainty of recursion may lead to a longer computational time, which leads to the real-time degradation of the positioning system. In order to solve the above problems, the dichotomy method is adopted to obtain the appropriate positioning interval quickly, and then the least-square method is used to determine the elevator node position.

When we acquire the positioning interval, the RSSI value of the end point may be in two cases. If the end point is an entity tag, then the value is read by the reader actually. Otherwise, when reference node is a virtual node and the location interval is [a, b], the elevator node currently projecting on the y-axis. Take a random point D_i (0, d_i) ($d_i \in$ [a, b]) as the D_{max} and calculate the virtual node signal power $RSSI_i$. Then, use the weighted

centroid algorithm to obtain the lift nodes (r_0, y_i). According to the principle of least squares, assume $Y = a+bx$, $\Delta = Yi - Yi$:

$$\varphi(a,b) = \sum_{i=1}^{n} \Delta^2, \quad (n \geq 20) \tag{4}$$

When $\varphi(a,b)$ is minimum, we can get the linear fitting relationship between the horizontal and vertical coordinates of the elevator node in the interval. Because the horizontal coordinate of the elevator node is r_0 in this case, the fitting relationship is a straight line parallel to the X-axis, that is, the equation of the line must be $Y = C$. Equation (4) can be simplified as:

$$\varphi(C) = \sum_{i=1}^{n} (y_i - C)^2 \tag{5}$$

When the above formula is used to obtain the minimum, that is, $C = C_{min}$, the location of the elevator is (r_0, C_{min}).

3.3 *Estimate the error tolerance δ and the location interval L*

In the previous two sections, we establish an algorithm model to calculate the position of the elevator node by inserting virtual labels between the two entity tags. In this section, we estimate the error tolerance (δ) and the interval distance (L) involved in the preceding model.

According to equation (2), when the distances from two reference nodes to lift are equal, the main cause of ΔRSSI $\neq 0$ is X_σ. In accordance with 3σ distribution law of normal distribution, the probability that $X_o \in [-\sigma, \sigma]$ is about 68%, and the value of σ can be calculated in the method provided by Cui-Cui Yu (Yu, 2015). Then, according to the size of the value of σ set δ, usually take $\sigma/3$.

Positioning distance (L) depends on the positioning accuracy requirements. As Figure 1 shows that the calculation error $\varepsilon < L$, the value of L must be taken less than positioning accuracy requirements.

4 EXPERIMENT AND SIMULATION

4.1 *Establish experimental system*

Establish a simulation system to simulate the movement of the elevator process in the laboratory environment. The RFID system hardware devices are shown in Table 1, and the entire system hardware deployment is shown in Figure 2.

RFID readers and laptops are placed on the top of the stroller bracket, the reader mode of work is set to active mode, the query interval is 100 ms, the transmission power is set to 30dBm, and the baud rate is 57600dps. The computer and the reader communicate through the RS232 serial port. The reader is connected to an external antenna through a TNC antenna interface. The antenna is suspended under the crosspiece of the trolley. The vertical distance to the ground is 0.5 m. The antenna plane is parallel to the ground. Passive tags are evenly spaced on the ground with a distance of 0.5 m between adjacent tags.

4.2 Results and analysis

4.2.1 Static measurement experiment

In the experiment, the displacement of equipment is fixed, and the center of the antenna is located as the lift node. Distances between lift node and two entity tags are 0.53 and 0.58 m. Then, 20 groups of RSSI data are collected and the position of lift is determined (setting $\delta = 4$dBm, $L = 0.1$ m). The error result is shown in Figure 3.

As shown in Figure 3, the improved positioning algorithm has obvious advantages compared with the weighted centroid algorithm, the error is reduced by nearly 50%, and the error is less than 0.1 m.

Table 1. RFID system equipment.

Reader	Brand model number	ZK_RFID406
	Frequency band	902–925 MHz
	Transmission power	30dBm
Tags	Brand model number	ZK-RFID606
	Mode standard	EPC C1G2
Antenna	Polarization mode	Linear polarization

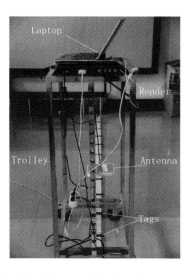

Figure 2. Simulation experiment system.

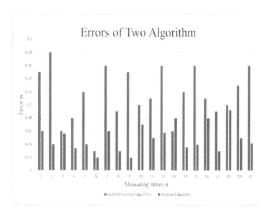

Figure 3. Errors of two algorithms.

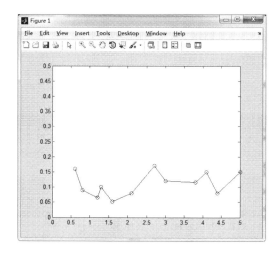

Figure 4. Errors of dynamic measurement experiment.

4.2.2 Dynamic measurement experiment

As the elevator is often in dynamic motion, it is necessary to verify the algorithm through dynamic experiments. The whole movement process of elevator is divided into three stages: constant speed, acceleration, and deceleration. Therefore, in the experiment process, we imitate the states of the movement, record the actual lift position, and the entity tags RSSI value, which constitutes the positioning interval. After the above processing, we process the data using MATLAB2010b, and the results are shown in Figure 4.

Figure 4 shows that when a car is in dynamic motion, its positioning error is higher than that in the static state, but the average error is about 0.1 m, which still has a high positioning accuracy.

5 CONCLUSION

In this paper, an improved algorithm is proposed on the basis of the weighted centroid algorithm. This is the first algorithm to use dichotomy reduced positioning interval and obtain position information of the lift range nodes, and finally through a linear least squares regression to estimate the node position of elevator. The feasibility of the experiment is verified. The experimental results show that the algorithm has a better positioning accuracy, meets the basic needs of elevator positioning system, and has a certain practical value.

REFERENCES

Cui-Cui Yu, The Design and Implementation of Location System Based on Active RFID, Computer Science Department, Beijing University of Posts and Telecommunications.

De Angelis A, Nilsson J, Skog I, et al, Metrology & Measurement System 17, 447–460 (2010).

Li-Rong Hui, Design and Realization of Elevator Safety Supervision System, Computer Science Department, Jilin University.

Tao Jiang, Guang-Jun Liu, Chinese Journal of Construction Machinery 13, 162–167 (2015).

Zhi-Feng Lin, RSSI Wireless positioning Technology Based on the Revised Environmental Factors, Applied Mathematics Department, Guangdong University of Technology.

Zhi-yong Tao, Sun Lu, Computer Applications and Software, 2014, 123–126.

Civil, Architecture and Environmental Engineering – Kao & Sung (Eds)
© *2017 Taylor & Francis Group, ISBN 978-1-138-02985-9*

Design of the omnidirectional mobile robot control system based on dSPACE

Huan-huan Liu, Chang-sheng Ai, Hong-hua Zhao & Xuan Sun
School of Mechanical Engineering, University of Jinan, Jinan, China

ABSTRACT: In recent years, the use of omnidirectional mobile robot has become a bright spot in the logistics scheme of equipment manufacturing industry. It has played a certain role in improving production efficiency, reducing costs, and improving product quality and management level. However, its application is limited due to its low flexibility, low efficiency, and difficulty of operation in complex environment. In order to realize the flexibility and maneuverability of this type of movement, the mobile structure and motion control system are designed. The structure design used mecanum wheel as its moving wheel. Its control system design used dSPACE real-time simulation and control platform, dSPACE simulation, and MATLAB/Simulink perfect combination. Through the establishment of the simulation model, the motor speed characteristics of the motor were tested. Therefore, it can realize the design and debugging of control system. The experiment shows that the control effect of dSPACE real-time simulation platform is good. The control system model can be built conveniently and quickly, and the development period of the control system can be shortened simultaneously.

1   INTRODUCTION

The omnidirectional mobile robot has the advantages of high self-planning, fast self-organization, self-adjustment, and so on (Wang, 2015). It is suitable for complex working environment, such as for indoor warehouse handling, space exploration, and security checking for automobile chassis, pipelines, and containers. Thus, the omnidirectional mobile robot will be widely used in factory automation, construction, mining, military, agricultural services, and so on. Therefore, more and more high requirements for its mobility and flexibility are required, especially for the performance requirements of the control system. There are many ways to design a control system for an omnidirectional mobile robot. For example, few scholars from the Islamic Azad University used PI-fuzzy control strategy to design the control system of four-wheeled omnidirectional mobile robot (Feng, 2010). Several researchers from Taiwan used the fuzzy logic controller and visualization to design the control system of the omnidirectional mobile robot (Ehsan, 2011). Although these methods can realize the control system design of the mobile platform, the design and development of the test cycle is long. Not only the development time is long, but also the cost is very high. To this end, the dSPACE real-time simulation software and hardware platform was used for the design of its control system. DSPACE is based on the development of

MATLAB/Simulink real-time simulation system software and hardware platform, which achieved a seamless combination with MATLAB/Simulink (Lang, 2013). ControlDesk interface can be used to provide various parameters of the window to adjust the parameters of the controller or real-time editing program so as to achieve rapid control prototype verification.

2   BASED ON THE PRINCIPLE OF DSPACE MOBILE ROBOT MOTION CONTROL SYSTEM DESIGN

The schematic diagram of control system of omnidirectional mobile robot based on dSPACE is shown in Figure 1. The figure shows the speed control system based on dSPACE, including the PC, dSPACE hardware, servo motor, servo motor driver, serial ISP download bus, and omnidirectional wheel. The PC computer is the bottom tool, which has the software of dSPACE of MATLAB/Simulink, RTW, RTI and ControlDesk installed in it. The PC has a serial port ISP download bus and dSPACE controller MicroAutoBox for signal transmission. When the block diagram is set up, the digital output channel and the port of the controller are set up. The MicroAutoBox slot of the controller is connected with the output port line corresponding to the controller, and is connected with the corresponding port of the servo driver to

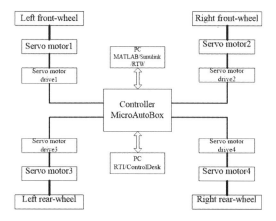

Figure 1. Schematic diagram of control system of omnidirectional mobile robot.

ensure the normal transmission of the signal. For the needs of different users, dSPACE provides a variety of alternative hardware systems, such as the single-board system and the standard-component system.

3 ANALYSIS OF THE FORCE APPLIED TO MECANUM WHEEL OF THE OMNIDIRECTIONAL MOBILE ROBOT

The Structure Design of Omnidirectional Moving Mechanism used Mecanum wheel (Gfrerrer, 2008). The shape of Mecanum wheel is like a helical gear. The tooth is a rotatable drum-shaped roller, and the roller axis and the wheel axis formed α angle. The roller has 3 degrees of freedom. It can rotate about itself and around the axle simultaneously and can rotate around a roller in contact with the ground point. The wheel itself also has 3 degrees of freedom: rotation around the axle and translation along the perpendicular direction of the roller axis and rotation around the contact point of the roller and ground. In this way, the driving wheel has the active driving capability in one direction, and the other direction has the movement characteristics of free movement. The wheel circumference is not composed of ordinary tires, but the distribution of many small rollers. The theory circle of these rolls contour line is coincident with the wheels, and the roller can rotate freely. During the motor rotation of the drive wheel, the wheel rotates normally forward along the direction perpendicular to the drive shaft, and the roller wheel rotates around the axis of rotation along their respective freedom. Accordingly, the deviation of the small roller is divided into left and right Mecanum wheels. Compared with the universal wheel, Mecanum wheel has the characteristics of flexibility, accuracy,

and efficiency, which is a kind of universal wheel control. Its mobile states are: forward, backward, left and right lateral and transverse shift left anterior oblique, right anterior oblique shift, left posterior oblique shift, right posterior oblique shift, rotation center with 0–360°, and the right after the origin of rotating omnidirectional mobile mode.

Among them, the omnidirectional mobile platform going forward, right lateral, right oblique shift, right rear wheel rotation, as the origin of omnidirectional mobile platform to center for the analysis of motion diagram of Figure 2 in the origin of 360° rotation and after rotation of the wheel center (a), (b), (c), (d), (e), and (f) shown. In Figure 2 (a), the torques of the four-wheeled mobile robot are T1, T2, T3, and T4. The X-direction decompositions of their moment are offset each other, and the Y-direction decompositions of the moment are pointing to the front; the synthesis of the moment is also pointing to the front so the omnidirectional mobile robot is moving forward. In Figure 2 (b), analysis of the omnidirectional mobile robot transverse direction, Y-decomposition torque offset each other, only the direction decomposition of X torque. Because the X-direction decomposition of the moment, the moments after the synthesis are pointing to the right and the omnidirectional mobile robot to traverse. In Figure 2 (c), the right front wheel and left rear wheel of the omnidirectional mobile robot have no input torque, and the left front wheel and the right rear wheel have torques T1 and T2, respectively. When T1 and T2 face the right front, the Y-direction decompositions of the moment are forward, X-direction decomposition of the moment fall to the right, when the omnidirectional mobile robot moves to the right front. In Figure 2 (d), the right front wheel and the right rear wheel of the omnidirectional mobile robot have no input torque, whereas the left front and left

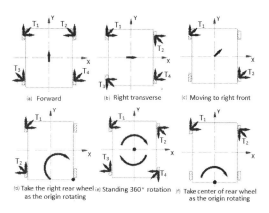

Figure 2. Graph of the analysis of the force applied to Mecanum wheel of the omnidirectional mobile robot.

rear wheel have torques T1 and T2, respectively. The left front wheel is subjected to the right front direction of the torque, the left rear wheel is left forward to the torque, when the omnidirectional mobile robot to the right rear wheel as the origin of rotation. In Figure 2 (e), the four wheels of an omnidirectional mobile robot are subject to torque. When the left front wheel torque is toward the right direction, the right front wheel toward right to the left rear wheel torque, torque is to the left front, right rear wheel is left to the torque, the clockwise rotation of an omnidirectional mobile robot. In Figure 2 (f), the left rear wheel and right rear wheel of the omnidirectional mobile robot have no input torque, and the left front wheel and the right front wheel have torques T1 and T2, respectively. When the left front wheel is subjected to the right front direction, the right front wheel is subjected to the right rear torque.

4 DESIGN OF FOUR-WHEEL DRIVE

Four-wheel drive (Xin, 2015), also called all wheel drive, refers to the front wheel and the rear wheel have driving force, and the engine output torque is distributed in all the wheels according to the different road conditions. Its main function is to effectively control the vehicle's lateral movement characteristics and improve the ability of the vehicle. Four-wheel drive relies on the four-wheel independent steering. All wheels can be turned around the same instantaneous center of rotation to achieve different steering radius and even zero turning radius so that it has a high degree of flexibility in the steering wheel. The narrow working space brings great challenge to the mobile robot. Therefore, the control system of the omnidirectional mobile robot adopted the four-wheel drive. The four-wheel drive system is mainly composed of a motor and its driver, driving power supply, and deceleration device. Among them, the servo motor is controlled by pulse and direction.

5 STRUCTURE OF DSPACE SYSTEM

DSPACE real-time simulation system was developed by German dSPACE company. It is a software and hardware platform for control system development, testing and hardware in-the-loop simulation based on MATLAB/Simulink. Its greatest feature achieved completely seamless connection with MATLAB/Simulink (Zhang, 2012). DSPACE real-time system mainly consists of two parts: software system and hardware system.

V-Cycle development process is a parallel development model, whose process is shown in Figure 3.

It is the software and hardware development for joint debugging. Automatically generated from the C Simulink code and downloaded to the controller (such as MicroAutoBox). The virtual instrument NG ControlDesk debugging is used to accelerate the design of the iteration loop process and greatly decelerate the control design process (Liu, 2014). In addition, the simulation can be carried out online or offline.

5.1 dSPACE software structures

The software system of dSPACE mainly consists of three parts. First, according to the needs, we can use MATLAB/Simulink to quickly and easily establish the simulation model. For the preparation of C and other procedures, the building block diagram is simple and easy to understand and accelerate the process of system development. Second, RTI (Real-Time Interface) is a bridge connecting the dSPACE real-time system and MATLAB/Simulink simulation (SHAO, 2012). According to the requirement of user selection model and the MATLAB/Simulink library RTI library to build the control model, we compile and generate C code and download the program to the controller. Finally, the ControlDesk interface provides users with a variety of parameters of the window, which can change the parameters of the controller in real time.

5.2 dSPACE hardware structure

DSPACE hardware provides a variety of components and a single-board system. DS1401/1512/1513 single-board system has two modules: 1512 modules and 1513 modules with 156 pins. Therefore, the function is very powerful. However, the 1512 module is much more complex than the 1513 module. Because it has two IP modules more than

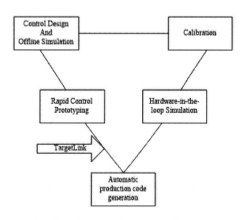

Figure 3. Flowchart of dSPACE V-cycle.

1513 modules, it is more complex to use. According to the motion control system design of mobile robot, the relative difficulty is not high, and considering the cost and other factors, the choice of 1513 can meet the design requirements. 1513 module I/O port resources are rich, including digital output, digital input, analog input, and analog output.

5.3 Establishment of the simulation platform of control system based on dSPACE

The motion control system of omnidirectional mobile robot adopted PWM pulse modulation (XU, 2012). Combined the MATLAB and the RTI module libraries provided by PC, we selected the 4 TP4 PWM DIO module. In the dSPACE real-time interface, we established the simulation model as shown in Figure 4. The model can output 4 PWM pulse signal. The signal output port are CH1, CH2, CH3, and CH4. The four-pulse signals can carry on the duty cycle and the start and stop of control system in real time. The PWM debugging of the servo motor is to change the pulse frequency and direction of the servo motor. The PWM module of the four corresponding output ports of CH1, CH2, CH3, and CH4 connected with the input port PUL1, PUL2, PUL3, and PUL4 of the four servo motor drives. The specific operation is to set the duty cycle T1, T2, T3, and T4 and the period cycle a, b, c, and d of the four PWM modules. Changing the cycle is to change the pulse frequency. Then, the corresponding four motors m1, m2, m3, and m4 will get the corresponding speeds n1, n2, n3, and n4. The omnidirectional mobile robot can achieve forward, right traverse, standing 360° rotation so as to realize the mobile control ability of the omnidirectional mobile robot.

After the model is built in MATLAB/Simulink, a key to compile and generate C code and save in the Sdf file is proposed. The project is set up and saved in ControlDesk. The hardware circuit and communication equipment are connected through the registration of the hardware platform to download the program to the Micro-AutoBoxII controller. In the ControlDesk interface, we establish a virtual instrument and the relevant parameters of the changes. As long as the program is downloaded to the controller, it also carried out off-line operation. In the ControlDesk interface, according to the variables in the column of relevant variables, such as Gain, Wave Sine, Sum, and constant, to establish the required virtual instruments, such as Plotter, Knob, and Bar, for the entire control system debugging. One of the great advantages of dSPACE is that the real time is very strong. We can change the program at the same time observe the running state of the actuator.

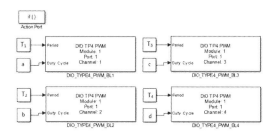

Figure 4.   Model of PWM debugging and simulation.

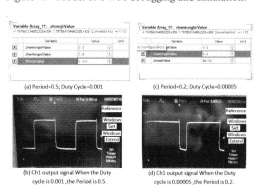

(a) Period=0.5; Duty Cycle=0.001   (c) Period=0.2; Duty Cycle=0.00005

(b) Ch1 output signal When the Duty   (d) Ch1 output signal When the Duty
cycle is 0.001 ,the Period is 0.5.   cycle is 0.00005 ,the Period is 0.2.

Figure 5.   PWM CH1 output pulse waveform.

In the motor PWM debugging, we establish an array of variables table. By changing the value of the array variable duty cycle and the cycle of two and so on simultaneously, the corresponding motor speed, the oscilloscope to detect the pulse signal waveform cycle, and pulse width are changed. Figure 5 shows the PWM port CH1 to the array of variable cycle and duty cycle to change and the corresponding pulse waveform changes. In Figure (a), the set period is 0.001 and duty cycle is 0.5. Figure (b) shows an oscilloscope to detect the corresponding output waveform. In Figure (c), the set period is 0.00005 and the duty cycle is 0.2. Figure (d) is an oscilloscope to detect the corresponding output waveform.

6   CONCLUSIONS

The motion control system of mobile robot is designed on the basis of dSPACE real-time simulation platform. The motion control model was built by using the RTI and MATLAB module library. A key to generate C code through the host computer will be downloaded to the controller. We use the ControlDesk user interface of the host computer to set the parameters. By rapid construction of motion control system simulation platform, mobile robot can realize the omnidirectional mobile and shorten the development period of the control system. DSPACE simulation platform is effective for the development of control system and the pre-experiment verification.

ACKNOWLEDGMENTS

This work was financially supported by the Shandong Province R&D Key Projects (2014GJJS0401) and Shandong Agricultural Equipment Research and Development Innovation Project 2015-09.

REFERENCES

A. Gfrerrer. Geometry and kinematics of the Mecanum wheel. J. CAGD. 25, 784–791 (2008).

Ehsan Hashemi. Model-based PI–fuzzy control of four-wheeled omni-directional mobile robots. J. Robot. 59, 930–942 (2011).

Hsuan-Ming Feng. Intelligent omni-directional vision based mobile robot fuzzy systems design and implementation. J. ESWA. 37, 4009–4019 (2010).

Lang pen fei. Research on hardware in loop simulation system based on dSPACE [D]. Guangzhou: South China University of Technology, 2013.

LIU Han, ZHANG Xian-min. Design of control system for high speed and high precision parallel robot based on dSPACE. J. EDE. 06, 22–11 (2014).

SHAO Xue-juan, ZHANG, Jing-gang, ZHAO. Design of experimental platform for motion control system based on dSPACE. J. EEE. 12, 34–06 (2012).

Wang Guan. Research on motion control technology of omni-directional mobile platform [D]. Beijing: Beijing Institute of Technology, 2015.

Xin xiao shuai. Study on the control strategy of multi drive mode of four wheel independent drive electric vehicle [D]. Chengdu: University of Electronic Science and technology, 2015.

XU Li-chuan, YANG Xiao-dong, CONG Pei-qiang. Experimental design of DC motor PWM based on dSPACE. J. EST. 08, 10–04 (2012).

Zhang qi. Research on control strategy of permanent magnet brushless DC motor based on dSPACE [D]. Hangzhou: Zhejiang University, 2012.

Civil, Architecture and Environmental Engineering – Kao & Sung (Eds)
© 2017 Taylor & Francis Group, ISBN 978-1-138-02985-9

An adapted NSGA-II approach to the optimization design of oil circuits in a hydraulic manifold block

Guang Li
Key Laboratory of Mechanism Theory and Equipment Design of Ministry of Education, Tianjin University, Tianjin, China
College of Packaging and Printing Engineering, Tianjin University of Science and Technology, Tianjin, China

Wentie Niu, Weiguo Gao & Dawei Zhang
Key Laboratory of Mechanism Theory and Equipment Design of Ministry of Education, Tianjin University, Tianjin, China

ABSTRACT: The optimization design of oil circuits in a Hydraulic Manifold Block (HMB) is a typical Multiobjective Optimization Problem (MOP), and typical optimization objectives include the number of cross-drills, the depth of working holes and cross-drills, and the length of oil circuits. MOP for HMB design is usually converted to a Single-objective Optimization Problem (SOP) by weighting method; however, the validity is not guaranteed. In this paper, maze algorithm—NSGA-II optimization method (MA-NSGA-II) is presented as improve NSGA-II. In this approach, a coordinate bunch format is proposed to encode the oil circuit parameters and the oil circuit connection; then, the maze algorithm is used to generate the initial population; further new operators of crossover and mutation are introduced and applied to adapt classical NSGA-II to suit the proposed encoding format. On the basis of this approach, an optimization design system for HMB is developed by using SolidWorks Application Programming Interface (API), VB.NET, and Microsoft Access. The system effectively realizes the automatic generation and optimization of oil circuits. A case study is conducted to verify the validity of the approach and the system.

1 INTRODUCTION

HMB is one of the key components in integrated hydraulic systems. However, the manual design of HMB is a complex and time-consuming process; and hence, it is highly desirable to generate oil circuits automatically. There are two types of holes in an HMB: working holes and cross-drills. Working holes are holes connected with hydraulic component ports while cross-drills are auxiliary holes. While designing an HMB, we have to consider the layout of hydraulic components, the connectivity of ports, and many other design constraints. For ports that cannot be connected directly, cross-drills are necessary.

In recent years, much research has been carried out to solve spatial pipe routing design and oil circuit design for HMB, for example, the design of pipeline layouts of aeroengines or ships. For the problem of pipe routing and oil circuit, a system utilizing AI and CAD techniques was described by Chambon (1991) to solve combined 3D component-placing and routing-design problems of HMB. Wong (1997) proposed an object-oriented approach to represent the behavior and the design knowledge of hydraulic circuit and component and presented a 3D circuit connection algorithm of HMB. A method was proposed by Feldmann (2006) for the generation and manipulation of bore chains of HMB, in which BOHRZUG objects are created between ports to be connected. When a new point is included in the chain, an existing bore is replaced by two new ones. Using the pattern match method, Park (2002) presented an automatic pipe-routing algorithm. For the problem of optimal design, Ito (1999) presented a genetic algorithm approach to support interactive planning of a piping route circuit. Fan (2007) presented a variable-length encoding format and corresponding genetic operators to ensure the association of genes chromosome. Guirardello (2005) presented an optimization approach for the design of a chemical plant geometric layout, and the task is decomposed into a sequence of subproblems, which are then solved by using Mixed Integer Linear Programming models (MILP). Ferreira (2010) presented a framework using a multidisciplinary design optimization methodology, which tackles the design of an injection mold. Liu (2008) studied the optimal method of oil circuit using genetic algorithm and simulated annealing algorithm. Although an MOP is converted to an SOP by using objective weight method above, the diversity of solutions cannot be reflected by the obtained results. Appropriate weights factors, which the objective weight method heavily relies on, are difficult to choose; meanwhile it is also time-consuming and inefficient to obtain an optimal solution. Hence, it is essential to investigate the application of multiobjective optimization methods, for example, NSGA-II proposed by Deb (2002), for HMB oil circuit.

On the contrary, the maze algorithm has been successfully applied to HMB design, but only little attention has been paid to enhance it with optimization methods. Lee (1961) proposed the maze algorithm to search the circuit of two points in a plane. Maze algorithm in HMB design is used by P.K. Wong (1997) to realize the oil circuit connection, but the strong connectivity of the algorithm leads to too many cross-drills.

In this paper, a novel approach termed as MA-NSGA-II is proposed by combining the maze algorithm and NSGA-II. The remainder of this paper is organized as follows. Section 2 introduces a multiobjective optimization model of HMB, explaining the design variables, objective functions, and design constraints. Section 3 presents the approach of MA-NSGA-II, and some new operators for genetic algorithm. Section 4 presents the implementation and case study of the approach. Finally, in Section 5, conclusion is drawn and future work is defined.

2 MULTIOBJECTIVE OPTIMIZED MODEL OF HMB

The oil circuit is utilized to realize the connective relation between ports of hydraulic components in HMB, and the relation is specified in the hydraulic scheme diagram. An oil circuit is composed of various oil holes. Because of the diversity of hydraulic components and oil circuit structure, the oil hole types and features are complex, and the related concepts are shown in Table 1.

Figure 1 shows an exemplary oil circuit connecting two hydraulic components, c_1 and c_2, and there are two working holes, h_1 and h_3, and one cross-drill, h_2, in which h_1 is a bottom hole and h_3 is a bottom center hole. The coordinate system of HMB is also shown in Figure 1.

An HMB is designed on the basis of its hydraulic scheme diagram. The design of HMB is to layout hydraulic components and to connect oil circuits, subject to some design objectives and design constraints. Parameters of hydraulic components,

Table 1. Oil hole types and features.

Oil hole type	Oil hole feature
Bottom hole	Connected with port of the plate valve
Bottom center hole	Connected with bottom center port of cartridge valve
Side hole	Connected with side port of cartridge valve
Cross-drill	Not connected with hydraulic components, but connected with holes that cannot be connected directly

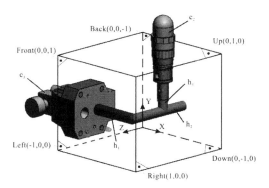

Figure 1. Coordinate system of HMB and flowpath model.

ports, and oil holes are involved in the design progress.

2.1 Design variables

To simplify the expression, the following matrices are used to represent related design variables.

2.1.1 Hydraulic components parameters matrix C
The basic information of hydraulic components is included in matrix C, and the ith row represents the parameters of the ith component, which is noted as $[c_i c_{vx_i} c_{vy_i} c_{vz_i} c_{x_i} c_{y_i} c_{z_i} c_{o_i}]$, where c_i is the ID of the hydraulic component HC_i, $c_{vx_i} / c_{vy_i} / c_{vz_i}$ is its installation surface vector, and the vectors of the surfaces of up, down, front, back, left, and right are listed in Figure 1, where $c_{x_i} / c_{y_i} / c_{z_i}$ is the benchmark point coordinate and c_{o_i} is the installation angle. If there are m hydraulic components, then the dimension of C is $m \times 8$.

2.1.2 Port parameters
The interface information of hydraulic components and oil circuits is described in matrix P, and the ith row $[p_i p_{x_i} p_{y_i} p_{z_i} p_{d_i} p_{s_i} c_j f_k]$ represents the parameters of the ith port, where p_i is the ID of port $port_i$, $p_{x_i} / p_{y_i} / p_{z_i}$ is its associate point coordinate, p_{d_i} is its diameter, p_{s_i} is its structure type, and

f_k is the ID of the owned oil circuit. If there are s ports, then the dimension of P is $s \times 8$.

2.1.3 Oil hole parameters matrix H

The information of all the holes is described in matrix H, where the ith row $[h_i h_{x_i} h_{y_i} h_{z_i} h_{vx_i} h_{vy_i} h_{vz_i} h_{d_i} h_{l_i} h_{s_i} f_j]$ represents the parameters of the ith working hole or cross-drill, where h_i is the ID of oil hole OH_i, $h_{x_i}/h_{y_i}/h_{z_i}$ is its starting point coordinate, $h_{vx_i}/h_{vy_i}/h_{vz_i}$ is its starting point vector, h_{d_i} is the hole diameter, h_{l_i} is the hole depth, h_{s_i} is the hole structure type (0/1/2/3: cross-drill/bottom hole/bottom center hole/side hole), and f_j is the ID of the oil circuit, to which the hole belongs. If there are p working holes and q cross-drills, then the dimension of H is $n \times 11$, where $n = p + q$.

2.2 Objective function

2.2.1 Shortest oil circuit length

There are three objective functions as shown in Equations (1)–(3) with respect to single oil circuit length, the length sum of all oil circuits, and the length of the longest hole, respectively:

$$\min L_{p_i}(C, P, H) = \sum_{j=1}^{M_i} h_{l_j} \quad i = 1, 2, \cdots, S \tag{1}$$

$$\min L_p(C, P, H) = \sum_{i=1}^{S} L_{p_i} \tag{2}$$

$$\min\left(\max\left(h_{l_1}, h_{l_2}, \cdots, h_{l_{M_1}}, \ldots, h_{l_1}, h_{l_2}, \cdots, h_{l_{MS}}\right)\right) = L_{\min(\max)} \tag{3}$$

where L_{p_i} is the length of oil circuit p_i, h_{l_j} is the depth of oil hole l_j of the oil circuit, M_i is the number of oil holes of p_i, L_p is the length sum of all oil circuits, and S is the number of oil circuits.

2.2.2 Minimize the number of cross-drills

There are three objective functions as shown in Equations (4)–(6) with respect to cross-drills number of single oil circuit, the sum of all cross-drills, and the most cross-drills number of one oil circuit, respectively.

$$\min N_{p_i}(C, P, H) \quad i = 1, 2, \cdots, S \tag{4}$$

$$\min N_p(C, P, H) = \sum_{i=1}^{S} N_{p_i} \tag{5}$$

$$\min\left(\max\left(N_{p_1}, N_{p_2}, \cdots, N_{pS}\right)\right) = N_{\min(\max)} \tag{6}$$

where N_{p_i} is the number of cross-drills in the oil circuit p_i and N_p is the number of cross-drills in all oil circuits.

2.3 Constraints

The oil holes and cross-drills must meet the following constraints:

1. The flow cross-sectional area (A_i) of holes intersection parts must be larger than the minimum flow cross-sectional area $(A_{\min})$, that is, $A_i \geq A_{\min}$.
2. The wall thickness of nonconnected flowpaths must be higher than the security wall thickness $(W_{\min})$, that is, $W_{ij} \geq W_{\min}$ and $W_{iF} \geq W_{\min}$, where W_{ij} is the wall thickness between holes and W_{iF} is the wall thickness between hole and surface of HMB.
3. A single oil hole can be processed within the depth scope, that is, $D_i / L_i \geq C_k$, where D_i is the hole diameter, L_i is the hole depth, and C_k is the hole machining depth index.

3 DESCRIPTION OF THE PROPOSED APPROACH (MA-NSGA-II)

NSGA-II is a fast algorithm to obtain a Pareto optimal solution set consisted of non-nominated solutions. NSGA-II uses a non-nominated sort algorithm to obtain Pareto front solutions. Moreover, a crowding distance is defined to preserve the diversity of solutions. The individuals are sorted according to the priority order defined by Equation (7):

$$i \prec_n j, \; if \, (i_{rank} < j_{rank}) \vee (i_{rank} = j_{rank})$$
$$\wedge (i_{dis\tan ce} > j_{dis\tan ce}) \tag{7}$$

where $\prec_n$ is the crowding comparison operator, $rank$ is the rank of non-nominated sort, and $distance$ is crowding distance.

NSGA-II has been widely applied in various fields because of its high efficiency and robustness. However, it has not been applied for the design of HMB because the encoding formats and operators in NSGA-II are not suitable for the complex oil circuit structure of HMB. Here, a coordinate bunch encoding format is proposed on the basis of grid concepts, and the maze algorithm is used to realize the oil circuit connection and to generate initial population. Furthermore, new crossover and mutation operators, which are suitable for the encoding format, are introduced. The flowchart of the approach is shown in Figure 2. Major concepts and steps of the flowchart are explained in the followed subsections.

3.1 Encoding method and initial population

When the maze algorithm is used to generate circuit and initial population, the quality of starting solutions can be improved.

3.1.1 Grids division and obstacles set

The layout space of HMB is divided into three-dimensional grids with a dimension of $M \times N \times L$, and each grid has a coordinate (x, y, z) based on the row, column, and floor in the space. The outer surface of the HMB, except places where ports lies, are denoted as obstacles, and the grids on existing oil circuit are denoted as obstacles, and their values are set to "#". All other grid's values are set to 0.

3.1.2 Circuit encoding

The encoding of circuit is described by using a chain table composed of linked nodes, such as [5 1 3;5 2 3;5 3 3;6 3 3;7 3 3;8 3 3;9 3 3;9 4 3;9 5 3;9 6 3;9 6 4;9 6 5;9 6 6;9 6 7;9 6 8;10 6 8], and a circuit also represents an individual of NSGA-II. In the chain table, the first and last coordinates represent ports. A circuit is a 3D polyline composed of a number of line segments from the start position to end position. In any two adjacent grids, the latter one must be an adjacent grid of the former one in six directions to ensure that the path is straight or orthogonal but not diagonal.

3.1.3 Circuit connection and population initialization

In the expansion process from one grid to the adjacent grid, a limit direction search strategy and the following rules are used: the node is specified by its tag value and direction, the value of the six neighbors of one grid is its value plus 1, and if one of the adjacent grid is the target port, then the extension stops.

By retracing from the target port to the source port, some circuits can be generated randomly, and these circuits are the initial population of NSGA-II.

3.2 Adapted NSGA-II

The existing encoding methods and genetic operators of NSGA-II are not suitable for the circuit encoding the format described above. Therefore, the crossover operator and mutation operator are redesigned as explained here.

3.2.1 Crossover operator

Crossover is the main way to generate new individuals in NSGA-II. The chromosomes generated after crossover or mutation must ensure the connective relationship of nodes to prevent producing illegal individuals. Here, a random point-cross strategy is described.

In a random point crossover, two different nodes are selected from two individuals of the parent generation, and then a sub-circuit is generated to connect the two nodes. The subcircuit will be inserted into the two individuals of the parent generation with correctitude or reverse, and the latter parts of crossover node are exchanged to generate two new individuals in the children's generation. Figure 3 shows an example of random point crossover. The path starts from point (5,1,3) and ends

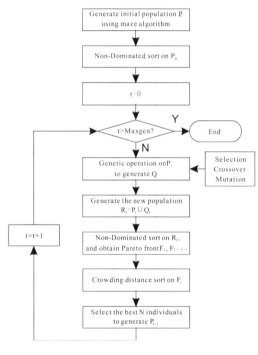

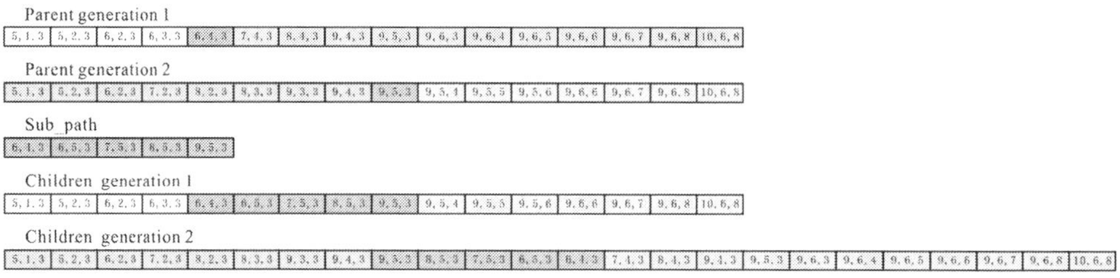

Figure 2. Flowchart of MA-NSGA-II.

Figure 3. Random point crossover.

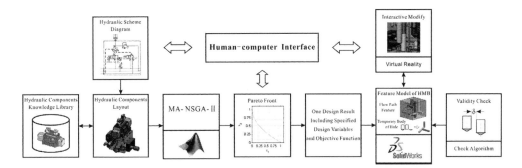

Figure 4.   Flowchart of optimal design system for HMB.

at point (10,6,8), and the random crossover points are (6,1,5) and (9,5,5).

3.2.2   *Mutation operator*

Mutation can maintain the diversity of population to prevent premature convergence. The mutation method is to delete one part of the chromosomes randomly and then add another part, which is a subcircuit. This heuristic mutation operation increases the chances to reduce turnings or bend around obstacles.

4   IMPLEMENTATION AND CASE STUDY

An optimal design system for HMB has been developed, and Figure 4 shows the flowchart. The basic design information is obtained from the hydraulic scheme diagram and knowledge library of hydraulic components. After the initial layout of hydraulic components by human–computer cooperation, the design parameters of hydraulic components and ports can be obtained. Then, the MA-NSGA-II approach will be operated to generate and optimize circuits by using MATLAB. The calculation result is a Pareto front set composed of population size solutions. The detailed results are the objective function valves, rank of non-nominated sort, and crowding distance of every solution. The optimal parameters of design variables will drive the system to generate 3D solid model of HMB with validity check in SolidWorks circumstance. The design results can be interactively modified until it is satisfying.

In the feature-modeling progress, the oil circuit feature is described by using temporary body model and B-Rep model. The temporary body model is created by solid modeling technology and used for representing feature semantics and showing the flowpath. The B-Rep model is used for saving the final results.

Figure 5 illustrates a hydraulic scheme diagram including nine hydraulic components and seven

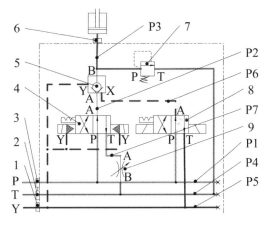

Figure 5.   Schematic diagram of one hydraulic system.

oil circuits. The hydraulic components, oil circuits, and connective relations of ports are listed in Table 2. These hydraulic components are the products of Bosch Rexroth company. The dimensions of HMB are determined according to the maximum edge length of hydraulic components, and the hydraulic components must have enough room for adjustment. In this case, the length, width, and height are 230, 212, and 250 mm, respectively, and the grids are divided into $23 \times 21 \times 25$.

In practical applications, the performances of MA-NSGA-II are dependent on some parameters of algorithms, but how to select these optimal parameters is not completely solved yet. In this paper, practical design experience of HMB and the orthogonal experiment method are combined to explore the optimal parameters. The running parameters are as follows: population size is 20, generation is 200, crossover probability is 0.8, mutation probability is 0.05, crossover distribution index is 20, and mutation distribution index is 20. In the optimization and validity check process, the constraints are as follows: the least-flow cross-sectional area (A_{min}) is 28 mm^2, the least security wall thickness (W_{min}) is 5 *mm*, and the oil hole machining depth index (C_k) is 0.045.

The Pareto set is constructed where each design has the best combination of objective values, so that improving one objective is impossible without sacrificing one or more of the other objectives. After the calculation, the system obtains 20 non-nominated sorted optimal solutions. An optimal design solution whose rank is 1 can be seen in Tables 3 and 4, where the values in Table 3 are the optimized design variables valves, and Table 4 shows the objective function values.

Table 2. Components, paths, and connective relation.

Components	Path ID	Connective relation
1:SNO-A8-SR3–8	P1	3:IO/4:P/8:P
2:VF3005	P2	4:A/5:A
3:VF3005	P3	5:B/7:P/6:IO
4:4 WEH32_6X	P4	2:IO/9:B/7:T
5:SL30P-4X_P1	P5	1:IO/5:Y/8:T/4:Y
6:VF3002	P6	5:X/8:A
7:DBDS10 K 1X	P7	9:A/4:T
8:_WE6_6XE_K4		
9:NDVP-30		

Table 3. Results of optimal design.

Hole ID	Starting point coordinate	Vector	Dia. (mm)	Depth (mm)	Path ID
4:P	159/212/134	0/–1/0	34	100	P1
3:IO	159/0/134.5	0/1/0	34	112	P1
8:P	0/16/134.5	1/0/0	7.6	159	P1
4:A	70/212/102.5	0/–1/0	38	100	P2
5:A	0/112/102.5	1/0/0	32	70	P2
5:B	0/112/153.3	1/0/0	20	100	P3
7:P	31/112/182.5	0/0/–1	10	29.2	P3
6:IO	100/112/250	0/0/–1	20	96.7	P3
2:IO	100/0/210	0/1/0	30	62	P4
9:B	230/62/156.5	–1/0/0	35	35	P4
7:T	31/0/209	0/1/0	10	112	P4
P4:T1	0/62/209	1/0/0	30	195	P4
P4:T2	195/62/250	0/0/–1	30	93.5	P4
1:IO	120/156/238	0/0/–1	14	103.5	P5
5:Y	0/156/145.3	1/0/0	6	120	P5
8:T	0/36.5/134.5	1/0/0	7.6	120	P5
P5:T1	230/155/188	–1/0/0	10	110	P5
P5:T2	120/212/134	0/–1/0	10	175.5	P5
4:Y	149.5/212/188	0/–1/0	10	57	P5
5:X	0/67.6/110.4	1/0/0	6	25	P6
8:A	0/26.2/125.7	1/0/0	7.6	25	P6
P6:T1	25/26/0	0/0/1	6	125.7	P6
P6:T2	25/0/110.4	0/1/0	6	67.6	P6
9:A	230/62/61.5	–1/0/0	35	71	P7
4:T	159/212/61.5	0/–1/0	38	150	P7

Table 4. Objective function values via MA-NSGA-II.

Length of circuit (P1–P7)	Total length	The deepest hole	Number of cross-drills
371/170/ 216.7/ 497.5/686/ 351.2/211	2513.4	P5:T2 175.5 mm	0/0/0/2/2/2/0

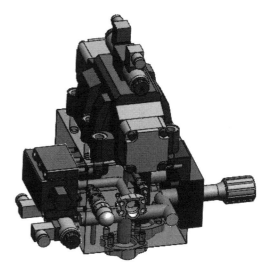

Figure 6. Optimal design plan model.

The choice of final HMB design should be made from the set of Pareto optimal designs, which may not be explicitly represented in the model by using additional customer preferences and criteria.

Figure 6 shows the optimal design plan model, and Table 3 lists its structure parameters. The layout of hydraulic components is compact and reasonable, the oil circuits can be connected with less cross-drills, and the flow cross-sectional area, wall thickness, and oil hole depth are all in the range of constrains. The results show the feasibility and validity of the approach and system.

5 CONCLUSIONS

In this paper, we presented a novel hybrid algorithm, MA-NSGA-II, aiming at the multiobjective design optimization of the oil circuits in HMB. This approach proposes a coordinate bunch encoding format for the oil circuit design of HMB. On the basis of the encoding format, the maze algorithm is proposed to set up the oil circuit connection and generate initial population, and new crossover and mutation operators are proposed to adapt NSGA-II to the encoding format. An HMB optimization design system is developed by using SolidWorks API and VB.NET, which can effectively realize the automatic generation and optimization of oil circuits of HMB.

To develop a more realistic model for HMB in future, it will be necessary to refine the optimization model by including all important variables. It is also important to integrate MATLAB with SolidWorks to generate some high-quality models, as well as CAD tools to visualize that the design solutions are also fundamental in a fully integrated HMB optimization system.

ACKNOWLEDGMENT

This work was financially supported by the National Natural Science Foundation of China (Grant Number: 51275340).

REFERENCES

Chambon, R. and M. Tollenaere, Computer-Aided Design. 23, 213 (1991).

Deb, K., A. Pratap, S. Agarwal and T. Meyarivan, IEEE Trans. Evol. Comput. 6, 182 (2002).

Fan, X., Y. Lin and Z. JI, Shipbuilding of china, 48, 82 (2007) (In Chinese).

Feldmann, D.G., Mechanika, 62, 54 (2006).

Ferreira, I., O.D. Weck and P. Saraiva, J. Cabral, Struct. Mutltidiscip. Opt. 41, 62 (2010).

Guirardello, R. and R.E. Swaney, Comput. Chem. Eng. 30, 99 (2005).

Ito, T., J. Intell. Manuf. 10, 103 (1999).

Lee, C.Y., IRE transactions on electronic computers. 9, 346 (1961).

Liu, W., S. Tian, C. Jia and Y. Cao, J. Shanghai Univ. (Engl. Ed.), 12, 261 (2008).

Park, J. and R.L. Storch, Expert Sys. Appl. 23, 299 (2002).

Wong, P.K., C.W. Chuen and T.P. Leung, Am. Soc. Mech. Eng. Fluid Syst. Technol. Div. Publ. FPST. 4,183 (1997).

Civil, Architecture and Environmental Engineering – Kao & Sung (Eds)
© 2017 Taylor & Francis Group, ISBN 978-1-138-02985-9

Quantification of single-event transients due to charge collection using dual-well CMOS technology

Zhun Zhang
Shenzhen Key Laboratory of Micro-nano Photonic Information Technology, Shenzhen University, Shenzhen, China

Wei He
College of Electronic Science and Technology, Shenzhen University, Shenzhen, China

Sheng Luo & Lingxiang He
Shenzhen Key Laboratory of Micro-nano Photonic Information Technology, Shenzhen University, Shenzhen, China

Qingyang Wu & Jianmin Cao
College of Electronic Science and Technology, Shenzhen University, Shenzhen, China

ABSTRACT: In this paper, we present the computation of single-event transients for calculating the charge collection in dual-well structure. A well gradient collapse drain injection for SRAM cell is demonstrated through TCAD modeling. Simulation presents that the deposited charge will induce parasitic bipolar amplification to broaden the transient widths for PMOS device under the effects of N-well region. The efficiency of charge collection is correlated to the distribution of SET pulse widths. Aiming at improving the reliability of devices, well potential contacts are expected to be established.

1 INTRODUCTION

Critical charge to represent a logic state in Integrated Circuits (ICs) is steadily decreasing with the shrinking device technology feature sizes. Single-Event Effects (SEEs) induced by heavy-ion irradiation has become a primary reliability issue for deep submicron processes because of the reduced nodal charge and space between devices (P.E. Dodd, 2010 & D. Munteanu, 2008). Traditionally, the close proximity of transistors means that the deposited charge cloud from an ion strike is expected to encompass multiple transistors on ICs, resulting in multiple-node charge collection (termed charge sharing) in devices (O.A. Amusan, 2006). Thus, it is imperative to characterize the single-event transient response to different design and layout practices. Dual-well bulk CMOS technology has been the subject of extensive studies on improving the reliability of manufacturing space and military electronic devices. The effects of dual-well structure were mainly analyzed and results showed that Single-Event Transients (SETs) using the dual-well technology were shorter than those in the triple-well technology (T. Roy, 2008). Perturbations in n-well and p-well potentials also have been shown to affect strongly the

charge collection, charge sharing, and parasitic bipolar transistor characteristics (N.J. Gaspard, 2011 & O.A. Amusan, 2006). A comparative analysis of heavy ion-induced upsets in dual-well bulk SRAMs shows that dual-well process is vulnerable to low-LET particles, whereas triple-well technology is more vulnerable to high-LET particles (I. Chatterjee, 2011). Despite these previous studies, response reported for dual-well and triple-well processes have been proposed to mitigate SEEs, single-event transients caused by heavy ions with high Linear Energy Transfer (LET) values are still inevitable to affect the stability and performance of the devices, resulting from the fact that much remains to be understood about the complex physics of carrier transport and enhancement in devices fabricated with well structures. Therefore, full three-dimensional physical models are conducive to the incorporation of device analyses into the behavioral abstractions of carrier motions and potential field perturbations in the wells.

This paper describes primarily the physical insights for accurately modeling single-event transient currents and charge collection using full three-dimensional Technology Computer-Aided Design (TCAD) simulations and explores the spatial extent and temporal characteristics of SEEs,

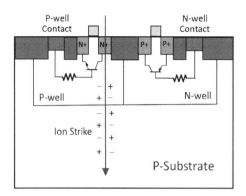

Figure 1. Schematic cross section view of NMOS and PMOS implemented with dual-well technology, showing parasitic elements.

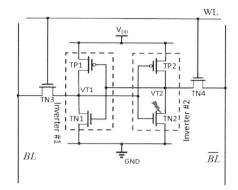

Figure 2. Conceptual schematic of a typical 6-T SRAM cell formed by two cross-coupled CMOS inverters. The blue dashed outline shows inverter #1, and the light red region represents inverter #2.

which are affected by dual-well structure. First, the schematic cross view of dual-well CMOS technology is illustrated in Figure 1. A particle strikes a sensitive region of a circuit module and deposits a dense track of electron–hole pairs, where the deposited charges will cause a potential gradient in the bulk from the strike location to the substrate, which could possibly activate the parasitic bipolar transistor of MOSFETs in the P-well and N-well, as well as affects the potential modulation process of multiple junctions.

2 TEST STRUCTURE AND EXPERIMENTAL SETUP

The modeling and simulation of the charge generation, transport, and collection processes are crucial issues in the computation subjected to ionizing radiation. In this work, Sentaurus H-2013 TCAD from Synopsys Corporation was adopted to perform structures and device simulations. For the dual-well structure, the typical 6-T SRAM cell as the storage unit was carried out by using 65 nm bulk CMOS technology. Figure 2 describes the conceptual schematic of a typical 6-Transistors SRAM cell. Two cross-coupled inverter modules, inverter #1 and inverter #2, are connect with each other and additional two access pass-gate transistors, and the output nodes VT1 and VT2 store opposite values. In the circuit, supply voltage is set to be 1.2 V.

For three-dimensional SRAM model calibration, the dual-well model was preliminarily referenced to the layout of target SRAM device cell with detailed designs, as illustrated in Figure 3. The ratio of width to length (W/L) in the NMOS is Wn/Ln = 0.30 /0.06 μm, whereas the W/L of the PMOS is Wp/Lp = 0.45 /0.06 μm, and they were generated with a 65 nm standard cell library in accordance with the Semiconductor Manufacturing International

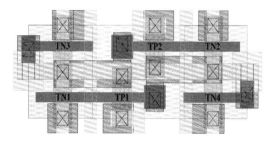

Figure 3. Layout of the target SRAM cell calibrated with 65 nm process design kit.

Corporation (SMIC) layout design rules. The I–V electrical characteristics (Id–Vd and Id–Vg curves) of NMOS and PMOS transistors, which are built into the dual-well CMOS structure, have obtained good agreements with the HSPICE models. Thus, the junction capacitance between drain/source and bulk agrees with reality.

As shown in Figure 4, a full 3-D SRAM cell structure is implemented on dual-well CMOS technology, N-channel transistors are placed in the P-type substrate, and all the P-channel transistors (TP1 and TP2) are generated in the implanted N-well inside the P-type substrate. An N-well contact band is placed near the two transistors with a distance of 0.3μm, the source and drain junctions of NMOS is 25 nm, and Light Doped Drain-source (LDD) are inserted between the channel and the source/drain, respectively. The thickness of the gate oxides are generated according to the spice model so that the gate capacitance agrees with actual process. Furthermore, the Shallow Trench Isolations (STI) are used to separate transistors for maintaining good electrical characteristics.

In the following 3-D TCAD simulations, the special-purpose supercomputer facilities are used

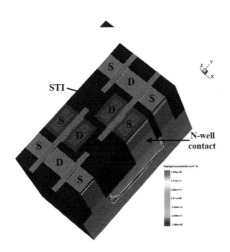

STI

S S D
D S
S S
D S
D
S
S D

N-well contact

Figure 4. Full 3-D SRAM model implemented on 65 nm CMOS process.

to shorten the work of extended numerical calculations at the MPIT laboratory in Shenzhen. Taking design parameters into account, Finite-Element Method (FEM) numerical simulations are performed to solve the Poisson and drift–diffusion current continuity equations, and the following physical models are incorporated during the whole simulations: 1) Firmi-dirac statistics; 2) bandgap narrowing effects; 3) doping-dependent Shockley–Read–Hall (SRH) recombination and Auger recombination; 4) mobility models, electric field, interface scattering effects, and carrier–carrier scattering impact on mobility; 5) density gradient quantization model; 6) a hydrodynamic model is used for carrier transport; and 7) heavy ions are modeled using a Gaussian distribution profile with a characteristic 1/e radius of 50 nm, and a Gaussian temporal profile with a 2 ps characteristic time.

To investigate the process of charge collection, heavy ions were used to strike the TCAD models, considering that the particle struck the drain region of TN2 transistor in inverter #2. Broad-beam heavy-ion experiments were performed with Linear Energy Transfers (LETs) ranging from 10 to 50 MeV-cm²/mg, and the LET suppression method was set as a stable step-up 10 MeV-cm²/mg iteration. Normally, the LET values were kept constant along the heavy-ion track, and every ion irradiation was performed at an angle of 90°; thus, the angular effects were ignored.

3 MATHEMATICAL FORMULATION

The physical mechanisms related to the production SETs in microelectronic devices consist in three

main successive steps: (1) the charge deposition by the energetic particle striking the sensitive region, (2) the transport of the released charge into the device, and (3) the charge collection in the sensitive region of the device. The mathematical formulations that affect the charge collection of SET in dual-well structure are explained in this section. When a heavy ion penetrated the SRAM cell, it loses energy and creates a trail of electron–hole pairs; the ion-striking deposited charge was collected by the drain and the transient currents are generated to turn on the output logic state of inverter #2. The device is assumed to consist of a collection of reverse-biased depletion regions and electrodes, and the device simulation considers the popular drift-diffusion transport equation by considering the methods of moments, and the electron (Jn) and hole (Jp) current densities are expressed as follows:

$$J_n = qn\mu nE + qDn\,grad(n)$$

$$J_p = qn\mu pE - qDp\,grad(p) \qquad (2)$$

where (1) and (2), n and p are the electron and hole densities, μ_n and μ_p are the electron and hole mobility, respectively, and E is the built-in electric field. The first term corresponds to the drift component, which is driven by the electric field and the second term to the diffusion component caused by the gradient of carrier concentrations. D_n and D_p are the diffusion coefficients corresponding to the carrier mobility via the Einstein equation (A. Grove, 1967):

$$D_{n,p} = \frac{k_B T}{q}\mu_{n,p} \qquad (3)$$

where k_B is the Boltzman constant and T is the carrier temperature equal to the lattice temperature as the carrier gas in the drift-diffusion approximation is assumed to be in thermal equilibrium. In the formulations, finite-element numerical simulations are performed to solve the Poisson and drift-diffusion current continuity equations. The transients due to particles striking the dual-well are computed by a double current pulse as follows:

$$I_d = \frac{Q}{\tau_\alpha - \tau_\beta}\left(e^{-t/\tau\alpha} - e^{-t/\tau\beta}\right) \qquad (4)$$

where I_d is the drain total current, whose unit is A, Q is the collected charge due to the LET values, τ_α and τ_β are the falling time and rising time constants of the current pulses to determine the amplitudes and widths respectively, and τ_α and τ_β can be defined as:

$$\tau_\alpha = \frac{4ds^2}{\pi^2 D_{n,\ p}} \quad \tau_\beta = \frac{4ds^2}{9\pi^2 D_{n,\ p}} \tag{5}$$

4 RESULTS AND DISCUSSIONS

When a charge column has been created in the dual-well SRAM structure by an ionizing particle, the physical phenomena in drift-diffusion, hydro-dynamic, and thermodynamic transport models are considered for characterizing the charges transported. Figure 5 shows the current pulses of the drain electrode of TN2 transistor, the amplitudes of parasitic transient currents have been decreased by varying time and sensitive duration to charge collection for maintaining the electrostatic potential gradients. The transients indicated that the drain total currents consist of a fast drift and a slower contribution from charge diffusion in the shallow P-well and N-well. This causes the electrons to drift into the n-well, leaving the holes behind in the p-well. When the P-well and its contact are held constant, NMOS TN2 transistor single-event transient currents are governed by the depressed drain contact voltage. After a short duration, current peaks corresponding to the channel conduction supported of drain voltage as well as most of deposited charges flow out of the struck transistor absorbed in P-substrate. Thus, the drain voltage recovers to its original value, and these current pulses decrease toward zero.

For quantifying the charge collection effects on dual-well SET performances, the cross-sectional view and simplified equivalent circuit models for TN2 and TP2 devices fabricated in the 6-T SRAM cell have been characterized as presented in Figure 6. These small signal models account for the series impedance (Z_{in}) between the source

(S_n, S_p) and drain (D_n, D_p) nodes of TN2 and TP2. The dual-well impedances (Z_{DW}) for TN2 and TP2 consist of back to back diodes for each transistor, which take into account the series impedance R3 in TN2 of the transition region between P-well and P-substrate. The channel impedance (R_{ch}) varies according to gate voltage biases (G_n and G_p). When a negative V_{Pwell} and a positive V_{Nwell} are applied to the shallow-well regions, and a negative voltage (V_{sub}) is applied to the P-substrate contact, the diode D2, D2', D4, and D4' junctions are reverse-biased, and only the electric yields of P-well to light P-doping region follow asymmetry conduction. SRAM soft-error rates depend on the critical charge of the circuit and the amount of charge collected by sensitive circuit nodes.

Because of the complex interaction between the heavy-ion charge track and the collection areas as well regions and resulting funnels, nearly simultaneous multiple nodes at the output voltage VT2 charge collection influenced by the SEE charge sharing effect were observed, as illustrated in Figure 7. Results present that the transient voltages are substantial steady with time, and after approximately 0.6 ns, both amplitude and width start to change the logic state. At the moment, PMOS devices are affected by the parasitic bipolar amplification effects, where the deposited charges are relative to the sensitive node and the resultant final collected charge. With the LETs increasing, a large amount of electronic charge within the well lead to the P-well and shallow N-well potential collapse, which results in parasitic bipolar conduction and the wider transients. Simulations verification of the effect of P-well contact together with the shallow N-well in reducing the charge collection lead

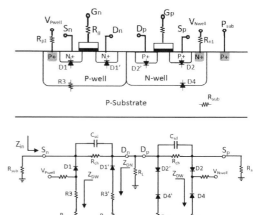

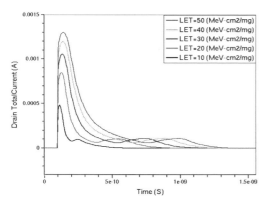

Figure 5. Current pulses of the drain electrode of TN2 transistor in dual-well structure.

Figure 6. (a) Cross-sectional view of dual-well process; (b) Simplified equivalent circuit model fabricated in dual-well process.

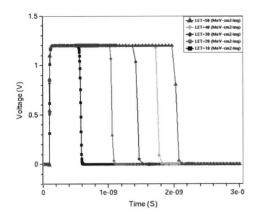

Figure 7. Transient voltages of VT2 node induced by the heavy-ion transient current duration.

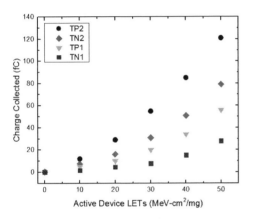

Figure 8. Charge collection with varying LETs for NMOS and PMOS devices. The significant increase of the charge collected with increasing LETs for PMOS device due to the parasitic bipolar amplification.

to the lower transient amplitudes than those of the TN2 transistor. Figure 8 shows the charge collection of NMOS and PMOS devices with varying LETs in the bulk CMOS technology, and PMOS transistors have significant increase of the charge collected with increasing LETs than NMOS transistors due to the parasitic bipolar amplification effects in the N-well region.

5 CONCLUSIONS

In this paper, the characterization of SET pulses among SRAM logic nodes was performed in 65 nm dual-well CMOS technology. Furthermore, we expatiated the pulse widths and amplitudes of NMOS and PMOS devices, simulation dates indicate that charge collection in N-well will activate the parasitic bipolar amplification for PMOS, and well contacts help maintain the well potential collapse, thereby reducing the charge collection and SET pulse width. This paper further provides referential analytical mechanisms for SETs fabricated in dual-well CMOS devices for ensuring good electrical reliability.

ACKNOWLEDGMENTS

The authors acknowledge the financial support provided in part by the Shenzhen Science and Technology Development Funds (JCYJ20140418095735 595, JCYJ20140418091413574, JCYJ201603080939471 32) and from the Open fund (MN201407) of Shenzhen Key Laboratory of Micro-nano Photonic Information Technology.

REFERENCES

Amusan, O.A., A.F. Witulski, and L.W. Massengill, J. IEEE Trans. Nucl. Sci. **53**, 6, (2006).

Amusan, O.A., A.F. Witulski, and L.W. Massengill, J. IEEE Trans. Nucl. Sci. **53**, 6, (2006).

Chatterjee, I., B. Narasimham, and N.N. Mahatme, J. IEEE Trans. Nucl. Sci. **58**, 7, (2011).

Dodd, P.E., M.R. Shaneyfelt, and J.R. Schwank, J. IEEE Trans. Nucl. Sci. **57**, 17, (2010).

Gaspard, N.J., A.F. Witulski, and N.M. Atkinson, J. IEEE Trans. Nucl. Sci. **58**, 7, (2011).

Grove, A., Physics and Technology of Semiconductor Devices. New York: Wiley, 201–204, (1967).

Munteanu, D., and J.L. Autran, J. IEEE Trans. Nucl. Sci. **55**, 25, (2008).

Roy, T., A.F. Witulski, and R.D. Schrimpf, J. IEEE Trans. Nucl. Sci. **55**, 9, (2008).

Research on missing value processing methods based on advanced community detection

Fuqiang Zhao, Guijun Yang & Xue Xu
Tianjin University of Finance and Economics, Tianjin, China

ABSTRACT: In complex network graph, connectivity of graph depends on the second smallest eigenvalue of the Laplacian matrix. Correlation coefficients are introduced to this cut model to resolve overlapping community detection by minimizing the algebraic connectivity of complex networks. In this paper, we define edge centrality for each edge by spectral analysis and propose an advanced algorithm of community detection on the basis of centrality measure and correlation coefficients. By the analysis of the algorithm and missing value processing, three methods of missing value processing are put forward. The study subject is classified into several groups by the algorithm of community detection on the basis of centrality measure. Then, it calculates the centrality of the study subject with missing value in the group and deletes the record with a low centrality; on the contrary, the traditional methods of missing value processing are adapted to process missing data in same group. This method is applied to evaluate a fast-food company with missing data. The results show that the missing value processing method based on community detected outperforms the traditional mean imputation method, multiple imputation method, and K-means algorithm. It offers a practical approach for missing value processing.

1 INTRODUCTION

With the rapid development of many social networks (Twitter, Google+, Facebook) in the last decade, the social network has a huge amount of data. The number of network nodes can reach millions or billions (Charu, 2011). The processing of huge data also promotes the development of complex network models and methods. Although some community detection algorithms can effectively identify communities in complex networks, the algorithm complexity is still high. The time complexity of Zhou & Lipowsky (2004) and Donetti & Muñoz (2005) is $O(n^3)$, and Clauset et al. (2004) time complexity is $O(nlog^2n)$. The time complexity of GN (Girvan, 2002) algorithm is $O(m^2n)$. The time complexity of sparse networks is $O(n^3)$. Duch & Arenas (2005) time complexity is $O(n^2logn)$. Fortunato et al. (2004) time complexity of is $O(nm^3)$. Eckmann & Moses (2002) time complexity is $O(m< k^2 >)$. Capocci et al. (2005) time complexity is $O(n^2)$. A cutting edge model based on the edge centrality measure is proposed in the Edge Centrality Cut Model (ECCM, Zhang, 2012). Although the model reduces the time complexity, it did not consider the overlapping community nodes. Its time complexity is $O(n + m)$.

There is a large amount of data in social network, but the experimental study or survey will inevitably cause missing data. The main reasons for this are:

1. The subjects are unwilling to provide the required survey information;
2. The uncontrollable factors cause the missing data;
3. The researcher or survey system does not collect complete information;
4. The error of information reporting summary cause the missing data.

The traditional methods of missing value processing mainly include deletion, imputation, and likelihood maximum (Ye, 2014). Imputation of missing values include imputation mean, random imputation method, and Multiple Imputation (MI) (Zhao, 2013; Yozgatligil, 2013). Multiple imputation method is widely applied in psychology, medicine, finance, climatology, pharmaceutics (Ji, 2013; Twisk, 2013; Donneau, 2015). Loss mechanism of missing data includes Missing Completely At Random (MCAR), Missing At Random (MAR), and Missing Not At Random (NMAR). Ye (2014) provides an inspection and identification method. Yang (2012) analyzed the effect of the full Bayesian and partial Bayesian methods on parameter estimation with different missing ratios. However, these studies do not deal with missing data based on the classification of data.

In summary, this paper presents an advanced community detecting algorithm, which effectively reduces time complexity. The missing data are processed in the same community. The centrality of survey subject decides whether missing data are deleted or not.

2 EDGE CENTRALITY MEASURE OF THE COMMUNITY DIVISION ALGORITHM WITH THE CORRELATION COEFFICIENT

Community detection algorithm can be divided into the following classic algorithms: Graph Partitioning (Kernighan, 1970), Agglomerative Clustering (Girvan, 2002), Divisive Algorithm (2004), Modularity Methods (J. Mei, 2009), Spectral Algorithm (Luxburg, 2007), and Markov Clustering (Dongen, 2000). The algorithms of community detection detect communities by optimizing an objective function. Among them, Graph Partitioning, Modularity Methods, and Spectral Algorithm optimize a specific objective function. The Advanced Edge Centrality Cut Algorithm (AECCA) with the correlation coefficient is introduced. The algorithm cuts graph edges and divides community into two graphs. The missing data are processed in the same community.

The Laplacian matrix, $L(G) = AA^T$, is independent of the orientation. In fact, $L(G) = D(G)-A(G)$, where $D(G)$ is the diagonal matrix of vertex degrees and $A(G)$ is the $(0, 1)$ adjacency matrix. The diagonal entry L_{ii} is the degree of node i, if $(i,j) \in E$ then $L_{ij} = -1$, otherwise $L_{ij} = 0$. $L(G)$ is a symmetric, positive semidefinite matrix (Merris, 1994).

The second smallest eigenvalue $\lambda_2(L)$ is called the algebraic connectivity of the graph G, and the corresponding normalized eigenvector is called the Fiedler vector. The algebraic connectivity is considered a measure how well-connected a graph is. That is, the more connected graph has the greater algebraic connectivity on the same vertex set. The magnitude of this value reflects how well connected the overall graph is. $\lambda_2(L) > 0$ if and only if G is connected. $\lambda_2(L)$ is monotone increasing in the edge set. The algebraic connectivity function of complex networks is a monotone convex function. If $G1 = (V, E1)$ and $G2 = (V, E2)$, $E_1 \subset E_2$, then $\lambda_2(L_1) \leq \lambda_2(L_2)$. The Courant–Fischer Minimax Principle implies (Mohar, 1991):

$$\lambda_2(L) = \min_{v \in 1^\perp} \left\{ \frac{v^T L v}{v^T v} \right\} \tag{1}$$

That is, we want to solve the problem that the following convex function is minimal:

$$\text{minimize} \quad \lambda_2 \left(L - \sum_{l=1}^{m} x_l a_l a_l^T \right)$$
$$\text{subject to} \quad 1^T x \leq k, \tag{2}$$
$$x \in [0,1]^m,$$

Where $x_l = 1$ if edge l belongs to the edge subset, otherwise, $x_l = 0$. l_{ij} is an edge connecting nodes i and j, $l \sim (i, j)$. Assign weight w_l to the every edges according to its importance, $0 \leq w_l \leq 1$. The weight is inversely proportional to the corresponding values in the Fiedler vector. The convex relaxation can be formulated as follows:

$$\text{minimize} \quad \lambda_2 \left(L - \sum_{l=1}^{m} x_l w_l a_l a_l^T \right)$$
$$\text{subject to} \quad 1^T x \leq k, \tag{3}$$
$$x \in [0,1]^m,$$

We choose k edges from candidate edges that lead to the greatest decrease in algebraic connectivity when these edges are cut from G. The gradient of the objective function is $w_l(v_i - v_j)^2$. The proof procedure is shown below. The objective function $\lambda_2 \left(L - \sum_{l=1}^{m} x_l w_l a_l a_l^T \right)$ is a monotone function. Let matrix $Y = \sum_{l=1}^{m} x_l w_l a_l a_l^T$ is the Laplacian matrix, which can be written as $\tilde{L} = L - Y$, such that

$$\lambda_2(\tilde{L}) \leq \frac{v^T(\tilde{L})v}{vv^T} \Rightarrow \lambda_2(\tilde{L})vv^T \leq v^T(\tilde{L})v \tag{4}$$

According to (4), we obtain:

$$\lambda_2(\tilde{L}) \leq \lambda_2(L) + \left\langle v^T v, (\tilde{L} - L) \right\rangle \tag{5}$$

where $\lambda_2(L)$ is an analytic function of L, and therefore of x. In this case, the supergradient is the gradient, that is,

$$\frac{\partial \lambda_2(L - w_l x_l a_l a_l^T)}{\partial x_l} = w_l v^T a_l a_l^T v \tag{6}$$

Definition 1. The edge that leads to the greatest decrease in the algebraic connectivity is defined as the center edge. The nature of this edge is called Edge Centrality.

Definition 2. The center node is the node in the core position of each community in the social network.

The cutting edge model based on edge centrality measures is different from the traditional community detection algorithm in social network. It is proposed to deal with the large-scale community structure based on the edge centrality measure. Nodes i and j correlation coefficients are calculated as:

$$r_{ij} = \frac{\langle x_i x_j \rangle - \langle x_i \rangle \langle x_j \rangle}{\left[\left(\langle x_i^2 \rangle - \langle x_i \rangle^2 \right) \left(\langle x_j^2 \rangle - \langle x_j \rangle^2 \right) \right]^{\frac{1}{2}}} \quad (7)$$

where the average $\langle . \rangle$ is over the first few nontrivial eigenvectors. The quantity r_{ij} measures the community closeness between nodes i and j.

ECCM with correlation coefficient is as follows:

1. Calculate the spectral centrality for each edge of graph G on the basis of the edge centrality function and sort them.
2. Find one edges with the highest spectral centrality, and then, delete them; renew the complex network to G^{new}. The choice of k is based on the edge sparse degree of complex networks. When G is a sparse graph, k is equal to one. Otherwise, k is more than one.
3. Calculate algebraic connectivity of a graph. If $\lambda_2 = 0$, the algorithm goes to step (4), otherwise it goes to step (1).
4. When the corresponding component value of the Fiedler vector of a node is less than the threshold α, we should analyze the necessity of computing the correlation coefficient. Moreover, we should calculate the correlation coefficient of the node if the difference in the number of edges that it connects to the two communities is less than 2.
5. Use formula (3) to calculate the correlation. By comparing the two correlation coefficients of the nodes in the two communities, we can determine to which community the node belongs to.
6. Update the complex network new graph G_1^{new} and G_2^{new}. The graph G is divided into two communities.

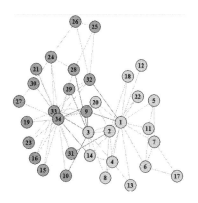

Figure 1. Karate network correct bisect result diagram.

The above six steps that described G are divided into two communities, G_1^{new} and G_2^{new}. If G_1^{new} and G_2^{new} need further partition, we need to repeat step (1) to step (6). Node 3 should have the same community with node 1 and 2. The result of community detection by AECCA is shown in Fig. 1.

3 MISSING VALUE PROCESSING METHOD BASED ON ADVANCED COMMUNITY DETECTION

Imputation mean is that missing data are replaced by the mean value of all the variables without missing values. Multiple imputation is to construct m estimates for each missing value (m > 1) and produce m a complete data set. K-means is the missing value that is replaced by clustering center value.

The missing value processing method with improved community partition algorithm is based on using the above cut edges algorithm. The steps are described in detail as follows:

1. Processing the survey data and constructing social network graph G based on the relationship between the subjects of social networks.
2. Cutting the graph G edges, calculating the spectral centrality for each edge of graph G based on the edge centrality function, and sorting them. Moreover, we should calculate the correlation coefficient of the node if the difference in the number of edges that it connects to the two communities is less than 2. The graph G is divided into two communities G_1^{new} and G_2^{new}.
3. Judging the centrality of the node in its community by $w_i(v_i - v_j)^2$.
4. If the centrality of node with missing data is less than a, its recording is deleted. Otherwise, the missing data are replaced by imputation mean (or K-means) in the same community.
5. Repeat step (3), traversing a community of all nodes and other communities.

The processing methods of missing value based on social network community detection include imputation mean, random imputation method, and multiple imputation with the community partition algorithm. Nodes (subjects) in the same community have the same preference and behavior; therefore, the missing value processing method is more accurate and effective.

4 EMPIRICAL ANALYSIS

Taking a fast-food company as the research object, the experiment collects the data of customer expectations, overall customer satisfaction, and corporate image by online surveys. The subject is RenRen user. A questionnaire link is sent to subjects. The subjects fill in and submit the questionnaire. The feedback results are stored in the database and complete the data collection. Every ID corresponds to specific questionnaire data. With the basic theory research of complex networks, R environment, Curl packages, and other related technologies, we crawl all the data on the RenRen and construct the actual network. According to the relationship between the subjects, using the community partition algorithm, the respondents can be divided into two groups.

Compared with K-means, the imputation mean with community detection is an improved one. Its deviation is smaller and closer to the original value. The number of iterations of multiple imputation with community detection is nine. Every community applies this method. The result is better than the traditional multiple imputation.

The overall expectation value density distribution is shown in Fig.2 with the missing value ratio of 30%. Fig.3 shows the overall expected missing rate and value distribution. The highest value of customer satisfaction is 10. Sample size is 112, blue indicates the observed value, red indicates the imputation value, and the number of iterations is nine. The imputation result is satisfactory.

Figure 4 indicates a comparison of missing data ratio and RMSE with different missing ratios (Zhao, 2013). Fig.4(a) shows respectively Mean Imputation, Classification Mean Imputation (C Mean Imputation), and Community Detection Mean Imputation (CD Mean Imputation). Fig.4(b) shows respectively K-means, Classification K-means, and Community Detection K-means (k = 5 or 10). Fig.4(c) shows respectively Multiple Imputation, Classification Multiple Imputation, and Multiple Imputation with Community Detection.

The results show that three kinds of missing data processing method exceed traditional

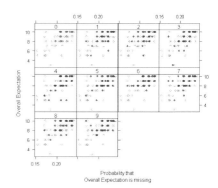

Figure 3. Overall expected missing rate and value distribution.

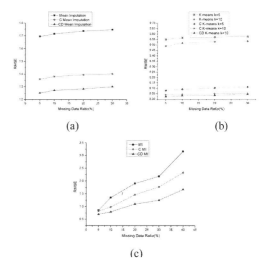

Figure 4. Missing data ratio and RMSE.

method. K-means method based on the community detection has the minimum RMSE. Multiple Imputation with Community Detection is the most important RMSE. Because there is a link between the subjects and larger correlation between datum in the social network, Imputation missing data in sub-community is closer to the original value than that in the whole community. The missing value processing method based on community detection is effective.

Although several EB (exabytes) of data are generated daily, missing data are relatively common in practical applications because of the problems of the existing software and hardware resources, data collection, processing, and storage. Accordingly, with the improvement of missing value processing precision and high speed, the missing value processing method has a partial use of space under certain conditions.

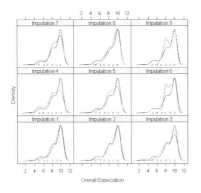

Figure 2. Overall expectation value density distribution.

5 CONCLUSION

Through analysis of community detection and missing value processing method, in this paper, we proposed the edge centrality measure of the community division algorithm with the correlation coefficient and three kinds of missing value processing method on the basis of advanced community detection (imputation mean, multiple imputation, and K-means with community partition algorithm). It takes a fast-food company as the research object and tests the three methods. The results show that our missing data processing method is superior to the traditional imputation methods. With the development of network, there are more and more nodes in a community. Therefore, future research should aim at decreasing the computational complexity of community detecting algorithms and applying faster cut edges method to process missing data.

ACKNOWLEDGMENTS

This paper was sponsored by the Natural Science Foundation of China (Project No. 11471239), Tianjin Natural Science Foundation of China (Project No. 15 JCYBJC16000), and Tianjin Philosophy and Social Science Research Program Foundation Project China (Project No. TJTJ15-002).

REFERENCES

AF Donneau, M Mauer, P Lambert, G Molenberghs, A Albert, Simulation-based study comparing multiple imputation methods for non-monotone missing ordinal data in longitudinal settings. Journal of Biopharmaceutical Statistics, 2015, 25(3): 570–601.

Capocci, A., V.D.P. Servedio, G. Caldarelli, F. Colaiori, Detecting communities in large networks, Physica A 352, 2005, 669–676.

Charu, C., Aggarwal. Social Network Data Analytics [M]. Springer Science + Business Media, LLC (2011).

Clauset, A., M.E.J. Newman, C. Moore, Finding community structure in very large networks, Phys. Rev. E 70 (6), 2004, 066111.

Donetti, L., M.A. Munoz, Improved spectral algorithm for the detection of network communities, in: P. Garrido, J. Maroo, M.A. Munoz (Eds.), Modeling Cooperative Behavior in the Social Sciences, in: American Institute of Physics Conference Series, vol. 779, 2005, 104–107.

Duch, J., A. Arenas, Community detection in complex networks using extremal optimization, Phys. Rev. E 72 (2), 2005, 027104.

Eckmann, J.-P., E. Moses, Curvature of co-links uncovers hidden thematic layers in the World Wide Web, Proc. Natl. Acad. Sci. USA 99 (2002) 5825–5829.

Fortunato, S., V. Latora, M. Marchiori, Method to find community structures based on information centrality, Phys. Rev. E 70 (5), 2004, 056104.

Girvan, M., M.E.J. Newman, Community structure in social and biological networks, Proc. Natl. Acad. Sci. USA 99 (12) (2002) 7821–7826.

Ji Jiachao, Wang Gang, Zhang Xiaoya, et al. The Analysis and Application of the Mixed-effect Pattern-mixture Model for Data with Missing Not at Random Mechanism. Chinese health statistics, 2013, 32(2): 221–225.

Kernighan, B.W., S. Lin, An efficient heuristic procedure for partitioning graphs, Bell Syst. Tech. J. 1970, 49(2): 291–307.

Mei, J., S. He, G. Shi, Z. Wang, W. Li, Revealing network communities through modularity maximization by a contraction-dilation method, New J. Phys. 2009, 11 (4), 043025.

Merris, Russell, Laplacian matrices of graphs: a survey. Linear algebra and its applications 197 (1994): 143–176.

Mohar, Bojan, and Y. Alavi, The Laplacian spectrum of graphs. Graph theory, combinatorics, and applications 2 (1991): 871–898.

Newman, M.E.J., M. Girvan, Finding and evaluating community structure in networks, Phys. Rev. E, 2004, 69 (2), 026113.

Twisk, J., MD Boer, WD Vente, M Heymans, Multiple imputation of missing values was not necessary before performing a longitudinal mixed-model analysis. Journal of Clinical Epidemiology, 2013, 66(9): 1022–1028.

Van S. Dongen, Graph Clustering by Flow Simulation. PhD thesis, University of Utrecht, 2000.

Von U. Luxburg, A tutorial on spectral clustering. Statistics and Computing, 2007, 17(4): 395–416.

Yang Linshan, Cao Yiwei, The Comparision of Two Approaches to Bayesian Method for Missing Data in Longitudinal Model-Growth Curve Model for Example. Journal of jiangxi normal university (natural science), 2012, 36(5), 461–465.

Ye Sujing, Tang Wenqing, Zhang Minqiang, Cao Weicong, Techniques for Missing Data in Longitudinal Studies and Its Application. Advances in Psychological Science, 2014, 22(12): 1985–1994.

Yozgatligil, C., S. Aslan, C. Iyigun, I. Batmaz, Comparison of missing value imputation methods in time series: the case of Turkish meteorological data. Theoretical & Applied Climatology, 2013, 112(1–2): 143–167.

Zhang Shuo, Community Detection of Complex Networks Based on the Algebraic Connectivity function [D]. Tianjin University (2012).

Zhao Fuqiang, Research on Missing Value Processing Methods in Measuring Customer Satisfaction Index [J]. Journal of statistics and decision, (6), 2013: 75–76.

Zhao Fuqiang, Zhao Shuo, He Li, Xing Enjun, Research of cutting edge Based on the algebraic connectivity in complex network [J]. Journal of computer engineering and Applications, 2014, 50 (11): 135–138.

Zhou, H., R. Lipowsky, Network brownian motion: A new method to measure vertex-vertex proximity and to identify communities and subcommunities, Lect. Notes Comput. Sci. 3038, 2004, 1062–1069.

Civil, Architecture and Environmental Engineering – Kao & Sung (Eds)
© 2017 Taylor & Francis Group, ISBN 978-1-138-02985-9

Research on the knowledge discovery system in intelligent manufacturing based on the big data

Jian Li, Xiangyang Liu, Shuming Jiang, Zhiqiang Wei & Shuai Wang
Information Research Institute, Shandong Academy of Sciences, Jinan, China

Jianfeng Zhang
Electrical Engineering and Automation of Tianjin University, Tianjin, China

ABSTRACT: Intelligent manufacturing has increasingly become the major trend of the future development of manufacturing and core content. It is an important performance measure of the comprehensive national strength. It is an important force to promote scientific and technological innovation, economic growth, and social stability, and become the opportunities of the various countries' development and transformation as well as form a new competitive battlefield. It is also an inevitable choice to build a new international competitive advantage under the new normal. However, in the development stage of the new things rapidly, accurately, and effectively in the first place to obtain new knowledge in the field of intelligent manufacturing new dynamic is the best way to lead enterprises an inevitable step in the market. Therefore, it is very important for the research of knowledge discovery system in the intelligent manufacturing under the Big Data. On the basis of the background of Big Data, this paper combines the knowledge discovery system and intelligent manufacturing field and designs the knowledge discovery system.

1 BIG DATA

In recent years, Big Data has attracted considerable attention. As early as the 1980s, an American social thinker Alvin Toffler praised the "Big Data" as "the third wave of the color movement" in the book The Third Wave.

Big Data is a combination of several old and new technologies. It can help companies to obtain meaningful data content, make in-depth analysis of users' needs, and even determine potential knowledge needs of users, revealing the relationship between information resources, to provide more accurate knowledge discovery services. Therefore, Big Data can be thought as the ability of management of vast amount irrelevant data at the right time and applying it to real-time analysis and response. Big Data has three typical features: volume (large amount of data), variety (variety), and velocity (fast and efficient) (Zhang, 2016).

In the era of Big Data, we can analyze more data, even sometimes deal with all the data associated with a particular phenomenon, rather than relying on random sampling. A well-known scholar at Harvard University suggested that the massive data sources had begun to quantify processes in all areas, whether academics, business, or government (Liu, 2014). With the rapid expansion of the amount of information, it has become the top

priority of information services how to effectively use a large number of structured, semi-structured, and unstructured complex data gradually.

2 INTELLIGENT MANUFACTURING AND BIG DATA

2.1 Key status of big data

The data have a huge value of innovation. We should not only pay attention to the actual amount of data, but also to the large data processing methods. With the increasing complexity of the data, the ability of transforming data into intelligence will be increasingly demanded.

How to achieve intelligent manufacturing from big data? A large number of entrepreneurs that come from excellent manufacturing have a general consensus. They realize intelligent manufacturing from digital transformation. Digital transformation not only means simply digital companies, but also the core driving force of intelligent manufacturing which needed to be used to integrate the industrial chain and value chain (Wang, 2008).

In the field of manufacturing, one type of data is derived from the human trajectory generated data, and the other is the machine automatically generated data. These two types of data constitute today's large multistructured data sources. In the field of industrial data, we have to pay more attention to the integration of machine data and industrial data with human behavior data in addition to continuing to care about human data or people-related data (Wang, 2010).

2.2 *Relationship between big data and intelligent manufacturing*

Enterprises can keep abreast of the processes, problems, and solutions in the production process and find new ways to create additional value on the basis of large data tools, data analysis, and mining.

Manufacturing industry can achieve business model changes, transform and enhance customer experience, improve the internal operational processes, and keep abreast of industry trends in order to market decision-making step ahead by using big data tools and thinking.

3 KNOWLEDGE DISCOVERY SYSTEM ARCHITECTURE

The data itself are worthless. The data will not let our manufacturing industry more advanced. It must be converted into information. The information will be valuable for industry (Wu, 2010). With the advent of the era of Big Data, users urgently need a simple and fast ideal platform that gains access to all the knowledge. Big Data is the basis of intelligent manufacturing; its core is the custom platform. If we do not invest in Big Data and data analysis, intelligent manufacturing pursuit of excellence operation will fall short. It is the Big Data analysis technology instead of Big Data itself that promotes intelligent manufacturing, namely the knowledge discovery system.

Knowledge discovery is a nontrivial process of identifying valid, novel, potentially useful, and eventually understandable patterns from the data set. The process of knowledge discovery translates information into knowledge, finds the data gold from the data mine, and contributes to knowledge innovation and economic development.

In the field of intelligent manufacturing, through cooperation with enterprises, we build knowledge discovery service platform. In the industrial technology information, talent discovery, knowledge recommendation, and other aspects of service enterprises in the benefit analysis, customer relations, and so on enhance the competitiveness of enterprises in all directions. The purpose of knowledge discovery is to break the limitations of the previous books directory, make full use of citation index and part of the literature, and provide users with comprehensive

and efficient knowledge mining and data analysis capabilities of the knowledge discovery system, in order to achieve the discovery from the resource to knowledge change.

Information organization supported by the technology of data mining carries out the knowledge association and the data analysis processing of books, further discovers large amount of hidden data, and then establishes a powerful new generation of academic resource discovery platform to help the information users to obtain the required knowledge or node more quickly.

3.1 *Data mining*

Data mining is a process that extracts implicit, unknown, and potential useful information from large amount of data that are incomplete, noisy, fuzzy, and random (Ouyang, 2001).

Data mining is also known as knowledge discovery in database. It creates model and finds out relationships, and then makes a decision and prediction among data from huge amounts of data by various methods and analysis tools.

As shown in Figure 1, the data mining process includes the following steps:

Understanding and defining the problem.

Data mining professionals work with domain experts to make an in-depth analysis of the problem in order to determine possible solutions and to evaluate the results of the study.

Related data collection and extraction.

Collect relevant data according to the definition of the problem. In the process of data extraction, database query function can be used to accelerate the data extraction.

Data exploration and cleaning.

Understand the meaning of the field in the database and its relationship with other fields. Check and clean out the data that are contained in the data extracted from the data.

Data engineering: reprocesses the data.

It mainly includes selecting related subsets of attributes to eliminate redundant attributes, and according to the knowledge discovery task to

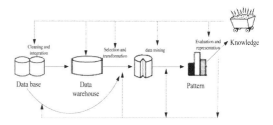

Figure 1. Data mining step in the process of knowledge discovery.

reduce the amount of learning, and transforming the way of expressing data to adapt the learning algorithm, in order to make the data and tasks achieve the best match. This step may be repeated several times.

Algorithm selection.

According to the data and the problem to be solved, select the appropriate data mining algorithm and decide how to use the algorithm on these data.

Run the data mining algorithm.

Using the selected data mining algorithm to extract the processed data pattern.

Evaluation of results.

The evaluation of learning results depends on the issues that need to be resolved. The novelty and effectiveness of the patterns will be evaluated by domain experts. Data mining is a basic step in the knowledge discovery process. It includes a specific mining algorithm that discovers patterns from the database. The knowledge discovery process uses data mining algorithms to extract or identify knowledge from a database on the basis of specific metrics and thresholds. This process includes preprocessing of the database, sample partitioning, and data transformation.

The task of data mining is finding models from data. It extracts useful and interesting knowledge and models from a large number of data by means of theory, methods, and tools related to the development. The enormous data in a database often contain high-level information or knowledge such as rules, laws, and assertion. We cannot acquire this information only through the query process. On the basis of the data provided by the database, data mining looks for some inner relationship between data to find potential and important roles for forecasting and decision-making behavior patterns through data analysis and reasoning and finally establishes a new business model to achieve the goal of helping decision makers to make the right decisions.

3.2 Data analysis

On the basis of the knowledge organization and presentation of existing structured and unstructured data by utilizing data mining and learning technology, reasoning knowledge discovery system could obtain a change direction and trend of literature resources. Automatic and intelligent analysis can help users to get dynamic, informative, and advanced knowledge of the literature. The primary role of data analysis is to help people to sample, extract, analyze operation process, and check the analysis results by setting the human–computer interaction interface and utilizing software environment. The potential regularity in huge amounts of information resources and its development

trend will be shown dynamically and intuitively by means of visualization technology. The data analysis flowchart is shown in Figure 2.

In the knowledge discovery system, data mining and data analysis make deeper development and application with information under cloud computing and Big Data environment. Data mining is the basis of data analysis; data analysis is the deepening of data mining. The content of data mining is the premise and guarantee for data analysis, data analysis systems reasoning, and the development trend and direction of knowledge based on the results of data mining. It thus puts forward ground-breaking and prospective prediction, verifies the depth and breadth of knowledge mining, and feeds back to the knowledge mining system for improvement. Therefore, data mining and data analysis, as two modules of knowledge discovery system, supplement and interact with each other.

3.3 System design principles

The design of knowledge discovery system should follow the basic principles of system development:

- Structural integrity

The knowledge discovery system gathers a large number of structured, semi-structured, and

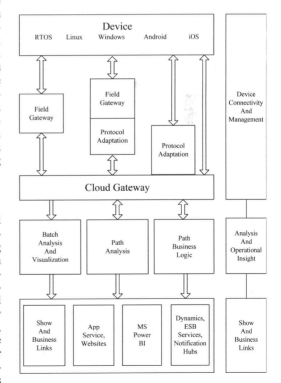

Figure 2. Data analysis flowchart.

unstructured complex data. It matches, analyzes, and mines the information agency and user needs and search behavior. Thus, it establishes a complete system architecture. And it widely and accurately reveals the multiple, three-dimensional relationship of the various types of data and the knowledge network link (Wang, 2010).

• Advanced technology

The knowledge discovery system innovates in the technical route. This system can not only dock with information service institution's original full-text information system, but also provide users with optimal results sorting and knowledge push through the advanced warehouse management technology, indexing technology, data mining technology, and so on.

• Platform security

Knowledge discovery system establishes a complete and effective system security, platform backup, and fault-handling mechanisms to ensure the stable operation of the platform. Job and synchronization system ensure safe and stable data operations. The design of hardware redundancy fully protects the daily data's security backup. Real-time monitoring module realizes the use of the database storage space and data access observation and monitoring, and it will alarm in case of emergency.

• System development

The knowledge discovery system adopts extensible development language and database type, such as distributed storage and distributed indexes. The performance of the system and its scalability have been greatly improved. The knowledge discovery system also provides OAI-DP services and standard interfaces. It allows seamless interfacing with other full-text access systems of information organizations. It can facilitate the docking of other platform on the system and the secondary use so that the utilization of information is improved and the system's life cycle is extended. The old and new systems of succession and development have been achieved.

• Interface friendliness

The knowledge discovery system obeys the basic principles of interaction design. It is user-centric interface designed. Its purpose is to plan and describe the mode that the users access knowledge, and then describe and communicate the knowledge information to the users efficiently. It provides users a friendly operating platform, and visualization results display and export services and gives a good user experience.

For the user, knowledge discovery system is a simple, easy-to-operate, and user-friendly platform.

3.4 Knowledge discovery system model

The knowledge discovery system extracts implicit, unknown, and potentially useful information from

a large amount of data that are incomplete, noisy, fuzzy, and random (Studer, 1998). Its purpose is to shield details for users, extract significant and succinct knowledge from original data, and directly report to users so that it can provide knowledge discovery service for business manager and information organization. In general, resource integration, knowledge discovery, and achievement exhibition are functional goals of knowledge discovery system.

The frame diagram of knowledge discovery system in the field of intelligent manufacturing is shown in Figure 3. It consists of four parts: data collection, data mining and analysis, data visualization, and knowledge service. Data collection module collects and receives original data, transports them to data mining and analysis module, in which important and interesting data are extracted, and the data are reduced for conversion to appropriate format. Finally, the generated knowledge model is evaluated, and valuable knowledge is integrated into corporate intelligent system.

This system combines data and knowledge discovery system and discovers the unknown correlation between them. It also makes it more possible to break information island Big Data used in operation and new data sources, such as social media, Internet of things, and so on, to analyze the solution capacity of Big Data. All these data are integrated, and knowledge discovery system can provide decision and prediction for management.

The firm boundary is becoming more and more vague in manufacturing industry. Subversive innovation is the most unpredictable external factor.

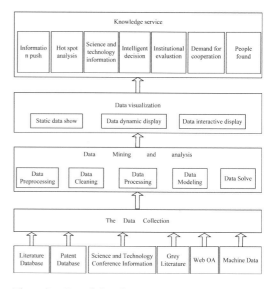

Figure 3. Knowledge discovery system framework diagram of intelligent manufacturing domain.

Interconnection entirely changed rules of the business and business managers need to realize cutting-edge technology and its correlations, make use of modern enterprise architecture to refine enterprise, and obtain more efficient, intelligent, and high-interest serving product through digital supply chain.

4 SUMMARY

The matching of knowledge discovery system and enterprise management is the key to the enterprise performance. In the era of knowledge economy, the competition between enterprises is the competition of knowledge, information acquisition, and their application ability. Knowledge discovery can discover useful knowledge from the mass data of the enterprise. On the basis of the mature information technology, this paper establishes an efficient search, discovery, accumulation, communication, sharing, and reuse of knowledge platform. It also makes the enterprise to realize the true meaning of the knowledge sharing and reuse in a wider range, promotes enterprises to achieve effective knowledge management so as to enhance the core competitiveness of enterprises.

The system architecture proposed in this paper provides a useful way for the development of knowledge discovery system. However, in the next step, we need to pay attention to the specific technical implementation details of the system architecture and its application effect.

REFERENCES

Liu Jiangling. Research on Knowledge Discovery System Faced in Big Data [J]. Information Science, 2014,03:90–92+101.

Ouyang Weimin, Zheng Cheng, Zhang Yan. International Survey of Knowledge Discovery and Data Mining Tools [J]. 2001, 28(3):101–108.

Shouhong Wang HW. Towards Innovative Design Research in Information System [J]. The Journal of Computer Information Systems, 2010, 51(1):11–18.

Studer R, et al. Knowledge Engineering: Principles and Methods [J]. IEEE Transactions on Data and Knowledge Engineering, 1998,25(1–2):161–197.

Wang Ning. Analysis of the Digital Library Knowledge Discovery System under the Big Data Background [J]. Library Work and Study, 2016,(4):58–61.EPL, 84 (2008).

Wu Sizhu. Analysis of Hot Spots in Field of Data Mining and Knowledge Discovery [J]. Journal of Intelligence, 2010, 29(7):18–24.

Zhang Lili. Research from Big Data to Intelligent Manufacturing [J]. China Industry Review, 2016,(7):66–71.

Civil, Architecture and Environmental Engineering – Kao & Sung (Eds)
© 2017 Taylor & Francis Group, ISBN 978-1-138-02985-9

Study on the positioning system in China's Advanced Broadcasting System-Satellite

Ming Yan & Jing-yi Yang
GxSOC Research Institute, Communication University of China, Beijing, China

Xin-gang Wang
Administrative Center for the DTH Service in China, Beijing, China

ABSTRACT: The Advanced Broadcasting System-Satellite (ABS-S) technical specification is an independent research and development satellite radio and television signal transmission technology in China. In this paper, we first introduce the development situation of satellite broadcasting technologies and then analyze the physical layer interface specifications of ABS-S. A positioning system in ABS-S is mainly studied because the system needs to know the user's position to insure they are in the rural area. Using the position management system, the security and stability of the total system have been improved to satisfy the demands in China.

1 INTRODUCTION

In the early 21st century, the State Administration of Radio Film and Television (SARFT) of China has begun to carry out the research on satellite system for radio and television (TV) broadcasting of Ku and Ka frequency band (Wang, 2015; Shi, 2008; Lin, 2007). For building a safer and more reliable satellite live broadcasting operating system, further standardizing the market of live satellite radio in China, and improving the safety of the satellite live broadcasting system, in July 2009, the SARFT approved the technical specification of radio, film, and television industry for the satellite live broadcasting: Advanced Broadcasting System-Satellite transmission system's frame structure, channel coding, and modulation: security mode (GD/JN 01-2009). On the basis of this specification, for building satellite live broadcasting system, in October 2009, the SARFT officially approved the technical specification for satellite live broadcasting security mode modulator: the technical requirements and measuring method for live broadcasting satellite safe mode modulator (GD/JN 02-2009).

As a supplementary technology of cable TV, the satellite live broadcasting system is forbidden to be used in the area with coverage of cable TV, such as metropolis, in China. Therefore, the positioning system is a key technology in ABS-S, which is used to locate the terminal receiver and judge the legality of the user. If the positioning system finds the terminal receiver located in illegal area, it will tell

the management system to lock the receiver (Cui, 2015; Liu, 2015).

The rest of this paper is organized as follows. In Section 2, we investigate the development situation of satellite live broadcasting technologies worldwide. In Section 3, we present the physical layer technologies and transmission parameters of ABS-S. In Section 4, we study the requirements of positioning system in ABS-S and design a positioning system, which can provide location service for ABS-S. Finally, the paper is concluded in Section 5.

2 DEVELOPMENT SITUATION OF SATELLITE LIVE BROADCASTING

The main feature of satellite live broadcasting system is that the TV program can directly transmit to the user with satellite, so the users can receive TV or radio programs by simply using a small satellite receiving antenna. Since the early 1990s, satellite live broadcasting television service, satellite mobile communication, and satellite digital audio broadcasting business had been undergoing a technical breakthrough firstly in the United States and stepping into industrialization and commercialization, and forming a new industry to promote the growth of the economy. At present, these three satellite communication and broadcasting businesses are rapidly expanding globally. Especially, the satellite live broadcasting radio and television industry is developed rapidly and achieved great success in operating (Wang, 2015; Shi, 2008; Zhao, 2015).

In 1993, US Hughes Corporation first developed and built a commercial television satellite system with digital video compression technologies and founded the DirecTV Company by a variety of financing channels in charge of the operation of the satellite system. After few years, the company dominated the market of satellite TV live broadcast in the United States and Canada and entered into the markets in Mexico, Latin America, and Japan. After the Hughes Corporation achieved success in developing satellite television broadcast, many large companies in the United States, Japan, and Europe also successively entered into the satellite digital TV market. Some traditional for-profit and non-profit satellite communications companies also began to enter into this market, trying to find their places in the world live satellite radio and television market.

The Europe SES Corporation owns about 10 communication broadcasting satellites, and the radio and TV programs has more than 1000 sets, which are transmitted by these satellites, with tens of millions of users. In the United States, there are many direct broadcast satellites in three satellite orbits, transmitting more than 600 programs. The Echostar Corporation owns nine direct broadcast satellites, with more than 1000 programs and 10 million users, and earned more than 8 billion dollars in 2004.

In China, the ABS-S technical specification is an independent research and development satellite radio and television signal transmission technology. The ABS-S has independent core technology and international advanced level in satellite broadcasting transmission technologies. It provides the interface specification of signal transmission in the physical layer, which is channel coding and modulation specification (Liu, 2008).

Compared with the Digital Video Broadcasting-Satellite (DVB-S2) technology, which is widely adopted worldwide currently, the ABS-S has obvious advantages. In the same transmission conditions, the ABS-S can provide greater signal transmission capacity; in other words, it can compete more poor transport conditions at the same transmitting efficiency. At the same time, the ABS-S also provides more transmission configuration choices. In addition, the ABS-S adopts different implementation methods in logical frame structure, physical frame structure, channel coding scheme, modulation system, pilots insertion, interleaving, scrambling, and signaling in band compared with DVB-S2 (Lan, 2015; Zhang, 2014).

3 PHYSICAL LAYER TRANSMISSION SYSTEM IN ABS-S

In the technology design, the various losses of satellite channel and their effects on signal receiving performance are considered adequately in the ABS-S system, such as the nonlinearity of satellite power amplifier, group delay characteristics, phase noise and frequency drift in the front-end of receiver, and thermal noise in uplink and downlink. At the same time, the ABS-S system takes advantage of the latest research findings of the satellite signal transmission technologies in the world and is improved and developed on the basis of the requirements of direct broadcast satellite in China. The ABS-S system adopts Low Density Parity Check (LDPC) code in channel coding, which leads to a completely different implementation model compared with the DVB-S2, in accordance with the current latest technology standards worldwide. The encoded frame length of the ABS-S is less than a quarter of the DVB-S2, and the ABS-S does not use the combination mode with LDPC and BCH. Therefore, the complexity degree of implementation of the ABS-S reduces significantly but has the same performance with the DVB-S2 (Zhou, 2006; Cheng, 2008).

3.1 Frame structure of physical layer

A more concise and more reasonable physical frame structure compared to the DVB-S2 is designed in the ABS-S system, and the ABS-S adopts a special inserting scheme of pilot training symbols, so it has better synchronization performance and shorter synchronization time than that of the DVB-S2. Meanwhile, it can merge Constant Coding and Modulation (CCB) and Adaptive Coding Modulation (ACM) together. Fig.1 shows the main functional block of the transmission system in ABS-S. The input data steam is formatted in baseband and then processed with Forward Error Correction (FEC). After bit mapping, the steam generates frame data in physical layer and then scrambles these symbols. The data go through a filter and are transformed from baseband to Radio Frequency (RF) data, which are Ku data transmitted to the receivers (Jiang, 2013; Yi, 2005).

The ABS-S is designed especially for broadcasting business application, considering the technological change and influence of the power capacity and repeater configuration in China's direct broadcast satellite adequately. In ABS-S system, two carrier modulation schemes, QPSK and 8 PSK, are adopted in satellite broadcast services. Combining with different channel coding rates, it can provide more transmission scheme solutions for the platform operation and take full advantage

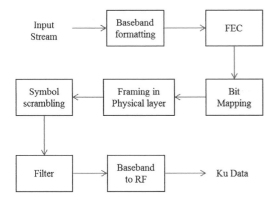

Figure 1. Functional block of the transmission system in ABS-S.

Table 1. Parameters of transmission coverage for ABS-S.

Parameters	Values
Repeater bandwidth	36 MHz
Repeater code rate	43.2 Mbps
MPEG-2 standard	2.677 Mbps
definition video bit rate	* 12 = 32.124 Mbps
(statistical multiplexing)	
TV audio bit rate	128 Kbps * 12 = 1.536 Mbps
SI bit rate	2 Mbps
Entitlement management	2 Mbps
message data rate	
System overhead data rate	2 Mbps

of the transmitting ability of the satellite platform. Meanwhile, it can resist the nonlinear impact of Travelling Wave Tube Amplifier (TWTA) power amplifier. In addition, the ABS-S can be extended to use higher-order modulation schemes, such as 16 APSK and 32 APSK, to satisfy the requirements of transmitting capacity enhancement and interactive business application in the satellite platform.

3.2 Video coding and parameters of transmission coverage

At present, the vast majority of digital television broadcast systems adopt Motion Picture Experts Group 2 (MPEG-2) as their video coding. In the application process of MPEG-2 standard, the technologies also develop continually, especially the application of noise-reducing filter, dynamic statistical multiplexing of multiple programs, so the compression efficiency increases greatly. Both the central and provincial satellite TV systems in China mainly adopt the compression bit rate from 4.2 to 6.9 Mbps. The China Central Television has tested the coding bit rate with the fixed rate of the encoding systems and found that the encoding rate with 3.2 Mbps can guarantee the video quality, which can completely meet the needs of radio and television broadcast system.

The key parameters applied in the ABS-S direct broadcast satellite platform are listed below:

a. Input signal: MPEG-TS bit stream with 188 bytes.
b. Coding scheme of FEC: LDPC coding, frame size is 15360 bits.
c. FEC code rate: 1/4–9/10, 14 combination modes with QPSK and 8 PSK.
d. Carrier modulation scheme: QPSK and 8 PSK.
e. Roll down factor of pulse shaping filter: 0.35, 0.25, 0.2.

f. Pilot frequency: optional, QPSK symbol.
g. Supporting variable code modulation and adaptive coded modulation.

Table 1 gives the parameters of transmission coverage for ABS-S.

4 POSITIONING SYSTEM IN ABS-S

4.1 Requirements of positioning system in ABS-S

In terms of positioning mechanism, although the mobile base stations can effectively control the moving range of the comprehensive decoding terminal, the base station cannot estimate whether the terminal locates in the legal installation area for the first installation, which results in illegal installation of decoding terminal. On the contrary, the built-in position lock module in the comprehensive decoding terminal of direct broadcast satellite is independent of the main module, so the positional information of the terminal can be modified through the replacement method of the built-in information in module, which also results in illegal moving of the terminal.

Aiming at the above technical questions, we improve and optimize the position management prototype system on the basis of the preliminary development, adding the function of collecting the geographical position information, and avoiding the illegal installation of direct broadcast satellite terminals in the area with cable TV coverage by using this geographical position information to confirm the installation location. A digital signature mechanism is used in the process of positioning information transmission between the position management module and the main module in the direct broadcast satellite terminals to prevent illegal module forging positioning information. Through the above improvement, the security and stability of the total system have been improved to satisfy the demands of the large-scale application

4.2 Design of position management system

The position management system takes advantage of the character that the geographic position of mobile communication base station is relative fixed, and the base station can provide effective return channel to lock the comprehensive decoding terminal with the base stations nearby the installing position when installing the terminal. The system uses the General Packet Radio Service (GPRS) uplink channel to return the base station information back to the position server, judging whether the current position is legal installation location by comparing with the information in the position server. When the decoder is working, the system can determine whether the position is moving by scanning the change of the surrounding base station (Kim, 2016).

The position management system is composed of position lock server, Short Messaging Service (SMS), Entitlement Management Message Generator (EMMG), signature validation server, decoding terminal with locating function, and geographical position information, which are shown in Fig. 2.

The position lock server is located in the direct broadcast satellite user management center. Its major functions are listed below:

a. Integrate the uploading data of the receiver and create the database of service stations.
b. Receive the base station information uploaded by the receiver and the user registration information from boss system simultaneously. Combining with the base stations list from the geographical position information block, complete the comparison of base station and return the comparing results to the boss system to determine whether to authorize the user's account.
c. Store all the location management information and status information of the receiver.
d. Output the combination statements.

The signature validation server verifies the signature validity of the location management information uploaded by the receiver and returns the verification result to the position management server.

The decoding terminal with locating function and geographical position information contains location management module, which can communicate with the server, supporting the two working modes of locking and unlocking. The key functions are listed below:

a. Collect the information of all retrievable base stations.
b. Transmit the base station information together with the encryption key of the receiver, smart

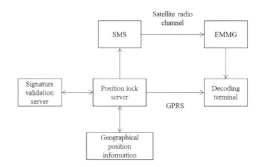

Figure 2. Position management system of ABS-S.

card serial number, and Interning Mobile Equipment Identity (IMEI) number to the direct broadcast satellite user management center.
c. Store the location management information in the smart card.
d. Obtain the new location management information when the terminal starts up every time and compare with the location management information stored in the smart card.
e. Realize location management function on the basis of the comparing result.

5 CONCLUSIONS

The ABS-S system is designed especially for broadcasting business application, considering the technological change and influence of the power capacity and repeater configuration in China's direct broadcast satellite adequately. In this paper, we studied a position management system, which takes advantage of the character that the geographic position of mobile communication base station is relatively fixed and the base station can provide effective return channel. This mechanism can lock the comprehensive decoding terminal with the base stations nearby the installing position when installing the terminal. Therefore, this system can judge whether the current position is legal installation location by comparing with the information in the position server.

REFERENCES

Cheng, X., Y. Wang, Z. Lu, Z. Zhou. Design and Implementation of DVB-S Multi- service Data Broadcast System, Television Technology, **32**, 23–25, (2008).
Cui, B. Discussion on the application technology of broadcast and TV satellite broadcasting, Journal of News Research, **1**, 79, (2015).
Jiang, W., J. Feng, Z. Cao. Radio and TV Transmission and Coverage Technology System in China, Television Technology, **38**, 18–21, (2014).

Kim, D., S. Lee, H. Bahn. An Energy-Efficient Positioning Scheme for Location-Based Services in a Smartphone, 2016 IEEE 22nd International Conference on Embedded and Real-Time Computing Systems and Applications, 139–148, (2016).

Lan, C. The Application Research of Broadband Satellite Communication Technology Based on DVB-S2, Master Thesis, Xidian University, (2015).

Lin, Z., L. Li. Technology Development and Applications of Satellite Communications, Modern Electronics Technique, **3**, 38–42, (2007).

Liu, C., Y. Shi. Design of ABS-S Modulator System Based on FPGA, Television Technology, **32**, 7–9, (2008).

Liu, Y. Research and Design Methods of ABS-S Signal Based on DSP, Master Thesis, Xidian University, (2015).

Shi, Y. M. Yang, J. Ma et al. A New Generation Satellite Broadcasting System in China: Advanced Broadcasting System-Satellite, 2008 4th International Conference on Wireless Communications, Networking and Mobile Computing, 1–4, (2008).

Wang, Q. Application and development of Ka broadband satellite communication market, Satellite Application, **8**, 49–51, (2015).

Yi, Z., S. Liu. An Implementation Scheme of Symbol-Processing in DVB-S Receiver, Bulletin of Science and Technology, **4**, 460–463, (2005).

Zhang, X. Analysis and Study on the Application of DVB-S2 in the System of Digital Satellite Earth Station, Master Thesis, Inner Mongolia University, (2014).

Zhao, Y. Probe into the Practical Coverage Effect of the DBS System, Journal of Shanxi radio & TV University, **4**, 102–106, (2015).

Zhou, R. Radio and television "village" construction: history, present and future, Modern Communication (Journal of Communication University of China), **5**, 45–50, (2006).

Civil, Architecture and Environmental Engineering – Kao & Sung (Eds)
© 2017 Taylor & Francis Group, ISBN 978-1-138-02985-9

Finite element simulation of compression on micropillars

G. Tang

Shenzhen Key Laboratory of Polymer Science and Technology, College of Materials Science and Engineering, Guangdong Research Center for Interfacial Engineering of Functional Materials, Nanshan District Key Lab for Biopolymers and Safety Evaluation, Shenzhen University, Shenzhen, P.R. China
Key Laboratory of Optoelectronic Devices and System of Ministry of Education and Guangdong Province, College of Optoelectronic Engineering, Shenzhen University, Shenzhen, P.R. China

Y.L. Shen

Department of Mechanical Engineering, University of New Mexico, Albuquerque, NM, USA

ABSTRACT: A numerical study was undertaken to investigate the mechanical properties of metal-ceramic nanolayered composites. We utilized Aluminum (Al)/Silicon Carbide (SiC) alternating layers with same thickness as a model system. Finite element modeling was employed to analyze the microcompression behavior on pillars which consist 41 Al/SiC multilayers. It deformed in a non-uniform way under compression, especially when a tapered side wall included in the numerical model. Then elastic modulus was obtained from stress-strain curve and compared with previously calculated modulus value of the composite. It was found that the base material connected to the pillar plays a significant role in the measured mechanical response. The simulation result were also used to rationalize some of the experimental observations.

1 INTRODUCTION

Nanomaterials are called "the future of materials". Nanotechnology is an emerging and rapidly growing field. Many of the devices and systems used in modern industry are already in the nano-scale domain. Nanomaterials may possess advantages of extremely high strength, fatigue resistance, thermal resistance, wear resistance and bio-compatibility, compared with traditional material. Nanomaterials are finding applications in area spanning from structural coatings to microelectronics. Synthetic and natural composite laminates have been shown to exhibit a combination of excellent strength and toughness (D.R. Lesuer, 1996). Composite laminates on the nano-scale with unique properties have been developed. These composites have been investigated in many different layered combinations: Metal-metal composites, metal-ceramic composites, and ceramic-ceramic composites. Metal-ceramic nanolaminate systems can exhibit a combination of high strength, high toughness, damage tolerance, as well as their potential applications in functional devices (T.C. Chou, 1992 & C.H. Liu, 1996).

Indentation technique has become the most popular approach to characterize the mechanical properties of nanomaterial. While in the past years, a new technique of microcompression on

free standing pillar-shaped materials was developed to investigate the mechanical behavior at the micro- and nano-scale (K.S. Ng, 2008; S.X. Song, 2009 & D. Kiener, 2008). The pillar samples with the size of a few hundred nanometers to several micrometers are prepared by Focused Ion Beam (FIB) milling. The compression tests are conducted using a modified nanoindentation device with a flat-punch indenter tip. The flat-punch indenter is usually produced by truncating the tip of a Berkovich indenter. The pillar is always attached to a base material (the original substrate). In order to avoid the buckling which usually happens when compressing on a thin and long pillar, a technique of tensile tests rather than compression on the pillar sample has also been developed. In most of the cases, the samples are cylindrical, although other type of pillar like a square cross section, does exist.

In this paper, a numerical model was constructed to simulate compression tests on the multilayered composite pillars. The experimental part of work has been performed by scientists at Arizona State University and the detail has been published in Acta Materialia (D.R.P. Singh, 2010). Nano-indentation simulations on the same structure (Al50/ SiC50 nanolaminates) have been investigated in the past by our authors (G. Tang, 2009; 2008 & 2010). The effective elastic modulus along perpendicular direction to the multilayer is found to be 117 GPa, which will be used as a guidance in this article.

2 MODEL

Microcompression can be treated as a special form of "indentation" test, where a flat-bottom indenter is used to press onto a rod-like specimen prepared by FIB-milling, shown schematically in Figure 1 (a). Two dimensional axisymmetric models were constructed for the analysis of the compression tests. Figure 1 (b) shows the schematics of a pillar consists 41-layer Al/SiC (same thickness) on

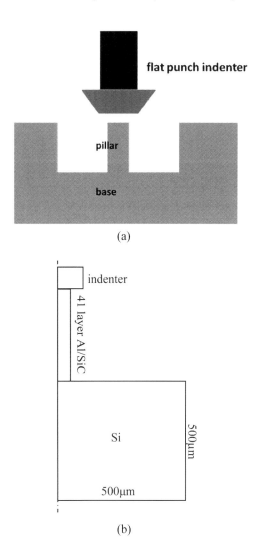

Figure 1. (a) Schematic showing microcompression on a pillar with base (b) Numerical model of the Al/SiC multilayer pillar on a Si base.

Si base; the compression axis is perpendicular to the layer directions. The geometry of the model is defined by the pillar height h, pillar cross section radius r (or cross section diameter d), base height H and base width W. The aspect ratio of the pillar is defined as the ratio of pillar height over pillar cross section diameter (h/d).

3 RESULTS AND DISCUSSION

3.1 Elastic modulus of the multilayer composite pillar

The stress-strain curve and apparent Young's modulus can be directly obtained from the finite element analysis. Table 1 lists the simulated Young's modulus for the model. The apparent modulus is 77 GPa, much below the true modulus value for the Al/SiC multilayers of 117 (GPa G. Tang, 2009). This is attributed to the compliance of the Si base and diamond indenter. The corrected pillar modulus can be obtained by subtracting the axial strains of the indenter and/or base from the total axial strain. Simulation shows that the contribution of the Si base is very significant and that of the indenter is moderate. When the compliances of both are accounted for, the true Al/SiC composite modulus of 117 GPa can be recovered. This finding suggests the importance of correcting the raw data by accounting for the base and "machine" deformation when conducting experimental studies.

We now focus on the stress-strain curves when the deformation is sufficiently large to cause significant plastic yielding. During deformation, the cross section area changes in a non-uniform manner. Here we use the cross section area in the middle of the pillar for calculating the stress. The stress-strain curve of the pillar structure, together with the stress-strain curve of the true composite, are plotted in Figure 2. The stress-strain curve of the true composite is plotted using the data from previous work (G. Tang, 2009; 2008 & 2010). The stress-strain curve of multilayer pillar is significantly below the curve of the true composite. One reason could be the compliance of the indenter and the base. At the nominal strain, the actual strain of

Table 1. Young's modulus of multilayered pillar.

Setting	Modulus (GPa)
FEM-with no correction	77.2
FEM-diamond compliance correction	80.7
FEM-Si compliance correction	110.0
FEM-diamond & Si compliance correction	117.3
True E_{22} of Al50SiC50 composite	117

the pillar itself is smaller than the nominal strain, so the stress is much lower than the stress of the true composite. Another reason could be the extrusion of the Al layer under compression which will be discussed in section 3.3.

3.2 *Elastic modulus of the multilayer composite pillar with taper*

In experiments, the pillar is produced by FIB milling. It is difficult to achieve a true vertical side wall from the process, so there is always an angle. To examine how this imperfect geometry may influence the test result, we constructed one set of models of the 41-layer pillar on a Si base with different taper angles of 1°, 2° and 4°. For comparison purposes, we also constructed another set of models of pure Al pillar on an Al base, with the same taper angles of 1°, 2° and 4°. For each model, we have conducted the compression simulation with a displacement up to 0.3 μm. The stress-strain curves of each model, together with the one with no taper, are plotted in Figure 3. When the strain is relatively

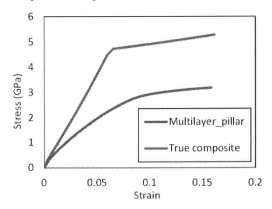

Figure 2.   Stress strain curves of multilayer pillar model and true composite.

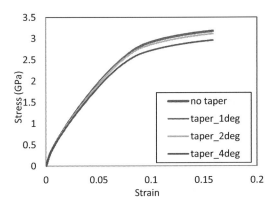

Figure 3.   Stress strain curves of multilayer pillar with taper angle of 1°, 2° and 4°.

small, all the curves are very close to one another. When the strain is large enough, the curves with taper are all below the curve of the non-tapered model. The difference increases with an increasing taper angle. For practical purpose, the effect may be ignored if taper angle is within 1–2°.

The Young's modulus of all the multilayer pillar models with different taper angels are calculated and listed in Table 2. The values in the left column are those obtained directly from finite element analysis, while the ones in the right column are the ones with the compliance correction of diamond and Si following the same approach as in the previous section. One can see that the taper does not have any significant effect on the modulus values.

3.3 *Stress and deformation evolutions*

Before examining the stress evolution of the multilayer pillar model, we first present the simulation results on the pure Al pillar with an Al base. Figure 4 (a) and (b) shows the equivalent plastic strain contours of the pure Al pillar with no taper, at the compression displacement of 100 nm and 325 nm respectively. Figure 4 (c) and (d) shows the equivalent plastic strain contours of the pure Al pillar with a 2° taper angle at 100 nm and 325 nm compression depths, respectively. The deformations for the tapered and non-tapered model are quite similar, with the deformation of the tapered model being a little stronger.

Figure 5 (a) and (b) show the Von Mises stress contours of the multilayer pillar, with no taper, at compression depths of 100 nm and 325 nm, respectively. It is evident that the Al layers deformed much more than the SiC layers. At greater depths, a significant portion of the soft Al layers was extruded out from the side. This extrusion of Al layer could be the reason that the stress-strain curve of the multilayer pillar is below the curve of true composite. When Al extruded out the overall stress will become smaller than the theoretical stress. Figure 5 (c) and (d) show the Von Mises stress contours of the multilayer pillar model with 2° taper angle at compression depths of 100 nm and 325 nm, respectively. The Al layer deformed much more than SiC layers. It is interesting to observe that, with only a small taper angle,

Table 2.   Young's modulus of multilayered pillar with taper.

Setting	FEM (GPa)	FEM-diamond & Si compliance correction (GPa)
No taper	77.2	117.0
1°	77.0	115.6
2°	76.8	115.4
4°	76.1	114.7

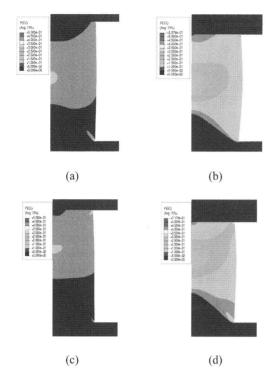

(a)   (b)

(c)   (d)

Figure 4. Equivalent plastic strain of the pure Al pillar (a) no taper at compression depth of 100 nm (b) no taper at 325 nm (c) taper angle of 2° at compression depth of 100 nm (d) taper angle of 2° at 325 nm.

the several Al layers near the top have undergone severe squeezing and were extruded much further out compared to the lower Al layers. It is worth pointing out that, in the case of a pure Al pillar, such type of deformation was not observed even with taper. The uneven extrusion of Al is caused by the multilayer arrangement.

3.4 *Comparison with experiments*

We now focus on the deformed configuration of the Al/SiC multilayer pillar. Figure 6 (a) and (b) show the experimental SEM pictures of the pillar before and after, respectively, the compression tests. The 2° taper can be seen in the as-processed specimen. Figure 6 (b) shows a dramatic view that the upper half of the pillar has been crushed, while the lower half remains relatively intact. Although the experimental picture at smaller displacement is not available, the severe localization of deformation in the upper portion can still be correlated with the simulated result in Figure 5 (d). This comparison, along with the other geometric features studied in the previous sections, serves to illustrate the versatility of applying finite element

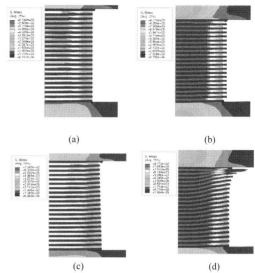

(a)   (b)

(c)   (d)

Figure 5. Von Mises stress contours of the Al/SiC multilayer pillar on a Si base (a) no taper at compression depth of 100 nm (b) no taper at 325 nm (c) taper angle of 2° at compression depth of 100 nm (d) taper angle of 2° at 325 nm.

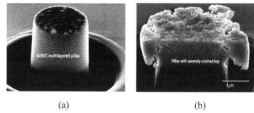

(a)   (b)

Figure 6. SEM images of multilayer pillar structure (a) before compression, and (b) after compression, Courtesy of N. Chawla and D.R.P. Singh.

modeling in studying the mechanical behavior of nano- and micro-scale materials.

4 CONCLUSIONS

A numerical study has been carried out on micro-compression of metal/ceramic multilayered pillars. The modulus obtained from simulation is compared with the true value of elastic modulus of the layered structure. Results from the compression tests on the micro-pillar structure showed a strong dependence on the specimen geometry. It is essential to correct the raw data by taking into account the compliance of the pillar base and the machine. In the case of Al/SiC multilayered pillar, a small degree of taper gives rise to severe extrusion of the Al layers out of the side boundary in the top portion of the pillar, leading to a highly non-uniform deformation configuration not seen in pillars made out of a homogeneous material. The simulated deformed contour plots serve well to rationalized deformation observed during experiments. Also the modulus value obtained from current work is found to be consistent with that of previous nano-indentation simulation. Thus microcompression, as another approach to characterize mechanical properties of nanomaterial, can be as important as nano-indentation.

ACKNOWLEDGEMENTS

The authors would like to thank D.R.P. Singh and N. Chawla in Department of Material Science and Engineering at Arizona State University, USA.

REFERENCES

Chou, T.C., T.G. Nieh, S.D. McAdams, G.M. Pharr, and W.C. Oliver, J. Mater. Res., 7, pp. 2774–2784 (1992).

Kiener, D., W. Grosinger, G. Dehm and R. Pippan, Acta Mater., 56, pp. 580–592 (2008).

Lesuer, D.R., C.K. Syn, O.D. Sherby, J. Wadsworth, J.J. Lewandowski, and W.H. Hunt, Int. Mater. Rev., 41, pp. 169–197 (1996).

Liu, C.H., W.Z. Li, and H.D. Li, J. Mater. Res., 11, pp. 2231–2235 (1996).

Ng, K.S. and A.H.W. Ngan, Acta Mater., 56, pp. 1712–1720 (2008).

Singh, D.R.P., N. Chawla, G. Tang and Y.-L. Shen, Acta Materialia, 58, 6628–6636 (2010).

Song, S.X., Y.H. Lai, J.C. Huang, and T.G. Nieh, Appl. Phys. Lett., 94, 061911 (2009).

Tang, G., D.R.P. Singh, Y.-L. Shen and N. Chawla, Materials Science and Engineering A, 502, 79–84 (2009).

Tang, G., Y.-L. Shen, D.R.P. Singh and N. Chawla, Acta Materialia, 58, 2033–2044 (2010).

Tang, G., Y.-L. Shen, D.R.P. Singh and N. Chawla, International Journal of Mechanics and Materials in Design, 4, 391–398 (2008).

Civil, Architecture and Environmental Engineering – Kao & Sung (Eds)
© 2017 Taylor & Francis Group, ISBN 978-1-138-02985-9

A neighborhood search algorithm for the unrelated parallel machine scheduling problem

Y. Zhan & Y.G. Zhong
School of Mechanical and Electrical Engineering, Harbin Engineering University, China

ABSTRACT: In this paper, the unrelated parallel machine scheduling problem with machine eligibility restrictions is studied with the objective of minimizing the maximum makespan. A new neighborhood search algorithm is proposed to find a near optimal solution, which is based on insertion and swap moves. An efficient method is developed to find a feasible swap move. We present computational results for a set of randomly generated instances. The results show that the proposed algorithm outperforms simple heuristics with respect to solution quality.

1 INTRODUCTION

Neighborhood Search is an important local search method that defines and explores neighborhoods to find a near-optimal solution. It can find high-quality solutions in the practical computation time in the cases where exact algorithms fail to return a solution. Therefore, it has been applied to a wide variety of NP-hard problems.

Parallel machines scheduling problem with machine eligibility restrictions is a well-known NP-hard optimization problem. Therefore, most algorithms are developed only for restricted versions of this problem. Ebenlendr et al. (2008) develop a 1.75-approximation algorithm for the problem where each job can be assigned to at most two machines. Lin and Liao (2008) proposed an exact algorithm for the situation, where machines and jobs can be classified into two levels: high and low levels. Some useful properties inherent in the problem are used in their algorithm, such as the impact of the number of high-level machines on solutions.

For general problems with machine eligibility restrictions, some two-phase algorithms are developed. Salem and Armacost (2002) presented a two-phase algorithm for the unrelated parallel machines scheduling problem. In phase 1, constructive heuristics are used to build an initial solution, followed by an improvement heuristic to improve the initial solution in phase 2. Eliiyi et al. (2009) dealt with the problem with time windows and eligibility constraints. They developed a constraint-graph-based construction algorithm for generating near-optimal solutions and then used a genetic algorithm to enhance the near-optimal solutions. For more detailed information about parallel machines scheduling problem with machine eligibility restrictions,

the reader is referred to the survey papers Leung and Li (2008) and Liao and Sheen (2008).

The purpose of this paper is to develop a new neighborhood search algorithm for the studied problem. The remainder of this paper is organized as follows. In Section 2, we give the formal introduction of the studied problem. The proposed neighborhood search algorithm is presented in Section 3. In Sections 4 and 5, we present the computational results and conclusion, respectively.

2 PROBLEM DEFINITION

The problem considered in this paper can be formally described as follows: a set $J = \{J_1, J_2, ..., J_n\}$ of n independent jobs are scheduled on m identical parallel machines $M = \{M_1, M_2, ..., M_m\}$. Each job has to be processed by exactly one machine. The processing time of job J_i is denoted by T_i. In identical parallel machine scheduling problem, each job has the same processing time regardless of the machine to which it is assigned. Machine eligibility restrictions are considered in this paper. That is, not all of the m parallel machines are capable of processing each job. For each job, it is eligible to be processed on certain machines only.

We select the minimization of the maximum completion time or makespan (C_{max}) as the optimization criterion. The makespan is determined by the maximum workload among parallel machines, where the workload of a machine is the sum of processing times of all jobs assigned to it. The goal of schedule is to distribute workload among parallel machines as equally as possible so as to minimize the makespan. According to the three-field classification scheme, the scheduling problem studied in this paper can be denoted by $P_m/M_j/C_{max}$.

There are many real problems that can be modeled as the $P_m/M/C_{max}$ problems, especially the production systems where one or more stages have several machines in parallel. Our research is motivated by some real-world scheduling problems arising in the shipbuilding company, where a fixed-position layout is used. The product is too large to move and requires machines and staff to bring it.

The following notations and definitions are used to define the studied problem.

Adjacency matrix $A_{n \times m}$: the machine eligibility restrictions can be represented by an adjacency matrix $A_{n \times m}$, where entry A_{ij} in row i and column j is:

$$A_{ij} = \begin{cases} 1 & \text{if job } J_i \text{ can be processed by machine } M_j \\ 0 & \text{otherwise} \end{cases}$$

C_j: the makespan of machine M_j.

C_{max}: the makespan of a feasible schedule.

x_{ij}: a binary variable which takes value 1 if J_i is assigned to M_j and 0 otherwise.

A straightforward Mixed Integer Linear Programming (MILP) formulation for the studied problem is as follows:

$$\min C_{max} \tag{1}$$

such that

$$C_{max} \geq C_j \quad \forall_j \tag{2}$$

$$C_j = \sum_{i=1}^{n} x_{ij} T_i A_{ij} \quad \forall_j \tag{3}$$

$$\sum_{j=1}^{m} x_{ij} A_{ij} = 1 \quad \forall_i \tag{4}$$

$$x_{ij} \in \{0,1\} \quad \forall_{i,j} \tag{5}$$

In the formulation described above, Equation (1) is the objective function. It concerns the minimization of the maximum completion time on the parallel machines. Equation (2) defines the maximum completion time, which is obviously as large as all completion times on each machine. Equation (3) is used to compute the completion time on each machine for a given schedule. Equation (4) indicates that each job is scheduled only once on an eligible machine. Finally, equation (5) indicates that x_{ij} is a binary variable.

3 NEIGHBORHOOD SEARCH ALGORITHM FOR $P_M/M_{Ji}C_{MAX}$ PROBLEM

In this section, we will describe four neighborhood structures: (Insert(i, j), Swap(i, j), Insert(num), and Swap(num)). And we proposed the neighborhood search algorithm for the $P_m/M/C_{max}$ problem.

3.1 Solution representation and initial solution

The studied problem includes one decision of assignment of jobs to machines. To represent a solution, we use the job-based encoding scheme. In the case of n different jobs, the schedule S is encoded into a string of digits. The length of the string is equal to the total number of jobs. The index of each digit represents the index of each job. The value of each digit represents the index of a machine used to process the job. Consider a problem with four jobs and three machines. Let $S = \{1, 3, 2, 2\}$ be a feasible solution. The decoding of S results in the following assignment: J_1 will be processed on M_1, J_2 on M_3, and the remaining two jobs on M_2.

The initial solution construction problem includes two decisions of job sequencing and machine selection. Job sequencing determines the arrangement of the jobs that is to be assigned. Machine selection finds which machine is used to process a given job.

In the process of building an initial solution, the order of jobs for assignment is determined by job sequencing at first; then, jobs are assigned in this order to machines determined by machine selection. Job sequencing is the major module of the initial solution construction algorithm. The heuristic is used in job sequencing, which is described below.

Shortest processing time first: select the currently available job with the shortest processing time to be processed first.

For machine selection, a greedy heuristic is used in this paper. For a given job, the eligible machine with the minimum workload is selected.

The initial solution construction algorithm is described as follows:

Input: J, M, T, A.
Output: initial solution S_i
Step 1: unfinished job set $J_w \leftarrow J$, $(S_i)_{n \times m} \leftarrow 0$.
Step 2: if $J_w = \Phi$, output F_i; otherwise, select a job J_x from J_w based on Shortest Processing First rule.
Step 3: select a machine M_y in U_x with the minimum workload.
Step 4: $(S_i)_{x,y} \leftarrow 1$.
Step 5: delete J_x from J_w, go to Step 2.

3.2 Neighborhood structure

The initial solution may be highly unbalanced. The following four neighborhood structures are used for searching a more balanced solution.

1. Insert(i, j): Select a job J_i randomly and then select a machine M_j on which J_i is not scheduled.

From the definition of Insert(i, j), we know that given a schedule S, its insertion neighborhood consists of all schedules that can be obtained by removing a job from the current machine in S and re-assigning it to another eligible machine.

2. Swap(i, j): Select two jobs, J_i and J_j, and then interchange the machine assignments of J_i and J_j.

From the definition of Swap(i, j), we know that given a schedule S, its swap neighborhood consists of all schedules that can be obtained by interchanging the assigned machines of J_i and J_j in S.

To avoid generating duplicates, J_i and J_j should be scheduled on different machines. Whether or not two jobs belong to two different machines, this can be represented by a undirected graph $G(V, E)$, where:

$V = \{V_1, V_2, ..., V_n\}$ is a set of vertices, which corresponds to the set of jobs. For example, V_i corresponds to job J_i.

E is the set of edges, which indicates that two jobs connected by an edge belong to two different machines.

On the basis of $G(V, E)$, we just need to select an edge in G randomly to generate a swap neighbor.

3. Insert(num): Perform independent Insert (i, j) num times simultaneously.

4. Swap(num): Perform independent Swap(i, j) num times simultaneously.

We now present a neighborhood search algorithm for the studied problem.

Input: initial solution S_i

Output: near optimal solution S_o

Step 1. $S_o \leftarrow S_i$, counter$\leftarrow 0$, input the value of maxIter and the value of maxCounter.

Step 2. While(counter < maxCounter) do
Nei: Select a neighborhood structure randomly and generate a neighbor S_n of S_i.
If S_n is better than S_o, then $S_o \leftarrow S_n$, counter$\leftarrow 0$. Otherwise, counter$\leftarrow$counter+1.

Step 3. Iter$\leftarrow$Iter+1. If(Iter > maxIter), output S_o. Otherwise, counter$\leftarrow 0$, $S_i \leftarrow S_o$, goto Step 2.

4 COMPUTATIONAL RESULTS

The neighborhood search algorithm presented in this paper is coded in Visual Studio 2010 C# and implemented on an i7@2.60GHz personal computer with 4G memory. The Central Processing Unit (CPU) time limit is set to 10 min.

The factors considered in this experiment are the number of jobs, the number of machines, the eligible machine set for each job, the processing time of each job, and the initial solution construction heuristics used in phase 1. The details of the above factors are as follows:

Table 1. Results of computational experiments.

m	n	T_s	S_i	S_o
2	10	10.18	268	253
2	20	98.39	652	608
4	10	100.04	188	154
4	20	489.63	224	217
6	10	326.34	126	99
6	20	517.11	227	192
8	10	458.31	121	91
8	20	598.74	184	156

1. Number of jobs (n): 10, 20.
2. Number of machines (m): 2, 4, 6, 8.
3. The machine eligibility restrictions are important factors in the studied problem, which have a strong impact on CPU time required to search the optimal solution. Obviously, CPU time will increase as the size of the eligible machine set increases, as the size of alternating search tree is much larger in this case. In order to evaluate the efficiency of the improvement algorithm in the worst cases, we consider the most difficult situation that every machine is eligible for every job.
4. The processing time of each job is uniformly distributed in the interval [50, 100].

Table 1 reports the average results of the computational experiments on 10 runs for each instance. T_s is the CPU time used by the proposed neighborhood search algorithm, which is in seconds.

As can be seen from Table 1, in each instance, the quality of S_o is better than that of S_i. Meanwhile, the solution times increase considerably with an increase in the number of jobs and the number of machines. In the worst case, the proposed algorithm can handle 8×20 size problem in reasonable time, which is less than 10 min.

5 CONCLUSIONS

In this study, the scheduling problem of minimizing the makespan on parallel machines with eligibility restrictions is studied. We developed a neighborhood search algorithm to find a near-optimal solution. A computational experiment is designed for evaluating the performance of the proposed algorithm. The results reveal that it can handle 8×20 size problem in reasonable time, which is less than 10 min.

ACKNOWLEDGMENTS

This work was financially supported by the Research Fund for the Doctoral Program of

Higher Education, China (20132304120021) and the National Natural Science Foundation of China (No. 51275104).

REFERENCES

Al Salem, A., & Armacost, R.L. (2002). Unrelated machines scheduling with machine eligibility restrictions. Qatar University.

Ebenlendr, T., Krčál, M., & Sgall, J. (2008). Graph balancing: a special case of scheduling unrelated parallel machines. In Proceedings of the nineteenth annual ACM-SIAM symposium on Discrete algorithms. 483–490.

Eliiyi, D.T., Korkmaz, A.G., & Çiçk, A.E. (2009). Operational variable job scheduling with eligibility constraints: A randomized constraint-graph-based approach. Technological and Economic Development of Economy. 15(2), 245–266.

Leung, J.Y.T., & Li, C.L. (2008). Scheduling with processing set restrictions: A survey. International Journal of Production Economics. 116(2), 251–262.

Liao, L.W., & Sheen, G.J. (2008). Parallel machine scheduling with machine availability and eligibility constraints. European Journal of Operational Research. 184(2), 458–467.

Lin, C.H., & Liao, C.J. (2008). Minimizing makespan on parallel machines with machine eligibility restrictions. Open Operational Research Journal. 2, 18–24.

Civil, Architecture and Environmental Engineering – Kao & Sung (Eds)
© *2017 Taylor & Francis Group, ISBN 978-1-138-02985-9*

A novel medical image enhancement algorithm based on ridgelet transform

Yunfeng Yang & Yuewen Yang
School of Mathematics and Statistics, Northeast Petroleum University, Daqing, China

ABSTRACT: A new medical image enhancement method based on ridgelet transform was proposed in this paper. Medical images enhancement is important in the clinical medicine. And the key work of enhancement methods based on ridgelet transform is how to enhance coefficients of ridgelet transform. The multiwavelet transform is used after the Radon transform in the process of ridgelet transform, and these multiwavelet coefficients are enhanced with weight factors. In order to reduce the effects of noise, the soft threshold method was selected to de-noise the coefficients of ridgelet transform. The piece wise histogram transform was used to stretch the range of gray level of the enhanced image for getting abundant gray levels. Experiments have shown that the method can not only enhance an image's details, but also hold the image's edge features effectively.

1 INTRODUCTION

With the development of information technology, medical images have been applied in the clinical medicine successfully. Medical images can provide more visual information about the disease in the clinical medicine. However, the quality of medical images are not high enough to use directly due to many reasons, such as low contrast and noise. Therefore, medical image enhancement technology is very important for clinical medicine. In addition, enhanced images can also be effective for the registration and segmentation of medical images and so on.

Wavelet transform has been widely used in image processing fields for that it is an effective time–frequency analysis tool developed in the 1980s. There are many medical image enhancement methods based on wavelet transform (Y. Yang et al. 2010, Bhutada 2011). After the wavelet transform, the ridglet transform is another important tool for multiscale analysis (Candes et al. 1998, Candes et al. 1999). It mainly includes two procedures: Radon transform and Wavelet Transform (WT). The ridgelet transform has been widely used in the field of image processing (Jiang Yuan et al. 2016, Deng Chengzhi et al. 2009). It is also widely used to enhance image quality (Li Hongbing et al. 2011, Qiu Ju et al. 2009).

Ridgelet transform is an effective processing method for an image. However there is also some high-frequency information hidden in high-frequency sub-images after wavelet transform of the image. Better enhancement results can be obtained if the information of high frequency is used effectively. On the basis of this thought, an improved enhancement method of a medical image based on ridgelet transform was proposed in this paper, in which multiwavelet transform is selected to replace wavelet transform after Radon transform.

2 MEDICAL IMAGE ENHANCEMENT BASED ON NEW RIDGELET TRANSFORM (NRT)

Ridgelet transform is a new representation method of an image, which mainly includes two procedures: Radon transform and Wavelet Transform (WT), as shown in Formulas (1) and (2). In this paper, the improvement of ridgelet transform is proposed for image enhancement:

$$R_f(\theta, t) = \int_{R^2} f(x)\delta(x_1 \cos(\theta) + x_2 \sin(\theta) - t)dx, \quad (1)$$

where $\theta \in [0, 2\pi], t \in R$ and δ is the Dirac function.

$$CRT_f(a, b, \theta) = \int_R \psi_{a,b}(t) R_f(\theta, t)dt, \quad (2)$$

where $\psi(x)$ is the wavelet function.

The proposed enhancement method based on ridgelet transform is described in Fig. 1. It mainly includes Radon transform, multiwavelet transform, enhancement of coefficients of ridgelet transform, inverse multiwavelet transform, inverse Radon transform, and histogram transform.

2.1 Improved ridgelet transform

The ridgelet transform proposed by Donoho is shown in Formulas (1) and (2). In this part, according to formula (2), the multiwavelet transform is provided to replace the wavelet after Radon transform. More high-frequency information can be obtained through multiwavelet transform. These more detail information can be used to enhance the features of an image. In order to watch clearly, the procedure of multi-wavelet transform of an image is shown in Figures 2 and 3.

2.2 Copying old text onto a new file and enhancement of ridgelet coefficients

In this part, the coefficients of ridgelet transform are de-noised by soft threshold method initially. After this, the coefficients are enhanced by enhancement factor shown in Formula (3):

$$CRT_f(a,b,\theta) = \omega_i \, CRT_f(a,b,\theta), \qquad (3)$$

where ω_i, (i = 1,2) denotes enhancement factor action on the coefficients of multiwavelet transform.

$$T_f(x,y)$$
$$= \begin{cases} f(x,y)M\Big/N, \\ \qquad f(x,y) \in [0,N] \\ \dfrac{(f(x,y)-N)(255-2M)}{(f_{max}-2N)} + M, \\ \qquad f(x,y) \in [N, f_{max}-N] \\ \dfrac{(f(x,y)-f_{max}+N)M}{N} + 255 - M, \\ \qquad f(x,y) \in [f_{max}-N, f_{max}] \end{cases} \qquad (4)$$

where $f(x,y)$ denotes the pixel of reconstructed image at position (x,y), f_{max} is the max value of the pixels, $M \in [0,255]$, and $N \in [0, f_{max}]$.

2.3 Histogram transform

The inverse multiwavelet transform and inverse Radon transform are executed on the basis of the

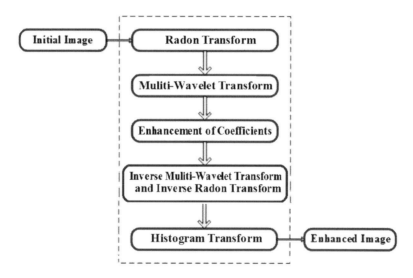

Figure 1. Enhancement procedure based on ridgelet transform.

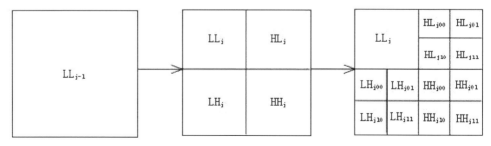

Figure 2. Multiwavelet transform of 2-D signal.

enhanced coefficients of ridgelet transform. So far, the reconstruction of the image is accomplished. However, gray level of the reconstructed image may be lower, which could result in a darker image. In order to change the problem, the piecewise histogram transform shown in Formula (4) is taken to obtain the more abundant information of an image's gray level.

3 EXPERIMENTS

In the experiments, we selected $M = \dfrac{255}{3}$, $N = \dfrac{f_{max}}{4}$, $\omega_1 = 1.5$, and $\omega_2 = 1.8$. The results of experiments based on the proposed enhancement method are shown in Figures 4 and 5, where panels (a) show the initial image, (b) show the

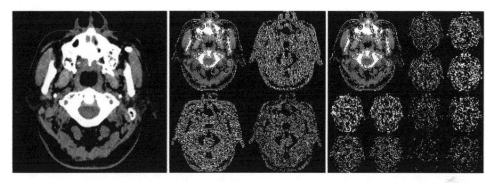

Figure 3.   Multiwavelet transform of an image.

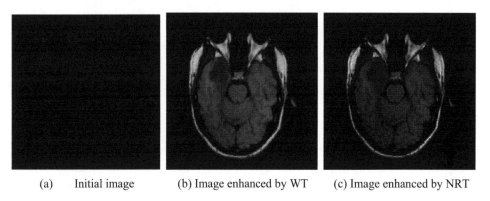

    (a)   Initial image   (b) Image enhanced by WT   (c) Image enhanced by NRT

Figure 4.   Image enhancement I.

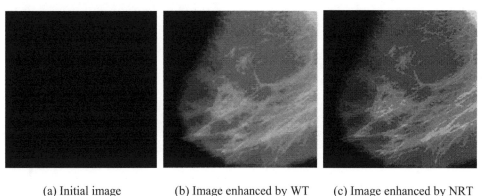

    (a) Initial image   (b) Image enhanced by WT   (c) Image enhanced by NRT

Figure 5.   Image enhancement II.

Table 1. Results of the experiments.

Image	Method	PSNR
Fig. 4(b)	Enhancement by WT	28.53
Fig. 4(c)	Enhancement by NRT	62.36
Fig. 5(b)	Enhancement by WT	27.19
Fig. 5(c)	Enhancement by NRT	52.31

enhancement result by wavelet transform, and (c) show the enhancement result by ridgelet transform proposed in the paper. The PSNR is shown in Table 1.

4 CONCLUSIONS

An important problem of medical image enhancement is how to use the coefficient of ridgelet transform effectively. In this paper, the two key steps are selected in the procedure of medical image enhancement. First, the multiwavelet transform is selected to replace wavelet transform after Radon transform. Second, different weights act on the ridgelet coefficients, that is, the weight factor ω_1 of enhancement is used to enhance the coefficients of multiwavelet transform, and the weight factor ω_2 is used to enhance the coefficients of obtained by inverse multiwavelet transform. Those techniques help acquire better enhancement results. Experiment results showed that better enhanced images could be obtained using the proposed method.

ACKNOWLEDGMENT

This work was financially supported by the Foundation of Northeast Petroleum University (XN2014106).

REFERENCES

Bhutada G. G., Anand R. S. & Saxena S. C. Edge preserved image enhancement using adaptive fusion of images denoised by wavelet and curvelet transform. *Digital Signal Processing*, 21(1), 118–130.

Candes E. J. (1998). Ridgelets: Theory and Applications, *Department of Statistics, Stanford University*.

Candes E. J. & Donoho D. L. (1999). Ridgolets: A key to higher-dimensional intermittency [J]. Philosophical Transactions of the Royal Society of London Series A.

Deng Chengzhi & Cao Hanqiang (2009). Construction of Multiscale Ridgelet Dictionary and Its Application for Image Coding [J]. Journal of Image and Graphics, 14(7), 1273–1278.

Jiang Yuan, Qu Changwen et al. (2016). SAS Image Denosing of Ship Wake Based on Ridgelet Transform. *Ship Electronic Engineering*, 36(1), 127–130.

Li Hongbing, Yu Chengbo et al. (2011). Study on finger vein image enhancement based on ridgelet transformation, 23(2), 224–230.

Qiu Ju, Zhang Zhongbo & Ma Siliang (2009). Enhancement Method Root Canal Image in Dental Film Based on Ridgelet Transform [J]. Journal of Jinli University (Science Edition), 47(4), 764–768.

Y. Yang, Z. Su & L. Sun (2010). Medical Image Enhancement Algorithm Based on Wavelet Transform. *Electronics Letters*, 46(2), 120–121.

Civil, Architecture and Environmental Engineering – Kao & Sung (Eds)
© 2017 Taylor & Francis Group, ISBN 978-1-138-02985-9

Significance of deep learning on big data analytics

Jilei Mao
Wuhan University of Technology, Hubei Province, China

Zijun Mao
Capital University of Economics and Business, China

ABSTRACT: Deep learning and big data analytics are two high foci of data science. With the huge volumes of data available today, big data brings us more opportunities and transformative potential; it also brings unprecedented challenges to address data and information. The data are too large to be analyzed using the traditional technology. Therefore, deep learning algorithm is becoming an important department in providing big data analytics a new way to get predictive analytics solutions. Deep learning gives us an important opportunity for completing neural network algorithms and models to deal with analytics problems related to big data compared with the traditional technology used to deal with data analytics problems.

1 INTRODUCTION

In the rapidly growing digital word, there are two hottest study trends: deep learning and big data. The concept, application, and relationship between big data and deep learning will be mainly introduced in this paper.

Big data refers to the digital data with the sheer growth and wide availability and is difficult or even impossible to be addressed and analyzed by traditional data analytics tools and technologies. The shapes and the size of data are growing at an astonishing rate.

While big data offers a great potential for revolutionizing all aspects of our society, getting useful information from big data is challenging. It requires the development of advanced technologies, and the work form virtuous tends experts working in close cooperation to get the large and rapidly growing size of information hidden in the astonishing volume of digital data.

As the size of data is becoming larger, deep learning is becoming an important department in providing big data predictive analytics solutions, particularly with the increased processing power and advances in graphics processors. Deep learning algorithm makes a better job of potentially providing a solution to solve the data analytics and learning problem found in massive volume of input data compared with another machine learning and feature engineering algorithms. More specifically, it aids in automatically extracting complex data representations from large volumes of unsupervised data. These features make deep learning a useful tool for the big data analytics of the data that are considered generally unsupervised and uncategorized. The multi-tier learning model and extraction of different complex levels and data abstractions in deep learning provide a certain degree of simplification for big data analytics tasks, especially for analyzing large volumes of data, semantic indexing, data tagging, and discriminative tasks such classification and prediction.

2 DEEP LEARNING ALGORITHM

The field of machine learning has made great progress especially in the algorithm. And we call it deep learning algorithm. In this part, a brief introduction about deep learning algorithm is given.

The deep learning algorithms are mostly influenced by the study about artificial intelligence, with the aim of emulating the human brain's ability to learn, analyze, observe, and make decisions, especially for complex problems related to large volumes of data. Work related to these complex challenges has been a key motivation behind deep learning algorithms, which strive to emulate the hierarchical learning approach of the human brain.

Deep learning algorithm is a learning model composed of multilayer neural networks, as shown in Figure 1. Neural network is a learning model consists of many logical units, which is organized in different levels. And the output of the first layer is the input variable of the next layer. The first layer is called the input layer, the last layer is called

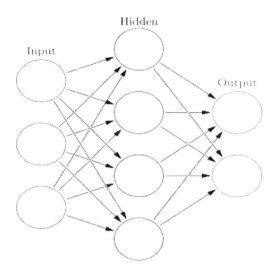

Figure 1. Three layers of neural network.

the output layer, and the middle layer is the hidden layer.

We can make a simple summary: Deep architecture of consecutive layers is the main idea of deep learning algorithm. A nonlinear transformation is set in the input of each layer, which provides processed data in its output.

The target of the algorithm is to learn a complex and abstract representation of the data in a hierarchical manner by passing the data through multiple transformation layers. The sensory data (such as an image) is fed to the first layer. The input layer transforms the features into another field, and the output features are the input of the next layer. The features of the data are transformed and passed by layers to learn and set the model of the data. Therefore, the concept of deep learning can be defined combined with data processing.

Deep learning is an algorithm that can learn a model from data and use the model to predict the new data. According to this feature, we cannot separate deep learning from data. The essence of learning is to build a machine learning model with several hidden layers and use large volumes of training data to learn more meaningful features so as to enhance the accuracy of classification or prediction. Therefore, "deep model" is a path and "feature learning" is the purpose of the algorithm.

Deep learning algorithm has two features: 1) it emphasizes the depth of the model structure, usually with one or two, or more hidden layers (as shown in Figure 2); 2) it highlights the importance of feature learning and transforms the features layer by layer. It transforms the samples features from the original space to a new feature space, thus making it easier to classify or predict.

Compared with the traditional way, deep learning algorithm is more accurate to portray the information inside the data.

3 TYPICAL APPLICATIONS OF DEEP LEARNING

Some basic concepts and principles of deep learning algorithm have been introduced. Examples of application are given in this part.

3.1 *Neural network machine translation system*

Google has released the Neural Machine Translation System (GNMT) recently. The improvement in speed and accuracy brings a better service to the users. The translation quality of English, France, and Spanish has reached about 90%. And the translation quality between Chinese and English can also reach 80%.

Figure 3 shows the scores of human translation, Neural Machine Translation System, and Phrase-based Machine Translation System, where 0 represents "totally meaningless translation" and 6 represents "perfect translation". Neural network Machine Translation System has a higher score than Phrase-based Machine Translation clearly.

3.2 *Mechanical arm*

Deep learning can teach robots how to use the robot arm to complete a task. Traditional mechanical arm, which is programmed by a programmer, cannot be seriously disturbed; otherwise, it cannot grasp the target accurately. Deep learning algorithm can perform much better than the traditional algorithm. Deep learning algorithm can train the mechanical arm to pick the object up. If failed in the first time, then re-learn and try again until the task is finished. Deep learning increases the ability of the robots to learn how to finish the task.

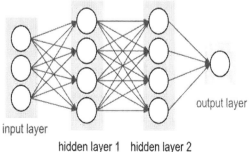

Figure 2. Neural network with two hidden layers.

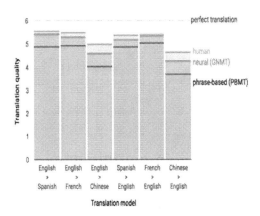

Figure 3. Translation model.

3.3 AlphaGo

In March 2016, AlphaGo won the world championship. Deep learning algorithm played a key role in this game.

In go game, each move has more than 300 types, and every chess has 200 steps on average. This question cannot be resolved in the traditional way. Deep learning algorithm can make machines have human intuition and predict the next step that the opponent will analyze, making the best decision. The deep learning algorithm is the reason for the success of Alphago.

3.4 Face recognition

By using deep learning algorithm in face recognition, more information can be obtained, such as the location of the face in the picture, name, gender, age, and facial expression of the person. The algorithm can even analyze the movement of the lips to know what people say. The speed and accuracy have been greatly improved compared with the technology before, such as support vector machines.

4 BIG DATA

Big data refers to the data with size beyond the traditional database system processing power. It requires high transfer speed. Its structure is not suitable for the original database system. This definition is intentionally subjective and incorporates a moving definition of how big a data set needs to be in order to be considered big data—the big data cannot be defined in terms of being larger than a certain number of terabytes (thousands of gigabytes). As technology advances over time, the data

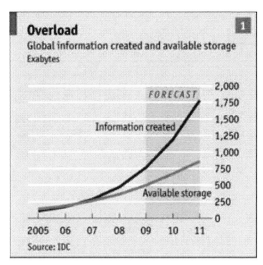

Figure 4. Global information created and storage.

sets that qualify as big data will also increase (as shown in Figure 4). The size of data in our world has been exploding. As an all-encompassing term, big data has a very vague definition. The input list of the big data is long, including the social network, satellite images, broadcast audio streams, bank transactions, rock music MP3, Web page content, government file scanning, GPS route, and financial market data. These data are not the same thing, and data sources are abundant.

5 CONCLUSIONS

Deep learning algorithm is mostly influenced by the study of artificial intelligence, which is aimed to emulate the human brain's ability by science analytics. It is a learning model composed of multilayer neural networks. Big data refers to the data whose size is beyond the traditional database system processing power.

As the data keep getting bigger, deep learning is coming to play a key role in providing big data predictive analytics solutions, particularly with the increased processing power and the advances in graphics processors. Compared with the traditional way, the deep learning algorithm is more accurate to portray the information inside the data and provide a solution to solve the data analysis and learning problems found in massive volumes of input data. The study about deep learning and big data is very important and significant and requires more attention from all fields. More efforts are necessarily taken to further apply deep learning to the problems associated with big data.

REFERENCES

Chen X W, Lin X, 2014, Big Data Deep Learning: Challenges and Perspectives [J]. IEEE Access, 2:514–525.

Greenspan H, Ginneken B V, Summers R M. Guest Editorial, 2016. Deep Learning in Medical Imaging: Overview and Future Promise of an Exciting New Technique [J]. IEEE Transactions on Medical Imaging, 35(5):1153–1159.

Mcafee A, Brynjolfsson E, 2012. Big data: the management revolution [J]. Harvard Business Review, 90(10):60–6, 68, 128.

Najafabadi M M, Villanustre F, Khoshgoftaar T M, et al, 2015. Deep learning applications and challenges in big data analytics [J]. Journal of Big Data, 2(1):1–21.

Rumelhart D E, Hinton G E, Williams R J, 2015. Learning In ternal Representations by Error Propagation[J]. Readings in Cognitive Science, 1:399–421.

Srinivas S, Babu R V, 2015 Deep Learning in Neural Net works: An Overview[J]. Computer Science.

Wu, Yonghui, Schuster, Mike, Chen, Zhifeng, et al, 2016. Google's Neural Machine Translation System: Bridging the Gap between Human and Machine Translation [J].

Civil, Architecture and Environmental Engineering – Kao & Sung (Eds)
© *2017 Taylor & Francis Group, ISBN 978-1-138-02985-9*

Branch pipe routing method based on a 3D network and improved genetic algorithm

Yaxiao Niu, Wentie Niu & Weiguo Gao
Key Laboratory of Mechanism Theory and Equipment Design of the State Education Ministry, Tianjin University, Tianjin, China

ABSTRACT: Pipe routing plays an important role in many industries, especially for the ship building design. In this paper, a rectilinear branch pipe routing method for automatic generation of the optimal branch pipe routes based on a 3D network and improved Genetic Algorithm (GA) for ships is proposed. By extending the 2D escape graph, a 3D network graph is constructed first. It can model the 3D constrained layout space and reduce the storage space effectively. Then, an improved genetic algorithm is employed to solve the rectilinear branch pipe routing optimization problem on the basis of the network graph. Fixed-length coding method, one-point crossover, and location mutation are adopted to improve the computational efficiency. Finally, a case study of pipe routing for a ship engine room is conducted to validate the performance of the proposed method.

1 INTRODUCTION

Pipe routing plays an important role in many industries, especially for the ship building design. Because of the large space of ship layout and various design constraints, it takes more than 50% of the man-hours in the detailed design phase by using the traditional pipeline design method. Therefore, an automatic pipeline layout approach might offer an attractive way to improve the design efficiency, leading to saving of time and money.

In the past decades, the two-terminal pipe routing problem has been widely studied. However, in practice, more than 70% of pipelines contain at least one branch pipe (Asmara & Nienhuis 2006). To solve the branch pipe routing problem, a series of research (Wu et al. 2008, Jiang et al. 2015, Sui & Niu 2016) has been conducted using a cell-generation method. According to computational geometry, the Steiner Minimal Tree with Obstacles (SMTO) problem can be formulated by constructing a shortest collision-free network interconnecting some given terminals while allowing for addition of auxiliary points called Steiner points (Liu & Wang 2012). By combining the Steiner tree theory with Particle Swarm Optimization (PSO) algorithm, rectilinear branch pipe routing method (Liu & Wang 2011) and nonrectilinear branch pipe routing approach (Liu & Wang 2012) were proposed to automatically generate the branch pipe routes of aero-engines.

In this paper, branch pipe routing method based on 3D network and improved genetic algorithm is

proposed for the rectilinear branch pipe routing problem. First, a 3D network graph is constructed by extending the 2D escape graph. Then, an improved genetic algorithm is employed to solve the rectilinear branch pipe routing optimization problem in the 3D network graph. Finally, a case study of pipe routing for ship engine room is conducted to validate the proposed method.

2 PROBLEM DESCRIPTION

Branch pipe routing problem can be formulated as finding the Rectilinear Steiner Minimal Trees with Obstacles (RSMTO) (Liu & Wang 2012). For the RSMTO problem, one of the important tasks is to construct a reasonable layout space model.

By extending the horizontal and vertical lines from the terminals and the obstacle vertices to any boundary of the design space and obstacles, 2D escape graph can be constructed. For the escape graph, if an instance of the RSMTO problem is solvable, there is an optimal solution composed only of escape segment in the escape graph (Ganley & Cohoon, 1994). The escape graph method is applied to construct the network graph model in the 3D layout space. By extending the 2D escape graph, 3D layout space can be constructed. The detailed steps are as follows:

Step 1. Extend lines from obstacle vertex in X, Y, and Z directions to any boundaries of the design space and obstacles and record the intersection points as vertexes.

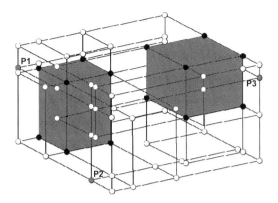

Figure 1. Example of 3D network graph.

Step 2. Extend lines from terminal in X, Y, and Z directions to any boundaries of the design space and obstacles and record the intersection points as vertexes.

Step 3. Combine all the lines and vertexes into the 3D network graph.

Figure 1 shows an example of 3D network graph with four obstacles and three terminals, in which black points denote the obstacle vertexes, white points denote the intersection points, and gray points denote the pipe terminals.

3 IMPROVED GENETIC ALGORITHM

Genetic algorithm is adopted to deal with the optimization problem for its global characteristic, robustness, and easy implementation. In this paper, the improved genetic algorithm is adopted to determine the number and position of the Steiner points on the 3D network graph. The flowchart of the improved genetic algorithm is shown in Figure 2.

3.1 *Fixed-length encoding*

The key problem of solving the Steiner minimal tree for a given graph $G = (V, E)$ with N connection points is to determine the number and positions of the Steiner points. The chromosome coding method of the improved genetic algorithm is presented referring to the PSO coding in the literature (Liu & Wang 2012). Assuming that V is a collection of all the points in the graph G, so $r = (V–N)$ is the collection of all the middle points. A chromosome is randomly generated by the fixed-length encoding method:

$$(Node(1),0),(Node(2),1),\cdots(Node(i),1)$$
$$\cdots(Node(r),0) \qquad (1)$$

where ($Node(2)$, 1) denotes that node 2 is the Steiner point and ($Node(1)$, 0) denotes that node 1 is the potential Steiner point, which may become Steiner point as the system iterates. For N-point connection problem, the number of Steiner points should be less than $(N – 2)$. Hence, the chromosomes that do not meet this requirement should to be deleted during the iteration. Figure 3 shows an example of a chromosome code and a Steiner tree on a network graph. As depicted in Figure 3, black points denote the pipe connection points, bold solid lines represent the Steiner tree, and points 4 and 10 denote the Steiner points.

3.2 *Genetic operator*

3.2.1 *One-point crossover*
The one-point crossover method is used in the improved genetic algorithm to generate the child chromosome. Take the network graph in Figure 3 as an example. First, randomly select two parent chromosomes P1 and P2 from the population. Then, a random number k $(0 < k <1)$ is generated.

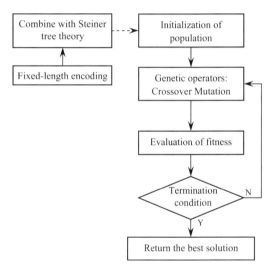

Figure 2. Improved genetic algorithm.

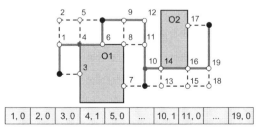

Figure 3. Example of chromosome code and a Steiner tree on a network graph.

Here, if $k \leq p_c$ (p_c denotes the crossover probability), the two parent chromosomes are recombined to produce two new children, as shown in Figure 4. There are four pipe connection points on the network graph, so the number of Steiner points must be less than two. The children with more than two Steiner points would be deleted.

3.2.2 Location mutation

Mutation operation can increase the population diversity and the search space. Location mutation is utilized in the genetic operation. The location mutation method is shown in Figure 5. In location mutation, a chromosome is randomly selected from the population first. Then, a random number k ($0 < k < 1$) is generated. If $k \leq p_m$ (p_m denotes the mutation probability), a random quantity of potential Steiner points are selected and transformed into Steiner points; accordingly, the corresponding quantity of Steiner points are turned

into potential Steiner points. Figure 5 shows the change of Steiner tree after the Steiner point 4 is substituted by point 5. Because the location mutation operation does not change the number of Steiner points, the filtering operation is not needed in the mutation operation.

3.3 Evaluating an individual

Considering practical engineering rules, the pipeline should be away from dangerous equipment, such as the electrical regions or the high temperature-areas, and close to some equipment that are easy to install the supports. To cope with the problems, the concept of potential energy is proposed by Ito (1999). By allocating diffident potential energy values to the cells around the equipment, different engineering constraints can be satisfied. On the basis of the concept of potential energy, a different value of potential energy

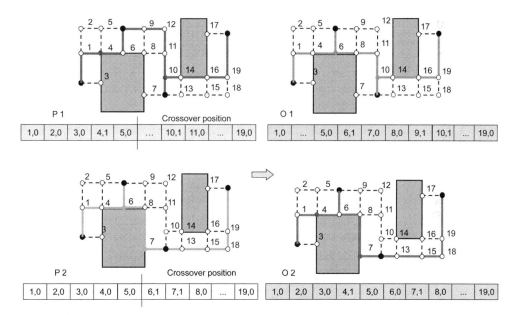

Figure 4.   One-point crossover method.

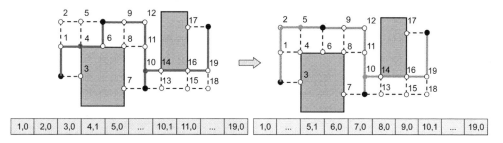

Figure 5.   Location mutation method.

(E_p) is assigned to each vertex in the 3D network graph for satisfying three engineering constraints.

A lower potential energy value means that the pipe route is close to equipment that are easy to install the supports, while a higher energy value means that the pipe route is close to the electric regions or high-temperature areas.

For example, as shown in Figure 3, assume that obstacle O1 is the equipment that may generate heat. Then, a potential energy value E_p is appended on the vertexes, which are on the boundary of the obstacle O1, such as vertexes 3, 4, 6, 7, and 8. The generalized length of pipeline segment between two vertexes can be calculated as follows:

$$L'_{seg} = L_{seg} + E_{p1} + E_{p2} \qquad (2)$$

where L_{seg} represents the length of the pipeline segment and E_{p1} and E_{p2} denote the potential energies of the two vertexes, respectively.

Another problem that needs to be considered is the number of pipe bends. The pipe routing cost increases with the increase of the number of bends. Considering the potential energy and the number of bends, the fitness function of a chromosome can be formulated as:

$$F = c_1 \times L_g + c_2 \times N_B \times B_{bend} \qquad (3)$$

where L_g represents the generalized length of the chromosome, B_{bend} denotes the equivalent length of a bend, N_B denotes the total number of bends

of the chromosome, and c_1 and c_2 denote the weighting factors associated with the objectives, respectively.

According to the aforementioned chromosome coding method and genetic operation, the fast algorithm for Steiner trees (Kou et al. 1989) is used to construct a minimal Steiner tree. The fitness value of a chromosome can be calculated as follows:

Step 1. Construct the point set of the chromosome, including the Steiner points and the connection points. There should be totally $(N + q)$ points in the set, where N is the number of connection points and q is the number of the Steiner points.

Step 2. The fast algorithm for Steiner trees is employed to solve the minimum Steiner tree problem on the network graph for connection of the $(N + q)$ points.

Step 3. According to Eqs. 2 and 3, calculate the Minimum Steiner Tree (MST) length, which is viewed as the fitness of the chromosome.

4 CASE STUDY

An equipment layout of a ship cabin is shown in Figure 6. The main equipment includes two fuel oil storage tanks, two marine main engines, two diesel generators, one steam boiler, one water boiler, and two oil pumps. Two fuel pipes are chosen as examples to validate the proposed method. The design information of these pipes is presented in Table 1.

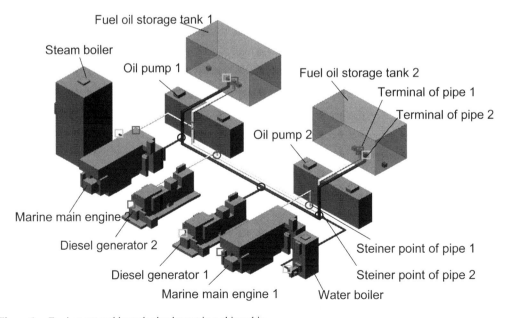

Figure 6. Equipment and branch pipe layout in a ship cabin.

For pipe 1, the 3D network graph contains 220 points and 800 edges. For pipe 2, the 3D network graph contains 260 points and 1060 edges, in which the routed pipe 1 is regarded as an obstacle. The parameters of the genetic algorithm run were the population size, 60; the number of generation, 100; the crossover ratio, 0.8, and the mutation ratio, 0.1; and the weighting factors $c_1 = 0.5$, $c_2 = 0.5$. As for electrical regions, potential energy value $E_p = 100$ is assigned to diesel generators 1 and 2.

The optimal results of the two pipes are shown in Figure 6. The total length of pipe 1 is 9526 cm, and it contains six bends and three T-joints. The total length of pipe 2 is 8722 cm, and it contains eight bends and three T-joints. In the optimization process of pipe 2, the parallel part of the pipe 1 is set as a shorter generalized pipe length, for the parallel part can share the supports. The convergence curves of two optimal branch pipelines are shown in Figure 7. The convergence generations of two optimal branch pipelines are about 22 and 44, respectively.

Table 1. Information of piping route path and connecting points.

Route path name	Included cabins and equipment with corresponding coordinates of connecting points
Path 1	Fuel oil storage tank 1: (100, 305, 318) Fuel oil storage tank 2: (100, 305, 823) Marine main engine 1: (695, 200, 335) Marine main engine 2: (695, 200, 744) Diesel generator 2: (905, 165, 940)
Path 2	Fuel oil storage tank 1: (100, 305, 280) Fuel oil storage tank 2: (100, 305, 793) Hot water boiler: (795, 115, 70) Steam boiler: (505, 65, 867) Diesel generator 1: (905, 165, 940)

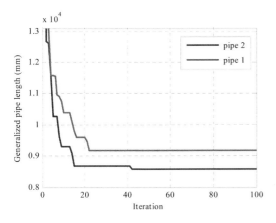

Figure 7. Convergence curves of two optimal branch pipelines.

CONCLUSIONS

In this paper, we presented an improved GA and 3D network-graph-based branch pipe routing approach, which mainly includes layout space construction, coding, fitness function, and specific genetic operation strategy.

A 3D network graph is first constructed by extending the 2D escape graph. Branch pipe routing problem can be formulated as finding the Rectilinear Steiner Minimal Trees with Obstacles. Branch pipes in the layout space are represented by the Steiner points and the connection points included in the chromosome. The fast algorithm for Steiner trees is employed to solve the minimum Steiner tree problem on the network graph for the connection of these points. On the basis of the 3D network graph and potential energy value, design goal and some engineering constraints are modeled. The genetic operator is then devised, in which one-point crossover and location mutation strategy are presented to improve the quality of chromosomes. A case study of a fuel piping system in a ship engine room demonstrates the feasibility and effectiveness of the proposed method.

Future research should focus on more engineering constraints, such as pipe grading, to avoid vibration and ensure the usage security of the pipeline.

ACKNOWLEDGMENT

This work was financially supported by the National Natural Science Foundation of China (Grant Number: 51275340).

REFERENCES

Asmara, A. & Nienhuis, U. 2006. Automatic piping system in ship. *Proceedings of 5th International Conference on Computer and It Application in the Maritime Industries* 269–280.

Ito, T. 1999. A genetic algorithm approach to piping route path planning. *Journal of Intelligent Manufacturing* 10(1): 103–114.

Jiang, W., Lin, Y., Chen, M. & Yu, Y. 2015. A co-evolutionary improved multi-ant colony optimization for ship multiple and branch pipe route design. *Ocean Engineering* 102: 63–70.

Kou, L., Markowsky, G. & Berman, L. 1981. A fast algorithm for Steiner trees. *Acta Informatica* 15(2): 141–145.

Lavinus, J.W. & Cohoon J.P. 1993. Routing a multi-terminal critical net: steiner tree construction in the presence of obstacles. *International Symposium on Circuits & Systems* 1: 113–116.

Liu, Q. & Wang, C. 2011. A discrete particle swarm optimization algorithm for rectilinear branch pipe routing. *Assembly Automation* 31(4): 363–368.

Liu, Q. & Wang, C. 2012. Multi-terminal pipe routing by Steiner minimal tree and particle swarm optimization. *Enterprise Information Systems* 6(3): 315–327.

Sui, H. & Niu, W. 2016. Branch-pipe-routing approach for ships using improved genetic algorithm. *Frontiers of Mechanical Engineering* 2016: 1–8.

Wu, J., Lin, Y., Ji, Z. & Fan, X. 2008. Optimal approach of ship branch pipe routing optimization based on co-evolutionary algorithm. *Ship & Ocean Engineering* 37(4): 135–138. (in Chinese)

Author index

1608

Wu, H. 67, 1185
Wu, J.Y. 331
Wu, J.Y. 1247
Wu, J.Y. 1521
Wu, L.G. 1021
Wu, Q.Y. 1559
Wu, T.B. 539
Wu, W.H. 1051
Wu, Y.-J. 1407
Wu, Y.W. 1233
Wu, Y.-W. 125
Wu, Y.Z. 3

Xia, C. 1101
Xia, W. 621
Xia, Y.X. 25
Xia, Y.Z. 597
Xiang, J. 693
Xiao, D.Q. 159
Xiao, H. 377
Xiao, J. 775
Xiao, M.Z. 985
Xiao, Z.L. 911
Xiaobin, W. 13
Xie, C.Y. 1147
Xie, D.P. 921
Xie, F.-J. 275
Xie, H.W. 861
Xie, H.Y. 939
Xie, J. 1253
Xie, J.G. 489
Xie, L. 135
Xie, L.Y. 883
Xie, Y. 3
Xing, Y.J. 549
Xiong, J. 843
Xiong, W. 991
Xiong, Y. 569, 577
Xu, C. 1129
Xu, D.F. 629
Xu, G.Y. 1121
Xu, H.B. 665
Xu, H.R. 1453
Xu, J. 205
Xu, J.J. 373
Xu, J.N. 991
Xu, L. 341
Xu, L. 345, 349
Xu, L. 559
Xu, M.D. 1511
Xu, N. 613
Xu, Q.Y. 363
Xu, R.Q. 381
Xu, S. 471
Xu, X. 1565
Xu, X.C. 159
Xu, X.T. 719

Xu, X.Y. 693, 1159
Xu, Y. 373
Xu, Y.D. 931, 945
Xu, Z.G. 975
Xuan, L. 833
Xuan, Z.-L. 959
Xue, C.Y. 1331
Xue, D. 145
Xue, K.D. 675
Xue, L. 629
Xue, S. 659
Xue, W.-F. 1367
Xue, W.J. 357
Xue, Z.X. 597

Yan, D.Y. 597
Yan, H. 1335
Yan, M. 1577
Yan, P. 971
Yan, Q. 381
Yan, R.Y. 927
Yan, S. 253
Yan, W. 823
Yan, X.J. 367
Yan, Z.-H. 979
Yang, B. 865
Yang, C.G. 1473
Yang, G.J. 1565
Yang, H. 1325
Yang, H. 1453
Yang, H.-C. 233, 447
Yang, J. 239
Yang, J.H. 1015
Yang, J.K. 1499
Yang, J.R. 1005
Yang, J.-Y. 1577
Yang, L.F. 295
Yang, L.L. 321
Yang, X.-B. 149
Yang, X.H. 581
Yang, Y.F. 1593
Yang, Y.J. 949
Yang, Y.Q. 195
Yang, Y.-S. 871
Yang, Y.W. 1593
Yang, Z. 581
Yao, J.-L. 495
Yao, T. 573
Ye, J. 719
Ye, J.M. 545
Ye, K. 1015
Ye, L. 49
Ye, Q.X. 651
Yi, Y. 289
Yin, H.W. 1223
You, A.J. 665
You, J.Y. 403

You, K.P. 403
Yu, D.M. 1129
Yu, G.L. 367
Yu, H.B. 377
Yu, H.Z. 1259
Yu, J. 1505
Yu, L. 465
Yu, Q. 377
Yu, S.-M. 411
Yu, S.N. 949
Yu, W. 1339
Yu, X. 995
Yu, Y.X. 1287, 1293
Yuan, J.Y. 1167
Yuan, L.Q. 265
Yuan, L.Y. 1445
Yuan, S.P. 1089
Yue, C.S. 387
Yue, G.Q. 589
Yue, M.-K. 1277
Yun, P. 517, 521, 525

Zeng, P.P. 1483, 1489
Zeng, R. 839
Zhai, J.Z. 1265
Zhan, C.H. 1159
Zhan, X.P. 609
Zhan, Y. 1589
Zhang, A.L. 75, 89, 97,
  215, 651
Zhang, A.Q. 19
Zhang, C.C. 861
Zhang, D.H. 1517
Zhang, D.W. 1551
Zhang, F. 621
Zhang, G.H. 377
Zhang, G.-J. 1095
Zhang, H. 585
Zhang, H. 823
Zhang, H.-W. 1367
Zhang, J. 939
Zhang, J.C. 177
Zhang, J.C. 341
Zhang, J.C. 377
Zhang, J.F. 1571
Zhang, J.J. 1073
Zhang, J.J. 1207, 1351
Zhang, J.S. 787
Zhang, J.X. 367
Zhang, J.Z. 589
Zhang, K. 353
Zhang, K. 659, 1147
Zhang, K.X. 1505
Zhang, L. 119
Zhang, L. 435
Zhang, L. 1089
Zhang, P.Y. 1347